3rd International Conference on Computer, Communications and Mechatronics Engineering (CCME 2021)

Journal of Physics: Conference Series Volume 2218

Online
17-18 December 2021

ISBN: 978-1-7138-5179-0
ISSN: 1742-6588

Printed from e-media with permission by:

Curran Associates, Inc.
57 Morehouse Lane
Red Hook, NY 12571

Some format issues inherent in the e-media version may also appear in this print version.

Institute of Physics
Dirac House, Temple Back
Bristol BS1 6BE UK

Phone: 44 1 17 929 7481
Fax: 44 1 17 920 0979

techtracking@iop.org

Additional copies of this publication are available from:

Curran Associates, Inc.
57 Morehouse Lane
Red Hook, NY 12571 USA
Phone: 845-758-0400
Fax: 845-758-2633
Email: curran@proceedings.com
Web: www.proceedings.com

TABLE OF CONTENTS

MECHANICAL DESIGN, MANUFACTURE AND AUTOMATION

Preface

2021 3rd International Conference on Computer, Communications and Mechatronics Engineering (CCME2021) was successfully held on December 17, 2021 in the form of Virtual Conference because of current epidemic prevention and travel restriction, which is a professional academic platform for all the participants. While improving the research motivation and academic literacy of all participants, CCME2021 will more strengthen safety protection, actively respond to the current epidemic prevention and cooperate with the normalization of epidemic situation.

CCME2021 virtual meeting mainly carries out academic sharing and exchanging in the forms of Opening ceremony, Keynote Speech, Oral presentations, Poster presentations and free discussions. In order to ensure this virtual meeting smooth running, CCME2021 organizing committee specially opened a test room before opening for experts and authors to test their participation conditions including video, audio and screen sharing, etc. in advance.

Thank you for all supports of Prof. Shuo Zhao, Dr. Hamed Taherdoost and all attendees. Prof. Shuo Zhao from Communication University of China, China made keynote speech about Corpus-involved E-learning and Education in European Universities, and Dr. Hamed Taherdoost from University Canada West, Canada shared his viewpoints on The Role of Information Security Awareness In Online Shopping Trends.

All anonymous submissions were reviewed by 2-3 experts in the relevant areas and based on their ranking, technical remarks and strict selection criteria the papers were chosen to the various tracks. And all selected papers were detected the instances of overlapping and similar text via iThenticate software, here are 88 papers were selected and divided into 2 parts in CCME2021 proceedings below:

1. Communications, Electronic and Information Engineering
2. Mechanical Design, Manufacture and Automation

Finally, CCME2021 are willing to express our sincerest appreciation to all the active authors with their latest academic research findings, to the sponsor of Advanced Science and Industry Research Center, Hong Kong, Journal of Physics: Conference Series with effective publication for all excellent paper submissions, all the members and reviewers in CCME2021 Technical Program Committee with careful reviewing and overwhelming response. Sincerely hope for more academic cooperation with all scholars and scientists of the related fields in the future.

CCME2021 Organization Committees

Committees

General Chairs
Prof. Hui-Ming Wee, Chung Yuan Christian University, Taiwan
Prof. Gherabi Noreddine, Sultan Moulay Slimane University, Morocco
Prof. Elżbieta Macioszek, Silesian University of Technology, Poland

General Co-Chairs
Prof. Dariusz Jacek Jakóbczak, Koszalin University, Poland
Dr. Yasin Polat, Nevsehir Haci Bektas Veli University, Turkey
Dr. Gianpaolo Di Bona, University of Cassino, Italy

Program Chairs
Prof. Andrey V. Brazhnikov, Siberian Federal University, Russia
Prof. Isidoros Perikos, University of Patras, Greece

Publication Chairs
Prof. Arnulfo Luévanos Rojas, Autonomous University of Coahuila, México
Dr. Bogdan Constantin Neagu, Gheorghe Asachi Technical University of Iasi, Romania
Dr. Nayef Abdulwahab Alduais, Universiti Tun Hussein Onn Malaysia, Malaysia

Publication Co-Chairs
Prof. Zine Ghemari, University Mohamed Boudiaf- M'sila, Algeria
Prof. Daschievici Luiza, "Dunarea de Jos" University of Galati, Romania
Dr. Gianpaolo Di Bona, University of Cassino, Italy

International Scientific Committees
Prof. Branko Karan, Danube Robotics Systems, Serbia
Prof. Valentin Mateev, Technical University of Sofia, Bulgaria
Prof. Wei Xie, Institute of Disaster Prevention, China
Dr. Shahram Babaie, Islamic Azad University, Iran
Prof. Alex Mathew, Bethany College, USA
Dr. S. Udhaya Kumar, Saveetha Institute of Medical and Technical Sciences, India
Dr. Pavan Kumar, VIT Bhopal University, India
Prof. Chandrani Singh, MCA Sinhgad Institute of Management, India
Dr. Kais Belwafi, King Saud University, Saudi Arabia
Prof. Zhiyu Jiang, Northwestern Polytechnical University, China
Prof. Lakhoua Najeh, Universty of Carthage, Tunisia
Dr. Daniela Litan, Hyperion University, Romania
Prof. Israa Shaker Tawfic, Ministry of Science and Technology, Iraq
Dr. Michele Fiorini, The Institution of Engineering and Technology (IET), Italy
Prof. Snježana Babić, Polytechnic of Rijeka, Croatia
Dr. Raghu N, Jain University, India
Dr. Shadi Abudalfa, University College of Applied Sciences, Palestine
Prof. Priyantha Wijayatunga, Umea University Sweden, Sweden
Dr. Ciortea Elisabeta Mihaela, "1 December 1918" University of Alba Iulia, Romania
Prof. Orchidea Maria Lecian, Sapienza University of Rome, Italy
Dr. Jan Kubicek, VSB-Technical University of Ostrava, Czech Republic
Dr. Jose Miguel Martínez Valle, University of Córdoba, Spain
Dr. José Joaquim de Moura Ramos, University of A Coruña, Spain
Dr. Mihaela Andrei, "Dunarea de Jos" University of Galati, Romania
Prof. S. Balakrishnan, Sri Krishna College of Engineering and Technology, Coimbatore
Dr. Pooja M R, Vidyavardhaka College of Engineering, India
Dr. Avisankar Roy, Haldia Institute of Technology, India
Dr. Jagdish Prasad Sharma, JK Lakshmipat University, India
Dr. M. Madhiarasan, Indian Institute of Technology, India
Prof. Mihailescu Marius Iulian, University of Spiru Haret, Romania
Prof. Arvind Kumar, VIT University, India

Peer Review Statement

All papers published in this volume have been reviewed through processes administered by the Editors. Reviews were conducted by expert referees to the professional and scientific standards expected of a proceedings journal published by IOP Publishing.

- **Type of peer review:** Double Anonymous
- **Conference submission management system:** Morressier
- **Number of submissions received:** 189
- **Number of submissions sent for review:** 171
- **Number of submissions accepted:** 88
- **Acceptance Rate (Submissions Accepted / Submissions Received × 100):** 46.6
- **Average number of reviews per paper:** 2
- **Total number of reviewers involved:** 86
- **Contact person for queries:**
  Name: Dawen Fang
  Email: dwfang@ser-center.org
  Affiliation: Science and Engineering Research Center, Hong Kong

A Single Channel End-to-End Speech Enhancement using Complex Operations

Jie Wu[1], Hongqing Liu[1,*], Lu Gan[2] and Yi Zhou[1]

[1]School of Communication and Information Engineering, Chongqing University of Posts and Telecommunications, Chongqing, China
[2]College of Engineering, Design and Physical Science, Brunel University, London, U.K

*Corresponding author email: hongqingliu@cqupt.edu.cn

Abstract. This paper investigates the possibility of using complex operations to perform speech enhancement task in time domain. To that end, first, the Hilbert transform is utilized to prepare the complex input in time domain. After that, the complex temporal convolutional network (CTCN) is developed to conduct complex convolutions. By cascading the TCN and the CTCN modules, the final proposed network form an encoder-decoder structure, which performs an end-to-end speech enhancement task. The results demonstrate that utilizing complex information in time domain indeed improves the enhancement performance. Compared to other approaches, the proposed network also demonstrates a superior performance in terms of objective evaluations.
Keywords: Speech Enhancement; Single-channel; Complex network; Time domain.

1. Introduction

Speech enhancement is one of highly expected tasks in modern speech applications. It aims to separate the target voice from mixture speech signal. In recent years, we have all witnessed that deep learning based approaches have outperformed traditional methods in speech enhancement. The noisy signals can be enhanced in both time-frequency domain and time domain. The time domain enhancement networks can be classified into two methods, which are direct regression approaches[1] and adaptive front-end methods[2]-[3]. The regression approach directly learns the regression function from the mixture to rebuild the target speech, and the adaptive front-end method usually uses convolutional encoder and decoder or a U-shaped network, which is similar to Fourier transform and its inversion.

As a different strategy, speech enhancement can also be conducted in frequency domain using spectrogram. The training methods in the frequency domain are mainly classified into two categories, namely mask-based and mapping-based approaches. For the masking methods, ideal binary masking (IBM)[4] and ideal ratio masking (IRM)[5] often use the amplitude between clean speech and mixed speech to perform enhancement. However, they only pay attention to amplitude information the magnitude spectrum and ignore the phase information because it was believed that the phase information was difficult to estimate. Due to that, the phase information was thus used to reconstruct target speech. This bounds the upper limit of network performance, for the phase deviation also increases interferences. Recently, several networks have adopted the phase reconstruction concept and seen certain improvements. For that, the deep complex U-Net[7] that simultaneously utilizes magnitude spectrum and phase spectrum based on a proposed variant of U-Net[8] was developed to process complex spectrogram. Soon after, the deep complex convolution recurrent network

(DCCRN)[9] adapted the complex multiplication rules to the LSTM part of the enhancement module on the basis of DCUNet.

Inspired by the complex network, this paper attempts to design a complex network in time domain, which means it is an end-to-end system. To do that, the real-valued time-domain signal is passed to Hilbert transform to produce the imaginary part of the corresponding real part. A complex TCN structure, termed as complex temporal convolutional network (CTCN), that uses the time modeling capability of TCN to process complex-valued information in the time domain, is developed.In our experiments, it is found that the addition of time-domain imaginary part information improves the speech quality of the reconstructed target signal and achieves a better performance than the other networks on the DNS1 Challenge dataset.

The rest of article is organized as follows. Section II introduces the descriptions of enhancement task. In Section III, we describe the proposed CTCN model. Section IV describes the experimental details and results. Section V summarizes the article.

2. Mathematical Background

2.1. Signal Model and Preparation

The single-channel speech denoising problem can be described as estimating the original clean source x(t), using the noisy mixture y(t), given by:

$$y(t) = x(t) + n(t) \tag{1}$$

where n(t) is the unwanted noise.

If the denoising network is conducted in frequency domain, one can transform the signal in(1) to spectrogram using short-time Fourier transform (STFT), and the resulting signal is complex that contains both real and imaginary parts. Based on that, the DCCRN network is developed that utilizes complex information[9]. Taking a different strategy, we attempt to perform speech enhancement in time domain by a use of complex network. To that aim, the imaginary signal needs to be generated from the real-valued time domain inputs.

According to Hilbert transform, an imaginary signal can be produced by its real-valued one, given by:

$$x_i(t) = H[x(t)] = x(t) * h(t) \tag{2}$$

where H[·] and * respectively represents the Hilbert transform and convolution, and subscript **i** indicates the imaginary part.

In (2), h(t) is the impulse response of the transform, given by:

$$h(t) = (\pi t)^{-1} \tag{3}$$

From the point of view of the frequency spectrum, this transform multiplies the positive frequency part of our original signal by -i, that is, while keeping the amplitude constant, the phase is shifted by -90 degree, and for negative frequency components is 90 degree.

Finally, performing Hilbert transform on both sides of (1) and putting the real and imaginary parts together, one obtains

$$y_{r,i}(t) = x_{r,i}(t) + n_{r,i}(t) \tag{4}$$

where **r** and **i** represent the real and imaginary parts of each signal after transformation. The real part represents the original time domain signal, and the imaginary part is the result of the real part signal shifted by a 90 degree.

2.2. Training Target

The model we propose runs in an end-to-end manner and it directly reconstructs the original signal to produce speech enhancement results. In this work, both the real part represented by the clean signal and the imaginary part of the signal after the Hilbert transform are used as training targets. The objective function uses a weighted scale-invariant source-to-noise ratio (SI-SNR). The SI-SNR is

$$\begin{cases} s_{target} := \langle \hat{s}, s \rangle s \times \|s\|^{-2} \\ e_{noise} := \hat{s} - s_{target} \\ \text{SI-SNR} := 10\log_{10} \|s_{target}\|^{2} \times \|e_{noise}\|^{-2} \end{cases} \tag{5}$$

where $\hat{s}$ and s are the estimator and original clean sources, and $\|s\|^2$ represents the energy of the signal. By taking the complex signal into account, the final weighted loss function is

$$J = a * L_{\text{SI-SNR}}(X_r) + (1-a) * L_{\text{SI-SNR}}(X_i) \tag{6}$$

Where **L** function is to calculate the SI-SNR of the estimator and the target. To ensure the real and imaginary parts play the equal importance, empirically, **a** is set to 0.5.

3. Proposed Time Domain Network

3.1. Complex Temporal Convolutional Module

The temporal convolutional network (TCN) module can be used to solve sequentially prediction[2]. Each TCN block includes a 1*1-conv operation, followed by a dilation convolution operation, and finally a skip connection path to avoid the problem of gradient disappearance, as shown in Figure 1. Despite the excellent performance of TCN, it only takes real signal as its input and real convolutions are conducted. To develop our time domain complex network, our building block is also TCN, but it takes complex signal as input and complex convolutions are performed, termed as complex temporal convolutional network module (CTCN), which is provided in Figure 2. Compared with original TCN, CTCN adds an imaginary branch to the ordinary real-valued TCN, and at the same time, introducing complex multiplication in the output part to simulate the correlation between magnitude and phase.

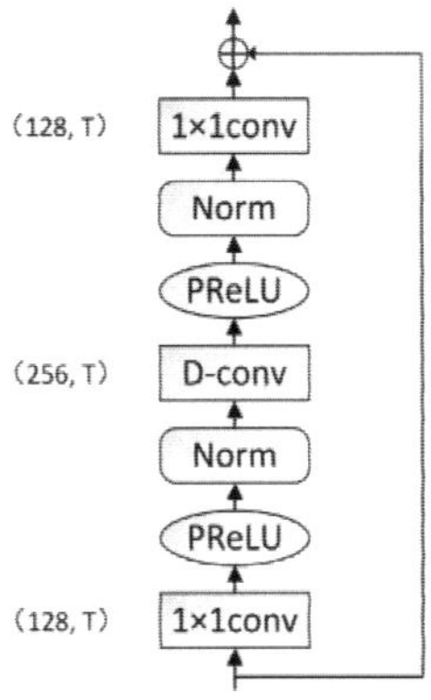

Figure 1. The diagram of each layer in original TCN

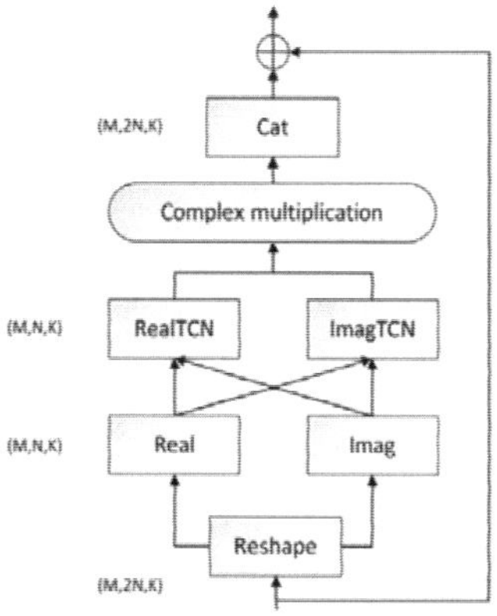

Figure 2. The diagram of each layer in the proposed CTCN

From Figure 2, the CTCN block simulates complex convolutions with real-value TCN, where CTCN consists of two real-value TCN operations, which controls the complex information from inputs.

3.2. Convolutional Time-domain Complex Speech Enhancement Network
The proposed network is an essentially causal convolutional architecture with a separation module. First, short segments of the input waveform are converted into intermediate high-dimensional features by an encoder that combines one-dimensional convolution and nonlinear activation functions. After that, the separation module stacked by 1-D dilated convolutional blocks is used to estimate the high-dimensional feature mask corresponding to the source. Finally, the decoder is used to transform the masking characteristics of the enhancement and reconstruct the time-domain waveform.
Specifically, the enhancements block is shown Figure 3. First, we perform Hilbert transform on the time domain signal to obtain the imaginary part information. The real and imaginary parts share the same one-dimensional convolutional encoder for feature modeling, and a speech enhancement module estimates the mask of the enhancement, where the last TCN module are replaced with CTCN module. The complex multiplication rule is used on the CTCN module to combine the real and imaginary parts. The network structure resembles the Conv-Tasnet and therefore, the proposed network is named as CTCN-Tasnet in this work。

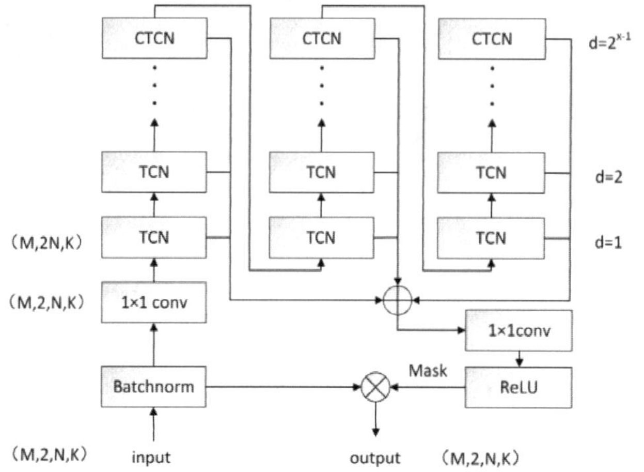

Figure 3. The structure of enhancement block

4. Evaluation Results and Comparisons

4.1. Datesets
In our experiments, we first evaluate the proposed model and several baselines on the Interspeech2020 DNS challenge dataset (DNS1)[9]. We use the script provided by DNS1 to generate training, validation, and evaluation sets, which are 40, 10, and 12.5 hours of utterance, respectively. The mixture in train and verification is generated by randomly selecting utterances from the speech set and noise set, and mixing them at a random signal-to-noise ratio (SNR) of -5 dB to 20 dB. An evaluation set is generated under 5 typical SNR of (0, 5, 10, 15, 20)dB.

4.2. Training Setup and Baselines
The time domain model cuts all audio into 2 ms segments, and the frequency domain model uses the original best configurations of each model.We adopt four state-of-the-art baselines for comparisons, and they are CRN [10], Conv-TasNet [2], DPRNN[3]. All the models are tested with their best configurations based on the suggestions in their papers.

4.3. Results and Analysis
We now compare the proposed model to different models with PESQ [11] and the results are provided in Table 1.

Table 1. PESQs with various numbers of CTCN on DNS1 dataset.

Model / SNR	0db	5db	10db	15db	Ave
Noisy	1.860	2.167	2.495	2.803	2.331
CRN	2.541	2.832	3.033	3.323	2.932
Conv-TasNet	2.623	2.974	3.151	3.432	3.043
DPRNN	2.682	3.022	3.237	3.504	3.115
DCCRN	2.752	3.043	3.277	3.513	3.144
CTCN-TasNet	**2.801**	**3.097**	**3.326**	**3.537**	**3.184**

From Table 1, It can be found that the performance of the CTCN-TasNet outperforms the baselines of CRN, Conv-TasNet, and DPRNN, which demosnrates the effectiveness of exploring the imaginary information in the signal. In addition, the performance of the CTCN-TasNet network is similar to DCCRN. In the low SNR case, the CTCN-TasNet is superior to the DCCRN, whereas in high SNR scenario, the DCCRN network is slightly better than the CTCN-TasNet. On average, however, they produce the same performance.

Table 2. PESQ under different future frames on the DNS datasets with SNR = 0 dB.

Model	Causality (look head)	DNS1 PESQ	DNS3 PESQ
Noisy		1.86	2.51
Conv-TasNet	nocausal	2.62	3.02
Conv-TasNet	21.25ms	2.55	2.96
Conv-TasNet	1.25ms	2.42	2.86
CTCN-TasNet	nocausal	**2.80**	**3.07**
CTCN-TasNet	21.25ms	**2.72**	**3.05**
CTCN-TasNet	1.25ms	**2.57**	**2.99**

To further show the performance improvement of the CTCN-TasNet over Conv-TasNet, we conduct the experiments on the DNS1 and DNS3 datasets at the same time allowing to access different future frames. For the TCN network, by applying causal convolution, the network can be easily modified to access limited future frames. We let the first few layers be non-causal, while the remaining layers are causal. In Conv-TasNet and CTCN-TasNet, when L = 40 (convolution kernel size) and N = 512 (channels of encoder and decoder) adopting 1/2 overlapping coding, the number of non-causal layer is three, and the number of causal layer is five and the rest of the hyperparameter settings are the same as the paper. By calculations, the future information accessed by the algorithm is (40*8+40/2)/16000=21.25 ms. Table 2 shows PESQ of different methods on DNS datasets with SNR = 0 dB.

As we can see from Table 2, the proposed CTCN-TasNet outperforms Conv-TasNet in all the cases considered. On the DNS1 dataset, CTCN-TasNet achieves an average improvement of 0.17 in terms of PESQ, whereas on the DNS3, CTCN-TasNet achieves an average of 0.09 in terms of PESQ. These results show that accessing future frame information can effectively promote the result of the network, but network still needs to be tuned for different scenarios.

5. Conclusion

This paper proposes a new way to process time-domain speech waveforms using complex operations. Given only the real waveform is available, Hilbert transform is utilized to construct the corresponding imaginary part. With both the real and imaginary signals in time domain, the CTCN module is developed to efficiently perform complex convolutions. The proposed CTCN-TasNet outperforms other networks in terms of PESQ for a smaller model size. It is of interest to point out that this transform is independent of the subsequent networks used and can be applied to different structures, which is our future work.

References

[1] S.-W. Fu, T.-W. Wang, Y. Tsao, X. Lu, and H. Kawai, "End-to-end waveform utterance enhancement for direct evaluation metrics optimization by fully convolutional neural networks," IEEE/ACM Transactions on Audio, Speech, and Language Processing, pp. 1570–1584, 2018.

[2] Y. Luo and N. Mesgarani, "Conv-tasnet: Surpassing ideal time-frequency magnitude masking for speech separation," IEEE Transactions on Audio, Speech, and Language Processing, pp. 1–1, 2019.

[3] L. Yi, C. Zhuo, and Y. Takuya, "Dual-path rnn: efficient long sequence modeling for time-domain single-channel speech separation," ICASSP, pp. 46–50, 2019.

[4] A. Narayanan and D. Wang, "Ideal ratio mask estimation using deep neural networks for robust speech recognition," Acoustics, Speech and Signal Processing, pp. 7092–7096, 2013.

[5] X. Li, J. Li, and Y. Yan, "Ideal ratio mask estimation using deep neural networks for monaural speech segregation in noisy reverberant conditions," INTERSPEECH, pp. 1203–1207, 2017.

[6] Y. Xu, J. Du, L.-R. Dai, and C.-H. Lee, "An experimental study on speech enhancement based on deep neural networks," IEEE Signal Processing Letters, pp. 65–68, 2014.

[7] O. Ronneberger, P. Fischer, and T. Brox, "U-net: Convolutional networks for biomedical image segmentation," MICCAI, 2015.

[8] Y. Hu, Y. Liu, S. Lv, M. Xing, S. Zhang, Y. Fu, J. Wu, B. Zhang, and L. Xie, "Dccrn: Deep complex convolution recurrent network for phaseaware speech enhancement," INTERSPEECH, pp. 2472–2476, 2020.

[9] C. K. R. A., G. Vishak, C. Ross, B. Ebrahim, C. Roger, D. Harishchandra, M. Sergiy, A. Robert, A. Ashkan, B. Sebastian, R. Puneet, S. Sriram, and G. Johannes, "The interspeech 2020 deep noise suppression challenge: Datasets, subjective testing framework, and challenge results," in INTERSPEECH, 2020, pp. 2492–2496.

[10] K. Tan and D. Wang, "A convolutional recurrent neural network for real-time speech enhancement," Interspeech, pp. 3229–3233, 2018.

[11] A. W. Rix, J. G. Beerends, M. P. Hollier, and A. P. Hekstra, "Perceptual evaluation of speech quality (pesq)-a new method for speech quality assessment of telephone networks and codecs," in 2001 IEEE International Conference on Acoustics, Speech, and Signal Processing. Proceedings (Cat. No. 01CH37221), vol. 2. IEEE, 2001, pp. 749–752.

Aircraft Engine Audio Signal Analysis in Assisting Maintenance Inspections

Yukino Sakai*

Dulwich College Beijing, Beijing, China

*Corresponding author email: yukino.chen22@stu.dulwich.org

Abstract. As the core component of modern commercial aircraft, turbofan engines have long been the center of focus in aircraft maintenance. Being subject to high temperatures and immense pressures causes problems for the engine components, such as the fan blades, as they are frequently burdened with the potential of overhaul and malfunction. Over many years, the industry has seen various methods of engine inspection and maintenance, ranging from manual inspection to computing large quantities of pre-existing data. Within, audio signal analysis has stood out as a productive, non-invasive method, with many alternate studies analyzing sound signals from components such as the combustion chamber. However, many of these methods, despite demonstrating good accuracy, are incredibly complex and require sophisticated apparatus. Therefore, this study begins by investigating the sound generation process of turbofan engines, especially how the features and form of the fan blade characterize its audio signals. This investigation proposes a solution that utilizes a fast Multi-Class Support Vector Machine (SVM) algorithm based on fan-blade-related audio signals from a perspective similar to the classification of music and images through supervised machine learning. Experimental results show that this fast Multi-Class SVM is more effective than traditional machine learning methods in its accuracy, F1-score, and other indicators.
Keywords: Aircraft turbofan; Engine inspection and maintenance; Audio signals; Fast Multi-Class SVM.

1. Introduction

From commercial aircraft to military-grade jets, our modern forms of aviation are heavily reliant on turbo engines for basic propulsion. As a crucial component of the aircraft's aerial capabilities, the engine must undergo regular inspection and maintenance procedures. Turbofan engines, which are turbomachines, are mechanical devices consisting of several blades in addition to stationary blades that allow a massive air volume flow rate and extract energy from the air to generate thrust [4]. Fig. 1 describes these aircraft noise sources visually. Modern aero engines are often characterized by their combustion and jet noises [3]. For instance, a study by M. J. Lighthill suggested that the noise from the exhaust of a jet engine reached values of the eighth power to its exhaust speed, which meant that an increase in the jet exhaust speed would exponentially increase the intensity level of generated noise [6]. The turbofan engine successfully reduced the noise generated by its jet exhaust through increased bypass ratios [1], but instead introduced a new source of characteristic fan noise; fan blades form a core component in the engine's composition and functionality, so the sound signatures generated by the fan blades alone are of the highest degree of interest for this paper. With the recent leaps in aviation technology, the possibility of system and component failures has dramatically decreased [12]. In the process of collecting fault detection data in aero-engine analysis, it is often found that there are far more normal samples than fault samples. Therefore, researchers often require fault detection classifiers with no or little fault data.

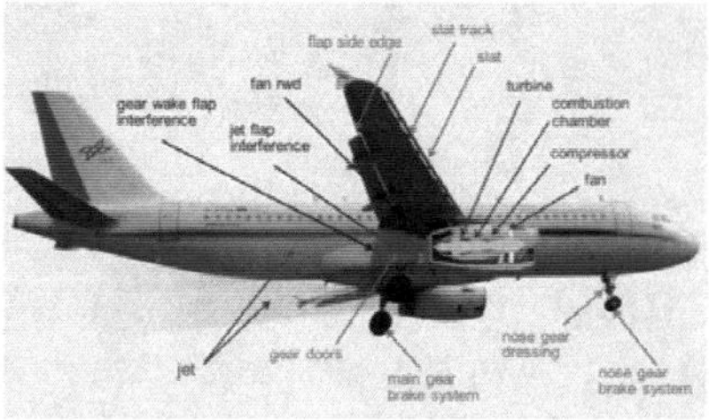

Figure 1. Overview of aircraft noise sources [1]

Most simply, professionals may be able to judge the state of the engine from experience via inspection or manually determine areas of error from observing changes in known parameters. We may compare measured parameters with known parameters of a 'healthy' engine via modeling. However, both methods are either largely subject to human error or risk yielding sub-optimal results, and therefore more advanced methods are necessary.

In recent years, engine fault detection through audio signal analysis has become an increasingly popular method of aircraft engine inspection, applicable for a wide variety of engines. In light aircraft, for example, fault detection of the aircraft engine can be conducted through analyzing the exhaust noise of piston engines. An exploration by Miljković claims that piston engine exhaust noise analysis is valuable, thereby generally illustrating the engine condition. This approach utilizes a statistical approach, including acoustic measurement, amplitude normalization, dividing audio signals into frames, extracting key features, and comparing the threshold of the distributive data. Overall, this method is advantageous due to its non-penetrating nature and short test time, which is typical of audio signal analysis. However, it comes short in being able to reveal other non-combustion-related engine problems [7].

Waligórski [11] investigates the acoustic characteristics typical of changes in the fuel mixture during combustion and claims that almost 80% of all failures of turbine aircraft engines are caused by the fuel system, compression chamber, and assembly of the compressor. However, this would require certain degrees of penetration into the engine system or an installation of inspection devices beforehand. Gharoun [5] innovatively utilizes data mining techniques applied to existing flight data and draws connections between the engine exhaust gas temperature and environmental conditions. The investigating claims to be successful at identifying faults and proposes that there would be no need for technical knowledge on the interior of the engine, and therefore theoretically simple to apply. These methods are generally practical, utilizing the comparison of raw data with theoretical models and then comparing the deviations of known parameters. Yet, they often require intricate modeling or sophisticated measuring equipment, and are still subject to assumptions, so that errors will only occur within specific frames or in specific combinations.

This paper proposes a fast multi-class SVM method for the identification and categorization of different forms of turbofan engine malfunction. The method follows the modern trends of audio signal analysis, but instead utilizes frequencies within the human hearing range. Following this, the paper will explore the Physical Analysis of Sound Generation in part II, propose an Improved Audio Signal Analysis Method in part III, and describe the experimental results in part IV before giving a conclusion.

2. Physical Analysis of Engine Sound Generation

Considering that turbofan engines are complex systems constantly subject to intense temperatures and pressures, it is inevitable that their functionality will undergo severe declines within observable time frames and eventually malfunction. For safety reasons, it is therefore crucial that they are thoroughly inspected at regular intervals for such failures. Of course, the causes of such failures span a diverse range of factors, each uniquely altering the sound that the engine generates. This section will explore the physics of sound generation by these fans and aims to determine how each different form of engine malfunction alters its sound signature.

Typically, the primary jet noise comes from the axial flow of air within the primary exhaust jet. The jet noises primarily dominate the noise composition during take-off, and the fan noises are characterized by high-pitched, discrete frequencies upon landing approach. During this time, the fan noise falls within the spectrum related to its rotational frequency and subsequent harmonics; its frequency generally falls in the region of 800 Hz and above (shown in Fig. 2), which is typically higher than the frequency of the jet-related noises, making the two relatively distinguishable.

When examining only the fan blade portion of a turbofan engine, its sound generation resembles that of any regular fan, and hence its overall noise is mainly composed of aerodynamic noise, magnetic noise, and mechanical noise; the latter two become negligible at low rotational speeds [9].

Aerodynamic noise features are mainly characterized by the type, dimensions, and geometry of the fan, as well as its rotational speed. Specifically, axial flow fans generate periodic fluctuations at harmonics of its rotational frequency. In turbofan engines, the fans reach supersonic speed in their rotations and generate shock waves that create the characteristic 'buzz-saw noise' illustrated by Fig. 3, which is the periodic fluctuation in this case [8].

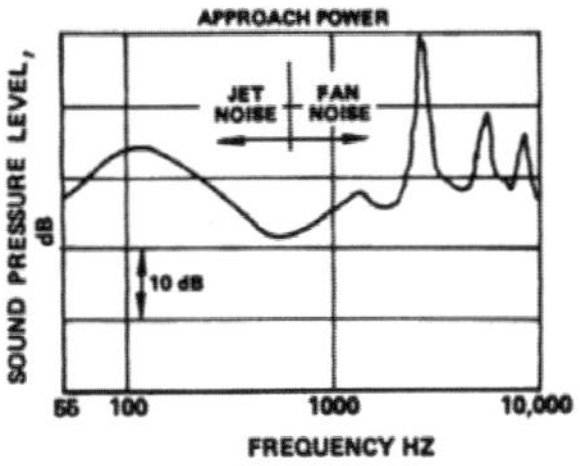

Figure 2. Sound pressure level of jet noises[10]

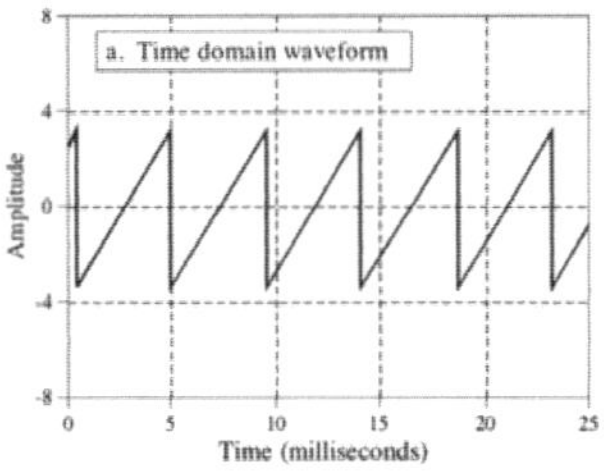

Figure 3. Spectrogram of the 'Buzz-Saw' Noise[8]

Nonetheless, as any mechanical component does, the fan blades of a turbofan engine will inevitably hit the end of its usage time as they are worn out [12] through damage via:

1) Domestic or foreign object debris (DOD and FOD), such as loose internal compartments that were rejected from their original positions due to improper assembly and intense vibrations, or bird-strike.
2) Fan blade corrosion due to external humidity and minerals, or internal minerals from fuel combustion
3) Build-up of substances such as byproducts of fuel combustion of ice on the fan blades
4) Abrasive substances such as passing air at immense velocities or external dust
5) Fuel leakages and subsequent substance build-up

Non-fuel-correlated turbofan engine failures occur in several ways, most often due to imbalances in its compression chamber or due to vibrations in the fan blade segment. Specifically, fan blades can lead to general malfunctions due to errors during assembly, blade distortion or deformation due to long-term use, air friction, and blade icing. Besides external causes of malfunction such as harsh environmental conditions, these blades can also be worn out over time, thereby changing their sound signature.

Certainly, any physical damage to the blades of a turbofan engine will directly cause its sound signals to deviate from its original signature, even before it hits a threshold of malfunction or complete destruction. For instance, when the input flow velocity of air into the plane of the fan blades is non-uniform (such as when a blade has been deformed or snapped, or when there is an obstacle), the amplitude of its original periodic sound fluctuation will experience deviations [9]. These fluctuations can be juxtaposed with the original sound signatures of healthy fan blade systems in the audio analysis stage to attempt categorization.

3. Improved Audio Signal Analysis Method

There are many methods for the fault detection of aircraft engines. For instance, professionals can judge the state of the engines via manual inspection or by observing changes in known parameters. The judgment of aircraft engine states can also be done by focusing on the combustion chambers and comparing measured parameters with known parameters from a 'healthy' engine. Both methods, however, are either largely subject to human error or unprecise, so more advanced methods are necessary. Thus, engine fault detection through the non-penetrating method of audio signal analysis has become an increasingly popular approach in recent years, applicable for a wide variety of engines.

Many current methods of aircraft engine fault detection focus on combustion-related malfunctions and omit the consideration of engine-blade-related malfunctions, which are incredibly dominant in the malfunctions of turbofan engines specifically. This can be done through utilizing artificial intelligence, or specifically machine learning.

As a branch of artificial intelligence, machine learning aims to categorize and 'learn' from data so that predictions and patterns can be made. For instance, backpropagation (BP) neural networks are the most popular feedforward networks (see Figure 4), which repeatedly conduct feed-forward computations and backpropagation to the output and hidden layers [2].

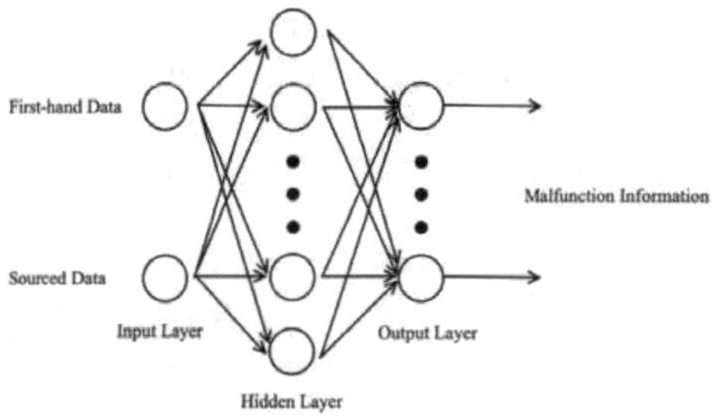

Figure 4. General Form of a BP Neural Network

The process uses a Sigmoid function as the activation function and can achieve high accuracies. At present, the most widely used BP network is a three-layer BP network. The linearization model of a set of Kalman filters to the actuator in the engine is:

$$\begin{aligned} \bar{x}^k &= \widetilde{\boldsymbol{A}}^k \tilde{x}^k + \widetilde{\boldsymbol{B}} u_{cmd} + w \\ y^k &= \widetilde{\boldsymbol{C}}^k \tilde{x}^k + \boldsymbol{D} u_{cmd} + v \end{aligned} \tag{1}$$

Where, k is an integer from 1 to p, which is the number of actuators, and b is the deviation of the kth actuator. If an actuator fails, then all filters may use the faulty signal. However, only one corresponding filter can accurately estimate the deviation, making the residual value very small; the remaining filters cannot correctly estimate the deviation, resulting in a large residual value. This paper proposes a fast multi-class support vector machine (SVM) algorithm to handle multi-class fault diagnosis problems of the turbofan engine where there are large samples. The approach is generalized in Figure 5.

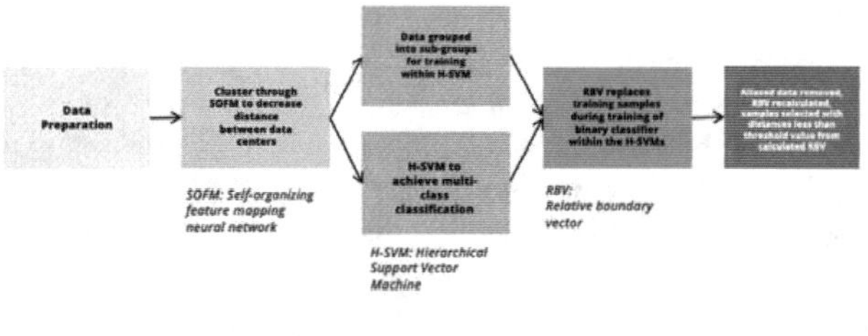

Figure 5. Classification Process Via the Fast Multi-Class SVM

First, data was prepared, and various data centers were used to represent the data and cluster through the self-organizing feature mapping neural network (SOFM) to decrease the distance between each center. A hierarchical SVM (H-SVM) was selected to achieve multi-class classification, whereby the data was grouped into sub-categories for training within an H-SVM hierarchy. Next, the relative boundary vector (RBV) replaced training samples during the training of the binary classifier within the H-SVM, greatly reducing the number of training samples necessary, significantly decreasing training time. As the number of support vectors was reduced, classification time is also shortened accordingly. In the case where data is severely aliased, the proposed algorithm removed the aliased data, recalculated the RBV, and then selected samples with distances less than a proposed threshold value from the calculated RBV; this prevented the loss of crucial data and thereby improved the accuracy of classification.

4. Experimental Results

We analyzed three main types of fan blade failures: cracking, deformation, and debris accumulation. These malfunctions were roughly approximated and modeled with miniature electric fans (as shown in Fig. 6 and Fig. 7), which adhere to the same sound generation principles as turbofan engine blades; the set-up was surrounded by sound-proofing foam in an otherwise quiet room of between 10 dB to 20 dB.

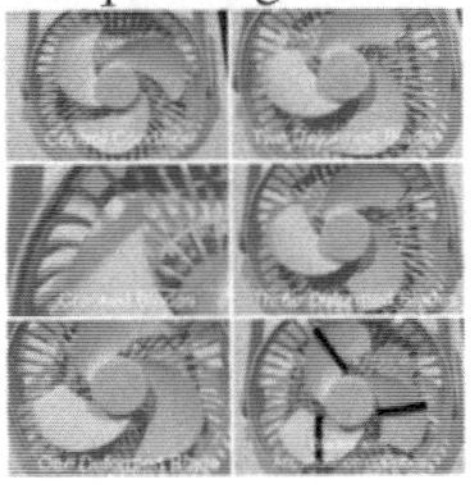

Figure 6. Simulated Conditions Using a Miniature Electric Fan

Figure 7. Approximated Experimental Set-up for the Collection of Simulated Audio Data

The main steps of this method are:

First, define the types of typical fan blade failures to simulate, and simulate the fan blade failures with physical models and record the sound generated from each for the duration of 1 minute and repeat three times for each recording. These recordings should automatically undergo analog-to-digital conversion and be saved in a .wav format. Then, generate spectrograms for each of these sets of audio signals and conduct Mel frequency extractions. After the root data is manipulated and input into a fast multi-class support vector machine for training, several test data recordings are generated, and the fast multi-class support vector machine is used to identify the type(s) of malfunction characteristics involved. This is further illustrated by Figure 8.

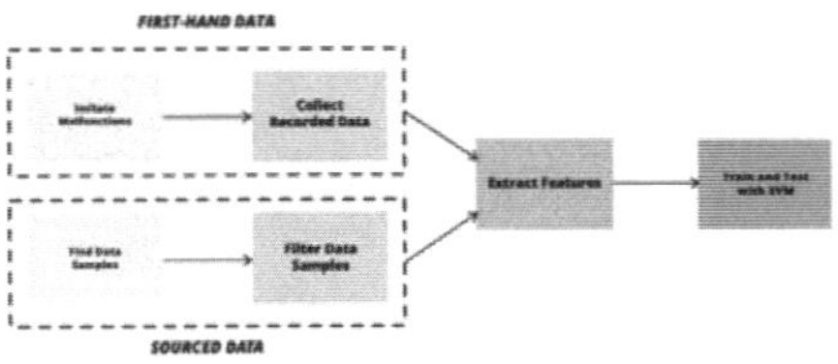

Figure 8. Experimental Process

Firstly, real turbofan engine sounds were collected from ontology and human-labeled datasets for audio events which include around 600 samples. After initial filtration, I obtained 324 .wav files containing audio samples from functional aircraft engines. Secondly, 120 audio samples of a miniature electric fan simulating each of the three malfunctioned conditions were collected and amplified via a professional recording software, then filtered in the same way. Finally, these audio samples were divided into a train dataset and test dataset with a 9:1 ratio respectively. The feature extraction process forms a critical portion of this investigation. Some captures of the spectrograms generated from my first-hand data are shown in Figure 9, which clearly demonstrate the differences in frequency distributions of noise generated by fans under various physical conditions.

Figure 9. Comparison of Spectrograms of First-hand Audio Samples with Fans Under Various Physical Conditions

The experimental results illustrate the exploration of two sets of data to relate the approximated model to real conditions and are detailed in Table 1.

Table 1. Experimental Results

Dataset	*Accuracy*	*AUC*	*F1*
Real Aircraft Engine	72%	0.53	0.70
Miniature Electric Fan	81%	0.65	0.72

From these results, it is evident that our approach performs relatively well for both real and simulated engine sounds in the judgement of aircraft engine malfunctions. The significance of these results lies in the new perspective provided to aircraft engine inspections that are highly accessible and largely simplified. Furthermore, the accuracy of my classification of sounds generated via the miniature electric fan was higher than for real aircraft engine noises, which is largely due to the presence of background noise for realistic scenarios. This raises the question of how background noise should be acknowledged and therefore reduced in future improvements of this proposed method.

5. Conclusion

Traditionally, aircraft engine detection was conducted manually, which was relatively inaccurate and at times destructive. Yet, many audio signal analysis methods proposed later were unable to perform with high efficiencies due to their complexity. Nonetheless, the correlation between the sound generation of turbofan engine blades and their physical conditions has led to many successful attempts in identifying and classifying aircraft engine faults through sound analysis. This paper proposes a fast Multi-Class SVM method to handle the connection between spectrogram and audio signal. Experimental results show that this method is suitable for analyzing engine failures related to fan blades using simulated data or real data. feasible. As the next step, we aim to predict these engine failures through the prediction of engine sound generation instead of sampled audio data.

References

[1] Bertsch, L., Snellen, M., Enghardt, L. et al. Aircraft noise generation and assessment: executive summary. CEAS Aeronaut J 10, 3–9 (2019).
[2] Cilimkovic M. Neural networks and back propagation algorithm[J]. Institute of Technology Blanchardstown, Blanchardstown Road North Dublin, 2015, 15: 1-12.
[3] Duran I, Moreau S, Nicoud F, et al. Combustion noise in modern aero-engines[J]. Aerospace Lab, 2014 (7): p. 1-11.
[4] Gerhart P M, Gerhart A L, Hochstein J I. Munson, Young and Okiishi's Fundamentals of Fluid Mechanics[M]. John Wiley & Sons, 2016.
[5] Gharoun H, Keramati A, Nasiri M M, et al. An integrated approach for aircraft turbofan engine fault detection based on data mining techniques[J]. Expert Systems, 2019, 36(2): e12370.
[6] Lighthill M J. On sound generated aerodynamically I. General theory[J]. Proceedings of the Royal Society of London. Series A. Mathematical and Physical Sciences, 1952, 211(1107): 564-587.
[7] Miljković D. Fault detection for aircraft piston engine by exhaust noise analysis[C]. 2011 Proceedings of the 34th International Convention MIPRO. IEEE, 2011: 756-761.
[8] Palleja Cabre S. Fan proximity acoustic treatments for improved noise suppression in turbofan engines[D]. University of Southampton, 2021.
[9] Roger M, Moreau S, Guedel A. Broadband Fan Noise Prediction Using Single-Airfoil Theory[J]. Noise control engineering journal, 2006, 54(1).
[10] Russell R E, Kester J D. Aircraft noise, its source and reduction[J]. SAE Transactions, 1971: 1295-1314.
[11] Waligórski M, Batura K, Kucal K, et al. Research on airplanes engines dynamic processes with modern acoustic methods for fast and accurate diagnostics and safety improvement[J]. Measurement, 2020, 154: 107460.
[12] Zhao Y P, Huang G, Hu Q K, et al. An improved weighted one class support vector machine for turboshaft engine fault detection[J]. Engineering Applications of Artificial Intelligence, 2020, 94: 103796.

Wireless Communication Strategy with BATS Codes for Butterfly Network

Shiheng Wang, Quan Zhou, Siyuan Yang, Chaoyuan Bai and Heng Liu*

Communication and Sensor Network for Modern Transportation International Cooperation Center (CSNMT), Southwest Jiaotong University, Chengdu, China

*Corresponding author email: hengliu@swjtu.cn

Abstract. In this paper, two transmission schemes for batched sparse(BATS) codes based on a multicast protocol in a butterfly network are studied. For both of the schemes, a source file is first segmented into several packets at the source node, which are then coded with outer coder of the BATS codes to generate potentially unlimited batches. Rather than the linear network, in which the packets are directly transmitted to the next node, the packets will be forwarded to multiple nodes in the butterfly network. All the intermediate nodes which receive the packets recode them with random linear network coding (RLNC). At the destination node receivers, the sink node decodes the packets transmitted from different links. The *Scheme I* requires the intermediate nodes to recode the packets until all of them are received. On the contrary, the intermidate nodes of *Scheme II* just recode the packets as soon as receiving them and forward the recoded packets to the next node immediately. The Belief Propagation(BP) decoders of these two schemes are studied and applied to *Scheme I* and *Scheme II* respectively. The simulation results show that *Scheme I* consumed fewer batches than *Scheme II*, which indicates *Scheme I* outperforms *Scheme II*.
Keywords: BATS codes; BP decoder; Multicast; Butterfly network.

1. Introduction

The BATS codes provide a two-step coding structure for the multi-hop wireless networks, in which the outer coder generates several potentially unlimited batches at the source node and the inner coder applies linear network coding on the packets belonging to the same batch at the intermediate network nodes[1]. The optimization and design issues of BATS codes based on linear network are studied by lots of previous works. [2] studies the normalized rank of BATS codes. [3] studies the degree distribution with better finite length performance. For the BATS code decoding, a BP decoder is presented in [4], in and excellent decoding performance with long code length is studied. The BATS protocols designed for a line network can be also used in a more complicated network with some modifications. [5] turns to a multicast network with one source node and multiple destination nodes, which is called the directed acyclic graph. It discusses multicast networks with homogeneous links and presents a multicast protocol, which can be applied to a general multicast network[6]. However, this protocol may be difficult to implement when the numbers of incoming and outgoing links are large. Therefore, [7] further presents another multicast protocol, which resolves the issues of previous protocol in [5]. BATS codes are also suitable for multi-path transmission to improve reliability and security. BATS codes in two-way relay networks have been studied in [8].

Although [7] presents a multicast protocol suitable for the butterfly network, the detailed scheme and BP decoder still need further discussion. In this paper, two transmission schemes with their BP decoding are designed for a multicast butterfly network presented in [7]. In the proposed schemes, for the intermediate nodes which have only one outgoing link and more than one incoming links, the *Scheme I* requires these nodes to receive all the packets from the same batch through different links. Then these intermediate nodes recode all the received packets and transmit the recoded packets to the following nodes. On the contrary, these nodes of *Scheme II* do not wait for all the packets to arrive. Instead they recode the packets immediately after receiving them from one incoming link, then forward the recoded packets to the next node at once. The simulation results show the *Scheme I* outperforms *Scheme II* with the same BP decoding.

The rest of this paper is organized as follows. BATS codes with BP decoder and a multicast protocol for BATS codes based on butterfly network is illustrated in section II. Section III proposes two transmission schemes with their the BP decoding are studied. The simulation results and performance comparisons between *Scheme I* and *Scheme II* are demonstrated in Section IV. Section V presents the conclusions.

2. Preliminaries and Related Work

2.1. BATS Codes

Assume that the source node needs to transmit a file divided into K packets represented by matrix $\mathbf{B} = [b_1, b_2, ..., b_K]$ to the sink node where the column vector b_j (j=1,2,...,K) is the j-th input packet which contains l bits. As a matrix form of fountain codes, the outer coder firstly generates several encoded packets in batches, each of which contains M encoded packets generated from a subset of the K input packets $\mathbf{B}$. The i-th batch $\mathbf{X}_i$ is generated from a subset $\mathbf{B}_i \subset \mathbf{B}$ of the input packets by the operation

$$\mathbf{X}_i = \mathbf{B}_i \mathbf{G}_i \tag{1}$$

in which $\mathbf{G}_i$ is the generator matrix of the i-th batch whose dimension is $dg_i \times M$ [7]. All the entries in $\mathbf{G}_i$ are i.i.d. and uniformly distributed over the base field. The degree dg_i is a random variable following the degree distribution $\mathbf{\Psi}$, which is a random variable and determines the number of packets contained in the i-th batch. Transmitted with $\mathbf{X}_i$, $\mathbf{A}_i$ is the index set of the packets in $\mathbf{B}_i$, which satisfies $dg_i = |\mathbf{A}_i|$.

The inner code of a BATS code is also called recoding, which applies the linear network coding at the intermediate nodes. The output packets $\mathbf{Y}_i$ can be defined as

$$\mathbf{Y}_i = \mathbf{X}_i \mathbf{H}_i = \mathbf{B}_i \mathbf{G}_i \mathbf{H}_i \tag{2}$$

where $\mathbf{H}_i$ represents the transfer matrix of the i-th batch $\mathbf{B}_i$, which is determined by the recoding, packet loss and the structure of network[7]. The column number of $\mathbf{H}_i$ is the number of recoding packets transmitted to the next node, which is usually no larger than M, and the number of rows is the batch size M.

2.2. BP Decoding of BATS Codes

BP(n) is used for limited batches which works with a given number n of batches and it stops as soon as the n batches consumed[3]. Suppose the time index is t, which starts at 0 and increases by 1 when each decoding step is over. For each batch i and time t, let $\mathbf{G}_i^{(t)}$, $A_i^{(t)}$ and $\mathbf{Y}_i^{(t)}$ be the values of $\mathbf{G}_i$, A_i and $\mathbf{Y}_i$ at time t respectively. Suppose that the i-th batch is decodable at time t if $\text{rk}(\mathbf{G}_i^{(t)}\mathbf{H}_i)=|A_i^{(t)}|$. then the decoded input packets b_j it contains can be substituted into the other batches they contribute to: For each batch i,

- If $j \in A_i^{(t)}$, then $A_i^{(t+1)} = A_i^{(t)} \setminus \{j\}$, $\mathbf{G}_i^{(t+1)}$ is generated by removing the row g of $\mathbf{G}_i^{(t)}$ corresponding to the j-th input packet b_j, and $\mathbf{Y}_i^{(t+1)} = \mathbf{Y}_i^{(t)} - b_j g \mathbf{H}_i$;
- If $j \notin A_i^{(t)}$, then $A_i^{(t+1)} = A_i^{(t)}$, $\mathbf{G}_i^{(t+1)} = \mathbf{G}_i^{(t)}$ and $\mathbf{Y}_i^{(t+1)} = \mathbf{Y}_i^{(t)}$.

2.3. A Butterfly Network and a Multicast Protocol

Linear networks not only have important applications, but also are the building blocks of many complicated networks such as the butterfly network in Figure. 1.

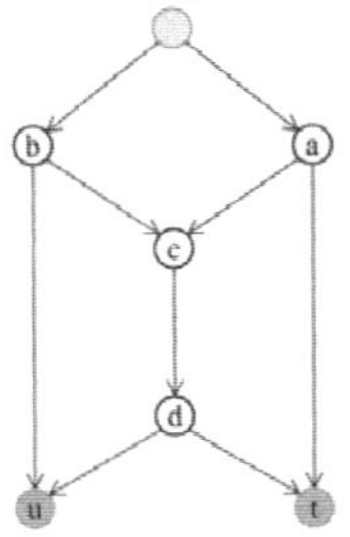

Figure 1. Butterfly network. Node Src represents the source node. Node u and t are the sink nodes.
All the network links in Figure. 1 have unit capacity (per use) and the same have packet loss rate ε, which are said to be homogeneous. The number of recoded packets is assumed to be the same as the batch size to simplify the discussion.

Multicast Protocol I can be illustrated by the butterfly network in Figure. 1. Suppose M is an even number and the source node generates batches which consists of M packets. The source node transmits $M/2$ packets of a generated batch on link (Src, b) and the other $M/2$ packets of the same batch on link (Src, a). For each batch received by node a, b or d, $M/2$ recoded packets are generated and transmitted on both outgoing links. Node c receives packets which belong to the same batch from both of its incoming edges, generates $M/2$ recoded packets, and transmits them on its outgoing edge. Only the packets from the same batch can be recoded by the network coding at the intermediate nodes.

The above protocol just provides a basic framework for the multicast scenario. More details should be considered to support the practical transmission. Therefore, based on this basic framework, two schemes are presented in the following.

3. A BP Decoder in Butterfly Network

This section proposes two transmission schemes and their decoding algorithms based on above multicast protocol framework in the butterfly network as shown in Figure. 1.

3.1. Scheme I

Since the network is symmetrical, the rank distributions of the sink nodes u and t are the same. To simplify the discussion, only the batches received by the sink node t are considered in the following.

When Src node generates a batch $\mathbf{X}_i$, the sink node t will receive the packets transmitted from three different links. The first link is a two-hop link, denoted with (Src, a, t). The remaining two links are four-hop link, denoted with (Src, a, c, d, t) and (Src, b, c, d, t) respectively.

For the proposed scheme, the source packets are encoded with the generated matrix $\mathbf{G}_i$ in (1) at the source node to generate the batches. The batch can be segemented into two parts, both of which contain $M/2$ packets and are transmitted on link (Src, a) and (Src, b) respectively. Therefore $\mathbf{G}_i$ can be rewritten as

$$\mathbf{G}_i = [\mathbf{G}_{1i} | \mathbf{G}_{2i}], \tag{3}$$

where the submatrices $\mathbf{G}_{1i}$ and $\mathbf{G}_{2i}$ have the same dimension of $dg_i \times M/2$. So the packets transmitted on link (Src, a) can be represented as

$$\mathbf{X}_{1i} = \mathbf{B}_i \mathbf{G}_{1i} \tag{4}$$

and the packets transmitted on link (Src, b) can be expressed as

$$\mathbf{X}_{2i} = \mathbf{B}_i \mathbf{G}_{2i}. \tag{5}$$

Consequently, the packets transmitted by the first link (Src, a, t) are

$$\mathbf{Y}_{1i} = \mathbf{B}_i \mathbf{G}_{1i} \mathbf{H}_{1i}, \tag{6}$$

where $\mathbf{H}_{1i}$ is a $M/2\times M/2$ tansfer matrix of the two-hop link (Src, a, t). For the immediate node c, it receives up to the M packets from both a and b, then it will recode them with RLNC to generate $M/2$ packets and deliver them to d. So the packets node t receives from d can be expressed as

$$[\mathbf{Y}_{2i} \mid \mathbf{Y}_{3i}] = \mathbf{B}_i[\mathbf{G}_{1i}\mathbf{E}_1\mathbf{\Phi}_a\mathbf{E}_2 \mid \mathbf{G}_{2i}\mathbf{E}_3\mathbf{\Phi}_b\mathbf{E}_4]\mathbf{\Phi}_c\mathbf{E}_5\mathbf{\Phi}_d\mathbf{E}_6, \tag{7}$$

where $\mathbf{E}_i$ (i = 1, 2, ..., 6) are independent and identically distributed (i.i.d.) random diagonal matrices and their diagonal components are 0 with probability ε and 1 with probability 1- ε. The upper triangular matrices $\mathbf{\Phi}_a$, $\mathbf{\Phi}_b$, $\mathbf{\Phi}_c$ and $\mathbf{\Phi}_d$ are the recoding generator matrices, where all entries are i.i.d. and uniformly distributed over the base field. (7) can be further simplified to

$$[\mathbf{Y}_{2i} \mid \mathbf{Y}_{3i}] = [\mathbf{B}_i\mathbf{G}_{1i}\mathbf{H}^*_{2i} \mid \mathbf{B}_i\mathbf{G}_{2i}\mathbf{H}^*_{3i}], \tag{8}$$

where $\mathbf{H}_{2i}$ and $\mathbf{H}_{3i}$ are the $M/2\times M/4$ tansfer matrix of the four-hop links (Src, a, c, d, t) and (Src, b, c, d, t) respectively. Therefore the packets which the destination node t received can be denoted with

$$\mathbf{Y}_i = [\mathbf{Y}_{i1} \mid \mathbf{Y}_{i2} \mid \mathbf{Y}_{i3}] = \mathbf{B}_i[\mathbf{G}_{1i}\mathbf{H}_{1i} \mid \mathbf{G}_{1i}\mathbf{H}_{2i} \mid \mathbf{G}_{2i}\mathbf{H}_{3i}]. \tag{9}$$

In the butterfly network, the sink node t receives packets from different links, all of which are originated from the same batch. So the BP decoder applied in linear networks is not suitable in the butterfly network. Referring to the BP decoding algorithm under linear network, the modified BP decoder for *Scheme I* in butterfly network is proposed in the following.
For the i-th batch and time t, let $A_i^{(t)}$, $\mathbf{G}_{1i}^{(t)}$, $\mathbf{G}_{2i}^{(t)}$, $\mathbf{Y}_{1i}^{(t)}$, $\mathbf{Y}_{2i}^{(t)}$, and $\mathbf{Y}_{3i}^{(t)}$ be the values of A_i, $\mathbf{G}_{1i}$, $\mathbf{G}_{2i}$, and $\mathbf{Y}_{1i}$, $\mathbf{Y}_{2i}$ and $\mathbf{Y}_{3i}$ at time t respectively. If $\mathrm{rk}([\mathbf{G}_{1i}^{(t)}\mathbf{H}_{1i} \mid \mathbf{G}_{1i}^{(t)}\mathbf{H}_{2i} \mid \mathbf{G}_{2i}^{(t)}\mathbf{H}_{3i}])=|A_i^{(t)}|$, the i-th batch can be decoded with Gaussian elimination at time t, then the decoded input packets b_j can be substituted into the undecodable batches which they contribute to. For undecodable batches, the update rules for *Scheme I*, which can be called UR-I, are shown as follows:

- If $j \in A_i^{(t)}$, then $A_i^{(t+1)} = A_i^{(t)} \setminus \{j\}$, $\mathbf{G}_{1i}^{(t+1)}$ is generated by removing the row g_1 of $\mathbf{G}_{1i}^{(t)}$ corresponding to the j-th input packet b_j. Meanwhile, $\mathbf{G}_{2i}^{(t+1)}$ is generated by removing the row g_2 of $\mathbf{G}_{2i}^{(t)}$ corresponding to the j-th input packet b_j, then $\mathbf{Y}_{1i}^{(t+1)} = \mathbf{Y}_{1i}^{(t)} - b_j g_1 \mathbf{H}_{1i}$, $\mathbf{Y}_{2i}^{(t+1)} = \mathbf{Y}_{2i}^{(t)} - b_j g_1 \mathbf{H}_{2i}$, and $\mathbf{Y}_{3i}^{(t+1)} = \mathbf{Y}_{3i}^{(t)} - b_j g_2 \mathbf{H}_{3i}$;
- $j \notin A_i^{(t)}$, then $A_i^{(t+1)} = A_i^{(t)}$, $\mathbf{G}_{1i}^{(t+1)} = \mathbf{G}_{1i}^{(t)}$, $\mathbf{G}_{2i}^{(t+1)} = \mathbf{G}_{2i}^{(t)}$, $\mathbf{Y}_{1i}^{(t+1)} = \mathbf{Y}_{1i}^{(t)}$, $\mathbf{Y}_{2i}^{(t+1)} = \mathbf{Y}_{2i}^{(t)}$ and $\mathbf{Y}_{3i}^{(t+1)} = \mathbf{Y}_{3i}^{(t)}$.

The update process does not stop until there are no decoded input packets. Meanwhile, the modified BP decoder stops if all the input packets are decoded.
The detail discription of the modified BP decoder in the butterfly network is given in **Algorithm 1**.

3.2. Scheme II
For the intermediate node c, it can recode the received packets and transmit them immediately instead of waiting for the M packets arriving before recoding them in one time. For the link (Src, a, t), the packets which the sink node t receives are the same as (6). The packets arrived at destination node t from the links (Src, a, c, d, t) and (Src, b, c, d, t) can be respectively reprenstend as

$$\mathbf{Y}_{2i} = \mathbf{B}_i\mathbf{G}_{1i}\mathbf{E}_1\mathbf{\Phi}_a\mathbf{E}_2\mathbf{\Phi}_{c1}\mathbf{E}_3\mathbf{\Phi}_{d1}\mathbf{E}_4 \tag{10}$$

and

$$\mathbf{Y}_{3i} = \mathbf{B}_i\mathbf{G}_{1i}\mathbf{E}_5\mathbf{\Phi}_a\mathbf{E}_6\mathbf{\Phi}_{c2}\mathbf{E}_7\mathbf{\Phi}_{d2}\mathbf{E}_8, \tag{11}$$

where $\mathbf{\Phi}_a$ and $\mathbf{\Phi}_b$ are the same as those in (7), $\mathbf{\Phi}_{c1}$ and $\mathbf{\Phi}_{c2}$ are the $M/2\times M/4$ recoding generator matrices, $\mathbf{\Phi}_{d1}$ and $\mathbf{\Phi}_{d2}$ are the $M/4\times M/4$ recoding generator matrices. (10) and (11) can be further rewritten as

$$[\mathbf{Y}_{2i} \mid \mathbf{Y}_{3i}] = [\mathbf{B}_i \mathbf{G}_{1i} \mathbf{H}^*_{2i} \mid \mathbf{B}_i \mathbf{G}_{2i} \mathbf{H}^*_{3i}] \tag{12}$$

where $\mathbf{H}^*_{2i}$ and $\mathbf{H}^*_{3i}$ are the $M/2 \times M/4$ tansfer matrices of the four-hop links (Src, *a*, *c*, *d*, *t*) and (Src, *b*, *c*, *d*, *t*). Therefore the packets which the destination node *t* received are

$$\mathbf{Y}_i = [\mathbf{Y}_{i1} \mid \mathbf{Y}_{i2} \mid \mathbf{Y}_{i3}] = \mathbf{B}_i [\mathbf{G}_{1i} \mathbf{H}_{1i} \mid \mathbf{G}_{1i} \mathbf{H}^*_{2i} \mid \mathbf{G}_{2i} \mathbf{H}^*_{3i}]. \tag{13}$$

The update rules for *Scheme II* based on above multicast protocol (UR-II) are similiar to UR-I, in which $\mathbf{H}_{2i}$ and $\mathbf{H}_{3i}$ should be replaced with $\mathbf{H}^*_{2i}$ and $\mathbf{H}^*_{3i}$.

Algorithm 1: the modified BP decoder in butterfly network

Input: $\mathbf{Y}_i$.
Initialization: $i = 0$.
1 **while** not all input packets are decoded **do**
2 fetch a new batch $\mathbf{Y}_i$;
3 **if** $\mathbf{Y}_i$ is decodable
4 Apply Gaussian elimination to $\mathbf{Y}_i$;
5 update the undecodable batches by UR-I;
6 **else**
7 Store $\mathbf{Y}_i$ temporarily;
8 **end**
9 $i = i+1$;
10 **end**
Output: Decoded packets.

4. Simulation Results

The performance of the proposed scheme are simulated, in which *Scheme I* and *Scheme II* are compared with each other. For the convenience of comparison, the degree dg_i is assumed to be fixed. The simulation parameters are listed in **Table 1**.

Table 1. Simulation Parameters

Simulation parameters	Explanation	Values
K	Input packets number	100
M	Batch number	32
l	Packet length	100
dg_i	Fixed degree value	19, 20, 21, 22
ε	Packet loss	0.14, 0.16, 0.18, 0.20, 0.22

Figure. 2 provides the numbers of batches consumed versus the packet loss rate when degree dg_i are 19 and 21. It can be seen from Figure. 2 that the number of batches consumed by *Scheme I* is less than that consumed by *Scheme II*. The gap between the two curves becomes larger when the value of ε increases under the same degree, which means that *Scheme I* outperforms *Scheme II*, espcially in the case of poor channel conditions.

Figure. 3 provides the numbers of batches consumed versus the packet loss rate when dg_i are 20 and 22. It can be observed from Figure. 3 that *Scheme I* still outperforms *Scheme II*. Moreover, the gap between the two curves becomes larger when the degree increases.

In summary, the modified BP decoder is suitable for *Scheme I* and *Scheme II*. The number of transmission times consumed by *Scheme I* is less than that of *Scheme II* under different degrees, it is more obvious when degree is large. Meanwhile, the interval between the two curves becomes larger under the same degree with the increase of ε. So *Scheme I* performs better than *Scheme II*, especially when degree and packet loss rate ε becomes larger

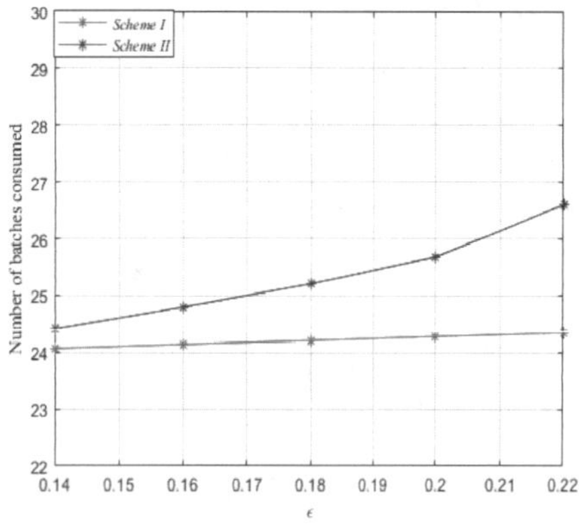

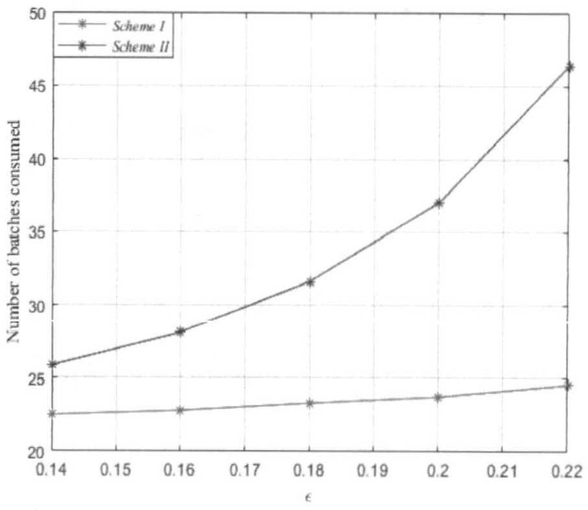

Figure 2. Consumed batches when dg_i=19 and 21. **Figure 3.** Consumed batches when dg_i=20 and 22.

5. Conclusions

This paper presented two transmission schemes for the batched sparse (BATS) codes for the multicast butterfly network over *GF(2)*, which are called *Scheme I* and *Scheme II* respectively. The modified BP decoders are further presented for these two schemes. The simulation results show that to transmit a complete file successfully, the number of batches consumed by *Scheme I* is less than that consumed by *Scheme II*, which indicates the *Scheme I* is more suitable than *Scheme II* in practical applications.

Acknowledgments

This research was supported by Sichuan Science and Technology Program (NO.2020YFG0303 and NO.2020YFH0111), and Chengdu Science and technology Program (2019-YF05-02657-SN).

References

[1] Tang L, Liu H, Yang L, Ma Z, and Xiao M 2020 Analysis for rank distribution of BATS codes under time-variant channels *IEEE 91st Vehicular Technology Conference (VTC2020-Spring)* pp 1-5.

[2] S Yang, R W Yeung, J. H. F. Cheung and H H F Yin BATS: network coding in action *2014 52nd Annual Allerton Conference on Communication, Control, and Computing (Allerton)* IEEE, 2014.

[3] S Yang, T C Ng and R W Yeung Finite-length analysis of BATS codes *IEEE Transactions on Information Theory*, vol 64, no 1, pp 322-348.

[4] S Yang and R W Yeung Coding for a network coded fountain *2011 IEEE International Symposium on Information Theory Proceedings*, pp 2647-2651.

[5] S Yang and R W Yeung 2014 Batched sparse codes *IEEE Transactions on Information Theory* vol 60, no 9, pp 5322-5346.

[6] X Xu, Y L Guan, Y Zeng and C C Chui 2017 Spatial-temporal network coding based on BATS code *IEEE Communications Letters*, vol. 21, no. 3, pp 620-623.

[7] S Yang and R W Yeung BATS Codes: Theory and Practice. *Synthesis Lectures on Communication Networks*, vol. 10, no. 2, pp 1-226.

[8] Q Huang, K Sun, X Li, and D O Wu 2014 Just fun: A joint fountain coding and network coding approach to loss-tolerant information spreading *In Proc. of the 15th ACM International Symposium on Mobile Ad Hoc Networking and Computing, MobiHoc'14, ACM*, pp 83–92.

Design of A Novel 1553B Data Communication Board Based on DSP and Integrated 1553B Terminal

Jinpeng Yang, Dong Liang, Shuaibing Wang and Tao Yang*

China Academy of Aerospace Aerodynamics, Beijing, 100074, China

*Corresponding author email: sdyangtao@qq.com

Abstract. In order to make the communication cables in airborne equipment not as complicated, large and heavy as before, the MIL-STD-1553B standard was released in the 1970s, thus reducing aircraft weight. Application of 1553B bus is becoming wider and wider. In recent years, UAV is a new force in modern battlefields and local conflicts. At the same time, the 1553B bus is prevalent in the domain of UAV for communication between different devices. A novel 1553B data communication board based on DSP and integrated 1553B terminal is designed in this paper. The principle of 1553B bus is introduced at first. And then it presents the hardware design as well as the software design. In the end, the conclusion part summarizes its applications. Undoubtedly, the 1553B board designed in this paper is convenient to be integrated to the whole system, providing a real-time, reliable and flexible communication solution for UAV.
Keywords: 1553B; UAV; DSP; Integrated terminal.

1. Introduction

Before the 1960s, there was no standard universal data bus used by airborne equipment and a huge amount of cables were needed for communication between different equipment. With the technology development, the airborne equipment is becoming more and more complex. The corresponding communication cables are large and heavy. At the same time, definitions and tests of the transmission cables are relatively complicated with high cost. In order to tackle this problem, in the 1970s, the MIL-STD-1553B standard was published, which substituted the massive devices of data communication between the transducer, actuator, computer and so on, which reduces aircraft weight a lot. Advantages of the 1553B bus are rapidity, flexibility, high communication efficiency as well as easy modification, expansion and maintenance. Because of excellent performance of the 1553B bus, along with the transformation and upgrading of weapons and equipment, not only is the 1553B bus used widely in aviation weapons and equipment, but also its application extends to weapons and military supplies of other corps such as army and navy forces, even to all kinds of the space aircrafts, ships, ground equipment and so on.

In modern wars and local conflicts, thanks to its advantages such as no casualty, little limitations, great concealment, high efficiency and cost ratio, military UAV (Unmanned Aerial Vehicle) is a force to be reckoned with in the air [1]. Military UAV features compact structure, convenient operation and strong maneuverability. It can carry different payloads for battlefield scout, electronic jamming and air strike with cluster bombs as well as guided missiles. In the future, various types of military UAV are capable to complete all kinds of strategic and tactical missions, including tracking and positioning, precision guidance, special operations, battlefield rescue and so on [2]. In order to secure great performance, communications between different devices are quite important. In this context, a novel 1553B data communication board is designed for UAV in this paper. A high performance DSP (Digital Signal

Processor) is selected as the control core, with an integrated 1553B terminal as the communication peripheral [3]. The board serves as an independent module and communicates with the flight mission control computer via RS422 interface.

2. The 1553B Bus

2.1. Bus Structure

The 1553B Bus mainly consists of two parts. One part is the terminal load in the data bus while the other is the transmission media. There are three kinds of terminals, namely, the BC (Bus Controller), RT (Remote Terminal) and MT (Monitor Terminal) [4]. The bus structure is shown in Figure 1.

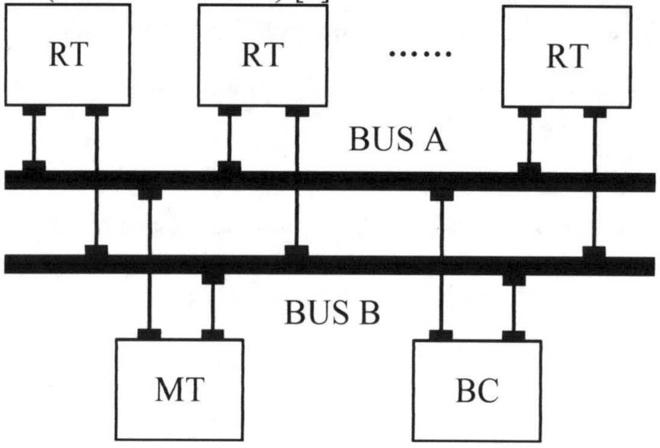

Figure 1. Bus Structure of the 1553B Bus

The dual-redundant 1553B bus consists of Bus A and Bus B to increase reliability of data transmission. In practice, one of the data buses is at work, while the other serves as hot backup. RT is the interface between each device hanging on the 1553B bus network, which is used for information transmission between the bus and the subsystem. Each RT has its own RT address, and there are 31 RTs at maximum in one certain network. BC is the control core of information transmission in the 1553B bus network. While it's at work, BC initiates the bus network and controls data stream. At the same time, according to BC command, RT begins to work, transmits information and reports working status back to BC. BM, a terminal that records information in the bus network, is also controlled by BC, which mainly monitors the current bus status online as well as collects remote data. All terminals in the 1553B bus network share one certain information path, and each of them plays a unique role. At any time, there is only one terminal transmitting data in the bus network while the other terminals are waiting for receiving. It is the terminal address that decides which terminal to receive. This network structure is pretty simple. System paralysis will not be caused by failure of any terminal. To summarize, advantages of the 1553B bus network are as follows [5]:

- Linear topology: easy maintenance, convenient nodes addition or deletion, improved system flexibility.
- Reliable transmission protocol: asynchronous operation in an instruction-to-response manner.
- Redundancy and fault tolerance: hot backup in the transmission lines and bus interfaces with increasing reliability.
- Real-time transmission: 1 Mb/s transmission rate with up to 32 words in each message.

2.2. Word Format and Communication Protocol

2.2.1. Word Format. The data flow in the 1553B bus consists of different messages. The message can be classified into three types with different types of words: command word, data word and status word. The word is a sequence of bits, and the length of command word, data word or status word is all 20 bits, including synchronization head (3 bits), valid data bit (16 bits) and parity bit (1 bit). Detailed definition is shown in Figure 2.

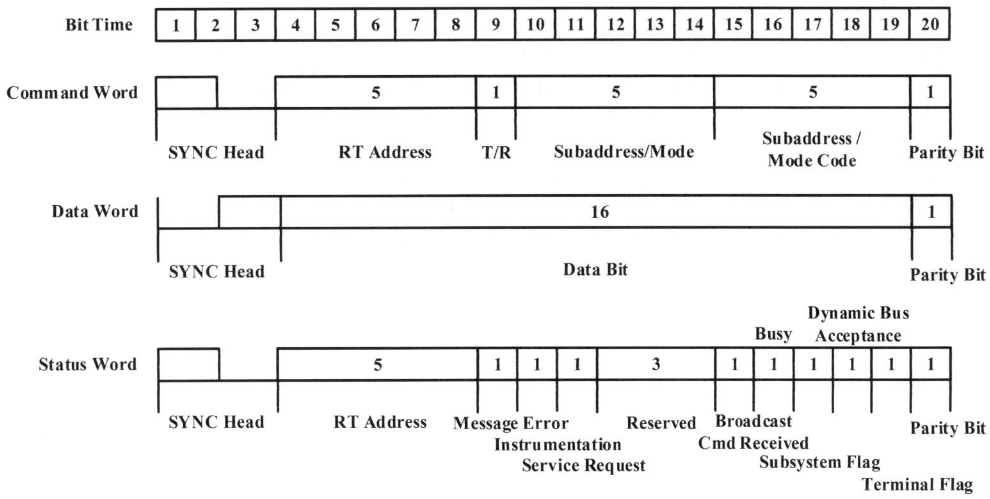

Figure 2. Word Definition of 1553B Bus

2.2.2. Communication Protocol. Each message in the 1553B bus network is composed of one command word, one status word and several data words, numbers of which vary from 0 to 32. These words construct a transmission sequence within status response intervals. In light of functions, there are 10 kinds of message formats, namely, BC to RT, RT to BC, RT to RT, mode instruction with data word (transmission), mode instruction with data word (reception), mode instruction without data word, broadcast from BC to RT (reception), broadcast from RT to RT, broadcast mode instruction with data word, broadcast mode instruction without data word. The first six messages are controlled by BC directly, requiring specific responses from accessed RT. The last four messages are broadcasts, allowing message transmission from BC or one certain RT to the other terminals without status confirmation.

3. Hardware Design

The hardware design of the 1553B data communication board is presented in the following contents one circuit by one circuit. There are the power circuit, the control circuit, the internal communication circuit, and the 1553B bus circuit. Power supply is provided by the power circuit. The digital control is performed by the control circuit. Communication between the control circuit and the 1553B bus circuit is conducted by the communication circuit. In the meanwhile, the communication circuit helps the 1553B board to communicate with other devices. The 1553B bus circuit contains the integrated 1553B terminal and its corresponding transformers.

3.1. Power Circuit

As shown in Figure 3, the power supply for different circuits in the 1553B data communication board is generated. Right after a TVS (Transient Voltage Suppressor) for suppression of any unexpected transient voltage over the limit which the components can tolerate in the board, the +5V power serves as the main power source of the whole board. And then it is sent to two regulators as input of the secondary power. +3.3V is transformed by a low dropout regulator for the control circuit and the internal communication circuit. Also, the +5V transforms into +1.8V as the power source of the other part of control circuit.

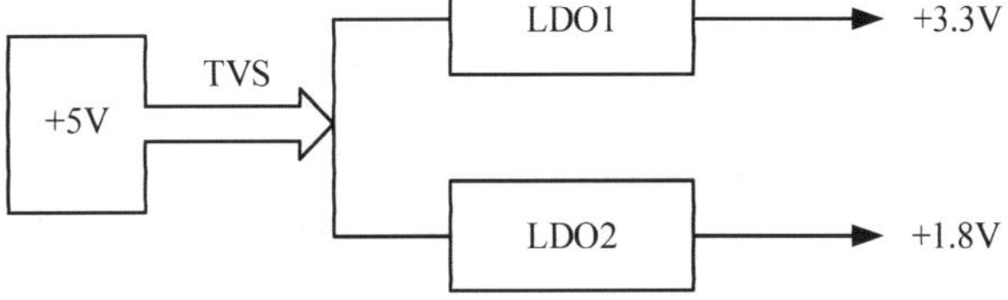

Figure 3. Power Circuit

3.2. Control Circuit

As shown in Figure 4, DSP is the core of the control circuit [6]. As for the design in this paper, TMS320F28335 is selected to conduct the digital control. TMS320F28335 is a high-performance 32-bit digital signal controller, with control frequency up to 150MHz [7]. It has an on-chip memory with 256K×16 Flash and 36K×16 SARAM, which makes software programming quite easy and convenient. There are several control and port peripherals, making it a perfect choice for digital control in this 1553B board design. The funtion modules used in DSP are XINTF (External Interface), GPIO (General-purpose Input and Output Multiplexer) and SCI (Serial Communication Interface). SCI module connects the RS422 transceiver for communication with other devices such as flight mission computer, control computer and so on. The control core and the 1553B circuit is linked via XINTF module with 16-bit address and 16-bit data, as well as GPIO module with input and output bits for control or status feedback from the 1553B circuit.

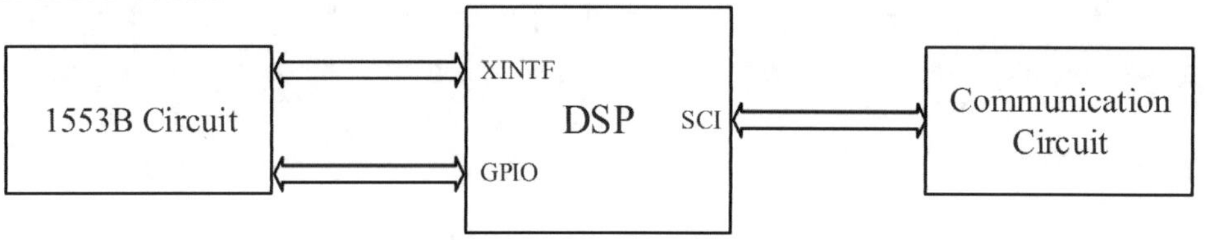

Figure 4. Control Circuit

3.3. Communication Circuit

The communication circuit conducts communication between the 1553B data communication board with other devices exteriorly, while serving as the medium between the control circuit and the 1553B circuit interiorly. The communication circuit is shown in Figure 5. In order to realize signal isolation, an RS422 transceiver featuring isolation is chosen as the interface chip. The internal communication between the control circuit and the 1553B circuit requires signal level shifts. Signals from the control circuit are of +3.3V level, while signals from the 1553B circuit are of +5V level. As a result, the level-shifting transceiver completes the task.

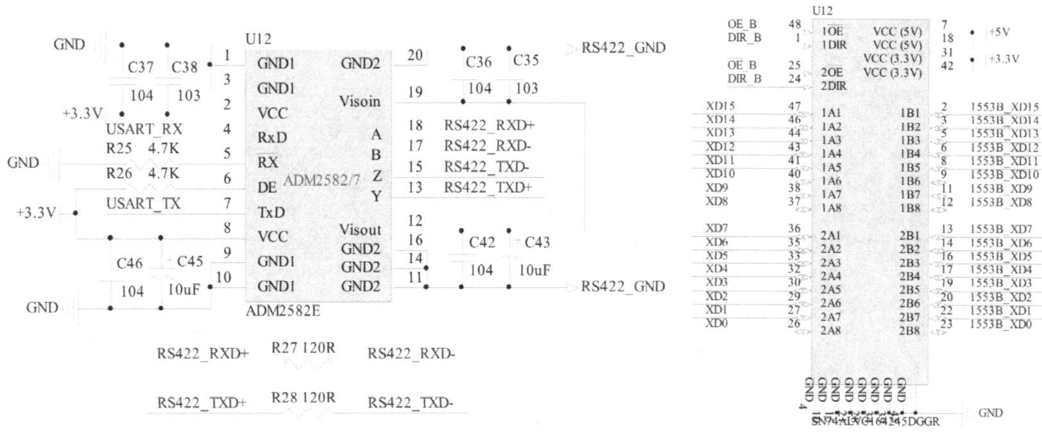

Figure 5. Communication Circuit

3.4. 1553B Bus Circuit

As shown in Figure 6, the 1553B bus circuit comprises the integrated 1553B terminal and two transformers, which are located in channel A and channel B respectively and the coupling voltage ratio is 1:2.5 [8,9]. The integrated 1553B terminal in this paper is BU-61580, a terminal widely used in the 1553B bus network design. It can be used as BC, RT or MT as required with extremely strong internal functions, flexible interface and easy control.

Interface between the integrated 1553B terminal and the external memory unit is relatively flexible. Not only can it communicate with 8-bit microprocessor, but also it is capable to work with 16-bit microprocessor, just requiring a few control logic circuits. There are six kinds of typical interfaces, including 16-bit buffering, 16-bit transparent, two-port RAM, 16-bit direct memory access, 16-bit DMA with extended logic, and 8-bit buffering. Among them, the 16-bit buffering interface is used to interact with DSP in this paper. It provides a buffering interface that shares RAM with DSP, in which the address and data buffer of the integrated terminal is isolated from DSP's address and data bus. The RAM address space is limited in the range of 4K RAM. As mentioned above, the integrated terminal communicates with DSP via 16-bit address and 16-bit data. In addition, there are some I/O signals for signal control, RT address selection and status feedback.

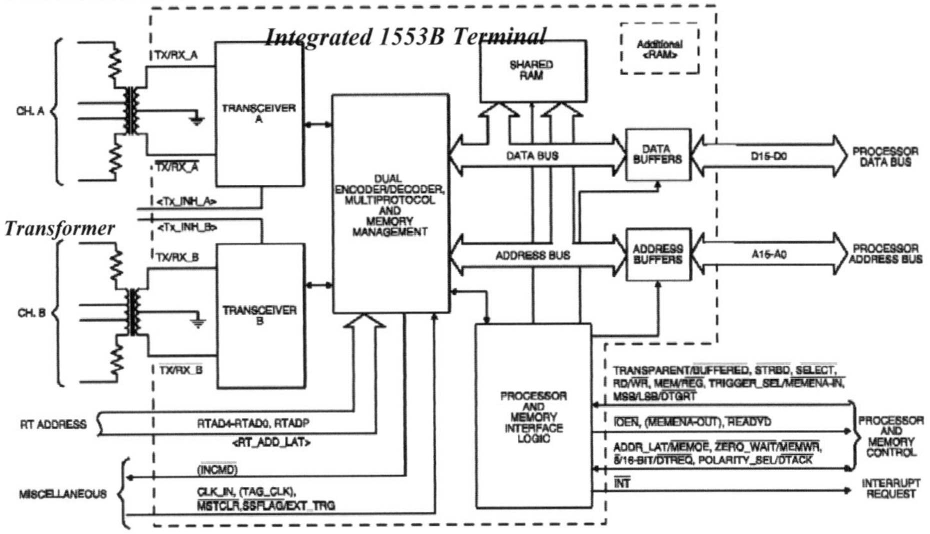

Figure 6. Structure of 1553B Bus Circuit

4. Software Interface Design

DSP interacts with the integrated 1553B terminal in the way of handshake and responds to the interrupt request. During DSP initialization after power up, certain configuration has to be made, which samples handshake signals, sets the synchronous mode and configures the external input triggered by falling edge. Taking the 1553B terminal's accessing sequence into consideration, the interface program uses NOP instruction to suspend DSP's continuous access in case of read or write failures [10]. Right after software reset, the 1553B terminal can enter BC, RT or MT mode. The 1553B board in this paper mainly works in the RT mode, and the software flow diagram is show in Figure 7.

4.1. Mode Initialization

As for RT mode, the mode initialization of the 1553B integrated terminal includes register configuration, initialization of sub-address query table and control word, and setup of illegal instruction area. The specific process is as follows [11]:

- Configure start/reset register.
- Configure configuration register #3 and enable enhanced mode.
- Configure interrupt mask register and enable sub-address control word interrupt.
- Configure configuration register #2 and enter enhanced RT memory management mode.
- Configure configuration register #3 and enable enhanced mode code handling.
- Initiate stack area and reset address space used for message status.
- Initiate mode code interrupt table and the sub-address query table.
- Configure address control word and assign interrupt condition as well as message storage pattern for each sub-address.
- Initiate the busy table of each sub-address and setup illegal instruction area.
- Configure configuration register #1 and enter RT mode.

After initialization, the 1553B terminal is online, ready to respond to command sent by BC controller, transmitting and receiving data in the 1553B bus network.

4.2. Data Reception Process

When the 1553B terminal receives messages from 1553B bus, the terminal requests for DSP interrupt if the corresponding interrupt can be triggered, and then DSP processes the messages. In the interrupt program, DSP reads interrupt status register of the 1553B terminal at first in order to judge whether the interrupt is triggered by interference or not. If not, DSP reads the status word of message blocks to decide whether it is an illegal command or not. If not, DSP reads the status word register to find whether the sub-address is busy or not. If not, DSP reads the T/R bit in the command of message description to judge whether it is a process of transmission or reception. If it is reception, according to the data block pointer in the message descriptor, the data reception block has to be found and data in it is sent to DSP. The number of reception data is increased by one and then the interrupt exits.

4.3. Data Transmission Process

Since RT is not able to initiate transmission of bus message, the format organization of messages with RT as data source, for instance, RT to BC message and RT to RT message, is more complex than that of receiving messages. The 1553B bus controller organizes two kinds of messages to be transmitted, which are classified into the periodic message and the non-periodic message. As for messages uploaded periodically by RT, DSP fills in the transmit buffer of the corresponding sub-address in the 1553B terminal with data. And then DSP sets the number of data packets for transmission, clears the "send /busy" bit, and waits for BC to take the data. When the transmission interrupt generates, the data is uploaded successfully with the number of transmission data reduced by one. As for non-periodic messages, BC adds the mode command which demands RT to send vector words during the intervals between different periodic messages.

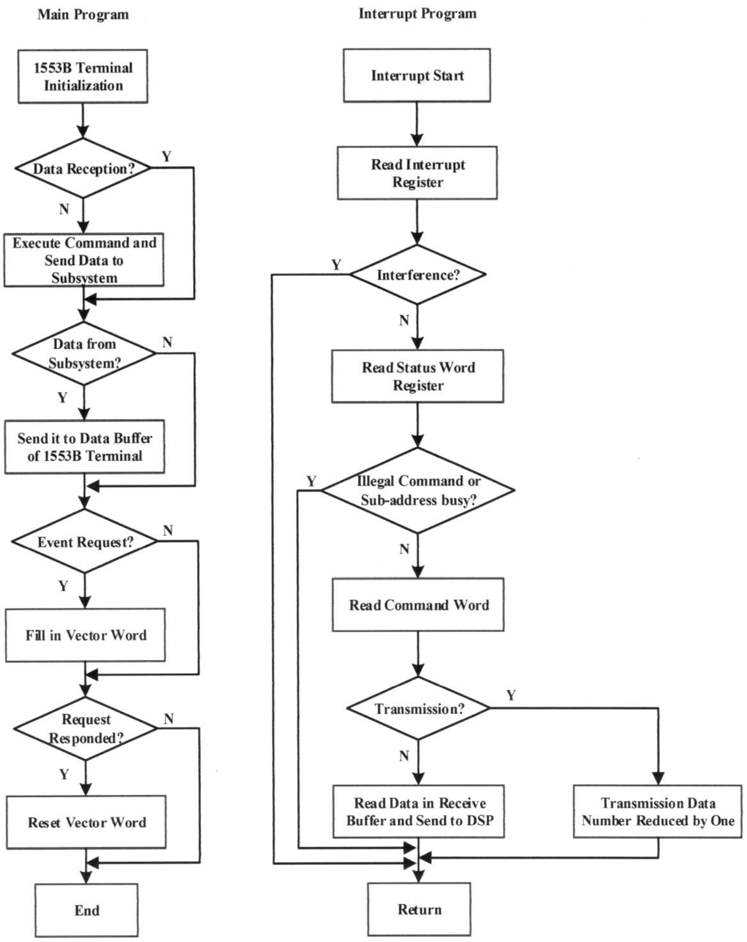

Figure 7. Flow chart of the embedded software

5. Experimental Test

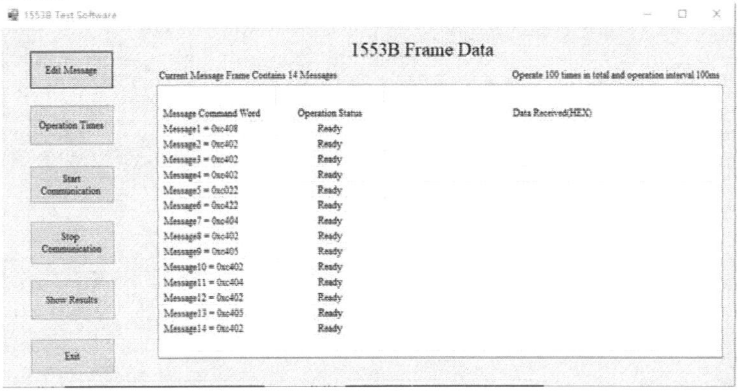

Figure 8. Upper Computer Software

In order to verify the 1553B data communication board designed in this paper, an upper computer software is developed as shown in Figure 8. The experimental tests include the command terminal control as well data transmission and reception. Results show the remote terminal receives and uploads data in real time immediately after responding to the control command. The tests are conducted 100 times with no failure. The success rate is 100%.

6. Conclusion

The 1553B standard is published in context of aircraft development. Instead of large and heavy communication cables, the 1553B bus network facilitates communication between different devices, with several advantages such as low cost, light weight, easy application and high efficiency. UAV is becoming more and more important in military fields. In order to complete complex flight missions, various devices collaborate with each other. The communication between them is quite significant. The 1553B bus network is a perfect choice in this application. A novel 1553B data communication board based on DSP and the integrated 1553B terminal is designed in this paper. The hardware circuit is introduced in detail and the corresponding software interface of RT mode is presented. The 1553B board is pretty convenient for communication with other devices as a module because of its easy integration. The main contribution of the 1553B data communication board is that it serves as a perfect substitute for the previous data bus in the UAV communication. In the future, it will be equipped in one certain UAV. And several flight tests are on the way to verify its performance.

References

[1] Office The Secretary of Defense 2005 *Unmanned aircraft systems roadmap* (Wash-ington, DC) pp 2005-2030

[2] Stefan Nitschke 2004 *Unmanned Aerial Vehicles-Capabilities And Limitations (Naval Forces* vol 2*)* pp 62

[3] Xiancheng Zheng, Guojun Zhang and Xiaobin Zhang 2008 *Design of 1553B Bus Interface Based on TMS320F2812 and DSP / BIOS* (*Measurement and Control Technology* vol 27) pp 53-58

[4] Feng Gao 2006 *Design of the MIL-STD-1553B RT Based on BU-61580 (Design Field* vol 8*)* pp 84-88

[5] Hongcai Zhang and Xuan Hao 2018 *Bus Competion Analyzing and Resolv Based on 1553B* (*Microcontroller and Embedded Systems* vol 10) pp 8-10

[6] Tao Liu, Chong Zhang and Zhihong Tan 2017 *Design and Realization of 1553B Bus System based on High- powered DSP* (*Aeronautical Computing Technique* vol 47) pp 109-112

[7] Zhang Jia 2020 *Design of MIL-STD-1553B Bus Test System Based on DSP* (*Automation and Instrumentation* vol 35) pp 61-66

[8] Hao Zhang, Weidong Ye and Lin Qi 2000 *Realizing the Connection from Processor to 1553 Bus by BU-61580* (*Measurement and Control Technology* vol 19) pp 58-59

[9] Yuanlin and Zhong Wu 2010 *Design of 1553B Bus Interface based on BU-61580 Computer Engineering and applications* (*Computer Engineering and Applications* vol 46) pp 65-68

[10] Xinliang Wang, Kai Chen, Qiqi Xue, Minzhou Dong and Minggang Wang 2020 *Design of Testand Simulation Platform for Flight Control Software Based on 1553B Bus* (*Computer Measuremnt and Contorl* vol 28) pp 12-15

[11] Shi Ran, Zhang Yongjie and Wang Jinfang 2018 *Design and Implementation of 1553B Bus with Double Buffering Mode* (*Navigation Positioning and Timing* vol 15) pp 83-89

Outage Probability Analysis for Two-Way Amplify-and-Forward Mobile Relay Assisted Device-to-Device Communication in Rayleigh Fading Channel

Weipeng Wang[1,*], Jihong Zhao[1,2] and Hua Qu[1]

[1] School of Electronic and Information Engineering, Xi'an Jiaotong University, Xi'an 710049, China
[2] School of Telecommunication and Information Engineering, Xi'an University of Posts and Telecommunications, Xi'an 710061, China

*Corresponding author email: wangweipeng88@stu.xjtu.edu.cn

Abstract. Device-to-Device (D2D) communication underlaying cellular network is potential to enhance the spectral efficiency, reduce the energy consumption and improve individual quality-of-service (QoS) experience. But intra-cell orthogonality is impaired due to spectral reusing and severe co-channel interference could be imposed between cellular and D2D links. Besides, the direct links between local communicating pairs may be unfavorable and suffer an unsatisfied QoS experience since the randomness of channel condition. So, we are focused on analyzing the outage probability of D2D links assisted by two-way relays on Rayleigh fading channel. Specifically, we consider an amplify-and-forward relay protocol at relay due to their limited processing ability, and devise power control at both D2D transmitter and relay to maintain the QoS requirement of prioritized cellular links. The simulation results verify the correctness of the outage probability analyses for D2D links, and indicate the performance gain when relays are introduced to facilitate D2D links.
Keywords: Device-to-Device (D2D); Spectral efficiency; Outage probability; Two-way relays.

1. Introduction

Recently, with the increasing demand for high-speed data services under short-distance communication, the resulting problems of data congestion and spectrum resource scarcity have stimulated research on ways to have higher spectrum reuse efficiency[1,2]. It is predicted that by 2022, the number of wireless connected devices will over 50 billions[3]. The overcrowded spectrum triggers research for new technologies for efficient utilization of the spectral resources[4]. D2D communication can improve spectrum efficiency and transmission rate since reusing radio spectrum resources without relay base station. In addition, it can also reduce terminal transmit power and shorten transmission time-delay[5-7]. However, D2D communication has some negative effects as follows: a) the QoS of cellular communication will be suffered sharply decreased since co-channel interference caused by spectrum reuse[5,8-10]; b) due to the randomness of channel condition and user location, the D2D pair link may be unfavorable and suffer an unsatisfied QoS experience when quality of channel is bad or the distance between D2D users is too far[11]. Especially, D2D communication has become an important communication way in 5G and the frequency band of future 5G network has expanded to 30-300GHz, which will weaken the ability of radio waves to bypass obstacles and penetrate solid matter[12].

In case a), the following methods have been proposed to reduce the co-channel interference in a cellular cell. In [13], the transmit power of D2D link users is controlled to minimize the interference to traditional cellular users by ensuring the minimum communication requirements of D2D link. In [14], the transmit power of D2D link users is controlled to maximize the performance of D2D link through QoS guarantee for the traditional cellular users. Both [13] and [14] are reuse uplink resource. In [15], in order to reduce interference, it allows a D2D transmitter to act as an in-band relay for a cellular link and at the same time transmit its own data by employing superposition coding in the downlink. However, all of them only consider the uplink resource or downlink resource separate reused and don't reuse these two resource simultaneously.

In case b), [16] uses the idle users as relay nodes to improve the link communication quality and assist the communication of D2D users. One-way amplify-and-forward (AF) and one-way decode-and-forward (DF) are the common relay protocols be used in the past and they need four time slots. At present, two-way amplify-and-forward and two-way decode-and-forward are the popular relay protocols, that because they only need two time slots which means the communication delay will be reduced[17,18]. Considering the relay has limited processing ability, we use two-way amplify-and-forward relay protocol in this paper.

In summary, through controlling co-channel interference and adopting two-way amplify-and-forward relay protocol can solve the case a) and b). In [19], it uses two-way DF relay protocol and analyzes the outage probability by dividing the distance from the relay to the user side into symmetric and asymmetric cases, but it doesn't resolve the co-channel interference which caused by D2D communication. [20] uses two-way AF relay protocol and computes the D2D link outage probability, but it also doesn't consider the co-channel interference. So we joint interference limitation and two-way AF relay protocol to study the outage probability of D2D link.

The contributions of this paper can be summarized as follows:

- Considering the limited processing ability of relay, the two-way AF relay protocol are applied to relay assist D2D link.
- Compared with conventional outage probability of D2D link methods, we propose a joint interference limitation and two-way AF relay protocol to analyze the outage probability of D2D link.
- The simulation results verify the correctness of our outage probability analyses for D2D link, and indicate the performance gain when relays are introduced to facilitate D2D links

The rest of this paper is organized as follows. Section 2 describes the system model. In Section 3, the power control at both D2D transmitters and relays are analyzed, and the outage probability of D2D link is derived. Furthermore, the outage probability analysis and verification have been simulated in Section 4. Finally, Section 5 concludes this paper.

2. System Model

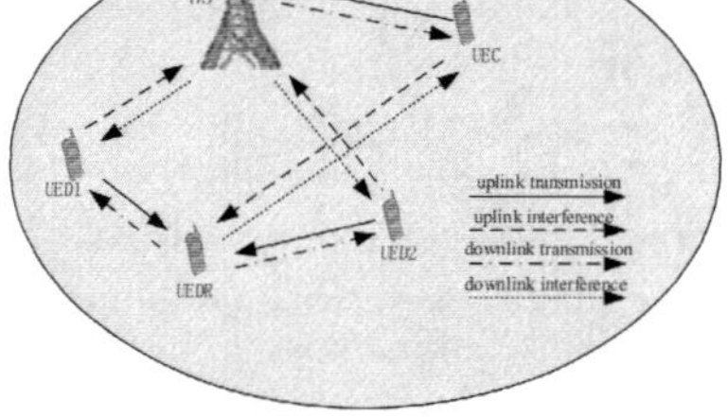

Figure 1. Two way AF relay assisted D2D communication model

The D2D communication under single cell of cellular is the matter that we focused on and the two-way AF relaying strategy is used in this paper. The system include one base station(BS), a pair D2D users (UED1 and UED2), relay nodes (UEDR) and cellular users (UEC) as shown in figure.1. We assume UED1, UED2, UEDR and BS all know the complete channel state information, and they can eliminate the self-interference. Different Rayleigh fading channels are i.i.d.

In this paper, we propose a D2D communication using a two-way AF relaying strategy by another D2D user. In particular, we consider an isolated cell scenario that includes one base station(BS), a pair of D2D users (UED1 and UED2), a relay D2D user (UEDR) and a cellular users (UEC) over a Rayleigh fading channel. As show in figure 1, UED1 and UED2 exchange the messages with each other via UEDR using the two-way AF protocol. We assume UED1, UED2, UEDR and BS know the complete channel state information (CSI) and all terminals can cancell self-interference.

We denote the transmitting power of cellular user i as p_i^C, send signal as s_i^c. The transmitting power of D2D link j are p_j^{D1} and p_j^{D2}, s_j^{D1} and s_j^{D2} denote their send signal. p_r^R is the transmitting power of relay r, s_j^R represents its send signal. The transmitting power of BS is p^{BS} and send signal is s^{BS}. It is assumed that the noise n_0 is an additive complex Gaussian noise $\sim\mathbb{N}(0,N_0)$. We assume all of the transmitting channel are Rayleigh fading channel, and the channel power gain can be expressed as $\left|h_{i,j,r}^{D1R}\right|^2=(\left|h_{i,j,r}^{D1R0}\right|^2\cdot(d_{j,r}^{D1R})^{-\alpha})/PL_0$,

$\left|h_{i,j,r}^{D2R}\right|^2=(\left|h_{i,j,r}^{D2R0}\right|^2\cdot(d_{j,r}^{D2R})^{-\alpha})/PL_0$, $\left|h_{i,j,r}^{CR}\right|^2=(\left|h_{i,j,r}^{CR0}\right|^2\cdot(d_{i,r}^{CR})^{-\alpha})/PL_0$, $\left|h_{i,j,r}^{D1BS}\right|^2=(\left|h_{i,j,r}^{D1BS0}\right|^2\cdot(d_j^{D1BS})^{-\alpha})/PL_0$,

$\left|h_{i,j,r}^{D2BS}\right|^2=(\left|h_{i,j,r}^{D2BS0}\right|^2\cdot(d_j^{D2BS})^{-\alpha})/PL_0$, where $h_{i,j,r}^{D1R}$, $h_{i,j,r}^{D2R}$, $h_{i,j,r}^{CR}$, $h_{i,j,r}^{D1BS}$, $h_{i,j,r}^{D2BS}$ denote the channel gain of UED1-UEDR, UED2-UEDR, UEDC-UER, UED1-BS, UED2-BS links respectively, PL_0 denotes the path-loss, $d_{j,r}^{D1R}$, $d_{j,r}^{D2R}$, $d_{i,r}^{CR}$, d_j^{D1BS}, d_j^{D2BS} are the distance between relevant links, and α is the coefficient of path-loss.

The process of two-way AF relaying D2D communication can divide into two time slots. We assume that D2D links reuse the uplink resource of cellular users to transmit signal at first slot and reuse the downlink resource of cellular users to transmit signal at second slot. The uplink and downlink channels are assumed reciprocal and invariant and are based on two consecutive equal time slots.

At first slot, UED1 and UED2 reuse the uplink resource of UEC to transmit signal to relay node UEDR simultaneously. As the UEC acts as interferences to the UED1-UED2 link, the receive signal of UEDR is

$$y_{i,j,r}^R=\sqrt{p_j^{D1}}h_{i,j,r}^{D1R}s_j^{D1}+\sqrt{p_j^{D2}}h_{i,j,r}^{D2R}s_j^{D2}+\sqrt{p_i^C}h_{i,j,r}^{CR}s_i^C+n_0 \tag{1}$$

After receiving signal, UEDR amplifies the signal to G times, where

$$G_{i,j,r}=\sqrt{p_r^R/(p_j^{D1}\left|h_{i,j,r}^{D1R}\right|^2+p_j^{D2}\left|h_{i,j,r}^{D2R}\right|^2+p_i^C\left|h_{i,j,r}^{CR}\right|^2+N_0)}$$

At second slot, UEDR reuse the downlink resource of UEC to transmit signal to UED1 and UED2 simultaneously. The receive signal of UED1 and UED2 can be expressed as

$$y_{i,j,r}^{D1}=G_{i,j,r}\sqrt{p_j^{D1}}(h_{i,j,r}^{D1R})^2s_j^{D1}+G_{i,j,r}\sqrt{p_j^{D2}}h_{i,j,r}^{D1R}h_{i,j,r}^{D2R}s_j^{D2}+G_{i,j,r}\sqrt{p_i^C}h_{i,j,r}^{D1R}h_{i,j,r}^{CR}s_i^C+G_{i,j,r}h_{i,j,r}^{D1R}n_0+\sqrt{p^{BS}}h_{i,j,r}^{D1BS}s^{BS}+n_0 \tag{2}$$

$$y_{i,j,r}^{D2}=G_{i,j,r}\sqrt{p_j^{D2}}(h_{i,j,r}^{D2R})^2s_j^{D2}+G_{i,j,r}\sqrt{p_j^{D1}}h_{i,j,r}^{D1R}h_{i,j,r}^{D2R}s_j^{D1}+G_{i,j,r}\sqrt{p_i^C}h_{i,j,r}^{D2R}h_{i,j,r}^{CR}s_i^C+G_{i,j,r}h_{i,j,r}^{D2R}n_0+\sqrt{p^{BS}}h_{i,j,r}^{D2BS}s^{BS}+n_0 . \tag{3}$$

The self-interference can be cancelled since each terminal knows its transmitted signal, we can get

$$y_{i,j,r}^{D1}=G_{i,j,r}\sqrt{p_j^{D2}}h_{i,j,r}^{D1R}h_{i,j,r}^{D2R}s_j^{D2}+G_{i,j,r}\sqrt{p_i^C}h_{i,j,r}^{D1R}h_{i,j,r}^{CR}s_i^C+G_{i,j,r}h_{i,j,r}^{D1R}n_0+\sqrt{p^{BS}}h_{i,j,r}^{D1BS}s^{BS}+n_0 , \tag{4}$$

$$y_{i,j,r}^{D2}=G_{i,j,r}\sqrt{p_j^{D1}}h_{i,j,r}^{D1R}h_{i,j,r}^{D2R}s_j^{D1}+G_{i,j,r}\sqrt{p_i^C}h_{i,j,r}^{D2R}h_{i,j,r}^{CR}s_i^C+G_{i,j,r}h_{i,j,r}^{D2R}n_0+\sqrt{p^{BS}}h_{i,j,r}^{D2BS}s^{BS}+n_0 . \tag{5}$$

The formula (4) denotes the receive signal of UED1 in UED2-UED1 link, and its SINR is written as

$$SINR_{i,j,r}^{D210} = \frac{p_j^{D2}\left|h_{i,j,r}^{D1R}\right|^2\left|h_{i,j,r}^{D2R}\right|^2 G_{i,j,r}^2}{p_i^C\left|h_{i,j,r}^{D1R}\right|^2\left|h_{i,j,r}^{CR}\right|^2 G_{i,j,r}^2 + \left|h_{i,j,r}^{D1R}\right|^2 N_0 G_{i,j,r}^2 + p^{BS}\left|h_{i,j,r}^{D1BS}\right|^2 + N_0} . \tag{6}$$

Assuming $\gamma_{i,j,r}^1 = p_j^{D1}\left|h_{i,j,r}^{D1R}\right|^2 / (p_i^C\left|h_{i,j,r}^{CR}\right|^2 + N_0)$, $\gamma_{i,j,r}^2 = p_j^{D2}\left|h_{i,j,r}^{D2R}\right|^2 / (p_i^C\left|h_{i,j,r}^{CR}\right|^2 + N_0)$, $\gamma_{i,j,r}^3 = p_r^R\left|h_{i,j,r}^{D1R}\right|^2 / (p^{BS}\left|h_{i,j,r}^{D1BS}\right|^2 + N_0)$, $\gamma_{i,j,r}^4 = p_r^R\left|h_{i,j,r}^{D2R}\right|^2 / (p^{BS}\left|h_{i,j,r}^{D2BS}\right|^2 + N_0)$, the $SINR_{i,j,r}^{D210}$ can be expressed by

$$SINR_{i,j,r}^{D210} = \gamma_{i,j,r}^2 \cdot \gamma_{i,j,r}^3 / (\gamma_{i,j,r}^1 + \gamma_{i,j,r}^2 + \gamma_{i,j,r}^3 + 1) . \tag{7}$$

Since $\gamma_{i,j,r}^1 + \gamma_{i,j,r}^2 + \gamma_{i,j,r}^3 \gg 1$, (7) can be simplified to

$$SINR_{i,j,r}^{D21} = \gamma_{i,j,r}^2 \cdot \gamma_{i,j,r}^3 / (\gamma_{i,j,r}^1 + \gamma_{i,j,r}^2 + \gamma_{i,j,r}^3) . \tag{8}$$

The formula (5) denotes the receive signal of UED2 in UED1-UED2 link, and we can get its SINR in a like manner

$$SINR_{i,j,r}^{D12} = \gamma_{i,j,r}^1 \cdot \gamma_{i,j,r}^4 / (\gamma_{i,j,r}^1 + \gamma_{i,j,r}^2 + \gamma_{i,j,r}^4) . \tag{9}$$

3. Power Control and Outage Probability Analysis

3.1. Power Control for D2D User

Since D2D communication reuse the resource of cellular networks, the co-channel interference is unavoidable. It's necessary to control the transmitting power of D2D users to guarantee the QoS of cellular users. Let $p_j^{D1}\left|h_{i,j,r}^{D1BS}\right|^2 < Q$, $p_j^{D2}\left|h_{i,j,r}^{D2BS}\right|^2 < Q$, the performance of two-way amplify-and-forward relaying is depended on the quality of worst link, so we can find a balance point of SINR to control the transmitting power of D2D users.
According to formula (8) and (9),

$$\begin{aligned} SINR_{i,j,r}^{D21} &= \gamma_{i,j,r}^2 \cdot \gamma_{i,j,r}^3 / (\gamma_{i,j,r}^1 + \gamma_{i,j,r}^2 + \gamma_{i,j,r}^3) \\ &< \gamma_{i,j,r}^2 \cdot \gamma_{i,j,r}^3 / (\gamma_{i,j,r}^2 + \gamma_{i,j,r}^3) \\ &< \min(\gamma_{i,j,r}^2, \gamma_{i,j,r}^3), \end{aligned} \tag{10}$$

$$\begin{aligned} SINR_{i,j,r}^{D12} &= \gamma_{i,j,r}^1 \cdot \gamma_{i,j,r}^4 / (\gamma_{i,j,r}^1 + \gamma_{i,j,r}^2 + \gamma_{i,j,r}^4) \\ &< \gamma_{i,j,r}^1 \cdot \gamma_{i,j,r}^4 / (\gamma_{i,j,r}^1 + \gamma_{i,j,r}^4) \\ &< \min(\gamma_{i,j,r}^1, \gamma_{i,j,r}^4). \end{aligned} \tag{11}$$

The outage probability of two-way amplify-and-forward relaying is given by

$$\begin{aligned} P_{i,j,r}^D &= \{\min(SINR_{i,j,r}^{D21}, SINR_{i,j,r}^{D12}) < \gamma_{th}\} \\ &= 1 - p\{SINR_{i,j,r}^{D21} > \gamma_{th}, SINR_{i,j,r}^{D12} > \gamma_{th}\} \\ &\approx 1 - p(\min(\gamma_{i,j,r}^2, \gamma_{i,j,r}^3) > \gamma_{th}, \min(\gamma_{i,j,r}^1, \gamma_{i,j,r}^4) > \gamma_{th}) \\ &= 1 - p(\min(\gamma_{i,j,r}^1, \gamma_{i,j,r}^2, \gamma_{i,j,r}^3, \gamma_{i,j,r}^4) > \gamma_{th}), \end{aligned} \tag{12}$$

while γ_{th} is the threshold of SINR that signal can be decoded. The value of outage probability is jointed decision by $\gamma_{i,j,r}^1, \gamma_{i,j,r}^2, \gamma_{i,j,r}^3, \gamma_{i,j,r}^4$, and the smaller one of $\gamma_{i,j,r}^1, \gamma_{i,j,r}^2$ will decide the performance of D2D link at first slot. So we can let $\gamma_{i,j,r}^1 = \gamma_{i,j,r}^2$ [21] and get the balance point of

SINR $p_j^{D1}\left|h_{i,j,r}^{D1R}\right|^2 = p_j^{D2}\left|h_{i,j,r}^{D2R}\right|^2$.

Depending on the power controlling of p_j^{D1} and p_j^{D2}, we have

$$p_j^{D1} < Q/\left|h_{i,j,r}^{D1BS}\right|^2 \Rightarrow p_j^{D2} < Q\cdot\left|h_{i,j,r}^{D1R}\right|^2/(\left|h_{i,j,r}^{D1BS}\right|^2\cdot\left|h_{i,j,r}^{D2R}\right|^2),$$

$$p_j^{D2} < Q/\left|h_{i,j,r}^{D2BS}\right|^2 \Rightarrow p_j^{D1} < Q\cdot\left|h_{i,j,r}^{D2R}\right|^2/(\left|h_{i,j,r}^{D1R}\right|^2\cdot\left|h_{i,j,r}^{D2BS}\right|^2).$$

In order to D2D link obtain the best performance, p_j^{D1} and p_j^{D2} are taken the maximum

$$p_j^{D1} = \min(Q/\left|h_{i,j,r}^{D1BS}\right|^2, Q\cdot\left|h_{i,j,r}^{D2R}\right|^2/(\left|h_{i,j,r}^{D1R}\right|^2\cdot\left|h_{i,j,r}^{D2BS}\right|^2)), \tag{13}$$

$$p_j^{D2} = \min(Q/\left|h_{i,j,r}^{D2BS}\right|^2, Q\cdot\left|h_{i,j,r}^{D1R}\right|^2/(\left|h_{i,j,r}^{D1BS}\right|^2\cdot\left|h_{i,j,r}^{D2R}\right|^2)). \tag{14}$$

Depending on formula (13) and (14), we can get

$$p_j^{D1}\left|h_{i,j,r}^{D1R}\right|^2 = p_j^{D2}\left|h_{i,j,r}^{D2R}\right|^2 = \min\{Q\left|h_{i,j,r}^{D1R}\right|^2/\left|h_{i,j,r}^{D1BS}\right|^2, Q\cdot\left|h_{i,j,r}^{D2R}\right|^2/\left|h_{i,j,r}^{D2BS}\right|^2\}. \tag{15}$$

(i) When $p_j^{D1}\left|h_{i,j,r}^{D1R}\right|^2 = p_j^{D2}\left|h_{i,j,r}^{D2R}\right|^2 = Q\cdot\left|h_{i,j,r}^{D1R}\right|^2/\left|h_{i,j,r}^{D1BS}\right|^2$, combining with (8)

$$\begin{aligned}
SINR_{i,j,r}^{D21} &= \frac{\gamma_{i,j,r}^2\cdot\gamma_{i,j,r}^3}{\gamma_{i,j,r}^1+\gamma_{i,j,r}^2+\gamma_{i,j,r}^3} = \frac{\gamma_{i,j,r}^1\cdot\gamma_{i,j,r}^3}{\gamma_{i,j,r}^3+2\gamma_{i,j,r}^1} \\
&= \frac{\frac{1}{2}\cdot 2\frac{Q\left|h_{i,j,r}^{D1R}\right|^2/\left|h_{i,j,r}^{D1BS}\right|^2}{p_i^C\left|h_{i,j,r}^{CR}\right|^2+N_0}\cdot\frac{p_r^R\left|h_{i,j,r}^{D1R}\right|^2}{p^{BS}\left|h_{i,j,r}^{D1BS}\right|^2+N_0}}{\frac{p_r^R\left|h_{i,j,r}^{D1R}\right|^2}{p^{BS}\left|h_{i,j,r}^{D1BS}\right|^2+N_0}+2\frac{Q\left|h_{i,j,r}^{D1R}\right|^2/\left|h_{i,j,r}^{D1BS}\right|^2}{p_i^C\left|h_{i,j,r}^{CR}\right|^2+N_0}} \\
&= \frac{1}{2}\cdot\frac{\left|h_{i,j,r}^{D1R}\right|^2}{\left|h_{i,j,r}^{D1BS}\right|^2}\cdot\frac{2Q/(p_i^C\left|h_{i,j,r}^{CR}\right|^2+N_0)\cdot\frac{p_r^R\left|h_{i,j,r}^{D1R}\right|^2}{p^{BS}\left|h_{i,j,r}^{D1BS}\right|^2+N_0}}{\frac{p_r^R\left|h_{i,j,r}^{D1R}\right|^2}{p^{BS}\left|h_{i,j,r}^{D1BS}\right|^2+N_0}+2Q/(p_i^C\left|h_{i,j,r}^{CR}\right|^2+N_0)} \\
&< \frac{1}{2}\cdot\min\{\frac{2Q\left|h_{i,j,r}^{D1R}\right|^2/\left|h_{i,j,r}^{D1BS}\right|^2}{p_i^C\left|h_{i,j,r}^{CR}\right|^2+N_0}, \frac{p_r^R\left|h_{i,j,r}^{D1R}\right|^2}{p^{BS}\left|h_{i,j,r}^{D1BS}\right|^2+N_0}\}.
\end{aligned} \tag{16}$$

(ii) When $p_j^{D1}\left|h_{i,j,r}^{D1R}\right|^2 = p_j^{D2}\left|h_{i,j,r}^{D2R}\right|^2 = Q\cdot\left|h_{i,j,r}^{D2R}\right|^2/\left|h_{i,j,r}^{D2BS}\right|^2$,

$$\begin{aligned}
SINR_{i,j,r}^{D21} &= \frac{\frac{p_j^{D1}\left|h_{i,j,r}^{D1R}\right|^2}{p_i^C\left|h_{i,j,r}^{CR}\right|^2+N_0}\cdot\frac{p_r^R\left|h_{i,j,r}^{D1R}\right|^2}{p^{BS}\left|h_{i,j,r}^{D1BS}\right|^2+N_0}}{\frac{p_r^R\left|h_{i,j,r}^{D1R}\right|^2}{p^{BS}\left|h_{i,j,r}^{D1BS}\right|^2+N_0}+2\cdot\frac{p_j^{D1}\left|h_{i,j,r}^{D1R}\right|^2}{p_i^C\left|h_{i,j,r}^{CR}\right|^2+N_0}} \\
&= \frac{\frac{1}{2}\cdot\frac{2Q\left|h_{i,j,r}^{D2R}\right|^2/\left|h_{i,j,r}^{D2BS}\right|^2}{p_i^C\left|h_{i,j,r}^{CR}\right|^2+N_0}\cdot\frac{p_r^R\left|h_{i,j,r}^{D1R}\right|^2}{p^{BS}\left|h_{i,j,r}^{D1BS}\right|^2+N_0}}{\frac{p_r^R\left|h_{i,j,r}^{D1R}\right|^2}{p^{BS}\left|h_{i,j,r}^{D1BS}\right|^2+N_0}+\frac{2Q\left|h_{i,j,r}^{D2R}\right|^2/\left|h_{i,j,r}^{D2BS}\right|^2}{p_i^C\left|h_{i,j,r}^{CR}\right|^2+N_0}} \\
&< \frac{1}{2}\cdot\min\{\frac{2Q\left|h_{i,j,r}^{D2R}\right|^2/\left|h_{i,j,r}^{D2BS}\right|^2}{p_i^C\left|h_{i,j,r}^{CR}\right|^2+N_0}, \frac{p_r^R\left|h_{i,j,r}^{D1R}\right|^2}{p^{BS}\left|h_{i,j,r}^{D1BS}\right|^2+N_0}\}.
\end{aligned} \tag{17}$$

(16) and (17) can be merged based on (15), we have

$$SINR_{i,j,r}^{D21} \approx \frac{1}{2}\cdot\min\{\frac{2Q\left|h_{i,j,r}^{D1R}\right|^2/\left|h_{i,j,r}^{D1BS}\right|^2}{p_i^C\left|h_{i,j,r}^{CR}\right|^2+N_0},\frac{2Q\left|h_{i,j,r}^{D2R}\right|^2/\left|h_{i,j,r}^{D2BS}\right|^2}{p_i^C\left|h_{i,j,r}^{CR}\right|^2+N_0},\frac{p_r^R\left|h_{i,j,r}^{D1R}\right|^2}{p^{BS}\left|h_{i,j,r}^{D1BS}\right|^2+N_0}\}. \tag{18}$$

In similar way, we can get

$$SINR_{i,j,r}^{D12} \approx \frac{1}{2}\cdot\min\{\frac{2Q\left|h_{i,j,r}^{D1R}\right|^2/\left|h_{i,j,r}^{D1BS}\right|^2}{p_i^C\left|h_{i,j,r}^{CR}\right|^2+N_0},\frac{2Q\left|h_{i,j,r}^{D2R}\right|^2/\left|h_{i,j,r}^{D2BS}\right|^2}{p_i^C\left|h_{i,j,r}^{CR}\right|^2+N_0},\frac{p_r^R\left|h_{i,j,r}^{D2R}\right|^2}{p^{BS}\left|h_{i,j,r}^{D2BS}\right|^2+N_0}\}. \tag{19}$$

3.2. Power Control for Relay Node

The co-channel interference will be generated when relay node UEDR reuse the downlink resource of cellular user at second slot. It's also necessary to control the transmitting power of relay node to ensure the QoS of cellular users. We need the outage probability of cellular user is not less than $P_{i,j,r}^{th}$ while cellular user get the minimum throughput $R_{i,j,r}^{low}$, and it can be denoted by

$$\Pr\{\log(1+p^{BS}\left|h_{i,j,r}^{CBS}\right|^2/(p_r^R\left|h_{i,j,r}^{CR}\right|^2+N_0))<R_{i,j,r}^{low}\}<P_{i,j,r}^{th}, \tag{20}$$

we have

$$\begin{aligned}&\Pr\{\log(1+p^{BS}\left|h_{i,j,r}^{CBS}\right|^2/(p_r^R\left|h_{i,j,r}^{CR}\right|^2+N_0))<R_{i,j,r}^{low}\}\\&=1-\Pr\{p^{BS}\left|h_{i,j,r}^{CBS}\right|^2/(p_r^R\left|h_{i,j,r}^{CR}\right|^2+N_0)>2^{R_{i,j,r}^{low}}-1\}<P_{i,j,r}^{th}.\end{aligned} \tag{21}$$

For simplification, we let $M=\left|h_{i,j,r}^{CBS0}\right|^2, f_M(m)=e^{-m}, m>0,\ N=p_r^R\left|h_{i,j,r}^{CR0}\right|^2\frac{(d_{i,r}^{CR0})^{-\alpha}}{PL_0}+N_0, f_N(n)=\frac{1}{p_r\chi}e^{-\frac{n}{p_r^R\chi}}e^{\frac{1}{p_r^R\chi}}, n>N_0$,

where $\chi=(d_{i,r}^{CR0})^{-\alpha}/PL_0$. The latter part of inequality (21) can be resolved as follow

$$\begin{aligned}\Pr\{\frac{p^{BS}\left|h_{i,j,r}^{CBS}\right|^2}{p_r^R\left|h_{i,j,r}^{CR}\right|^2+N_0}>2^{R_{i,j,r}^{low}}-1\}&=\Pr\{\frac{p^{BS}\left|h_{i,j,r}^{CBS0}\right|^2(d_i^{CBS})^{-\alpha}/PL_0}{p_r^R\left|h_{i,j,r}^{CR0}\right|^2(d_{i,r}^{CR0})^{-\alpha}/PL_0+N_0}>2^{R_{i,j,r}^{low}}-1\}\\&=\Pr\{\frac{p^{BS}\left|h_{i,j,r}^{CBS0}\right|^2}{p_r^R\left|h_{i,j,r}^{CR0}\right|^2(d_{i,r}^{CR0})^{-\alpha}/PL_0+N_0}>\frac{(2^{R_{i,j,r}^{low}}-1)PL_0}{(d_i^{CBS})^{-\alpha}}\}\\&=\Pr\{\frac{p^{BS}\left|h_{i,j,r}^{CBS0}\right|^2}{p_r^R\left|h_{i,j,r}^{CR0}\right|^2\chi+N_0}>(2^{R_{i,j,r}^{low}}-1)\delta\}\\&=\int_{(2^{R_{i,j,r}^{low}}-1)\delta/p^{BS}}^{\infty}\int_{N_0}^{\infty}n\frac{1}{p_r^R\chi}e^{-\frac{n}{p_r^R\chi}}e^{\frac{N_0}{p_r^R\chi}}e^{-mn}\,dndm\\&=\frac{1}{p_r^R\chi}e^{\frac{N_0}{p_r\chi}}\int_{N_0}^{\infty}e^{-(\frac{1}{p_r^R\chi}+\frac{2^{R_{i,j,r}^{low}}-1}{p^{BS}}\delta)n}\,dn\\&=\frac{p^{BS}}{p^{BS}+(2^{R_{i,j,r}^{low}}-1)\chi\delta p_r^R}e^{-\frac{2^{R_{i,j,r}^{low}}-1}{p^{BS}}\delta N_0},\end{aligned} \tag{22}$$

where $\delta=PL_0/(d_i^{CBS})^{-\alpha}$. After simplification, we can get

$$p_r^R < (p^{BS} / (1 - P_{i,j,r}^{th}) e^{-(2^{R_{i,j,r}^{low}} - 1)\delta N_0 / p^{BS}} - p^{BS}) / ((2^{R_{i,j,r}^{low}} - 1)\chi\delta) \quad . \tag{23}$$

To achieve the best performance of D2D link, it can make p_r^R be the maximum value

$$p_r^R = \min\{p_j^{\max}, \frac{p^{BS} / (1 - P_{i,j,r}^{th}) e^{-(2^{R_{i,j,r}^{low}} - 1)\delta N_0 / p^{BS}} - p^{BS}}{(2^{R_{i,j,r}^{low}} - 1)\chi\delta}\} \quad , \tag{24}$$

where $p_j^{\max}$ is the maximum transmitting power.

4. Outage Probability Analysis

There are two kinds of links under cellular network in this paper, one is cellular communication and another is D2D communication. We have do some work to limit the co-channel interference in former part and the QoS of cellular communication can be promised. The performance of D2D link is the next problem that we should consider, then we will analysis the outage probability of D2D link. The two-way transmission processing should be considered when we compute the outage probability. We have

$$\begin{aligned} P_{i,j,r}^{D} &= \Pr(\min\{SINR_{i,j,r}^{D21}, SINR_{i,j,r}^{D12}\} < \gamma_{th}) \\ &= 1 - \Pr(SINR_{i,j,r}^{D21} > \gamma_{th}, SINR_{i,j,r}^{D12} > \gamma_{th}) \\ &= 1 - \Pr(\frac{1}{2}\min\{\frac{2Q|h_{i,j,r}^{D1R}|^2 / |h_{i,j,r}^{D1BS}|^2}{p_i^C |h_{i,j,r}^{CR}|^2 + N_0}, \frac{2Q|h_{i,j,r}^{D2R}|^2 / |h_{i,j,r}^{D2BS}|^2}{p_i^C |h_{i,j,r}^{CR}|^2 + N_0}, \frac{p_r^R |h_{i,j,r}^{D1R}|^2}{p^{BS}|h_{i,j,r}^{D1BS}|^2 + N_0}\} \\ &> \gamma_{th}, \frac{1}{2}\min\{\frac{2Q|h_{i,j,r}^{D1R}|^2 / |h_{i,j,r}^{D1BS}|^2}{p_i^C |h_{i,j,r}^{CR}|^2 + N_0}, \frac{2Q|h_{i,j,r}^{D2R}|^2 / |h_{i,j,r}^{D2BS}|^2}{p_i^C |h_{i,j,r}^{CR}|^2 + N_0}, \frac{p_r^R |h_{i,j,r}^{D2R}|^2}{p^{BS}|h_{i,j,r}^{D2BS}|^2 + N_0}\} > \gamma_{th}) \\ &= 1 - \Pr(\frac{Q|h_{i,j,r}^{D1R}|^2 / |h_{i,j,r}^{D1BS}|^2}{p_i^C |h_{i,j,r}^{CR}|^2 + N_0} > \gamma_{th}, \frac{Q|h_{i,j,r}^{D2R}|^2 / |h_{i,j,r}^{D2BS}|^2}{p_i^C |h_{i,j,r}^{CR}|^2 + N_0} > \gamma_{th}, \\ &\quad \frac{p_r^R |h_{i,j,r}^{D1R}|^2}{p^{BS}|h_{i,j,r}^{D1BS}|^2 + N_0} > 2\cdot\gamma_{th}, \frac{p_r^R |h_{i,j,r}^{D2R}|^2}{p^{BS}|h_{i,j,r}^{D2BS}|^2 + N_0} > 2\cdot\gamma_{th}). \end{aligned} \tag{25}$$

Since $N_0 \ll p^{BS}|h_{i,j,r}^{D1BS}|^2$, $N_0 \ll p^{BS}|h_{i,j,r}^{D2BS}|^2$, the formula (25) can be denoted by

$$\begin{aligned} P_{i,j,r}^{D} = 1 - \Pr(&\frac{Q|h_{i,j,r}^{D1R}|^2 / |h_{i,j,r}^{D1BS}|^2}{p_i^C |h_{i,j,r}^{CR}|^2 + N_0} > \gamma_{th}, \frac{Q|h_{i,j,r}^{D2R}|^2 / |h_{i,j,r}^{D2BS}|^2}{p_i^C |h_{i,j,r}^{CR}|^2 + N_0} > \gamma_{th} \\ &, \frac{p_r^R |h_{i,j,r}^{D1R}|^2}{p^{BS}|h_{i,j,r}^{D1BS}|^2} > 2\cdot\gamma_{th}, \frac{p_r^R |h_{i,j,r}^{D2R}|^2}{p^{BS}|h_{i,j,r}^{D2BS}|^2} > 2\cdot\gamma_{th}). \end{aligned} \tag{26}$$

For simplifying, we let $\varepsilon_{j,r} = (d_{j,r}^{D1R} / d_j^{D1BS})^{-\alpha}$, $f_{i,r} = (d_{i,r}^{CR})^{-\alpha} / PL_0$, $\varphi_{j,r} = (d_{j,r}^{D2R} / d_j^{D2BS})^{-\alpha}$, $X_{i,j,r} = |h_{i,j,r}^{D1R0}|^2 / |h_{i,j,r}^{D1BS0}|^2$, $Y_{i,j,r} = p_i^C |h_{i,j,r}^{CR0}|^2 \cdot f_{i,r} + N_0$, $Z_{i,j,r} = |h_{i,j,r}^{D2R}|^2 / |h_{i,j,r}^{D2BS}|^2$ and we have

$$
\begin{aligned}
&\Pr\left(\frac{Q\left|h_{i,j,r}^{D1R}\right|^2 / \left|h_{i,j,r}^{D1BS}\right|^2}{p_i^C\left|h_{i,j,r}^{CR}\right|^2+N_0}>\gamma_{th}, \frac{Q\left|h_{i,j,r}^{D2R}\right|^2 / \left|h_{i,j,r}^{D2BS}\right|^2}{p_i^C\left|h_{i,j,r}^{CR}\right|^2+N_0}>\gamma_{th}, \frac{p_r^R\left|h_{i,j,r}^{D1R}\right|^2}{p^{BS}\left|h_{i,j,r}^{D1BS}\right|^2+N_0}>2\cdot\gamma_{th}, \frac{p_r^R\left|h_{i,j,r}^{D2R}\right|^2}{p^{BS}\left|h_{i,j,r}^{D2BS}\right|^2+N_0}>2\cdot\gamma_{th}\right)\\
&=\Pr\left(X_{i,j,r}>\frac{\gamma_{th}Y_{i,j,r}}{Q\varepsilon_{j,r}}, Z_{i,j,r}>\frac{\gamma_{th}Y_{i,j,r}}{Q\varphi_{i,r}}, Y_{i,j,r}>\frac{2Qp^{BS}}{p_r^R}\right)+\Pr\left(X_{i,j,r}>\frac{2\gamma_{th}p^{BS}}{p_r^R\varepsilon_{j,r}}, Z_{i,j,r}>\frac{2\gamma_{th}p^{BS}}{p_r^R\varphi_{i,r}}, N_0<Y_{i,j,r}<\frac{2Qp^{BS}}{p_r^R}\right)\\
&=\int_{\frac{2Qp^{BS}}{p_r^R}}^{\infty}\int_{\frac{\gamma_{th}Y_{i,j,r}}{Q\varphi_{i,r}}}^{\infty}\int_{\frac{\gamma_{th}Y_{i,j,r}}{Q\varepsilon_{j,r}}}^{\infty}\frac{1}{(1+x)^2}\frac{1}{(1+z)^2}e^{-\frac{y}{p_i^C f_{i,r}}}\frac{1}{p_i^C f_{i,r}}e^{\frac{N_0}{p_i^C f_{i,r}}}dxdzdy+\int_{N_0}^{\frac{2Qp^{BS}}{p_r^R}}\int_{\frac{2\gamma_{th}p^{BS}}{p_r^R\varphi_{i,r}}}^{\infty}\int_{\frac{2\gamma_{th}p^{BS}}{p_r^R\varepsilon_{j,r}}}^{\infty}\frac{1}{(1+x)^2}\frac{1}{(1+z)^2}e^{-\frac{y}{p_i^C f_{i,r}}}\frac{1}{p_i^C f_{i,r}}e^{\frac{N_0}{p_i^C f_{i,r}}}dxdzdy\\
&=\frac{Q\varepsilon_{j,r}\varphi_{i,r}}{(\varphi_{i,r}-\varepsilon_{j,r})p_i^C f_{i,r}\gamma_{th}}\left(Ei\left(-\frac{p_r^R\varphi_{i,r}Q+2p^{BS}\gamma_{th}Q}{\gamma_{th}p_r^R p_i^C f_{i,r}}\right)e^{\frac{Q\varphi_{i,r}+\gamma_{th}N_0}{\gamma_{th}p_i^C f_{i,r}}}-Ei\left(-\frac{p_r^R\varepsilon_{j,r}Q+2p^{BS}\gamma_{th}Q}{\gamma_{th}p_r^R p_i^C f_{i,r}}\right)e^{\frac{Q\varepsilon_{j,r}+\gamma_{th}N_0}{\gamma_{th}p_i^C f_{i,r}}}\right)\\
&\quad+\frac{(p_r^R)^2\varepsilon_{j,r}\varphi_{i,r}}{(p_r^R\varepsilon_{j,r}+2\gamma_{th}p^{BS})(p_r^R\varphi_{i,r}+2\gamma_{th}p^{BS})}\left(1-e^{-\frac{2Qp^{BS}-p_r^R N_0}{p_i^C p_r^R f_{i,r}}}\right).
\end{aligned}
\tag{27}
$$

The outage probability of D2D link can be expressed as

$$
\begin{aligned}
P_{i,j,r}^D=1-&\frac{Q\varepsilon_{j,r}\varphi_{i,r}}{(\varphi_{i,r}-\varepsilon_{j,r})p_i^C f_{i,r}\gamma_{th}}\left(Ei\left(-\frac{p_r^R\varphi_{i,r}Q+2p^{BS}\gamma_{th}Q}{\gamma_{th}p_r^R p_i^C f_{i,r}}\right)\cdot e^{\frac{Q\varphi_{i,r}+\gamma_{th}N_0}{\gamma_{th}p_i^C f_{i,r}}}-Ei\left(-\frac{p_r^R\varepsilon_{j,r}Q+2p^{BS}\gamma_{th}Q}{\gamma_{th}p_r^R p_i^C f_{i,r}}\right)e^{\frac{Q\varepsilon_{j,r}+\gamma_{th}N_0}{\gamma_{th}p_i^C f_{i,r}}}\right)-\\
&\frac{(p_r^R)^2\varepsilon_{j,r}\varphi_{i,r}}{(p_r^R\varepsilon_{j,r}+2\gamma_{th}p^{BS})(p_r^R\varphi_{i,r}+2\gamma_{th}p^{BS})}\left(1-e^{-\frac{2Qp^{BS}-p_r^R N_0}{p_i^C p_r^R f_{i,r}}}\right).
\end{aligned}
\tag{28}
$$

5. Simulation Results

We consider a circular cell with a radius of $200m$ in the simulation. It assumes channel obeys Rayleigh distribution, and each channel bandwidth is set as $150KHz$. The white Gaussian noise power spectral density is $-174dBm/Hz$. The noise coefficient of BS and user are $5dB$, $9dB$ respectively. We set $PL_0=10^{2.8}$, $\alpha=4$, $p^{BS}=30dBm$, $p_j^{\max}=21dBm$. We simulate 5000 times for different Rayleigh fading channel and achieve the average value to make sure the result is accuracy.

First, we verify the consistency of simulation results and analysis results of D2D link outage probability, and analyze the influence of different parameters on D2D link outage probability. As show in figure 2 and figure 3, the analysis value is highly consistent with the simulation value, it means our analysis results is correct.

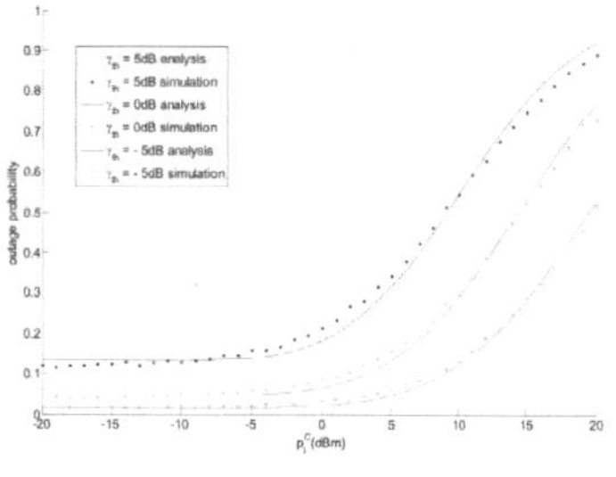

Figure 2. The relationship between p_i^C and outage probability on different γ_{th}

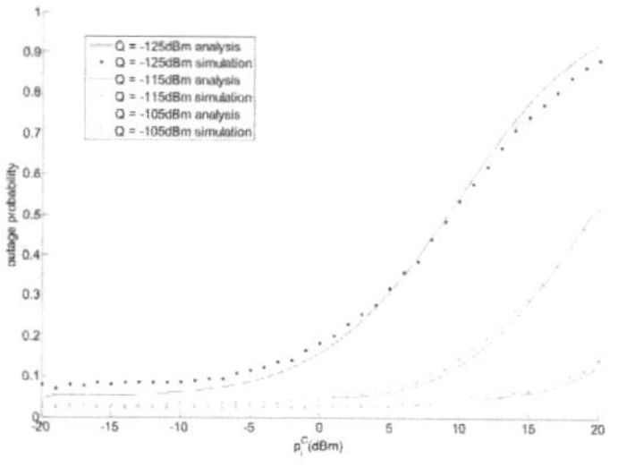

Figure 3. The relationship between p_i^C and outage probability on different Q

Figure 2 depicts the impact of γ_{th} on D2D link outage probability. We set $Q=-115dBm$ for different SINR threshold of D2D links in this part. With p_i^C increasing, the D2D link outage probability increased in figure 2. That is because when p_i^C increases, the co-channel interference which caused by D2D user will increase, leading to the performance of communication network reduced. In addition, the bigger γ_{th} means the D2D users need higher quality of link to decode they have received signal, which also cause the D2D link outage probability increased. Then, the outage probability changed much smaller when $p_i^C<-5dBm$, that reason is too small p_i^C will improve the performance of D2D link in first time slot, and make the holistic performance of D2D link is depended on the second time slot. Therefore, the effect of p_i^C on the performance of D2D link system is reduced at this time.

Figure 3 depicts the impact of Q on D2D link outage probability. We set $\gamma_{th}=0dB$. With Q increasing, the power constraint on the D2D link is weakened, and the performance of the D2D link is improved in figure 3. When $Q=-105dBm$, it will further reduce the transmit power limitation of D2D link users, enhance the received signal strength, and weaken the impact of interference on the network system performance.

Finally, in order to ascertain whether relay can improve the communication performance of D2D link, the influence of relay with and without on D2D link outage probability is analyzed in figure 4. Obviously, the outage probability of D2D link with relay is less than without relay, and the performance increased by an average of 32.3%.

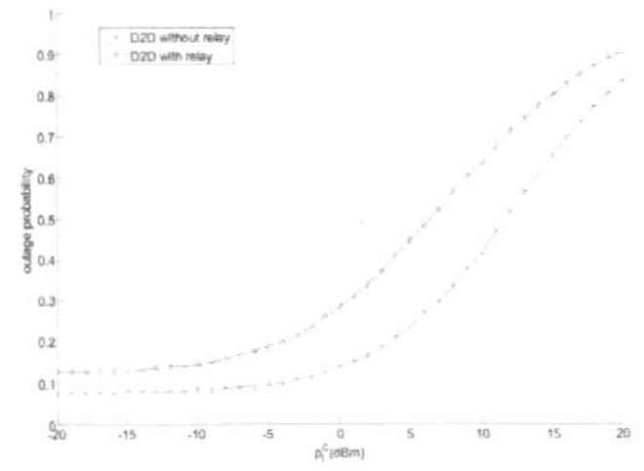

Figure 4. The influence on outage probability of D2D with and without relay

6. Conclusion

Considering intra-cell orthogonality is impaired due to spectral reusing and severe co-channel interference could be imposed between cellular and D2D links. We use two-way AF relay to assist D2D link communication first. Then, in order to limit the interference of spectrum multiplexing to existing cellular users and maintain the QoS requirement of cellular users, the power control at both D2D transmitters and relays are proposed. Finally, we analyze the outage probability of D2D links assisted by two-way AF relays on Rayleigh fading channel. The simulation results verify the correctness of the outage probability analyses for D2D links, and indicate the performance gain when relays are introduced to facilitate D2D links.

Acknowledgments

This work has been supported by National Key Research and Development under Grants No.2018YFB180024.

Reference

[1] Ge X, Cheng H, Guizani M, Han T. 5G wireless backhaul networks: Challenges and research advances. *IEEE Network* 2014; **28**(6): 6–11. DOI: 10.1109/MNET.2014.6963798.

[2] Zhu X, Chen S, Hu H, Su X, Shi Y. TDD-based mobile communication solutions for high-speed railway scenarios. *IEEE Wireless Communications* 2013; **20**(6): 22–29. DOI: 10.1109/MWC.2013.6704470.

[3] Gandotra P, Kumar Jha R, Jain S. A survey on device-to-device (D2D) communication: Architecture and security issues. *Journal of Network and Computer Applications* 2017; **78**(July 2016): 9–29. DOI: 10.1016/j.jnca.2016.11.002.

[4] Hu RQ, Qian Y. An energy efficient and spectrum efficient wireless heterogeneous network framework for 5G systems. *IEEE Communications Magazine* 2014; **52**(5): 94–101. DOI: 10.1109/MCOM.2014.6815898.

[5] Ma C, Sun G, Tian X, Ying K, Yu H, Wang X. Cooperative relaying schemes for device-to-device communication underlaying cellular networks. *GLOBECOM - IEEE Global Telecommunications Conference* 2013: 3890–3895. DOI: 10.1109/GLOCOM.2013.6831680.

[6] ElSawy H, Hossain E. Analytical Modeling of Mode Selection and Power Control for Underlay D2D Communication in Cellular Networks. 2014; **62**(11): 4147–4161. DOI: 10.1109/TCOMM.2014.2363849.

[7] Lin X, Andrews JG, Ghosh A. Spectrum Sharing for Device-to-Device Communication in Cellular Networks. *IEEE Transactions on Wireless Communications* 2014; **13**(12): 6727–6740. DOI: 10.1109/TWC.2014.2360202.

[8] Rigazzi G, Chiti F, Fantacci R, Carlini C. Multi-hop D2D networking and resource management scheme for M2M communications over LTE-A systems. *IWCMC 2014 - 10th International Wireless Communications and Mobile Computing Conference* 2014: 973–978. DOI: 10.1109/IWCMC.2014.6906487.

[9] Mustafa HAU, Imran MA, Shakir MZ, Imran A, Tafazolli R. Separation Framework: An Enabler for Cooperative and D2D Communication for Future 5G Networks. *IEEE Communications Surveys & Tutorials* 2016; **18**(1): 419–445. DOI: 10.1109/COMST.2015.2459596.

[10] Sexton C, Bodinier Q, Farhang A, Marchetti N, Bader F, DaSilva LA. Coexistence of OFDM and FBMC for Underlay D2D Communication in 5G Networks. In *2016 IEEE Globecom Workshops (GC Wkshps)*, IEEE, 2016; 1–7. DOI: 10.1109/GLOCOMW.2016.7848863.

[11] Nguyen H-S, Nguyen T-S, Voznak M. Wireless powered D2D communications underlying cellular networks: design and performance of the extended coverage. *Automatika* 2017; **58**(4): 391–399. DOI: 10.1080/00051144.2018.1455016.

[12] Qiao J, Shen X, Mark J, Shen Q, He Y, Lei L. Enabling device-to-device communications in millimeter-wave 5G cellular networks. *IEEE Communications Magazine* 2015; **53**(1): 209–215. DOI: 10.1109/MCOM.2015.7010536.

[13] Ma X, Yin R, Yu G, Zhang Z. A distributed relay selection method for relay assisted Device-to-Device communication system. *IEEE International Symposium on Personal, Indoor and Mobile Radio Communications, PIMRC* 2012: 1020–1024. DOI: 10.1109/PIMRC.2012.6362495.

[14] Wang L, Peng T, Yang Y, Wang W. Interference Constrained D2D Communication with Relay Underlaying Cellular Networks. In *2013 IEEE 78th Vehicular Technology Conference (VTC Fall)*, IEEE, 2013; 1–5. DOI: 10.1109/VTCFall.2013.6692247.

[15] Shalmashi S, Ben Slimane S. Cooperative device-to-device communications in the downlink of cellular networks. *IEEE Wireless Communications and Networking Conference, WCNC* 2014; **3**: 2265–2270. DOI: 10.1109/WCNC.2014.6952682.

[16] Xia M, Aissa S. Cooperative AF relaying in spectrum-sharing systems: Outage probability analysis under co-channel interferences and relay selection. *IEEE Transactions on Communications* 2012; **60**(11): 3252–3262. DOI: 10.1109/TCOMM.2012.082712.120108.

[17] Louie RHY, Li Y, Vucetic B. Practical physical layer network coding for two-way relay channels: Performance analysis and comparison. *IEEE Transactions on Wireless Communications* 2010; **9**(2): 764–777. DOI: 10.1109/TWC.2010.02.090314.

[18] Cui Q, Yang X, Hämäläinen J, Tao X, Zhang P. Optimal energy-efficient relay deployment for the bidirectional relay transmission schemes. *IEEE Transactions on Vehicular Technology* 2014; **63**(6): 2625–2641. DOI: 10.1109/TVT.2013.2295413.

[19] Ni Y, Jin S, Ran Tian, Wong K-K, Zhu H, Shao S. Outage analysis for device-to-device communication assisted by two-way decode-and-forward relaying. In *2013 International Conference on Wireless Communications and Signal Processing*, IEEE, 2013; 1–6. DOI: 10.1109/WCSP.2013.6677242.

[20] Ni Y, Jin S, Wong KK, Zhu H, Shao S. Outage performances for device-to-device communication assisted by two-way amplify-and-forward relay protocol. *IEEE Wireless Communications and Networking Conference, WCNC* 2014; **1**: 502–507. DOI: 10.1109/WCNC.2014.6952079.

[21] Havary-Nassab V, Shahbazpanahi S, Grami A. Optimal distributed beamforming for two-way relay networks. *IEEE Transactions on Signal Processing* 2010; **58**(3 PART 1): 1238–1250. DOI: 10.1109/TSP.2009.2026067.

Frequency Increments Optimization Based on IPSO to Extend Beam Dwell Time for Frequency Diverse Array

Yuwen Pan[1], Yuanwang Yang[1,2,*] and Libin Jiao[2]

[1] Information and Communication Engineering, University of Electronic Science and Technology of China, Chengdu, 611731, China
[2] Science and Technology on Communication Networks Laboratory, Shijiazhuang, China

*Corresponding author email: yuanwangyang@uestc.edu.cn

Abstract. Compared with traditional phased array radar, frequency diverse array (FDA) can form a range-dependent beampattern and overcome the shortcoming that the traditional phased array radar cannot effectively control the range direction of the transmit beam. Nonuniform linear frequency increments has been widely applied in FDA to promote the performance of FDA. However, few methods can effectively extend the beam dwell time. In this paper, we propose an improved particle swarm optimization algorithm (IPSO) that can adjust the frequency increment of every array element to extend the dwell time of the beam within a period of time. Numerical simulation results show that our method can outperform other existing methods, which implies that a more stable and flexible FDA communication system can be achieved for the proposed method.
Keywords: Beam dwell time; Particle swarm optimization; Frequency diverse array; Frequency increment.

1. Introduction

Compared with phased array, frequency diverse array (FDA) applies the frequency increment of each array element, which can result in range, angle, and time-dependent beampattern. It can overcome many limitations of phased array such as scan loss due to the low dynamic range of phase-shifters and the beam direction is constant in the distance direction (the beam direction is independent of the distance) [1], [2]. Nevertheless, compared with phased array, the beam dwell time of a standard FDA (using linearly increasing frequency increments) is too short to satisfy a stable and flexible communication between two targets [3]. In other words, the dwell time of beam is still not ideal in the applications. Hence, an effective method is necessary to decide the feasible frequency increments to extend the dwell time of beam.

Selection of the suitable frequency increment of each array element of FDA is one of the key challenges and there are several researches in this field. A standard FDA usually uses the linear frequency increments of every array element. Nevertheless, it cannot unambiguously localize targets [2]. Non-linear frequency increments can generate a wider beam, which addresses this problem to a certain extent [2], [4]. W. Khan et al. [1] proposed a logarithmically increasing frequency increments to achieve a beampattern with a single maximum at the target location. In [2], J. Xiong et al. proved that logarithmically increasing frequency increments will cause the poor performance in the range and angle dimension and proposed a frequency increments strategy optimized by genetic algorithm to solve this problem. And there are many other methods to decide the frequency increment of every array element to decouple the angle-range beampattern [5]. However, the above researches only consider the static beampattern and ignore the beam dwell time, which is a key factor that has an influence on the stability and flexibility of communication in the applications [3]. Most existing researches have discussed the beam dwell time for phased array [6]. However, for FDA, few studies have paid enough attention to beam dwell time. Therefore, an effective algorithm is essential to optimize the frequency increment of every array element to achieve longer beam dwell time and more stable and flexible communication system.

In this paper, a far-field application is assumed and an improved particle swarm optimization algorithm (IPSO) is proposed to select the optimal frequency increments for FDA. It focuses on changing the frequency increment of every array element to maximize the integral of square of absolute value of array-factor (AF) over a desired time period at the target location, which can extend the beam dwell time by a large margin. Numerical simulation results show that our proposed algorithm (IPSO) can outperform the existing method, including linear frequency increments (standard FDA) and logarithmical frequency increments in [1].

2. Basic Theory of FDA

In the FDA with M array elements, supposing the carrier frequency is f_c, then, the frequency increment of mth array element can be expressed as Δf_m. The radiated frequency of mth array element is

$$f_m = f_c + \Delta f_m \tag{1}$$

where $0 \leq m \leq M-1$. The model of FDA transmitter is presented in Figure 1.

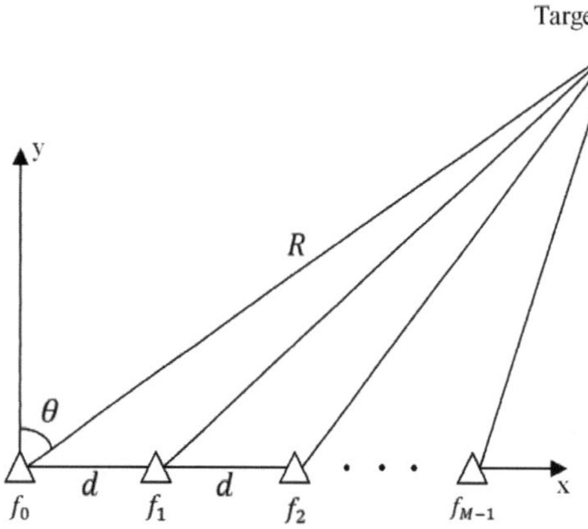

Figure 1. Model of FDA radar transmitter

In the far-field condition, the signal transmitted by mth array element can be denoted as [1]

$$s_m(t) = \omega_m e^{-j2\pi f_m t} \tag{2}$$

where ω_m is a coefficient for mth array element. If the target location is R meters away from FDA and the distance between two adjacent array elements is d meter (shown in Figure 1), then, the signals superposition of M array elements can be expressed as

$$s_m(\theta, R, t) = \sum_{m=0}^{M-1} \omega_m e^{-j2\pi f_m (t - \frac{R - m d \sin(\theta)}{c})} \tag{3}$$

where c is the speed of light. Then, the AF can be written as [4]

$$AF(\theta, R, t) = \sum_{m=0}^{M-1} \omega_m e^{-j2\pi[\Delta f_m \left(t - \frac{R}{c}\right) + m f_c \frac{d \sin(\theta)}{c} + \frac{m \Delta f_m d \sin(\theta)}{c}]} \tag{4}$$

Considering that the frequency increments Δf_m is kept much lower than the carrier frequency f_c, so $m f_c \frac{d\sin(\theta)}{c} \gg \frac{m\Delta f_m d\sin(\theta)}{c}$ and $\frac{m\Delta f_m d\sin(\theta)}{c}$ can be ignored in equation (4). Hence, equation (4) can be expressed in more compact form as

$$AF(\theta,R,t)=\sum_{m=0}^{M-1}\omega_m e^{-j2\pi[\Delta f_m\left(t-\frac{R}{c}\right)+mf_c\frac{d\sin(\theta)}{c}]} \tag{5}$$

Supposing $\omega_m=e^{-jm\varphi}$ [4], the final expression of AF can be written as

$$AF(\theta,R,t)=\sum_{m=0}^{M-1}e^{-j2\pi[\Delta f_m\left(t-\frac{R}{c}\right)+mf_c\frac{d\sin(\theta)}{c}+\frac{m\varphi}{2\pi}]} \tag{6}$$

3. Frequency Increments Optimization Based on IPSO

To extend the beam dwell time in the desired time period (suppose it is t_1 to t_2), frequency increment of every array element will be adjusted and $E_{AF}(t_1,t_2)$ is considered as the optimization target, where the expression of $E_{AF}(t_1,t_2)$ can be written as

$$E_{AF}(t_1,t_2)=\int_{t_1}^{t_2}|AF(\theta,R,t)|^2dt \tag{7}$$

Here, $E_{AF}(t_1,t_2)$ represents the energy in the time period t_1 to t_2. So, maximizing $E_{AF}(t_1,t_2)$ is to extend the beam dwell time from t_1 to t_2. Then, this problem can be recast as

$$\begin{cases}\max[E_{AF}(t_1,t_2)]\\ s.t.\begin{cases}\Delta f_{min}<\Delta f_0<\Delta f_{max}\\ \Delta f_{min}<\Delta f_1<\Delta f_{max}\\ \cdots\\ \Delta f_{min}<\Delta f_M<\Delta f_{max}\end{cases}\end{cases} \tag{8}$$

where Δf_{min} represents the minimum value of frequency increments and Δf_{max} is the maximum value of frequency increments. To find the maximum value of $E_{AF}(t_1,t_2)$ by mathematical derivation is relatively difficult. Therefore, in this paper, we choose heuristic algorithm (particle swarm optimization algorithm) to solve this problem. Furthermore, an improved particle swarm optimization algorithm (IPSO) is proposed to promote the efficiency and applied in the optimization process. The process of frequency increments optimization base on IPSO will be introduced below.

Supposing the number of array elements is M and the population size is N. Then, we use a matrix P to represent the population which consists of N individuals that present the frequency increments.

$$P=\begin{bmatrix}\Delta f_{11} & \Delta f_{12} & \cdots & \Delta f_{1M}\\ \Delta f_{21} & \Delta f_{22} & \cdots & \Delta f_{2M}\\ \cdots & \cdots & \cdots & \cdots\\ \Delta f_{N1} & \Delta f_{N2} & \cdots & \Delta f_{NM}\end{bmatrix} \tag{9}$$

Here, Δf_{ij} $(0<i\le N, 0<j\le M)$ represents the frequency increment of jth array element of ith individual. And another matrix V is given to express the velocity of individuals.

$$V=\begin{bmatrix}v_{11} & v_{12} & \cdots & v_{1M}\\ v_{21} & v_{22} & \cdots & v_{2M}\\ \cdots & \cdots & \cdots & \cdots\\ v_{N1} & v_{N2} & \cdots & v_{NM}\end{bmatrix} \tag{10}$$

$V_i=[v_{i1},v_{i2},\ldots,v_{iM}]$ is the velocity of ith individual. Then, combined with equation (7), fitness of all individuals in population P can be calculated as

$$\rho_i=\int_{t_1}^{t_2}\left|\sum_{k=1}^{M}e^{-j2\pi[\Delta f_{ik}\left(t-\frac{R}{c}\right)+kf_c\frac{d\sin(\theta)}{c}+\frac{k\varphi}{2\pi}]}\right|^2dt \tag{11}$$

where ρ_i is the fitness value of ith individual. And the greater value of ρ, the better the individual.

Based on the above theory, we give the process of frequency optimization based on the IPSO. First, initialize population P and velocities of all individuals V. Second, calculate the fitness of all individuals by the equation (11). Then, update V based on the equation (12).

$$V_i\leftarrow\alpha V_i+c_1r_1(\eta_{besti}-P_i)+c_2r_2(\zeta_{best}-P_i) \tag{12}$$

Here, η_{besti} and ζ_{best} denote the historical best position (position of ith individual is $P_i=[\Delta f_{i1}\ \ \Delta f_{i2}\ \ \cdots\ \ \Delta f_{iM}]$) of ith individual and optimal position of all individuals. c_1 and c_2 are the coefficients, r_1 and r_2 are the random number between 0 and 1. If $V_i > V_{max}$ or $V_i < V_{min}$, set $V_i = V_{max}$ or $V_i = V_{min}$, where V_{max} and V_{min} denote the maximum and minimum velocity of individuals. Third, apply elite strategy [7] to keep the best individual unchanged, other individuals' positions will be updated through equation (13).

$$P_i \leftarrow P_i + V_i \tag{13}$$

If $P_i > \Delta f_{max}$ or $P_i < \Delta f_{min}$, set $P_i = \Delta f_{max}$ or $P_i = \Delta f_{min}$, where Δf_{max} and Δf_{min} represent the maximum and minimum value of frequency increments of every array element (maximum and minimum position boundaries). Then, calculate the new fitness for all individuals and repeat the above steps (update velocity and positions). If the fitness of ith individual is unchanged for γ iterations, we reinitialize this individual (include its position and velocity) to prevent it falling into local optimal solution (this part is the improvement for PSO). If there are no changes for the best fitness in ε iterations, the algorithm is terminated. The process of this algorithm is summarized as follow.

Algorithm 1: Frequency increments optimization based on IPSO

Input: Desired time period $t_1 \sim t_2$

Output: Best frequency increments ζ_{best}

1: Initialize population P (all individuals) and their velocities V, where $V \in (V_{min}, V_{max})$ and $P \in (\Delta f_{min}, \Delta f_{max})$

2: Initialize all coefficients ($\alpha = 0.9$, $c_1 = 2.05$, $c_2 = 2.05$, $M = 75$, $\gamma = 10$, $\varepsilon = 500$).

3: Calculate fitness for all individuals by equation (11).

4: Update velocity and position of all individuals by equation (12) and (13).

(1): If velocity of individuals is out of range, then, keep it in the range $[V_{min}, V_{max}]$.

(2): If positions of individuals are out of range, then, keep it in the range $[\Delta f_{min}, \Delta f_{max}]$.

5: Repeat **3**~**4** until **5**.(2) is executed.

(1): If the fitness of ith individual is unchanged for γ iterations, reinitializing this individual.

(2): If the best fitness has no change for ε iterations, terminating algorithm.

4. Simulation Results

In this paper, we set the distance between FDA and target location $R = 300$ km and $\theta = \frac{1}{6}\pi$. And the number of array elements M is set to 20. Other parameters are set as $f_c = 5$ GHZ (carrier frequency), $d = 0.03$ m (distance between two adjacent array elements) and $\varphi = \frac{1}{4}\pi$. Referring to [4], we set the minimum value of frequency increment is $\Delta f_{min} = 0$ HZ and the maximum value of frequency increment is $\Delta f_{max} = 250$ HZ. The minimum value of velocity is $V_{min} = -\frac{1}{15}\Delta f_{max}$ and the maximum value of velocity is $V_{max} = \frac{1}{15}\Delta f_{max}$.

In the simulations, we applied three methods include linear frequency increments (method 1), logarithmical frequency increments (method 2) and frequency increments based on IPSO (our method) to form a contrast. Here, for linear frequency increments, in equation (6), we set $\Delta f_m = m \times f_o$, where $f_o = 100\ HZ$. And for logarithmical frequency increments, in equation (6), we set $\Delta f_m = \log(m + 1) \times f_l$, where $f_l = 3000$ HZ. Then, we fix R, θ, f_c, d and select the desired time period $t_1 \sim t_2 =$ 0,0145 s~0.0185 s. Using method 1, method 2 and our method to obtain the beam dwell time. Results with method 1, method 2 and our method are shown in Figure 2 and Table 1.

Table 1. Comparison of beam dwell time with method 1, method 2 and our method.

Method	Beam dwell time (3 dB)
Method 1	0.44 ms
Method 2	0 ms
Our method	7.11 ms

From Figure 2 and Table 1, we can notice that the 3dB beam dwell time [5] with method 1, method 2 and our method are 0.44 ms, 0 ms and 7.11 ms (the maximum value of $|AF|^2$ obtained by method 2 is 90, which cannot reach to $\frac{400}{2} = 200$, so the beam dwell time is 0 ms.). Therefore, we can see that the beam dwell time can be effectively extended by our method.

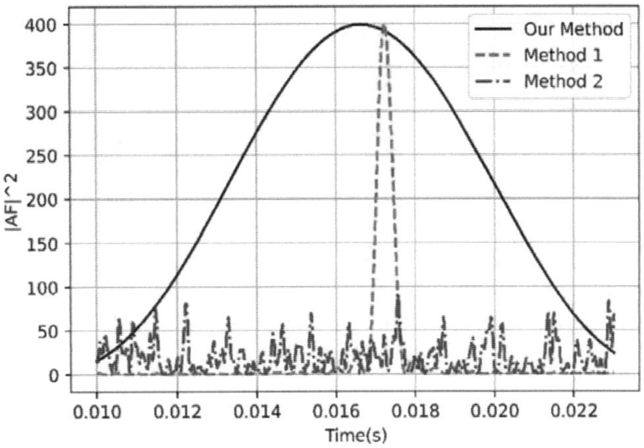

Figure 2. Beam dwell time with method 1, method 2 and our method.

5. Conclusion

The feasible frequency increments of FDA will affect the beam dwell time. However, many researches did not take this factor into consideration. And existing methods cannot effectively resolve this problem, which will lead to an unstable and inflexible communication system. In this paper, we proposed frequency increments optimization for FDA based on IPSO to extend the beam dwell time. Comparative simulation results show that our method can outperform other existing methods to obtain longer beam dwell time for FDA.

Acknowledgments

This work was supported in part by the Science and Technology on Communication Networks Laboratory Foundation Project under Grant 6142104190202.

Reference

[1] W. Khan, I. M. Qureshi, and S. Saeed, "Frequency diverse array radar with logarithmically increasing frequency offset," *IEEE Antennas Wireless Propagation Letters*, vol. 14, pp. 499–502, 2015.

[2] J. Xiong, W. Wang, H. Shao and H. Chen, "Frequency Diverse Array Transmit Beampattern Optimization With Genetic Algorithm," in *IEEE Antennas and Wireless Propagation Letters*, vol. 16, pp. 469-472, 2017.

[3] Chen, Hui, Zi - Hao Li, and Wen - Qin Wang. "Analysis of beampattern dwell time for planar frequency diverse array." *IET Signal Processing* 15.1 (2021): 40-45.

[4] M. Zubair, S. Ahmed and M. Alouini, "Frequency Diverse Array Radar: New Results and Discrete Fourier Transform Based Beampattern," in *IEEE Transactions on Signal Processing*, vol. 68, pp. 2670-2681, 2020.

[5] K. Gao, H. Chen, H. Shao and J. Cai, "A two-dimensional low-sidelobe transmit beampattern synthesis for linear frequency diverse array," *2015 IEEE China Summit and International Conference on Signal and Information Processing (ChinaSIP)*, Chengdu, China, 2015.

[6] Y. Zhang, Y. Yuan, J. Wang and H. Xiang, "Improved adaptive scheduling algorithm for real-time dwells in multifunction phased array radars," *2014 12th International Conference on Signal Processing (ICSP)*, Hangzhou, China, 2014, pp. 2018-2021.

[7] Y. Pan, Y. Yang and W. Li, "A Deep Learning Trained by Genetic Algorithm to Improve the Efficiency of Path Planning for Data Collection With Multi-UAV," in *IEEE Access*, vol. 9, pp. 7994-8005, 2021.

A Knowledge Base-based Method for Line Loss Abnormality Diagnosis in Station Area and Automatic Generation of Loss Reduction Measures

Linna Ni*, Li Yao, Qi Ding and Jiangming Zhang

State Grid Zhejiang Marketing Service Center, Hangzhou 310000, China

*Corresponding author email: nilinna918@qq.com

Abstract. In view of the current low efficiency of line loss abnormality analysis and the inability to automatically generate loss reduction strategies, this paper proposes a knowledge base-based method for the abnormality diagnosis of line loss and automatic generation of loss reduction strategies. In this study, the research on line loss in station area at home and abroad is summarized, the classification of line loss abnormality is sorted out. Combined with the operation status of low-voltage distribution network and the current status of line loss, the intelligent diagnosis method of line loss abnormality is proposed and line loss reduction strategies is automatically generated according to the classification of line loss abnormality, which provide a decision-making support for improving the lean level of line loss control in the station area. Through application system development and display, the effectiveness of this method is verified.
Keywords: Knowledge base; Line loss in station area; Abnormality diagnosis; Loss reduction measures; Automatic generation.

1. Introduction

At present, the line loss control in the station area still mainly relies on formulating relevant indicators based on the statistical function of the information system. The line loss control method of station area based on the electricity consumption information collection system is proposed, the factors that affect station area line loss control are analyzed in literature [1-4] to solve the problems that arise in the line loss control of the station area. The current problems in line loss control is analyzed in literature [5], and corresponding control measures are proposed in terms of detailed responsibilities, rewards and punishments, and professional collaboration. The paths and methods of line loss control in low-voltage station area from six aspects are put forward in literature [6], such as improving the control network, establishing control mechanism and quantifying the quality of checking and receiving work, with examples to prove the application of these control methods. Problems in the lean control of low-voltage station areas are analyzed, and an analysis of the line loss control mode in low-voltage station areas is proposed in literature [7].

The knowledge base is an important part of the knowledge organization theory, and plays an important role in traditional industries. It is prominently manifested in the aspects of knowledge acquisition, knowledge representation and knowledge reasoning. Regarding the line loss reduction measures, the usage of the knowledge base also focuses on the knowledge acquisition, knowledge representation and knowledge reasoning.

In order to better improve the line loss control level, a knowledge operation system with independent "cognitive ability" is established, which can quickly and intelligently provide corresponding loss reduction strategies. All kinds of line loss-related information and data are combined meaningfully, and then become knowledge after interpretation and evaluation through contextual thinking models. The knowledge and loss reduction measures activities are linked to generate problem-solving solutions based on changes to form a knowledge base of loss reduction strategies.

2. Line Loss Abnormality Diagnosis in Station Areas

2.1. Diagnosis Process of Line Loss Abnormalities

The line loss abnormality diagnosis mainly include: determining whether the line loss in the station area is normal, diagnosis of the line loss abnormalities in the station area, diagnosis suggestions and on-site treatment. The specific process is shown in Figure 1:

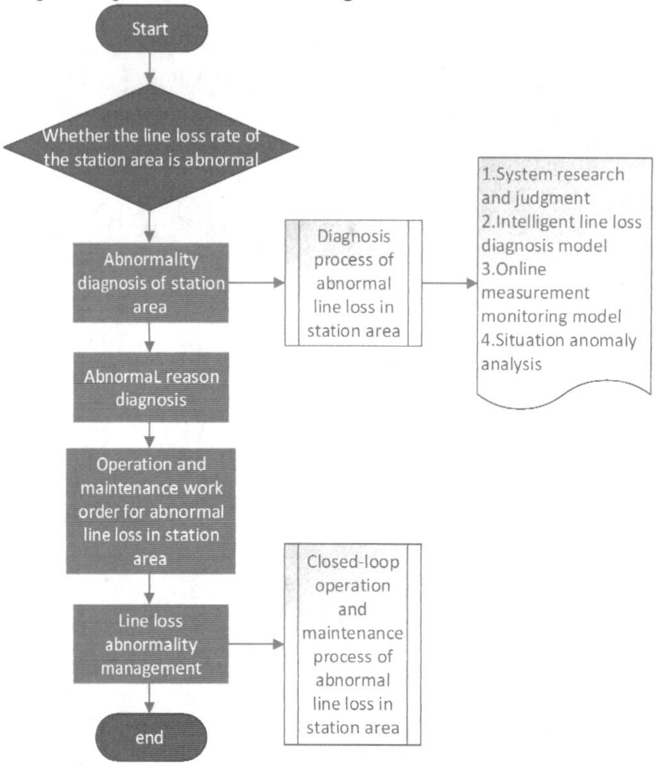

Figure 1. The abnormality diagnosis of line loss in station areas.

The diagnosis process of line loss abnormality in the station area is based on whether the line loss rate in the station area is qualified or not. The system diagnoses the causes of abnormality in the diagnostic station area every day, gives specific diagnosis results and treatment suggestions, and generates a closed-loop work order for defect processing. On-site personnel conduct on-site treatment and feedback according to the work orders and recommended treatments, and complete the work order filing.

2.2. The Classification of Abnormal Causes of Line Loss

There are many factors related to the line loss in the station area. The main reasons include the operation status of the power system, the user files modifications, the power system transformation, the circulation lag of data flow between different systems, the acquisition equipment failure, measurement imprecision, electricity consumption abnormality, etc. In this paper, the line loss abnormalities in the station area can be divided into technical factors and management factors, and also into 10 categories and 46 subcategories. The line loss abnormalities are classified as follows:

Table 1. The classification of line loss abnormalities

The technical factors	Unreasonable topology	*a*) Excessive power supply radius; *b*) The load concentrated at the end; *c*) unbalanced voltage and current
	Distributed household photovoltaic	*a*) Distributed household photovoltaic installation location and capacity
	Reasons for the equipment	*a*) The smaller wire diameter; *b*) The aging line; *c*) The lightning arrester is damaged; *d*) The electricity leakage caused by line grounded
	Reasons for load	*a*) Frequent overload
The management factors	Abnormal collection and communication	*a*) The collection and communication failure
	File abnormality	*a*) CT magnification error; *b*) User file changes; *c*) The household relationship change
	Abnormal measurement	*a*) The meter stops; *b*) Abnormal deviation of metering data; *c*) The measurement value is uneven; *d*) Abnormal communication; *e*) The unbalanced current; *f*) Phase failure; *g*) The unbalanced voltage; *h*) Abnormal power factor; *i*) Abnormal voltage loss; *j*) Overcapacity of the measurement equipment; *k*) Long-term loss of current; *l*) Abnormal current loss
	Photovoltaic abnormalities	*a*) Abnormal measurement equipment gateway; *b*) Abnormal photovoltaic power generation; *c*) Abnormal measurement area; *d*) Abnormal reverse power; *e*) No photovoltaic collection
	Abnormal power consumption	*a*) kWh meter; *b*) Abnormality relationship between power and line loss; *c*) Continuous 0 kWh meter; *d*) Abnormal power consumption change; *e*) Big power users; *f*) Abnormal power fluctuation; *g*) Abnormal power level; *h*) Abnormal peak-to-valley ratio; *i*) Abnormal power percentage
	Breach of contract/theft of electricity	*a*) Suspected electricity theft; *b*) Users without measurement equipment

3. Automatic Generation of Loss Reduction Strategies Based on the Knowledge Base

3.1. In-depth Correlation between Knowledge Base Structure and Loss Reduction Measures

The structure of knowledge base is analyzed based on the knowledge organization theory to make it deeply accommodated to the decision-making needs of loss reduction events. The deep correlation between the knowledge base structure and the loss reduction measures can be abstracted as shown in Figure 2. Based on such a structural system, line loss event factors are regarded as inputs, loss reduction strategies are regarded as outputs in a big data environment, , and a knowledge map is constituted by the knowledge base connected to the input and output ends.

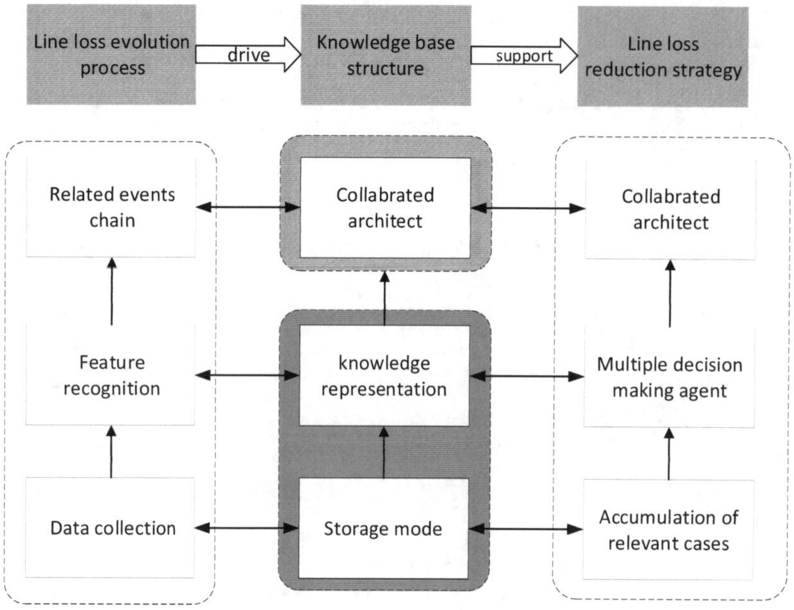

Figure 2. The deep correlation between the knowledge base structure system and the line loss reduction measures.

3.2. The Relationship between Knowledge Representation and Loss Reduction Events

Based on the collection of online loss event data, it is more necessary to gather and integrate data from multiple sources around the occurrence of line loss to match the characteristics of previous loss reduction information. As a result, the knowledge base is required to carry out effective knowledge representation, which is conducive to the stored knowledge to accurately describe the characteristics of line loss event factors. The line loss factor information collected from different information sources and through different channels are analyzed and processed, and disorderly and unstructured data are transformed into orderly, structured, and usable knowledge. A reasonable knowledge base structure is not only limited to the storage of massive knowledge, but also easier to form the correlation between knowledges, which helps to infer new knowledge.

First of all, non-clean data needs to be cleaned. From the input of data to the output of knowledge, the ultimate goal of the knowledge base is to provide accurate and timely knowledge for loss reduction decision-making, so the data must be cleaned at the beginning. Through data cleaning, the scope of knowledge collection in the knowledge base is narrowed, and the noise is removed, so the accuracy of line loss event feature description is improved. On this basis, the association between multiple information sources and the knowledge base is further studied to reveal one-to-one correspondence between the line loss information source and the knowledge base.

In actual work, when storing line loss event data, according to the different application goals of technical line loss and management line loss, the loss reduction measures information of different fields can be correspondingly stored in knowledge bases of different line loss factor. On this basis, the integration between these sub-event knowledge bases is realized by means of modern information technology (Figure 3). According to the principle of high cohesion and low coupling, this integration is mainly realized through interfaces.

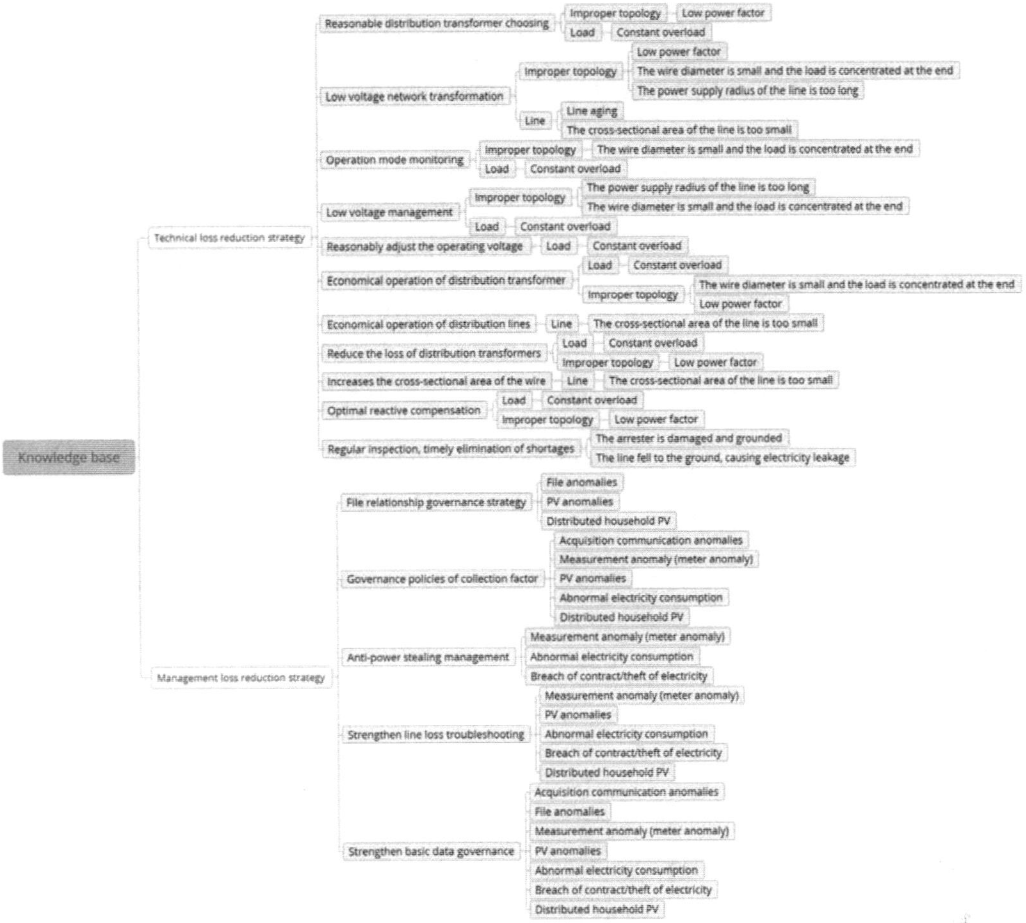

Figure 3. Line loss reduction strategies.

3.3. Display of Line Loss Abnormality

The system is developed based on the above theory, and the station area abnormality and the user abnormality are displayed as examples.

The station area abnormality take the station area as an object, displaying data such as the main cause of the line loss abnormality in the station area, the number of abnormal users and the abnormal time of line loss in the station area. As shown in Table 2, it is recommended whether the station area needs on-site treatment through the abnormal type of the station area and the influence degree on the line loss.

Table 2. Line loss abnormal details in the station area

Electricity supply unit	Station area NO.	Line loss rate	Condition	Abnormal type	Abnormal number of users	Abnormal level
Anji Bureau	0001227126	2.86	nomal	File anomalies	1	The third level
Anji Bureau	0002102865	2.3	nomal	File anomalies	1	The third level
Anji Bureau	0002272821	4.34	nomal	File anomalies	1	The third level
Anji Bureau	0000588395	1.55	nomal	File anomalies	1	The third level

Anji Bureau	0000589305	0.92	nomal	PV anomalies	1	The third level
Anji Bureau	0002462567	5.9	nomal	File anomalies	2	The third level
Anji Bureau	0005893033	5.06	nomal	Meter anomalies	1	The third level
Anji Bureau	0002435244	5.44	nomal	File anomalies	2	The third level
Anji Bureau	0001432832	4.61	nomal	File anomalies	1	The third level
Anji Bureau	0000588432	1.19	nomal	File anomalies	1	The third level
Anji Bureau	0001966369	1.69	nomal	File anomalies	1	The third level
Anji Bureau	0001318310	7.82	Extra large	File anomalies, electric larceny	2, 5	The third level
Anji Bureau	0000587489	3	nomal	PV anomalies	1	The third level

User abnormalities take a single user as a subject, displaying the user information, the abnormal type, the abnormal duration and other data. The data can support further verification of the abnormal cause.

Table 3. Line loss abnormal details of users

Electricity supply unit	Station area NO.	User NO.	Electricity consumptio n	Abnormal number	Abnormal type	Abnormal duration
Anji Bureau	000058843 2	37180547 8	34	1	Abnormal power level	1
Anji Bureau	000058843 2	37200659 6	10	1	electric larceny	1
Anji Bureau	000058843 2	37203684 2	0	1	electric larceny	1
Anji Bureau	000058843 2	37201195 2	0	1	electric larceny	1
Anji Bureau	000058843 2	37200186 4	0	1	electric larceny	1
Anji Bureau	000058843 2	37162504 4	0	1	electric larceny	1
Anji Bureau	000058843 2	37160256 3	0	1	User relationship error	1
Anji Bureau	000058843 2	27200984 5	45	1	User relationship error /Abnormal power level	2

4. Conclusion

The line loss research status of the low voltage distribution networks at home and abroad is summarized, and the classification of line loss abnormalities are sorted out firstly. Subsequently, the intelligent line loss diagnosis method is proposed according to the classification of line loss abnormalities. Based on the knowledge base theory, line loss reduction strategies are automatically generated to provide decision-making support for improving the lean control level of the line loss in the station area. Finally, the decision making system is developed to verify the proposed method. The research in this paper can provide theoretical guidance for the low voltage line loss reduction, the efficiency improvement of the power enterprises, and have significant importance for the lean control of the low voltage power network and promotion of the productivity and economy benefits.

Acknowledgments
This work is supported by the China State Grid Corporation Sci.& Tech. Project 5600-201919168A-0-0-00.

References
[1] T. Song, Q. Wang, S. Zhang, et al, "New line-loss management technology of distribution area based on electricity information collection system," *J. Eng.*, no. 3, pp. 1-4, Mar. 2019.
[2] Z. Ma, X. Zhou, Y. Shang, et al, "Form and development trend of future distribution system," *Proc. CSEE*, vol. 35, no. 6, pp. 1289-1298, Mar. 2015.
[3] C. Samende, S. M. Bhagavathy, M. McCulloch, " Distributed state of charge-based droop control algorithm for reducing power losses in multi-port converter-enabled solar DC nano-grids," *IEEE Trans. Smart Grid*, vol. 12, no. 6, pp. 4584-4594, Nov. 2021.
[4] Z. Ma, T. An, Y. Shang, et al, "State of the art and development trends of power distribution technologies," *Proc. CSEE*, vol. 36, no. 6, pp. 1552-1567, Mar. 2016.
[5] Y. Cheng, H. Yang, J. Xiao, et al, "Operation errors analysis of electric energy metering device and state evaluation method research," *Adv. Technol. Elec. Eng. Energy*, vol. 33, no. 5, pp. 76-80, May 2014.
[6] J. Wang, C. Liu, J. Qin, et al, "Multidimensional design ideas of reducing loss and increasing benefit based on ubiquitous power Internet of Things," *2020 IEEE International Conference on Power, Intelligent Computing and Systems* (*ICPICS*), 2020, pp. 63-66.
[7] W. Hu, Q. Guo, Y. Liu, et al, "Intelligent line losses management system for secondary power distribution network based on SSM framework," *2020 IEEE 4th Conference on Energy Internet and Energy System Integration* (*EI2*), 2020, pp. 2815-2820.

Research on Super-resolution Location Technology of Leakage Source Based on Ultrasonic Array Sensor

Xu Bian[1,*], Biwan Tian[2] and Shijiu Jin[2]

[1]School of Information and Intelligence Engineering, Tianjin Renai College, Tianjin, China.
[2]State Key Laboratory of Precision Measurement Technology and Instrument, Tianjin University, Tianjin, China.

*Corresponding author email: bx332@tju.edu.cn

Abstract. Leakage in the pressure structure is a problem that must be paid attention to, especially in some special applications, such as spacecraft, which will have a serious impact on the entire system. This paper presents a new method of leak source location based on a single sensor array. The elastic wave excited by the leakage is collected by an array sensor placed on the wall, and then the frequency-space characteristics of these signals are analyzed to obtain the spatial point that meets these characteristics, that is, the spatial location of the leakage source. At the same time, combining the super-resolution technology and the randomness of the signal, integrating the distribution maps of the positioning results in different time periods, and finally obtaining a high-precision estimation of the location of the leakage source. The experimental results show that, under the conditions considered in this paper, the average value of the absolute positioning error on a square meter board is about 2 cm.
Keywords: Leakage; Locate; Sensor array; Super-resolution.

1. Introduction

Leakage is a noteworthy problem in pressured structures[1]. Once the leakage occurs, such as spacecraft, the consequences could be fatal. Thus, an effective real-time detection method, which can quickly confirm the leakage source is very necessary. Meanwhile, the situation that we need to study should not only include the source of leakage that occurs suddenly (such as impact), but also the source of leakage that has already leaked.

After a leak occurs, the source of the leak will generate elastic waves in the wall. Based on this phenomenon, many previous studies are based on the time-domain information at the time when the elastic wave is generated[2]. The typical are Time Difference of Arrival (TDOA) technique[3], time reversal approach technique[4] and so on. However, because the leakage will continue to produce broadband noise signals, when the elastic wave excited by it propagates in the wall for a period of time, the signal we obtain has no obvious time-domain characteristics[5], it is more difficult for us to locate the continuous leakage source, thus, it is needed to be further explored. Combined with super-resolution technology, Ashida, Yuto et al. [6] have given the maximum likelihood estimation of the location of multiple transmitters through theoretical derivation, and a method that can simultaneously locate overlapping sound sources is proposed. W. Djerir et al.[7] have researched on super-resolution in hydrophone measurements of pulsed ultrasonic fields to deconvolve the spatial effects of the receiver hydrophone so as to reconstruct the pulsed ultrasonic field with a better spatial resolution. Under the circumstance of a single snapshot and coherent sources, Anupama Govinda Raj et al.[8] have resolved the problem of search-free direction of arrival (DOA) estimation for sensor arrays of arbitrary geometry, and extend a method of search-free super-resolution beamforming. An, Yun-Kyu et al. [9] have studied the propagation law of ultrasonic waves in stiffeners and proposed a Lamb wave line sensing technology for crack detection in welded stiffeners. James S Hall and Jennifer E Michaels [10] have proposed a multipath guided wave imaging algorithm to successfully detect and locate damage in geometrically complex structures by exploiting the echoes and reverberations present in ultrasonic waves propagating in the structure. Bolotina et al.[11] have advanced a passive leak detection method for underwater pipelines by the use of phased antenna arrays. To detect multiple leaks in water-filled pipes, XunWang et al. [12] have proposed a machine learning algorithm based on an iterative process. Numerical simulations show that the method can accurately identify up to 5 leak points and correctly estimate the number of leak points. Using the ultrasonic sensor array, we have presented a method based on frequency compensation and time domain superposition (SITD) to realize real-time location of the leak source[13]. The maximum positioning error is generally 10mm. However, this method uses two or more L-shaped ultrasonic sensor arrays placed in different positions to locate the leak source. The detection range of an array is 90°, which limits its wide application.

For container leakage, this article proposes a new method to locate the source of leakage in real time, which uses one ultrasonic sensor array place on the wall of the vessel. Different from the previous method[13], this method has widely applicability because of it only need one array and the coverage angle is 360°. The experimental results show that the method can locate the leak source accurately.

2. Location Method

Based on the theoretical analysis[13], The elastic wave generated by the leakage source propagates in the vessel wall, so an ultrasonic sensor array is placed on the vessel wall to sample the signal. The schematic diagram of leakage source location is shown in Figure 1:

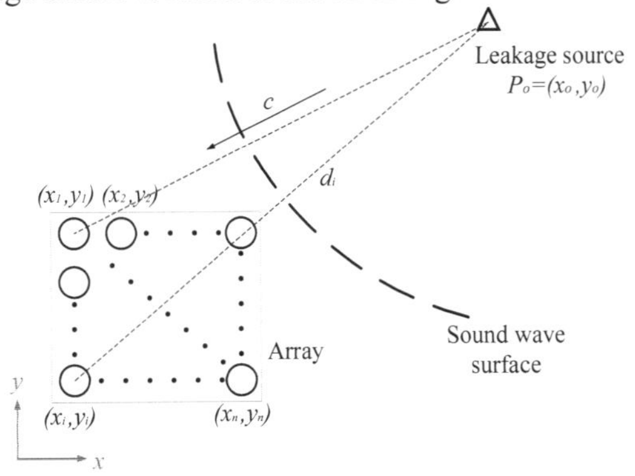

Figure 1. Schematic diagram of leakage source location

As show in Figure 1, c represents the velocity of elastic wave propagation in the medium, P_o represents the real location of the leak source, so, the distance between the i-th sensor and P_o is:

$$d_{o-i}=\sqrt{(x_o-x_i)^2+(y_o-y_i)^2} \tag{1}$$

Let $S(t)$ represents the signal excited by the leakage. Theoretically, the signal obtained by the i-th sensor S_i can be estimated as shown:

$$S_i=\alpha_i\cdot S\left(t+\frac{d_{o-i}}{c}\right)+R\left(t_R\right) \tag{2}$$

In equation 2, α_i represents the influence factor caused by different positions and frequency response property of the i-th sensor. Meanwhile, $R\left(t_R\right)$ represents the reflected wave, which can be written as:

$$R(t_R)=\sum_{l=1}^{L}\beta_l \cdot S\left(t+\frac{d_{l-i}}{c}\right) \tag{3}$$

In equation 3, d_{l-i} represents l numbers of different propagation paths of the reflected wave which collected by the i-th sensor, and β_l represents the influence factor which is mainly affected by the d_{l-i}, signal frequency and reflection condition.

In practice, P_o is unknown, so $P_k=(x_k,y_k)$ is selected instead of P_o in coverage area, then, the distance between the i-th sensor and P_k is:

$$d_i(k)=\sqrt{(x_k-x_i)^2+(y_k-y_i)^2} \tag{4}$$

The experimental results show that the elastic wave generated by the leakage is broadband in the frequency domain, and according to the Lamb wave theory, it has significant dispersion when propagating in the thin plate[14]. That is, the propagation speed of elastic wave (c) is related to its frequency (f). If the leak source is P_k, let c_f represent the sound velocity at frequency f, which can get from the dispersion curves[13], according the equation 1-4, the signal $p_i(x_k,y_k)$ excited by P_k which is calculated by the data of the i-th sensor S_i can be expressed as:

$$p_{i,f}(x_k,y_k)=\alpha_i(f)\cdot S_f\left[t+\frac{d_{o-i}-d_i(k)}{c_f}\right]+R\left[t_R-\frac{d_i(k)}{c_f}\right] \tag{5}$$

Normally, $d_i(k)$ is much longer than the size of the array and the properties of each sensor in the array are similar, so the difference of α_i can be ignored. Let $\alpha(f)=\alpha_i(f)$ $i=1,2,\ldots,n$ and $p_{\Sigma,f}(x_k,y_k)$ represents the sum of $p_{i,f}(x_k,y_k)$, which can be written as:

$$p_{\Sigma,f}(x_k,y_k)=\alpha(f)\cdot\sum_{i=1}^{n}S_f\left[t+\frac{d_{o-i}-d_i(k)}{c_f}\right]+\sum_{i=1}^{n}R\left[t_R-\frac{d_i(k)}{c_f}\right] \tag{6}$$

Let $E(k)$ represents the energy of $p_{\Sigma,f}(x_k,y_k)$ in time range $[0,T]$ and frequency range (f_0, f_m):

$$E(x_k,y_k)=\int_{f_0}^{f_m}\int_0^T\left|p_{\Sigma,f}(x_k,y_k)\right|^2 dtdf \tag{7}$$

Let **E** denote all the values of $E(x_k,y_k)$ in the range of x and y, Taking into account the energy attenuation of the elastic wave propagating in the wall, the proportion of signals in $p_{\Sigma,f}(x_k,y_k)$ which transmitted directly from the leak source to the sensor is relatively large, that is, it accounts for a large proportion in the value of **E** in any time segment. Thus, there is only one-point P_k can make $E(x_k,y_k)$ reaches the maximum in theory, that is $P_k=P_o$. Meanwhile, due to the randomness of the signal $S(t)$ excited by the leak hole in time domain, **E** has different distribution in different time segments. Based on sensor position, combined with the super-resolution method, the **E** on different time segments have been integrated and reconstructed to solve the problem of the limitation of sensor array aperture for locating accuracy. The locating method has implemented with MATLAB®, and the method flow chart is shown in Figure 2.

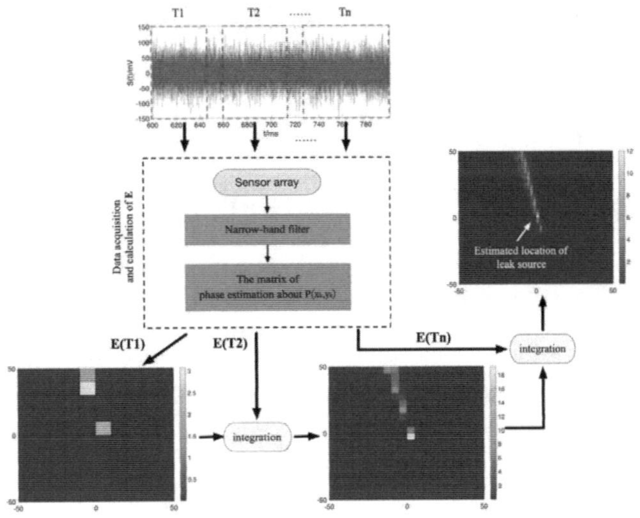

Figure 2. The method flow chart

Furthermore, on the basis of the above algorithm, through reasonable iterative calculation, the calculation time and scale can be further reduced, and higher positioning accuracy can be obtained at the same time.

3. Experiments and Discusses

302 stainless steel is selected as the test medium, and a plate with a size of 1000mm×1000mm×2.5mm is made. Some circular holes with diameters (0.8-2mm) are made in advance at different positions of the board to simulate leak holes. The leakage pressure is provided by a vacuum pump, which is connected to the leakage hole through a vacuum nozzle. With a vacuum pump running, air flows continuously from the atmosphere side to the vacuum side through the hole to simulate a leak. According to the principle of aeroacoustics, the elastic wave excited by the leakage propagates in the medium, the sensor array is fixed on the stainless-steel plate, and the data is obtained from the array by a data acquisition system (16-channel, 3 MHz sampling rate) and sent to computer. The experimental setup is shown in Figure 3.

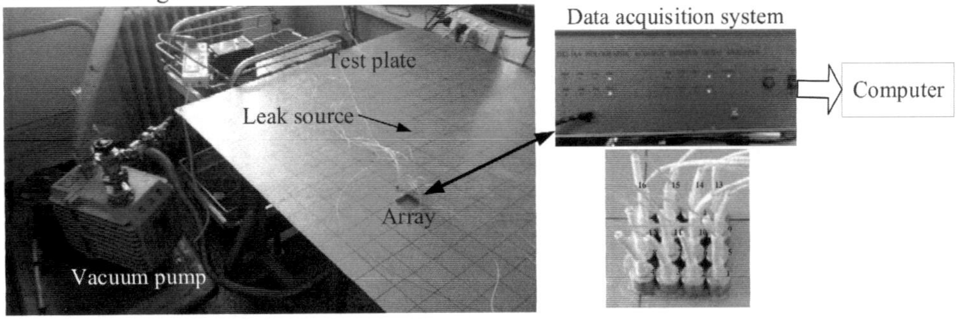

Figure 3. The schematic diagram of experimental set-up

Considering that the frequency response characteristics are stable and consistent within the range (100kHz, 400kHz), the sensor array uses Nano-30 sensors produced by PAC (America Physical Acoustics Co) to form a square. When the leak exists, the sensor array collects the wall vibration data and is applied by the method proposed in this article to obtain the location of the leak source. Theoretically, the number of sensors in the array has an impact on the positioning accuracy. A series of experiments have been designed to discuss this, and the experimental results are summarized in Table 1.

Table 1. The relationship between the number of sensors in the array and the locating accuracy

The sensor number for use	Average errors/cm	Variance
1-5	12.1	139.2
1-6	9.1	102.2
1-7	6.8	73.0
1-8	4.8	29.4
1-9	4.4	14.7
1-10	6.4	30.6
1-11	7.2	77.6
1-12	10.0	135.4
1-13	9.1	91.6
1-14	10.7	142.5
1-15	11.3	121.3
1-16	11.9	214.1

According to the Table 1, with the increasing number of sensors, the average error and the variance of the locating results decrease firstly and then increase. By comparison, the results are better when the number of sensors is from 7 to 11. This phenomenon can be explained as the increase of the sensor leads to the increase of the array area and the decrease of the consistency, which leads to the non-establishment of $\alpha(f)=\alpha_i(f) \quad i=1,2,\ldots,n$.

In theory, the smaller the number of sensors, the higher the computational efficiency. Therefore, according to Table 1, taking the 8 sensors in the array as a standard, the experiments are all based on the array of different shapes formed by the standard. Taking an L-shaped array experiment as an example, the actual coordinates of the leakage are (16.5, 28), which is estimated to be (16, 28). A large number of experiments were designed and conducted to cover all the spatial distribution between the leak source and the array. Taking the L-shaped array as an example, the statistical results of positioning errors are shown in Figure 4. D (position error) is defined as the distance between the estimated leakage position and the real hole, and r is the ratio of D to the distance between the real leakage hole and the array.

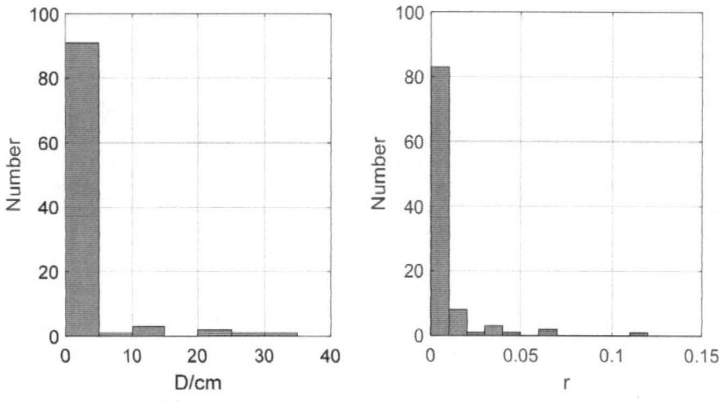

(a) (b)

Figure 4. the statistical results of positioning errors

As shown in Figure 4, the total sample size was 100. The percentage of $D \le 5$ is about 91% of the total sample, and the percentage of $D \ge 20$ is 4%. Meanwhile, the percentage of $r \le 0.01$ is about 83% of the total sample, and the percentage of $r \ge 0.1$ is 1%. That means, the method we proposed can obtain the accurate and stable results of leakage location. Moreover, the locating accuracy can be further improved, if there are two or more arrays placed in different positions, and through calculating their data, the mean of D is about 1.2cm and maximum of D is less than 2cm.

4. Conclusions

This article introduces a new method of leak source location based on ultrasonic sensor array. The following conclusions can be drawn from the research results:

1) The more sensors in the array are not always better, too many sensors will introduce too much interference, such as larger echo interference, worse performance consistency, etc., which will make the positioning result Getting worse. Under our experimental conditions, 8 sensor arrays are the best choice.

2) Aiming at container leakage, this paper proposes a high-precision real-time leakage source location method. The experimental results show that the method has low complexity and wide application. Under the experimental conditions, the average positioning error is 2.07cm.

Acknowledgments

This paper is supported by a grant from the Science & Technology Development Fund of Tianjin Education Commission for Higher Education. (No. 2019KJ150).

References

[1] Murvay, P.S.; Silea, I. A Survey on Gas Leak Detection and Localization Techniques. J. Loss Prevent. Process Ind. 2012, 25, 966–973.
[2] Ramadevi R, Jaiganesh J, Krishnamoorthy N R. Leak Detection Methods—A Technical Review. International Conference on Communications and Cyber Physical Engineering 2018. Springer, Singapore, 2018: 125-139.
[3] Dong, Linxi, et al. "The Gas Leak Detection Based on a Wireless Monitoring System." IEEE Transactions on Industrial Informatics 2019.
[4] F Ciampa and M Meo. Impact detection in anisotropic materials using a time reversal approach. Structural Health Monitoring, January 2012; vol. 11, 1: pp. 43-49.
[5] Wang T, Wang X, Hong M. Gas Leak Location Detection Based on Data Fusion with Time Difference of Arrival and Energy Decay Using an Ultrasonic Sensor Array. Sensors, 2018, 18(9): 2985.
[6] Ashida Y, Ueda M. Precise multi-emitter localization method for fast super-resolution imaging[J]. Optics Letters, 2016, 41(1):72.
[7] Djerir W, Boutkedjirt T . Simulation study of super-resolution in hydrophone measurements of pulsed ultrasonic fields[J]. Canadian Acoustics - Acoustique Canadienne, 2018, 46(3):19-28. Raj
[8] A G, Mcclellan J H . Single Snapshot Super-Resolution DOA Estimation for Arbitrary Array Geometries[J]. IEEE Signal Processing Letters, 2018, 26(1):119-123.
[9] An, Y.K.; Kim, J.H.; Yim, H.J. Lamb wave line sensing for crack detection in a welded stiffener. Sensors 2014, 14, 12871–12884.
[10] James S Hall and Jennifer E Michaels. Multipath ultrasonic guided wave imaging in complex structures. Structural Health Monitoring. July 2015; vol. 14, 4: pp. 345-358.
[11] Bolotina, I.; Borikov, V.; Ivanova, V.; Mertins, K.; Uchaikin, S. Application of phased antenna arrays for pipeline leak detection. J. Petrol. Sci. Eng. 2018, 161, 497–505.
[12] Wang X , Ghidaoui M S . Identification of multiple leaks in pipeline II: Iterative beamforming and leak number estimation[J]. Mechanical Systems and Signal Processing, 2019, 119:346-362.
[13] Bian, X.; Zhang, Y.; Li, Y.; Gong, X.; Jin, S. A new method of using sensor arrays for gas leakage location based on correlation of the time-space domain of continuous ultrasound. Sensors 2015, 4, 8266–8283.
[14] Rose, J.L. Ultrasonic Waves in Solid Media; Cambridge University Press: Cambridge, UK, 2004.

Anti-asphyxiation System for Children in Vehicle Based on Multi-sensor Fusion

Shaolong Peng[1], Bokang Yang[2], Zhihui Yang[2] and Shaoyue Shi[2*]

[1] Department of navigation, Wuhan University of Technology, Wuhan city, Hubei province, China
[2] School of Automation, Wuhan University of Technology, Wuhan city, Hubei province, China

*Corresponding author email: 1194159927@qq.com

Abstract. With the rapid development of automobile industry, cars have been widely used in thousands of households. But the car brings convenience to people's life, at the same time, the safety hidden trouble it brings is also more prominent and serious. In recent years, more and more accidents caused by asphyxiation of children in cars have prompted the scientific community to pay attention to and study this phenomenon. Therefore, we design a vehicle anti-asphyxiation system based on multi-sensor fusion.
Keywords: Multi-sensor fusion; Children asphyxia; Temperature and CO2 detection.

1. Introduction

At present, the frequent death cases of children stranded in cars at home and abroad have aroused widespread concern of public opinion. A study by US scientists found that when the temperature reached 35 °C and the sun shone for 15 minutes, the temperature in a closed carriage could rise to 65°C. In a test conducted by the Royal Automobile Club of Queensland, Australia, when the car is parked in the sun for 1 minute, the temperature inside the car will rise from 19°C to 30°C, and after 7 minutes, the temperature will reach more than 40°C.

In such a "tank car", it only takes less than half an hour for a child left alone in the car to encounter fatal danger, and it will not take long for excessive accumulation of heat in the body to cause damage to nerve organs, until death [3].

According to the survey data from Kids And Cars, a well-known child safety organization in the United States, at least 944 children have died from being trapped in Cars in the United States since 1990, And an average of 39 children die from being locked in Cars every year (1 every 9 days)[4]

Therefore, we design a vehicle anti-asphyxiation system based on multi-sensor fusion.

2. Existing Means of Protection

The protection measures that have been proposed include active devices and passive devices. Active safety devices detect children with cameras and sensors, and then ventilate and cool them down by controlling doors and Windows and air conditioning. The passive safety device is in the detection of children through the means of alarm protection. The passive safety device is in the detection of children through the means of alarm protection.

The former because of the need to be designed at the factory, so it is difficult to late modification, so the cost is too high. And the latter because of the detection means is too single, resulting in insufficient sensitivity and intelligence. [1]

In view of these problems, we finally decided to make a children's anti-asphyxiation system with multi-sensor fusion, and to achieve the integration of active and passive protection to a certain extent.

3. Hardware System Design

The system consists of main controller, comprehensive alarm module, personnel detection module, environment awareness module and CAN bus communication system. The detailed block diagram is as follows:

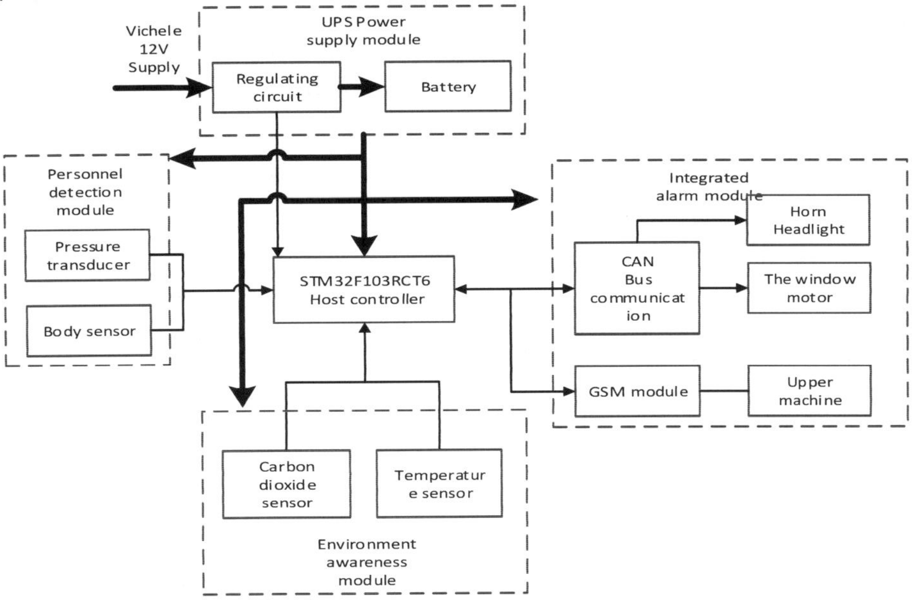

Figure 1. Hardware system design.

3.1. Main Control Module

The main control module needs to sense the occupant status in real time according to the transmitted data. The hardware system uses STM32F103RCT6 chip as the main control, which has strong performance and abundant peripherals. It has 48K RAM, multiple timers and DMA controller, and comes with SPI, IIC and serial port for easy communication. At the same time, the chip has its own watchdog, with strong anti-interference ability.[1]

3.2. Personnel Detection Module

The pressure sensor adopts thin film contact sensor, and the sensor's exit point is evenly distributed on the surface of the seat. When the seat is subjected to external pressure, a trigger signal is generated to judge whether the driver is in the seat. The passive safety device is in the detection of children through the means of alarm protection. The human body sensor uses PIR infrared pyroelectric induction switch probe, under normal circumstances, the human body temperature is generally maintained at 36.5 °C or so, will send out a specific wavelength of 10 μm or so infrared, used to judge whether there are children missing back row.

3.3. Integrated Alarm Module

When the device detects that a child has been stranded in the car, it is necessary to broadcast and alarm the outside world so that the child can be rescued in time. The passive safety device is in the detection of children through the means of alarm protection.

GSM transmission part adopts SIM-800C module, which is widely used, with support RS232 serial port and TTL serial port, and with hardware flow control, support 5V~24V ultra-wide working range, so that the module can be very convenient to connect. To provide including voice, SMS, MMS, Bluetooth and GPRS data transmission and other functions. [5] It is used to send SMS messages to the computer directly for reminding. [2]
CAN bus is used for communication with vehicles. CAN bus is A multi-master bus system. It is different from traditional bus systems such as USB or Ethernet, which realizes the transmission of large amounts of data from node A to node B under the coordination of bus controller. The message of CAN network is broadcast, that is, the data detected by all nodes on the network is consistent at the same time. Therefore, the vehicle system can maintain a high degree of consistency of information.
When a child is stranded, the SIM-800C module will send an alarm message to the car owner's mobile phone. If the car owner does not respond for a long time, the car communication system will lower the window to protect the child from suffocation, and turn on headlights and horns to alert the surrounding crowd.

3.4. Environment Awareness Module
The temperature sensor adopts DS18B21 to detect the real-time temperature inside the vehicle and sets 35°C as the alarm limit. The passive safety device is in the detection of children through the means of alarm protection. When the temperature is higher than 35°C and there are people in the car, the response system opens the skylight. Voice reminder device in the car plays prompt voice to open doors and Windows for ventilation and cooling; If the high temperature in the car does not change three minutes later, the SIM900 module sends a text message to the owner to remind him of the abnormal situation in the car. [4]
Carbon dioxide sensor by MC8ll carbon dioxide sensor detection of carbon dioxide concentration to solve. MC8ll is used to detect CO2 concentration in the vehicle. MC8ll carbon dioxide sensors usually carbon dioxide by infrared irradiation at room temperature, carbon dioxide gas to specific wavelengths of infrared radiation absorption effect, and by measuring the infrared radiation energy wanes, often can clear understanding to the concentration of carbon dioxide, the principle of using a certain photoelectric technology, Therefore, it has higher accuracy. Set the alarm limit of CO2 concentration to 3000ppm. When CO2 concentration is detected higher than 3000ppm, it will be a threat to children's breath, and the response system will open the skylight [4].

3.5. USP Power Module
The module is powered by the vehicle's 12V power supply, and has a spare battery designed at the same time. The 18650 lithium battery can work continuously in the state of power failure to ensure the stable operation of the system. The passive safety device is in the detection of children through the means of alarm protection.

4. Software System Design
On the basis of the above hardware system, the supporting software structure is developed to realize the accurate judgment of the situation in the car and the alarm in case of emergency, so as to realize the protection of children. The specific software process is as follows: The passive safety device is in the detection of children through the means of alarm protection.

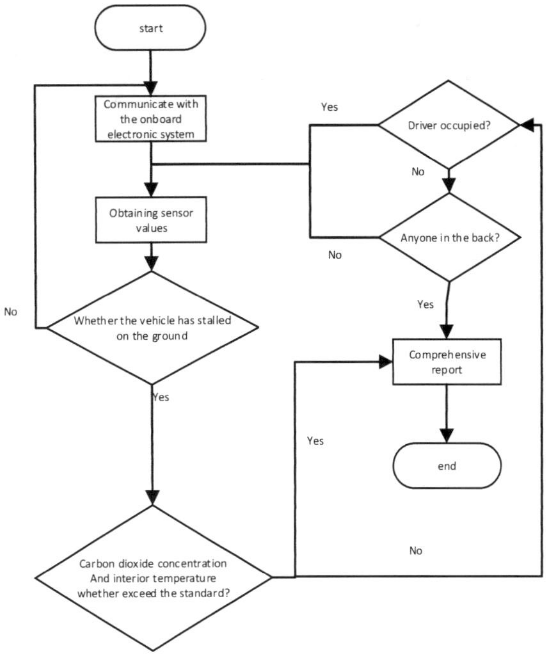

Figure 2. Software system block diagram.

After the system is started, it communicates with the car system to determine whether the car is flameout, and then the main controller obtains the state of the car through various sensors. The passive safety device is in the detection of children through the means of alarm protection. When carbon dioxide concentration or temperature exceeds the limit, direct alarm. If not, the system will judge whether the car flameout. If not, the system will return to the state of obtaining information inside the car. If it has been stalled by pressure sensor to determine whether a driver someone, when the system identification to the driver without the person, we are using the human body sensor to judge whether the back someone, when judging someone back, will include honking flash, immediately up machine send text messages, automatically lowered the window, the integrated alarm [2], and to protect the safety of the children in the car.

5. Experimental Analysis

In order to verify the reliability and rationality of hardware system and software system, test is set up to test the system. The passive safety device is in the detection of children through the means of alarm protection. In the experiment, we verify the reliability of each module in reality.

Table 1. Reliability of environment awareness module and alarm module.

CO_2 concentration	Whether the alarm	Whether the false positives
1000PPM	No	No
1500PPM	No	No
2000PPM	No	No
2500PPM	No	No
2800PPM	No	No
3000PPM	Yes	No
3200PPM	Yes	No
3400PPM	Yes	No

Table 2. CO_2 does not exceed the pressure and body sensor reliability.

Type of activity	Number of experiments	Number of alarms	False positive rate
Front seat occupied Back seat empty	30	0	0%
Front seat occupied Back seat occupied	30	0	0%
Front seat empty Back seat occupied	121	120	0.82%
Front seat empty Back seat empty	30	0	0%

As can be seen from the table, there is almost no misjudgement of the vehicle. The carbon dioxide detection module, alarm module and human body detection module both maintain good sensitivity and accuracy, and can adapt to the needs of daily life The passive safety device is in the detection of children through the means of alarm protection.

6. Conclusion

In this paper, a children's anti-asphyxia system based on multi-sensor fusion is studied. The passive safety device is in the detection of children through the means of alarm protection. At the same time, CAN bus communication keeps the high consistency of information between modules. By communicating alarm information to the computer and the environment around the car, it provides valuable rescue time for unexpected situations and provides better safety for children.

References

[1] Zhu Zhaopei, Li Hui, Wang Jiabing. The passive safety device is in the detection of children through the means of alarm protection. Instrument Technology,2021(05):38-40. (in Chinese)

[2] Liang Yushi, Zhu Dong, Sun Xuelian, Liu Guokai, Zhang Feng, Liang Xinxin. Design of emergency Escape Device for Stranded Children in Car [J]. China New Technology and New Product,2021(04):143-145.

[3] Zhu Hang, Ma Zhixiong, Dong Liping, Wu Xiuying. Journal of Jiamusi[J] university (natural science edition),2014,32(02):246-248.

[4] Li Yuying, Li Zesen, Cheng Junfeng, Wang Tian, Zhang Yingjie. Analysis and Preventive measures of hidden danger of Children stranded in Cars [J]. Time Automobile,2020(18):192-193.

[5] Chen Weixing, LUO Jiangpeng, He Yingxin. Design of Anti-Child Suffocating Alarm System in Car Based on GSM [J]. Computer Products and Circulation,2020,(06):118.Chen Peng, Tu Yaqing, Tong Junping, Zhao Yunyong.Design of Wearable Fall Detection Intelligent System[J].Transducer and Microsystem,2017,36(02):114-116.

Bayesian Regression Analysis of Spatio-temporal Location Information of Ship Electromagnetic Signal

Min Wang*, Chong Sun, Rangcheng Wu and Xiangqian Shi

Mathematics Department, Army Academy of Artillery and Air Defense, China

*Corresponding author email: wangminhf@163.com

Abstract. For spatio-temporal location information of electromagnetic signals according to maritime ship targets, the regression model based on Bayesian statistics is established based on the Bayesian estimation method, and some examples are analyzed. The results show that the model reduces the error, improves the accuracy of the sample data, significantly improves the data pretreatment effect, and effectively facilitates the further modeling and analysis of the target identification problem, so as to increase the credibility of the ship target identification.
Keywords: Ridge regression; Bayesian statistics; Target identification.

1. Introduction

Based on the needs of navigation, positioning and communication, a ship emits or reflects multiple types of electromagnetic signals at different points in time during the voyage at sea. With the advancement of science and technology, a variety of sensors are often used to investigate, monitor and identify maritime targets. In the important technical means of sea management, improving the identification accuracy of target positioning is the current problems encountered [1].

The electromagnetic signals of ships often include spatio-temporal position information and signal characteristic information, which are received by sensors and are the main basis for identifying the target identity. In the process of emitting, transmission and reception, signals affected by complex electromagnetic environment will produce certain error.Sometimes the model is necessary to be established only for the spatio-temporal location information to reduce the inevitable error caused by the sensor due to noise, interference and sensor calibration in the process of acquiring signal and data preprocessing, improve the accuracy of target identification, and thus improve the reliability of the identification result [2].

Although the regression analysis method can help to identify the relationship or correlation between the spatio-temporal location information and the ship trajectory, the error is often very large, even sometimes the result does not conform to the reality. When the regression model based on Bayesian statistics is established, spatio-temporal data can be preprocessed through the Bayesian inference process, resulting in a higher precision of the regression coefficient, which can help identify the essential differences between spatio-temporal location information, thus helps to identify which signals come from which ships, and improves the accuracy of ship target identification.

2. Analysis of the Spatio-temporal Location Information Characteristics

The affections and the causes of the spatio-temporal location information of maritime ship are different from other information such as signal characteristics, and they are greatly affected by ship movement. The movement of the same ship will make the spatio-temporal position information of the signal source constitute a movement trajectory, while the changes of longitude and latitude are probably monitored

by different sensors. Due to the multiple types of signals sent by the ship, and the same ship may have multiple movement trajectories. Nevertheless, the formed trajectories have some characteristic features, and within some error range, the two trajectories obtained after data processing can be regarded as generated by the same target. Therefore, it is necessary to model the error reduction first, and then use the motion trajectories generated by the spatio-temporal location information to verify the effectiveness of the model [3].Before this, assumptions need to be assumed:
(1) Different maritime ships work separately, and there is no close and parallel driving;
(2) No shelter to block on the way of target signal transmission;
(3) The ship is far enough from the land, and the sea surface environment is relatively stable;
(4) Various types of sensors work stable, and the error is within a controllable range.
Based on the above assumptions, a certain target ship with known electromagnetic signal characteristics is selected, which may produce four types of signal sources: R, A, L1, and L2. At the same time, the corresponding four types of sensors have received the spatio-temporal position information of the signal source batch number 241,379 and 524 generated by the movement of the ship, mainly including longitude and latitude data, some of which showed in table 1. The spatio-temporal position information of the signal source batch number 658 does not belong to the ship.

Table 1. Data of the signal source location record section.

time	longitude	latitude	Sensor type	Signal source batch number
2018042911165273	124.334621	26.1452947	L1	241
2018042911254861	124.28891	26.0940492	L1	241
2018042910394327	124.530224	26.35488	L1	379
2018042910413931	124.52033	26.3437849	L1	379
2018042910381903	124.543676	26.3602248	L1	524
2018042910410371	124.529617	26.3444802	L1	524
2018042911191426	123.112725	27.2452745	L1	658
2018042911230635	123.127072	27.2759208	L1	658

Figures 1 and 2, respectively, are the spatio-temporal location information of the lesser signal source error and the actual motion trajectory. Through the comparison and observation, the agreement verified by the movement trajectory can be high.

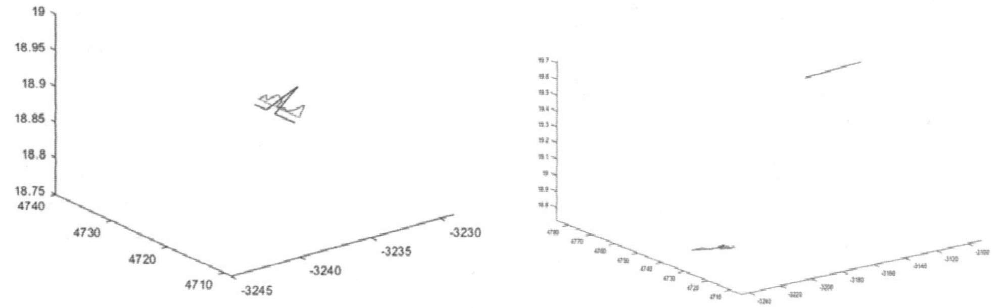

Figure 1. The spatio-temporal position information of the signal source batch number 241,379,524 and 241,379,524,658.

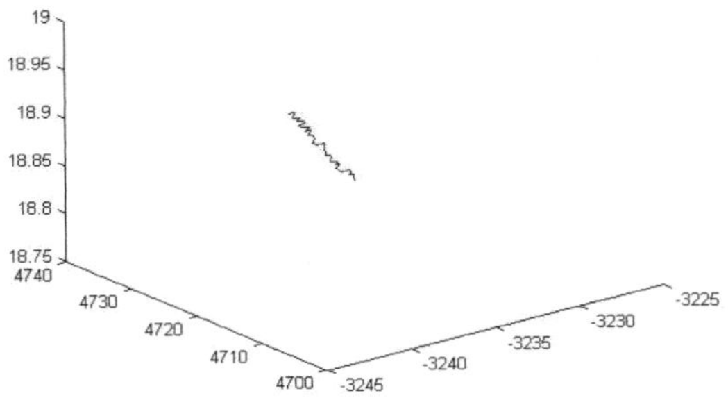

Figure 2. Actual movement trajectory of the target ship.

3. Rridge Regression Analysis Based on Classical Statistics

A univariate linear regression model study the interrelationship between a single dependent variable and an individual independent variable, with the model of:

$$b_i = \alpha_1 + \alpha_2 a_i + \varepsilon_i \quad i = 1,2,\cdots,n \tag{1}$$

Where: a_i represents the ith observation of the independent variable, b_i represents the ith observation of the dependent variable under the independent variable a_i, ε_i represents the ith error, α_1、α_2 are unknown parameters. Here ε_i are assumed to obey a normal distribution $N(0,\sigma^2)$ and are independent of each other, and α_1、α_2、σ^2 are also independent of each other.

Ridge regression analysis is the least squares regression that increases the canonical term, a modified least squares method specifically for biased estimation of collinar data analysis. When there is multiple linearity in time, longitude, latitude[5], using ridge regression to estimate parameters can reduce the mean square error by abandonning partial accuracy and the unbias of the least square, achieve the minimum variance of parameter estimation, so as to obtain regression models with slightly poor effect but better stability, which has the functions of independent variable screening and anti-collinearity [6].Ridge regression estimate the regression coefficients by:

$$\alpha(t) = (M'M + tI_m)^{-1}M'N \tag{2}$$

I_m is the m-order unit matrix, $t > 0$is the ridge parameter, when $t = 0$ it is the least squares estimation ofα .At this time, Some values near zero in the eigenvalue of $M'M + tI_m$ will improve accordingly, thus breaking the complex collinearity of the original matrix $M'M$, and the mean square error of the ridge estimate is smaller than the least squares estimate [7].

First, 54 groups of raw data were selected for the spatio-temporal location information received by class L1 sensors, and the time, longitude and latitude variables were standardized and centralized. After the operations of regression-optimal scale-regularized in the SPSS 19, the ridge regression analysis results based on classical statistics were calculated as follows.

Table 2. Residual statistics.

	minima	maxima	mean	Standard, deviation	N
predicted value	123.9878082275	124.1952285767	124.0860489195	.07471641670	54
residual	-1.25998687744	.36852827668	.00000000083	.59169058510	54
normal expected value	-1.315	1.461	.000	1.000	54
Standard, residuals	-2.109	.617	.000	.991	54

Table 3. Anova.

model		quadratic sum	df	mean square	F	Sig.
1	regression	.296	1	.296	.829	.367
	residual	18.555	52	.357		
	amount to	18.851	53			

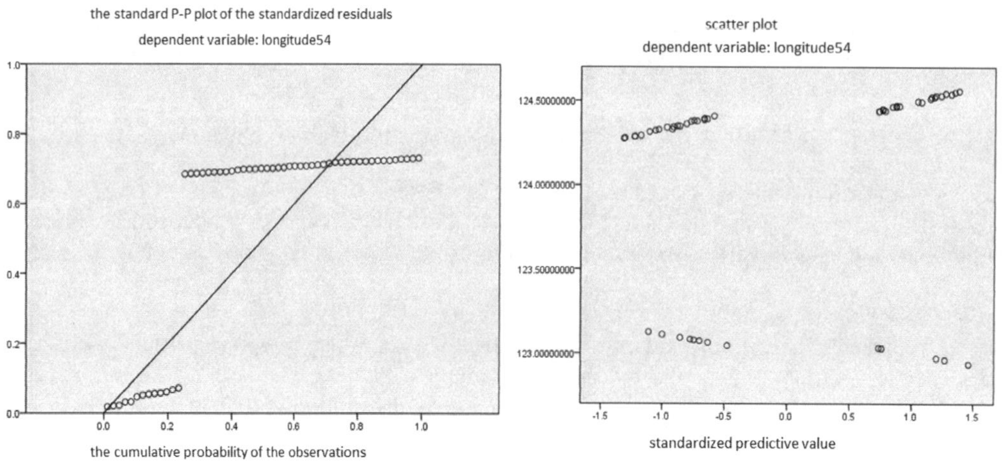

Figure 3. The cumulative probability of expectation and scatter plots of the 54 group data.

Tables 2 and 3 show that the mean square error of the ridge regression based on classical statistics is not large, but has not reached the expected smaller values. The large value of F indicates that the regression results are of some significance, and from the cumulative probability of Figure 3, we can basically see the small difference between the two signal source batch numbers 241,379,524 and 658. However, the Sig value is greater than 0.05, the difference between groups is not significant.Especially, the scatter plot of Figure 3 can not significantly show the class difference of batch No., that is, one can not determine which batch numbers are from the electromagnetic signal of the ship. To more clearly identify the limitations of the ridge regression analysis when processing spatio-temporal location information, the 1425 sets of data received by class R sensors were reselected for a similar ridge regression analysis and the following results were obtained .

Table 4. Residual statistics.

	minima	maxima	mean	Standard, deviation	N
predicted value	126.2997131348	126.3023452759	126.3015456585	.00059551448	1425
residual	-4.12265491486	2.11978507042	-.00000000016	1.72541631147	1425
normal expected value	-3.074	1.349	.000	1.000	1425
Standard, residuals	-2.389	1.228	.000	1.000	1425

Table 5. Anova.

model		quadratic sum	df	mean square	F	Sig.
1	regression	.001	1	.001	.000	.990
	residual	4239.336	1423	2.979		
	amount to	4239.336	1424			

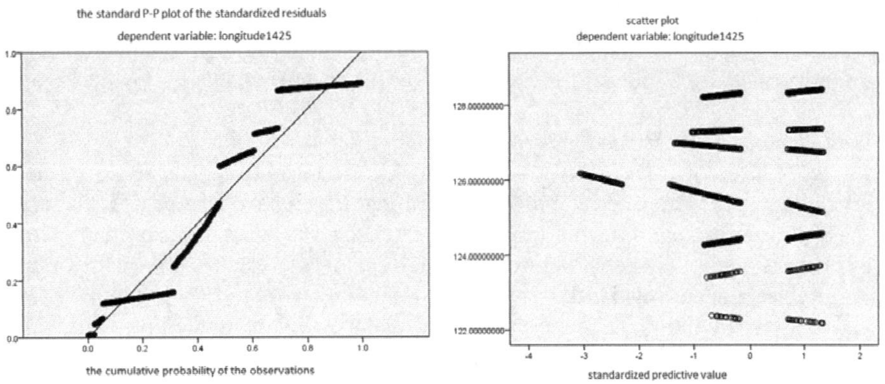

Figure 4. The cumulative probability of expectation and scatter plots of the 1425 group data.

At this time, the F value in Table 5 is very small and the sig value is very large, indicating that the ridge regression analysis has no significance. Not only the difference of the signal source batch number is not significant, it is also difficult to identify which are electromagnetic signals from the target ship according to the spatio-temporal location information distribution of the signal source and the longitude residual cumulative probability map and scatter map, combined with the actual movement trajectory of the ship. Thus this greatly reducing the credibility of the identification results. The fundamental reason is that the pretreatment effect of more data is not ideal in the process of ridge regression analysis, and a large error appears. Therefore, this error needs to be overcomed or reduced to improve the credibility of signal source identification. And the ridge regression analysis based on Bayesian statistics can achieve this goal.

4. Ridge Regression Analysis Based on Bayesian Statistics

4.1. Advantages of the Bayesian Statistics

The inference made by classical statistics is based on frequency, and the statistical properties of estimates such as regression coefficients are based on a large number of repeated experiments, and cannot give explicit answers and contain more randomness when applied. In general, the historical and empirical information before obtaining the sample is usually referred to as prior information. Bayesian statistics is an intuitive way of viewing the world, when comprehensively using the prior information and sample information, the advantage of Bayesian statistics is highlighted, which reduces the randomness of sampling, thus making the estimation results more accurate [8].In this respect, Bayesian inference is a practical alternative to the corresponding frequency inference based on the classical statistical.

Specifically, Bayesian estimates are obtained from the posterior distribution [9]. Assuming the prior density is $\pi(\theta)$, for unknown parameter θ, there is generally insufficient information to help determine $\pi(\theta)$. Bayes proposes that it can be assumed as a uniform distribution, and this principle of determining the prior distribution is also known as the Bayesian assumption, expressed as $\pi(\theta) \propto$constant. When the observation amount x is obtained, the posterior density $\pi(\theta|x)$ is a new understanding of θ after the test sample was obtained, and a starting point of statistical inference. It is given by the density function form of the Bayesian formula:

$$\pi(\theta|x) = \frac{f(x|\theta)\pi(\theta)}{\int_{\Theta} f(x|\theta)\pi(\theta)d\theta} \tag{3}$$

Where: $f(x|\theta)$ is the probability density function of the sample when x is given, also known as the likelihood function. The denominator is independent of θ and only dependent on x, and the molecule is related to the likelihood function and prior distribution, so equation (3) can also be equally expressed as $\pi(\theta|x) \propto f(x|\theta)\pi(\theta)$, and the posterior distribution of the Bayesian parameter is also considered as a product of the likelihood function and prior distributions.

Based on the advantages of the Bayesian estimation method, the ridge regression analysis based on Bayesian estimation is used to obtain the ridge regression coefficient solving the problem that the ridge regression cannot effectively reduce the data error of the ship electromagnetic signal.

4.2. Bayesian Estimates of the Ridge Regression Coefficients

Bayesian regression uses of experimental data repeated, introduces canonical terms in the estimation process.It is adaptive to the data and can prevent overfitting. If you have limited data that is known or if you want to use prior information in your model, you can use a Bayesian linear regression method. The Bayesian estimation method is constructing an initial estimate, and then constantly improving the estimation as more data is collected [10].

The Bayesian analysis of the ridge regression model mainly aims to reduce the error of the spatio-temporal location information of the ship's electromagnetic signal. The signal data needs precision processing, and then the new sample data of the spatio-temporal location information is got to be processed regression. The new data is more accurate sample value, which will help to the comprehensively analysis of the difference of sensor types and the signal source.Then it will be easier to use other methods to determine the movement trajectory of the ship target, and obtain accurate target identification results.

First, observe the results of general maximum likelihood estimation.

Without losing generality, verifying training samples that the spatio-temporal location information is subject to an independent identical distribution[11], the likelihood function of the sample set can be expressed as

$$(\alpha_1, \alpha_2, \omega) = \left(\frac{1}{2\pi\sigma^2}\right)^{\frac{j}{2}} \exp\left\{-\frac{1}{2\sigma^2}\sum_{i=1}^{j}(b_i - \alpha_1 - \alpha_2 a_i)^2\right\}, -\infty < \alpha_1, \alpha_2 < +\infty, 0 < \sigma < +\infty \quad (4)$$

Takes the log of equation (4) and the partial derivative respectively of α_1、α_2. Let the partial derivatives be 0 ,and the maximum likelihood estimates of α_1、α_2 are

$$\hat{\alpha}_1 = \bar{B} - \hat{\alpha}_2\bar{A}, \quad \hat{\alpha}_2 = \frac{\sum_{i=1}^{j}(a_i - \bar{A})(b_i - \bar{B})}{\sum_{i=1}^{j}(a_i - \bar{A})^2} \quad (5)$$

Where: $A = (a_1, \cdots, a_j)', B = (b_1, \cdots, b_j)', \bar{A} = \frac{1}{j}\sum_{i=1}^{j} a_i, \bar{B} = \frac{1}{j}\sum_{i=1}^{j} b_i$.

Next consider the Bayesian estimates of the ridge regression coefficient α_1、α_2.

Here consider the uninformative prior distribution [12], i.e., assuming that the prior distribution of the parameters α_1、α_2、σ are respectively $\pi(\alpha_1) \propto 1$ 、$\pi(\alpha_2) \propto 1$、$\pi(\sigma) \propto \frac{1}{\sigma}, \sigma > 0$. At this moment, the likelihood function $L(\alpha_1, \alpha_2, \sigma|A, B)$ in equation (4) is

$$L(\alpha_1, \alpha_2, \sigma|A, B) = \left(\frac{1}{2\pi\sigma^2}\right)^{\frac{j}{2}} \exp\left\{-\frac{1}{2\sigma^2}[(j-2)S^2 + C(\alpha_1, \alpha_2)]\right\} \quad (6)$$

Where: $S^2 = \frac{1}{j-2}\sum_{i=1}^{j}(b_i - \hat{\alpha}_1 - \hat{\alpha}_2 a_i)^2$, $C(\alpha_1, \alpha_2) = j(\alpha_1 - \hat{\alpha}_1)^2 + (\alpha_2 - \hat{\alpha}_2)^2 \sum_{i=1}^{j} a_i^2 + 2(\alpha_1 - \hat{\alpha}_1)(\alpha_2 - \hat{\alpha}_2)\sum_{i=1}^{j} a_i$.

According to the Bayesian formula, the posterior density of the parameter $(\alpha_1、\alpha_2、\sigma)$ is proportional to the product of the likelihood function $L(\alpha_1、\alpha_2、\sigma)$ and the prior density $\pi(\alpha_1、\alpha_2、\sigma)$, so that the joint posterior density function of the parameter $(\alpha_1、\alpha_2、\sigma)$ is

$$\pi(\alpha_1, \alpha_2, \sigma|A, B) \propto \pi(\alpha_1, \alpha_2, \sigma|A, B)\pi(\alpha_1, \alpha_2, \sigma) \propto \frac{1}{\sigma^{j+1}}\exp\left\{-\frac{1}{2\sigma^2}[(j-2)S^2 + C(\alpha_1, \alpha_2)]\right\} \quad (7)$$

From equation (7), integrating σ over interval $(0, +\infty)$, and the posterior edge density function of the regression coefficient (α_1, α_2) is

$$\pi(\alpha_1, \alpha_2|A, B) \propto \int_{R^+} \frac{1}{\omega^{j+1}}\exp\left\{-\frac{1}{2\omega^2}[(j-2)S^2 + C(\alpha_1, \alpha_2)]\right\} d\omega \propto \frac{1}{[(j-2)S^2 + C(\alpha_1, \alpha_2)]^{\frac{j}{2}}} \quad (8)$$

When A、B is given, the parameter (α_1, α_2) obeys the binary t-distribution [13], whose posterior edge density function is

$$\pi(\alpha_1, \alpha_2|A, B) = \frac{(j-2)^{(j-2)/2}\Gamma[j/2]|(X^T X)/S^2|^{-1/2}}{\pi\Gamma[(j-2)/2]}\left[1+\frac{1}{(j-2)S^2}C(\alpha_1, \alpha_2)\right]^{-j/2} \quad (9)$$

And as known from the nature of the binary t-distribution, the posterior expectations of the parameter α_1、α_2 are $E(\alpha_1|A, \mathrm{B}) = \hat{\alpha}_1, E(\alpha_2|A, \mathrm{B}) = \hat{\alpha}_2$, and the Bayesian estimates of α_1、α_2 under the quadratic loss function[14] are

$$\hat{\alpha}_1 = \bar{B} - \hat{\alpha}_2\bar{A}, \quad \hat{\alpha}_2 = \frac{\sum_{i=1}^{j}(a_i-\bar{A})(b_i-\bar{B})}{\sum_{i=1}^{j}(a_i-\bar{A})^2} \quad (10)$$

Formally, the Bayesian estimate of the parameter α_1、α_2 in formula (10) is exactly the same maximum likelihood estimate in Equation (5), but the meaning are fundamentally different: here $(\alpha_1$、$\alpha_2)$ is a random variable, and the posterior expectations $\hat{\alpha}_1, \hat{\alpha}_2$ are a specific number; while in classical statistics, $(\alpha_1$、$\alpha_2)$ is only the unknown parameter and not random, and the maximum likelihood estimate $(\hat{\alpha}_1, \hat{\alpha}_2)$ is a random variable.
With the above comparative analysis, it is clear that the regression coefficient α_1、α_2 based on the Bayesian statistics will reduce the randomness of the samples than the classical statistics, thus making the estimation results more accurate and more credible.

4.3. Results of a Bayesian Regression Analysis of Spatio-temporal Location Information

Bayesian estimates for the 54 data are as follows.

Table 6. Residual statistics.

	minima	maxima	mean	standard deviations	N
predicted value	122.9170	124.6116	124.0860	.54816	54
residual	-.55325	.39941	.00000	.23494	54
normal expected value	-2.133	.959	.000	1.000	54
Standard residuals	-2.332	1.684	.000	.991	54

Table 7. Anova.

model		quadratic sum	df	mean square	F	Sig
	regression	15.926	1	15.926	283.069	.000
1	residual	2.926	52	.056		
	amount to	18.851	53			

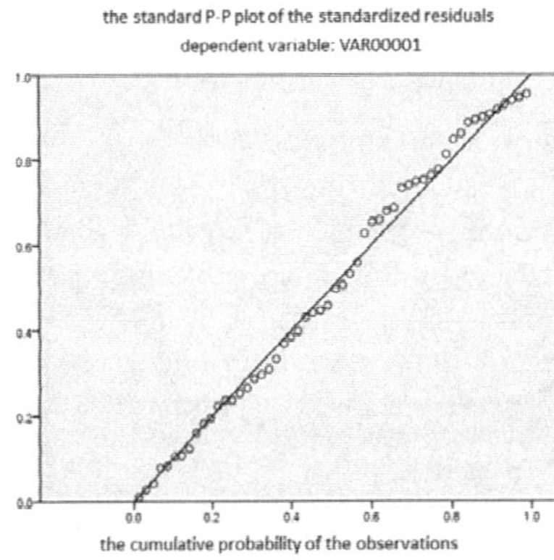

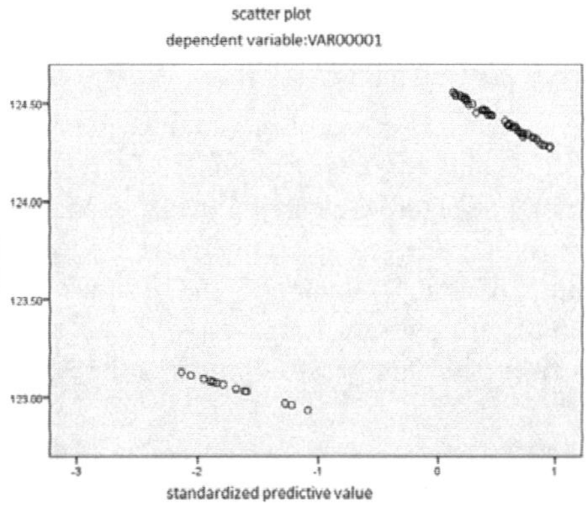

Figure 5. The cumulative probability and scatter plots based on Bayesian statistics.

Tables 6 and 7 show that the ridge regression based on Bayesian statistics has very large F value and very small Sig value, indicating that the regression works very well. The cumulative probability plot of Figure 5 can also see the good regression results of the data. The scatter plot illustrates the obvious difference between the two types of signal sources batch numbers 241,379,524 and 658. The signal sources of the first three batch numbers can be significantly classified as one, and they are from the ship can be quickly judged, while the other signal source is not an electromagnetic signal sent by the same ship. To further identify the ship by the spatio-temporal location information of the electromagnetic signal, a ridge regression analysis of the 1425 sets of data received by class R sensors was finally performed, and the following results were obtained.

Table 8. Residual statistics.

	minima	maxima	mean	standard deviations	N
predicted value	124.3409	127.5018	126.3015	1.14276	1425
residual	-4.45881	1.61063	.00000	1.29273	1425
normal expected value	-1.716	1.050	.000	1.000	1425
Standard residuals	-3.448	1.245	.000	1.000	1425

Table 9. Anova.

model		quadratic sum	df	mean square	F	Sig
1	regression	1859.614	1	1859.614	1111.992	.000
	residual	2379.722	1423	1.672		
	amount to	4239.336	1424			

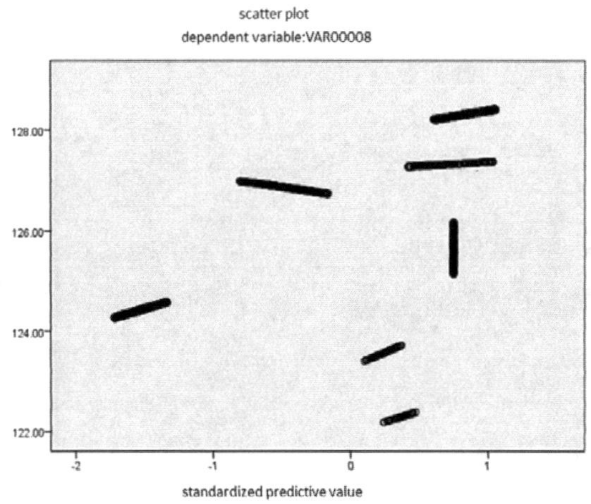

Figure 6. Longitude scatter map based on Bayesian statistics.

Table 9 has the same characteristics as Table 7, the F value is very large and the Sig value is very small. The scatter plot of Figure 6 clearly shows that some signal sources are very concentrated, thus obviously are the electromagnetic signals sent from the same ship received by the sensor. Compared with Figure 4, some data are more concentrated and are more likely from the same trajectory. The ridge regression analysis based on Bayesian statistics well reduces the error of the data, which makes the regression results very accurate, and greatly improves the credibility of maritime ship target identification.

5. Conclusion

Using Bayesian regression analysis reduces the data error in the spatio-temporal location information of ship electromagnetic signals. For different examples, the regression based on Bayesian estimate accurate samples when processing spatio-temporal data, which significantly improved the preprocessing effect of the data compared with the ridge regression analysis based on classical statistics, and increased the credibility of target identification. Usually, after a successful error reduction in obtaining new spatio-temporal location information, the new sample data is applied to the SVM [15] sometimes, and samples were classified based on the signal characteristic information. Or it is necessary to determine which target ship the different movement trajectories belong to.There are some other methods such as calculating sample entropy to obtain the number of target ships and the corresponding motion trajectory. Further, the obtained data can be compared with the identification database of the target ship. For example, using the decision tree algorithm or the K-L divergence to exact range of data. Finally, various target identification information is obtained, such as matching signal source number, the number and type of ships, target ship matching each signal source, and ship matching motion trajectory. To achieve these goals, the inevitable error of the electromagnetic signal data must first be reduced. And on that basis, the role of quantile regression analysis in data processing can be further explored[16].

References

[1] Bisai Bao Jianrong Wu Xiaojun Lou and Haitao Liu 2012 *Model and algorithm fusion of wireless sensor network* vol 31 Sensor and Microsystems, pp 43-46
[2] Wenjuan Zhou Mingfeng Zhang and Guangfa Lin 2017 *Lost personnel space-time information fuzzy matching model* vol 19 Journal of Earth Information Sciences pp 886-894
[3] Hongying Zhang and Zheng Hu 2013 *Motor target detection combining CenSurE features and spatio-temporal information* vol21 Optical Precision Engineering pp 2452-2463
[4] ChanLin Xu and Lili Wei 2009 *Bayesian analysis of a uniary linear regression model vol 23* Journal of Chongqing Institute of Technology (Natural Science Edition) pp 98-100
[5] Huiming Zhu Yuqi Han and Zhenggang Wu 2005 *A Bayesian forecast analysis of multiple linear regression models* vol 14 Operations Research and management pp 44-48
[6] Zhitao Zhang Haifeng Wang KAR NIELI Arnon Junying Chen and Wenting Han 2018 *Hyperspectral inversion study of soil water content based on ridge regression* vol 5 Agricultural Mechanical Journal pp 240-248
[7] Bo Zhang 2016 *Estimation of Ridge regression parameters in econometric models* Tianjin University of Finance and Economics
[8] Juan Peng 2012 *The Bayesian regression approach estimates the cost-effect acceptable curves* vol 6 Drug Economy pp 10-12
[9] Yan Ao and Chenwu Xu 2005 *The Bayesian regression analysis method and its application in QTL mapping* vol 26 Journal of Yangzhou University pp 44-49
[10] Lequn Su and Aimin Feng 2015 *Abnormal detection based on sparse Bayesian regression* vol 233 Computer and Modernization pp 57-60
[11] Qun Li 2017 *Research on survival analysis regression model and its variable selection under Bayesian framework* Changchun University of Technology
[12] Hongli Li 2015 *The study of Bayesian imputation techniques in the regression analysis* Jiangnan University
[13] Qun Li Xiaogang Dong Chunjie Wang and Bo Zhao 2016 *Bayesian regression analysis under the generalized exponential distribution vol 37* Journal of Changchun Polytechnic University pp 597-602
[14] Huanhuan Huang 2014 *Bayesian regression under asymmetric square loss function* Zhejiang University
[15] Chang Gu 2013 *Supports the relationship between vector machine classification and regression algorithm* Liaoning Normal University
[16] Hengming Gu Liangping Hu 2018 *Comparison of Bayesian and classical statistics in the quantile regression analysis* vol 48 Military Medicine pp 149-153

Diagnosis of Interturn Short Circuit of Permanent Magnet Synchronous Motor Based on Stacked Normalized Sparse Autoencoder

Yunfeng Chen[1] and Yuanjiang Li[2,*]

[1]School of Electronic Information, Jiangsu University of Science and Technology, Zhenjiang, China
[2]Suzhou Shijing Technology Co. , Ltd, China

*Corresponding author email: liyuanjiang@just.edu.cn

Abstract. Regarding the problems such as the traditional auto-encoders' tendency to learn the similar features during the feature extraction and limited capability of feature learning in the shallow network models, the paper puts forward an interturn short circuit diagnosis method for Permanent Magnet Synchronous Motor (PMSM) based on the Stacked Normalized Sparse Autoencoder (SNSAE). The method is to guarantee the representativeness of the extracted features, acquire stronger anti-interference capability, and is more suitable for training diversified and large-capacity mechanical equipment failure data. Four types of Inter-Turn Short-Circuit (ITSC) fault data were collected on the PMSM fault diagnosis experimental platform for testing. The experimental outcome suggests that the SNSAE algorithm can effectively extract the fault features, and the fault diagnosis effect is better than the traditional intelligent diagnosis method, and its diagnosis accuracy is as high as 98.91%.
Keywords: Inter Turn Short Circuit; Stacked Normalized Sparse Autoencoder; Permanent Magnet Synchronous Motor; Fault diagnosis.

1. Introduction

PMSM has been extensively used in the industrial production and daily life. But it is prone to faults such as ITSC. If faults can be found and repaired in a timely manner, efficiency can be improved and losses reduced.

The nature of deep learning [1] is to construct a multi-layered non-linear neural network. Cheng [2] proposed a new type of CNN-based intelligent fault-tolerant control system for turbofan engines, which uses a CNN-based fault diagnosis module to improve the feature learning ability of data. Liu [3] adopted the stacked sparse auto-encoder (SSAE), realize the unsupervised learning and the extraction of such feature-related information as vibration signal layer by layer, and use such information in the malfunction diagnosis for the rolling bearings. Liu [4] proposed a novel partial adversarial domain adaptation (SPADA) model based on stacked autoencoders to solve the problem of fault diagnosis in the case of PDA. Jia [5] put forward the normalized sparse auto-encoder (NSAE), which uses standard orthogonal constraints to improve NSAE's capability of identifying different features and avoid the extraction of excessive similar features.

In combination with the backgrounds above, the paper puts forward an interturn short circuit diagnosis technique for PMSM based on the SNSAE to guarantee representativeness of extracted features and improve the diagnosis precision of interturn short circuit malfunctions greatly.

2. Deep Learning

2.1. Normalized Sparse Autoencoder

Compared with traditional autoencoders, NSAE [5] adopts standard orthogonal constraints to improve the ability of learning and identifying different features and ensure the importance and difference of network feature learning.

The following are the primary differences between NSAE and regular SAE:

(1) Compared with other activation functions, ReLU can train the network more effectively.

(2) Bias is not used in NSAE.

(3) The sparse solution of autoencoder is obtained by replacing KL divergence function with L1 norm.

(4) Soft orthonormal constraint is introduced into the NSAE cost function to force NSAE to learn different features.

Given the label data $\{X_m, K_m\}_{m=1}^{M}$, where M is the sample size; K_m is label that indicates the degree to which an inter-turn short circuit is severe. The loss function of NSAE is shown in Equation (1).

$$
\begin{aligned}
J_{NSAE} &= \frac{1}{2M}\sum_{m=1}^{M}\left\|W^T\sigma_r(Wx_m)-x_m\right\|_2^2 \\
&+\lambda\sum_{m=1}^{M}\left\|\sigma_r(Wx_m)\right\|_1 \\
s.t\ \sigma_r(z) &= \begin{cases} 0\ , & if\ z<0 \\ z\ , & if\ z\geq 0 \end{cases}
\end{aligned}
\tag{1}
$$

Where λ is the NSAE regularization coefficient and W is the weight matrix. Since NSAE encourages Wx_m to be positive, the value of $\sigma_r(Wx_m)$ is always greater than zero. Therefore, the value of $W^T\sigma_r(Wx_m)-x_m$ can be represented as $W^TWx_m - x_m$ During network training, when the value of the first term of the loss function is close to zero, that is, the outcome of W^TW will become I when the quantity of $W^TWx_m - x_m$ is near to zero.

NSAE helps extract the multi-dimensional malfunction features, and with a more transparent weight matrix, which helps avoid learning excessive similar features.

2.2. Stacked Normalized Sparse Autoencoder

The SNSAE network is a deep network structure formed by stacking multiple NSAE. The features extracted from the previous layer are input to the latter layer, and the data features at different scales are deeply learned. In iteration, a softmax classifier is superimposed on the rightmost layer to classify different health conditions according to the mined features. Compared with shallow neural networks, deeper stacked networks [6] are better in feature extraction and have obvious advantages in processing high-dimensional data. The stacked network structure is shown in figure 1.

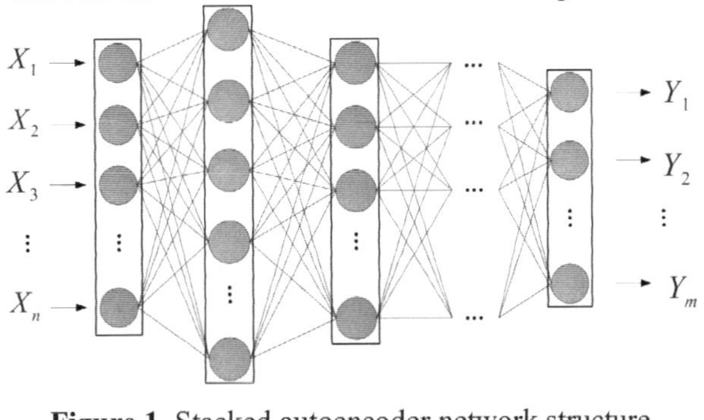

Figure 1. Stacked autoencoder network structure.

The fault diagnosis process of the SNSAE method is shown in figure 2. First, the data set is randomly divided into two groups, one group is used for training, and the other group is used for testing the network model. Second, use the training data set to pre-train and fine-tune the stacked network. Pre-training is used for weight initialization, and greedy strategy is used to train each layer of the network in turn. Then select the appropriate optimization algorithm to fine-tune the network to get the best parameters. Finally, the diagnostic performance of the stacked network model is verified through the test data set.

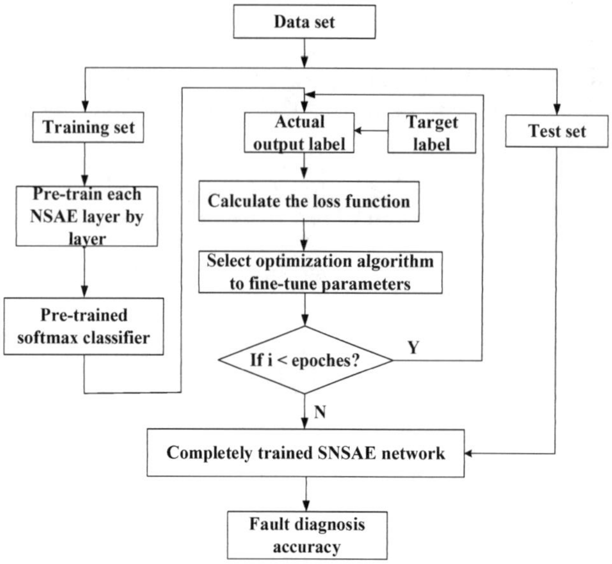

Figure 2. Fault diagnosis flowchart.

3. The ITSC's Characteristics Analysis

3.1. Negative Sequence Current

Negative sequence current [7] is a common basis for PMSM state detection. It is assumed that the A-phase stator windings of PMSM have ITSC fault. The fault diagram is shown in figure 3.

As shown in figure 3, the A-phase winding is split into sections R_1 and R_2 by the short-circuit resistance R_f. In particular, $R_s = R_1 + R_2$, R_s is the stator resistance. Assume that the amount of faulty turns is n and N ($n < N$) is the total number of turns, the ratio of short-circuit turns is $k = n / N$. At this point, short-circuit current i_f can be expressed as Equation (2):

$$i_f = -\frac{k(V_a - V_n)}{k(1-k)R_s + R_f} \tag{2}$$

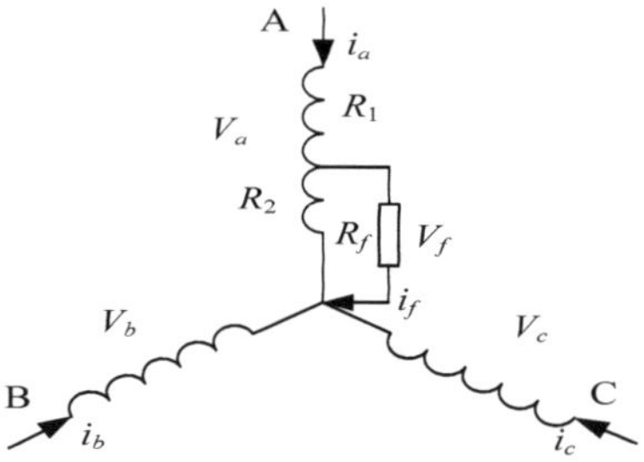

Figure 3. Schematic diagram of ITSC fault.

Where V_a represent A-phase voltage, and V_n represents neutral voltage. The negative sequence component will increase as the short-circuit current increases if the PMSM has an asymmetric short-circuit defect [8]. f represents the degree of ITSC, then the negative sequence current i_{f-} could be expressed as Equation (3):

$$i_{f-} = \frac{1}{3}(i_{af} + a^2 i_{bf} + a i_{cf}) \tag{3}$$

Where i_{af}, i_{bf} and i_{cf} respectively represent the three-phase current after inter-turn short-circuit fault occurs.

According to Equation (3), the negative sequence current is strongly related to the three-phase current. The three-phase current will be warped to variable degrees if the A-phase short-circuit occurs in the PMSM. As a result, using the negative sequence current to diagnose a short-circuit defect is useful.

3.2. Electromagnetic Torque

Electromagnetic torque [9] is also a very important indicator of permanent magnet synchronous motor, which is related to the working efficiency, durability and safety factor of the motor. When the motor fails, the electromagnetic torque and instantaneous phase current can be expressed as Equations (4) and (5) respectively:

$$T_e = T_u + T_r = \sum_{i \in s_n} K_i I_i + \sum_{j \in s_f} K_j I_j \tag{4}$$

$$I_{j_n} = (T_e - T_r)\frac{K_{j_n}}{5K_m^2 - \sum_{j_f \in S_f} K_{j_f}^2} \tag{5}$$

In the above formula, T_u is controllable electromagnetic torque, T_r is uncontrolled electromagnetic torque, s_n is normal phase winding, s_f is fault phase winding, K_j / K_i is electromotive force coefficient of j / i phase, I_j / I_i is phase current of j / i phase, K_m is the peak value of the phase electromotive force coefficient.

When PMSM encounters ITSC fault, the short-circuit current is limited to the rated current. According to Equation (5), when the motor is in the fault state, the instantaneous phase current will have high-order harmonic components, so the electromagnetic torque of PMSM in inter-turn short circuit is:

$$T_r = \sum_{i \in S_f} \frac{K_m I_m}{2} \sin(2\theta_i) \tag{6}$$

where I_m is the peak value of the phase current and θ_i is the phase angle.

When an inter-turn short circuit defect develops, the electromagnetic torque variation is obvious. If the number of short-circuit turns increases, the fluctuation becomes larger and larger, and even deviates from the normal curve. Therefore, electromagnetic torque can be used to make up for the problem that it is difficult to distinguish ITSC with single characteristic.

4. Experiment Analysis

4.1. The model of PMSM

AnsysMaxwell electromagnetic simulation software was used to build the motor model of PMSM in this paper. Table 1 shows the values of the model motor's individual parameters.

Mark the collected motor data according to the state of the PMSM. ITSC fault labels are shown in table 2. In the table 2, fault degree A and fault degree B respectively indicate that PMSM may have minor or severe inter-turn short circuit within the allowable range of unbalanced voltage. Fault degree C indicates that PMSM is in a damaged state.

Table 1. The PMSM model's parameter values.

The parameter	The value
The rated power/W	550
The rated voltage/V	380
The rotated frequency/Hz	100
The number of pole pairs	4
The parallel winding turns number	257

Table 2. The sample labels.

Name of the label	Value of the label
Health status	0
Fault degree A	1
Fault degree B	2
Fault degree C	3

A total of 4500 samples were collected through the PMSM simulation platform as training data set, among which the ratio of the samples in the four states of "health status", "Fault degree A", "Fault degree B" and "Fault degree C" was 2500:1200:55:250. In addition, a 1000 sample size test data set was created. in which the ratio of the number of different samples was 480:250:150:120.

4.2. The Parameters Optimization of Network

The number of hidden layers, learning rate, sparsity rate and optimization algorithm are important factors affecting feature learning and fault diagnosis. This experiment uses the same sample to evaluate various hyperparameters in the SNSAE network.

It can be seen from table 3 that it is not that the more and more stacked layers, the greater the diagnostic accuracy. Setting an appropriate number of hidden layers can improve the network training effect. Comparing the accuracy of fault diagnosis when the hidden layers are 2, 3, and 4, the results show that stacking two NSAE has the best diagnosis effect. The learning rate determines the amplitude of each update of the parameter. If the amplitude is too large, it may cause the parameter to move back and forth on both sides of the optimal value. On the contrary, if the learning rate is too small, the convergence can be guaranteed, but it will greatly reduce the optimization speed. Comparing the diagnostic accuracy under different learning rates, the best performance is when the learning rate is 0.005. The sparse rate is too high or too low will affect the data information, resulting in the inability to obtain better fault diagnosis capabilities. From table 3, it can be seen that the diagnostic accuracy when the sparse rate is 0.5 is higher than that when the sparse rate is 0.1 and 0.9. In addition, according to the diagnosis effect, this article chooses Adam as the network optimization algorithm.

Table 3. The SNSAE's parameter optimization.

Hidden Layers	Learning Rate	Sparse Rate	Optimizer	Accuracy (%)
2	0.005	0.5	Adam	98.91
3	0.005	0.5	Adam	92.39
4	0.005	0.5	Adam	85.86
2	0.005	0.5	Adam	98.91
2	0.01	0.5	Adam	96.73
2	0.05	0.5	Adam	94.56
2	0.005	0.5	SGD	87.35
2	0.005	0.5	Adam	98.91
2	0.005	0.5	Adadelta	79.34
2	0.005	0.1	Adam	96.73
2	0.005	0.5	Adam	98.91
2	0.005	0.9	Adam	97.14

Under this network structure, the training effect of algorithm model in this paper is shown in the figure 4.

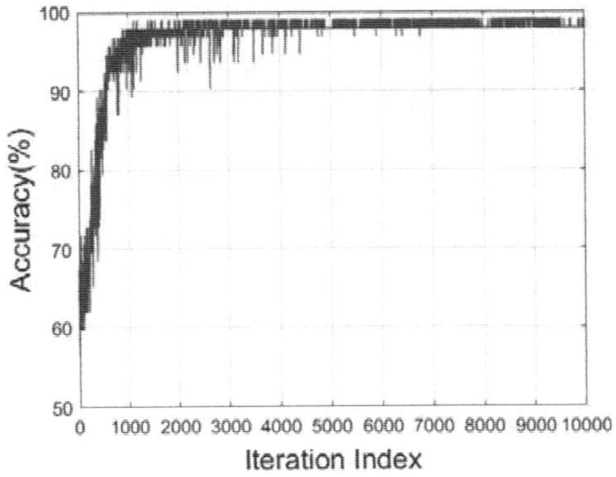

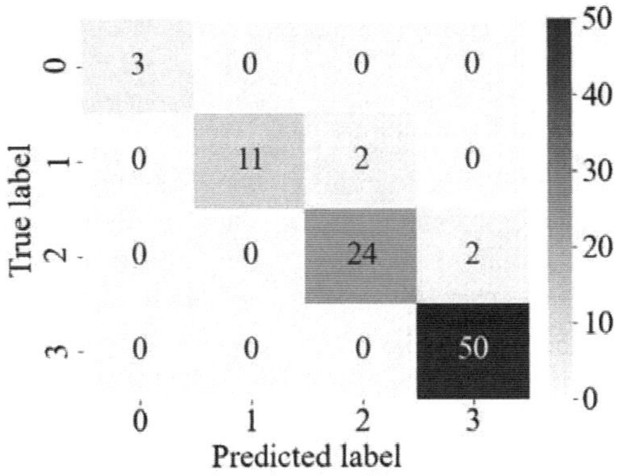

Figure 4. Network diagnosis accuracy. **Figure 5.** The confusion matrix of the diagnosis results.

Figure 4 shows that the network's diagnosis accuracy eventually converges to around 98.91%, which proves the feasibility of this method. In order to describe the fault classification results more intuitively, the confusion matrix of this method for each type of diagnosis is shown in figure 5.

Figure 5 shows that the SNSAE algorithm can well overcome the impact of unbalanced data, and has effective diagnostic capability for motor turn-to-turn short-circuit faults.

4.3. The Comparison of Different Methods

This experiment contrasts the old method with the SNSAE in order to better reflect the benefits of the SNSAE algorithm. Table 4 shows the results of the experiment.

Table 4. Diagnostic results of different methods.

methods	accuracy(%)
BP	86.21
RNN	86.15
SDAE+SOFTMAX	93.52
PCA+SVM	66.45
SAE +SOFTMAX	89.13
The proposed method	98.91

It can be seen from table 4 that compared with other methods such as BP [10], RNN [11], SDAE [12] and PCA+SVM [13], SNSAE algorithm has a higher fault diagnosis accuracy, reaching 98.91%.

5. Conclusion

An ITSC defect diagnosis approach based on the SNSAE network is presented in this paper. By stacking two NSAE networks, this strategy retrieves hidden abstract features in fault data, prevents the network from learning too many similar features, efficiently distinguishes different types of failures, and improves the network's generalization ability and classification accuracy. Experiments were carried out by gathering ITSC data from various sorts of faults. The results reveal that the method extracts fault features well and has a greater diagnosis accuracy than existing methods. It is suitable for accurate diagnosis of ITSC faults in PMSM.

References

[1] M. Azamfar, X. Li and J. Lee. Deep Learning-Based Domain Adaptation Method for Fault Diagnosis in Semiconductor Manufacturing [J]. IEEE Transactions on Semiconductor Manufacturing, 2020, 33(3): 445-453.

[2] D. Cheng, L. Liu and Z. Yu. Fault Diagnosis for Rotating Machinery Using Multiple Sensors and Convolutional Neural Networks[J]. IEEE Access, 2021, 9: 28122-28139.

[3] Z. Liu, Q. Li, B. Yan, et al. Application of Depth Neural Network Algorithm with Stacked Sparse

Autoencoder in Rolling Bearing Fault Diagnosis[J]. Machine Tool &Hydraulics, 2020, 48(23): 208-213.

[4] Z. -H. Liu, B. -L. Lu, H. -L. Wei, et al. A Stacked Auto-Encoder Based Partial Adversarial Domain Adaptation Model for Intelligent Fault Diagnosis of Rotating Machines [J]. IEEE Transactions on Industrial Informatics, 2021, 17(10): 6798-6809.

[5] F. Jia, Y. G. Lei, L. Guo, et al. A neural network constructed by deep learning technique and its application to intelligent fault diagnosis of machines[J]. Neurocomputing, 2018: 6772-6786.

[6] D. Zhu, X. Cheng, L. Yang, et al. Information Fusion Fault Diagnosis Method for Deep-Sea Human Occupied Vehicle Thruster Based on Deep Belief Network [J]. IEEE Transactions on Cybernetics, 2021.

[7] Y. J Li, Y. B Wang, Y. Zhang, et al. Diagnosis of Inter-turn Short Circuit of Permanent Magnet Synchronous Motor Based on Deep learning and Small Fault Samples[J]. Neurocomputing, 2021, 442: 348-358.

[8] X. Y Zuo. Fault Diagnosis of Rolling Bearings Based on the Fusion of Gaussian Dimensionality Reduction and Hybrid Kernel SVM [D]. Wuhan University of Technology, 2017.

[9] Y. Li, Z. Zhang, M. Li. Fault Diagnosis Method of Interturn Short Circuit of Permanent Magnet Synchronous Motor Using Deep Learning [J]. Journal of Electrical Engineering and Control, 2020, 24(09): 173-180.

[10] Q. Lu, R. Yang, M. Zhong. An Improved Fault Diagnosis Method of Rotating Machinery Using Sensitive Features and RLS-BP Neural Network[J]. IEEE Transactions on Instrumentation and Measurement, 2020, 69(4): 1585-1593.

[11] J. Zhang, Y. Wang, K. Zhu, et al. Diagnosis of Inter-Turn Short Circuit Faults in Permanent Magnet Synchronous Motors Based on Few-Shot Learning Under a Federated Learning Framework[J]. IEEE Transactions on Industrial Informatics, 2021.

[12] R. Liu, S. Feng, Y. Cai, et al. State Assessment and Fault Prediction Method of Distribution Terminal Based on SDAE and Hierarchical Bayesian [C]. IEEE Sustainable Power and Energy Conference (iSPEC), 2019, 2783-2787.

[13] B. Li, D. Cheng, Y. Tian, et al. GIS Insulation Defect Diagnosis Method Based on Improved MFCC and PCA-SVM Model [C]. 2020 5th International Conference on Communication, Image and Signal Processing (CCISP), 2020, 33-37.

Analysis of the Impact of Substation Switching Operations on 5G Base Station Antenna Feeder Ports

Zhihui Wang[1], Shulei Liu[3], Ying Zhang[2], Yue Hu[1,*] and Sicheng Zhu[1]

[1]China Electric Power Research Institute Company Limited
[2]Nanjing NARI Information & Communication Technology Company Limited
[3]Jinan Power Supply Company, State Shandong Electric Power Corporation, P.R China

*Corresponding author email: freya1025@163.com

Abstract. With the rapid development of the construction and application of 5G communication networks in the power grid, more and more 5G base stations need to be built in substations. 5G base stations have antennas placed outside in a complex electromagnetic environment, and they also become sensitive ports that induce external electromagnetic interference while sending and receiving signals. Especially the Electrical Fast Transient/Burst generated during the switch operation in the substation through the cable connecting the antenna and the base station equipment, at the antenna feeder port, generates electromagnetic disturbance that poses a potential threat to the operation of the equipment. This paper proposes an analysis method of an electromagnetic disturbance at the antenna feeder port of a 5G base station under the condition of switching operation of a substation. A 500kV substation is used to calculate the impact size, and the minimum distance between the antenna of the 5G base station and the switch operation device is determined. When the distance between the cable and the ground is two meters, the cable disturbance voltage reaches the maximum.
Keywords: Substation; 5G Base Station; Electro Magnetic Compatibility; Radiation emission; Switch operation; Electrical Fast Transient/Burst.

1. Introduction

With the development of power 5G communication network construction and application, as well as the needs of telecom operators network construction and power business applications [1-3], especially the promotion of 5G applications in substations, more and more 5G base stations need to be built in substations. The electromagnetic interference intensity, protection methods, and immunity limits of the communication equipment in the 5G base station under the complex environment in the substation have become important issues that need to be recognized and resolved [4], so as to improve the communication between the secondary equipment in the substation and 5G. Studies on the electromagnetic environment and electromagnetic interference of power systems in foreign countries began in the 1960s [5]. At present, studies in all aspects have been formed, such as the electromagnetic environment and electromagnetic interference characteristics of substation [6], proposing disturbance immunity requirements of electronic equipment in the station such as secondary equipment [7], and improving the electromagnetic compatibility performance of sensitive equipment ports [8]. There is a lot of research on the electromagnetic environment of substations after 2000 in China. At present, studies in all aspects have been formed, such as the electromagnetic environment and electromagnetic interference characteristics of the substation. However, transient electromagnetic disturbances such as switching operations and lightning surges in substations will generate conduction disturbances through cables, which are the main disturbances of electronic equipment in substations [9]. For this aspect of the problem, the current research is still insufficient.

In order to recognize the influence of complex electromagnetic environment in the substation on 5G base station and provide a reference for site selection and electromagnetic harassment protection of 5G base station, this paper proposes an analysis method of electromagnetic harassment in antenna port of 5G base station under the condition of substation switch operation and calculates the influence size of a 500kV substation.

2. Influencing Factors and Pathways of Substations on 5G Base Stations

As shown in Figure 1, the complex electromagnetic environment in the substation can be directly radiated to the communication facilities in the 5G base station through space radiation, and it can also be affected by conducted disturbances through cables, antennas, and other connection ports with the base station.

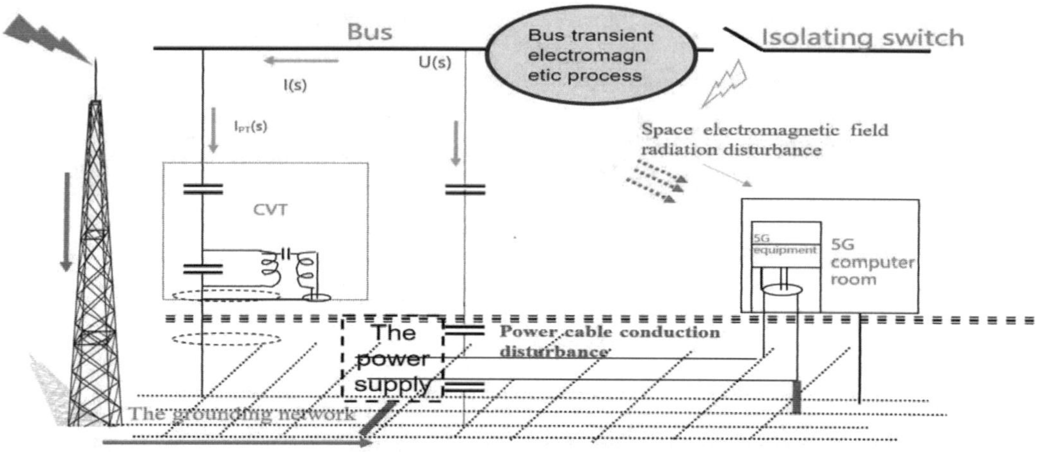

Figure 1. The complex electromagnetic environment affects 5G base stations in substations.

As the main facility for the installation of 5G communication equipment, 5G base stations are subjected to radiated and conducted disturbances in the complex electromagnetic environment in the substation to affect the ports of communication equipment through electromagnetic interference from enclosure ports and Communication ports of communication equipment.

3. Calculation Method of Antenna-fed port Disturbance Voltage during Switching Operation

3.1. Calculation Method of the Space Radiation Field

In the case of switch operation, the current passing through the switch, overhead wire, and other metal conductors will generate radiation electromagnetic fields in the adjacent space. Due to the existence of a large number of metal structures in the substation, when calculating the radiation disturbance near the 5G base station, it is necessary to establish all the metal structures in the substation in addition to the conductors of the current channel, and consider the electromagnetic scattering of all the metal conductors [12].

3.2. TDIE Solves the Radiation Field in Space

Calculate the space radiation field under switch operation. Firstly, the time domain integral equation of the fine conductor is established. On the surface of a conductor with a thin wire structure, the tangential component of the electric field intensity of the incident field is equal to that of the scattering field [10]

$$E_x^i = \frac{\partial A_x(x,t)}{\partial t} + \partial\varphi(x,t)/\partial x\,(\mathrm{V/m}) \tag{1}$$

In the formula:

E_x^i is the tangential component of electric field strength in the incident field in V/m ;

A_x is the dynamic vector of electric current on the surface of a conductor in wb/m;

φ is the dynamic scalar potential generated by the conductor axis current on the surface in V.

In infinite free space, the relation between dynamic vector potential and dynamic scalar potential and surface current and the surface charge is:

$$A_x(x,t) = \frac{\mu}{4\pi}\int_{x'}\frac{I(x,x',t-R/c)}{R}\mathrm{d}x' \quad \text{(wb/m)} \tag{2}$$

$$\varphi(x,t) = \frac{1}{4\pi\varepsilon}\int_{x'}\frac{q(x,x',t-R/c)}{R}\mathrm{d}x' \quad \text{(V)} \tag{3}$$

In the formula:

μ,ε are respectively the permeability and permittivity of the space where the wire is located;

$\boldsymbol{R} = |\boldsymbol{r} - \boldsymbol{r}'|$ Represents the distance from the field point to the source point in m.

$I(x,x',t-R/c)$ in A and $q(x,x',t-R/c)$ in F/m are line current and line charge at time t =R/c at x prime in s.

According to the law of conservation of charge, and make $\psi = \partial\varphi(x,t)/\partial t$ Equation(3) can representable:

$$\varphi = \int_t \left(-\frac{1}{4\pi\varepsilon}\int_{x'}\frac{1}{R}\frac{\partial I(x,x',t-R/c)}{\partial x'}\mathrm{d}x' \right) dt \tag{4}$$

By substituting Equation (2) and (4) into Equation (1), the time-domain integral equation of the axis current in free space can be obtained. The horizontal electric field can be regarded as the incident field of the wireline, so the incident field on the left side of the equation can be regarded as known [11].

The conductor axis is divided into N segments and a piecewise uniform function (impulse function) is chosen as the basis function[12]. The current at each point is called: I_k, k=1,2…N+1.

The section is calculated by constructing the field on the surface of the conductor. The point matching method was used for weighting operation, and the weight function was chosen as $\omega_i(x) = \delta(x - x_i)$, you can get:

$$V^s(t) = \frac{\partial \Phi(t)}{\partial t} + \varphi(x_i,t) - \varphi(x_{i-1},t) \tag{5}$$

In the formula:

$$\int_{x_i} (\partial\varphi/\partial x)\, dx = \varphi(x_i,t) - \varphi(x_{i-1},t) \tag{6}$$

Φ is the scalar potential $\boldsymbol{A}$ the result of the weighted integral operation;

The time step method is applied to discretize equation (5) in each calculation segment, and the weighting function is chosen as:

$$\omega_j(t) = \delta(t - t_j) \tag{7}$$

The following difference relation can be formed:

$$\Phi_{i,j} = \sum_{k=1}^{N}\frac{\mu}{4\pi}\int_{x_i}\int_{x'}\frac{I_{k,j-n}}{R}\mathrm{d}x'dx \tag{8}$$

Define two variables:

$$K_{i,k}^{a}=\frac{\mu_0}{4\pi}\int_{x_i}\int_{x_k'}\frac{1}{R}dx'dx \tag{9}$$

$$K_{i,k'}^{b}=\frac{1}{4\pi\varepsilon}\int_{x_k'}\frac{1}{R}dx' \tag{10}$$

These two variables are independent of time and can be taken separately to obtain the integral value. In the Equation(8)and Equation (9), Reformulated as follows:

$$\Phi_{i,j}=I_{k,j-n}\sum_{k=1}^{N+1}K_{i,k}^{a} \tag{11}$$

$$\psi_{i,j}=-\sum_{k'=1}^{N}\left(\frac{I_{k'+1,j-n}-I_{k',j-n}}{\Delta x'}\right)K_{i,k'}^{b} \tag{12}$$

The distance difference between the segment on the conductor surface and the k segment on the conductor axis can be approximated by the distance difference R between the midpoints of the two segments [13].

Conductor surface section and x_i section of conductor axis x_k' it doesn't mean the same period, when $n\neq 0$ define:

$$\Phi_{i,j}^{D}=I_{k,j-n}\sum_{k=1,n\neq 0}^{N+1}\frac{\mu}{4\pi}\int_{x_i}\int_{x_k'}\frac{1}{R}dx'dx \tag{13}$$

In the Equation (12) it can be reformulated as follows:

$$\frac{\Phi_{i,j}^{0}}{\Delta t}=V_{i,j}^{S}-\frac{\Phi_{i,j}^{D}-\Phi_{i,j-1}}{\Delta t}-(\varphi_{i,j-1}-\varphi_{i-1,j-1}) \\ -(\psi_{i,j}-\psi_{i-1,j})\Delta t \tag{14}$$

In the Equation (13), the horizontal electric field induced by a lightning strike at a point at the current moment is known; $\varphi_{j,n-1},\varphi_{j-1,n-1}$ is the dynamic scalar bit obtained after the calculation of the previous moment, which is also equivalent to the known quantity when calculating at the current moment; Φ and ψ can use Equation (11)and Equation (12)calculation.

The radiation field of the thin conductors in space can be obtained by the above calculation method. After the space radiation field is obtained, the transmission line method can be used to calculate the disturbance voltage of the antenna feeder cable port. The calculation model is shown in Figure 2. Agrawal transmission line model under excitation of external radiation field can be described by the following frequency-domain telegraph equation:

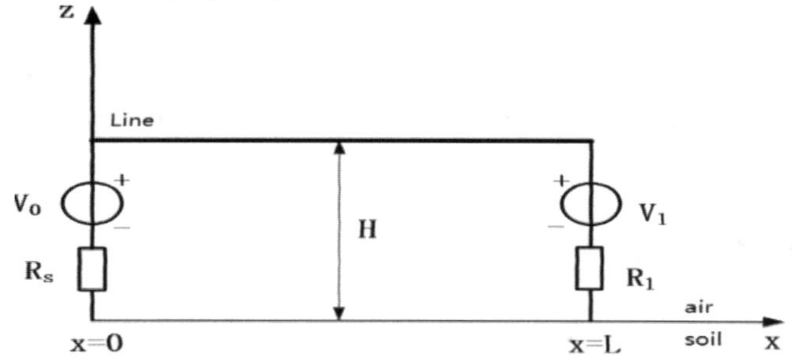

Figure 2. Agrawal Transmission line model

$$\frac{dI(x)}{dx} + j\omega CV^s(x) = 0 \tag{15}$$

$$\frac{dV^s(x)}{dx} + j\omega LI(x) = V_F^s(x) \tag{16}$$

In the formula, $I(x)$ and $V^s(x)$ are respectively the cable surface current and voltage to be solved, L and C are respectively the cable surface current and voltage to be solved, $V_F^s(x)$ is the equivalent distributed voltage source of the space radiation field on the cable surface of the transmission line calculated in the previous section, which can be calculated by the following formula:

$$V_F^s(x) = E_x^i(x,0,h) + E_x^r(x,0,h) \tag{17}$$

In the formula, E_x^i and E_x^r Respectively is a tangential component of incident wave electric field and reflected the wave electric field at wire. By solving Equation (15) and Equation (16), the disturbance voltage of the cable port can be obtained.

3. Algorithm Validation

In order to verify its correctness, this section compares the induced lightning overvoltage of overhead line obtained by the above method and the method in reference [Ref] under the same condition [14]. When the line height h=10m, Wire radius 0.5cm, the length of a single line is 1000m. The distance between the lightning strike point and both ends of the line is the same, and the vertical distance between the lightning strike point and the line is 50m, The propagation speed of lightning current along the channel is 1.3×108m/s, The peak wave of lightning current is 12kA, The maximum time derivative is 40kA/μs. Figure 3 shows the voltage waveform at the front end of the line.

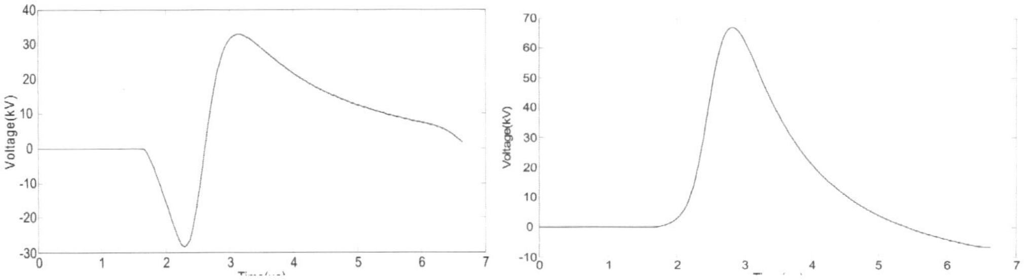

Figure 3. When the earth conductivity is respectively σ=0.001m/s and Lightning induction overvoltage at infinity

The results obtained by the method in this paper have a little deviation from the literature [14] method, so the correctness of the method in this paper is verified.

4. Analysis of the Influence of Switch Operation on 5G Base Station

In order to master the influencing factors of the complex electromagnetic environment in the substation on 5G base station, this paper takes a 500kV interval in a 500kV substation as the object and establishes a simplified analysis model [15]. Combined with the method mentioned above, the electromagnetic disturbance analysis method and installation location selection method of the 5G base station are analyzed in detail. Based on the substation layout and section drawing, the calculation model of the 500kV substation is established. As shown in Figure 4.

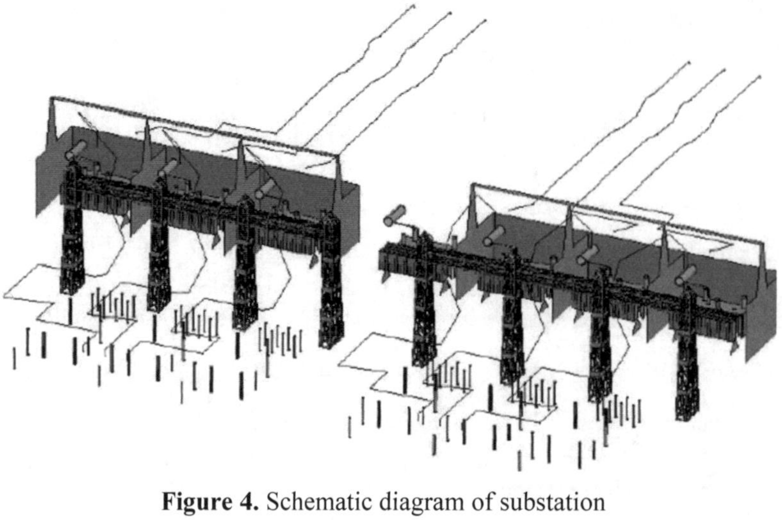

Figure 4. Schematic diagram of substation

The radiation field generated in space during switch operation in the substation is coupled to antenna and antenna feed cable port of communication equipment. Figure 5 shows the conducted disturbance voltage at different lengths and distances. The results show that the cable shielding mode, cable length, and distance from closing current all affect the disturbance voltage of the antenna feed port, and the value is relatively high.

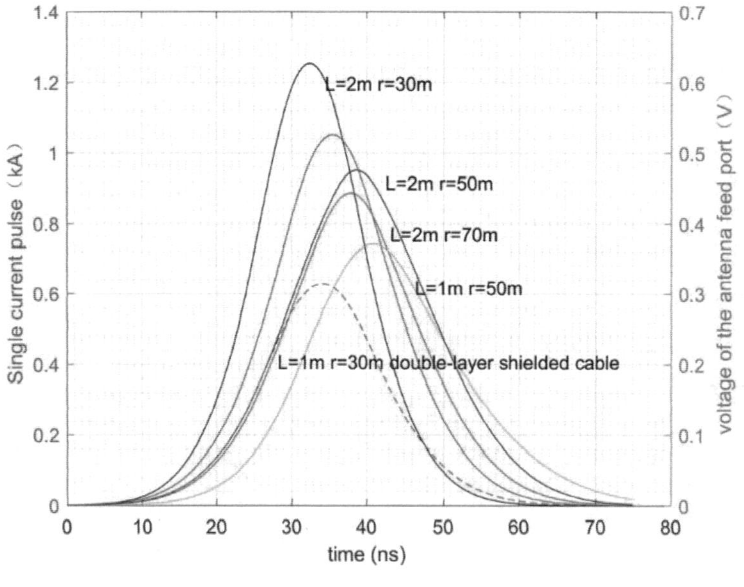

Figure 5. Voltage between core wire and shield layer of antenna feeder cable port

Electrical transient bursts of influence are closely related to the cable length and cable location, and 5G base station without electrical transient bursts to specialized for tolerance of protection device, so by reasonable location to avoid 5G base station power supply cable electrical transient bursts of coupled conduction harassment over the limit, is the key factor in 5G base station location.

5. Conclusion

As the electromagnetic disturbance in the complex electromagnetic environment in the substation cannot be suppressed from the disturbance source, the 5G communication equipment in the substation needs to master the main disturbance factors and determine the installation position of the 5G base station while reducing the coupling path and improving the anti-interference ability to improve the electromagnetic protection ability in future. In this paper, the calculation method of electromagnetic disturbance caused by switching operation of substation on antenna feed cable of 5G base station is analyzed, and the influence of electric fast transient pulse group generated during switching operation is analyzed with a simplified model of 500kV substation. The calculation results show that the port disturbance voltage of the ANTENNA feed cable of the 5G base station can be avoided by increasing the distance between the port and the switch on the one hand, and the influence can be reduced by using a double-layer shielded cable.

Acknowledgments
This work is supported by the project Research on 5G and Satellite Integrated Network Architecture for Power Transmission and Distribution Scenario (WBS: 5442XX210005, Project code of assignment: 5700-202155325a-0-0-00).

Reference

[1] Xiaohui You, Zhiwen Pan, Xiqi Gao, Shumin Cao and Hequan Wu 2014 *The 5G mobile communication: the development trends and its emerging key techniques* vol 44 (Science China Press: Information Science) No.5 pp 551-563

[2] Pei Yong, Ning Zhang, Song Ci and Chongqing Kang 2021 *5G Communication Base Stations Participating in Demand Response: Key Technologies and Prospects* vol 41 (Proceedings of the CSEE) No.16 pp 5541-5552

[3] Yi Wang, Qixin Chen, Ning Zhang, Cheng Feng, Fei Teng, Mingyang Sun and Chongqing Kang 2019 *Fusion of the 5G Communication and the Ubiquitous Electric Internet of Things: Application Analysis and Research Prospects* vol 43 (Power System Technology) No.5 pp 1575-1585

[4] Haun Li, Yue Xu, Fanbo Meng, Shuai Ren, Haijun Li, Xiaojing Pang 2021 *Modeling and analysis of 5G Terminal Communication Channel in Substation*(STUDY ON OPTICAL COMMUNICATIONS) No.223 pp 63-66

[5] C. M. Wiggins, D. E. Thomas, F.S. Nickel,etc. 1987 *Development and validation of a model to predict EMI effects of switching transient on protective circuit in high voltage substations* vol 3D (IEEE Symposium on EMC) PP 209-220

[6] The state bureau of quality and technical supervision 1999 GB/T17626.1-1998 *Electromagnetic compatibility Testing and measurement techniques Overview of immunity tests* (Beijing: Standards Press of China)

[7] Jiahui He, Dandan Zhang, JunQin Chen, Xin Tong, Yongqin Wang, Lu Zhang 2020 *Research on Coupling Mechanism of Substation Electromagnetic Interference Source to Secondary System* vol 44 (HUBEI ELECTRICPOWER) No.4 pp 14-19

[8] Zhenguang Liang 2008 *Coupling Path Analysis of Electromagnetic Disturbances in a Substation* vol 34(High Voltage Engineering) No.11 pp 2423-2427

[9] Ge Yan, Xixiu Wu, Yun Tian, Huan Zhang 2014 *Review on the Research of Electromagnetic Transient Interference Caused by Arc Discharge Due to Switchgear Operation* vol 50 (High Voltage Apparatus) No.2 pp 119-130

[10] Abdenbi Mimouni, Zin E. Azzouz, Boualem Ghemri 2007 *Lightning Induced Overvoltages on Overhead Lines: Modelling And Experimental Validation.* (Electrical Engineering) pp 131-146.

[11] The state bureau of quality and technical supervision 2019 YD/T 2583.17-2019 *Requirement and Measurement Methods of Electromagnetic Compatibility for Mobile Telecommunications Equipment -- Part 17: 5G Base Station and Ancillary Equipment* (Beijing: Standards Press of China)

[12] The state bureau of quality and technical supervision 2017 GB/Z 17799.6-2017 *Electromagnetic compatibility-Generic standards-Immunity for power station and substation environments* (Beijing: Standards Press of China)

[13] Xiong Wu, Yansheng Ding 2008 *Status and Strategies for Electromagnetic Environment of 500kV Power Network in China* vol 34 (High Voltage Engineering) No.11 pp 2408-2411

[14] Federico Delfino, Renato Procopio, Mansueto Rossi 2017 *An Algorithm for the Exact*

Evaluation of the Underground Lightning Electromagnetic Fields vol 49 (IEEE Transactions on Electromagnetic Compatibility) No.2 pp 401-411
[15] Yang Liu 2015 *Time-Domain Characteristics of Corona Current Pulses on DC Conductord* (Beijing: North China Electric Power University) pp 42-43

Research on Multi Charging Path Planning for Network Layout

Shan Huang, Yahui Zuo and Xiaolu Liu*

Department of Operations Research and Mission Planning, National University of Defense Technology, Changsha, Hunan, China

*Corresponding author email: lxl_sunny_nudt@live.cn

Abstract. With the development of Internet of things, Rechargeable Wireless Sensor Network (WRSN) has been widely used. In order to keep the WRSN working, the mobile charger needs to be charged regularly. This paper studies the practical problems of WRSN charging. Aiming at the problem of path and battery capacity planning under complex constraints of multiple chargers, using the traditional MTSP model for reference, the minimum energy consumption path and battery capacity planning model under multiple chargers are established, and the equilibrium factor is creatively designed and applied. Considering the NP-hard characteristics of the problem, an improved genetic algorithm based on Balance degree is designed to solve it. Finally, four cases are tested and compared with the traditional genetic algorithm. The results show that the improved genetic algorithm saves 31.56% of the total distance on average compared with the traditional algorithm, and has better stability and better optimization ability. The results are reasonable and have wide application value.
Keywords: WRSN; MTSP; path planning; Genetic algorithm; Balance degree.

1. Introduction

1.1. Background

With the rapid development of the Internet of things, wireless sensor network (WSN) is more and more widely used in life. One of the most important factors affecting the life cycle of WSN is energy. One way to provide energy is energy harvesting[1], which maintains its operation by using environmental energy such as solar or wind energy. However, because of environmental uncertainty, this is an unstable mode of energy supply. Another way to provide energy is to introduce wireless charging technology and mobile charging devices[2]. Use Mobile Charger (MC) to replenish the battery of the sensor regularly to make it operate stably. The network that supplies power in this way is also called wireless rechargeable sensor network (WRSN)

For a variety of constraints, Shi Yi, Xie Liguang and Hou YT [3]proposed the concept of periodic charging planning for the first time , and proved that MD charging sensor nodes along the shortest Hamiltonian loop can ensure the continuous operation of sensor networks. In addition, there are research precedents for the restriction of total energy of electrical appliances[4] and charging distance[5].

In the case of one-to-one service between sensor and MC, Zhao et al[6] optimized MC scheduling, sensor charging time allocation and node rate control to improve network performance. Zhao et al[7] also proposed a mixed integer optimization model for charging scheduling and charging time allocation with the goal of maximizing charging efficiency. An off-line algorithm is designed to optimize the charging task and allocate charging time. Huong et al[8], aiming at minimizing the number of starving nodes, proposed a charging scheduling scheme based on genetic algorithm.

One to one charging generally adopts contact charging. This charging method has high charging efficiency, but the moving distance of MC is long, which is suitable for the situation of scattered sensor distribution. When the sensor nodes in the network are densely deployed, one to many charging is used, and non-contact charging technology is adopted. MC can supplement energy for multiple nodes in the range at the same time, but more energy will lost.[9] [10][11]

This paper comprehensively considers the sensor energy consumption, battery capacity and distributed multi MC path planning. Adopt the strategy of one-to-one service between MC and sensor. This kind of problem can usually be equivalent to multi traveling salesman problem (MTSP). Intelligent algorithms are widely used in such problems, such as genetic algorithm (GA) [12], simulated annealing algorithm (SA)[13], ant colony algorithm (ACA)[14] and the combination of the above algorithms.

2. Problem Description and Models

2.1. Problem Description

The following symbols in table 1 will appear below or in the model:

Table 1. Main notations.

Symbols	Definition	Symbols	Definition	Symbols	Definition
S	Sensors	d_{ij}	Distance between sensors I and j	M	MC
x_{ijk}	Decision variable (whether MC_k is from sensors I to J)	D_k	Total distance of MC_k charging cycle	T_k	Total time of MC_k charging cycle
v	MC movement speed	r	Charging speed for sensor	t_0	Charging time of charger in data center
c_i	Power consumption of sensors I	f	Sensor minimum power of Sensor operation	t_i	Time for Sensor I wait for charging
F_i	Battery capacity of Sensor I	m	The number of MC	n	The number of Sensors

There is a WRSN distributed in an area, including data center DC (Data Center), n scattered sensors $S\{S_1, S_2, S_3... S_n\}$). The battery capacity of Si is F_i, the minimum power required for operation is f, and the power consumption speed is Ci. MC $\{M_1, M_2, M_3... M_m\}$ charge the sensors, MC is sensors, and the contact charging rate is r. Suppose that the data center and several sensors are distributed in a two-dimensional space, and each node is connected with each other. Based on the existing research work, this paper will design the charging problem planning model to obtain an energy-saving and efficient scheme, and can solve the minimum capacity of sensor battery.

3. Planning Model of Periodic Charging Problem Based on Balance Degree

In the process of MC charging the sensor, higher efficiency needs to evenly distribute the charging task to different MC as much as possible. The goal of energy saving puts forward requirements for the total length of the planned path. There is a certain conflict between the two objectives. If we blindly consider the demand for the minimum path consumption of mobile chargers, the workload of each mobile charger will be extremely unbalanced (for example, when only one charger is working and other chargers are idle). Obviously, in the context of practical application, this is not a good method.

Balance degree is an improvement of the traditional multi traveling salesman problem (MTSP) model in this paper. The concept of job balance is proposed and applied to the model, which can improve the balance of charger tasks and make the solution consistent with the actual application. This concept will be introduced while building the model.
Objective function:

$$\min z=\sum_{i=0}\sum_{j=0}\sum_{k=1}^{m}x_{ijk}d_{ij}+\frac{\mathrm{K}m}{\prod_{k=1}^{m}\left(\sum_{i=0}\sum_{j=0}x_{ijk}\right)} \tag{1}$$

The objective function consists of two parts. $\sum_{i=0}\sum_{j=0}\sum_{k=1}^{m}x_{ijk}d_{ij}$ represents the total path of all mobile chargers. x_{ijk} is the decision variable, indicating whether the No. 7 charger is from No. I sensor to No. j sensor. The expression is as the equation (2):

$$\mathrm{x}_{ijk}\begin{cases}1 & \text{via path(i,j)}\\ 0 & \text{not via path(i,j)}\end{cases} \tag{2}$$

The other part of the objective function is the balance degree operator. The numerator M represents the number of mobile chargers and the denominator represents the product of the number of sensors each charger is responsible for. Only when the workload is balanced, the equilibrium operator will be small. K is the equilibrium coefficient, and its value is determined according to the specific scale of the problem.

$$\sum_{i=1}^{n}\sum_{s=1}^{m}x_{ijs}=1 \quad j=1......n \tag{3}$$

$$\sum_{j=1}^{n}\sum_{k=1}^{m}x_{ijk}=1 \quad i=1......n \tag{4}$$

$$\sum_{k=1}^{m}\sum_{j=1}^{n}x_{ijk}=4 \quad i=0 \tag{5}$$

$$\sum_{k=1}^{m}\sum_{i=1}^{n}x_{ijk}=4 \quad \mathrm{i}=0 \tag{6}$$

$$\sum_{i=1}^{n}\sum_{j=1}^{n}x_{ijk}\leq m-1 \quad k=1......m \tag{7}$$

Equation (3) And equation (4) restrict that each sensor can be charged by only one mobile charger.
Equation (5) And equation (6) ensure that the starting point and ending point of the four mobile chargers are the data center.
Equation (7) Ensure that there are no sub circuits in the mobile charger.
After the MC is charged for a period of time according to the same route, the residual energy of the sensor node will show a certain periodicity in a time cycle [4], that is, the total power consumed by the sensor in each charging cycle is equal to the total power supplemented by the mobile charger. Through periodicity, the energy consumed by each sensor in each cycle can be known, and then the minimum capacity required to maintain the normal operation of the system can be obtained.

$$T=\frac{\sum c_i T}{r}+\frac{D}{v}+t_0 \tag{8}$$

$$t_i = T - \frac{c_i T}{r} \tag{9}$$

$$\mathrm{F}_i - c_i t_i \geq f \tag{10}$$

$$(\mathrm{i} = 1, 2 \ldots n)$$

Equation (8) Represents a charging cycle time: including the time for charging the sensor, the total time consumed on the road and the charging time of the mobile charger.
Equation (9) Indicates that the waiting time of the ith sensor for charging is the total time of a cycle minus its own charging time.
Equation (10) That is, the power of each sensor before charging is required to be greater than the required minimum power.
The formula of the minimum electric quantity of each sensor can be obtained by solving the above formula:

$$\mathrm{F}_i = \frac{\mathrm{c}_i (D + vt_0)(r - c_i)}{v(r - \sum c_i)} + f \tag{11}$$

4. Improved Genetic Algorithm Based on Balance Degree

Considering the NP hard characteristics of the problem, the improved genetic algorithm based on Balance degree is used to solve the problem. The design of genetic algorithm will be introduced below

4.1. Gene Coding

The algorithm adopts symbol coding and takes the sequence number of the mobile charger passing through each sensor as the gene. Set n-1 more gene as segmentation points. These segmentation points are used to divide the original gene into N parts to represent the paths of different chargers. The distance between them is infinite

4.2. Fitness Function

In addition to considering the minimum total energy consumption, the algorithm also considers the Balance degree , so the fitness function is designed as follows

$$\xi(x) = K\alpha + \frac{1}{D} \tag{12}$$

α is the Balance degree and *K* is the coefficient, which is used to balance the conflict relationship between the shortest total distance and the Balance degree, so that the influence between the two goals can be compromised.

4.3. Crossover

Considering that the path cannot be repeated, the characteristics of order crossover and multi-section path are combined to design multi-section sequential intersection:
Firstly, two genes, parent1 and parent2, were selected by roulette. Copy the gene representing the segmentation points of each charger path in parent1 into offspring.

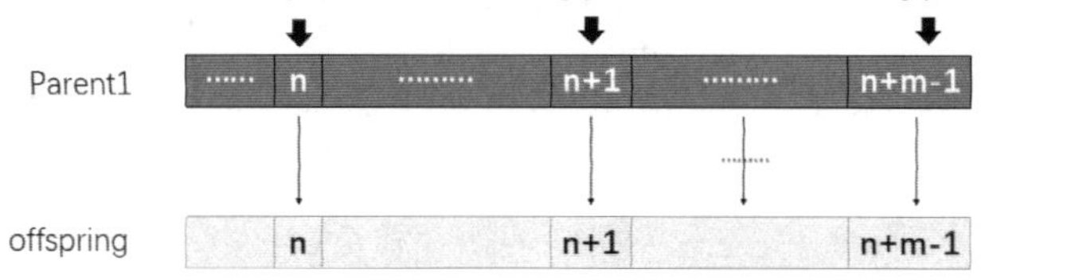

Figure 1. Cross - keep segment points.

Journal of Physics: Conference Series **2218** (2022) 012013 doi:10.1088/1742-6596/2218/1/012013

Then randomly select a segment in parent2 (ignoring the segmentation point) and copy it to the corresponding position of offspring. When encountering the segmentation point in offspring during replication, skip this gene and continue to copy from the next gene.

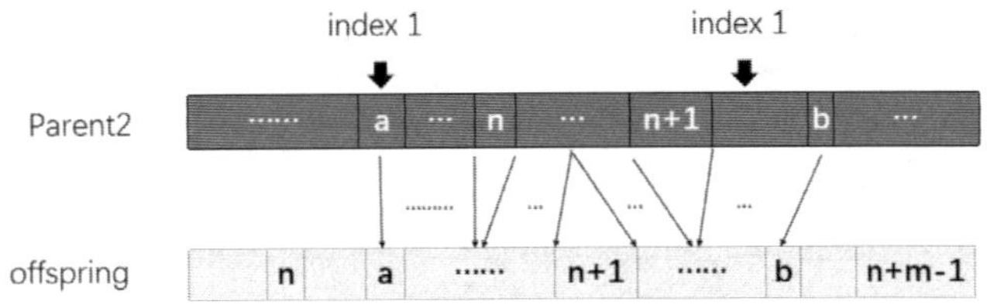

Figure 2. Cross - copy the selected segment.

Finally, the genes in parent1 that did not appear in the fragment from parent2 were sequentially filled into the blank of offspring.

4.4. Other

The selection strategy adopts proportional route wheel selection, and the variation adopts 2-exchange. The program is terminated by specifying the genetic times, and the elite retention strategy is adopted to retain the optimal individual of each generation.

5. Simulation and Data Analysis

5.1. Solution

The following will use specific examples for experimental verification, and illustrate the effectiveness of the improvement by comparing with the classical genetic algorithm:
In the experiment, the sensors and data center will be randomly distributed on the sea surface between 120.689662 and 120.7130185 east longitude and 36.3624281 and 36.3874186 north latitude to monitor the water area data, and conduct one-to-one charging through four MCs to make the WRSN work continuously. For the number of sensors, two situations are set: 29 sensors (small scale type) or 49 sensors (large scale type) ; Two situations are also set for the location of the data center: the data center is located in the central (central type) or the edge (edge type). A total of four calculation cases are obtained through the cross combination of two groups of cases. The location distribution of data center and sensor in the case is shown in figure 3-6, and the location of data center in the figure is marked in Black:

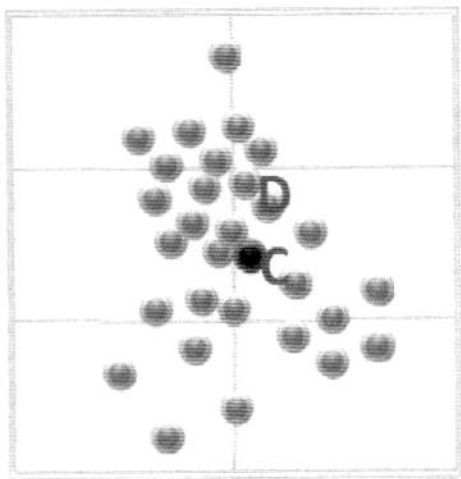

Figure 3. Small scale - central type (case 1).

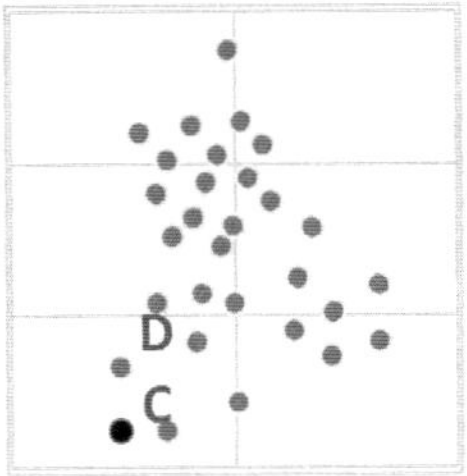

Figure 4. Small scale - edge type (case 2).

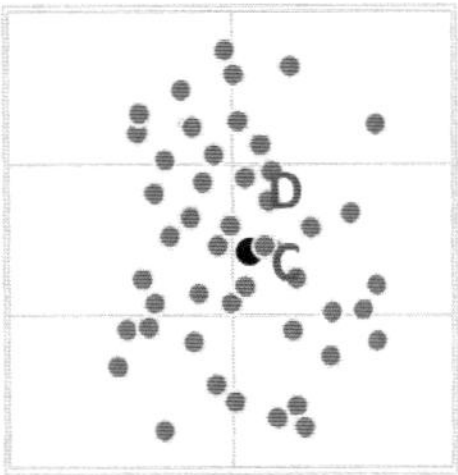

Figure 5. Large scale - central type (case 3).

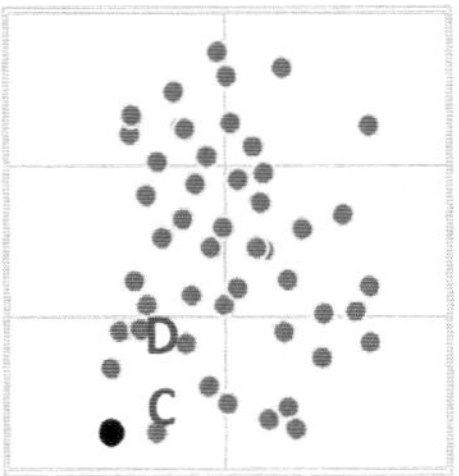

Figure 6. Large scale - edge type (case 4).

The experiment uses classical genetic algorithm and improved genetic algorithm based on Balance degree to plan the path of the above four cases. The maximum algebra of classical genetic algorithm is 5000 generations, the crossover probability is 70%, and the mutation probability is 5%; Based on the improved genetic algorithm of Balance degree, the K value is set to 0.0002, the designed crossover mode is adopted and the Balance degree is considered.

According to the above cases and parameters, the optimal path results calculated by the improved genetic algorithm based on Balance degree are shown in figures 7, and the calculation results of classical genetic algorithm are shown in figures 8:

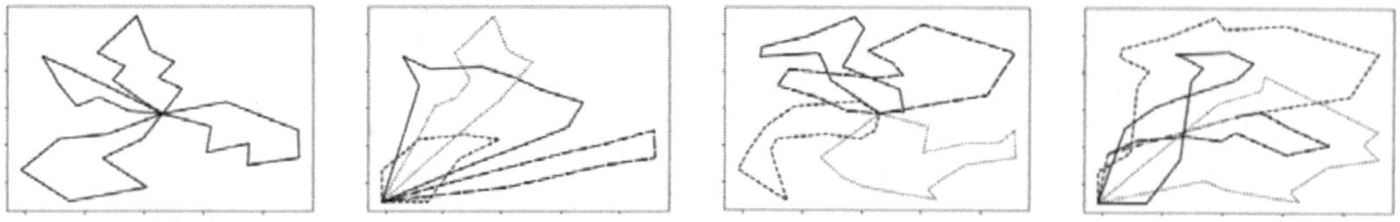

(a) (b) (c) (d)

Figure 7. Improved algorithm results of case 1, 2, 3 and 4.

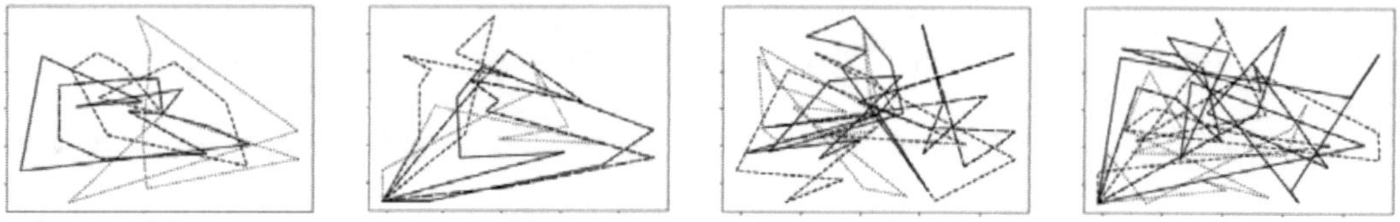

(a) (b) (c) (d)

Figure 8. Traditional algorithm results of case 1, 2, 3 and 4.

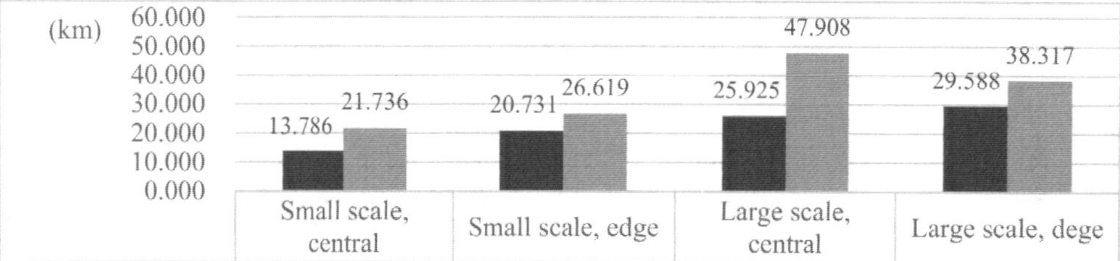

	Small scale, central	Small scale, edge	Large scale, central	Large scale, dege
■ Improved GA	13.786	20.731	25.925	29.588
■ Traditional GA	21.736	26.619	47.908	38.317

Figure 9. Comparison of results between traditional and improved algorithm.

In terms of the total distance of the results, the improved genetic algorithm based on equilibrium saves 31.56% of the total distance on average compared with the traditional algorithm, and compared with edge type, central type can give better play to the advantages of the improved algorithm. In the case of more sensors, the improved algorithm will perform better than the traditional algorithm.

For each case, the comparison of the algorithm convergence process is shown in the figure. The ordinate represents the total distance value and the abscissa represents the genetic times.

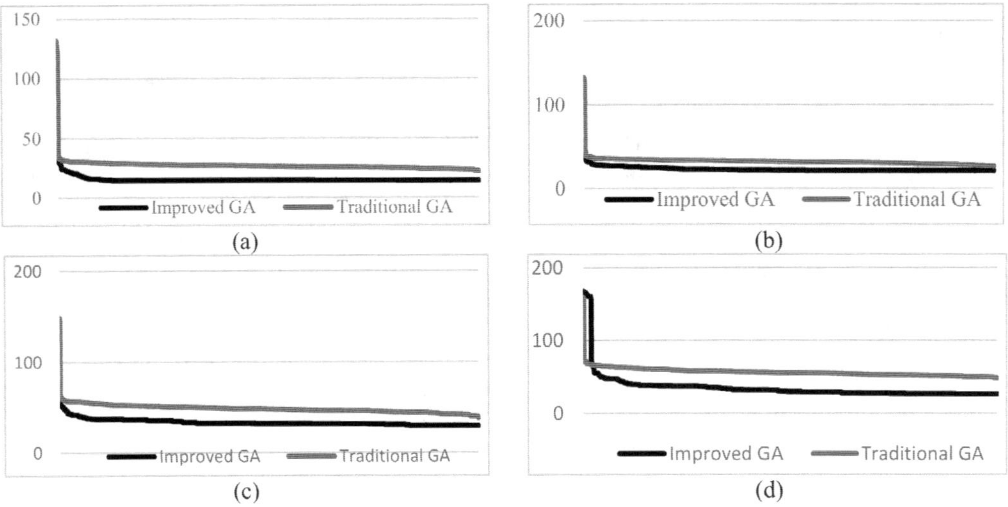

Figure 10. The comparison of algorithm convergence process in case 1, 2, 3 and 4.

By analyzing the above figure, it can be found that when the number of iterations is about 760, the fluctuation is small, and when the number of iterations is 2000 to 3000, it tends to be stable. It shows that the improved algorithm has good stability. Although the traditional genetic algorithm can converge effectively in the initial stage, the convergence ability in the middle and later stage is obviously not as good as the genetic algorithm based on the improvement of Balance degree. This shows that the improved algorithm can find a better solution more easily.

6. Conclusion

In this paper, the practical problems of WRSN charging are studied. Aiming at the problem of path and battery capacity planning under complex constraints of multiple chargers, considering a variety of constraints and practical factors, using the traditional MTSP problem model for reference, the equilibrium factor is creatively designed and applied, and the minimum energy consumption path and battery capacity planning model under multiple chargers is established. Considering the NP hard characteristics of the problem, an improved genetic algorithm based on Balance degree is designed to solve it. Finally, four cases are tested and compared with the traditional genetic algorithm. The results show that the improved genetic algorithm saves 31.56% of the total distance on average compared with the traditional algorithm, and has better stability and better optimization ability. It shows that the results have better applicability and high reliability, the designed model and algorithm are practical and have wide application value.

Reference

[1] Adnan Harb, 2010. Energy harvesting: State-of-the-art. Renewable Energy, 36(10), p.2641-2654.
[2] L. Xie, Y. Shi, Y. T. Hou and A. Lou, 2013. Wireless power transfer and applications to sensor networks, IEEE Wireless Communications, 20(4), p.140-145
[3] Y. Shi, L. Xie, Y. T. Hou and H. D. Sherali, 2011. On renewable sensor networks with wireless energy transfer. 2011 Proceedings IEEE INFOCOM, p.1350-1358.
[4] TOMAR A,KASWAN A,and JANA P K, 2018. On-demand energy provisioning in wireless sensor networks with capacity-constrained mobile chargers. The 11th International Conference on Contemporary Computing, p.1-6.
[5] Haipeng, CHEN Guihai, XU Lijie, et al, 2015. Effective algorithm for placement of directional wireless chargers. Journal of Software. 26(7), p. 1711-1729.
[6] ZHAO Chuanxin, ZHANG Xin, WU Changzhi, et al, 2020. Design of optimal utility of wireless rechargeable sensor networks via joint spatiotemporal scheduling. Applied Mathematical

Modelling. 86, pp. 54-73.

[7] ZHAO Chuanxin,ZHANG Hengjing,CHEN Fulong,et al (2020). Spatiotemporal charging scheduling in wireless rechargeable sensor networks. Computer Communications. 152, p.155-170.

[8] Huong T T,Le Nguyen P,Binh H T T,et al, 2020. Genetic algorithm-based periodic charging scheme for energy depletion avoidance in WRSNs. 2020 IEEE Wireless Communications and Networking Conference. p.1-6.

[9] LIU Tang,WU Baijun,ZHANG Shihao,et al (2020). An effective multi-node charging scheme for wireless rechargeable sensor networks. 2020 IEEE Conference on Computer Communications pp.2026–2035.

[10] ZHANG Fan,ZHANG Jie,and QIAN Yujie, 2018. A multi-node rechargeable algorithm via wireless charging vehicle with optimal traveling path in wireless rechargeable sensor networks.The 10th International Conference on Ubiquitous and Future Networks. p.531–536.

[11] LV Zengwei,WEI Zhenchun,HAN Jianghong,et al, 2019. A mobile charging and data collecting algorithm based on multi-objective optimization. Journal of Electronics&Information Technology p. 1877–1884.

[12] F. Liu 2020, A Genetic Algorithm Based Scheduling Scheme for WRSN. 2020 IEEE 3rd International Conference on Electronics Technology. p.751-755

[13] W. Chien, H. Cho, H. Chao and T. K. Shih, 2016. Enhanced SA-based charging algorithm for WRSN. 2016 International Wireless Communications and Mobile Computing Conference. p. 1012-1017.

[14] Chen Zhenhao,Xiang Shuwen and Zhang Li, 2020. Charging Route Planning for Sensors Based on Ant Colony Algorithm—Consider the Presence or Absence of Obstacles. Operations Research and Fuzziology. 10, p. 269-277.

A Image Segmentation Method Based on Dual Convolutions

Xiaoyu Dong and Xichuan Hu*

College of Information Engineering, Shanghai Maritime University, Shanghai, China

*Corresponding author email: huxc@shmtu.edu.cn

Abstract. Fundus retinal vascular image has high medical diagnostic value. Its segmentation image can help assess diseases that cause retinopathy, such as cerebral small vessel disease, coronary artery disease, glaucoma, and diabetes. Classical encoder-decoder segmentation methods like Unet cannot extract subtle details as well in the retinal vessel due to its complex structure. In this paper, we propose an improved Unet algorithm based on dual convolutions and integrate MultiRes feature extraction module. The MultiRes module improves the feature learning ability of micro vessels, which can collect further subtle details. The dual convolutions module utilizes dual feature maps with varying scales and channels, exchanges feature information with each other, thus can capture retinal vasculatures with better performance. Experiment based on the DRIVE dataset indicates that the accuracy, sensitivity, specificity, AUC(area under curve), and Dice coefficient of the proposed model are 96.13%, 78.57%, 98.17%, 97.29% and 80.87%. Compared with the existing advanced algorithms, our proposed algorithm shows its effectiveness.
Keywords: Retinal vessels; Dual convolutions; Unet; MultiRes; DRIVE.

1. Introduction

In modern medicine, fundus retinal vascular segmentation images can provide important supporting information for clinical diagnosis. Regular observation of fundus retinal vascular images to measure the growth status of blood vessels can provide diagnostic information [1][2]. The traditional method of manual segmentation of fundus retinal vessels relies on experienced clinicians, and it is time-consuming. The application of computerized images for fundus retinal vessel segmentation has shown significant advantages.

Retinal vascular segmentation is mainly divided into unsupervised and supervised. Unsupervised algorithms rely on the raw information of vessels for segmentation, such as region growth [3], thresholding [4], clustering [5], etc. Supervised learning relies on manually labeled standard data and iterates to the best results by training the model. Deep learning has gained widespread attention in supervised learning due to the rapid improvement of computer hardware, and it can give good results by leveraging massive amounts of data and various training models. Convolutional neural network CNN shows significant achievements in deep learning. In 2014 Long [6] et al. proposed the fully convolutional networks that can accept input images of arbitrary size. The FCN extracts feature information of the image by convolutional downsampling, then uses deconvolution to upsample the last downsampled feature map, ensuring that there will be a prediction for each pixel. The shortcoming of FCN is that it does not consider global contextual information. Ronneberger [7] et al. proposed an encoder-decoder type of deep learning algorithm Unet. The down-sampling part of Unet is similar to FCN, and in the decoding part, each up-sampling is connected with a layer of the same size as the previous down-sampling part, and then the up-sampling continues. Unet introduces the concept of jump connection, which makes full use of contextual information and improves the accuracy of image segmentation

For more complex medical images Unet algorithm cannot fully recover some details of the original image. Based on the Unet algorithm many improved algorithms have been proposed later. Zhou [8] et al. proposed the Unet++ model. Unet++ consists of Unets with different depths and its decoder is a redesigned dense jump connection with the same resolution. Different scales of feature maps are provided at the decoder nodes so that the aggregation layer can decide how to fuse the various feature maps carried in the jump connections with the decoder feature maps, leading to more robust segmentation results. Huang [9] et al proposed the DenseNet model to enhance the feature propagation by reusing the features through dense connections. Simon [10] et al used the dense convolution in DenseNet instead of the convolution in Unet for segmentation to make the model more accurate. Lv [11] et al. added the dilated convolution to Unet. The dilated convolution is added to the standard convolution with spacing to make the convolution operation provide more receptive field. Xiao [12] et al. proposed the Unet algorithm with a weighted attention mechanism. The attention mechanism improves the accuracy by providing the channel weights of the feature maps during the model training. To address the problems of low accuracy and difficulty of retinal image vessel segmentation, we designed a retinal vessel segmentation model based on the improved Unet model. By introducing dual convolution structure, we take advantage of the different attention to low-frequency and high-frequency information in the retinal vessels, exchange feature information through the cross fusion block, and integrate the MultiRes [13] network block instead of the convolution block in the Unet network, enable the model to achieve better segmentation performance.

2. Methodology

2.1. MultiRes Block

In standard convolution, increasing its width or depth can bring better performance. But it also brings excessive parameters, which commands higher hardware requirements. The model itself may also suffer from network degradation problem. In solving the depth problem, He [14] et al. proposed the Resnet model. Residual network obtains the difference between the observed and estimated values by shortcut mechanism, solves the network degradation. In solving the width problem, Szegedy [15] et al. proposed the Inception network structure. Multiple disparate convolutional kernels are used to extract features of different scales in parallel, then connected to improve the learning efficiency. MultiRes module combined the advantages of both Inception and Resnet. Figure 1 illustrates the architecture. The MultiRes module reuses convolution kernels to extract various scale features from the feature map with reduced parameters. Finally, add to the 1*1 convolutional map to complete the residual connection. In comparison with standard convolution, a model with different proportional features brought by multi convolutional kernels can generally achieve better performance. We integrate the CBAM dual attention mechanism based on the MultiRes convolution module. CBAM consists of channel [16] and spatial [17] attention combined in parallel. The channel attention module obtains the weights by squeeze and excitation operation, enhances the useful channels and suppresses the useless. The spatial attention module obtains the global maximum and average pooling, enhances the spatial features by multiplying the connection of two pooling layers with the original feature map. Figure 2 shows the MultiRes block with CBAM module integration.

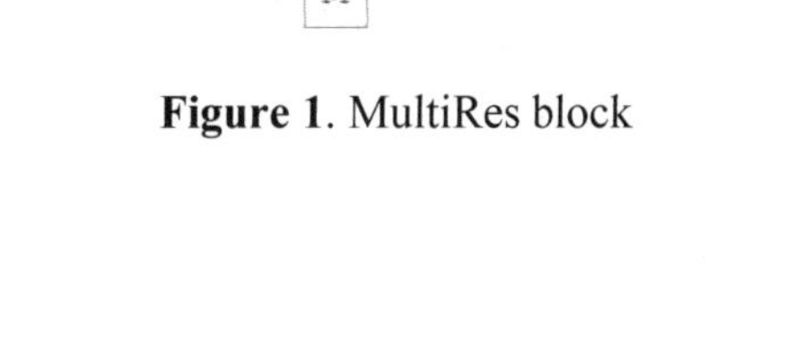

Figure 1. MultiRes block

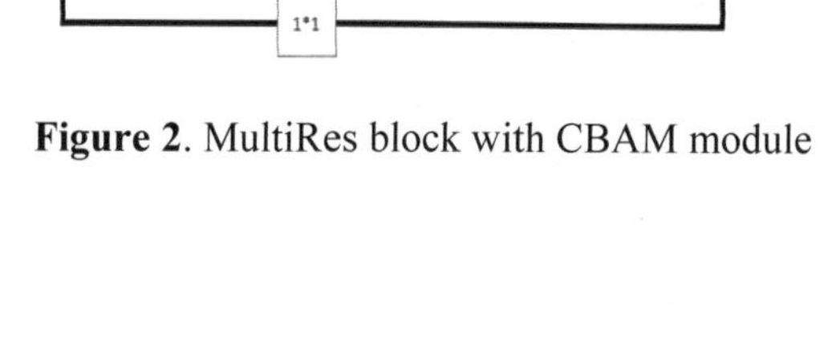

Figure 2. MultiRes block with CBAM module

2.2. Overview Model Structure

In the process of convolutional downsampling, the upper layer network has smaller receptive field, richer geometric information, while the lower layer network has a larger receptive field and provides more spatial information compared with the upper layer network. Retinal vessels have multi-branch dendritic vessels and intricate structures. The information provided by the upper and lower layers in downsampling generally cannot provide accurate spatial and geometric information at the same time when jumping connections, for which we propose a dual convolutional architecture. Figure 3 shows the proposed model architecture. Pair MultiRes convolutional modules are applied to downsample simultaneously so that pair feature maps of different sizes and channels are obtained in the same encoding layer, one focusing on fine structure and the other focusing more on spatial location information. Then they are both fed to the CFB (Cross Fusion Block), which is shown in Figure 4, to exchange spatial and geometric information. The cross fusion block downsamples the high-level feature map, adds it to the low-level feature maps, upsamples the low-level feature map, adds it to the high-level feature map. Thus completes the feature fusion. Pair outputs are used as the inputs of the next encoding layer. After four down samplings, the bottom feature fusion map starts to upsample. Before each upsampling, the feature map from decoding layers is connected with the feature maps of the encoding layer with the same size as itself. It can provide more spatial and geometric information concurrently as a contrast to Unet, and provides more reliable performance support for the segmentation of retinal vessels. After the upsampling, the final end-to-end output is obtained by a 1*1 convolution block and Sigmoid function.

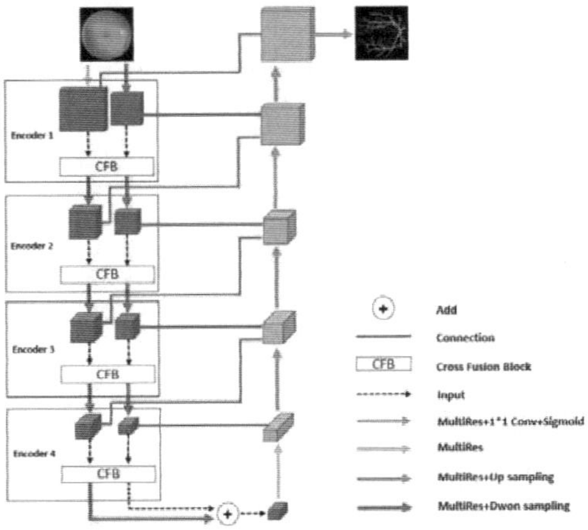

Figure 3. Overview architecture of proposed model

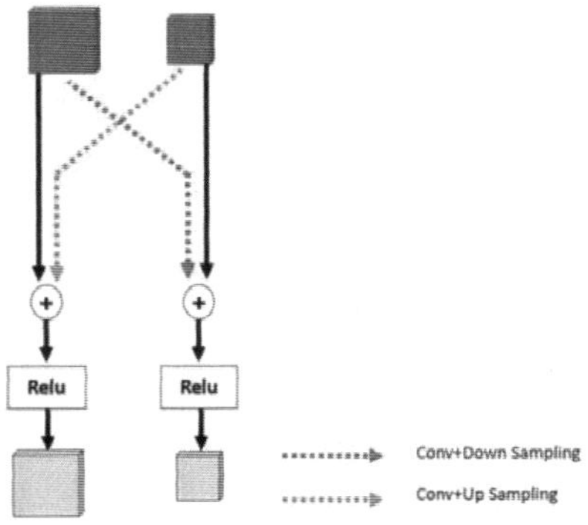

Figure 4. Cross fusion block

3. Experiment Preparation

3.1. Loss Function

To solve the problem of unbalanced category occupancy in retinal vessels, this model uses a weighted cross-entropy loss function. The following equation (1) defines it.

$$L = \sum_{n=1}^{m} (w_{pos} y_n \log(\hat{y}_n) + (1 - y_n)\log(1 - \hat{y}_n)) \tag{1}$$

where $w_{pos} = p_{pos} / p_{neg}$, p_{pos} represents the ratio of the positive pixel count, p_{neg} represents the ratio of the negative pixel count, w_{pos} represents the positive class weight, m denotes the number of samples in each batch, y_n denotes the ground truth of the nth sample, and $\hat{y}_n$ denotes the model predicted output value for the nth sample.

3.2. Dataset and Dataset Pre-processing

The dataset used for the experiment is the DRIVE [18]. DRIVE contains 40 color fundus images. 20 fundus images for training and the other 20 for testing. Each fundus image has been manually segmented, and the resolution of each image is 584*565. To give the image a more significant vascular structure, we pre-processed those images. Firstly, those images are converted into grayscale maps to facilitate the subsequent processing. Then histogram equalization with restricted contrast is used to improve the local contrast. Finally, we normalized those gray images to facilitate model calculation. The DRIVE dataset has an insufficient amount of data while the training model requires large samples, so we processed it by using chunk cropping. In the original DRIVE training set, 20 fundus images and retinal vascular true value images are cropped into 5000 small blocks of the size of 96*96 for training. The remaining 20 images in the test set are used for evaluating models. Figure 5 illustrates the details of cropped sample images;

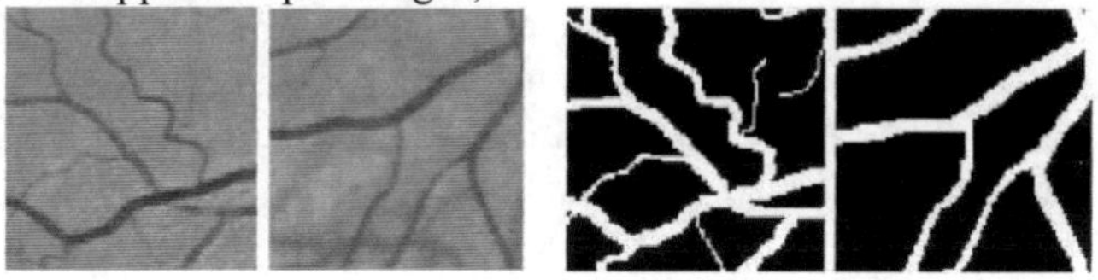

(a) (b)

Figure 5. Cropped sample; (a) Cropped images (b) Cropped masks

3.3. Evaluation Metrics

The following metrics are selected: Accuracy; Sensitivity; Specificity; Dice. The definitions are as following equations (2) (3) (4) (5).

$$Accuracy = \frac{TP + TN}{TP + TN + FP + FN} \tag{2}$$

$$Sensitivity = \frac{TP}{TP + FN} \tag{3}$$

$$Specificity = \frac{TN}{TN + FP} \tag{4}$$

$$Dice = \frac{2TP}{2TP + FP + FN} \tag{5}$$

We also calculated the AUC (Area under curve) whose horizontal coordinate is the specificity and vertical coordinate is the sensitivity. The value closer to 1 indicates the better model segmentation performance.

3.4. Experiment Setup and Configuration

The experiments are based on Windows 10 and the Keras framework, accelerated by Nvidia RTX3060. Epoch is set to 50, batch size is 6, learning rate 0.0001, and Adam optimizer is used at the same time.

4. Experiment

Our quantitative results on DRIVE dataset show excellent segmentation performance, and Figure 6 shows the comparative segmentation samples of our model and Unet.

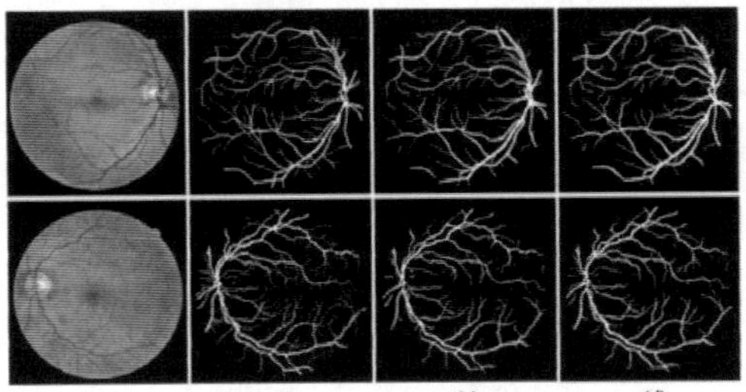

Figure 6. Segmentation diagram (a) Grayscale sample (b)Ground truth (c)Unet (d)Ours

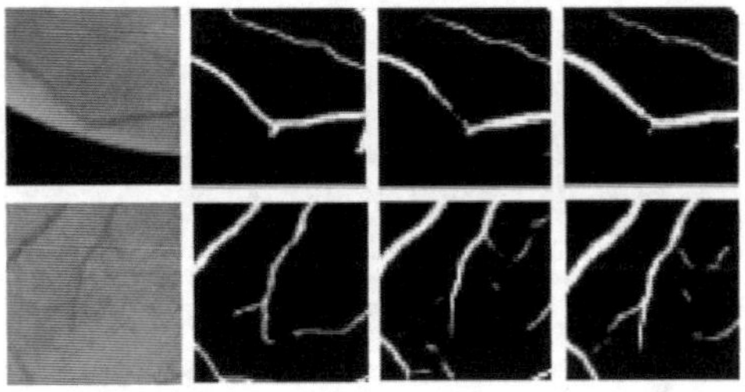

Figure.7 Local segmentation diagram (a) Grayscale sample (b)Ground truth (c)Unet (d)Ours

Figure 7 zooms in on detailed parts of Figure 6 to better see the differences. The following problems exist in Unet network segmentation, ① the problem of discontinuity of segmented vessels, as in the first case; ② the problem of the disappearing vessel ends, as in the second case. ③ Unet is prone to the problem of mis-segmentation when segmenting less obvious features, as in the second case.

Our model solves the discontinuity of blood vessels segmented by Unet while moderating the disappearance of blood vessel ends to some extent, and achieves better segmentation results by minimizing mis-segmentation.

Table 1 shows the values of Accuracy, Sensitivity, Specificity, and AUC for the different models in the DRIVE dataset. Our model outperforms other models in Accuracy, Specificity, AUC by providing more useful information in jump junction due to dual parallel convolutional structure. AA-Unet integrates attention guide and atrous convolution, which is better than ours in SE and Dice evaluation. Our model does not conflict with its advanced idea. Attention guide and atrous convolution can also be integrated into our model for more accurate segmentation results.

Table 1. Comparison of different models on DRIVE dataset

Model	Acc	SE	SP	AUC	Dice
Unet	95.23%	75.42%	97.86%	96.68%	78.75%
Unet++	95.51%	78.37%	97.89%	96.90%	80.97%
DenseUnet	95.44%	77.63%	97.72%	96.87%	79.44%
AA-Unet	95.56%	79.29%	97.77%	97.16%	81.03%
Res-Unet	95.60%	78.45%	97.84%	96.98%	80.46%
Ours	96.13%	78.57%	98.17%	97.29%	80.87%

The above comparison of segmentation results shows that our model is able to perform more accurate segmentation in complex retinal vascular maps compares with other models.

5. Conclusion

To solve the problems of difficult segmentation of retinal vessels in clinical diagnosis, we proposed an end-to-end Unet architecture-liked model. The dual convolutional module and MultiRes module are combined to extract more details of different scales in the encoding stage. Cross fusion block exchanges the spatial and geometric information of dual feature maps in different sizes and channels and preserves more vascular details while suppressing background noise. Experimenting on the DRIVE dataset shows that our method improves the segmentation performance efficiently.

In the future, we plan to optimize the network model, reduce its number of parameters caused by dual convolution and MultiRes block, and consider finding a balance point between accuracy and model parameters to integrate attention gate and atrous convolution for more accurate results.

References

[1] Budai A, Bock R, Maier A, et al. Robust Vessel Segmentation in Fundus Images[J].

International Journal of Biomedical Imaging, 2013, 2013(6):154860.

[2] Zhuang Q X, Zhang H. The application of fundus fluorescein angiography and optical coherence tomography in the examination of diabetic retinopathy[J].Practical Journal of Clinical Medicine,2020,17(05):117-120.

[3] Montoya M G, Gil C, Garca I. The load unbalancing problem for region growing image segmentation algorithms[J]. Journal of Parallel & Distributed Computing, 2003, 63(4):387-395.

[4] Queiroz R, Fan Z, Tran T D. Optimizing block-threshold segmentation for MRC compression[C]// Image Processing, 2000. Proceedings. 2000 International Conference on. IEEE, 2000.

[5] Xu Y B, Zhou Y, et al .Retinal vessel segmentation based on super pixel affine propagation clustering[J].Acta Optica Sinica, 2020,40(02):60-70.

[6] J. Long, E. Shelhamer and T. Darrell, "Fully convolutional networks for semantic segmentation," 2015 IEEE Conference on Computer Vision and Pattern Recognition (CVPR), Boston, MA, USA, 2015,pp.3431

[7] Ronneberger O, Fischer P , Brox T . U-Net: Convolutional Networks for Biomedical Image Segmentation[J]. 2015.

[8] Z. Zhou, M. M. R. Siddiquee, N. Tajbakhsh and J. Liang, "UNet++: Redesigning Skip Connections to Exploit Multiscale Features in Image Segmentation," in IEEE Transactions on Medical Imaging, vol. 39, no. 6,pp.1856-1867,June2020,doi:10.1109/TMI.2019.2959609.

[9] Huang G, Liu Z, Laurens V, et al. Densely Connected Convolutional Networks[J]. IEEE Computer Society, 2016.

[10] S Jégou, Drozdzal M, D Vazquez, et al. The One Hundred Layers Tiramisu: Fully Convolutional DenseNets for Semantic Segmentation[C]// 2017 IEEE Conference on Computer Vision and Pattern Recognition Workshops (CVPRW). IEEE, 2016.

[11] Y Lv, Ma H, Li J, et al. Attention Guided U-Net With Atrous Convolution for Accurate Retinal Vessels Segmentation[J]. IEEE Access, 2020, 8:32826-32839.

[12] X. Xiao, S. Lian, Z. Luo and S. Li, "Weighted Res-UNet for High-Quality Retina Vessel Segmentation," 2018 9th International Conference on Information Technology in Medicine and Education (ITME), 2018, pp. 327-331, doi: 10.1109/ITME.2018.00080.

[13] Ibtehaz N, Rahman M S. MultiResUNet: Rethinking the U-Net architecture for multimodal biomedical image segmentation[J]. Neural Networks, 2019, 121.

[14] He K, Zhang X, Ren S, et al. Deep Residual Learning for Image Recognition[J]. IEEE, 2016.

[15] Szegedy C, Liu W, Jia Y, et al. Going Deeper with Convolutions[J]. IEEE Computer Society, 2014.

[16] Jie H, Li S, Gang S, et al. Squeeze-and-Excitation Networks[J]. IEEE Transactions on Pattern Analysis and Machine Intelligence, 2017, PP(99).

[17] Woo S, Park J, Lee J Y, et al. CBAM: Convolutional Block Attention Module[C]// European Conference on Computer Vision. Springer, Cham, 2018.

[18] Staal J, Abramoff, M.D, Niemeijer M , et al. Ridge-based vessel segmentation in color images of the retina.[J]. IEEE Transactions on Medical Imaging, 2004, 23(4):501-509.

Influence of Mesh Density on Repeatability of FSW Process Simulation Based on DEFORM

Xiaohong Lu[1], Jinhui Qiao[1,*], Yihan Luan[1], Zesheng Han[1] and Steven Y. Liang[2]

[1] Key Laboratory for Precision and Non-traditional Machining Technology of Ministry of Education, Dalian University of Technology, Dalian, China
[2] The George W. Woodruff School of Mechanical Engineering, Georgia Institute of Technology, Atlanta, GA 30332-0405, USA

*Corresponding author email: qiaojinhui1997@mail.dlut.edu.cn

Abstract. To solve the problem of the relative error caused by mesh density is difficult to quantitatively evaluate, a process simulation model of FSW (friction stir welding) 2219 aluminum alloy is built based on DEFORM. Process simulation models with different mesh densities are solved, the influence of mesh density on simulation repeatability and time-consuming is analyzed, and the mesh density suitable for the process simulation model is determined. The evaluation method proposed offers a theoretical reference for the determination of local optimal mesh density of the FSW process simulation model based on DEFORM. This study provides a basis for high precision prediction and characterization of the temperature distribution of FSW on heavy-lift launch vehicle fuel tank.
Keywords: Friction stir welding; Repeatability; Simulation; Mesh density; DEFORM.

1. Introduction

Friction stir welding (FSW) is a solid-phase jointing technology invented by TWI in 1991 [1]. The temperature distribution of FSW directly determines the quality of joint. Finite element method has become an important means to analyze the temperature distribution of FSW. DEFORM is a metal plastic analysis software, and a professional process simulation system based on finite element method.

Scholars have used DEFORM to establish process simulation models of the FSW. Zhang et al. [2] established a model of FSW 10mm thick LY12 aluminum alloy, refined the meshes at the welding seam of the workpiece, studied the temperature distribution of the welding process. Hu et al. [3] qualitatively analyzed the influencing factors of simulation accuracy of the simulation model, and proposed that mesh size would affect the simulation accuracy. But they didn't perform quantitative analysis.

Due to the strong mesh refinement function of DEFORM, most of scholars adopted the mesh refinement function in the simulation of FSW process based on DEFORM. Empirical method is often used to determine the mesh density in process simulation model, and the effect of mesh density on the repeatability of the process simulation model which will affect the prediction accuracy of the simulation has not been quantitatively studied.

To solve the problem of the relative error caused by the mesh density is difficult to quantitatively evaluate, a process simulation model of FSW 2219 aluminum alloy is established based on DEFORM. Simulation models with different mesh densities are established to simulate the process of plunging stage. The temperature of feature points is extracted, and the influence of mesh density on the repeatability of the process simulation model is studied.

2. Simulation of FSW 2219 Aluminum Alloy Based on DEFORM

The tool shoulder diameter is 32mm, the pin diameter is 15mm, and the pin length is 17.8mm. The size of the weldment is $100 \times 150 \times 18mm^3$. The diameter of mesh window 1 is set as 38mm, and mesh window 2 is set as 20mm. The mesh density is set as 1mm. JMatPro software is used to obtain the material parameters of 2219 aluminum alloy by importing chemical composition, as shown in figure 1.

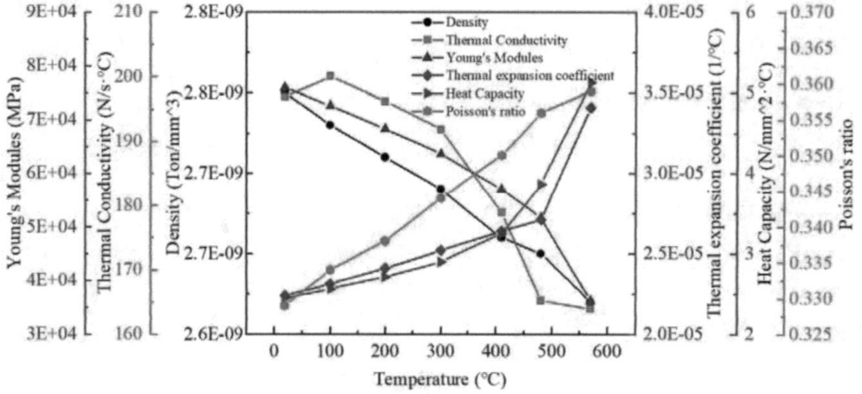

Figure 1. Material parameters of 2219 aluminum alloy.

Johnson-Cook constitutive model is used to describe the plastic behavior of materials. Restrict the movement freedom of each surface of the weldment to avoid the displacement of the weldment during welding. The setting of thermal boundary conditions is shown in [4]. To simplify the complexity of the process simulation model, the friction coefficient is assumed to be temperature-dependent. The friction coefficients are show in [5].

Finally, the FSW process simulation model based on DEFORM is established. When the rotational speed is 400r/min and the welding speed is 100mm/min, the temperature diagram of each phases is shown in figure 2.

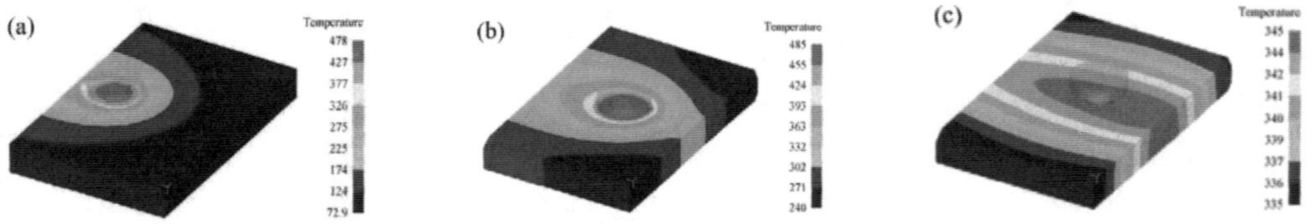

Figure 2. Temperature diagram: (a) the plunging and dwelling phase, (b) the welding phase, (c) the tool withdrawal phase.

3. Establishment of Simulation Models with Different Mesh Densities

The mesh density affects computation time, as well as has a great impact on the repeatability accuracy of the process simulation model. Therefore, it is necessary to comprehensively consider the repeatability accuracy and computation time of the simulation to select the appropriate mesh density. Five simulation models of different mesh densities as shown in figure 3 have been considered.

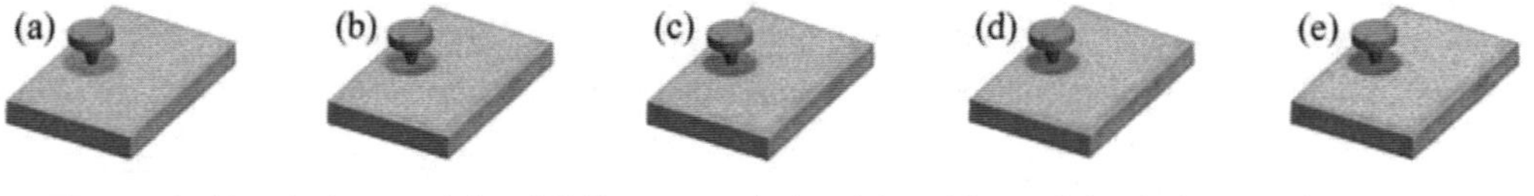

Figure 3. Simulation models of different mesh densities: (a) Model 1 (1.3mm), (b) Model 2 (1.2mm), (c) Model 3 (1.1mm), (d) Model 4 (1.0mm), and (e) Model 5 (0.9mm).

In the simulation process, the tool tilt angle is 2.5°, the rotational speed is 400r/min, and the plunging traverse speed is 20mm/min. To quantitatively study the repeatability accuracy of the process simulation model, three groups of simulations are performed for each mesh size model.

4. Results and Discussion

4.1. Selection of Feature Points

According to the research on the microstructure and tensile test by related scholars [6, 7], during the FSW process, the intense flow range of the weldment material is located at approximately 2mm on the side of the conical stir pin, and the fracture position of the FSW joint tensile specimen is at TMAZ near NZ or TMAZ near HAZ.

The data points on the weldment at AS and RS section and at 2mm on the side of the conical stir pin (Y = 25mm, XOZ section) are selected as feature points, where each feature point is separated by 0.92mm. Figure 4(a) shows the schematic diagram of feature points selection on AS and RS sections. The data points on the weldment along the welding seam direction and at 2mm on the side of the conical stir pin (X = 75mm, YOZ section) are selected as feature points, where each feature point is separated by 0.92mm. Figure 4(b) shows the schematic diagram of feature points selection on the welding seam direction.

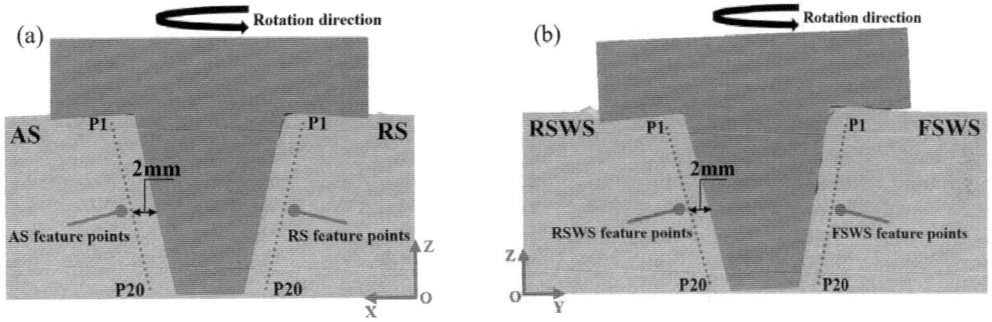

Figure 4. Diagrammatic drawing of feature points selection (a) AS and RS section (Y = 25 mm, XOZ section) (b) welding seam direction (X = 75 mm, YOZ section).

4.2. Error Analysis

Maximum relative percentage error (δ_{MRPE}) is used to describe the repeatability error of the process simulation model. The calculation equations are as follows:

$$\overline{y}_i = \frac{\sum_{j=1}^{3} y_{ij}}{3} \quad (1)$$

$$\delta_{MRPE} = \max\left(\frac{1}{3}\cdot\sum_{j=1}^{3}\frac{\left|y_{ij}-\overline{y}_i\right|}{\overline{y}_i}\right)\times 100\% \quad (2)$$

Where y_{ij} is the temperature of the i feature point of the j group model, i=1-20, j=1-3; $\overline{y}_i$ is the average value of the three groups of temperatures at the i feature point.

The calculation results of the δ_{MRPE} of the feature point temperature of the simulation models with different mesh densities are shown in table 1.

Table 1. The influence of mesh density on simulation repeatability.

Model	AS	RS	RSWS	FSWS
Model 1 (1.3mm)	4.10%	3.55%	2.41%	4.00%
Model 2 (1.2mm)	2.94%	3.17%	2.64%	5.62%
Model 3 (1.1mm)	2.84%	2.00%	2.40%	3.17%
Model 4 (1.0mm)	1.20%	1.97%	2.11%	2.93%
Model 5 (0.9mm)	2.35%	2.64%	1.95%	2.32%

The computer for simulation is equipped with Intel® Core (TM) i7-10700 CPU @ 2.90 GHz, 16 GB RAM. The computation time for each group of the five models and the average computation time of the five models are shown in table 2.

Table 2. Computation time for each group of models (min).

	Model 1	Model 2	Model 3	Model 4	Model 5
Group 1	188	270	312	457	781
Group 2	187	243	334	451	846
Group 3	207	264	318	436	816
Average computation time	194	259	321	448	814

The maximum relative percentage error of each feature point temperature and the average computation time of the simulation models are represented in figure 5.

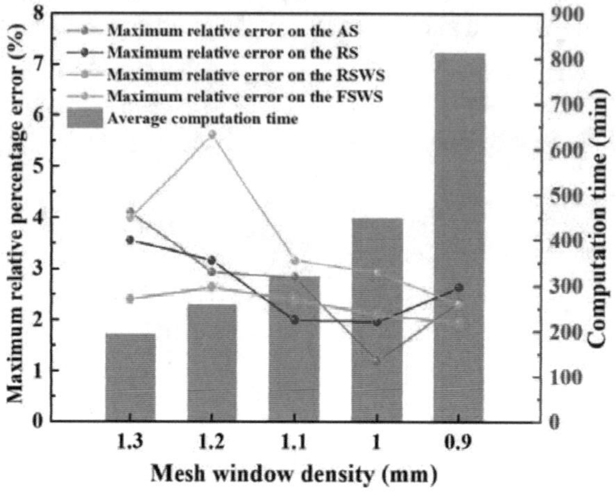

Figure 5. Comparison of the error of each feature point temperature and average computation time of each model.

Figure 5 shows that the accuracy of the feature points on the AS is ranked from high to low: model 4 > model 5 > model 3 > model 2 > model 1. The accuracy of the feature points on the RS is ranked from high to low: model 4 > model 3 > model 5 > model 2 > model 1. The accuracy of the feature points on the RSWS is ranked from high to low: model 5 > model 4 > model 3 > model 2 > model 1. The accuracy of the feature points on the FSWS is ranked from high to low: model 5 > model 4 > model 3 > model 1 > model 2. The larger error on the FSWS is mainly because the tool has a tilt angle during the plunging stage, which causes the tool to not fully contact the material on the FSWS, resulting in a large change in the heat generation on the FSWS.

Based on the above analysis, mesh sizes of 1.0mm and 0.9mm seem to be the optimum mesh sizes, but the computation time of model 5 is nearly double that of model 4. Considering the simulation accuracy and computation time, mesh size of 1.0mm is a more suitable mesh density.

5. Conclusion

A process simulation model of FSW is established, and the repeatability error is analyzed. The results prove that even when the condition including the tool size parameters, the welding process parameters, the geometrical model, the material properties, the boundary conditions, the frictional model, and the mesh density are unchanged, the temperature of the feature points in the core area of the process simulation model output is still different. This is because although the mesh density is set to the same value in the simulation models, the meshes layout generated by automatic meshing method of DEFORM is different each time, which leads to the difference of simulation results and computation time.

Through the comparison of the simulation results of models with different mesh densities, it is found that the smaller the mesh size in the mesh window, the better the repeatability of the process simulation model. Considering the simulation accuracy and computation time comprehensively, the mesh density of 1.0mm is more suitable for the process simulation model. The evaluation method proposed offers a theoretical reference for the determination of local optimal mesh density of the FSW process simulation model based on DEFORM.

Acknowledgments
This work was supported by the National Key Research and Development Program of China (Grant No. 2019YFA0709003), the Dalian Science and Technology Innovation Fund (Grant No. 2020JJ26GX041), and the Fundamental Research Funds for the Central Universities (Grant No. DUT20ZD204).

References

[1] Li Y, Trillo E A and Murr L E 2000 Friction-stir welding of aluminum alloy 2024 to silver *Journal of Materials Science Letters* vol 19 no 12 pp 1047-1051
[2] Zhang Z K, Peng J and Wang X J 2015 Numerical Simulation of the Dynamic Stress Field Based on DEFORM in the Friction Stir Welding of Al Alloy *Advanced Materials Research* vol 3857 no 1095 pp 781-785
[3] Hu J J, Xu H B and Jin Y 2009 Approach for increasing precision of numerical simulation of plastic forming *Forging and Stamping Technology* vol 34 no 2 pp 149-151
[4] Jain R, Pal S K and Singh S B 2019 Investigation on effect of pin shapes on temperature, material flow and forces during friction stir welding: A simulation study *Proceedings of the Institution of Mechanical Engineers, Part B: Journal of Engineering Manufacture* vol 233 no 9 pp 1980-1992
[5] Xiong F, Qian W and Liu G 2015 Frictions Stir Welding Process for 2219 Aluminum Alloy *Journal of University of Shanghai for Science and Technology* vol 37 no 1 pp 79-83
[6] Sun G, Niu J, Wang D and Chen S 2015 Fatigue experimental analysis and numerical simulation of FSW joints for 2219 Al-Cu alloy *Fatigue and Fracture of Engineering Materials and Structures* vol 38 no 4 pp 445-455
[7] Wen Q, Li W Y, Patel V, Gao Y J and Vairis A 2020 Investigation on the Effects of Welding Speed on Bobbin Tool Friction Stir Welding of 2219 Aluminum Alloy *Metals and Materials International* vol 26 pp 1830-1840

A Pipeline-based Low Complexity Dual Thresholds and Multiresolution Mouth Detection Method

Jingli Lin, Changwei Li*, Xi Lan and Jinge Li

Xihua University, College of Electrical and Information Engineering, Chengdu, China

*Corresponding author email: LCW@stu.xhu.edu.cn

Abstract. Mouth detection is a basic step in the automatic diagnosis of buccal diseases. This paper proposes a method of using FPGA(Field Prog ram mable Gate Array) to detect the human mouth region in images, which adopts a pipelined design for high-speed operating. Thus, it can be applied to mouth motion detection as well. After an original image is processed regularly by median filtering, color conversion is performed to obtain its corresponding H component, which is used to create a partitioned binary image. Each constituent region of the binary image uses two thresholds for regional processing separately, which results in two binary images: one comprises the mouth and some other regions with a color close to that of the mouth, and the other indicates the position of the mouth. A logical AND is performed on these two binary images to calculate the border of mouth region with reduced resolution, which is then superimposed on the buffered original image to confirm the mouth region eventually. An FPGA-based scheme is presented, which shows the method could detect the whole mouth properly, with a maximum clock speed of 150MHz. In addition, processing an image of 200*200 only consumes 2178 Slice LUTs(Look Up Table), which is a relatively low resource consumption.
Keywords: Mouth detection; FPGA; Pipelined Architecture; Dual Thresholds; Multiresolution.

1. Introduction

Mouth recognition, like the recognition of other human body parts, is essential in the image-based biomedical analysis of human features. The recognition methods can be generally classified into two categories: 1) unsupervised or semi-supervised feature deep learning algorithms [1]; 2) traditional pattern recognition algorithms [2].

In general, deep learning algorithms have high accuracy. However, they require massive annotated data to train neural networks, which results in high cost and long development cycles. In addition, the neural networks based on deep learning are more like black boxes, in which a challenge of properly adjusting some algorithm parameters exists. On the other hand, the traditional pattern recognition algorithms are mainly based on the processing of statistical signals to determine an adaptive/solid decision boundary.

In practical applications, the capacity of high-speed processing is sometimes required for real-time operation. To meet these requirements, FPGA[3], DSP[4],GPU and cloud computing are some of the main platforms employed[5]. Neural networks have pipelined architectures naturally, which is efficient and effective for an FPGA-based design. However, neural networks based on deep learning have a large number of hidden layers and neurons. Thus, they usually run with GPU or cloud computing, which means higher cost. Compared with DSP, FPGA and GPU have complete parallel architectures. Therefore, they have an advantage over DSP in processing speed. Taking various factors into consideration, including the oral feature, processing speed, complexity, and low-power consumption, a dual thresholds mouth recognition method is proposed in this paper, which has the following features: 1) high accuracy, 2) low complexity, 3)high speed.

2. The Detection Model

2.1. Color Model Conversion

In a facial region, the mouth region has more obvious color characteristics than other parts, and thus a detection method based on color characteristics can be an option. Common color models include RGB, HSV, YCbCr, and HSI. Because lips are redder than other regions of the face, their H value is around zero degree. Therefore, this paper adopts the HSV color model for the preprocessing of face images. Since the images captured by a common camera are in RGB format, the image format should be converted first. In this paper, only color features are considered, i.e., only the H conversion is required. Assume that all colors have been normalized to be within the range [0, 1]. Among the three components of RGB, the maximum value is MAX, and the minimum value is MIN. The conversion model from RGB to H [6-7] is as follows:

$$H = \begin{cases} \frac{G-B}{MAX-MIN} \times 60^\circ \text{ , } (R = MAX) \\ \left(2 + \frac{B-R}{MAX-MIN}\right) \times 60^\circ, (G = MAX) \\ \left(4 + \frac{R-G}{MAX-MIN}\right) \times 60^\circ, (B = MAX) \end{cases} \tag{1}$$

2.2. Thresholding Function to Create Binary Image

After RGB to HSV conversion, the image will be further converted to a binary image: each pixel containing more red components is with value 1, and the others are with value 0. The thresholding function is:

$$Value_{pixel} = \begin{cases} 1, \ a \cdot H_{max} < H \ or \ b \cdot H_{min} > H \\ 0, \ otherwise \end{cases} \tag{2}$$

where 0<a<1, and b>1.

An RGB image of size 200*200 pixels, its corresponding H component model and binary image are shown in Figure 1, where a is 0.9, and b is 2. As shown in Figure 1, due to the presence of red component. Not only the mouth is 1-valued, but also some other parts of the face, the clothing, and backgrounds, etc. Besides, if all binary pixels are used for subsequent processing, resource consumption would be a challenge.

2.3. Detection with Dual thresholds and Multiresolution

Since mouth has more red component relatively, the parts in the binary image containing more 1s could be mouth more possibly. We divide a binary image of size $M * N$ pixels denoted by I into $U * V$ pixel blocks, and each block contains $S * T$ pixels, then $M = U * S$, $N = V * T$. Each pixel block in I is defined as I_{ij}, where $0 \le i < U$, $0 \le j < V$. Each element $a_{ij\_t}(0 \le t < S * T)$ of I_{ij} is stored in memory sequentially, which can be represented as $A_{ij} = [a_{ij\_0}, a_{ij\_1}, \cdots, a_{ij\_S*T-1}]$. Then, the total number of 1s in I_{ij} can be calculated by:

$$X_{ij} = A_{ij} * O' \tag{3}$$

where $O = \underbrace{[1,1,\cdots,1]}_{S*T}$, and$'$ denotes transposition operator.

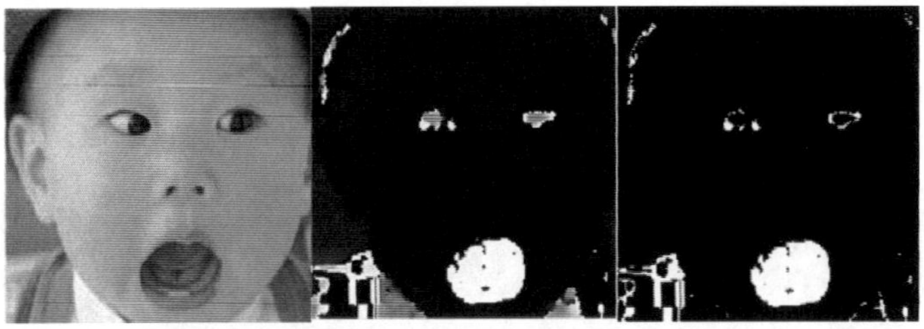

Figure 1. (a) Original image (b) H component model (c) binary image.

When a threshold t_b is set, where $t\_b < S * T/2$. For a binary image I_1 of size $U * V$ pixels, the value of each element is I_{1_ij}:

$$I_{1\_ij} = \begin{cases} 1, if\ X_{ij} \geq t\_d \\ 0, otherwise \end{cases} \quad (4)$$

Figure 2 shows an I_1 obtained from Figure 1 by setting $U = 8,\ V = 8,\ S = 25,\ T = 25,\ t\_d = 20$. As shown in Figure 2, the mouth part is kept maximally in I_1, however, some other non-target regions are also included.

Figure 2. Binary image $\boldsymbol{I_1}$ under the circumstances of $\boldsymbol{U} = 8,\ \boldsymbol{V} = 8,\ \boldsymbol{S} = 25,\ \boldsymbol{T} = 25,\ \boldsymbol{t_d} = 20$.

Figure 3. Binary image $\boldsymbol{I_2}$ under the circumstances of $\boldsymbol{U} = 8,\ \boldsymbol{V} = 8,\ \boldsymbol{S} = 25,\ \boldsymbol{T} = 25,\ \boldsymbol{t\_u} = 360$.

In order to remove the interference from the non-target regions, similarly, a threshold t_u is set for a binary image I_2 of size $U * V$ pixels whose value of each element is I_{2_ij}:

$$I_{2\_ij} = \begin{cases} 1, if\ X_{ij} \geq t\_u \\ 0, otherwise \end{cases} \quad (5)$$

where $S * T > t\_u > t\_b$.

Figure 3 shows an I_2 obtained from Figure 1 by setting $U = 8,\ V = 8,\ S = 25,\ T = 25,\ t\_u = 360$. As shown in Figure 3, except for the mouth part, the interference regions are completely removed because the red component in these regions is not concentrated enough, but the mouth region is also narrowed significantly.

Let $R(A)$ be the set of pixels with value 1 in a binary image A. According to Figure 2 and Figure 3, both the binary images I_1 and I_2 reflect the features of the mouth, but $\mathrm{R}(I_1)$ not only includes the mouth but also other backgrounds. In contrast, $\mathrm{R}(I_2)$only include a part of the mouth. Neither of them can accurately and completely determine the mouth region. Therefore, we divide I into $P \times Q$ pixel blocks again, and each block contains $G * H$ pixels, then $S = P * G,\ T = Q * H$. The value of each pixel block in I_3 is defined as I_{3_mn}, where $0 \leq m < P,\ 0 \leq n < Q$. Then let

$$I_{3\_mn} = \begin{cases} 1, if (I_{2\_ij}=1) \cap ((m*G,(m+1)*G) \cap (i*S,(i+1)*S) \neq \emptyset) \cap ((n*H,(n+1)*H) \cap (j*T,(j+1)*T) \neq \emptyset) \\ 0, otherwise \end{cases} \quad (6)$$

Where $\emptyset$ is an empty set.

Create a binary image I_4 of size $U * V$ pixels based on I_3 and the value of each element in I_4 is:

$$I_{4\_ij} = \begin{cases} 1, if (I_{3\_mn}=1) \cap (((m*G,(m+1)*G) \cap (i*S,(i+1)*S) \neq \emptyset) \cap ((n*H,(n+1)*H) \cap (j*T,(j+1)*T) \neq \emptyset) \\ 0, otherwise \end{cases} \quad (7)$$

In this way, I_4 can maximally expand the mouth region, but it introduces some other backgrounds near the mouth inevitably. Figure 4 shows an I_4 which results from the binary image I_2 being resampled twice in Figure 3.

Figure 4. $\boldsymbol{I_4}$:The result of $\boldsymbol{I_2}$ is resampled twice.

Figure 5. $\boldsymbol{I_5}$: logical AND result of $\boldsymbol{I_1}$ and $\boldsymbol{I_4}$.

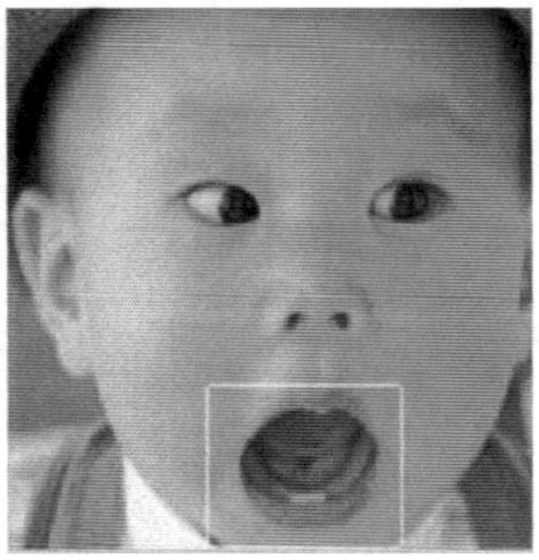

Figure 6. The confirmed mouth region.

Comparing Figure 2 with Figure 4, we notice that both $\mathrm{R}(I_1)$ in Figure 2 and $\mathrm{R}(I_4)$ in Figure 4 contain the whole mouth region. However, $\mathrm{R}(I_1)$ includes some other backgrounds unconnected to the mouth whereas $\mathrm{R}(I_4)$ includes some other backgrounds connected to the mouth region. Therefore, the interference regions would be maximally removed through a logical AND of I_1 and I_4. In this way, a binary image I_5 of size $U * V$ pixels can be obtained.

Finally, each pixel of I_5 is interpolated to a block of size $S \times T$ pixels by pixel replication. Based on the coordinates of the four vertices in the mouth region, the mouth region in the original picture can be determined. The confirmed mouth region in the original image is shown in Figure 6.

3. Pipelined Architecture

As described in Section II, the utilization of the original image data is twofold. One is processed to confirm the mouth region, and the other one is stored for the final mouth display. The architecture is shown in Figure 7.

3.1. Pipelined Dual Thresholds Architecture

In general, an original image of size $M \times N$ pixels is stored in memory as a vector of size $1 \times MN$ which is formed by letting the first N elements of the vector be the first row of the image, the next N elements be the second row, and so on. Therefore, the data is stored sequentially after H conversion on the second processing path in Figure 7, which is shown in Figure 8.

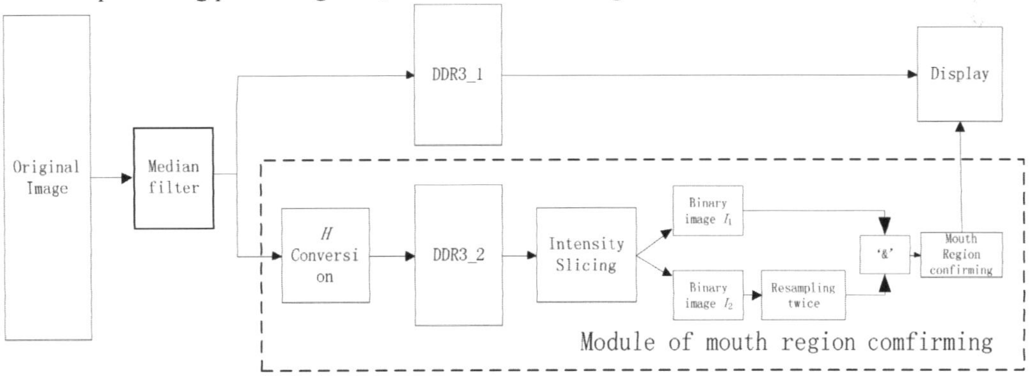

Figure 7. System block diagram.

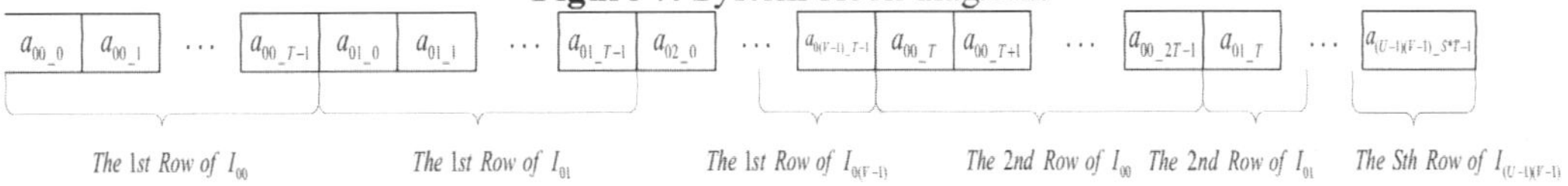

Figure 8. Data storage format after $\boldsymbol{H}$ conversion.

In Figure 8, the output of H conversion circuit is synchronized with a global clock signal. An accumulator ACC_{ij} and two threshold comparators $C1_{ij}$ and $C2_{ij}$ are applied to each I_{ij}, respectively. ACC_{ij} is used to generate X_{ij} ; $C1_{ij}$ and $C2_{ij}$ are used to generate I_1 *and* I_2 , respectively. When the first row of I_{ij} is calculated, ACC_{ij} is cleared. ACC_{ij} checks the current pixel value at the rising edge of the clock signal. If the value is 1, ACC_{ij} adds 1 to the current value of its register. If the current processing row is the last row of I_{ij}, then the value of ACC_{ij} is X_{ij}, which is compared with thresholds T_{ij_d} and T_{ij_u} to determine the pixel values I_{1_ij} and I_{2_ij} of I_1 and I_2, respectively. The generation architecture of each pixel of I_1 and I_2 is shown in Figure 9, where $AND1_{ij}$ and $AND2_{ij}$ are AND gates; $M1_{ij}$ is a 2-to-1 multiplexer. When its selection signal is true, the input a is routed to the output to clear ACC_{ij}. Otherwise, the data connected to port b is passed to ACC_{ij}; $M2_{ij}$ is a 1-to-2 multiplexer. When its selection signal is true, the input is routed to the output b to calculate I_{1_ij} and I_{2_ij}. Otherwise, the input is routed to the output a; And $Index(R)_{ij}$ denotes the current processing row index of I_{ij}, therefore, $0 \leq Index(R)_{ij} \leq S-1$.

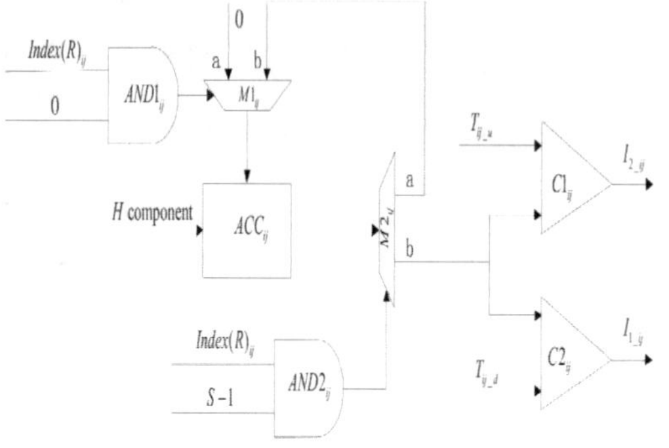

Figure 9. Pipelined Dual Thresholds Architecture.

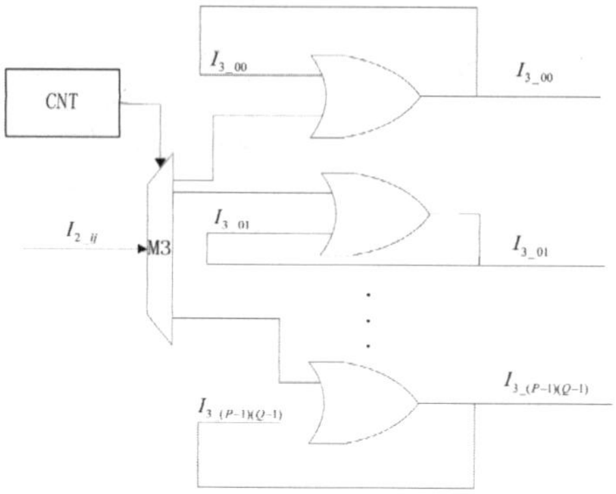

Figure 10. Pipelined $\boldsymbol{I_3}$ Creating Architecture.

According to Figure 9, as the H component of the first video frame is generated, I_{1_00} and I_{2_00} of the frame can be obtained after $T(V*S-V+1)$ clock cycle, and following I_{1_0j} and I_{2_0j} $(0<j<V)$ can be obtained at T clock cycle intervals. Then, after another $T(V*S-V+1)$ clock cycles, I_{1_10} and I_{2_10} can be obtained..., and so on. In particular, after I_{1_00} and I_{2_00} are obtained, ACC_{00} is available to the next frame image processing(i.e., the interval between two consecutive frames is only $T(V*S-V+1)$ clock cycles), which greatly improves the processing speed compared to the routine $U*V*S*T$ clock cycles.

3.2. Pipelined Binary Image of Size $\boldsymbol{P \times Q}$ *Pixels Creating Architecture*

A binary image I_3 of size $P*Q$ pixels is obtained on the basis of I_2 in Figure 9. According to Figure 9, I_2 is generated row-by-row. Therefore, the pipelined I_3 creating architecture is shown in Figure 10 with an initialization of $I_{3\_mn}=0$ for all the elements. In Figure 10, one OR gate is used for each pixel of I_3, respectively. M3 is a 1-to-$P*Q$ multiplexer whose selection signal comes from a counter denoted as CNT. When $(m = floor\left(i*\frac{S}{G}\right)$ or $floor\left((i+1)*\frac{S}{G}\right))$ && $(n = foor\left(j*\frac{T}{H}\right)$ or $floor\left((j+1)*\frac{T}{H}\right))$, where the notation $floor(\cdot)$ denotes the operator of rounding down to nearest integer, the input of M3 routes to the next output port. Due to the different resolution, the clock speed in Figure 10 is $1/(S*T)$ of that in Figure 9.

4. Conclusion

The key point of the mouth region detection algorithm in this paper is the multiresolution binary images creating based on a dual thresholds scheme. The algorithm is simulated with Matlab and its pipelined architectures are designed. The result shows that the processing has high efficiency with low resource consumption, so it can be used for mouth motion detection. Similarly, the method can also be applied to the detection of other image features with relatively specific colors. In addition, the simulation shows when we create I_3 based on I_2, even I_3 having a higher resolution than that of I_2, the same result can be obtained. However, taking complexity into consideration, an I_3 with lower resolution is preferred. In the case of an original image of size 200*200 pixels, the processor is implemented in a XILINX XC6SLX15[8] FPGA. It costs 1138 Slice Registers and 2178 Slice LUTs in total which are 6% and 24% of the whole corresponding resources, respectively. The maximum global clock speed is 150MHz.

Acknowledgments
We thank the associate editor and the reviewers for their useful feedback that improved this paper. In addition, we acknowledge the support received from the Chunhui project(Z201405)

References
[1] R. Pai, A. Dubey and N. Mangaonkar, "Real Time Eye Monitoring System Using CNN for Drowsiness and Attentiveness System," 2021 Asian Conference on Innovation in Technology (ASIANCON), 2021, pp. 1-4
[2] F. Wen, M. Qin, P. Gratz and N. Reddy, "An FPGA-based Hybrid Memory Emulation System," 2021 31st International Conference on Field-Programmable Logic and Applications (FPL), 2021, pp. 190-196
[3] J. Kim and L. Wei, "Performance Analysis of Machine Learning-based Face Detection Algorithms in Face Image Transmission over AWGN and Fading Channels," 2021 IEEE International Symposium on Broadband Multimedia Systems and Broadcasting (BMSB), 2021, pp. 1-5
[4] Gary Storey, Richard Jiang, et al., "3DPalsyNet: A Facial Palsy Grading and Motion Recognition Framework Using Fully 3D Convolutional Neural Networks", IEEE Access, vol. 7, pp. 121655-121664, 2019.
[5] S. Swetha, P. Kamali, B. Swathi, R. Vanithamani and E. Karolinekersin, "Oral Disease Detection using Neural Network," 2020 9th International Conference System Modeling and Advancement in Research Trends (SMART), 2020, pp. 339-342.
[6] U. I. Minhas, R. Woods, D. S. Nikolopoulos and G. Karakonstantis, "Efficient, Dynamic Multi-Task Execution on FPGA-Based Computing Systems," in IEEE Transactions on Parallel and Distributed Systems, vol. 33, no. 3, pp. 710-722, 1 March 20
[7] R. Ranjan et al., "A Fast and Accurate System for Face Detection, Identification, and Verification", IEEE Transactions on Biometrics, Behavior, and Identity Science, vol. 1, no. 2, pp. 82-96, April 2019.
[8] D. Arruda and N. H. Madhavji, "QualiBD: A Tool for Modelling Quality Requirements for Big Data Applications," 2019 IEEE International Conference on Big Data (Big Data), 2019, pp. 5977-5979

Research on Network Routing Control Algorithm Based on OpenFlow and IGP

Xuejun Tian[1], Lizi Wen[1], Xingyan Yang[1], Liangzhe Chen[1], Gang Min[2] and Zhong Shu[1,2,*]

[1] School of Electronic Information Engineering, Jingchu University of Technology, Jingmen, China
[2] Jingmen Mobile Media Co. Ltd, Jingmen, China

*Corresponding author email: 421934337@qq.com

Abstract. Given the coexistence of heterogeneous hybrid networks such as IP network, 4G network and Software Defined Network (SDN) in the application of digital news media system, this paper designs a routing control algorithm based on SDN network architecture, which can realize the compatibility of internal gateway protocol and OpenFlow communication strategies. Through the analysis of OpenFlow communication protocol in SDN network architecture, the relationship between Interior Gateway Protocol (IGP) and OpenFlow routing rules is found out. Based on the existing hardware structure of the OpenFlow router, the configuration parameters of a port table, port address table and routing table are improved, and the adjacent routing relation table and information transmission port are used to realize the connection between routers are designed. The attention of users is considered It can reduce the burden of system operation and is compatible with the forwarding control of route forwarding and routing filtering. The adjacent routing relation table is designed to control the process of data collection, storage and forwarding, and the input and output ports between adjacent routers are defined. The open router designed in this paper integrates Simple Network Management Protocol (SNMP) agent and SDN agent mechanism organically to achieve the goal of SDN network management; the routing control module focuses on the management of key data flow, realizes the control of routing state and data flow at the same time, and provides the visualization function of routing state and routing state data flow. By constructing the experimental platform of routing transmission and control functions based on routing control and interior gateway protocol, the effect of routing state control of the two methods is compared and analysed from three aspects of packet transmission state, packet processing state and packet receiving delay in routing control process. The experimental results verify the correctness of the design results.
Keywords: Interior Gateway Protocol (IGP); OpenFlow; Open router; Routing control.

1. Introduction

In a heterogeneous hybrid network, the construction of a routing information control model dominated by SDN technology faces several problems [1-4]. Firstly, different information communication strategies are adopted in routing control. In the Internet environment where distributed routing mechanism is widely used, unified routing control is directly realized by using OpenFlow communication protocol, but it is difficult for the configured routing control server to realize routing control in the control layer of the SDN model. Then, the OpenFlow communication policy has no direct control over the IP network. IGP and Border Gateway Protocol (BGP) communication policies are the main actors of IP network routing control. To realize the convergence and co-governance of

OpenFlow, IGP and BGP communication policies, another control platform needs to be designed, and its stability is questioned. In addition, the bandwidth provided by the transmission channel of routing control information is insufficient. When multiple communication policies are started at the same time, it is necessary to provide network status information exchange between multiple routing control servers and SDN master control servers, and information loss is inevitable [5-8]. Hence, it's meaningful to solve the above problem.

In this paper, the heterogeneous hybrid networks that consist of IP and SDN network system is employed as the network platform, which is mainly for a digital media platform to provide graphic information collection, processing, and publishing services. Besides, the SDN network management strategy is introduced, and the OpenFlow communication and fusion of IGP communication protocol coexistence is set as the breakthrough point. On basis of the OpenFlow communication protocol, a new algorithm is designed to realize the routing control of heterogeneous hybrid network systems on the Internet, which will provide support for the design of visual routing control system. Compared with some current algorithms, the algorithm proposed in this paper improves the communication transparency between different networks and makes a new attempt on load balancing control of routing policy. In the construction of the SDN interdomain model, boundary switches can be directly controlled to forward routing control information, saving a lot of information exchange in intermediate links.

2. Routing and Forwarding Model Construction and Function Implementation

2.1. Open Router design

Data forwarding layer of the main equipment for a router (can also be a data switch), the present OpenFlow router usually provides the data flow and function modules of the control of two parts, therefore, need not tinkering with existing router hardware, only need to increase in routing control layer to support the RC protocol communication port, The connection to the data forwarding layer can be realized. In the network routing control policy shown in figure 1, the connection mode between the data forwarding layer and the routing control layer, the internal design of the routing control layer and the transmission process of the data forwarding layer are defined. In a traditional IP network using SNMP network management strategy [9], an open router is designed.

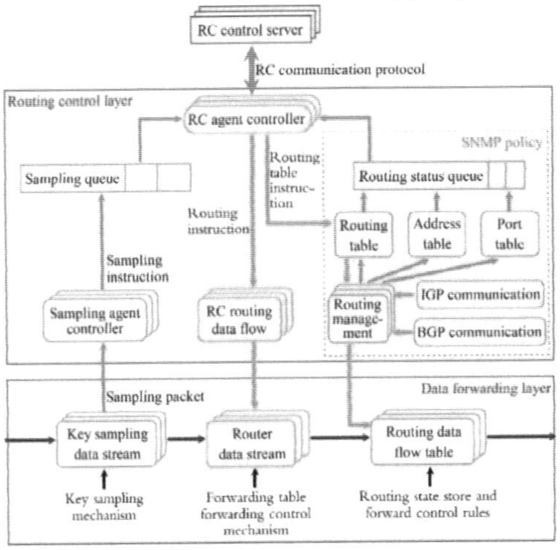

Figure 1. The internal structure of an open router.

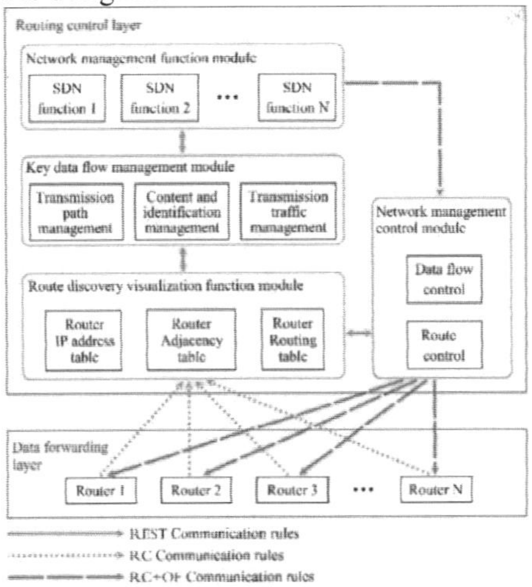

Figure 2. Internal structure diagram of RC routing control layer.

Data forwarding layer open router within the physical hardware part is divided into the data flow in a sequence sampling, access control, and data stream processing three parts, data packets sent to a router before, through the three hardware parts processing, to construct the routing table of the next-hop routing path involved (namely before a followed by a router, the router is connected, Of course, it can also be directly the destination host of final access). Routing control layer by all routing protocols (including RC communication protocols, OF communication protocol, IGP communication protocol, etc.), the routing table management (routing table includes the source IP address, destination IP address, the source host port, the destination port, the next-hop routing path, etc.), the key data flow sampling (to reduce the operation OF the master server load, At the same time, it is compatible with SNMP network management agent policy, data flow sampling can also use the proxy mechanism), RC routing control (also can introduce the proxy mechanism to reduce the load of RC routing control server) and other modules. In the RC routing control module, the routing status data flow table and information transmission cache queue are introduced, the RC routing control agent relates to the RC routing master server, and the standard OF communication protocol is used to transmit the routing status information and data flow information, which provides conditions for realizing the visualization OF route discovery. The information transmission control rules of the RC routing master server are defined, and these executable standard rules are all implanted into an open router to ensure that the router feedback standard routing status information and data flow information.

2.2. RC Control Module Design

The RC control module (the main body is the RC routing control layer) mainly consists of three functional modules: visualization of route discovery, data flow management and network management and control. The internal structure of the RC routing control layer is shown in figure 2.

To specify the IP address of the destination host and generate the next-hop routing path, a router needs to define the port table (used for data packet storage and forwarding), port address table, and internal information of the routing table, and provide the corresponding three ports. If the set of all ports is defined as the set of all ports is (Formula 1) :

$$P(r,i) = [p(r,1), \ldots, p(r,i), \ldots, p(r,n)] \tag{1}$$

Formula 1, r is the set of open routers, which can be defined by identifying each router. such as $r(1),\ldots,r(i),\ldots,r(n)$; i is a certain router among all routers, and its assigned value range is $(1,n)$; n is the total number of routers; $P(r,i)$ is the port of a router. For a router, there must be two input and o utput ports. The input port can be defined as $P(r,i,\leftarrow)$ the output port can be defined as $P(r,i,\rightarrow)$.

After the IP address of a router is loaded on a port, the connection relationship between at least two ro uters should be included. Then, the connection between the IP address loaded on the port and the port can be expressed as $P(r) \leftrightarrow D(r)$, D is the IP address loaded by the port, n is the total number of rout ers, then the set of $P(r) \leftrightarrow D(r)$ is (Formula 2) :

$$P(r) \leftrightarrow D(r) = [(p(r,1), d(r,1)), \ldots, p(r,i), d(r,i)), \ldots, p(r,n), d(r,n))] \tag{2}$$

In this case, the routing table (that is the routing rule) can be defined as $R(r)$, where $R(r)$ contains $P(r) \leftrightarrow D(r)$ and the next hop $S(r,i)$ (the relevant port information of the next router), the set of $R(r)$ is (Formula 3) :

$$R(r) = [(D(r,1), S(r,1)), \ldots, (D(r,i), S(r,i)), \ldots, (D(r,n), S(r,n))] \tag{3}$$

In Formula 3, the next-hop $S(r,i)$ mainly includes the output port of the ith router and the input port inf ormation of the j TH router. j is the number of all neighbouring routers of i. $\rightarrow$ is the output port inf ormation, $\leftarrow$ is the input port information, then the next-hop information can be expressed as (Formul a 4) :

$$S(r,i) = (p(r,i,\rightarrow), p(r,j,\leftarrow)) \tag{4}$$

Once you have defined the port table, port address table and routing table, you can start defining the ne ighbour router relationship table (that is defining neighbouring routers). The relationship table of adjac ent routers includes router set r (all routers) and the output port $P(r,i,\rightarrow)$ of a certain router i, neighb our router $R(r,r(i+1),\ldots,r(i+j))$, input ports $P(r,r(i+1),\ldots,r(i+j),\leftarrow)$ of neighbouring routers, then r outer $R(i)$ neighbouring router relationship table $T_i(r,r(i+1),\ldots,r(i+j))$ Can be defined as (Formula 5) :

$$T_i(r,r(i+1),\ldots,r(i+j))=[r,P(r,i,\rightarrow);R(r,r(i+1),\ldots,r(i+j));P(r,r(i+1),\ldots,r(i+j),\leftarrow)] \quad (5)$$

Formula 5 describes the relationship between adjacent routing table is only one router, all their corresponding relationship between adjacent routing table n router description, it should be described according to the formula 5 for four kinds of information, starting from the first router until the end of the first n routers, the relationship between adjacent to the first router's routing table is defined.

The design process of the relationship table of adjacent routers is as follows:(1) Within the IP address of the next-hop router defined by $R(i)$ in the routing table of a certain router $R(i)$, the IP address of all adjacent routers $R(i+1)$ $R(i+2)$, …, $R(i+j)$, clearly defined; The routing information defined in each router must contain information about the output ports that determine the output ports of each router to its neighbors. It should be noted that related routers can be identified by defining only identifiers. The next-hop router must be generated among all neighboring routers. (2) Among neighboring routers of a router $R(i)$, adjacent routing relation table $T_i(r,r(i+1),r(i+2),\ldots,r(i+j))$. The information is contained in the routing table of the neighboring router $R(r(i+1),r(i+2),\ldots,r(i+j))$. The key is to define the input port $P(r(i+1),r(i+2),\ldots,r(i+j),\leftarrow)$ of neighboring routers, Output port $P(r(i),\rightarrow)$ with router $R(i)$ As long as the input port information of the neighboring router is determined as the output port of $R(i)$, the normal connection between the two ports can be realized.

When a data stream (usually multiple packets) enters the network system, the acquisition and monitoring of the transmission path need to complete the capture of the transmission path information of this group of packets. The main method of capture is the feedback information after receiving the RC control module instruction through the open router. The packet enters the router through the input port. The router analyzes the IP address of the destination host in the packet and finds the IP address of the next-hop router (or the destination host) in the routing table to forward the packet. The key to obtaining and monitoring the transmission path is to include the IP address of the input port and destination host in the input information of each router. If the network transmission path set involves the source host, router and destination host is defined as $H(w,r,v)$ and $W(d,p,\rightarrow)$ is defined. Is the IP address and output port information of the source host, $V(d,p)$ is the IP address and access port information of the destination host, and $H(r,d,p,i,\rightarrow)$ is the next-hop transmission path of a router in the network transmission channel (including the IP address and output port information of the router indicated by the next hop), $H(r,d,p,n,\rightarrow)$. Is the router transmission path of the last access host, then the set of the entire data flow transmission path is (Formula 6):

$$H(w,r,v)=[W(d,p,\rightarrow),H(r,d,p,1,\rightarrow),\ldots,H(r,d,p,i,\rightarrow),\ldots,H(r,d,p,n,\rightarrow),V(d,p)] \quad (6)$$

3. Verification and Discussion

3.1. Packet Transmission Status Detection

Packet transmission state detection experiments mainly through the strategy of using SNMP agent and RC agent to compare the two ways of routing control method, compare and evaluate the main indexes of main from two aspects of resource-intensive and packet transmission capacity, resources refer to the packets during transmission using the number of router Packet transmission capacity refers to the number of packets transmitted. The experimental results are shown in Table 1.

Table 1. Comparison table of data packet transmission results under two communication strategies.

Data packets (DPs)	the SNMP agent policy							the RC agent policy						
	0	10	20	30	40	50	60	0	10	20	30	40	50	60
Labeled and key	0	50	100	150	200	250	300	0	50	100	150	200	250	300
Actual transmission	0	307	503	728	906	1082	1198	0	50	100	150	200	250	300
Actual transmission of the key DPs	0	28	73	96	107	126	159	0	0	0	0	0	0	0

The data listed in Table 1 show that the RC routing control algorithm proposed in this paper fully meets the target requirements of algorithm design in terms of routing information acquisition and transmission.

3.2. Packet Processing Status Detection

In this experiment, the processing of data packets mainly refers to the formulation and operation process of routing rules, and finally, the visualization result of the routing process can be mentioned. The packet processing status detection experiment mainly compares the processing efficiency of the server CPU of the two communication policies in a certain period. The main index comparison and evaluation is the CPU usage of SNMP agent and RC agent server in a unit time. The experimental results are shown in Table 2.

Table 2. Comparison table of packet processing results under two communication strategies.

CPU utility (%)	Processing times						
	0	100	200	300	400	500	600
SNMP	0	49	54	46	58	62	42
RC	0	0	18	29	0	0	24

The data listed in Table 2 show that the RC routing control algorithm proposed in this paper does not occupy too many CPU resources in the process of routing information processing. The experiment not only verifies the accuracy of the algorithm proposed in this paper in capturing data packets but also verifies that the process of routing information processing consumes fewer resources.

3.3. The Packet Receiving Delay Detection

The experiment of packet receiving delay detection mainly compares the speed of receiving routing information by a proxy server in a certain time with two communication strategies. The main indicator for comparison and evaluation is the delay of the SNMP agent and RC agent server receiving router feedback information in unit time. The experimental results are shown in Table 3.

Table 3. Comparison table of receiving packet delay results under two communication strategies.

Processing times (s)	SNMP							RC			
	0	100	200	300	400	500	600	0	1-10	11-20	21-100-600
Delay (s)	0	0.28	0.4	0.39	0.61	0.7	1	0	0.01-0.05	0.06-1	0

Under the RC proxy policy, when the delay of the proxy server receiving information is 1 second, the processing time of the system is only 20 seconds. With the increase of the system running time, the increase of the delay of receiving information basically does not show a big jump. The data listed in Table 3 show that the RC routing control algorithm proposed in this paper, when receiving routing information, the system runs faster, the server receives routing status information later, and the system runs more stably. This experiment also verified the accuracy of the first two test results.

4. Conclusion

Under the Routing Control mechanism OF the coexistence OF IGP and OF (OpenFlow) communication protocols, this paper designs the RC (Routing Control) communication protocol based on the communication strategy and the existing OF router and constructs the Routing Control system with the open router as the breakthrough. It provides a guarantee for visualization of the routing control process. The main conclusions of this study are as follows:

(1) The SDN network system provides a mechanism to coexist with the SNMP network management agent policy in the IP network system to ensure that the IGP internal gateway protocol and the RC protocol proposed in this paper can realize communication connection.

(2) The design of an open router needs to consider the initial synchronization of routing state, focus on the sampling of data flow, the storage and forwarding routing state information, and the compatibility of information forwarding modes.

(3) The key to the implementation of the RC routing control strategy is to control both routing state and data flow by establishing the relationship table of adjacent routers.

(4) The acquisition and monitoring of route status information transmission path and packet transmission flow are the keys to directly realizing route discovery and visualization of data flow transmission.

Through constructing the experimental platform of routing information transmission and control based on RC and IGP, the routing control ability of traditional SNMP and RC network management strategy in this paper is tested from the perspectives of packet transmission state, packet processing state and packet receiving delay in the routing control process. The experimental results show that the fewer routers occupy, and the fewer packets are transmitted, the stronger the system is in obtaining, transmitting, and controlling routing information during packet transmission. In the process of data processing, the fewer CPU resources occupied by the proxy server, the higher the processing efficiency, and the more accurate the key packets obtained. In the data receiving process, the less time the system runs, the shorter the delay of the server receiving the routing status information feedback from the router, and the more stable the system runs.

Acknowledgements

The authors are grateful for the financial support of the Scientific Research Project of Hubei Education Department (Grant No. Q20204306), Jingmen Science and Technology Project (Grant No. 2021YFZD076 and 2020YFYB049), Scientific Research Project and Team of Jingchu University of Technology (Grant No. YY202102 and TD202101).

References

[1] CAO Y, WANG J and LUO Q 2018 *Optical Communication Technology* vol 42, pp 21.
[2] YANG H, LIANG Y, YAO Q, et al. 2019 *China Communications*, vol 16, pp 42.
[3] Khaled A and Robin B 2020 *Communications and Network*, vol 12, pp 28.
[4] KOPONEN T AMIDON K BALLAND P et al. 2014 *Proceedings of the 11th USENIX Conference on Networked Systems Design and Implementation* pp 203.
[5] TILMANS O VISSICCHIO S 2014 Proceedings of 2014 10th International Conference on Network and Service Management (CNSM), Rio de Janeiro Brazil: CNSM014 pp 127.
[6] NASCIMENTO M ROTHENBERG C SALVADOR M, et al. 2011 *ACM SIGCOMM Computer Communication Review* vol 41, pp 441.
[7] VISSICCHIO S, CITTADINI L, BONAVENTURE O, et al.2015 *IEEE Conference on Computer Communications, Kowloon Hong Kong: IEEE INFOCOM* pp 469.
[8] ZUO Q, CHEN M, ZHAO G, et al. 2013 *Journal of Software* vol 24, pp 1078.
[9] Jakma P, Lamparter D 2014 *IEEE Network* vol 28, pp 42.

Research on Architecture of Local Domain Application Based on Radio Frequency Identification

Tingshun Li[1,*], Zesan Liu[2], Dadi Wang[3] and Suling Ma[3]

[1] School of Control and Computer Engineering, North China Electric Power University, Beijing, China
[2] R & D, State Grid Information & Telecommunication Group Co., Beijing 102211, China
[3] Beijing China Power Information Technology CO. Ltd, China

*Corresponding author email: lits@ncepu.edu.cn

Abstract. Along with the speedy improvement of sensing systems, wireless transmission technology, and faster and faster chip, the scope of application of the Internet of things (IoT) has substantially and rapidly increased, and research and innovation that integrate artificial intelligence. At same time, campus application construction as the foundation of intelligence education is being on the march, which demands newly emerging computer technologies support, such as IoT, AI, Big data and so on. In this study, we provide an architecture of smart campus based on Radio Frequency Identification (RFID) which can offer new solutions to solve some problems such as position, identification, store information and so on because of its strong anti-interference capability and non-intrusiveness, as core technology of IoT. Then based on this architecture, a kind of intelligent campus application system scheme is proposed, which can realize the function of automatic identity recognition, book reading, internal consumption, student positions. Moreover, for some special exact location demands, this paper proposes a method to precisely measure RFID tag location based on triangulation principle, which is embedded in the architecture. The application of Internet of things technology can better promote some origination or company efficiency, such as more efficient service for teachers' and students' work, study and life.
Keywords: IoT; Smart campus; RFID; Sensor technology.

1. Introduction

After IBM proposed “smart planet” in 2018, smart campus in China is being constructed, spreading out from universities and colleges to secondary schools, even primary schools (Geng Zhen, 2019) [1]. The rapid development of modern network communication technology and the rapid popularization of computer related hardware equipment have promoted the pace of the construction of University information, at present, have built different scale information digital campus, which has realized the integration of information resources, electronic payment and financial system and other information management systems to a certain extent, and has brought a lot of convenience to the work, study and life of teachers and students. For example, the application of Campus IC card system has provided a convenient way for students in school dining, shopping, library and so on. In 1994, Schilit and Theimer first proposed the concept of the context awareness (Schilit, et al. 1994) [2]. With the deepening of research, Dey introduced the concept of adaptive (Dey, et al.1998) [3-4]. Salber introduced a new definition (Salber, et al. 1998) [5]. Dey and Abowd further expanded the field of application of context awareness (Dey, et al. 1999) [6]. Tewari developed a comprehensive intelligent decision support system

(Tewari, et al. 2012) [7]. Tseng provided green smart campus monitoring and detection using LoRa (Tseng KuoHsiung, et al. 2021) [8].All of these have contributed to the development of smart campus. However, there are many problems in the current digital campus, such as the information system integration is difficult, user information collection and update are not convenient, the IC card is difficult to retrieve and so on. At the same time, with the constant introduction of all kinds of institutions and bank information cards, each student often has more than one card, which brings a lot of inconvenience to the daily use of students. Therefore, based on the analysis of the relevant research of the Internet of things technology, combined with the application requirements of Internet of things technology based on Intelligent campus, a set of intelligent campus system based on Internet of things was proposed in this paper.

2. State of the IoT

2.1. Definition of the Internet of Things

The IoT is an important part of the new generation of information technology. One word in its name is "Internet+", so the Internet of things can be understood as "the material objects connected to the Internet" (Ma, et al. 2011) [9]. From the name of the Internet of things, it can be seen that there are two layers of meaning: Firstly, the core and the foundation of the Internet of things is still the Internet, it is the extension and expansion of the network based on the internet; Secondly, its user side extends and expands to any items to carry on the information exchange and the communication. Internet of things is the integration application based on the intelligent sensing, recognition technology and pervasive computing, ubiquitous network, it is known as the world's third wave of information industry development following the computer and the Internet. Internet of things is the application development of the Internet. Therefore, the application of innovation is the core of the development of the Internet of things, the innovation 2.0 is the soul of the development of the Internet of things with the user experience as the core.

The definition of the Internet of things was originally proposed in 1999, the Internet of things was defined as: It refers to a kind of network which can connect any items with the Internet, and carry out the information exchange and communication through the radio frequency identification (RFID), infrared sensors, global positioning system, laser scanner, gas sensors and other information sensing equipment, so as to realize the intelligent recognition, positioning, tracking and monitoring and management (AR Al-Sudani, et al. 2018) [10]. In short, the Internet of things is "the material objects connected to the Internet".

In China, the Internet of things is defined as the combination of almost all the technologies and computer, Internet technology by Chinese Internet of things Enterprise Alliance, it realizes the real-time sharing of the environment and the state information among the objects and the intelligent collection, transmission, processing and execution. Broadly speaking, the application of information technology can be included in the scope of the Internet of things (Qianwen Miao, et al. 2020) [11].

2.2. Characteristics of the Internet of Things

Connectivity

Connectivity is the basis of the Internet of things. Network, wireless, wired or sensing objects must reflect the "connection" state. The International Telecommunications Union believes that the connectivity of the Internet of things includes four aspects: the first is that it can carry on the connection at any time: the second is that it can carry on the connection in any place; the three is that it can be connected to any object, the four is that it can be connected to any person (Angelis, et al. 2015) [12].

Physical object association

In the Internet, people use computers to communicate with each other, while the Internet of things through the sensor, radio frequency identification technology, global positioning system and other technology to achieve the exchange of things and objects. Internet achieves the long-distance communication between the people and people, and the Internet of things achieves a real-time interaction between people and objects, objects and things, so as to achieve the transition from the virtual world to the real world (Angelis, et al. 2015) [12].

Intelligence
Intelligence refers to carry out the intelligent control for each object by the use of computer technology, sensor technology and control technology and intelligent monitoring system, from this point of view, the Internet is an important part of the Internet of things, which makes that the objects in the world not only can be associated by the way of sensing, but the way of intelligent. Internet of things has the intelligent perception, it can sense the environment, and maximize the support of people to better insight into the use of a variety of resources, so as to make a more accurate judgment (He-quan Wu. 2017) [13]. Therefore, the Internet of things has been widely used in other fields, and has produced the industry chain. At present, some colleges and universities in China have set up the Internet of things, and have researched the intelligent sensor network, which has laid the foundation for the further development of the Internet of things technology. However, the Internet of things technology has not been widely used in the digital campus construction. Therefore, based on the actual needs of the wisdom of the campus, and combined with the Internet of things RFID technology and sensor technology research, the smart campus system based on the application of Internet of things technology was designed in this paper (Ruttala, et al. 2015) [14].

2.3. Requirement Analysis of the Construction of Smart Campus
Internet of things industry chain can be divided into: identification, perception, information processing and transmission of 4 links, the key technologies of each link are RFID, sensor, smart chip and the wireless transmission network of the telecommuter operators. In view of the problems that the domestic Internet of things industry foundation and the key link were relatively weak, the standardization work was relatively backward and so on, the national organization started the Internet of things technology research and development and industrialization projects, launched a national networking application demonstration project, planned to promote the development and industrialization of key technologies and industrial, standard system and public service platform, and focused on the breakthroughs in core technologies, and the improvement of the industrial chain, so as to provide an effective support for the application of key areas of the Internet of things.
With the development of information technology, the higher education institutions have adopted information technology to strengthen management according to their own needs. Campus "one card" has become a common choice for the development of colleges and universities. Campus "one card", the wisdom of the campus and network information technology applications and other related industries are rapidly developing. The goal of "one card" is to realize the integration and sharing of data management through the common identity authentication mechanism, the main management functions of which are: logistics management system, attendance management system, computer management system, library management system and access control management system. Combined with the use of the situation, in view of the problems in the digital campus management, the non-contact RFID electronic tag sensor card was used to achieve the "one card" management, according to the typical application of "one card" system, a design scheme of intelligent campus based on Internet of things technology was proposed in this paper.

3. Methodology
According to the basic idea of the effective integration of various functions of digital campus and the user's convenience, based on the Internet of things technology, this paper presents a design scheme of intelligent campus management system, which can realize the functions of user management and automatic identification, library borrowing, internal consumption, student information management and so on.

3.1. Architecture
According to the demand of the automatic identification, application data management integration and sharing of teachers and students in the era of Internet of things intelligent digital campus, the intelligent digital campus based on Internet of things adopts the three-tier system, as shown in Figure 2, which mainly includes the application data service layer, network communication layer and terminal user layer.

Among them, the application data layer includes user information database and user information management server and database management. The user information database is mainly used for information storage and management. The user information management server is mainly used for the related data and real-time communication, and sending and receiving SMS intelligent digital campus information exchange network communication layer. The user layer includes smart mobile phones, IC cards, RFID cards, a tag identification and so on, which is mainly used for real-time data acquisition and storage of user information in the mobile phone SIM card. In this way, students can be convenient to realize efficient management of all kinds of data and information through the mobile phone SIM card unified user identity recognition mechanism, the intelligent campus information management system can realize the office automation system integration.

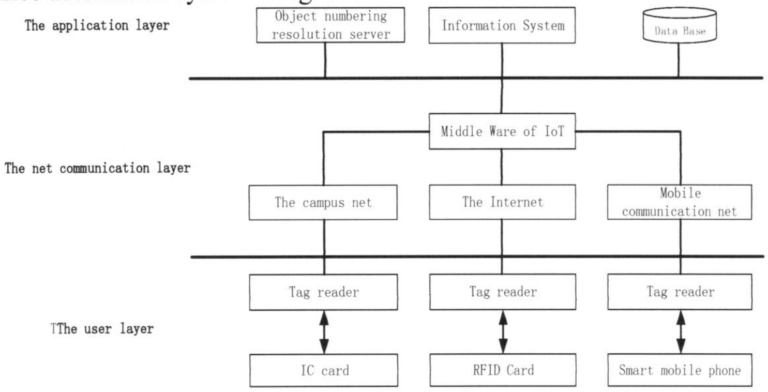

Figure 1. Intelligent campus architecture based on IoT

3.2. System Function Module

Based on the unified user identity recognition mechanism, the intelligent digital campus system based on Internet of things mainly consists of user identity management, conference management, student attendance management, school vehicle management of four function modules.

User identity management module: it through the sensing technology and RFID technology to achieve personnel in and out of control in library and other important places, in addition, it also comes with recording function, which can count the passengers and time through the signal.

School vehicle management module: it uses RFID technology to achieve rapid, real-time, accurate acquisition and processing for the information, besides, it is regarded as the upgrade and supplement of the traditional vehicle management methods to meet the requirements of intelligent campus vehicle management.

4. System Design Based on RFID Technology

4.1. RFID Meeting Non-obstacle Recognition System Design

The organization and management of large-scale conference are complex, and an important issue of the organization and management of the meeting is to identify and count the participants effectively. Traditional conference management methods lead to a long waiting time, especially the staff attendance is more concentrated, the problem is more serious. At present, the conference attendance and identification generally use the manual check documents and credit card and other means. These methods exist the following problems: Paper and pen sign is easy to cause a large area of congestion, which can't guarantee the correct participants data; The efficiency of manual verification documents is low, which can't guarantee the accuracy, and it is not easy to statistics. Card needs to be represented by the designated locations to produce documents and line card, which lacks of human. According to the needs of the meeting organization, the more ideal sign and identification program must have the following characteristics: Security features (representative certificate can't be forged), the identification of the auxiliary portrait and the leakage of the situation to the police (Yang Ping, et al. 2019) [15].

The function of the design scheme based on the long-distance RFID electronic tag technology is realized by using the digital camera and the infrared sensor technology. Through the occurrence of infrared sensor, the system will start the camera to capture the face image, if it is effective, the host will transfer the photos and the original data, and the manager will verify the image, the whole process is fast, convenient and accurate.

Central algorithm: which consists of a sensor network in a reference node and a common node. The position of the reference node or coordinate is called (X_i, Y_i), which estimates its position or coordinates by the receiving position or coordinates of a common node:

$$(X_{test}, Y_{test}) = [(X_1+X_2+...+X_K) / k, (Y_1+Y_2+...+Y_K)/k] \quad (1)$$

Through the process, there will be two radio frequency identification areas and two-time cutting infrared, which avoids the blind read, infrared cut can also determine the role of participants to enter or leave. Of course, the primary target infrared identification is a warning after the radio frequency identification.

4.2. Design of Fast Channel System without Obstacles

University library access control system often uses the design of the channel, in the actual use, it is prone to mechanical failure, and will lead that the certified personnel can't normally pass. The barrier free fast track is used to achieve personnel in and out of control, which also is a more advanced management methods. Application goal is to establish a fast channel based on RFID technology in library management system, so as to achieve the effective management of personnel access control system. The system doesn't have a physical barrier to facilitate the rapid access of personnel, and the no-credit card can be achieved, so as to send a warning to the undocumented personnel (KANG, et al. 2012) [16]. If it is necessary, the facial features can be captured by the camera to compare with the reader photo library of educational administration system, so as to realize the scientific and humane management. The system is composed of an induction device, a card reader, an antenna and a communication module(Zhou Z, et al. 2020) [17].

4.3. Design of Open Active RFID Recognition System

The class attendance of the students is particularly difficult for the teachers. The teachers usually take time for class to carry out the pen and paper statistics, and often meet the impostor situation. In view of the needs of the parents and school communication, the school communication system and other systems provide effective means for communication between schools and parents. But most of the systems are still not perfect, which need to be carried on the artificial intervention. Based on the combination of the 2.4 GHz active RFID technology, computer technology and wireless communication technology, the learning and information communication platform is provided for parents through the 2.4 GHz radio frequency identification system, Internet and communication technology, RFID card.

RFID card reader is installed at the school gate, the stairs and the door of the classroom, so as to achieve real-time positioning of the students. A bus type network is connected to the controller through the connection of the card reader. Through the original data of the students, the attendance of students is analyzed. Attendance of each class will be sent to the teacher's mobile phone to check. At the end of the statistical work, the educational management system will automatically statistics the attendance of each student, so as to solve the problem of teachers' attendance.

4.4. Remote Vehicle Identification and License Plate Recognition System

The efficiency of the traditional vehicle management system is low, and is prone to congestion and so on. Artificial credit card and automatic lift rod will be forgotten with magnetic card, magnetic card without electricity and other situation, which causes the release rate slower.

Here the three side of the measurement principle is needed. Three side measurement principle is to determine the coordinates of the node according to the 3 known coordinates of the distance of the unknown node of the node. Here, A, B, C are the coordinates of the 3 nodes (X_A, Y_A), (X_B, Y_B), (X_C, Y_C), it is assumed that the node of the program D is (X, Y), then the formula can be obtained by calculating:

$$\begin{bmatrix} x \\ y \end{bmatrix} = \begin{bmatrix} 2(x_a - x_c) & 2(y_a - y_c) \\ 2(x_b - x_c) & 2(y_b - y_c) \end{bmatrix}^{-1} \times \begin{bmatrix} x_a^2 - x_c^2 + y_a^2 - y_c^2 + d_c^2 - d_a^2 \\ x_b^2 - x_c^2 + y_b^2 - y_c^2 + d_c^2 - d_a^2 \end{bmatrix} \tag{2}$$

As a new technology for rapid, real-time, accurate collection and processing of information, radio frequency identification (RFID) tag is taking as the upgrade and supplement of the traditional vehicle management methods, so as to meet the requirements of today's vehicle management. The vehicle electronic tag is mounted on the windshield of the vehicle or the campus "one card", and the RFID reader is installed in the vehicle passage or the parking lot. In the normal operation, when the vehicle passes, the reader makes a prompt tone, and indicates the success of automatic identification, the reader reads the information transmitted to the microprocessor of the access control machine, realizes the collection of the vehicle information and the payment of the temporary vehicle, and completes the channel control by controlling the channel gate and the like.

Triangle measurement principle is based on the relative angle of the unknown node of the node's 3 known coordinates to determine the node coordinates. Here, A, B, C are the coordinates of 3 nodes: (X_A, Y_A), (X_B, Y_B), (X_C, Y_C). For the nodes A, C, D, if the arc AC is in the triangle ABC , then only a circle can be determined, and as below.

$$\begin{cases} \sqrt{(x_{o1} - x_a)^2 + (y_{o1} - y_a)^2} = r_1 \\ \sqrt{(x_{o1} - x_c)^2 + (y_{o1} - y_c)^2} = r_2 \\ (x_a - x_c)^2 + (y_a - y_c)^2 = 2r_1^2 - 2r_2^2 \cos\alpha \end{cases} \tag{3}$$

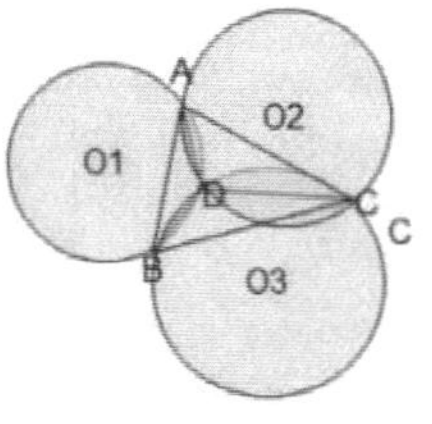

Figure 2. Triangulation principle. Note: R1 is the radius of the circle $O1(x_{o1}, y_{o1})$.

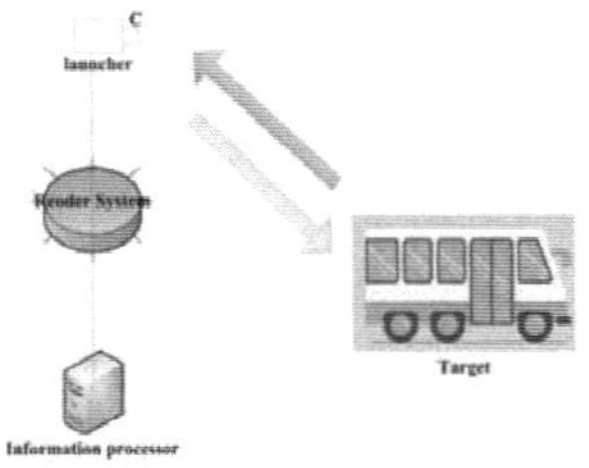

Figure 3. Principle of license plate recognition

RFID system is an automatic identification technology that can realize the information acquisition and data exchange by the use of induction, radio wave or microwave energy to carry out non-contact bidirectional communication. The reader sends out the microwave inquiry signal through the antenna to the electronic tag, and the electronic tag is activated by the microwave energy, after receiving the microwave signal, the echo signal with the vehicle data message is received and sent. After the reader receives the electronic tag signal of the RFID vehicle, the decoder will parse out the information which is related to the vehicle, so as to establish the communication between the reader and the RFID tag, the vehicle information can be processed by computer to realize the automatic identification of the vehicle, as shown in Figure 3. For foreign vehicles, through the camera, the license plate scanning and camera archive are carried out, and the temporary vehicle information data is established to be a payment document to leave.

5. Conclusion

The smart campus achieves intelligent campus services and management through the Internet of things to, it can achieve the interconnection of any person, anything, any information carrier, any time, any place through the Internet of things, besides, the polymerization of the vast amounts of information in the Internet of things platform provides smart business and service model for the majority of teachers and students.

There is no doubt that the smart campus has brought great convenience to teachers and students, but also has faced many challenges. Firstly, there is the issue of cost, the construction of the smart campus must have a unified infrastructure platform, including wired and wireless dual network coverage of the network environment; The followed is the teachers and students privacy issues, since the Internet of things can track the user's actions, habits and preferences, therefore, how to protect the information resources and the privacy of teachers and students becomes the urgent problem to be solved in creating intelligent campus; Finally, the management mechanism is not complete, how to maintain, manage and use the Internet of things is a big problem. At present, the research of the smart campus is still in the exploratory stage, the outlook is not clear, but in the long run, because the Internet of things has a broad application prospect, so the smart campus will continue to improve and progress in the process of its development.

References

[1] Geng Zhen J 2019 *A new stage of educational Informatization: from "digital campus" to "smart compus"* (Computer Era) p 76-79+83
[2] Schiliti J 1994 *Disseminating active map information to mobile host* vol 5(IEEE Network) p 22-32
[3] Dey A.K C 1998 *Context-Aware Computing: The Cyber Desk Project In AAAI* (Spring Symposium on intelligent environments technical report) p51-54.
[4] Dey K. Anind and Abowd C 1999 *Gregory towards Better Understanding of Context and Context Awareness Technical report* (GIT-GVU-99-22 Georgia Institute of Technology)
[5] Salber D. Dey A.K. and Abowd C 1998 *Ubiquitous computing: defining and HCI research agenda for an emerging paradigm* (Georgia Tech GVU Technical Report GIT-GVU-98-01)
[6] Dey A.K J 1999 *Cyber Desk: A framework for providing self-integrating context-aware services* vol 1 ed Abowd GD, Wood A (Knowledge-Based Systems) p3-13
[7] Tewari J 2012 *Personalized location-based brokering using an agent-based intermediary architecture* vol 34 ed G; Youll J; Maes, P (Decision support systems) p127-137
[8] Tseng KuoHsiung J 2021 *Green Smart Campus Monitoring and Detection Using LoRa* vol 21 ed Chung MengYun, Chen LiHsien, Chang PeiYao (Sensors) p6582-6582
[9] Ma D H J 2011 *Internet of Things: Objectives and Scientific Challenges* vol 11 (Journal of Computer Science & Technology) p15.
[10] Al-Sudani A R, Zhou W, Wen S, et al. *Hiding RFID in the Image Matching Based Access Control to a Smart Building*[C]// 2018 IEEE SmartWorld, Ubiquitous Intelligence & Computing, Advanced & Trusted Computing, Scalable Computing & Communications, Cloud & Big Data Computing, Internet of People and Smart City Innovation (SmartWorld/SCALCOM/UIC/ATC/CBDCofm/IOP/SCI). IEEE, 2018.
[11] Qianwen Miao J 2020 *Smart Attendance System Based on Frequency Distribution Algorithm with Passive RFID Tags* vol 25 Fu Xiao, Haiping Huang, et al. (Tsinghua Science and Technology) p217-226
[12] De Angelis E J 2015 *The Brescia Smart Campus Demonstrator. Renovation toward a zero energy classroom building* vol 118 Ciribini A L C and Tagliabue L C, et al. (Procedia Engineering) p735-743
[13] He-quan Wu 2017 *Technology and application progress on Internet of Things* (Chinese Journal on Internet of Things) p1-6
[14] Ruttala U K, J 2015 *NFC based Smart Campus Payment System* vol 8 Balamurugan M S and Chakravarthi M K (Indian Journal of Science and Technology)
[15] Yang Ping J 2019 *Research Review of Wisdom Campus Construction* vol 29 Yao Yuxiang, Shi Beibei, et al. (Modern Educational Technology) p18-24
[16] Kang S O J 2012 *The implement of smart campus through future-oriented infrastructure and services* (Network Times) p159-161
[17] Zhou Z, Yu H, Shi H 2020 *Optimization of Wireless Video Surveillance System for Smart Campus Based on Internet of Things* (IEEE Access) p1-23

Allocation Structure Optimization of Distributed Power Supply Based on Improved Self-adapting Cuckoo Algorithm

Lei Li[1,*] and Guige Gao[2]

[1]School of Electrical Engineering, Shanghai DianJi University. Shanghai 201306, China
[2]Shanghai DianJi University, China

*Corresponding author email: 206001010312@st.sdju.edu.cn

Abstract. Aiming at the problems of environmental protection and economy caused by the integration of distributed power (DG) into distribution network, an optimal allocation model of distributed power was established with generation cost, network loss cost and DG environmental cost as the optimization objectives. In order to solve the model, an improved adaptive cuckoo algorithm was introduced, which combined the discovery probability and scaling factor of the adaptive strategy. Through the example analysis of IEEE-33 bus, the simulation results of different algorithms are compared, for which, in distributed power allocation structure, the availability and advantages of this improved algorithm optimization are verified.
Keywords: Cuckoo search algorithm; Adaptive strategy; Distributed power supply;

1. Introduction

With the opening of the intelligent era, people gradually pay attention to the related problems of Distributed Generation (DG) and power network connection. Due to the randomness and intermittency of distributed power supply, a large number of grid connections will change the original structure of power network and affect its stability. If the location and capacity of the power network are different, the voltage stability, network loss and power flow direction of the power system will be significantly affected [1-3]. To sum up, reasonable planning of the access location and capacity of distributed power supply is of great significance to reduce the loss of active power network, reduce voltage deviation and improve the safety, economy and stability of power network. However, if the location and capacity of the DG are not properly configured, the power system will be burdened. At present, many scholars study the optimal structural allocation of DG from different perspectives.

A novel Adaptive Cuckoo Search (IACS) algorithm based on Improved Adaptive Cuckoo Search (IACS) is proposed to improve the convergence speed and accuracy of the algorithm. Taking IEEE-33 node as an example, IACS is used to calculate the optimal configuration of DG, and the results are compared with those of Cuckoo Search (CS).

2. Construction of Optimal Distribution Model of Distributed Power Supply

2.1. Objective Function

The solution of distribution network related problems often involves nonlinearity and multi - objective. Different objective functions can be selected according to actual requirements. The DG model in this paper aims to minimize the annual DG comprehensive cost, which includes power generation cost, network loss cost and DG environmental protection cost. The object function is shown in formula (1) below:

$$\min C_{total} = C_i + C_{loss} + C_{co2} \tag{1}$$

Where, C_i is the annual cost of distributed power generation; C_{loss} refers to distribution network loss cost; C_{co2} is the carbon emission cost of distribution network.

$$C_i = \left(\frac{r(1+r)^n}{(1+r)^n - 1} \right) C_{DG} S_{DG} + C_{OM} \tag{2}$$

Where, r is the annual rate of return, in %; N is the service period, in years. C_{DG} is the investment cost of unit capacity DG, 1500 yuan /kW; S_{DG} is the total capacity of the installed DG, in kW. C_{OM} is the running cost of DG (S_{DG}*500 yuan).

$$C_{loss} = C_p T_{\max} P_{loss} \tag{3}$$

Where, C_{loss} is the cost of network consumption. C_P is unit electricity price; T_{max} is the annual maximum load utilization hours; P_{loss} refers to the active power loss.

$$C_{CO_2} = \frac{0.96AWg}{1000} \tag{4}$$

Where, A is the environmental penalty, 0.97 yuan /t, W_g is the generating capacity, the unit is kW*h.

2.2. Constraints

(1) Node voltage constraint expression is:

$$U_{\min} \le U_i \le U_{\max} \tag{5}$$

Where, $U_{max}$and U_{min} are the minimum and maximum values of node voltage respectively. U_i is the node voltage of node I.

(2) The constraint expression of the maximum power of the branch is

$$|P_i| \le P_{i.\max} \tag{6}$$

P_i is the transmission power of branch I, and the upper limit of transmission power of branch I of $P_{i.max}$.

(3) Constraints on DG installed capacity

$$0 \le P_{DG.\mathrm{sum}} \le P_{DG.\max} \tag{7}$$

Where, $P_{DG.sum}$ is the total installed active capacity of the distributed power supply, and $P_{DG.max}$ is the maximum active capacity of the distributed power supply that is allowed to access.

(4) Node power flow constraint expression is:

$$\begin{cases} P_{Gi} + P_{DGi} - P_{Li} = U_i \sum_{j=1}^{N} U_j \left(G_{ij} \cos\delta_{ij} + B_{ij} \sin\delta_{ij} \right) \\ Q_{Gi} + Q_{DGi} - Q_{Li} = U_i \sum^{N} U_j \left(G_{ij} \sin\delta_{ij} - B_{ij} \cos\delta_{ij} \right) \end{cases} \tag{8}$$

Where, P_{Gi} is the injected active power of node I; Q_{Gi} is the injected reactive power of the node. P_{DGi} is the active power of DG; P_{Li} is active load; Q_{DGi} is the reactive power of DG; Q_{Li} is reactive load; U_i and U_j are the potential difference of node I and j respectively. G_{ij} and B_{ij} are the admittances of the lines between node I and j. δ_{ij} is the power factor Angle.

3. IACS Algorithm Analysis

In the basic cuckoo algorithm, the random step size is generated by levy flight. When searching, the step size increases and decreases, and the larger the step size is, the better the global optimization is, but the lower the search precision, and sometimes the oscillation phenomenon occurs. If the step size is smaller, the result is opposite. Therefore, in view of the problem that the step size generated by Levy flight is random but not adaptive, generally the adaptive step size cuckoo search algorithm only improves the step size, and its discovery probability Pa is fixed at 0.25, which is not conducive to maintaining the balance between global and local random search of the algorithm. Its adaptive strategies are as follows:

$$b = (f_i - f_{\max}) / (f_{\min} - f_{\max}) \tag{9}$$

$$alpha1 = (alpha_{\max} - alpha_{\min}) * b + alpha_{\min} \tag{10}$$

$$alpha = alpha1.*(pop(ranperm(n),:) - pop(randperm(n),:)) \tag{11}$$

Where, $f_i$represents the ith nest position, f_{max} represents the optimal state of the nest position at this time, (f_{min}-f_{max}) represents the maximum distance between the optimal position and all other nest positions, and $alpha_{max}$ and $alpha_{min}$represent the upper and lower limits of step size respectively. According to Equations (9) and (10), the step size can be dynamically adjusted adaptively. Formula (11) is used to generate random solutions to avoid premature development.

At the same time, the scale factor r in this algorithm is uniformly distributed random number in the interval of [0,1], which does not produce large variation for poor individuals, which is not conducive to the generation of new solutions. Therefore, if the scaling factor R and discovery probability can be well controlled to make the poor individuals produce large variation, the solution accuracy of the algorithm will be greatly improved.

To overcome the above problems, this paper proposes a cuckoo search algorithm (IACS) based on the improved adaptive Step Size Cuckoo search algorithm (ASCS)[4-5], and introduces the adaptive dynamic adjustment strategy of discovery probability and scaling factor, whose scaling factor and discovery probability are respectively adjusted as follows:

$$r_i^t = r_{\min} + (r_{\max} - r_{\min}) \frac{f_i^t - f_{best}}{f_{worst} - f_{best}} \tag{12}$$

$$p_a^t = p_{a\max} + (p_{a\max} - p_{aanin}) \left(1 - \frac{f_{best}}{f_i^t}\right) \tag{13}$$

Where r_i^t is the scaling factor of the individual in the t generation population; f_i^t, f_{best}.however, f_{worst} are the fitness of the individual, the best individual and the worst individual in the t generation population, respectively. Pa_t: probability that the individual in the t generation population is discovered and a new solution is regenerated; Pa_{max} and Pa_{min}: upper and lower limits of Pa_t; r_{max} and r_{min}: ceiling and floor .

3.3 Steps to improve the algorithm

Step 1 Initial solution: n nests were randomly generated, the dimension of search space was D, the maximum iteration was [iter] _max, the ceiling and floor of step size alpha, the ceiling and floor of probability Pa and scaling factor R were found, and the fitness of all nests was calculated.

Step 2 Retain the optimal nesting position of the previous generation and update other nests according to Equations (9) and (10). Equation (11) is used to eliminate prematurity and improve random probability. If the new nesting fitness is higher after calculation, the old solution is replaced.

Step 3 Use dynamic scaling factor formula (12) and dynamic discovery probability formula (13) to eliminate some poor solutions, and use preference random generation formula (10) to generate new solutions with the same number of eliminated solutions.

Step 4 Calculate the updated nest fitness and output the current optimal solution.

Step 5 When the algorithm terminates, the result is obtained; otherwise, return to Step 2.

4. Simulation Verification

Using ieee-33 nodes as an example, the reference capacity of the system is 10MVA, the total load capacity is 4369.65kVA, and the total active and reactive power is 3175kW and 2300kvar respectively. The total capacity of the distributed generation (DG) connected to the system does not exceed 20% of the total capacity of the system load. The DISTRIBUTED generation (DG) adopts the PQ node type and the power factor is 0.9 [17-20]. Standard system diagram of IEEE-33 node is shown in Figure 1:

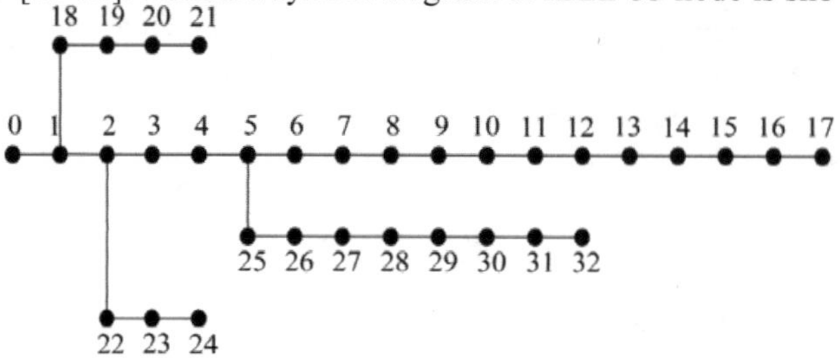

Figure 1. IEEE-33 node power distribution system

4.1. Example Analysis

The IACS algorithm proposed in this paper is applied to the configuration structure optimization of distributed power supply, and verified by ieEE-33 system. By comparing the genetic algorithm (GA), cuckoo (CS) algorithm and adaptive genetic algorithm (AGA) in the same premise conditions of convergence results, verify the IACS algorithm is proposed in this paper can not only better approximate optimal solution set, and the stability of the iterative convergence is very good, to give full expression to the algorithm in DG configuration optimization advantages. The comparison of optimization iterative convergence process of various intelligent algorithms is shown in Figure 2:

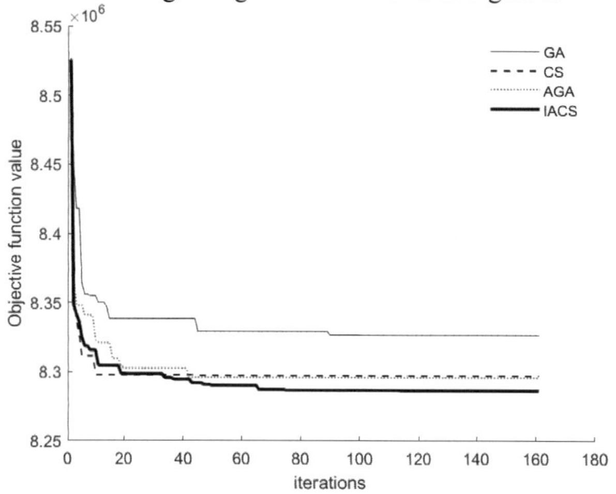

Figure 2. Comparison of convergence curves of the two algorithms

In order to avoid the contingency of the algorithm, CS algorithm and MOCS algorithm were used in MATLAB/Simulink to simulate the IEEE 33 node system, and each algorithm was simulated 10 times. 10 experiments with stable data ratio were selected for statistics. Table 1 shows the compromise solution after optimization of the two algorithms. The optimal configuration scheme is shown in Table 2:

Table 1. Optimization results of CS and IACS

algorithm	running costs (ten thousand yuan)	Network loss (ten thousand yuan)	Environmental costs (ten thousand yuan)
CS	55.72	17.79	12.71
IACS	55.00	16.06	11.50

Table 2. Comparison of optimal configuration schemes

Algorithm	DG Access	capacity of a-Node (MVA)
CS	4	48
	9	36
	15	100
	31	71
	32	350
IACS	6	100
	15	115
	30	90
	31	75
	32	200

As can be seen from Table 1, IACS algorithm has low investment and operating cost in active power network loss, and its environmental cost is lower than the optimal configuration result of distributed power obtained by CS algorithm, which can provide more optimization space for decision makers. It can be seen from the optimal configuration scheme in Table 2 that the capacity optimal configuration scheme obtained by the IACS algorithm proposed in this paper is more evenly and reasonably distributed.

5. Conclusion

In this article, an allocation distribution model of distributed power supply with the objective function of minimizing comprehensive investment cost is established, and the objective function is calculated by improved IACS and CS. Taking IEEE-33 node as an example, the following conclusions are obtained:
(1) The access of DG to distribution network can cut the loss of active power network and raise the node voltage level. It is beneficial to the safe and stable operation of distribution network.
(2) The adaptive step size cuckoo algorithm is improved by introducing adaptive strategies for discovery probability and contraction factor. The simulation results show that IACS algorithm has better searching performance than other algorithms, and can improve convergence speed and solving accuracy.
(3) By comparing the genetic algorithm (GA), cuckoo (CS) algorithm and adaptive genetic algorithm (AGA) in the same premise conditions of convergence results .IACS algorithm has more advantages in reducing the network loss of system active power and optimizing the environment, and can obtain a better optimal configuration scheme.

References

[1] Hanoun S, Creighton D, Nahavandi S. A hybrid cuckoo search and variable neighborhood descent for single and multiobjective scheduling problems[J]. The International Journal of Advanced Manufacturing Technology. 2014, 75(9-12): 1501-1516.
[2] Qiu Haifeng, Zhao Bo, Lin Da, et al. Regional coordination scheduling of AC-DC hybrid microgrid considering energy storage loss and converter cost [J]. Automation of Electric Power Systems, 2017, 41(23): 29-37.
[3] YANG X S,DEB S. Cuckoo search via Lévy flights [C]//Proceedings of World Congress on Nature and Biologically Inspired Computing. India: IEEE Publications, 2009:210-214. [J]. Automation of Electric Power Systems, 2018, 42 (21) : 36-44.
[4] Li Rongyu, Dai Ruiwen. Adaptive step size cuckoo search algorithm [J]. Computer science,2017,44(05):235-240.
[5] Xie Hai-bo, WEI Ming-lei, Chen Zhuo-you, XIE Xu-quan, Lin Xiang. Ningxia Electric Power,2016,(05):6-10+48.
[6] Li Na, He Xing Shi. Cuckoo search algorithm based on Particle Swarm Optimization [J]. Journal of basic science of textile universities, 2014,27 (3) : 374-379.

Hierarchical Attention Networks for Multimodal Machine Learning

Haotian Liang and Zhanqing Wang*

School of Automation, Beijing Institute of Technology, China

*Corresponding author email: wangzq@bit.edu.cn

Abstract. The Visual Question Answering (VQA) task is to infer the correct answer to a free-form question based on the given image. This task is challenging because it requires model handling both visual and textual information. Most successful attempts on VQA task have been achieved by using attention mechanism which can capture inter-modal and intra-modal dependencies. In this paper, we raise a new attention-based model to solve VQA. We use question information to guide model concentrate on special regions and attribute and hierarchically reason the answer. We also propose multi-modal fusion strategy based on co-attention method to fuse both visual and textual information. Under the same experimental conditions, extensive experiments on VQA-v2.0 dataset illustrate our method performance exceeds the performance of some state-of-the-art methods of the same experimental conditions.
Keywords: Guided attention; Visual question answering; Self-attention; Co-attention.

1. Introduction

In recent years, due to the development of deep learning, great progress has been made in the fields of computer vision (CV) and natural language processing (NLP). As the two most important modes of human interaction, vision and natural language play a key role in advancing the development of AI. Since the tasks, such as image captioning [1], visual dialogue [2][3] and visual question answering [4][5], include both visual and language modal inputs, these tasks have attracted more and more attention. Compared with other multi-modal tasks, visual question answering is a much more challenging task. The goal of the task is to answer questions based on given images. This task requires the model to be able to simultaneously represent the input of both visual and language information, and perform inference to generate the correct answer.

For VQA task, early methods mostly used CNN to extract image features and RNN to encode questions. Then, simple methods such as addition and multiplication were used to fuse two modal information to reason about answers. In order to avoid being limited to the dataset, [6] uses an external knowledge base to process VQA task, so that the model can have the same common sense as humans, and it has also improved the generalization ability of the model. With the successful application of attention mechanism in the domains of computer vision [7]and natural language processing [8], some models based on attention mechanism have been proposed. Yang et al proposed the Stacked Attention Networks (SAN) [9] through the multi-step attention mechanism, gradually find the target area. Ilievski et al used the Focused Dynamic Attention matches the keywords (nouns) in the question with the candidate regions in the image, and simulates the attention shifting process when people see question and image. Lu et al proposed co-attention method [10] which applied attention mechanism is to image and question at the same time.

However, most of these attention-based methods are limited to spatial attention, the visual features extracted by convolution neural network contain a large number of attributes, such as color, size, shape and so on. These attributes do not play a role in question reasoning. Although spatial attention mechanism filters out spatial noise input, it will still bring attribute noise to the model. In this paper we use two basic units Question-Guided Spatial Attention (QGSA) and Question-Guided Attribute Attention (QGAA) to construct the spatial and attribute reasoning module to focus on specific regions and capture attributes which are related to question.

Contributions. Our contributions are:

- We use QGSA and QGAA units form a Hierarchical Reasoning Networks (HRN). HRN can simulate human spatial and attribute reasoning when answering questions.
- Detailed experimental results illustrate that our model can effectively focus on the most critical information on both modes for the given task.
- Under the same experimental conditions, our model can outperform some state-of-the-art methods on VQA-v2.0 dataset.

2. Related Works

2.1. Visual Question Answering

The visual question answering task requires the model to infer the correct answer based on the given image. The most intuitional solution is to extract the global features of the picture through CNN, use RNN to represent the semantic features of the problem, and then use the method of concatenation or element-wise multiplication to fuse the two modal features to get the joint information of the picture and the problem, finally it will be sent to the classification, in which the most likely answer is inferred from the predefined candidate answer set, or sent to the cyclic neural network to directly generate the answer, but the effect is inferior to the classification method at present. This processing method inputs all visual feature information into the model, but not all visual features are helpful to solve the problem due to the complexity of content in images in most cases. Therefore, attention mechanism is introduced into VQA task. Stack attention networks (SAN) [9] learns visual spatial attention through multi steps. Dual attention networks (DAN) [11] learns the attention information of visual and question modes separately. As the result of the semantic gap between visual features and question representation, bilinear models have more performance and expressiveness than the linear model which linearly integrates the two modes. However, because the representations of the two modes are high-dimensional, the use of outer product will lead to a very high dimension of the output. Hadamard product for low rank bilinear pooling (MLB) [12] proposed a low rank bilinear pooling to effectively deal with this problem. Subsequently, Jin Hwa Kim et al. proposed the bilinear attention network (BAN), which learned the co-attention map of vision and question.

2.2. Attention Mechanism

Attention mechanism first succeeded in complex image classification tasks by Mnih V et al. [13], and it also widely used in other single-modal task fields such as visual, language, speech.

In summary, the attention mechanism endues the model the ability to distinguishing and opens the black box of deep learning for us, providing a window. Generally, it is divided into soft attention, hard attention, and self-attention widely used in NLP. Soft attention can be divided into Item-wise Soft Attention and Location-wise Soft Attention. The same hard attention can also be divided into Item-wise Hard Attention and Location Hard Attention. Self-attention is a variant of the attention mechanism, which reduces the dependence on external information and can effectively capture the internal correlation of data or features.

The attention mechanism has also been successfully applied to multi-modal tasks like VQA, image captioning and image classification. These studies show that learning the attention of visual and language modalities helps to enhance the fine-grained representation of problems and images, thereby effectively improving the accuracy of the model.

3. Our Approach

In this section, we introduce our Hierarchical Reasoning Networks (HRN). First, we introduce the way to deal with images and question in section 3.1, and then illustrate the two most important components: Question-Guided Spatial Attention (QGSA) unit and Question-Guided Attribute (QGAA) unit in section 3.2 and section 3.3 respectively. In section 3.4, we discuss the different combination modes of reasoning module using the two units and fusion strategy of our model, followed by section 3.5, we show our overall architecture of our method.

3.1. Preliminaries

For a given question and picture, we need to convert it into the required form. We use the Faster R-CNN to detect the objects in the image, so we can convert the input image into features $V = (v_1, v_2, \dots, v_n), v_i \in R^D$ which D is the dimension of each detected object's feature and n represent the number of detected objects. For a given question Q, the length of question is limited to m, for question with different lengths will be expanded or cut. Then we use GloVe to embedding it into a vector, and use LSTM to encode them, we only take the hidden layer output from the LSTM which can be represented as $Q = (q_1, q_2, \dots, q_m), q_m \in R^C$, where C is the feature dimension of hidden state. In the VQA-v2.0 dataset, most answers are not composed of multi-words, just one word, so under the common practice, we regard it as a classification problem rather than a generative type. From the training set, we count the k answers with the highest frequency $A = (a_1, a_2, \dots, a_k)$ which is used as the candidate answer. It uses k-dimensional one-hot encoding, and the output of the network is set to the probability that each candidate answer is the correct answer. Use y to represent the true label, and $\hat{y}$ to represent the answer predicted by the model.

3.2. Question-Guided Spatial Attention Unit

In most cases, only one or a few areas in the picture are helpful to answer the question. The model need to ignore areas that are irrelevant to the question, and focus our visual attention on the question-related areas in the image. So we propose QGSA unit as shown in the figure 1, we first input visual features and question representation into the Bilinear Model (BM) module, calculate the relationship between each object in the image and the question, and then use $sigmoid$ to output the spatial attention map as follows:

$$s_i = BM(v_i, q_m) \tag{1}$$

$$S = [s_1, s_2, \dots, s_n] \tag{2}$$

$$att = sigmoid(WS) \tag{3}$$

$$v_i^s = att_i \circ v_i + v_i \tag{4}$$

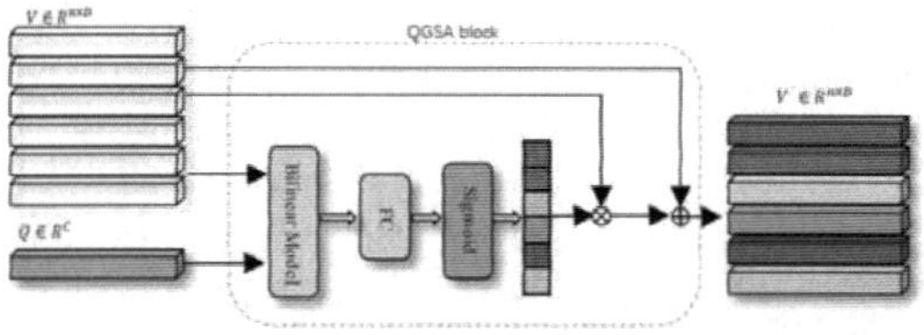

Figure 1. Question-Guided Spatial Attention unit

Where s_i represents relevance between the question and the i-th target in the image, it is worth mentioning that only the last hidden layer output from the LSTM is used in this unit. $att \in R^n$ is a vector represents the spatial attention distribution of the question to the image. The $\circ$ in the formula denotes element-wise multiplication which used broadcast mechanism and $v_i' \in R^D$ represents the i-th weighted visual feature.

3.3. Question-Guided Attribute Attention Unit

For the VQA problem, the existing methods all consider the use of spatial attention which calculating the relevance between the question and the target in the image. But although the model focuses on a specific spatial area, the visual features of the area include redundant information of different kinds of attribute, such as color, size and shape. Inspired by SCA-CNN, we proposed the QGAA unit to obtain the attribute attention distribution map, and then highlight the problem-related attributes in the visual features. Specifically, it includes three steps: 1) Obtaining attribute distribution statistics in visual features. 2) Calculate the correlation between each attribute and the problem. 3) Re-weight visual features.

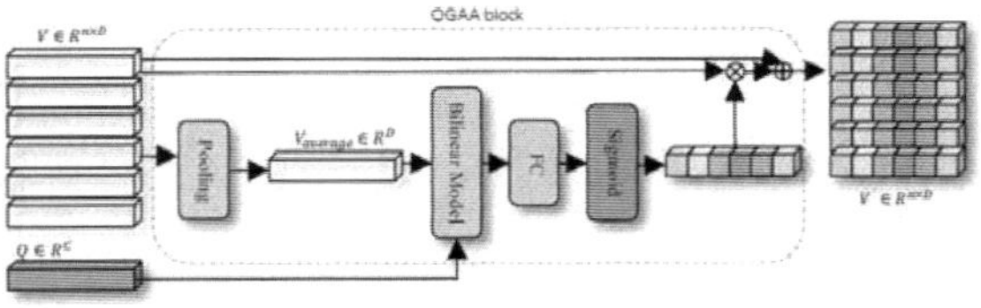

Figure 2. Question-Guided Attribute Attention unit

We use an average pooling layer to generate features-wise descriptors of visual features, which describes the attribute distribution of visual features:

$$V_{average} = \frac{1}{K}\sum_{i=1}^{K} V_i \tag{5}$$

Where $V_{average} \in R^D$ denotes the average statistics of image features with D attribute. By averaging the pooling layer, features can be compressed at the same time, so the parameters are reduced, and then reduced the risk of overfitting.

We further use the Bilinear Model module to calculate the relevance of each attribute of the image to the problem, and then decide which attributes are important.

$$Z = BilinearModel(V_{average}, Q) \tag{6}$$

Where Z denotes the relevance of each attribute of the image to the problem.

Then we use a single-layer fully connected layer and an activation function $sigmoid$ to generate the attribute-attention map, and re-weight the visual features:

$$A = sigmoid(W_a Z) \tag{7}$$

$$V_A = A \circ V \tag{8}$$

Where A denotes the attribute-wise attention vector, $W_a \in R^{D\times D}$ is a parameter matrix, $\circ$ denotes element-wise multiplication which used broadcast mechanism.

3.4. Reasoning Module and Fusion Strategy

Based on the two basic units described before, we design two types of reasoning module Encoder-Decoder and Stacking as shown in figure 3 Both QGSA and QGAA can be cascaded, such that the outputs of current QGSA or QGAA can be fed into the next directly. To cascade these two basic units, we test two types of construction: stacking strategy and encoder-decoder strategy.

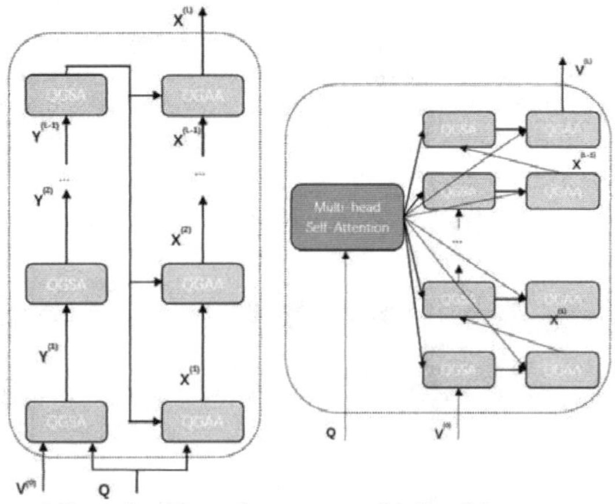

a) Encoder-Decoder b) Stacking

Figure 3. Reasoning modules

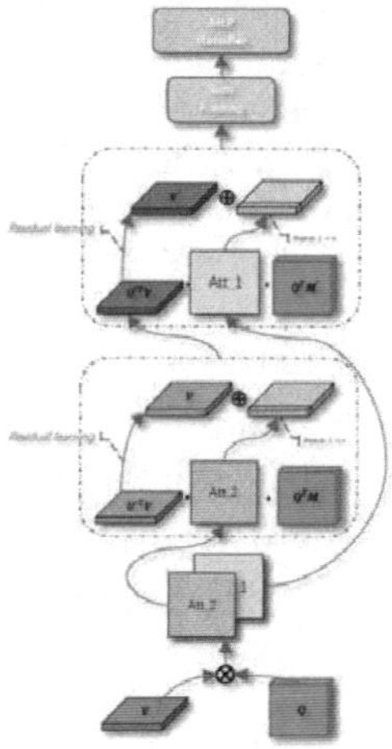

Figure 4. Fusion module

After passing through the reasoning module, the spatial and attribute attention visual features can be obtained, but they still have a different distribution compared with question representation. Inspired by some co-attention [10] based models, we use co-attention layers as fusion strategy. The attention map $\mathcal{A}$ is obtained using Hadamard product and matrix-matrix multiplication:

$$A = [att_1, att_2, ..., att_g] \tag{9}$$

$$att_g = soft\max(((1 \cdot p_g^T) \circ VU)MQ^T) \tag{10}$$

Where $\mathbb{1} \in R^D, p_g \in R^K, U \in R^{D\times K}, M \in R^{C\times K}, att_i \in R^{n\times m}$, $V \in R^{n\times D}, Q \in R^{m\times C}$ represent visual features and question features, g denotes the index of glimpses. Different from the Q used in the previous section, the Q used in this section is all the hidden states of LSTM.

Figure 4. shows the fusion module structure when $g = 2$. After getting $\mathcal{A}$, use it to fuse the two modal inputs V and Q as:

$$f_k' = (VW_V)_K^T A(QW_Q)_k \tag{11}$$

Where $W_V \in R^{D\times K}, W_Q \in R^{C\times K}$, and the subscript k denotes the k-th column of matrices. The joint representation is $f = P^T f'$, where $f \in R^n, P \in R^{K\times n}, f \in R^K$, So the whole fusion process can be expressed as:

$$f_i = FM(f_{i-1}, Q; A) \tag{12}$$

Special, when $i = 0, f_0 = V$.

3.5. Hierarchical Reasoning Networks

We describe the architecture of our Hierarchical Reasoning Networks in this section.

First, we use GloVe and LSTM to represent question features from input question, and use bottom-up attention visual features which used Faster R-CNN to detect object. Second, both visual and question features will be sent to reasoning module followed by co-attention layers in order to fuse the visual features and question features. Finally, the fused features will be sent into a multi-label classifier to generate answer. The flowchart of the whole model is illustrated in figure 5.

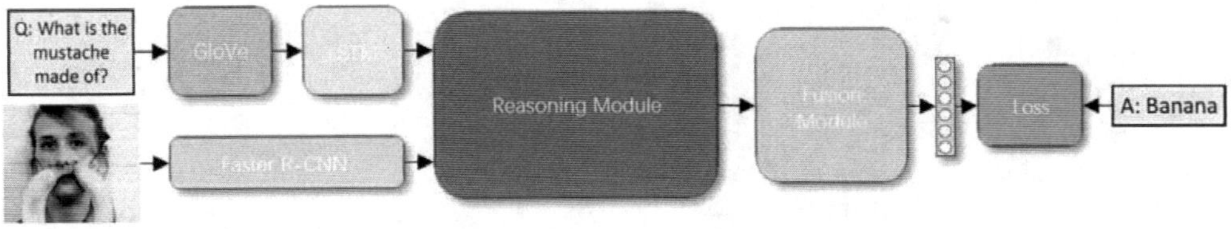

Figure 5. Overall flowchart of Hierarchical Reasoning Network (HRN)

4. Experimental Analysis

4.1. Dataset

We evaluate our model on the VQA-v2.0 dataset, which is proposed to mitigate the language bias. Compared with VQA1.0, VQA2.0 contains over 204K images, 1 million questions and 11 million answers. The overall size is twice that of the VQA1.0 dataset and the VQA2.0 dataset is more balanced. For each question in the dataset, it is a question of two similar images, so there will be two different answers corresponding to it, emphasizing the understanding of visual images and then answering the questions, thus effectively mitigate the problem in the VQA1.0. The VQA2.0 dataset is divided into training set and validation set and test set according to the ratio of 2:1:2. It contains 444,000 questions for 80,000 images, 214,000 questions for 40,000 images, and 448,000 questions for 80,000 images respectively. The training sets and validation sets as open parts, researchers can use them to train the model. The images and questions of the test set are open, but the answer labels are not visible. You can test the performance of the model by submitting the output answers of the model. Questions are classified according to the type of answers, which can be roughly divided into three types: "yes/no", “number” and “other” types.

4.2. Implementation Details

We used the same hyper-parameters in all the state-of-the-art models and our model experiments. The input image features we used in experiments are pretrained by BUTD And each image represented in 36 features which is of 2048 dimensions. The dimensionality of input question features is 512. Candidate answer vocabulary size N is set to 3129 dimensions using the strategy of D Teney et al. and the loss function in the process of model training is binary cross-entropy:

$$L = -\frac{1}{N}\sum_{i=1}^{N} y_i log\left(p(\hat{y_i})\right) + (1-\hat{y_i})log\left(1-p(\hat{y_i})\right) \tag{13}$$

In the process of model training, all the models used the Adamx solver. All the models in the experiment are trained for 20 epochs, and the batch size will be set as 32. Both train and val splits of VQA-v2.0 are used for training. And the number of co-attention layers in our experiments is g = {1,2,4,8} which represent the depth of co-attention layers. It is worth mentioning that the experiment is implemented on the validation set.

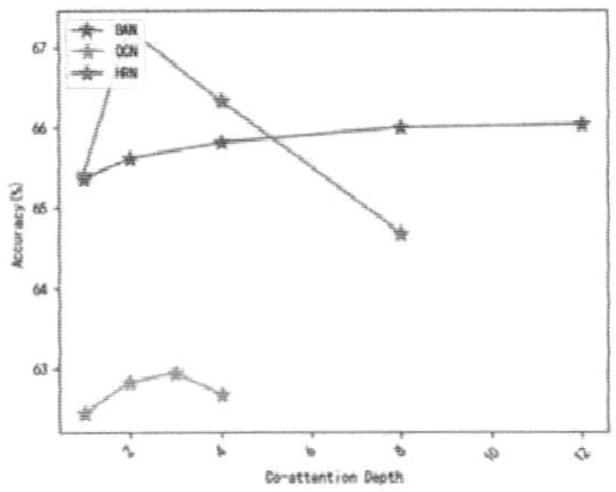

Figure 6. Accuracies of models with different co-attention depth.

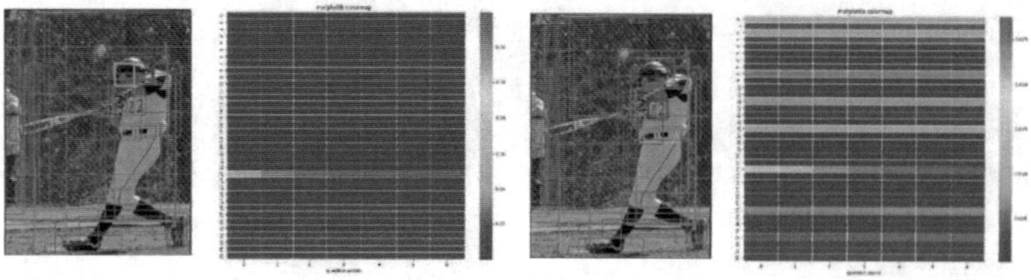

Figure 7. Visualization of the QGSA unit's attention maps and heatmaps of the co-attention layer. The red box(helmet) and yellow box in the first picture match the 23-th and 34-th regions in the second picture, respectively.

4.3. Ablation Studies
A series of ablation in our experiments, investigating the reasons why our model is effective. As mentioned earlier, we use a co-attention layer to calculate the interaction between image features and question features, we show the performance of three different models with different co-attention depth in figure 6.
Compared with DCN and BAN, our model significantly surpass the other leading methods with co-attention models, and reaches its best when $g = 2$, and shows declined trend when co-attention depth deepening.
In Table 1, we illustrate the performances of our model with different reasoning modules. With the deepening of reasoning modules, the performance improved quickly and reached the top at L=4;

Table 1. Experimental results with different depth

L	**Stacking**	**Encoder-Decoder**
1	62.67	62.54
2	64.30	64.92
3	65.42	62.99
4	**65.77**	65.05
5	65.40	65.46

4.4. Qualitative Analysis
We also did some qualitative analysis, but we only shown two examples from the same image but different questions. As shown in figure 7, the first and third picture are the visualization of the attention maps from QGSA unit, and the other two are the heatmaps of attention maps from co-attention layer. From the results, we have the following observations.
Figure 7. illustrate the powerful visual reasoning ability of QGSA unit. The model can focus on image correctly.

Table 2. Experimental results with state-of-the-art method in VQA v2.0 test-dev

Methods	**VQA v2.0 test-dev**			
	All	**Y/N**	**Number**	**Other**
Bottom-up	59.69	75.61	42.24	50.05
BAN	64.20	78.73	42.18	55.86
Counter	64.09	79.14	47.62	54.97
BAN+Counter	64.57	78.96	45.35	**56.67**
MFH	62.12	-	-	-
MFH+Bottom-up	64.76	80.27	45.56	55.89
HRN (our)	**65.76**	**81.07**	**48.37**	56.64

4.5. Comparison with State-of-the-Art
As shown in Table 2, we compare our best single model (g = 2, L=4) with strong benchmarks using

identical visual features. Our method significantly outperforms these leading methods in terms of **ALL**, **Y/N** and **Number** accuracy.

Table 3. Experimental results with state-of-the-art method in VQA v2.0 test-standard

Methods	**VQA v2.0 test-standard**			
	All	**Y/N**	**Number**	**Other**
Bottom-up	59.98	75.89	42.33	50.37
BAN	64.52	79.04	42.10	56.21
BAN+Counter	64.99	79.30	45.33	**57.03**
Counter	64.41	79.53	47.59	55.42
HRN (our)	**66.13**	**81.48**	**48.44**	56.82

5. Conclusion

In this paper, for visual question answering task we present a new a Hierarchical Reasoning Networks (HRN) which consists of a deep reasoning module based on two basic units, QGSA and QGAA and use co-attention layer to fuse the visual and textual information. By stacking reasoning module and co-attention module, our method outperforms some leading methods for VQA.

Acknowledgments

This work was supported by the National Natural Science. Foundation of China under grant Nos. 61621063, 62076028.

References

[1] You Q, Jin H, Wang Z , et al. Image Captioning with Semantic Attention[J]. IEEE Computer Society, 2016.
[2] Abhishek Das, Satwik Kottur, et al. Visual dialog. In Proceedings of the IEEE Conference on Computer Vision and Pattern Recognition, volume 2, 2017.
[3] Dalu Guo, Chang Xu, et al. Image-questionanswer synergistic network for visual dialog. In Proceedings of the IEEE Conference on Computer Vision and Pattern Recognition, pages 1043410443, 2019.
[4] Stanislaw Antol, Aishwarya Agrawal, et al. Vqa: Visual question answering. In Proceedings of the IEEE international conference on computer vision, pages 2425 2433, 2015.
[5] Zhou Yu, Jun Yu, et al. Deep modular co-attention networks for visual question answering. In Proceedings of the IEEE Conference on Computer Vision and Pattern Recognition, pages 62816290, 2019.
[6] Q. Wu, C. Shen, et al. Wang, A.R. Dick, Image captioning and vi- sual question answering based on attributes and their related external knowl- edge, CoRR abs/1603.02814 (2016b).
[7] A Voulodimos, N Doulamis, A Doulamis, et al. Deep Learning for Computer Vision: A Brief Review.
[8] Dzmitry Bahdanau, Kyunghyun Cho, et al. Neural machine translation by jointly learning to align and translate. arXiv preprint arXiv:1409.0473, 2014.
[9] Yang Z, He X, et al. Stacked Attention Networks for Image Question Answering[J]. 2015.
[10] Lu J, Yang J, et al. Hierarchical Question-Image Co-Attention for Visual Question Answering[J]. 2016.
[11] Nam H, Ha J W, et al. Dual attention networks for multimodal reasoning and matching[C]//Proceedings of the IEEE conference on computer vision and pattern recognition. 2017: 299-307.
[12] Kim J H, On K W, et al. Hadamard product for low-rank bilinear pooling[J]. arXiv preprint arXiv:1610.04325, 2016.
[13] Mnih V, Heess N, et al. Recurrent models of visual attention[C]//Advances in neural information processing systems. 2014: 2204-2212.

Digital Analysis of Movements on Characters Based on OpenPose and Dlib from Video

Kai Ye[1,2], Jianmin Dong[1,2,*] and Lijun Zhang[1,2]

[1]School of Information Engineering, Xizang Minzu University, Shanxi, China
[2]Key Laboratory of Optical Information Processing and Visualization Technology of Tibet Autonomous Region, Xizang Minzu University, Shanxi, China

*Corresponding author email: jmdong@xzmu.edu.cn

Abstract. In order to avoid the disadvantages of complex operation and expensive equipment for the extraction method of motion capture systems and other equipment , a method which can extract key points of characters movements was proposed based on video of Tibetan drama with OpenPose and Dlib. And, the movement trajectory and movement fluctuation amplitude which were obtained by digital analysis from characters' movement were used as indicators to provide a quantitative plan for movement essentials of Tibetan drama's performing arts. The digital analysis was implemented by tracking key points' two-dimensional coordinate data with the form of line chart and reflecting movement fluctuation amplitude with absolute difference of the same key point's two-dimensional coordinate data belonged to two frame of video. Experiments showed that this method was suitable for video and was correct and effective in the image coordinate system, which provided a reliable foundation for the digital analysis.
Keywords: OpenPose; Dlib; Key point; Digital analysis.

1. Introduction

Digital technology is an important means for the protection and promotion of national culture such as traditional dances and dramas. Due to many national culture rely on traditional inheritance methods and the rapid development of economy and information technology in nowadays, more and more young people in the location of national culture go to big cities for employment, so many national culture are facing a crisis of loss[1] .Therefore, in order to protect and promote the national culture, it is necessary to accomplish the digital collection of character movements through digital technology and obtain the digital characteristics of the characters' movements based on the collected data.

There are many researches on digital technology for national culture in China. Wang G J et al.[2] proposed to use inertial motion capture technology combined with Kinect equipment to digitally capture the movements of Huangmei Drama. Ye W Z [3] proposed to use Vicon optical capture system to obtain Tibetan Guozhuang dance movement data, and Yao J D [4] proposed to use 24 sets of high-definition lenses optical capture system to obtain the Mongolian dance skeletal motion files. Tang C Q et.al[5] proposed to use the Kinect device to extract human bone data to match the shadow puppet movement. These methods are complicated, the equipment is complex with the cost is high.

As Tibetan drama is the most dazzling pearl in Tibetan culture, Tibetan drama fully embodies the cultural and artistic views of the Tibetan nation and has extremely high artistic value[6]. Therefore, this paper takes Tibetan drama as example to conduct a digital analysis to obtain the digital characteristics of the characters in Tibetan drama.

In view of disadvantages of complex operation and expensive equipment for the extraction method of motion capture systems, a method which can extract key points of characters movements is proposed based on video of Tibetan drama with OpenPose and Dlib which only need to import video as data. And, the movement trajectory and movement fluctuation amplitude which are obtained by digital analysis from characters' movement are used as indicators to provide a quantitative plan for movement essentials of Tibetan drama's performing arts. The digital analysis is implemented by tracking key points' data with the form of line chart and reflecting movement fluctuation amplitude with absolute difference of the same key point's data belonged to two frame of video. Experiments showed that this method is suitable for video.

2. OpenPose and Dlib Toolkit

2.1. OpenPose Libarary

OpenPose is the world's first deep learning-based real-time multi-person pose estimation application based on the real-time multi-person gesture recognition using Part Affinity Fields (PAFs) proposed by Zhe C et al.[7]. It innovatively increases the affinity domain vector field based on the Part Confidence Maps(CMP), which can predict the position and direction of key points at the same time so that Openpose is both robust and has extremely high processing speed[8] [9] .

OpenPose processing process in short: Firstly, the original feature map F is extracted by the VGG-19 network from an image. Secondly, F is entered into two branches which composed of the same multiple Convolutional Neural Networks (CNN) to acquire Part Confidence Maps and Part Affinity Fields. Finally, in addition to the first stage, the result of the previous stage is combined with the original feature map F for multiple iterations. The iteration formula is as follows:

$$S^t = \rho^t(F, S^{t-1}, L^{t-1}) \forall \mathrm{t} \geq 2 \tag{1}$$

$$L^t = \phi^t(F, S^{t-1}, L^{t-1}) \forall t \geq 2 \tag{2}$$

Where F is the original feature map, ρ is the predicted key point confidence map, ϕ is the predicted key point confidence map, S is the key point confidence map set, L is the key point affinity field set, superscript 1 represents the first stage, t represents the t stage.

After iteration, the final Part Confidence Map and Part Affinity Field are obtained that the position information of all key points and the direction vector of the connected key points are known. Finally, human joints are connected based on these data with bipartite matching [10]

2.2. Dlib Library

The Dlib library is an open source library, which has a large number of commonly used image processing algorithms and trained models, so it is widely used in the field of face detection [11] .The face model of Dlib library used in our works is based on the Ensemble of Regression Tress (ERT)algorithm, which is proposed by Kazime et al. [12] .The principle of the ERT algorithm: The face gradually returns to its true shape by building a Cascaded Residual Regression Tree (GBDT). Each leaf node of each residual regression tree stores a residual regression value. When data is entered to each leaf node, residuals of each leaf are added to data until all the residuals are added together which has completed the extraction of the key points of the face[13].

Among them, the ERT algorithm builds GBDT to gradually return the face to its true shape through multiple iterations. The formula is as follows:

$$S^{(t+1)} = S^{(t)} + r_t(I, S^{(t)}) \tag{3}$$

r_t is the prediction residual based on the feature, $S^{(t)}$ represents the shape of the feature point predicted in the previous iteration, $r_t(I, S^{(t)})$ represents the residual calculated in the current layer, $S^{(t+1)}$ represents the result of the tth iteration.

3. Key Points Extraction of Characters

3.1. Pose Key Points Extraction of Tibetan Characters

Nowadays, motion capture system which contains many expensive and complicated equipment are often used to perform digital collection of motion. An optical motion capture system which shown as the picture below built by Cao J H[14] .

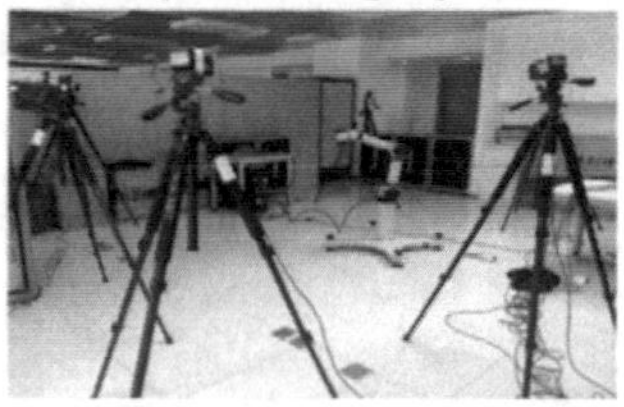

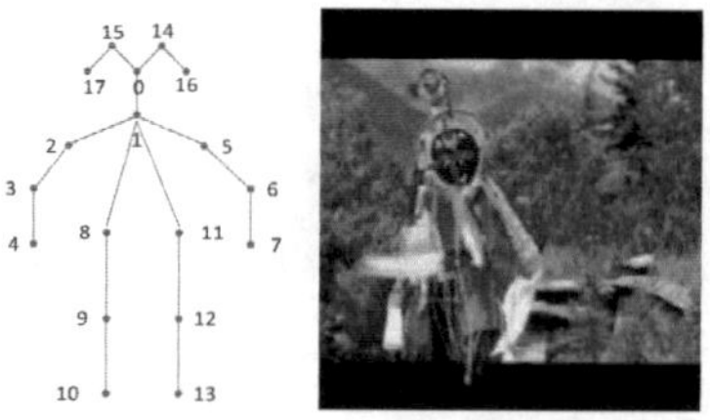

Figure 1. An optical motion capture system **Figure 2.** OpenPose Pose Model

Comparing with it, OpenPose is convenient and simple that it only need to import video as data without any expensive and complicated equipment.

OpenPose is an open libaray with built-in well-trained models, which contains the pose model of the character. After importing the Tibetan drama video, the pose model will recognize the posture of the Tibetan drama character, and store the 18 key points'data.

The detailed acquisition steps are as follows:

Step 1: Import the video into the VGG19 network trained by OpenPose to obtain the original feature map F, and input F into two multi-layer Convolutional Neural Networks to respectively predict Part Confidence Map C and Part Affinity Field P.

Step 2: Part Confidence Map C is used to determine key points' detection C', and the Part Affinity Field P is used to determine joint's direction V of key points which belongs to one joint.

Step 3: Use C' and V to connect joints through bipartite matching which get the data set S.

Step 4: Export data set S.

Table 1. Key Points of OpenPose Pose Model

Key Point	Body Part
0	Nose
1	Neck
2	Right Shoulder
3	Right Elbow
4	Right Wrist
5	Left Shoulder
6	Left Elbow
7	Left Wrist
8	Right Hip
9	Right Knee
10	Right Ankle
11	Left Hip
12	Left Knee
13	Left Ankle
14	Left Eye
15	Right Eye
16	Left Ear
17	Right Ear

3.2. Facial Key Points Extraction of Tibetan Face Masks

The face model in OpenPose uses 70 key points to return the shape and position of face. But, the face model in OpenPose cannot recognize the facial shape of the face mask when it detects the Tibetan drama mask. Aiming at the disadvantages of the face model in OpenPose, considering the uniqueness of the Tibetan drama face mask, after comparison, it is proposed to use the Dlib library face model to obtain the key points of the Tibetan drama face mask.

(a) OpnePose (b) Dlib

Figure 3. Comparsion of OpenPose and Dlib face models to obtain key points

The comparison of the key points obtained by the OpenPose and Dlib face models is showed on figure 3. The result picture on figure 3(a) is the key points obtained by the OpenPose face model. The key points of the obtained face mask are seriously damaged. The result picture on figure 3(b) is the key points obtained by the Dlib face model. The key points of the face are extracted completely.

The detailed acquisition steps are as follows:

Step 1: Import the video into the feature point detector generated by the ERT cascade algorithm to obtain the position information of 68 key points of the face masks on Tibetan drama characters which is the face key point data set T.

Step 2: Export data set T.

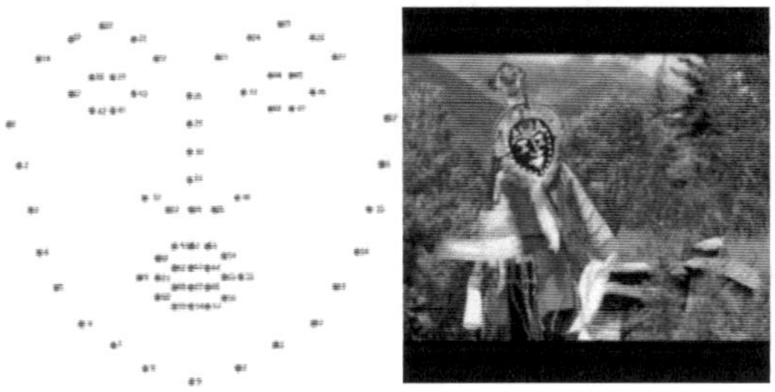

Figure 4. Dlib Face Model

Table 2. Key Points of Dlib Face Model

Key Point	Face Part
1-17	Facial contour
18-22	Right Eyebrow
23-27	Left Eyebrow
28-36	Nose
37-42	Right Eye
43-48	Left Eye
49-68	Mouse

According to figure 4, the distance between points 37 and 40 of the face model is the distance of the left eye. Let the x and y coordinates of point 37 be x_1 、 y_1, the x and y coordinates of point 40 are x_2, y_2, the distance is:

$$d_1 = \sqrt{(x_2 - x_1)^2 + (y_2 - y_1)^2} \tag{4}$$

Let the distance between points 43 and 46 be the distance of the right eye. Let the x and y coordinates of point 40 be x_3, y_3, let the x and y coordinates of point 43 be x_4, y_4. Then the right eye distance is as follows:

$$d_2 = \sqrt{(x_4 - x_3)^2 + (y_4 - y_3)^2} \tag{5}$$

Let d be the correct rate of the distance between the two eyes, so

$$d = 1 - \frac{|d_2 - d_1|}{(d_1 + d_2)} \tag{6}$$

4. Digital Features of Movements on Characters

4.1. Movement Data Accuracy on Method

Tibetan drama is selected as experimental data. The characters will be performed according to the view, and the movements must be complicated and changeable. Therefore, the movement trajectory of the characters in Tibetan drama is visualized by line graph. The fragment of Nob Sangpo's farewell to his wife is selected as an example to explain three movements of Nob Sangpo, "Sleeve Waving", "Sideways", and "Head Up" in the fragment. "Sleeve Swing" is related to key points of shoulder, selecting left shoulder key point 2 and right shoulder key point 5 in figure 2; "Sideways" is related to the head key point, selecting nose key point 0 and neck key point 1 in figure 2; "Head up" is related to the key points of the face, selecting the key point 40 of the left eye and the key point 43 of the right eye in figure 4.

The pose data of movement extracted by OpenPose can be expressed by CMP which represents that the pose data extracted is good or bad. The face data of movement extracted by Dlib can be expressed by d which represents that face data extracted is good or bad.

When the character is in front, the accuracy of the data obtained is high. When the character moves quickly or turns around, the accuracy of the data obtained is low.

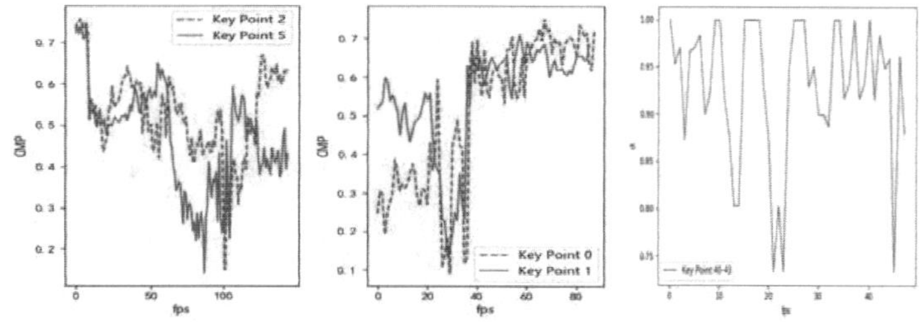

Figure 5. Movement data accuracy extracted by OpenPose and Dlib

4.2. Movement Trajectories of Characters

On the "Sleeve Swing" movement, Nob Sangbo leaned up sideways firstly that the X and Y coordinates of key points on shoulders rose rapidly. Then, he gradually raised his hands and waved his right hand to right. As his hands were raised, his shoulders were raised that the X and Y coordinates of key points on shoulders showed an upward trend. After that, when he leaned over that his shoulders were laid out that the X and Y coordinates of key points on shoulders fell. Finally, when he straightened his body that his shoulders were raised the X and Y coordinates of key points on shoulders rose again.

On the "Sideways" movement, Nob Sangbo swung his sleeves and his body turned right firstly in which his head also raised and turned right that the X coordinate of key points on nose showed an upward trend. Then, he bowed his head and moved his head to right that the X coordinate of key points on nose showed an upward trend. On this movement, the character's neck had changed small in Y coordinate that the Y coordinate of key points on neck had almost never changed.

On the "Head Up" movement, Nob Sangbo turned his head right firstly the X coordinate of key points on eyes rose quickly and the Y coordinate of key points on eyes fell. Then, he bowed his head and moved his head to the left that the X and Y coordinates of key points on eyes showed an downward trend. Finally, he raised his head and moved slowly down that the X and Y coordinates of key points on eyes rose and fell.

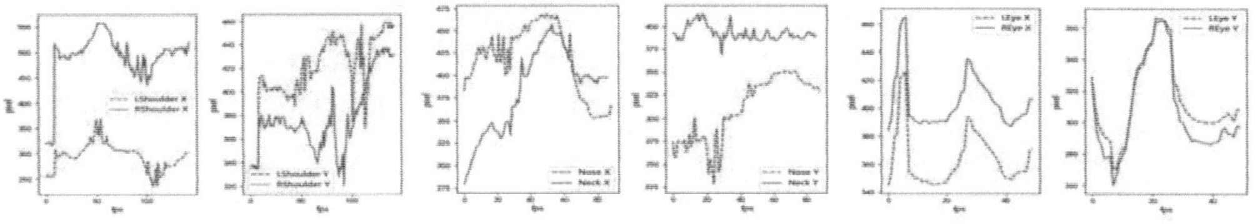

(a) "Sleeve Swing" (b) "Sideways" (c) "Head Up"

Figure 6. The Movement Trajectory of Key Points

4.3. Movement Fluctuation Amplitude of Characters

Let the movement fluctuation amplitude be K. Let the one frame of key point's coordinate be (X_{now}, Y_{now}) and the next frame of key point's coordinate be (X_{next}, Y_{next}). The fluctuation amplitude can be expressed by the following formula:

$$K_X = |X_{next} - X_{now}| \quad (7)$$

$$K_Y = |Y_{next} - Y_{now}| \quad (8)$$

On the "Sleeve Swing" movement, the movement of shoulders changed quickly and greatly that the movement fluctuation amplitude of shoulders also changed quickly and greatly which showed in figure 5(a) that the left picture represented movement fluctuation amplitude of right shoulder and the right picture represented movement fluctuation amplitude of left shoulder.

On the "Sideways" movement, the movement of neck and nose had small change fluctuation amplitude but performed complicated that the movement fluctuation amplitude had complex changes in figure 5(b) in which the left picture showed the movement fluctuation amplitude of neck and the right picture showed the movement fluctuation amplitude of nose.

On the "Head Up" movement, the movement of eyes changed quickly at the beginning and then changed relatively steady that the movement fluctuation amplitude of eyes also changed greatly at the beginning and then had almost never change which showed in figure5(c) in which the left picture showed the movement fluctuation amplitude of right eye and the right picture showed the movement fluctuation amplitude of left eye.

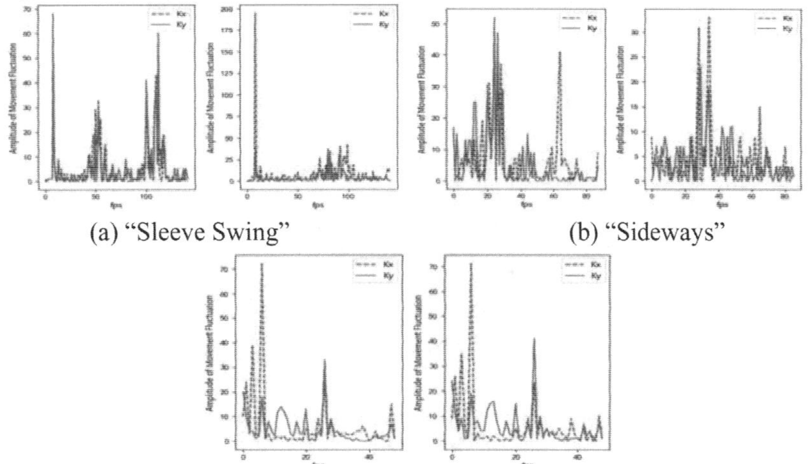

(a) "Sleeve Swing" (b) "Sideways"

(c) "Head Up"

Figure 7. The Amplitude of Movement Fluctuations

5. Conclusion

In order to avoid the disadvantages of the digital method of human body movements by motion capture systems and other equipment, a method which can extract Spatial digital features of the characters' movements is proposed based on the video data of Tibetan dramas with OpenPose and Dlib. The method acquires digital characteristics of movement trajectory and movement fluctuation amplitude by tracking and analyzing Spatial digital features of the characters' movements. The movement trajectory and movement fluctuation amplitude are used as indicators to provide a quantitative plan for movement essentials of Tibetan drama's performing arts. Experiments show that the method in this article is correct and effective in the image coordinate system, which provides a reliable foundation for the digital analysis of this traditional art. In addition, our works use Tibetan drama as the experimental data. In the future, other traditional dramas and dances will be used as experimental data to realize the digital analysis, and obtain the digital characteristics of characters' movements in other traditional dramas and dances.

Acknowledgments
This work is supported by National Key R&D Program of China (No.2017YFB1402101) and the Key Natural Science Foundation of Tibetan Autonomous Region (No.2016-ZR-15-14).

References

[1] Xiao W Q 2021 *Research on Digital Protection of Intangible Cultural Heritage in Ganzi Tibetan Autonomous Prefecture [J]* (Hubei: JinGu Creative Literature)p 72-74
[2] Wang G J et.al 2016 *Digital protection of Huangmei Drama actions based on inertial motion capture system [J]* (Anhui: Journal of Anqing Teachers College)p 77-80+110
[3] Ye W Z 2020 *Research on digital method of Tibetan Guozhuang Dance based on motion capture technology [J]* (Beijing: China New Communications)p 235-236
[4] Yao J D 2018 *Development and application of Mongolian dance digital dance simulation action resource library [J]* (Beijing: Educational Science Edition)p 116-120
[5] Tang C Q et.al 2021 *Research on real-time matching of shadow puppet action based on Kinect [J/OL]* http://kns.cnki.net/kcms/detail/11.3092.v.20210104.1909.022.html.(Beijing: Journal of System Simulation)
[6] Hu X Y 2020 *A Narrative Study of Eight Traditional Tibetan Dramas [D]* (Yunnan: Yunnan Normal University)
[7] Zhe C et.al 2017 *Realtime Multi-person 2D Pose Estimation Using Part Affinity Fields[C]//* IEEE Conference on Computer Vision and Pattern Recognition (CVPR).
[8] Yang J et.al 2021 *Comparative research on human action recognition based on OpenPose [J]* (Hei Longjiang: Sensors and Microsystems)p 5-8
[9] Zhang X X et.al 2020 *Application research of intelligent escalator monitoring system based on OpenPose [J]* (Hei Longjiang: Journal of Heilongjiang Institute of Technology)p 49-55
[10] Ma J 2018 *Research and Implementation of Behavior Recognition Method Based on Posture and Skeleton Information [D]* (Shandong: Shan Dong University)
[11] Liu Z F 2020 *Application of Dlib in face recognition technology [J]* (Beijing: Electronic production)p 39-41+7
[12] Kazemi V 2014 *One Millisecond Face Alignment with an Ensemble of Regression Trees[C]//* IEEE Conference on Computer Vision & Pattern Recognition.
[13] Rui C 2019 *Feature Point Extraction of Dog Face Based on ERT Algorithm [D]* (Shanghai: Shanghai Normal University)
[14] Cai J H 2021 *Research on a Method of Object Motion Data Capture in Optical Motion Capture System [J]* (Beijing: Computer Measurement & Contorl)

A Text Location Method for Sign Based on Pixel-level Gradient Ray Images and Edge Contour Mapping

Jiani Du, Jun Bai and Guanglu Zhou*

School of computer science and technology, Harbin Institute of Technology, Weihai, China

*Corresponding author email: estellanini@163.com

Abstract. Compared with deep learning methods, traditional image processing methods have lower computational costs and have advantages for typical problems such as sign text, this paper proposes a sign text location method based on pixel-level gradient ray map and edge contour mapping. First, based on image pre-processing, the pixel-level gradient ray map is obtained by using the Stroke Width Transform. Then, using the Maximally Stable Extremal Region method to locate the irregular text candidate area in the ray map. At the same time, add a mask layer to filter the irregular candidate area, and then regularize the candidate area and perform Non-Maximum Suppression processing. Finally, the background non-text candidate area is processed by extracting the mapping of the closed edge contour of the image, and the remaining text area positioning candidate frame coordinates are mapped back to the original picture. The experimental results show that the method is simple and clear, the effect is remarkable, and it has good robustness to the sign of different sizes.
Keywords: Text location; Maximally stable extremal regions; Pixel-level gradient ray; Edge contour mapping.

1. Introduction

Nowadays, the field of image processing and computer vision has developed rapidly, and text detection and recognition are the basis of many types of research. With the emergence of new research fields such as intelligent transportation and intelligent driving assistance, the extraction of sign information from scene pictures has an important meaning. Text area location technology is to locate the input scene picture according to a certain method and extract the text part for subsequent processing again. At present, the text area location technology of scene pictures is mainly divided into traditional methods and methods based on deep learning. Since traditional image processing methods have lower computational costs than deep learning methods, they have advantages for typical problems such as sign text. Therefore, our research direction is mainly to use traditional image processing methods for text positioning to obtain sign information and to provide basic research for intelligent driving assistance.

Traditional text area positioning methods generally use manual feature extraction for text location. Commonly used are methods based on edge features[1,2], texture features[3,4], or based on stable regions[5,6]. When using the edge feature to locate the text area in the scene image, you need to rely on the use of operators, such as Canny[7], Sobel, Laplacian. Firstly, edge detection is performed on the scene image, and the gradient image of the edge of the picture is obtained through the operator, and then the detection result is filtered to filter the non-text area, and finally, the remaining dense edges are merged into the same area to realize the positioning of the text area. When using texture features to locate text in a scene picture, the text area in the scene picture and the texture feature of the background need to be regarded as opposite features, and the candidate frame in the picture is obtained through the positioning method of the sliding window. Finally, through the wavelet transform or other algorithms to distinguish between text areas and non-text areas. Due to the computational complexity and low efficiency of this method, it has not been widely used in text area locations. The most common positioning methods based on stable regions include Stroke Width Transform, Maximum Stable Extremal Region, and so on. B.Epshtein et al.[8] proposed a stroke width transform algorithm. Since the text has a fixed stroke width, this feature is used to locate the feature in other elements of the picture. First, find the stroke width transformation of the image, that is, assign a stroke width to each pixel, and then use flexible geometric reasoning to aggregate the positions with similar stroke widths into words. J. Matas et al.[9] proposed an algorithm for image text location and recognition based on the largest extreme value stable area, clustering analysis of pixels based on the extreme value attributes of the area[10], while excluding non-text areas, and then according to the structural attributes of the stable connected areas. Position the text area. M.Jiang et al.[11] introduced frequency tuning (FT) visual saliency to remove non-text candidate regions after adopting the SWT algorithm. Azmi Can et al.[12] used the Tesseract optical character recognition engine to eliminate non-text groups after traditionally extracting text candidate frames.

In this paper, the method we propose is based on the stroke width transform method to provide a pixel-level gradient ray map to filter the most irrelevant backgrounds. In the ray map, the method of maximum stable extreme value area is used to locate the irregular text candidate area. At the same time, a mask layer is added to filter the individual rays, so that the image to be processed contains only the irregular candidate area, and then the candidate area is regularized and Non-maximum suppression processing solves the nesting problem. Finally, the background non-text candidate area is processed by extracting the mapping of the closed edge contour of the image, and the remaining text location area coordinates are mapped back to the original picture.

2. Related Work

2.1. Edge Detection

There are many edge extraction algorithms, such as Canny, Sobel and Laplacian. This paper chooses to use Canny to achieve edge extraction because the Canny method is not susceptible to noise interference and can detect real weak edges. The advantage is that two different thresholds are used to detect strong and weak edges respectively. The Sobel method has a better effect on image processing with more gray gradients and noise, but Sobel is not very accurate in edge positioning, and the edge of the image is more than one pixel. When the accuracy requirements are not high, Sobel is a more commonly used edge detection method. The Laplacian method is more sensitive to noise, so this operator is rarely used to detect edges. Instead, it is used to judge whether edge pixels are regarded as bright or dark areas of the image. In the research content proposed in this paper, the closed contour after edge extraction needs to be mapped, so an edge extraction method that is not interfered with by noise and accurate edge positioning is required, so Canny is selected for edge extraction.

2.2. Stroke Width Transform

The main idea of the algorithm[8] is to extract edges and gradients, and then search for edges in the opposite direction through the gradient direction. The path length recorded during the search process is the stroke width corresponding to all pixels on the path. Choose an edge pixel point S and search along its gradient direction (d_p). If an edge pixel point E with the opposite direction (d_q) and the gradient absolute value difference is less than $\frac{\pi}{6}$ can be found, then the two points are considered to be one pixel pair, otherwise discard the pixel. Then we need to calculate the distance from S to E, and assign the stroke width of all pixels between these two points to this distance value. The SWT-based text detection method can detect text areas of multiple scales and multiple arrangements. At the same time, the detection speed is fast and the obtained candidate areas are also fewer. However, this method heavily relies on the results of edge detection, when the environment is complex and the contrast is low or under poor lighting conditions, more missed inspections are likely to occur. Moreover, text-like objects in the background such as tree branches, railings, will also detect similar stroke widths and certain errors will occur.

2.3. Maximum Stable Extremal Region

The algorithm[9,10] uses the gray value in the gray image to operate. In this process, the image is understood as a topographic map, and the gray value is regarded as the altitude. The pixel with a gray value of 0 represents the lowest point and pixels with the gray value of 255 represent the highest point. The process of finding the area with the largest stable extreme value can be regarded as the process of injecting water into the topographic map: the area with the smallest altitude will be filled first, and then the area with the second smallest altitude will be filled. In the process of continuous water injection, the terrain will be completely submerged. When the water surface continues to rise, the area of these submerged areas no longer changes, and areas like this are called MSER areas. The idea of this watershed algorithm is transferred to the image: threshold changes are made to the gray value in a certain order, and a binary image is obtained for each threshold. In the process of change, some connected areas have little or no change, and these areas are MSER areas.

In the process of threshold change, the judgment of the speed of the area change is based on Eq. (1). Among them, Q_i represents the area of a connected region when the threshold value is i, Δ is the threshold change amount, and $q(i)$ represents the area change rate of the region. When it takes the local minimum value, the maximum stable extreme value region is obtained.

$$q(i) = \frac{|Q_{i+\Delta} + Q_{i-\Delta}|}{|Q_i|} \tag{1}$$

3. Proposed Method

In this section, we will introduce the specific steps of the sign text location method based on pixel-level gradient ray images and edge contour mapping.

3.1. Pre-processing

Before the usual image processing operations, the original pictures need to be pre-processed. The image to be processed is gray-scaled, and the image contrast is enhanced by gray-scale linear stretching to improve the image quality and prepare for the subsequent text location. Grayscale linear stretching is also called a piecewise linear transformation. If the grayscale of an image is concentrated in a darker area and the image is dark, we can use the grayscale stretching function to stretch the grayscale range of the object to improve the image. If the image gray is concentrated in the brighter area and the image becomes brighter, we can also use the grayscale stretch function to compress the grayscale interval of the object to improve the image quality. For brighter pictures, the histogram averaging method can also be used to increase the contrast. Pre-processing comparison results are shown in Figure 1.

3.2. Obtain a Pixel-level Gradient Ray Image

After the pre-processed image is obtained, the pixel-level gradient ray map is obtained through the stroke width algorithm[8]. The process is as follows: the edge of the image is extracted by the operator. In this paper, the dual thresholds 185 and 255 are selected. The larger threshold is used to detect the obvious edges in the image. In generally, the detection effect is not ideal, and the edge detection is intermittent. At this time, the smaller first threshold is used to connect these discontinuous edges. We calculate the derivative of the image in the horizontal and vertical directions, and jointly calculate the gradient of the edge pixels of the image. For each pixel, the gradient is the best tangent direction, and the rays drawn through the gradient direction can best simulate the width direction of the text stroke. The X gradient map and the Y gradient map are obtained through the Sobel operator. For each pixel at the edge of the image, the arctangent angle in the x and y directions is obtained, and the ray is drawn along the gradient direction θ. If it encounters other image edge pixels, the ray stops. Otherwise, the rays will be extended on the image. Record the stroke width data so that each pixel in the image corresponds to the stroke width value of each pixel. If a pixel belongs to multiple edges, the median value is used to coordinate the difference. Use the K-D tree to find similar stroke shapes and calculate the text angle between two characters at the same time. Re-set a representative label for each connected tree. Finally, the pixel-level gradient ray map is obtained. Processing results are shown in Figure 2.

Figure 1. Pre-processing process: (a) original image. (b) gray image. (c) contrast-enhanced image. (d), (e), (f) are corresponding contour extraction images.

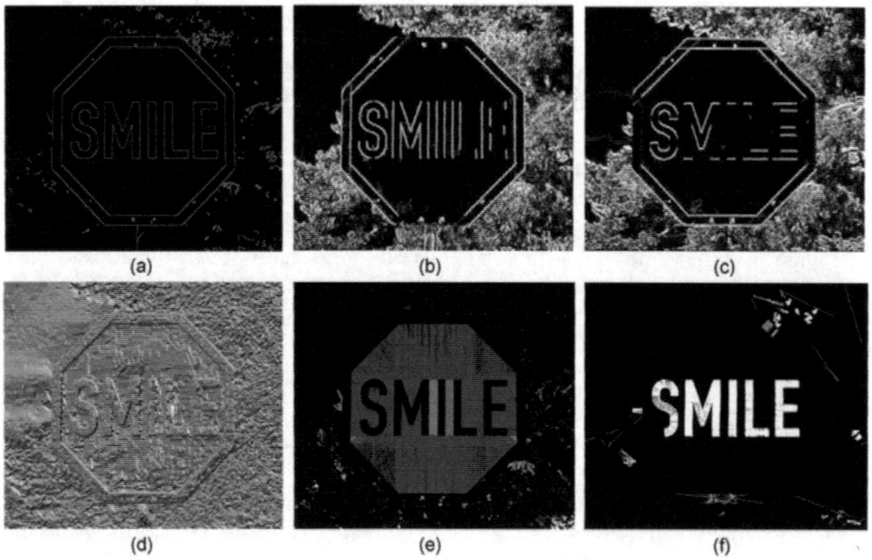

Figure 2. Pixel-level gradient ray image process: (a) operator contour map. (b) sobel X gradient map. (c) sonbel Y gradient map. (d) theta map. (e) label map. (f) pixel-level gradient ray image.

3.3. Processing Extreme Value Areas through the Mask Layer

The gray value is changed so that the maximum stable extreme value area starts to be gradually generated, and a corresponding binary image is generated for each gray value change. We record the number and area of connected regions contained in each generated binary image are recorded. According to formula (1), determine whether the area belongs to the MSER area. Abandon the area and the rate of change is too large or too small. The extracted area will have certain angles, rotation differences, and distortions. The irregular areas are ellipticity and adjusted to regular affine invariant areas. The pixels in the same text in a natural scene usually have the same gray value, so the MSER algorithm can be used to extract the text area.

The MSER algorithm can find areas that are indeterminate within a certain threshold range. Through the comparison images, it is found that the MSER processing on the pixel-level gradient ray map will filter out many cluttered areas, and the processing effect will be better. At this time, NMS processing can be carried out to solve the nesting problem of candidate frames. The idea of the NMS algorithm is as follows: First, we need to set the confidence threshold of the target box, and the common value of the threshold is 0.3-0.5. Then the list of candidate boxes is arranged in descending order according to the degree of confidence. Push the top of the sorted candidate box list to the output list and delete it from the candidate box list at the same time. We need to calculate the Intersection over Union (IoU) value of this candidate box and all items in the candidate box list, and delete candidate boxes whose intersection ratio is greater than the threshold. Repeat the above process until the candidate box list items are decremented to empty, and then return the output list. Finally, we add a black mask layer, leaving only the candidate area after MSER. At this time, most of the non-text areas in the background have been successfully filtered, you can proceed to the next step. Processing results are shown in Figure 3.

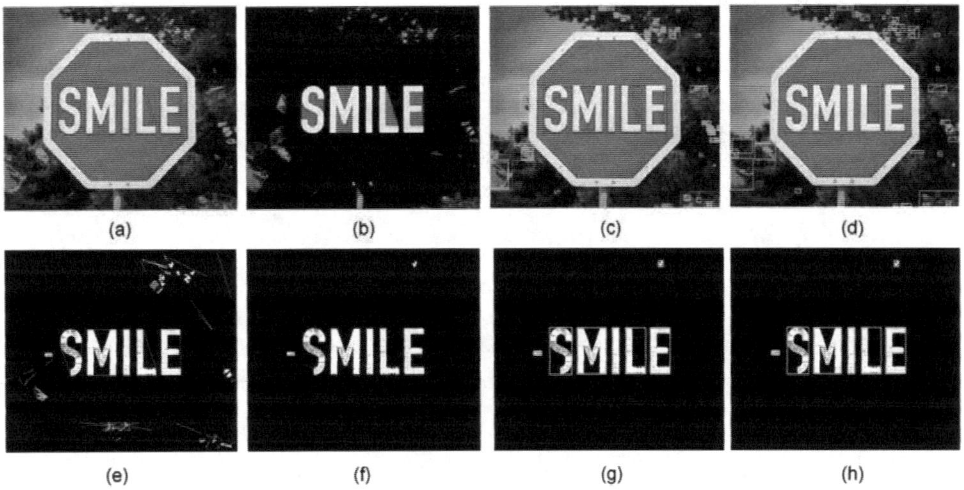

Figure 3. MSER and mask layer processing comparison: (a), (b), (c), (d) are the MSER image, mask layer processing image, circumscribed rectangle processing image, and NMS processing image of the original image. (e), (f), (g), (h) are the results corresponding to the pixel-level gradient ray image.

3.4. Processing the Mapping of Closed Edge Contours and Result Mapping

The closed edge contour of the picture is obtained by the canny operator, and the closed edge is mapped to the black mask image in the previous step. Only the candidate boxes within the closed edge contour are retained. In this step, using the feature that the edge contour of the sign picture to be processed is obvious, and the smaller noise candidate frame in the picture to be processed is processed. A better effect can be achieved. Finally, we merge the remaining text boxes to construct the smallest neighbor pair, aggregate each nearest neighbor pair, scan each nearest neighbor pair, and aggregate to obtain the smallest bounding rectangle of the area. At last, the coordinates of the text area positioning area are mapped back to the original picture. Processing results are shown in Figure 4.

Figure 4. Mapping image and final result image.

4. Experiments

The performance of this method is in the self-built data set and the standard data set ICDAR2013. The ICDAR2013 data set contains 229 training pictures and 233 test pictures. We used 233 images in the test set and 95 images in the self-built data set for performance evaluation. Some experimental results are shown in Figure 5. The recall rate (R), precision rate (P), and F-measure (F) are calculated by the following formulas. Among them, E is the number of all candidate frames located by this method, C is the number of text areas in the area located by this method, and T is the exact number of text areas of the actual data to be tested. The parameter in F is usually set to 0.5. Experiments show that the effect is better than existing methods.

$$R = C/T \tag{2}$$

$$P = C/E \tag{3}$$

$$F = (\alpha/P + (1-\alpha)/R)^{-1} \tag{4}$$

Figure 5. Example of experimental results.

5. Conclusion

In this paper, we propose a sign text location method based on pixel-level gradient ray map and edge contour mapping. This method mainly uses the pixel-level gradient ray map to extract the maximum stable extreme value region, instead of directly using the original image to extract the maximum stable extreme value region. Then, the candidate frame is processed by the mapping of the closed edge contour. Experiments show that our method achieves better performance on the self-built data set and ICDAR 2013 data set. However, this method still has certain limitations. The advantages of this method are more obvious in the context of the horizontal text of the sign with obvious outlines. For curved and slanted text, the candidate box we locate will be larger and may contain some noisy pixels. If the difference between the sign outline and the background environment is small, the edge extraction effect of the operator will not good, which will eventually lead to a poor text positioning effect. We will continue to improve in the later period.

References

[1] X.Liu, J.Samarabandu. Multisccale edge-based text extraction from complex images[C]//International Conference on Multimedia and Expo, IEEE, 2006:1721–1724.

[2] Q. Ye, J. Jiao, J. Huang, H. Yu. Text detection and restoration in natural scene images [J]. Journal of Visual Communication and Image Representation, 2007, 18(6) 504-513.

[3] G. Zhou, Y. Liu, Q. Meng, Y. Zhang. Detecting multilingual text in natural scene [C]//International Symposium on Access Spaces, ISAS, IEEE, 2011:116–120.

[4] Y.F. Pan, C.L. Liu, X. Hou. Fast scene text localization by learning-based filtering and verification [C]// 17th IEEE International Conference on Image Processing, ICIP, IEEE, 2010:2269-2272.

[5] Y. Pan, X. Hou, C. Liu. A hybrid approach to detect and localize texts in natural scene images [J]. IEEE Transactions on Image Processing, 2011,20:800-813.

[6] H. Zhang, C. Liu, C. Yang, X. Ding, et al. An improved scene text extraction method using conditional random field and optical character recognition [C]//International Conference on Document Analysis and Recognition, ICDAR, IEEE, 2011:708-712.

[7] A computational approach to edge detection[J]. Readings in Computer Vision, 1987:184-203.

[8] Epshtein B, Ofek E, Wexler Y. Detecting Text in Natural Scenes with Stroke Width Transform[C]// Computer Vision & Pattern Recognition. IEEE, 2010.

[9] Matas J, Chum O, Urban M, et al. Robust wide-baseline stereo from maximally stable extremal regions[J]. Image and Vision Computing, 2004, 22(10): p.761-767.

[10] Maximally Stable Extremal Regions[M]. Betascript Publishing, 2010.

[11] Jiang M, Cheng J, Chen M, et al. An Improved Text Localization Method for Natural Scene Images[C]// 0.

[12] Ozgen A C, Fasounaki M, Ekenel H K. Text detection in natural and computer-generated images[C]// 2018 26th Signal Processing and Communications Applications Conference (SIU). 2018.

A S-transform Based Algorithm for Doppler Frequency Rate-of-Change Estimation from Coherent Pulse Train

Zhangfu Lan, Huafeng Peng* and Peiyan Zhao

National Key Laboratory of Science and Technology on Blind Signal Processing, Chengdu 610041, China

*Corresponding author email: hualong913@sina.com

Abstract. The traditional Fourier correlation method has the problem of poor accumulation effect, which leads to the low estimation accuracy of Doppler frequency change rate coherent pulse signal under the condition of low signal-to-noise ratio or short pulse duration. To solve this problem, an estimation algorithm based on S-transform is proposed to estimate the Doppler frequency change rate. By means of higher time-frequency aggregation of S transform, the coherent accumulation of the coherent pulse signal is better realized; Then the high accuracy Doppler frequency change rate is obtained by using the minimum variance criterion in the phase domain. Theoretical derivation and simulation results show that the algorithm in this paper can improve the accuracy of about 1 times compared with the traditional algorithm. It lays a solid foundation for the estimation of the high-precision Doppler frequency change rate in the practical application of subsequent projects, and improves the accuracy of target positioning and tracking.
Keywords: Frequency estimation; Coherent pulse train; Coherent accumulation; Doppler frequency rate; S-transform.

1. Introduction

In interferometry, due to the use of phase information for delay estimation, it is particularly important to estimate the consistency of frequency and Doppler frequency change rate between the received signals of two stations. The difference of target motion relative to two receiving stations will lead to the different Doppler frequency change rate in signals received by two stations, so it is necessary to study a high precision Doppler frequency change rate estimation algorithm. In reality, the Doppler frequency change rate is generally small. For example, for a moving target 100km away from the observation platform, if the tangential velocity is 200m/s, its radial acceleration is 0.5m/s^2, and when the carrier frequency is 2GHz, the Doppler frequency change rate is only 3.3Hz/s. And the duration of a single pulse of a pulse system radar is extremely short, generally on the order of μs. Because the Doppler frequency change rate is small and the pulse duration is short, using a single pulse to achieve high-precision signal Doppler frequency change rate estimation is difficult. Consider that the new generation of radar mostly adopts the signal form of coherent carrier wave, where the coherent characteristic of this type of signal can be utilized. Hence, the intra-pulse accumulation is firstly carried out to accumulate N sampling points within a pulse into a single point, and then the pulse train is equivalent to a new pulse signal, thus prolongating the observation time of the signal. In this way, the duration of the signal is the duration of the whole pulse train, which also means tha the gap between pulses is used. Thus, the measurement accuracy of Doppler frequency change rate can be improved by extending the signal observation time[1].

The Doppler frequency change rate can be obtained indirectly by measuring the frequency change of the received signal or directly by measuring the phase of the received signal. In recent years, many scholars have studied high-precision frequency estimation or Doppler frequency change rate estimation based on coherent radar pulse train[2]. Calculating the phase difference between each pulse and the reference pulse ensures a longer time accumulation, and effectively improves the frequency estimation accuracy of the pulse signal. But when the accuracy of the rough measurement frequency is less than the ambiguity period, the phase defuzzification cannot be completed, so that the frequency cannot be accurately estimated[3]. A frequency estimation algorithm based on two-dimensional DFT variation was proposed in [4]. It realizes high-precision frequency estimation without periodic fuzzy through two steps of rough estimation and fine estimation. However, when the fast-time frequency rough estimation error exceeds half the pulse repetition frequency(PRF), the defuzzification will be wrong, resulting in large frequency estimation error. A new method of accurately measuring carrier frequency using FFT and data extraction technology was proposed in [5], which can greatly reduce the amount of calculation, but the carrier frequency estimation error is large when the real value of frequency deviates from discrete frequency points more. In conclusion, most of the existing methods have strict requirements on the accuracy of frequency rough estimation or large amount of calculation
S transform is a time-frequency analysis method that can adaptively adjust the analysis time width according to the change of frequency and provide intuitive time-frequency characteristics, and does not need to choose the window function domain analysis scale. Hence, S-transform has been widely used in high resolution spectrum estimation in recent years[6,7]. In this paper, S-transform is applied to the estimation of the Doppler frequency rate-of-change of the coherent pulse signal, and the S-transform of the coherent pulse signal is derived in detail. Through the higher time-frequency aggregation of S-transform, the coherent pulse signal can be accumulated better. Then the high accuracy Doppler frequency change rate is obtained by using the minimum variance criterion in the phase domain. The method presented in this paper has the advantages of stable performance, good accumulation effect and low requirement on the accuracy of coarse frequency estimation. Theoretical derivation and simulation results show that the algorithm in this paper can improve the accuracy of about 1 times compared with the traditional algorithm. It lays a solid foundation for the estimation of the high-precision Doppler frequency change rate in the practical application of subsequent projects, and improves the accuracy of target positioning and tracking.

2. Signal Model and S transform Basic Theory

Assuming that the moving target transmits a signal of constant amplitude and carrier frequency of f_0, the expression of the PTH pulse after mixing the received phase-pulse impulse signal with frequency conversion is [8]:

$$s_p(t) = A_p \exp\left[j\left(2\pi f_I t + \pi\alpha t^2 + \varphi_0\right)\right] + w(t) \quad t \in \left[T_p, T_p + \tau_p\right] \tag{1}$$

where A_p is the PTH pulse's amplitude , f_I is the signal frequency after down-conversion, α is the Doppler frequency change rate of the signal, φ_0 is the signal's initial phase , T_p is the PTH pulse's initial moment , τ_p is the pulse width, and $w(t)$ is complex Gaussian white ,its means noise is zero and variance is σ^2.

Let $t_p = t - T_p$, then (1) can be rewritten as:

$$\begin{aligned} s_p(t_p) &= A_p \exp\left[j\left(2\pi f_I \left(t_p + T_p\right) + \pi\alpha \left(t_p + T_p\right)^2 + \varphi_0\right)\right] + w(t_p) \\ &= A_p \exp\left[j\left(2\pi f_I t_p + 2\pi f_I T_p + \pi\alpha t_p^{\,2} + \pi\alpha T_p^{\,2} + 2\pi\alpha t_p T_p + \varphi_0\right)\right] + w(t_p) \quad t_p \in \left[0, \tau_p\right] \\ &= A_p \exp\left[j\left(\omega_p t_p + \pi\alpha t_p^{\,2} + \varphi_p\right)\right] + w(t_p) \end{aligned} \tag{2}$$

where ω_p , φ_p and are the initial frequency and initial phase of the Pth pulse respectively, and their expressions are as follows:

$$\begin{cases} \omega_p = 2\pi\left(f_I + \alpha T_p\right) \\ \varphi_p = \varphi_0 + 2\pi f_I T_p + \pi\alpha T_p^{\ 2} \end{cases} \tag{3}$$

For any function or signal $s(t)$, its S-transform is defined as[9]:

$$S(\tau, f) = \int_{-\infty}^{+\infty} s(t) w(\tau - t, f) \exp(-j2\pi ft) \mathrm{d}t \tag{4}$$

where

$$w(t, f) = \frac{|f|}{\sqrt{2\pi}} \exp\left(-\frac{t^2 f^2}{2}\right) \tag{5}$$

where $w(t, f)$ is the Gaussian window function, $S(\tau, f)$ representing the result of S-transformation, τ and t is time, controlling the position of the Gaussian window function on the time axis, $s(t)$ is the signal to be analyzed, f and is the frequency.

In (4), when $f = 0$, $S(t,0) = 0$ was established. Therefore, $S(t,0) = 0$ will be defined as follows[10] :

$$S(t,0) = \lim_{T\to\infty} \frac{1}{T} \int_{-\frac{T}{2}}^{\frac{T}{2}} x(t) \mathrm{d}t \tag{6}$$

Where T is a cycle time of signal $x(t)$.

3. Estimation Method of Doppler Frequency Rate-of-change Based on S- transform

For the interconnected pulse impulse series signal $s_p(t_p)$ defined in equation (2), substitute it into Equation (4) to obtain the S-transformation of the PTH pulse:

$$S_p(\tau, f) = \int_0^{\tau_p} \frac{A_p|f|}{\sqrt{2\pi}} \exp\left(j\left(\omega_p t + \pi\alpha t^2 + \varphi_p\right) - \frac{(\tau - t)^2 f^2}{2} - j\omega t \right) \mathrm{d}t + \varepsilon(\tau, f) \tag{7}$$

where: $\varepsilon(\tau, f)$ is the complex Gaussian white noise with zero means and variance σ^2. α is extremely small, and the α product term in Equation (6) is the square of time, which makes $\pi\alpha t^2$ extremely small relative to the phase value of the signal and has little influence on the phase change. Therefore, (6) can be simplified as:

$$S_p(\tau, f) = \int_0^{\tau_p} \frac{A_p|f|}{\sqrt{2\pi}} \exp\left(j\left(\omega_p t + \varphi_p\right) - \frac{(\tau - t)^2 f^2}{2} - j\omega t \right) \mathrm{d}t + \varepsilon(\tau, f) \tag{8}$$

We rewrite (7) as:

$$S_p(\tau,f)=\frac{A_p}{\sqrt{2\pi}}\exp\begin{bmatrix}-\frac{1}{2f^2}(\omega_p-\omega)^2+\\ j(\varphi_p+\tau\omega-\tau\omega_p)\end{bmatrix}\int_0^{\tau_p}\exp\left\{-\frac{1}{2}\begin{bmatrix}(\tau-t)f+\\ j\frac{1}{f}(\omega_p-\omega)\end{bmatrix}^2\right\}dt+\varepsilon(\tau,f) \quad (9)$$

Take the time shift factor $\tau=\tau_p/2$, and according to the properties of Γ function $\Gamma(0.5)=\int_{-\infty}^{+\infty}\exp(-x^2)dx=\sqrt{\pi}$, equation (8) can be simplified as:

$$S_p(\tau,f)=A_p\exp\left[-\frac{1}{2f^2}(\omega_p-\omega)^2+j\left(\varphi_p+\frac{\tau_p}{2}\omega-\frac{\tau_p}{2}\omega_p\right)\right]+\varepsilon(\tau,f) \quad (10)$$

In formula (9), when $\omega=\omega_p$, the modulus $|S_p(\tau,f)|$ of S-transformation has a maximum value and is independent of the time-shift factor τ. In Equation (1), since the Doppler frequency rate-of-change α is greatly small, item $\pi\alpha t^2$ in (1) can be ignored. Therefore, the frequency estimation algorithm of sinusoidal signal can be used to estimate the frequency of P pulses, and then the arithmetic average of the estimated values of P frequencies can be ω_p [11]. When the SNR is higher than 10 dB[12], additive noise $\varepsilon(\tau,f)$ can be approximately equivalent to phase noise $u(\tau)$. Therefore, when the SNR is large enough, make $\omega=\omega_p$, the phase of $S_p(\tau,f)$ can be approximately expressed as:

$$\arg[S_p(\tau,f)]=\varphi_p+\frac{\tau_p}{2}\omega-\frac{\tau_p}{2}\omega_p+u(t)=\varphi_0+2\pi f_I T_p+\pi\alpha T_p^{\,2}+u(T_p) \quad (11)$$

Due to the limitation of the value range of the inverse trig function, the phase is folded between $(-\pi,\pi]$, which means that the phase is wound. The winding phase will cause a large deviation in the estimation of α in the phase domain. Therefore, the differential method is used for phase unwinding:

$$Q(p)=S_p\left(S_{p+1}^{\,2}\right)^* S_{p+2}\exp\left(-j2\pi f_p \mathrm{ET}_p\right) \quad (12)$$

where: $\mathrm{ET}_p=(T_p+\tau_p/2)+(T_{p+2}+\tau_p/2)-2(T_{p+1}+\tau_p/2)$ is approximately the repeated jitter between pulses when the pulse repetition frequency is constant. So the phase of $Q(p)$ can be written as:

$$\varphi_Q(p)=2\pi(f_l-f_I)\mathrm{ET}_p+\pi\alpha\begin{bmatrix}(T_p+\tau_p/2)^2+\\ (T_{p+2}+\tau_p/2)^2-\\ 2(T_{p+1}+\tau_p/2)^2\end{bmatrix}+\mathrm{MA}(p) \quad (13)$$

where: $\mathrm{MA}(p)=u(p)-2u(p+1)+u(p+2)$ is a MA (moving average) process. Pulse repetition frequency in general radar is relatively stable, so the value of equation (12) can generally be guaranteed to fall in $(-\pi, \pi]$.

Equation (12) can be rewritten in matrix form:

$$\mathbf{Z}=\mathbf{H}\boldsymbol{\theta}+\mathbf{MA} \quad (14)$$

where

$$\mathbf{Z} = [\varphi_Q(1), \varphi_Q(2), \cdots, \varphi_Q(P-2)]^T \tag{15}$$

$$\boldsymbol{\theta} = \left[\left(f_I - f_I \right), \alpha \right]^T \tag{16}$$

where H is the $\left[(p-2) \times 2 \right]$-dimensional time dimension matrix. The least square method can be used to obtain the estimated value of $\boldsymbol{\theta}$:

$$\hat{\boldsymbol{\theta}} = \left(\mathbf{H}^{\mathrm{T}} \mathbf{R}_{\mathbf{u}} \mathbf{H} \right)^{-1} \mathbf{H}^{\mathrm{T}} \mathbf{R}_{\mathbf{u}} \mathbf{Z} \tag{17}$$

where: $R_u = Cov(MA) = \sigma_u^2 R$, R is equivalent to the covariance matrix of the moving average noise formed by gaussian white noise with zero means and variance 1. The composition coefficients of MA noise are 1, -2 and 1:

$$\mathbf{R} = \begin{bmatrix} 6 & -4 & 1 & 0 & \cdots & 0 \\ -4 & 6 & -4 & 1 & \ddots & \vdots \\ 1 & -4 & 6 & \ddots & \ddots & 0 \\ 0 & 1 & \ddots & \ddots & -4 & 1 \\ \vdots & \ddots & \ddots & -4 & 6 & -4 \\ 0 & \cdots & 0 & 1 & -4 & 6 \end{bmatrix}_{(p-2)\times(p-2)} \tag{18}$$

In this case, equation (14) can be rewritten as:

$$\hat{\boldsymbol{\theta}} = \left(\mathbf{H}^{\mathrm{T}} \mathbf{R} \mathbf{H} \right)^{-1} \mathbf{H}^{\mathrm{T}} \mathbf{R} \mathbf{Z} \tag{19}$$

Thus an estimate of the rate of change of Doppler frequency change rate α can be obtained.

4. Algorithm Flow and CRLB

The specific algorithm steps are shown in Table 1.

Table 1. Algorithm steps

Step 1 Calculate the S-transform of each pulse of the collected signal according to Equation (6).
Step 2 Take the time shift factor $\tau = \tau_p / 2$, the frequency $\omega = \omega_p$, so that the modulus $\left| S_p(\tau, f) \right|$ of S-transformation has a maximum value.
Step 3 Multiply the complex numbers of $S_p(\tau, f)$ and extract the phase spectrum.
Step 4 The phase spectrum is expressed as a matrix, and the parameters are estimated according to the minimum variance criterion.

The CRLB of parameter estimation of $A\exp\left[j\left(a_0 + a_1 t + a_2 t^2 + \cdots + a_N t^N \right) \right]$ was derived in reference [13]. Referring to the formula in the literature, the CRLB expression of α can be obtained as follows:

$$\mathrm{var}(\alpha) = \frac{\mathrm{var}(\hat{a}_2)}{\pi^2} \geq \frac{90\sigma^2}{\pi^2 A^2 N^5 T_s^4} = \frac{90}{\pi^2 \mathrm{SNR}_{in} N T^4} \tag{20}$$

where: SNR_{in} is the input SNR, $\mathrm{SNR}_{in} = A^2 / \sigma^2$, N is the total number of sample points of pulse string signals, T_s is the sampling interval, and $T = NT_s$ is the time length of pulse string signals.

5. The Simulation Analysis

The performance of the proposed algorithm is verified by experiments on different factors that affect Doppler frequency change rate estimation. The Root Mean Square Error (RMSE) of Doppler frequency change rate estimation can be expressed as[14] :

$$\mathrm{RMSE} = \sqrt{\frac{1}{\mathrm{MC}} \sum_{mc=1}^{\mathrm{MC}} (\hat{\alpha} - \alpha)^2} \tag{21}$$

where: MC is the number of Monte Carlo experiments.

Simulation experiment 1 assumes that the center frequency $f_I = 50MHz$, sampling frequency $f_s = 200MHz$, pulse repetition frequency $f_{PRF} = 5KHz$, single pulse pulse width $5\mu s$, the true value of Doppler frequency change rate is 50Hz/s, and the number of Monte Carlo experiments is 100 under the condition of signal-to-noise(SNR) ratio of 20dB. The statistical simulation results are shown in table 2 and figure 1.

Simulation experiment 2 has the same parameters as simulation experiment 1 except signal duration. Under the condition of signal duration of 60ms, the number of Monte Carlo experiments is 100. The statistical simulation results are shown in table 3 and figure 2.

Table 2. Experimental results

Time(ms)	20	30	40	50	60	70	80	90	100
DFT(Hz/s)	3.791	2.212	1.485	0.712	0.492	0.285	0.146	0.082	0.054
S-transform(Hz/s)	2.271	1.223	0.903	0.354	0.232	0.141	0.081	0.024	0.025
CRLB(Hz/s)	0.243	0.132	0.072	0.043	0.021	0.01	0.007	0.004	0.002

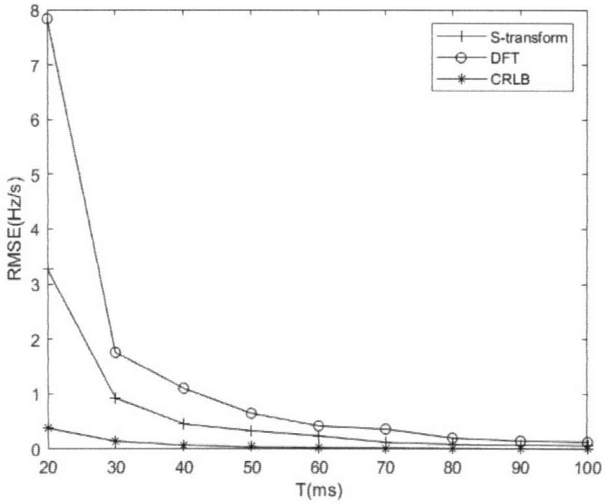

Figure 1. Change curve of root mean square error with observation time

Table 3. Experimental results

SNR(dB)	0	5	10	15	20	25	30	35	40
DFT(Hz/s)	3.799	2.219	1.483	0.713	0.499	0.286	0.146	0.082	0.053
S-transform(Hz/s)	2.271	1.224	0.904	0.356	0.235	0.149	0.084	0.024	0.016
CRLB(Hz/s)	0.242	0.136	0.076	0.043	0.024	0.013	0.007	0.004	0.002

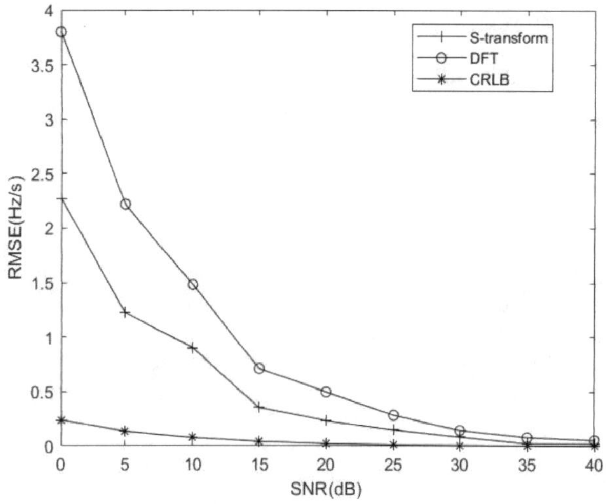

Figure 2. Change curve of root mean square error with SNR

As table 2 and figure 1 shown, under the condition of signal duration less than 30ms, estimation precision of proposed method(or S-Transform) has a big gap between the CRLB. It is known that the estimation accuracy of Doppler frequency rate-of-change is inversely proportional to the fourth power of time. When signal duration is shorter, the estimation precision is strongly influenced by noise, and estimation error is amplified by the fourth power. Therefore, the estimation accuracy of the Doppler frequency rate-of-change is low.

As seen from the table 3 and figure 2, the SNR estimation precision under the condition of less than 15 dB has larger difference with CRLB. The reason is the method accumulated in the pulse accumulation effect are sensitive to the signal-to-noise ratio. When SNR is low, the frequency estimation precision of a single pulse is low, which leads to the poor accumulation effect, and make the Doppler frequency rate-of-change estimation precision is low.

As can be seen from the above picture and table that the estimation accuracy of Doppler frequency rate-of-change is significantly affected by observation time and SNR. The estimation accuracy of this algorithm is better than that of DFT. Under the condition of the same estimation accuracy, the SNR required by the proposed algorithm is about 5-10dB lower than that of the existing algorithms, and in the middle and high SNR region (SNR > 15dB) and longer signal duration (T < 30ms) The estimation accuracy of the proposed algorithm is within 1Hz/s, and it can well approximate CRLB.

6. Conclusion

In this paper, the method of Doppler frequency rate-of-change of phase-connected pulse signal based on S-transform is proposed, and the s-transform form of phase-connected pulse signal is derived in detail. The energy accumulation between signal pulses is realized by S-transform, and the estimation of Doppler frequency rate-of-change is realized by the least square method according to the observation value of phase. The proposed algorithm has higher parameter estimation accuracy than the DFT algorithm. Under the same estimation accuracy, the SNR required by the proposed algorithm is about 5~10dB lower than that of the existing algorithms. In the middle and high SNR region (SNR > 15dB) and longer signal duration (T > 30ms) The estimation accuracy of the proposed algorithm is within 1Hz/s, and it can well approximate CRLB. It lays a solid foundation for the estimation of the high-precision Doppler frequency change rate in the practical application of subsequent projects, and improves the accuracy of target positioning and tracking.

References

[1] YU Chunlai. Research on key technologies of single-station passive positioning and tracking using spatial frequency domain information[D]. National University of Defense Technology, 2008.(in Chinese)

[2] ZHANG Gangbing, LIU Yu, DENG Zhenmiao. Frequency Estimation for Staggered Coherent Pulse Train[J]. Journal of Data Acquisition & Processing, 2001,26(2): 135-139. (in Chinese)

[3] XIONG Bo, FENG Bing, ZHANG Wei. High Precision Frequency Estimation of Coherent Pulse Signal[J]. Electronic Information Warfare Technology, 2019, 34(03): 19-23. (in Chinese)

[4] YI Zhouwei, LI qiqin.Frequency Estimation of the Coherent Pulses Train Based on 2-Dimension DFT[J]. Electronic Information Warfare Technology, 2020, 35(04): 34-38. (in Chinese)

[5] ZHANG Wei, LV Jingqing, DUAN Junqi, A Down-Sampling Based Method for Estimation of Carrier Frequency of a Coherent Pulse Train[J]. Measurement & Control Technology, 2010, 29(11): 30-32. (in Chinese)

[6] LI F, YANG C, XIAY, et al. An Adaptive S-Method to Analyze Micro-Doppler Signals for Human Activity Classification[J]. Sensors, 2017, 17(12): 2769-2772.

[7] BABU P R, DASH P K, SWAIN S K. et al. A new fast discrete S-transform and decision tree for the classification and monitoring of power quality disturbance waveforms[J]. International Transactions on Electrical Energy Systems, 2014, 24(9): 1279-1300.

[8] Chen Xi, Deng Zhenmiao, Fu Maozhong. Doppler Rate Estimation of Coherent LFM Pulse Train[J]. 2014, 36(03): 29-33. (in Chinese)

[9] Stockwell R G. "S-transform analysis of gravity wave activity from a small scale network of airglow imagers"[J]. Thesis, 1999:4662.

[10] Feng Yongxin. Multi-component LFM signal parameter estimation based on S transform[D]. Changchun University of Science and Technology, 2020. (in Chinese)

[11] Zhang Gangbing. Research on Key Technologies of Single Station Passive Positioning and Tracking[D]. Nanjing University of Aeronautics and Astronautics, 2010. (in Chinese)

[12] Tretter S A. Estimating the Frequency of a Noisy Sinusoid by Linear Regression [J]. IEEE Transactions on Informations on Theory, 1985, 32(6): 832-835.

[13] Ristic B, Boashash B. Comments on "the cramer-rao lower bound for signals with constant amplitude and polynomial phase"[J]. IEEE Tarns. On Signal Proc, 1998, (6): 1708-1709.

[14] Zhao Peiyan, Ouyang Xinxin, Peng Huafeng. A Phase Delay Estimation Algorithm of Frequency Hopping Signal Based on Chinese Reminder Theore[J]. Journal of Electronics & Information Technology, 2018, 40 (03): 656-662. (in Chinese)

The Echo Model and Simulation Method of Impulse Radar Fuze Based on Attribute Scattering Center Model

Ming Kong[1,2], Yuzhao Li[2,3], Jichi Chen[1], Xu Wang[1,*] and Guohua Wei[1]

[1]School of Information and Electronics, Beijing Institute of Technology, Beijing 100081, China
[2]Science and Technology on Electromechanical Dynamic Control Laboratory, China
[3]Beijing Institute of Remote Sensing Equipment, Beijing 100854, China

*Corresponding author email: wangxu@bit.edu.cn

Abstract. This article focuses on solving the problem of performance evaluation and algorithm verification of ultra-bandwidth impulse radar fuze. A modeling and simulation method for the scattering echo of an impulse radar target is proposed. We distinguish the scattering center of the extended target and construct its echo based on the attribute scattering center model. Besides, the position variation characteristics of the scattering center, and the azimuth dependence of the scattering amplitude are fully considered. At the same time, the motion information of the target is modulated to obtain the target scattering echo of the impulse radar. We generate the scattering echo of the typical streamlined moving target at different observation angles by calculation. The changes of the echo position and amplitude of each scattering center are consistent with the scattering echo model. We use the generated echo data to image, and the imaging results are basically consistent with the shape of the target, which verifies the effectiveness of the proposed method.
Keywords: Signal and information processing; Ultra-bandwidth impulse radar fuze; Moving target modeling; Extended target modeling; Simulation echo generation; Attributed scattering center model.

1. Introduction

To accurately and effectively perform performance evaluation and system verification of the radar system, the development process of the radio fuze requires a large number of simulation tests[1]. The target echo signal simulators have been improved to conduct the radio fuze simulations of hardware-in-the-loop(HIL)[2]-[3]. However, the estimated velocity Doppler information and the power information are obtained in the condition that the target is regarded as a simple point target to generate the echo signal of the target simulation. In the simulation of the above literature, the shape feature of the detected target is ignored and regarded as a point target, which is no longer applicable to existing high resolution radar. Ultra-bandwidth radar has advantages in high precision, strong anti-jamming, strong recognition, anti-stealth with low power consumption, which can be used for all kinds of high-precision strike ammunition[4] and has been applied to radio fuze[5]. However, the echo signal generated from a simple point target can neither reflect the characteristics of the target motion in the detection process nor reflect the characteristics of time-varying and space-varying in the echo signal. The method of simply treating the target as an extended target composed of multiple scattering points cannot reflect the actual scattering characteristics[6][7]. Therefore, the echo generation methods mentioned above cannot meet the test requirements of modern ultra-bandwidth impulse radar fuze.

The parameters of the attributed scattering center model can well describe the shape, attitude, and position information of the scattering center of the target, which has been widely studied and implemented[8]. The scattering echo of the moving target can be quickly obtained by distinguishing the scattering centers and calculating their echo respectively. At the same time, the motion information of the target is modulated based on the attribute scattering center model. The scattering echo contains refined structure information, which reflects the features of target in scattering, time-varying and space-varying that can be used to evaluate the signal processing algorithm of ultra-wideband impulse fuze.

In this paper, we introduce the attributed scattering center model with its classification and propose a novel method for echo generation of moving target in ultra-bandwidth impulse fuze. Furthermore, we verify the effectiveness of the proposed method by simulation calculation and echo imaging.

2. Attribute Scattering Center Model

According to the azimuth dependence of scattering center position and echo amplitude, the scattering centers of complex targets can be divided into three categories: Local Scattering Center (LSC), Distributed Scattering Center (DSC), and Sliding Scattering Center (SSC).

2.1. LSC

LSC is usually generated by tip diffraction, corner diffraction, edge diffraction of a tip, smooth bulge, or smooth top structure with small relative incident wavelength, and scattering of the discontinuity target surface, where the position is relatively fixed. The expression of the LSC scattering model is shown in equation (1):

$$E(f,\xi)=A(f,\xi)\cdot\exp\left(-j\frac{4\pi f}{c}\mathbf{r}_i\cdot\hat{\mathbf{r}}_{\mathbf{los}}\right) \tag{1}$$

f represents the incident wave frequency, $\xi(\theta,\varphi)$ represents the spatial angle of radar line-of-sight, θ is the pitch angle, φ is the azimuth angle; $\mathbf{r}_i$ represents the position vector of the scattering center in the target-centered coordinate system, and $\hat{\mathbf{r}}_{\mathbf{los}}$ is the unit vector of the radar view direction; $A(\cdot)$ represents the amplitude term, which is related to the incident wave frequency f and radar line-of-sight angle $\xi(\theta,\varphi)$. It can be seen from equation (1) that in the frequency-domain, the magnitude of the scattering field is affected by frequency and radar line-of-sight angle, while the magnitude of the phase is affected by the frequency and position vector.

For ultra-bandwidth impulse radar, Doppler and phase information cannot be obtained[9], it is often processed in time domain. Since the azimuth dependence of LSC is relatively small, its azimuth dependence is taken as $A_{azm}=1$. Moreover, the physical meaning of this item is the change of the echo amplitude caused by the electromagnetic scattering mode, so the loss of this term can be directly equivalent to an efficient term for calculating the echo amplitude, so the amplitude term $A(\cdot)$ can be estimated according to the target scattering loss formula[10] as shown in equation (2):

$$A_\sigma=\Gamma\sqrt{\sigma} \tag{2}$$

$\Gamma_{n,n-1}=(\eta_n-\eta_{n-1})/(\eta_n+\eta_{n-1}),\eta_n=\sqrt{\mu/\varepsilon_n}$ is the intrinsic impedance of the n-th layer medium of the magnetic permeability and permittivity. σ is the radar cross-section of the structure corresponding to the scattering center.

2.2. DSC

DSC is usually generated by specular reflection or edge diffraction of structures such as planes, cylinders, cones, and straight edges. Its positions are distributed on the surface or edge; the scattering amplitude is greatly affected by observation angle, and it can be observed only at some angles. sinc function is often used to express the azimuth dependence of scattering amplitude. The expression of the DSC scattering model is shown in equation (3).

$$E(f,\xi) = A_\sigma \cdot A_{azm}(\xi) \cdot \exp(-j \cdot 4\pi f \cdot \mathbf{r}_i \cdot \hat{\mathbf{r}}_{\mathbf{los}} / c) \tag{3}$$

$A_{azm}(\xi)=\text{sinc}\left[2\pi f_0 L \sin(\xi - \xi_i)/c\right]$, L is the length parameter of DSC, f_0 is the center frequency of signal, $\xi_i(\theta_i, \varphi_i)$ is the effective observation angle, and the calculation method of the amplitude term A_σ is the same as the method shown in equation (2).

2.3. SSC

The position of SSC slides on the surface or edge, but its scattering structure is mainly quadric surface, curved edge, etc. Since the surface normal vector changes with the incident direction, the position vector of the scattering center will also slide with the incident direction. For a quadric surface with constant curvature, $A_{azm}=1$, while the scattering amplitude of a quadric surface with varying curvature is affected by the curvature, as shown in equation (4).

$$A_{azm}(\xi) = \sqrt{r_1(\xi) r_2(\xi)} \tag{4}$$

$r_1(\xi)$、 $r_2(\xi)$ is the radius of curvature in the orthogonal direction.
The SSC scattering model is shown in equation (5).

$$E(f,\xi) = A_\sigma \cdot A_{azm}(\xi) \cdot \exp\left[-j\frac{4\pi f}{c}\mathbf{r}_\mathbf{i}(x_i, y_i, z_i) \cdot \hat{\mathbf{r}}_{\mathbf{los}}\right] \tag{5}$$

The calculation method of A_σ is as shown in equation (2), it needs to select an appropriate calculation method according to the target structure to substitute $A_{azm}(\xi)$, $\mathbf{r}_\mathbf{i}(x_i, y_i, z_i)$ is the position vector of the SSC, which is a function of the radar viewing direction, which can be obtained by the system of equations set shown in equation (6).

$$\begin{cases} F(x_i, y_i, z_i) = 0 \\ \hat{\mathbf{n}}(x_i, y_i, z_i) \times \hat{\mathbf{r}}_{\mathbf{los}} = 0 \end{cases} \tag{6}$$

$F(\cdot)$ is the quadric surface equation, and $\hat{\mathbf{n}}(x_i, y_i, z_i)$ is the normal vector of a point on the surface. Only when the radar line-of-sight direction coincides with the normal vector of the curved surface, the position of the point be the SSC position, which can be substituted into equation (5).

3. Echo Simulation Generation Method

Ultra-bandwidth impulse radar transmits Nano-Second Grade Gaussian pulses or its first-order or high-order differential, and the signal bandwidth can reach several GHz. The expression of the Gaussian pulse signal is shown in (7).

$$s(t) = A\exp\left\{-4\pi\left[(t - \Delta T)/\Delta T\right]^2\right\} \tag{7}$$

A is the peak amplitude of the transmitted signal, t is the time variable, and ΔT is the effective pulse width. The autocorrelation function and Fourier transform of Gaussian signals are both in Gaussian form[11]. However, it is not easy to implement engineering because Gaussian pulse has a DC component, so in practical applications, the first-order Gaussian differential signal is often used as the transmitted signal, so the form of the transmitted signal in this article is

$$s(t) = AA_0\frac{t - \Delta T}{\Delta T}\exp\left\{-4\pi\left[(t - \Delta T)/\Delta T\right]^2\right\} \tag{8}$$

A is the signal amplitude, $A_0 = \exp(-1/16\pi)$, t is the time variable, and ΔT is the effective pulse width. The signal shape is shown in Figure 1(a).

The shape of the scattered echo signal of a complex target is the differential form of the transmitted signal or the Reversed-phase transmitted signal, and different types of scattering centers have different differential forms[12]. However, each differential form of Gaussian signal has multiple peaks and negative peaks[13], which interferes with echo signal simulation and processing. NGUYEN L proposed a method to obtain the envelope of echo signal by Hilbert transform, which can solve the multi-peak problem[14]. In order to facilitate the distinction between the types of scattering centers, taking LSC as the second-order differential of the transmitted signal, DSC is the first-order differential of the transmitted signal, and SSC is the first-order differential of the inverted transmission signal. The signal shape of each scattering center is shown in Figure 1(b)-(d).

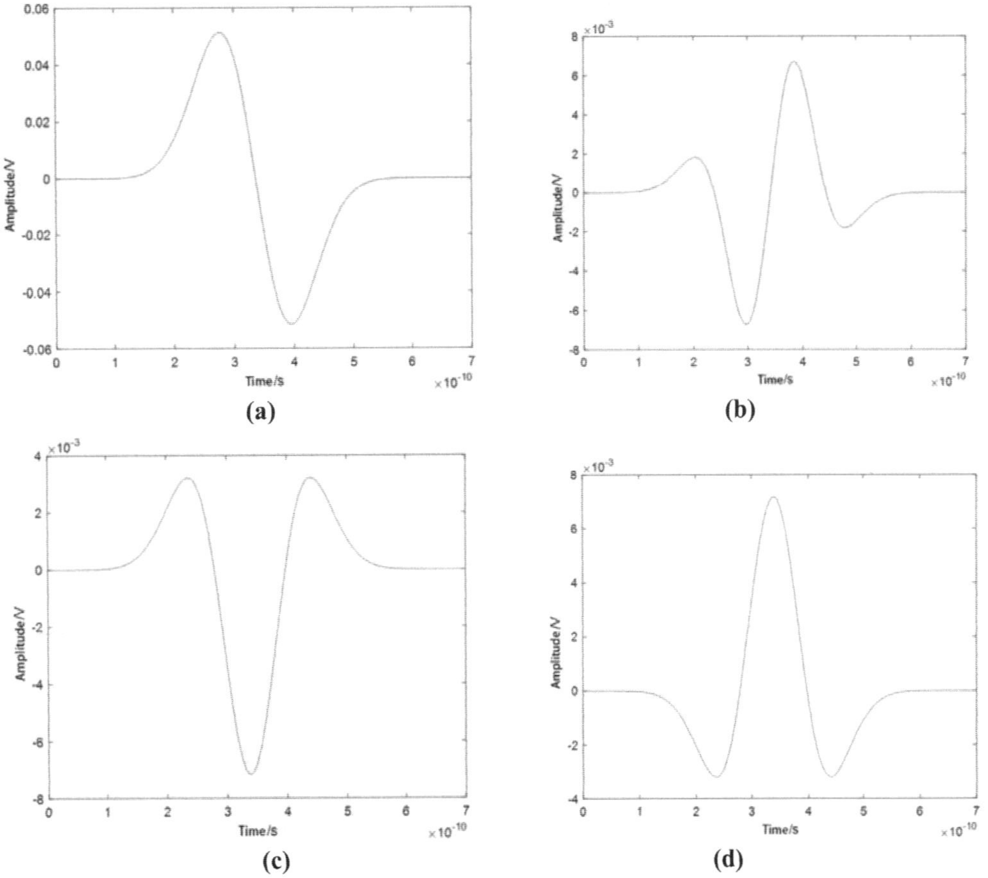

Figure 1. Signal waveform diagrams. **(a)** Transmitting signal waveform diagram. **(b)** The echo diagram of LSC. **(c)** The echo diagram of DSC. **(d)** The echo diagram of SSC.

Due to its high-resolution characteristics, it is necessary to modulate the delay information of each scattering center in the scattered echo. Therefore, the echo simulation generation method is summarized in Figure 2, and the received echo expression is shown in equation (9).

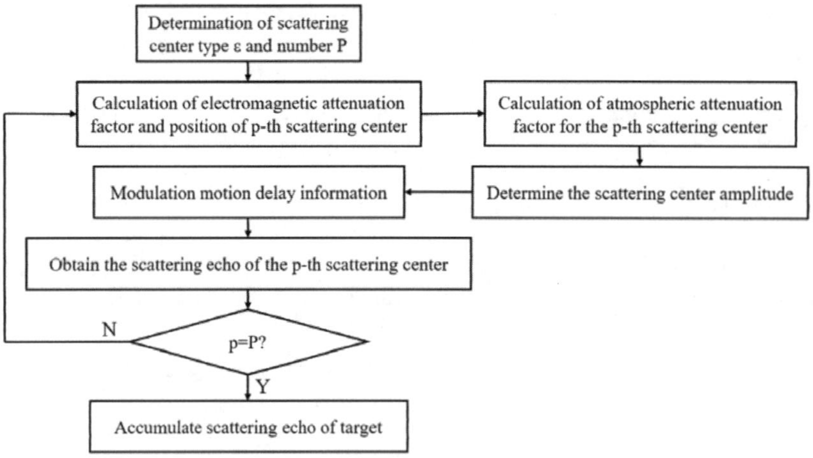

Figure 2. Diagram of echo simulation method.

$$Y(t,\xi,r)=\sum_{p=1}^{P}T_p(r)G_p(\xi,\varepsilon)s^{(\delta)}(t-\tau_p) \tag{9}$$

Firstly, determine the type of each scattering center according to the shape of the target and its scattering characteristics; then, calculate the electromagnetic attenuation factor $G_p(\xi,\varepsilon)$ and position of the p-th scattering center according to the method in Section 2. It should be noted that since the transmitted signal is a carrier-free signal, there is no phase information, and only the scattering amplitude determined by the type of scattering center and the radar viewing angle is calculated as the electromagnetic attenuation factor. The calculation of $G_p(\xi,\varepsilon)$ is shown in equation (10).

$$G_p(\xi,\varepsilon)=A_\sigma\cdot A_{azm}(\xi,\varepsilon) \tag{10}$$

ε represents different types of scattering centers.
The calculation of the atmospheric attenuation factor $T_p(r)$ of the p-th scattering center is shown in equation (11).

$$T_p(r)=(G_tA_{er})^{1/2}(4\pi r^2)^{-1} \tag{11}$$

G_t is the gain of the transmit antenna, A_{er} is the effective aperture of the receive antenna, and r is the distance from the radio fuze to the scattering center.
Since the target's speed is much lower than the speed of the incident wave, the "Stop and Go" model can be used. When calculating the time delay information at each sampling time, the target is regarded as the position stationary at this time, and the time delay of each scattering center position is calculated. Then the time delay of the p-th scattering center is:

$$\tau_p=2r/c \tag{12}$$

r is the distance from the radio fuze to the scattering center, and c is the velocity of light.
Calculate the electromagnetic attenuation factor, position, atmospheric attenuation factor, and time delay of each scattering center to get the echo of each scattering center, and accumulate all the scattering center echoes to get the target scattering echo at the current sampling time shown in formula (9).

4. Simulation Examples

4.1. Simulation Parameters Design

The cross-sections, distributions, and scattering center serial numbers of the typical streamline target model are shown in Fig. 3. This target is the target of the rotating body, and the upper half of the side bus is divided into semi-elliptical curves with $h_1 = 2\text{m}$ in height and a radius of $R_1 = 0.5\text{m}$. The middle part is a straight line with $h_2 = 2\text{m}$ in height. The height of bottom part is $h_3 = 0.4\text{m}$, and a radius of $R_3 = 0.65\text{m}$. The upper half of the target will form SSCs because of its quadratic surface structure, and its position calculation method is shown in formula (6). The top of the upper part will form LSCs due to the tip diffraction. The middle part is a cylinder surface that can form DSCs distributing on the cylinder surface. The side of the bottom part is conical that can form DSCs on the conical surface. The scattering at the discontinuity between the bottom and the middle part forms LSC, and two LSC are formed at the bottom edge.

The target moves horizontally on xoz planes along its Z-axis, and the relative trajectory scene is shown in Figure 4 with a speed of 500 m / s. The fuze is located on xoz surface of the target coordinate system, and the starting coordinate is $\left(25,0,25\sqrt{3}\right)\text{m}$. The parameters of the transmitted signal are set as peak amplitude $A=1$, effective pulse width $\Delta T = 1/3\,\text{ns}$, center frequency is $f_0 = 3\text{GHz}$, incident azimuth $\varphi = 0^\circ$. As the target moves, the incident pitch angle ranges from 30° 150°, and the distance from the fuze to the starting position of the target center is $(0,0,0)\text{m}$ at 50m, and the pulse sampling interval is $\text{PRI} = 0.1\text{ms}$.

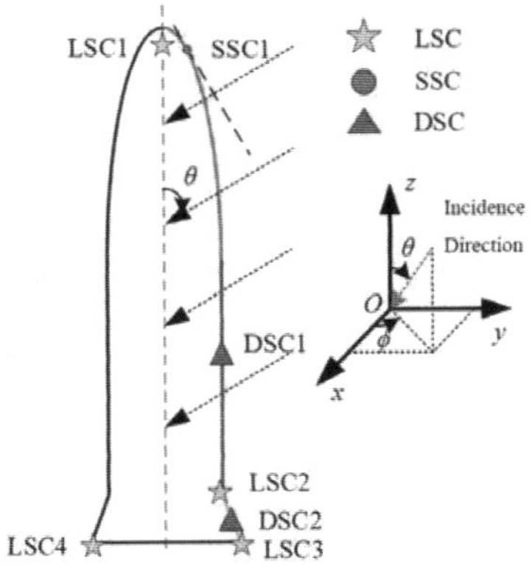

Figure 3. The streamlined target model and its scattering centers distribution.

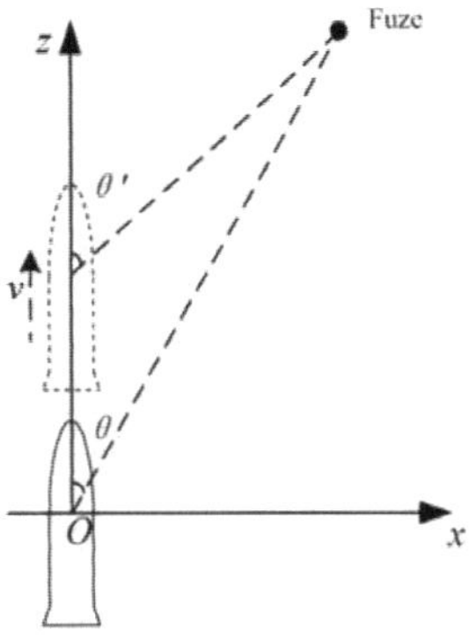

Figure 4. Scene graph of missile relative motion.

4.2. Results and Analysis

In the target motion process, choose the following pitch angles $\theta = 30^\circ、69^\circ、90^\circ、100^\circ、120^\circ$ to generate the simulation echo of the target. The scattering echo results are shown in Fig5. The corresponding relationship between the scattering echo and each scattering center are marked in the figure. The position of each scattering center in the target coordinate system is shown in Table 1. It should be noted that the values of $G_t A_{er}$ and $\Gamma_{n,n-1}$ should be 1 in the calculation process. Since the above parameters are the multiplier items in the calculation of each scattering center, their values do not affect the relative amplitude of each scattering center. In the practical application of this model for echo generation, the above parameters can be calculated according to the actual situation of the hardware system and the environment, which can be directly multiplied by the scattering echo calculated when the value is 1.

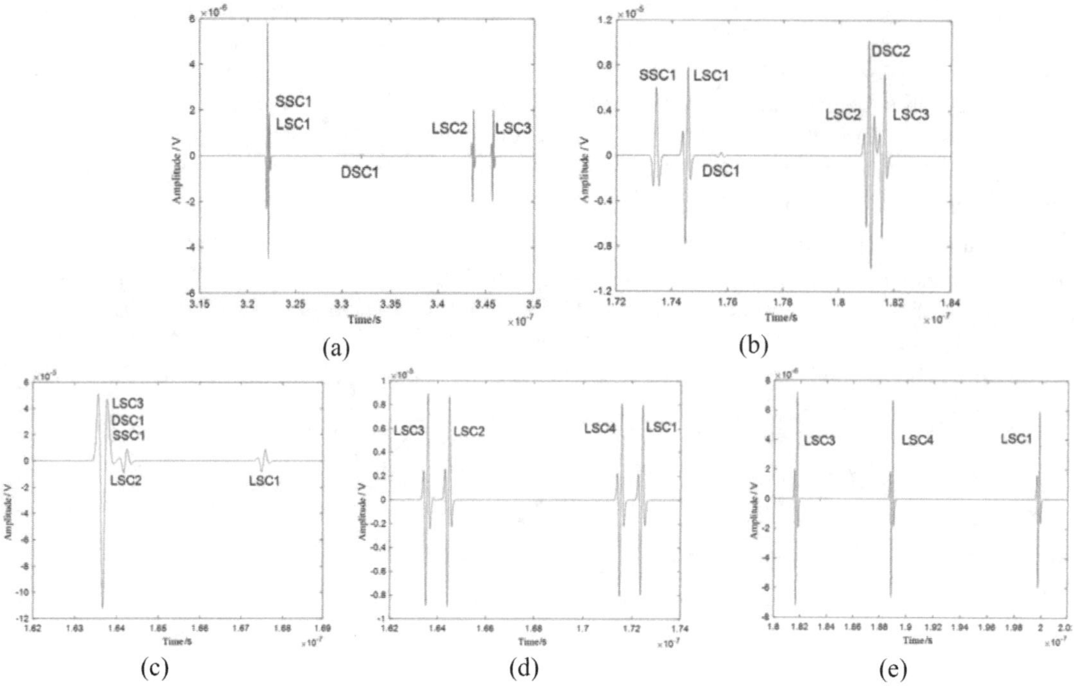

Figure 5. Scattering echo results at different observation angles. **(a)** Scattering echo when $\theta = 30°$. **(b)** Scattering echo when $\theta = 69°$. **(c)** Scattering echo when $\theta = 90°$. **(d)** Scattering echo when $\theta = 100°$. **(e)** Scattering echo when $\theta = 120°$.

Table 1. Position of scattering centers at different observation angles (moments) in target coordinate system.

Scattering Center	Position Coordinate/m				
	$\theta = 30°$	$\theta = 69°$	$\theta = 90°$	$\theta = 100°$	$\theta = 120°$
SSC1	(0.07,0,1.98)	(0.27,0,1.68)	(0.5,0,0)	--	--
DSC1	(0.5,0,-1)	(0.5,0,-1)	(0.5,0,-1)	--	--
DSC2	--	(0.58,0,-2.2)	--	--	--
LSC1	(0,0,2)	(0,0,2)	(0,0,2)	(0,0,2)	(0,0,2)
LSC2	(0.5,0,-2)	(0.5,0,-2)	(0.5,0,-2)	(0.5,0,-2)	--
LSC3	(0.65,0,-2.4)	(0.65,0,-2.4)	(0.65,0,-2.4)	(0.65,0,-2.4)	(0.65,0,-2.4)
LSC4	--	--	--	(-0.65,0,-2.4)	(-0.65,0,-2.4)

The results above, showing that with the movement of the target, the distance between the scattering center and the fuze change constantly as well as the scattering intensity, and the scattering echo of multiple scattering centers can be observed at each time which reflects the high-resolution characteristics of ultra-bandwidth impulse fuze.

DSC is most seriously affected by azimuth, and the scattering amplitude varies significantly at different observation angles. The scattering echo simulation results of DSC1 also show that the scattering echo amplitude is the strongest at the effective observation angle ($\xi_1(\theta) = 90^{\circ}$), especially when the radial distance of the scattering center is almost coincident with other scattering centers. Besides, the scattering echo shape is almost unaffected, and the scattering echo amplitude is much stronger than other scattering centers. In other observation angles, the scattering echo can be almost ignored. The scattering echo simulation results of DSC2 also show that the echo amplitude is the strongest at the observation angle ($\xi_2(\theta) = 69^{\circ}$), and because of its small length, the scattering amplitude is also smaller than that of DSC1 and can hardly be observed at other angles.

SSC can be observed at more observation angles ($\theta = 30^{\circ}$、69°、90°)with strong scattering echo amplitude. Its position also slides with the change of observation angle. However, when the radar sight cannot coincide with any point of normal vector ($\theta = 100^{\circ}$、120°), SSC will not be observed. LSC can be observed at almost all angles. However, it can be seen from the figures that LSC4 is occluded at certain pitch angles ($\theta = 30^{\circ}$、69°、90°), and LSC2 is occluded at a pitch angle of120°. Thus, part of LSC cannot be observed at the above angles. The results above are consistent with the theoretical analysis.

In order to further verify the effectiveness of the echo generation method, a circular array is used to observe the target. Sixteen self-receiving antennas are distributed on the circle of plane $x = 25\text{m}$. The circle has a center of $\left(25,0,25\sqrt{3}\right)$m with a radius of 10m. The target's initial position is $(0,0,0)$, and the velocity is still along the z-axis, and the velocity is 500m/s. The scattering center imaging results of the target using the calculated echo signal are shown in Fig. 6. The LSC imaging results are obtained by the back projection (BP) algorithm[15]and the motion compensation. The dark blue, bright spot in the figure is the LSC position. The energy at LSC1, LSC2, and LSC3 is focused and displayed, while LSC4 fails to display in the image due to its too few observable positions and too little energy accumulation. SSC1, DSC1 imaging results can be obtained by the range point migration (RPM) method[16], corresponding to the green lines in Figure 6; DSC2 cannot be presented because too few positions are observed for imaging. The imaging results of the scattering center in Fig 6 are consistent with the shape of the detected target, which verifies the effectiveness of the scattering echo calculation method in this paper.

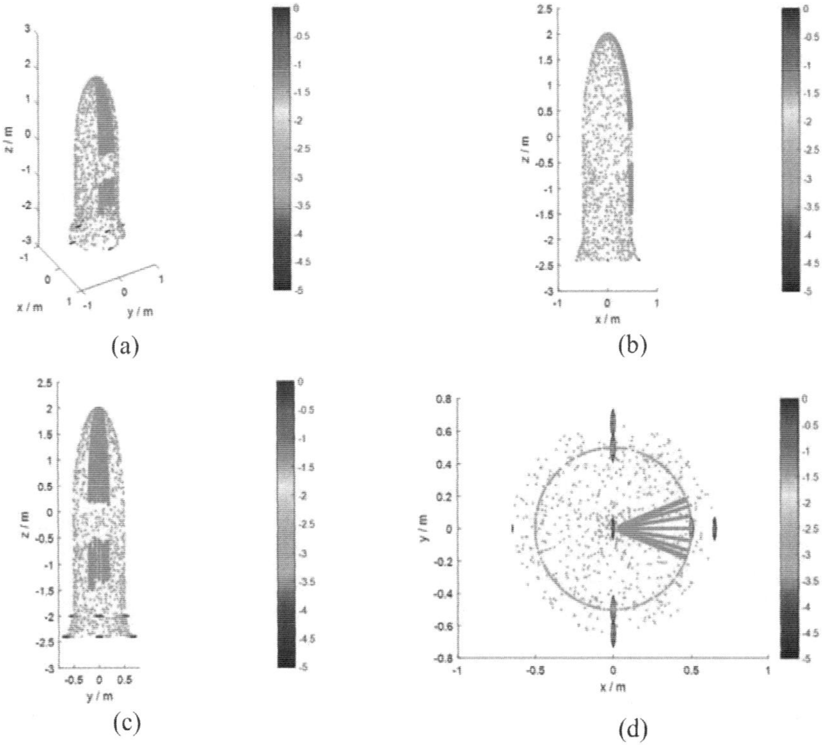

Figure 6. Imaging result diagrams. **(a)** 3D imaging result diagram. **(b)** XZ view. **(c)** YZ view. **(d)** XY view.

5. Conclusion

Ultra-bandwidth impulse radar fuze has various advantages, especially high range resolution. The scattering echo obtained during the detection of moving targets contains more structure and motion information of the target, which will help outputting the ignition signal to trigger the warhead accurately.

Compared with the complex target echo signal generation method based on electromagnetic calculation, our model has the advantages of small computation, less time-consuming and low hardware requirements. Compared with the traditional point echo model and extended echo model, the method we proposed can obtain a more realistic description of ultra-bandwidth impulse radar fuze echo and reflect the main scattering characteristics of the target and provides data for the evaluation of radio fuze signal processing algorithm. In this paper, the scattering centers of the target are distinguished, and the scattering echo model and its generation method, which are suitable for each scattering center and reflect its scattering characteristics, are established based on the attribute electromagnetic scattering center model.

The echo signal we obtained can reflect the main scattering characteristics of the target and provides data for the evaluation of radio fuze signal processing algorithm.

Acknowledgements
This work was supported by the National Nature Science Foundation of China (Grants nos. 61671059).

References

[1] FENG J, CAO J, FEI Y C, et al 2008 Missile radio fuze echo simulation based on bright spot model(Modern Defence Technology no. 5) pp 32-36
[2] CAI S J, ZUO D G, ZHU X F 2019 Influence of motion factors on Doppler frequency of radar fuze echo(Electronics Optics & Control vol 26, no. 1) pp 55-60
[3] ZHANG H Q, LI X, SHEN X 2020 Simulation of LFMCW fuze based on Simulink(Aero Weaponry vol 27,no. 2) pp 53-58
[4] TAO L B, FU HONGWEI, ZHOU G A, et al 2011 Anti-jamming performance analysis and simulation of ultra-wideband fuze(Journal of Projectiles, Rockets, Missiles and Guidance vol 31, no. 4) pp 89-91+98
[5] WANG X, FU H W, XIANG Z Y 2006 Simulation design of ultra-wideband fuze pulse source (Journal of Projectiles, Rockets, Missiles and Guidance vol 26, no. 3) pp 266-268
[6] LU Z Z, LIU Y J, YIN Y Z 2013 Construction and implementation of dynamic scene echo simulator for radio fuze (Computer science vol 40, no. 8) pp 53-58
[7] YU W G, WU C S, ZHANG D X 2020 Design of a multifunctional wideband radar target and jamming simulator (Modern defense technology vol 48, no. 5) pp 105-112
[8] GUO K Y, XIAO G L, ZHAI Y, et al 2018 Angular Glint Error Simulation Using Attributed Scattering Center Models(IEEE Access vol 6) pp 35194-35205
[9] LV T G, LU Z L, SU Y, et al 2000 The analysis of the azimuth resolution of impulse signal SAR(Acta Electronica Sinica vol 28, no. 6) pp 36-39
[10] NGUYEN C, HAN J 2014 Time-Domain Ultra-Wideband Radar, Sensor and Components: Theory, Analysis and Design(New York: Springer-Verlag Press)
[11] HUSSAIN M G M 1989 Principles of high-resolution radar based on nonsinusoidal waves. I. Signal representation and pulse compression(IEEE Transactions on Electromagnetic Compatibility vol 31, no. 4) pp 359-368
[12] SHENG H, FU H W, W X 2008 Optimization and implementation of ultra-wideband impulse radar receiver(Telecommunication Engineering vol 48, no. 12) pp 71-74
[13] FEI Y C 2010 Ultra-wideband radar theory and technology(Beijing: National Defense Industry Press
[14] NGUYEN L 2009 Signal and image processing algorithms for the U.S. Army Research Laboratory ultra-wideband (UWB) Synchronous Impulse Reconstruction (SIRE) radar(Adelphi: Army Research Laboratory)
[15] ZHOU X P, WEI G H, WU S L, et al 2014 Overview of imaging algorithm based on impulse radar(Systems Engineering and Electronics vol 36, no. 6) pp 1081-1087
[16] KIDERA S, KIRIMOTO T 2013 Efficient three-dimensional imaging method based on enhanced range point migration for UWB radars(IEEE Geoscience and Remote Sensing Letters vol 99) pp 1-5

Design of Large Monitoring System Based on Heterogeneous Network

Shuai Liu*, Xubo Tang and Chunhua Zang

[1]Department of Electronic Information Engineering, Nanjing University of Aeronautics and Astronautics, Nanjing, China
[2]Department of Integrated Circuits, Nanjing University of Aeronautics and Astronautics, Nanjing, China

*Corresponding author email: sxsmliushuai@163.com

Abstract. In order to ensure the quality of tobacco products, it is necessary to strictly control the environment of tobacco during storage. In order to save manpower and material resources and realize accurate monitoring of the storage environment, a large-scale tobacco storage monitoring system based on heterogeneous network was designed. Combining the cross-regional characteristics of NB-IoT and low-frequency characteristics of LoRa, NB-IoT was used as a wide area network and LoRa as a subordinate local area network to form a two-layer network structure. Its structure includes sensor node and gateway node, and the gateway node is composed of LoRa gateway and NB-IoT module. This paper introduces the composition and working mode of each part in detail.

1. Introduction

Tobacco is an important source of economic income for the country. The tobacco industry chain is very long and tobacco leaves are the basis of the tobacco industry. The tobacco leaves must go through a series of production processes before they can be put into the market. After the tobacco leaves are picked, they must be processed by primary roasting, redrying, storage and mellowing before they can be provided for cigarette companies. Therefore, the storage period is long, generally 1-3 year[1]. And the storage method is after packing, the packages will be stacked in order. Therefore, the storage area is large and the surrounding environment is complicated. In order to ensure the quality of the final tobacco products, manufacturers need to strictly control the stability of the storage environment of tobacco leaves.

The quality of tobacco leaves is not only related to the variety of tobacco leaves, but also affected by the surrounding environmental factors during storage. However, it is obviously unrealistic to keep the storage environment in a stable state through manpower. A more practical solution is to determine whether the current state is suitable for continued storage of tobacco by regularly testing various factors of the storage environment. If not, measures will be taken immediately. With the advancement of intelligence, Internet of Things technology is more and more widely applied to production and life, so it is the best solution to realize the remote monitoring of tobacco storage environment through Internet of Things.

NB-IoT, LoRa, WiFi, Bluetooth, and Zigbee are all commonly used wireless transmission schemes. The characteristics of these schemes are compared as shown in the table 1[8].

Four important features of the NB-IoT follow: wide signal coverage and high transmission gain, allowing for good signal support indoors, even in basements and sewers; As the service does not have a very high requirement for delay, theoretically 100K devices can be accessed within a signal sector. Low power consumption, theoretical calculation shows that a AAA battery can meet the requirements of the NB-IoT terminal can be standby for ten years; The cost is not high either. Currently, the module costs less than \$5, and monthly network data charges are lower than what 4G operators charge. However, LoRa also has these advantages. The biggest difference between NB-IoT and LoRa is that NB-IoT can be networked regardless of region. LoRa is only suitable for forming a LAN in a certain area, while NB-IoT can form a wide area network with great coverage by setting NB-IoT nodes across regions. Therefore, it is feasible to monitor tobacco storage environment data through NB-IoT network. However, there is still a problem, that is, tobacco itself has a great attenuation effect on communication signals, which will seriously affect the communication quality, and the higher the signal frequency is, the greater the impact will be. Therefore, the communication quality within the reservoir area cannot be guaranteed if only NB-IoT is used as the networking. Therefore, LoRa has the lowest working frequency band of 433MHz, and LoRa is used as a subordinate subnet to take charge of the data set inside the reservoir area, and the signals of the two will not interfere with each other due to the difference of communication frequency band. The LoRa of the subnet adopts a star structure, and each sensor node is independent of each other. Data collected by sensors is sent to the gateway. LoRa gateway and NB-IoT module together constitute gateway nodes connected to the public network. Based on the above, this paper finally designed a heterogeneous storage environment monitoring network formed by using NB-IoT as a wan and LoRa as a LAN.

Table 1. Comparison of different communication schemes

	NB-IoT	LoRa	Zigbee	WiFi	Bluetooth
Networking method	Cellular network	LoRa gateway	Zigbee gateway	Wireless router	Bluetooth mesh
Transmission distance	More than 10km	More than 10km	10-100m	50m	10m
Nodes capacity	200000	500-5000	200-500	50	60000
Frequency band	Operator frequency band	Sub-GHz 433, 868, 915MHz	2.4GHz	2.4GHz, 5GHz	2.4GHz

2. The Hardware Design

In order to meet the requirements, an NB-IoT networking system needs to be designed. The entire architecture is shown in Figure 1, which is mainly composed of sensor nodes, gateway nodes combined with LoRa gateway and NB-IoT module, and cloud servers.

The sensor node is composed of temperature and humidity sensor, LoRa module and MCU. The temperature and humidity sensor uses DHT11 digital low-power temperature and humidity sensor, whose measurement range is 0-50℃ and 20-90%RH, which is the most common temperature and humidity sensor. MCU of the sensor node and gateway node uses STM32F103C8T6, which is responsible for regularly obtaining the current ambient data from the temperature and humidity sensor and driving LoRa module for communication between the node and gateway. LoRa uses a communication module with a core chip of SX1278. It communicates over long distances by modulating gain rather than enhancing transmission power, so its power consumption is lower. Due to the selection of low power devices, so the use of lithium battery power can be.

The gateway node is composed of LoRa gateway and NB-IoT module. The NB-IoT module adopts the remote BC26 module, which can push data to Ali Cloud server through built-in MQTT. In terms of performance, the power consumption is low, and the working environment temperature range is large. The SX1278 module of LoRa gateway is responsible for receiving the temperature and humidity data sent by the sensor node in the current reservoir area, and then MCU sends these data to the operator base station through the NB-IoT module, and the base station transmits the data to the server. Similarly, the data sent by the server is also transmitted to the gateway node through the operator. The NB-IoT network formed by MCU in this way realizes the timely transmission of multi-point detection data to the server, and the server obtains the data of each node and saves it in the database for further analysis and display on the web page, so as to achieve the purpose of networking.

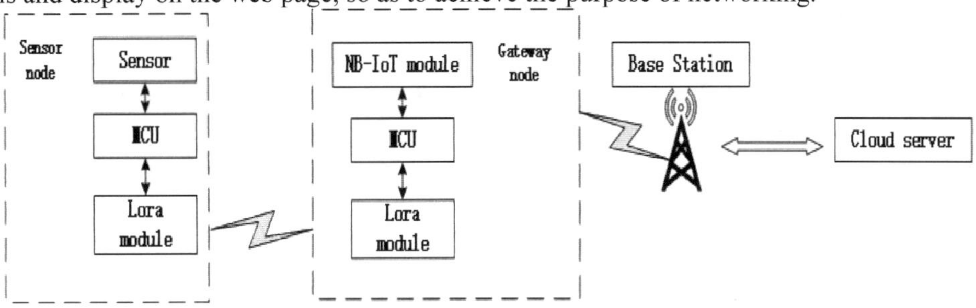

Figure 1. The structure of the system

3. The Software Design

The software mainly includes the work of sensor node and gateway node, and the gateway node includes the communication between LoRa gateway and sensor node and the communication between NB-IoT and base station.

3.1. Sensor Node Software Flow

The software flow of sensor node is shown in Figure 2. When the node is waiting for the query command from the gateway after initialization, LoRa can use the wake up technology to wake up the sensor node by setting the duration of the leading code of the query command sent from the gateway to a little longer than the sleep period of the sensor node. After the sensor node receives the query command, MCU obtains the current temperature and humidity data through the temperature and humidity sensor, and then sends the data to the gateway through the LoRa module, and waits for the gateway ACK. After receiving an ACK, the device enters the sleep state again and waits for the next query command. If no ACK is received within the waiting time, the device resends data. If no response is received after the retransmission, the device directly enters the sleep state.

3.2. Gateway Node Software Flow

As shown in figure 3, The LoRa module of the gateway node is mainly responsible for communication with the LoRa module of the sensor node. Based on the polling mechanism, the gateway sends query commands to the first sensor node. After receiving the data from the sensor node, it will parse the information contained in the signal and issue ACK response. Then it turns to the next node and polls all the sensor nodes covered by the gateway in turn. If no data is received, the query command is resend. If no data is received, the node is considered damaged and the node is directly transferred to the next node. After the data of all sensor node in the current reservoir area covered by the gateway has been reported once, the data will be packaged and uploaded. MCU will first check the status of the NB-IoT module, including judging whether the module is attached to the base station and checking whether the network status is normal. If the status is normal, the data will be uploaded to the server. If the status of the module is abnormal or no ACK signal is received, the data will be cached and sent again after the network status is normal, ensuring the stability of communication.

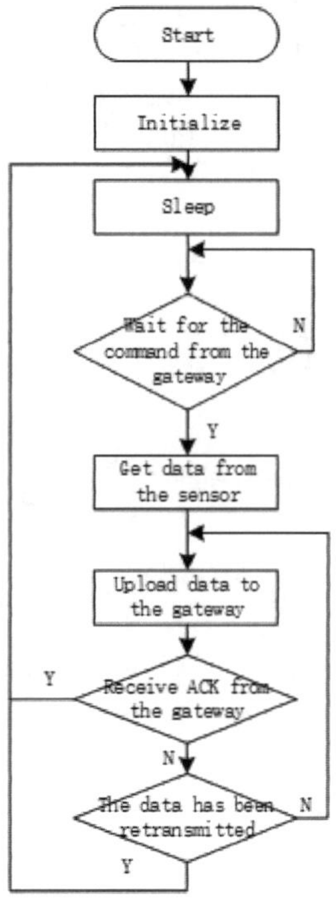

Figure 2. The software flow of sensor node

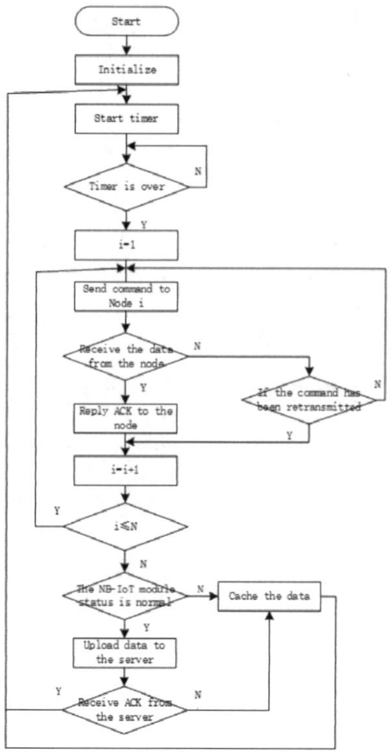

Figure 3. The software flow of gateway node

4. Conclusion

This article firstly introduces the current status of tobacco leaf and its storage. Secondly, it analyzes the advantages of NB-IoT networking and why LoRa is used to form a heterogeneous network with it. Finally, the tobacco storage monitoring system is proposed and from the point of view of hardware and software are given details.

References

[1] Jiang Yousheng, Liu Yong, Zhang Pingde, Li Qidong, Cheng Xueqing, Wang Lei. Storage and moisture control management of flue-cured tobacco leaves [J]. Modern Agricultural Science and Technology, 2014(05):74-75.

[2] Chen Qingli, Xiao Xi, Jiang Xiaobin, Zhu Yuan. Characteristic analysis of electromagnetic wave attenuation coefficient [J]. Journal of Oil and Gas Technology, 2014,36(08):43-45+51+4.

[3] Paredes M, Bertoldo S, Carosso L, et al. Propagation Measurements for a LoRa network in an Urban Environment[J]. Journal of Electromagnetic Waves and Applications, 2019(10):1-15

[4] Vangelista L, Centenaro M, Magrin D. Performance Evaluation of LoRa networks in a Smart City Scenario[C]. 2017 IEEE International Conference on Communications (ICC). IEEE, 2017

[5] Zhu Tianpei. Distributed Monitoring System of Phosphine in Tobacco Warehouse Based on NB-IoT[D]. Nanjing University of Aeronautics and Astronautics, 2018.

[6] Baronti P, Paolo P, Chook V, et al. Wireless Sensor Networks:a Survey on the State of the Art and the 802.15.4 and Zigbee Standards[J].Computer Communications,2007,**30(7)**:1665-1695

[7] Nie Zutian, Qiao Juying. Design of Smart Library Monitoring and Management System Based on LoRa Wireless Network[C]. Proceedings of the 17th International Conference on Innovation and Management (ICIM2020). 2020.

[8] Wang Ting. Review of the Research Results of Smart Library in China[D]. Northeast Normal University, 2019.

Research and Design of Remote Online Supervision System of Coal Mine Electromechanical Equipment

Yunqi Chen[1,*] and Jin Xu[2]

[1] China Coal Technology Engineering Group Chongqing Research Institute, No. 6 Kecheng Road, Jiulongpo District, Chongqing, China
[2] School of Big Data & Software Engineering, Chongqing University, Chongqing 400000, China

*Corresponding author email: chenyunqee@foxmail.com

Abstract. Increasing the supervision of coal mine electromechanical equipment is an urgent need to ensure the safe production of coal mines. This paper designs and develops a coal mine electromechanical equipment supervision system, which includes four parts: mine-side monitoring subsystem, mine-side data acquisition and transmission module, local-side data analysis and storage module, and coal mine electromechanical equipment supervision platform. It realizes coal mine ventilator, drainage pump, air compressors, hoists and other major electromechanical equipment online monitoring and equipment failure analysis. The system satisfies the remote online inspection and management of equipment by multi-level departments, reduces the possibility of enterprise accidents, innovates the supervision mode, and improves the efficiency of supervision.
Keywords: Coal mine; Electromechanical equipment; Supervision; Online monitoring; Failure analysis.

1. Introduction

Coal is China's fundamental energy guarantee and key chemical raw material. In 2020, the country's total energy consumption is 4.98 billion tons of standard coal, of which coal consumption accounts for 56.8%[1]. Therefore, safe production and stable supply of coal are important to national security cornerstone.

In recent years, with the rapid development and application of new technologies such as the Internet of Things and intelligence, the level of coal mine mechanization, automation and informatization has become higher and higher, and the number of accidents and deaths in coal mines has decreased year by year. In 2020, there were 122 coal accidents nationwide, with 255 deaths and a death rate of 0.059[2] per million tons of coal, which dropped to the lowest level in previous years. The overall situation of coal safety production is improving.

With the widespread use of mechanical and electrical equipment and the frequent occurrence of equipment failures, the number of electromechanical accidents and the number of deaths have increased compared with 2019. Therefore, the use of information technology to realize online supervision of major electromechanical equipment in coal mines, reduce the occurrence of electromechanical accidents, and improve coal mine safety production is an urgent need for coal mine enterprises and supervisory units.

This paper designs a real-time monitoring, collection and uploading analysis of equipment working condition parameters from bottom to top, discovering equipment abnormalities and fault alarms in time, realizing remote supervision of coal mine electromechanical equipment, and preventing equipment irregular use, and reducing accidents possibility of occurrence to ensure safe production in coal mines.

2. System Architecture Design

The design of the system is carried out of logic architecture, deployment architecture, and technical architecture to meet the requirements for rapid data collection and transmission, and to ensure the functional performance indicators of the web platform.

2.1. Logic Architecture

The logic architecture[3] of system is divided into three levels, namely the mine subsystem layer, the data collection and transmission layer and the business application layer, as shown in the figure 1. The mine subsystem layer is composed of equipment, IoT sensors and host computer monitoring system to realize online monitoring and control of main fans, hoists, compressors and other equipment. The data collection and transmission layer consists of the collection software responsible for collecting monitoring data from the mine monitoring system and the data transmission software responsible for uploading the data to the higher-level data center. The business application layer consists of the data center, website and mobile app, and is responsible for providing online monitoring services of equipment to coal mine staffs and supervisors.

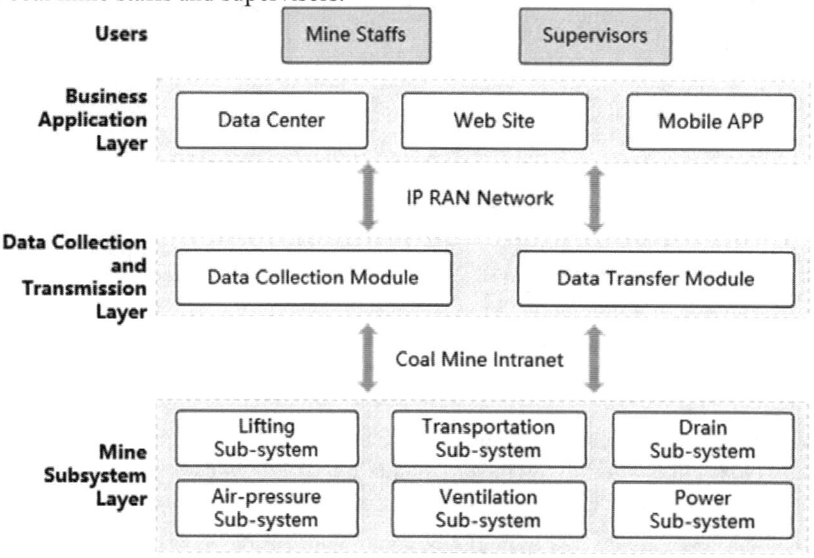

Figure 1. System logical architecture.

2.2. Deployment Architecture

The system is deployed at two levels, the coal mine side and the supervisory side, and its architecture[4] is shown in Figure 2. On the coal mine side, use IoT sensors such as temperature, vibration, pressure, speed, voltage, current, etc. to collect the operating status of electromechanical equipment and various working condition parameters in real time, and store and display this information in real time through the monitoring subsystem. The system development data collection module collects these data from the subsystem, and uploads the data to the data receiving platform of the supervisory end through the data transmission module. Both the data collection module and the data transmission module are deployed at the coal mine side.

On the supervision side, deploy data analysis modules, supervision platform website, and app software that provides mobile supervision services. In this way, supervisors can use the Internet to implement online supervision anytime, anywhere.

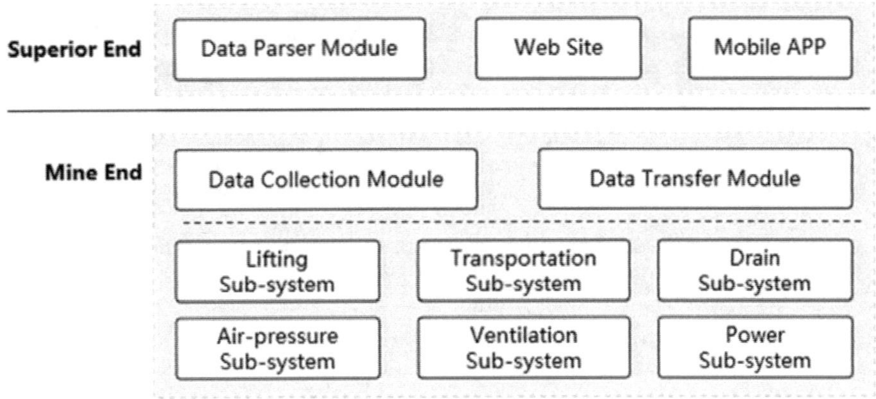

Figure 2. System deployment architecture.

2.3. Technical Architecture

The system adopts a layered architecture design method[5], which is divided into an access layer, a collection and transmission layer, a data storage layer and a platform application layer. In the software module design, C/S, B/S and mobile app coexist. The website is developed based on the ASP.NET MVC 5.0 framework[6] and third-party components and technologies such as jQuery, ECharts[7], HTML5, and SVG. The app software based on the Android platform[8] is developed using Java.

The platform uses a combination of Redis[9], SQL Server and HBase[10] to store real-time monitoring data, alarm and statistical data, and massive historical sampling data of equipment to meet the needs of different businesses. In the data access, the OPC UA/DA[11] protocol is used to realize the data collection and access of the monitoring subsystem of the mine-end equipment; the FTP protocol or message queue transmission technology[12] is used to realize the transmission of the data to the platform.

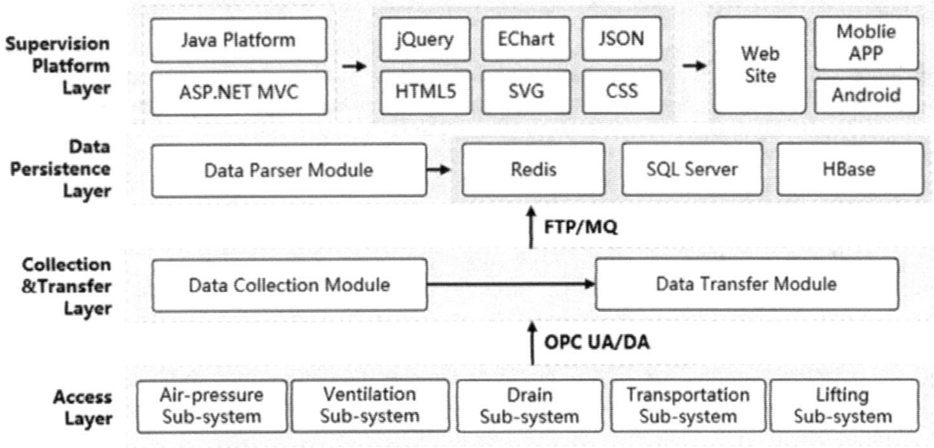

Figure 3. System technical architecture.

3. Main Functions

The platform is simultaneously open to coal mine managers and supervisors. Coal mine managers use the platform to fill in and improve the basic and maintenance information of mining equipment. Supervisors can use the information, as well as equipment real-time monitoring data, alarm fault data, etc., to understand the use of equipment, and it is online and remote supervision. The main functional modules of the platform are shown in Figure 4.

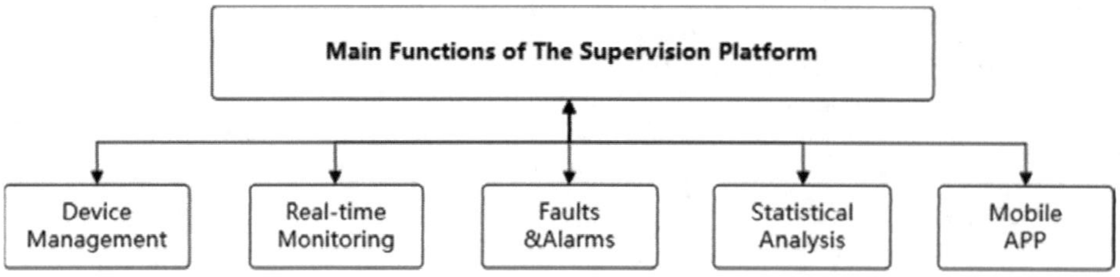

Figure 4. The main functions of the supervision platform.

3.1. Device Management
The module realizes online management of basic equipment information, maintenance records, and inspection reports. Mine users fill in basic information such as the model, type, production date, manufacturer, safety standard information, installation location, etc. of the equipment, and timely report the maintenance and delivery record information of the equipment, including maintenance time, personnel, content, process records and costs. For equipment that requires regular inspection, it need to be uploaded or updated in time. For those that are not inspected on time, the system must promptly send out warning or alarm prompts.

3.2. Real-time Monitoring
Real-time monitoring module realizes online monitoring of the real-time operating status of the equipment. The monitoring content includes equipment operating status, current, voltage, vibration, temperature, speed and other different information. Such as monitoring the lifting height of the hoist and brake oil pressure and other parameters, monitoring the wind speed, air volume and other parameters of the main fan, monitoring the water level of the main drainage pump and the water flow rate and other parameters, monitoring the running speed of the belt transportation and other parameters.

3.3. Faults & Alarms
This module realizes real-time alarm prompts for abnormal alarms and fault information generated during the use of the equipment, and queries and displays historical fault alarms. The module displays equipment failure alarm information through pop-up windows, curve and column diagrams, report charts, etc.

3.4. Statistical Analysis
The module reveals the safe operation status and trend of the equipment through mufti-dimensional statistical analysis of the basic information and monitoring data. For example, by analyzing the completeness of the reported equipment information, we can evaluate the degree of importance that each coal mine attaches to the equipment information management; by analyzing the number and duration of equipment abnormal alarms, we can grasp the operating status and use efficiency of the equipment in the jurisdiction; by analyzing the equipment frequency of maintenance, timely discover potential hidden dangers of equipment that have not been repaired and maintenance due, remind the coal mine to deal with it in time.

3.5. Mobile APP
The APP software is developed based on the Android platform, using XML and HTML5 mixed interface design simulation, using SQLite and Preference cache mechanism. It reduces memory consumption and improves software performance and user experience. The app is installed in smart phones of supervisors, the basic information, real-time monitoring data, alarm fault records, operation records and other information of major electromechanical equipment can be viewed anytime and anywhere. When equipment alarms and abnormal information are generated, the software can automatically pop up prompt messages to actively inform supervisors, which greatly improves supervisory efficiency.

4. Conclusion

This system breaks through the limitation of time and space, and uses information technology to realize the collection, upload, analysis and display of the working condition parameters of the main electromechanical equipment, realizes remote online supervision, and effectively promotes the normal operation of equipment, the timely handling of faults and the correct maintenance. This will help reduce the occurrence of electromechanical accidents in coal mines and improve the level of safety production. At present, the system has been promoted and used in some provinces and cities, which has improved the efficiency of supervisory departments.

Acknowledgement

This work was supported by funds from the Major Special Project of CCTEG Chongqing Research Institute (2021ZDZX01).

References

[1] National Bureau of Statistics. volume of major energy products[EB/OL]. [2021-12-06].https://data.stats.gov.cn/easyquery.htm?cn=C01
[2] Zhang Shilong, ZhangMinbo, Zhu Renhao, et al. Optimization Analysis of the characteristics of China's mine accidents in the past five years and countermeasures for prevention and control[J].Coal and Chemical Industry, 2021,44(8):101-106,109.
[3] Dehraj, P., Sharma, A. An approach to design and develop generic integrated architecture for autonomic software system. Int J Syst Assur Eng Manag 11, 690–703 (2020).
[4] Adrian P. Woźniak. Optimization of Deployment Architecture in SOA systems[J].Procedia Manufacturing,2020,44(5):543-550.
[5] ZHANG Guo-sheng. Microservice Software Modeling Based on Domain Driven Design and C4 Layered Architecture Model[J].Journal of CAEIT, 2021, 16(2):119-126.
[6] Wang Zhenyi. Design and Implementation of Teaching Diagnosis System based on ASP.NET MVC[D]. Zhejiang University of Technology, 2020.
[7] Zhou Weiyi. The enterprise marketing analysis based on ECharts[J].Application of Electronic Technique, 2019, 45(12):101-105.
[8] HUANG Xue-Qian, TAN Jun-Yuan, LI Xin, et al. Optimization Application of Marine Ecological Environmental Data Display Based on Android[J]. Computer Systems & Applications, 2019,28(2):213−218.
[9] CHEN Qing. Networking architecture of large-scale electromechanical equipment in coal mine based on Redis[J].Industry and Mine Automation,2020,46(10):109-113.
[10] Wen Zhenhui, Fan Yongsheng, Yu Hongying. Storage mechanism optimization of HBase data based on Thrift[J]. Science Technology and Engineering,2019,19(6) : 185-189.
[11] YU Xin-yi,YIN Hui-wu, SHI Tian-feng,et al.Data Acquisition System of Industrial Equipment Based on OPC UA[J].Computer Science,2020,47(S2):609-614.
[12] Kang Yunchuan, Zhong Jing, Li Ruofeng. Irrigation system based on message queuing telemetry transport protocol[J]. Science Technology and Engineering,2020,20(8):3109-3116.

An IoT Assisted Early Warning System for Smart Grid

Jinglong Xie[1,*], Xinxiu Xiao[2], Yao Xu[3] and Bing Jin[3]

[1] Security Supervision Center, State Grid Hubei Electric Power Co., Ltd, Hubei, China
[2] State Grid Hubei Electric Power Co., Ltd, Hubei, China
[3] Hubei Huazhong Electric Power Technology Development Co., Ltd, Hubei, China

*Corresponding author email: 1853026491@qq.com

Abstract. Safety monitoring is an important component in early-warning system, and it is crucial to ensure safe operations in various industries. However, the majority of the existing safety monitoring system suffer some limitations, either they cannot cover the whole working area or their intelligence is inadequate to provide effective early warnings. In order to ensure safe operations in smart grid, we propose an IoT assisted early-warning system. The system employs some latest technologies, such as electronic fence, video surveillance, UWB positioning and the SVM model, to achieve real-time accurate positioning of operators. Moreover, it can detect the distance between operators and dangerous areas, and the system can alert the operators when they are approaching the dangerous areas. Our proposed approach helps to make the early warning system more intelligent and more automated, improving safety supervision and providing effective security management for smart grid.
Keywords: Internet of Things; Electronic fence; Intelligent early warning.

1. Introduction

Smart grid integrates modern sensors into the traditional power distribution system. It can collect and analyse various data information in order to reduce costs and promote efficiency. Moreover, it can improve the reliability and optimize the operation and management of the entire grid [1]. However, the operational areas within smart grid are normally hazardous, and manual management is insufficient as it suffers various problems, such as inaccurate operation, non-standard inspections, accidental entry dangerous areas, and accidental touch of risky equipment [2][3]. Although electronic fences are helpful to prevent illegal intrusions, and video surveillance can detect the situation on the job site, they need to be properly integrated in order to provide an effective early warning system for dangerous behaviours.

In order to fulfil the above requirements, this paper introduces an Internet-of-Things (IoTs) assisted early warning system for smart grid. Our proposed system integrates various latest information techniques, such as electronic fence, video surveillance, UWB positioning, etc., to deliver an intelligent management system, ensuring workers' safety in operation as well as simplifying the supervision processes for the management staffs.

1.1. Related works

Recently, many works have explored how to design an effective early warning system for smart grid. Pan et al. [4] have proposed a management system using impulse electronic fence. The system realises perimeter intrusion prevention, i.e. it can prevent illegal entity from entering the operational area. But this system still has some limitations. For example, the probability of false alarms is very high, and it is difficult for the management staffs to locate and track the intruder. Kang et al. [5] have proposed a

monitoring system using video surveillance, electronic fence and intrusion alarm. The system can alert when someone tries to cross the electronic fence. In addition, the management staffs can locate the intruder through video surveillance. But the system lacks the ability to warn dangerous activities, and it requires the staffs to be online for monitoring. Therefore, it is still not intelligent enough. Liu et al. [6] have designed a similar system based on GIS, which uses GPS positioning technology to obtain the real-time position of operators. But the system lacks the ability to ensure the safety of indoor operators. Xu et al. [7] have proposed a novel monitoring system using UWB positioning, electronic fence and monitoring technology, which realises high-precision and real-time positioning and early warning for dangers, i.e. when entities enter the restricted area, the system can send an warning signal. But the restricted area not only limits the size of the operation area, but also may cause false alarms. Recently, NIST has issued a report discussion the security requirements for smart grid [8], pointing out the management system should achieve perimeter prevention, intelligent early warning and automatic supervision. Based on these requirements, Mitchell et al. [9] have proposed an evaluation framework for early warning in smart grid.
In this paper, we propose an IoT assisted early warning system for smart grid, integrating electronic fence, UWB positioning and SVM model to provide real-time monitoring. In our system, accidents caused by non-standard operations can be prevented by employing the position and video monitoring techniques, and it is more intelligent thanks to the use of the SVM model. Compared with the existing works, our proposed system satisfies more desirable security features. Therefore, it helps to make the early warning system more intelligent and more automatic, improving safety supervision and providing effective security management for smart grid.

2. Building Blocks

2.1. Internet of Things
The Internet of Things (IoTs) have three main functionalities: real-time perception, communication, and intelligent processing.

- **Intelligent perception:** It uses various technologies, such as QR code, sensors, cameras, and RFID to collect information.
- **Communication:** It uses communication networks to ensure the stability and reliability of message exchange.
- **Intelligent processing:** It uses a series of intelligent computing technologies to complete scientific analysis and related data processing tasks, which is conducive to subsequent decision-making and management.

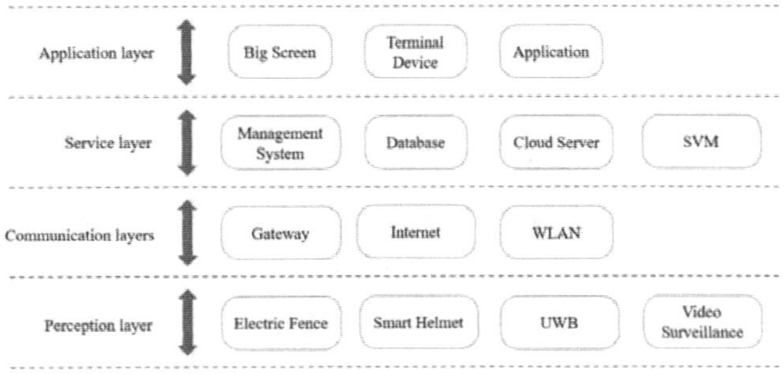

Figure 1. System architecture for IoT.

2.2. Electronic Fence
1) Physical electronic fence

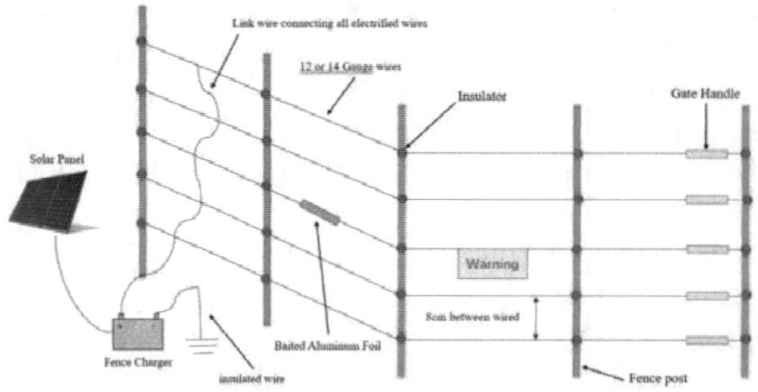

Figure 2. An illustration of physical electronic fence.

As shown in figure 2, the physical electronic fence is a widely used anti-theft alarm system, which is generally composed of three parts: front-end accessories, host and back-end control equipment. In general, physical electronic fences can be classified into three categories, infrared through-beam electronic fence, high-voltage pulse electronic fence and vibrating fiber. All of them can provide accurate and safe protection after being laid at the perimeter guard site.

2) Virtual fence

The virtual electronic fence is an important technology in the geographic information system. It realises real-time monitoring of people and vehicles by drawing a graphic area on the map. When an object enters or leaves the drawing area, the system makes corresponding processing. In smart grid, electronic fence can be used to provide early warning.

2.3. Video Surveillance

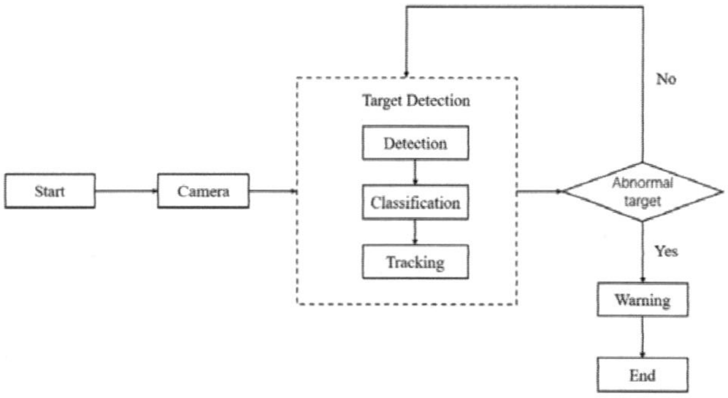

Figure 3. The basic structure of video surveillance.

Video surveillance [13] consists of five parts: camera, control, transmission, display and recording. It mainly contains two functions. One is to control the on-site video capture equipment and receive the digital video signals. The other is to identify, analyse, and process the collected digital video signals.

The specific working process of the scene is shown in figure 3. First, the surveillance camera collects the image and video information of the scene. Then, the system performs target detection on the collected images or videos. If suspicious behaviour is found, the video surveillance system will track the target and send the video or image to the analysis centre for analysis. Finally, if the target is classified as having dangerous or abnormal, an alarm is triggered and the information is reported to the monitoring centre.

2.4. Positioning Technology

Positioning has various applications in many surveillance and security systems. The mainstream positioning technologies include Beidou positioning, infrared positioning, ultrasonic positioning, Bluetooth positioning, Wi-Fi positioning, and ZigBee positioning. In this paper, we will use Ultra-Wide Band (UWB, Ultra-Wide Band) positioning technology 0, because of its advantages, such as low power consumption, anti-multipath, high security, low system complexity, and can provide accurate positioning.

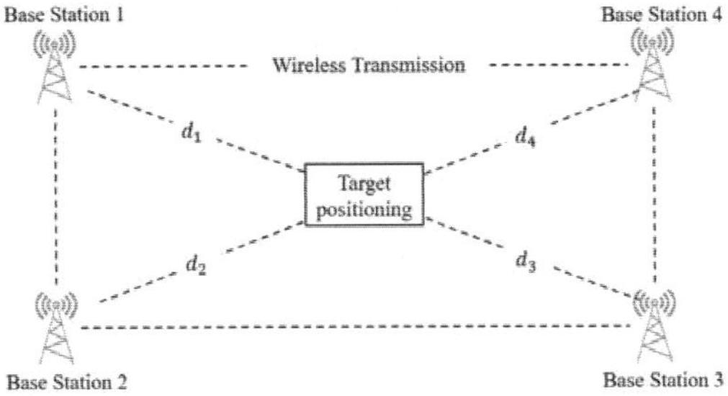

Figure 4. Basic principles of UWB positioning technology.

UWB locates the target by measuring the direction and distance between the base station and the target. Generally, the measurement algorithm includes Angle of Arrival (AOA), Received Signal Strength (RSS) and Time Difference of Arrival (TOA/TDOA).

As shown in figure 4, TOA uses 4 fixed base stations to transmit signals to mobile tags. The tag calculates $d_i = (t_i - t_0)c$ (where c is the speed of light) according to the signal receiving time t_i and the time t_0 of each base station transmitting the signal. Finally, the system can calculate the specific location of the tag through the distance between the four base stations and the tag (d_1, d_2, d_3, d_4).

2.5. Support Vector Machines (SVM)

Support Vector Machines (SVM) is a two-classification model, whose purpose is to find a hyperplane with the largest geometric interval to divide the sample, and finally, transform it into a convex quadratic programming problem to solve. As shown in figure 5, the hyperplane is expressed as $\omega x + b = 0$. For a linearly separable data set, there are infinite many such hyperplanes, but the hyperplane with the largest geometric interval is unique.

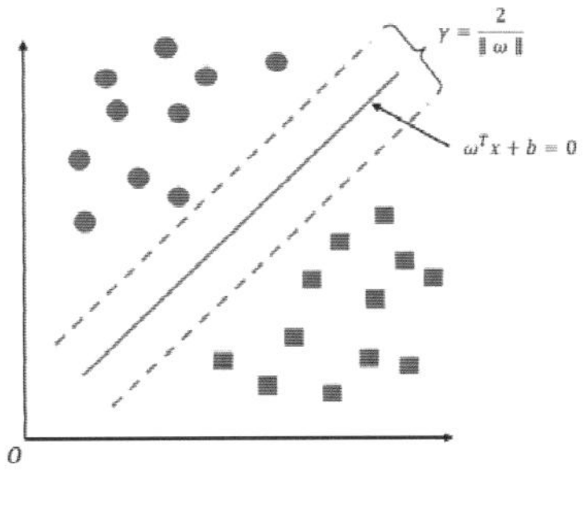

Figure 5. Support vector and interval.

The basic principle of the SVM model is to apply the kernel function expansion theorem to map the sample data to a high-dimensional feature space, so that a linear algorithm can be used to analyse the sample. The key of the SVM model is to find the optimal classification surface in the case of linear separability. First, one needs to collect a training set of user's movement trajectories. Then, the system selects the radial basis function (RBF) function K to realize the nonlinear transformation, and calculates the Lagrangian multiplier a_j^*, the weight vector ω and the classification offset b^* and other parameters. Finally, one solves the classification interface according to formula (1), and predicts whether the operator has a tendency to cross the boundary.

$$f(x) = sgn(\sum_{j=1}^{l} a_j^* y_j K(x_j \cdot x) + b^*) \tag{1}$$

Among them, $f(x)$ is the best classification function, x is the unknown vector, and y_j is the parameter obtained by model optimization.

3. System Function Design

The main functions of our proposed early warning system include perimeter intrusion detection, operation area delimitation, operation site management, early warning for risks, trajectory playback, query statistics, etc. As shown in figure 6, the system first realises intrusion detection through video surveillance and electronic fence installed in the operation area. Then, the system uses UWB positioning to determine the location of entities within the field, the insulation distance between entities and the surrounding equipment, and the distance between non-operating entities and the work area. Finally, the system uses the SVM model to realize intelligent risk warning.

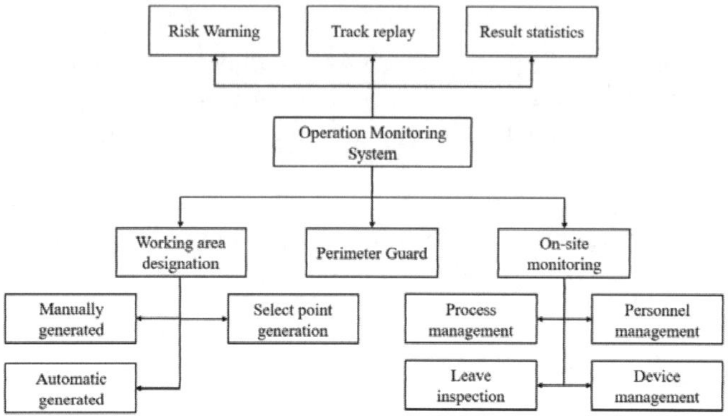

Figure 6. System structure.

3.1. Perimeter Intrusion Detection

Perimeter intrusion detection prevents the outsiders from crossing the border and entering the inside of the power station. As shown in figure 7, the system uses physical electronic fences and surveillance cameras. It installs physical electronic fences at the border. If an abnormal situation occurs, the alarm device will immediately issue an alarm and report the abnormal behaviour to the monitoring centre. At the same time, the camera will record in real time and track the abnormal targets. By integrating the surveillance camera with the electronic fence, the management staffs can effectively receive the alarm signal, locate the intrusion area and track the intruder, and immediately rush to the scene to deal with the abnormal situation.

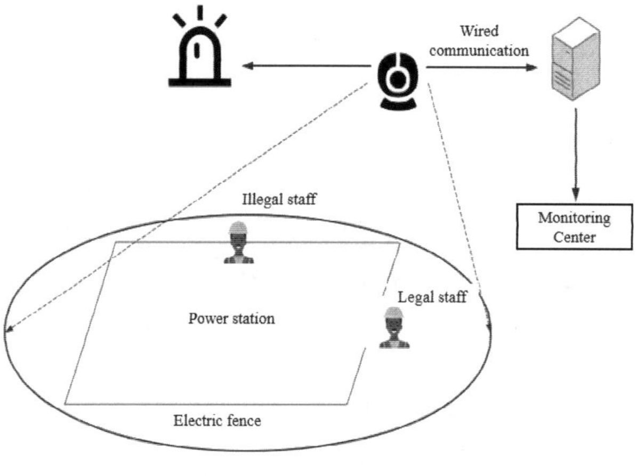

Figure 7. Schematic diagram of preventing perimeter intrusion.

3.2. Automatic Division of Work Area

The system is based on UWB positioning and virtual electronic fences to realize the function of dividing the work area. The work ticket clearly specifies the area of the power operation. The operator can enter the work area with the work ticket and can only work in the area. Non-operating entities cannot enter the work area.

The process of automatically generating the work area by the system is shown in Figure 8. Based on the information provided in the work ticket, the system divides the work area on the substation plan and sets up the entrances and exits for electronic fence. The operation manager can check and confirm the automatically generated operation area. If it does not match the area given in the work ticket, it can be manually modified it.

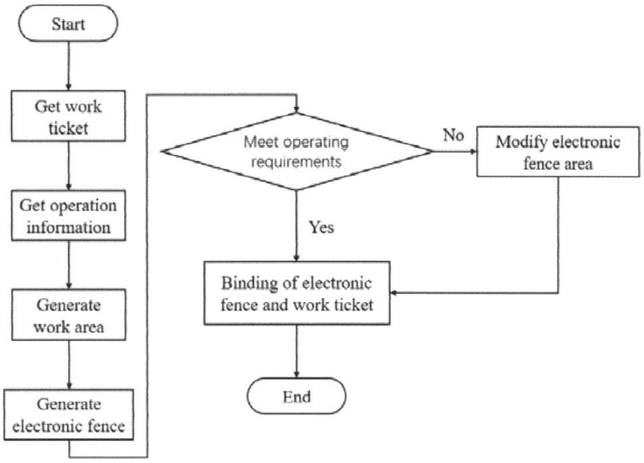

Figure 8. A flowchart about automatically dividing the work area.

3.3. Work Site Control

Safety management for the work area includes three components: personnel and equipment management, real-time monitoring of the operation process, and inspection of entity join and leave. The purpose is to ensure smooth operation and operators' safety. Through the establishment of a virtual electronic fence in the operation area, combined with UWB positioning and video surveillance, the overall safety management for the operation site can be realized.

3.3.1. Management of operating personnel and operating equipment. It includes identity verification for personnel and equipment, access inspection, records of entering and leaving the work area, and positioning.

1) **Identity verification for personnel and equipment**: The system first uses UWB positioning tags to store the key for personnel or equipment. When detecting personnel or equipment entering the work area, the system checks it with the database. Therefore, personnel and equipment with operating qualifications can complete the check, but entering the operating area without a permission or without operating qualifications will trigger an alarm.
2) **Personnel and equipment admission detection**: The system first judges whether the worker is wearing a smart helmet through a surveillance camera. It then confirms whether the personnel have the authority to enter the operation area and whether the operation equipment has the qualification to use the personnel and equipment information.
3) **Monitoring of personnel and equipment entering and leaving the work area**: UWB tag reading terminals are installed at the entrance and exit of the work area. The terminal can automatically read the coding information of the UWB positioning tag within the effective communication range, and record the time of personnel entering and exiting the work area.

4) **Positioning of personnel and equipment**: Install positioning tags in the smart helmets and equipment of the operators, and use UWB positioning technology to monitor the positions of personnel and equipment in real time.

3.3.2. Real time monitoring. The monitoring system collects the position and movement track information of the operators in real time and stores them in the database. When operators such as inspections, maintenance, visits, etc. enter the substation, the positioning base station captures the signal from the positioning tag in the helmet. Therefore, the system can establish communication, measure the RSSI value, and complete the positioning of the operator. An alarm will be triggered when the operator's activity range exceeds the safety range set by the electronic fence. In addition, based on the movement trajectory information, the system will automatically remind the inspectors when it determines that there is no inspection or repeated inspections.

3.3.3. Inspection. After the operation is completed, the system will track and confirm the departure of the operators and equipment based on the acceptance checklist, surveillance video and smart wearable equipment.

3.4. Risk Warning Based on SVM Model

The system first uses UWB positioning to collect the movement trajectory information of the workers. It then uses the SVM model to predict the tendency of entity approaching or leaving and issues early warnings in time. Finally, the managers can set the time limit for staying in important areas according to their needs.

The detection process of Support Vector Machine (SVM) is to select the features that can fully describe the behaviours from a large amount of training data. And then, it trains the classification model. The early warning system can judge violations based on the SVM classification model. Therefore, the system has the advantages of high detection accuracy, no manual involvement, strong generalization, and predictability of violation trends.

3.5. Track Playback

The trajectory information generated by the operator in the work is stored in the database. Therefore, managers can accurately determine the location, time and related information of the violation.

3.6. Query and Statistics

The system supports querying the number of violation alarms, the number of operations, the number of operations, and the duration of operations in a certain period or within a certain period of time. It is convenient for the management staffs to establish a safety and quality evaluation system, and strengthen the work areas that were frequently being violated.

4. System Analysis

This section provides brief analyses of the proposed system. Table 1 compares the functionality of our system with some existing ones, including the thermal power plant security monitoring system [2], the GIS-based distribution field operation system [3], and the UWB positioning-based operation monitoring System [4]. Since the systems [2] and [3] only use electronic fence and low-precision positioning technology, they can only provide simple perimeter protection. The system [4] adopts high-precision positioning, electronic fence, and video surveillance technology, which has high practicability in perimeter prevention and job site management. But it has a low degree of intelligence. Our system integrates various latest technologies, such as electronic fence, video surveillance, UWB positioning and SVM model. It has good practicability in perimeter prevention, job site management and risk warning. The use of SVM model for early warning of dangerous behaviours effectively guarantees operators' safety, enhances the intelligence and automation of the system, and improves the management level of smart grid. Therefore, it achieves more desirable security requirements.

Table 1. Function comparison.

Scheme	Perimeter prevention	Work site management	Intelligent early warning	Unattended	Degree of intelligence
[1]	√	×	×	×	×
[2]	√	√	×	×	×
[3]	√	√	×	×	low
[4]	√	√	√	×	low
Our system	√	√	√	√	√

5. Conclusion
This paper proposes an IoT assisted early warning system for smart grid. Thanks to the use of electronic fences, video surveillance and UWB positioning, the system can effectively monitor the infrastructure, operators and abnormal conditions. Moreover, the system uses wireless networks for communication, and it combines the SVM model to perform early warning and intelligent decision-making. Security analyses demonstrate that the system can improve safety for smart grid. Hence, it can make the early warning system more intelligent and more automatic, improving safety supervision and providing effective security management for smart grid. The proposed system has been implemented and used in our daily operations for three years. Data is collected continuously over time, and in the future, we plan to apply the big data and artificial intelligence techniques to further improve its intelligence and automation.

References
[1] Chen Shuyong, Song Shufang, Li Lanxin, Shen Jie. Survey on Smart Grid Technology[J]. Power System Technology, 2009, 33(8):1-7.

[2] Ma Jianwei. Research and Application of the Substation Safety Operation Monitoring based on the Internet [D]. North China Electric Power University, 2014.

[3] Metke A R, Ekl R L. Security technology for smart grid networks[J]. IEEE Transactions on Smart Grid, 2010, 1(1): 99-107.

[4] Pan Linzhou. Application of pulse electronic technology in substation safety protection [J]. Changjiang Information &Communications ,2014(09):286.

[5] Kang Xuan, Cao Xiangbin. Application of Security Monitoring System in Thermal Power Plant [J]. Electric Drive Automation, 2016,38(05):32-36.

[6] Liu Zhijun, Sun Jun, Yang Yushan, Liang Zhijun. Distribution Field Operation System Based on Mobile GIS [J]. Electric Age, 2010(01):92-93.

[7] Xu Baojun, Li Xinhai, Zeng Lingcheng, et al. Research on 3D Visual Operation Monitoring System of Substation Based on UWB Positioning [J]. Electric Power Information and Communication Technology, 2021,19(02):29-35.

[8] NIST technical report, Cybersecurity for smart grid system. nist.gov/programs-projects/cybersecurity-smart-grid-systems

[9] Robert Mitchell, Ing-Ray Chen. Behavior-role based intrusion detection systems for safety critical smart grid applications. IEEE Transactions on smart grid 4(3):1254-1263, 2013.

[10] Zhang Quan, Zhang Zijing. Pulse Electronic Fence Perimeter Alarm System [J]. Science & Technology Vision, 2012(08):137-138.

[11] Guan X, Lu L I, Tiantian Y I, et al. Research on the Application of Smart City Electronic Fence[J]. Information and Communications Technology and Policy, 2019.

[12] Chen ziye. Design and Realization of Operation Monitoring and Management System Based on the Beidou for Marshal1ing Stations [D]. China Academy of Railway Sciences, 2018.

[13] Yang Fu. Rescarch and Application on the Vidco Surveillance System for Unattended Transformer Substation [D]. Hunan University, 2018.

[14] Zhou Gang. Study on Power Substation Operation Safety Monitoring Technology Based on UWB Location [J]. Journal of Safety Science and Technology, 2016,12(08):125-129.

An Anti-noise Fault Detection Algorithm for Distribution Network

Ming He[1,*], Yi Li[1], Wan Zou[1], Xiangxi Duan[2], Jie Gao[1] and Rui He[1]

[1]State Grid Sichuan Electric Power Company, Chengdu, China
[2]Electric Power Research Institute, State Grid Sichuan Electric Power Company, Chengdu, China

*Corresponding author email: a1333303@163.com

Abstract. Power fault directly threatens the safety of the power system. Aiming at the recognition ability of power line fault detection disturbed by noise, A fault detection method based on attentional mechanism combined with autoencoder neural network is proposed. The fault signal is decomposed by wavelet and the subband energy is used as the input of the neural network, which is composed of autoencoder and fault recognition network. The input signal is compressed by an encoder, represented by a small number of features, and transmitted to the fault detection network. By extracting features and training fault signals, the network structure can effectively identify fault categories. Simulation experiments based on IEEE34 show that the fault recognition rate of this method is over 98%, and it can detect distribution network faults effectively under the condition of noise, and has high anti-interference performance. It provides a new idea for fault detection of distribution networks and lays a foundation for a high robustness detection method.
Keywords: Fault detection; Wavelet; Autoencoder; The neural network.

1. Introduction

With the development of the power system, the improvement of voltage levels, and the increasing scale of transmission lines, the possibility of line failure is increased. Therefore, power fault detection is very necessary. Conventional power fault detection algorithms have accumulated in years of research, but the effect of conventional power fault detection algorithms is not satisfactory in the face of fault information containing noise. Therefore, how to detect power fault under noise interference is a very important research direction at present.

Fault detection and classification of distribution lines can be roughly divided into traditional algorithm and deep learning method. Y.hang[1] used S transform to extract low-frequency fault current according to the difference in the variation trend of fault current at both ends of the line when a fault occurred, and used F test to carry out trend test on it, to identify in-zone and out-zone faults. D.Fang[2] used a new method for detecting HIF in high-impedance transmission lines by wavelet transform, and adopted the absolute value of coefficients in multi-resolution signal decomposition based on discrete wavelet transform, combined with fault indicator and fault criterion to detect HIF in transmission lines. X.Wu[3] proposed a multi-support vector machine model diagnosis method based on electrical volume fault information features. Firstly, the symmetric component method was used to extract fault information features and establish feature sets. Secondly, a genetic algorithm is used to optimize the parameters of the support vector machine model to build a diagnosis model. Finally, the d-S evidence theory fusion method is used to fuse the diagnosis results of different support vector machine models to obtain the final fault diagnosis results. There are many traditional fault detection algorithms, including methods based on fault current and voltage amplitude [4], Kalman filter [5], etc. But the conventional algorithm is simple, poor robustness and sensitive to noise. In recent years, with the development of deep learning, more and more methods have been applied to the classification of distribution network faults, such as target detection method [6], image-based detection method [7]. However, image-based methods can only identify external faults, cannot detect internal electrical faults, and conventional neural network methods are susceptible to noise. To improve the robustness of the algorithm and alleviate the influence of noise on the algorithm, a power fault detection method based on a self-coding neural network is adopted in this paper. The feature vector of the signal is extracted by the self-coding neural network, which can effectively filter out noise, retain key information and improve the robustness of the algorithm.

2. Technical Work Preparation

DNN neural network is composed of the hidden layer and the input and output layer. Through a large amount of data to train the model and mine the characteristic relationship between the data, the network structure can fit the mathematical model of the problem, and achieve classification recognition and other functions.

The basic method of neural network fault detection is to decompose the signal by wavelet transform, and take the decomposed information as the input of neural network, and minimize the output and calibration categories by fitting neural network to achieve fault detection.

First of all, the fault signal is broken down by the orthogonal wavelet function, and the wavelet coefficient is obtained, as shown in equation 1:

$$\begin{cases} d_l^{j-1,2n} = \sum_{k\in Z} d_l^{j,n} g_{k-2l} \\ d_l^{j-1,2n+1} = \sum_{k\in Z} d_l^{j,n} h_{k-2l} \end{cases} \tag{1}$$

Among them, the $d_l^{j,n}$ is the coefficient of the nth wavelet base corresponding to the decomposition of the jth wavelet, and g and h are the high-pass filter coefficient and the low-pass filter coefficient of the wavelet function and the scale function. Using subband energy as a secondary feature of a cable failure:

$$E = f\left\{\left|d_l^{j,n}\right|^2\right\} \tag{2}$$

Where f is regular.

The output vector of the network is Y, and T is the fault marker category, where Tj represents the jth fault type and T is the c-type cable fault collection. The distance between the identified cable fault feature and the Class j fault signal is:

$$d(X,T_j) = |Y - T_j| = \sqrt{\sum_{i=1}^{n} |y_i - t_i|^2} \tag{3}$$

For the initial fault feature input vector, satisfy $d(X,T_j) = min\, d(X,T_q), q = 1,2,\dots\dots,c$, X∈Tj.

Autoencoder is an unsupervised learning algorithm that compresses the input and restores the output. autoencoder includes encoders and decoders. Assuming that there is an input data x, The compressed information after encoding is calculated by encoder F, and the output information after decoding is obtained by decoder G. In the training process of the autoencoder device, the refactoring error L is used as the target function, and the mean error is used as the reconstruction function and. The cross-entropy loss function can also be used as the reconstruction error of the binary input vector, and the autoencoder can be trained by minimizing the objective function. After training, the coded signal can replace the original input as a classification feature and usually has better noise resistance than the original signal. To prevent the autoencoder's output from being equal to the input, which would make it meaningless, we can add some restrictions or rules to the autoencoder unit.

3. Network Model

The autoencoder extracts the input into a feature vector with less information, which can represent the characteristics of the input. By representing input by compressed feature information, redundant information can be effectively reduced and the efficiency of network identification can be improved. The specific structure of the network is shown in Figure 1 and is divided into the autoencoder section and the fault identification part. Since the encoding network compresses the input and then restores it, by restoring the input as much as possible by a small number of features, and by constraining the similarity of the input-output, the compressed small number of features can be the main information of the input. The compressed information is used as the input of the fault identification part, and finally, the fault type is output by softmax, wherein the model is trained as part A, i.e. the whole network, and the fault detection is part B, that is, the decoder part does not participate in fault detection.

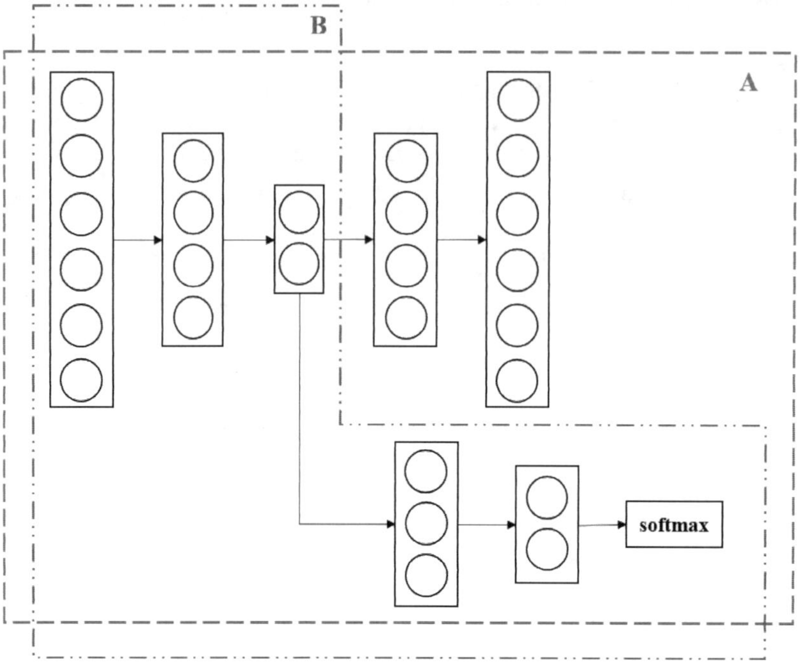

Figure 1. Network structure diagram

The loss function of the autoencoder uses L1 distance to constrain the coding result, as shown in the equation (4):

$$L_a = \|F(x) - G(F(x))\|_1 \tag{4}$$

The loss function of the fault identification section uses cross-entropy loss, as shown in equation (5):

$$Lc = -\sum_{i=1}^{K} y_i \log(p_i) \tag{5}$$

Therefore, the total loss function of the network can be expressed as an equation (6):

$$LOSS = L_a + Lc \tag{6}$$

4. Simulation Experiments

Simulation experiments are conducted based on the IEEE34 node model, and the system structure is shown in Figure 2.

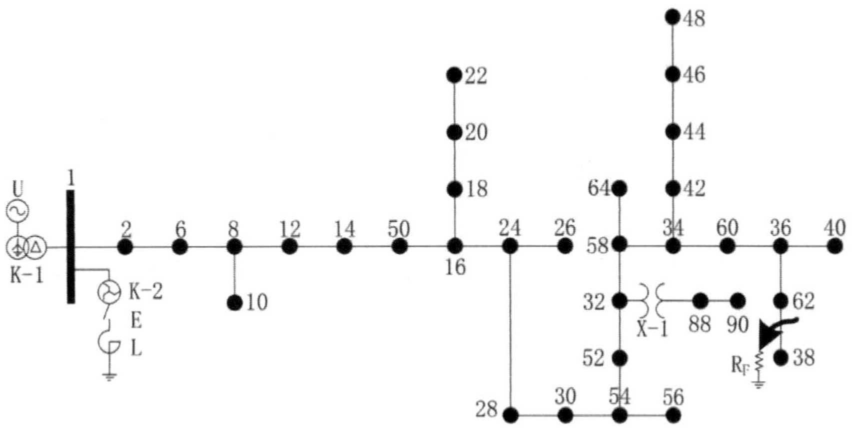

Figure 2. System simulation structure diagram

The training sample is the fault sample of the neutral point ungrounded system when E is on. Ten fault nodes were selected to test the initial phase Angle and transition resistance of six fault types, and the number of training sets was 3600. The test sample contains more data types than the training set. The transition resistance exceeds the range of the training sample (0 ~ 1kω) and reaches 1.2 kω. The specific information of the training set is shown in Table 1.

Table 1. Sample information

	The fault node	Transition resistance	Fault initial phase Angle
Train set	806,810,814,822,826, 838,840,844,856,864	0,50,100, 200,500,1000	0,18,36, 90,126,162
Test set	802,812,816,818,830, 832,834,848,860,862	20,70,150, 300,600,1200	18,54,72, 90,108,144

In order to verify the effect of the algorithm in the case of noise, the fault recognition rate of wavelet coefficient [8], BP neural network [9] and self-coding neural network algorithm are compared, and the test is carried out in the case of without noise and with noise. The experimental results are shown in Figure 3. As can be seen from FIG. 3, among the 10 types of faults, two-phase short circuit has the highest fault identification rate. Without noise, the recognition rates for two-phase faults are 91.95%, 94.6% and 97.4%, respectively, which are 15.45%, 10.35% and 1.46% higher than those for three-phase faults, respectively. In addition, in the case of noise, the average recognition rate of the three algorithms has a certain degree of decline. It can be seen that the overall recognition rate of the wavelet coefficient method is lower than the other two algorithms, and is greatly affected by noise. BP neural network is slightly better than wavelet coefficient method, but it is also affected by noise because the whole input is computed. Based on the autoencoder neural network algorithm, the main features of the data are extracted to the hidden layer through the encoder, which is used as the basis of fault recognition and can effectively deal with the influence of noise. Experiments show that the algorithm based on autoencoder neural network has the best effect and strong anti-interference ability.

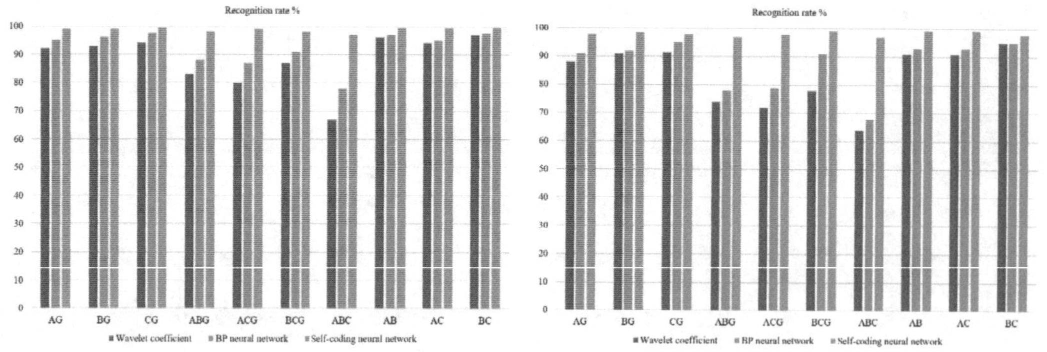

(a) without noise (b) with noise

Figure 3. Comparison of detection algorithms

The attention mechanism is a common optimization method used in the codec. Attention module is added into the network model, and the autoencoder part is shown in Figure 4. The attention module can improve the influence of key information on the calculation results by assigning different weights. For distribution network fault detection, the attention module strengthens the weight of useful information in the feature map and weakens the weight of irrelevant information, to reduce noise interference. The data types that did not participate in the training before and after adding the attention module were compared experimentally, and the results were shown in Figure 5. It can be seen that the network with added attention module has better anti-interference and generalization ability based on the autoencoder network.

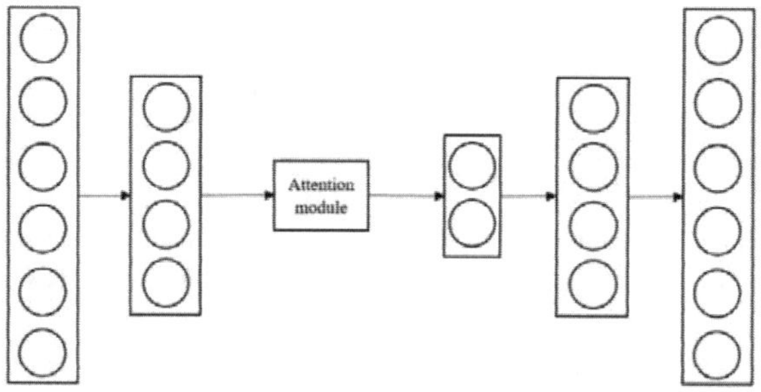

Figure 4. Autoencoder structure with attentional module

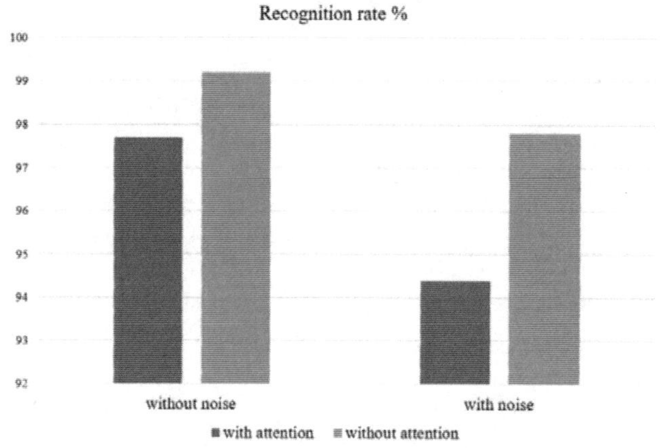

Figure 5. Comparison of recognition rate of added attention module

5. Conclusion

In order to improve the identification ability and anti-interference ability of distribution network fault detection, an anti-noise fault detection method is proposed. The experimental results show that the features obtained by self-encoder compression can effectively represent the input information, reduce the noise interference, and effectively detect the power fault of distribution network. However, the current research methods remain in the recognition of known fault types, and how to detect and deal with unknown fault types is a major direction of future research. It is very important for the stable and safe operation of the power grid to identify the fault information before the distribution network fault occurs, and it can recover the inestimable loss in some serious cases. This method has some contribution both in economic level and technical level.

Acknowledgments

This work was supported by State Grid Sichuan Electric Power Company Science and Technology Project under Project Numbers SGSC0000DDJS2000106.

References

[1] Y. Zhang, D. Zhang, C. Wu, et al. "A New Principle of HVDC Line Backup Protection based on Low-frequency fault Component F Test" China Southern Power Grid Technology, 2021,15(07),pp. 26-33.
[2] D. Fang, F. Zhang, W. Hou , et al "Research on Transmission line fault Detection Based on Discrete Wavelet Transform" Information Technology,2019(10),pp. 110-114+120.
[3] X. Wu, W. Cao, D. Wang, et al. "Transmission line fault diagnosis method based on multi-support vector machine model" High voltage technology,2020,46(03),pp. 957-963.
[4] L. Song, X. Han, M. Yang, et al. "Fault detection and protection in a meshed MMCHVDC grid based on bus-voltage change rate and fault component current" Electric Power Systems Research,2021(201),pp. 1-10.
[5] Z. Liu, Z. Wang, Z. Ji, " Fault diagnosis method of power converter based on reverse kalman filter " Power system protection and control,2019,47(09),pp. 19-26.
[6] J. Cui, X. Hou. "YOLOv4 Transmission line Fault Detection Algorithm Based on attention mechanism"Foreign Electronic Measurement Technology, 2021,40(07),pp. 24-29.
[7] S. Hao, R. Ma, X. Zhao, et al. "Fault Detection method for YOLOv3 transmission line based on Convolutional Block attention Model " Power Grid Technology, 2021,45(08),pp. 2979-2987.
[8] H. Liu, Q. Cheng, "Microgrid fault detection method based on dq transform and wavelet packet transform" Water and power energy science,2020,38(08),pp. 188-191+157.
[9] J. Chen, Y. Gao, Y. JIN. "Microgrid Fault Diagnosis Method Based on MODWT and BP Neural Network" Journal of Shanghai Electric Power University, 2021,37(01),pp. 57-60+77.
[10] W. Li, Y. Li, J. Shi, et al. "Power grid fault detection method based on cloud platform and improved isolated forest" 2021 IEEE 5th Advanced Information Technology, Electronic and Automation Control Conference (IAEAC), 2021, pp. 864-867.
[11] J. Xu and H. Yue. "Research on Fault Diagnosis Method of Power Grid Based on Artificial Intelligence" 2020 IEEE Conference on Telecommunications, Optics and Computer Science (TOCS), 2020, pp. 113-116.
[12] F. Mohammadi, G. Nazri and M. Saif. "A Fast Fault Detection and Identification Approach in Power Distribution Systems" 2019 International Conference on Power Generation Systems and Renewable Energy Technologies (PGSRET), 2019, pp. 1-4.
[13] X. Cui, B. Wang, X. Dong, et al. "Experimental Research on Typical Single-Phase High Resistance Grounding Fault Detection Methods in Resonant Grounding Distribution System" 2019 IEEE 8th International Conference on Advanced Power System Automation and Protection (APAP), 2019, pp. 432-436.

Accurate Measurement of DC Electric Energy in Power Plant

Dan Gong[1], Chuang Zhang[2], Jian Ma[2], Congyu Zhang[2], Xinyu Liu[2] and Shuai Guo[2,*]

[1]State Grid Jiangsu Electric Power co., LTD Marketing Service Center, China
[2]Heilongjiang Electrical Instrument Engineering Research Center Co., Ltd., China

[*]Corresponding author email: skyard2010@126.com

Abstract. With the development of electric vehicles, the corresponding charging facilities are gradually accelerating the construction. Aiming at the corresponding problems existing in the domestic electric vehicle charging facilities, such as management and control, operation and maintenance and testing, this paper studies the integrated test system for AC / DC charging facilities, and introduces the design concept, hardware scheme, software scheme and processor platform of the system in detail. The system is based on GB/T18487.1-2015, NB/T33008-2018 and other test standards, and can be used for type experiments and performance tests of charging piles. The research results in this paper are conducive to the development of charging inspection facilities in the direction of integration, digitization, and automation. It is of great significance for charging facilities to reduce operation and maintenance costs and improve charging service level. It is conducive to promoting and ensuring the sustainable development of the domestic electric vehicle industry and the needs of implementing China's long-term development plan and energy conservation and emission reduction policies.

1. Introduction

Energy crisis and environmental problems have been on the focus list of countries around the world in recent decades. Traditional fuel vehicles are currently facing the depletion of Earth's oil resources, and the environmental pollution problems have been increasingly valued. The emergence of new energy technology provides a solution to the problem of future energy sources, also flourished in the automotive industry. The electric vehicles have been vigorously promoted in the world, as a pollution-free representative. The mainstream direction of the future development of the automotive industry, electric vehicles can effectively reduce a series of environmental pollution problems caused by traditional fuel vehicles.For achieve sustainable development of electric vehicles and being favored by the society and the market, in addition to the continuous exploration and advances in electric vehicle technology, charging facilities as a foundation are also very critical.

This paper mainly studied and designed the integrated test system of AC / DC charging facilities. Based on the system design concept, the hardware scheme, the software scheme and the processor platform, this paper introduces the system composition principle in detail. The system can simulate the real load situation of the charging pile, and test the power supply of the charging pile and the electrical parameters of the charging output.The system is conducive to the development of charging testing facilities to integration, digitalization and automation. This article is of great significance for reducing the operation and maintenance cost and improving the charging service level of charging facilities. The research results of this paper are conducive to promoting and ensuring the sustainable development of the domestic electric vehicle industry, and to meet the needs of national long-term development plans and energy conservation and emission reduction policies.

2. Design Solutions

2.1. Design of Integrated AC/DC Charging Facility Inspection and Testing System

Figure 1 shows the total system topology, which shows that the system mainly consists of a mobile platform, an integrated test system for AC charging piles, an integrated test system for DC chargers, three-phase adjustable voltage, and safety regulation equipment.

The schematic diagram of the integrated AC/DC charging facility inspection and test system is shown in Figure 2, which consists of an integrated AC/DC charging facility electrical performance test device, a three-phase AC programmable power supply, an electronic load, a safety test device, a power analysis device, a DC standard power meter, an AC standard power meter, etc.

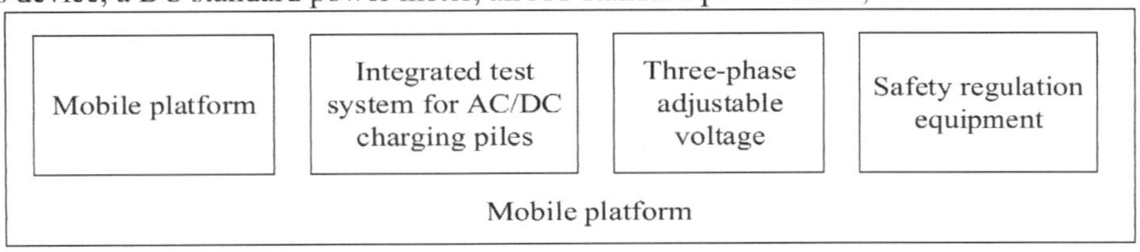

Figure 1. General system topology diagram.

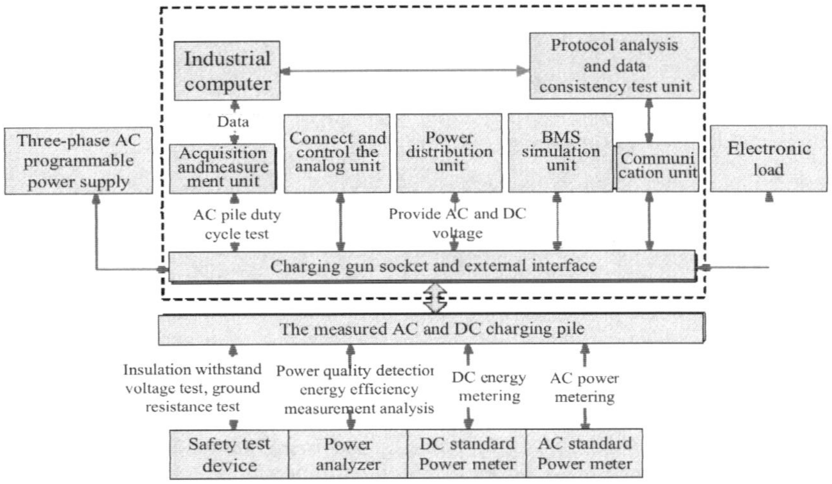

Figure 2. Integrated AC/DC charging facility inspection and test system.

In particular, the DC charger system consists of a three-phase adjustable power supply, a DC charger laboratory calibrator, a power pair interface and current and voltage sensors, back-end calibration software, and a DC load box.

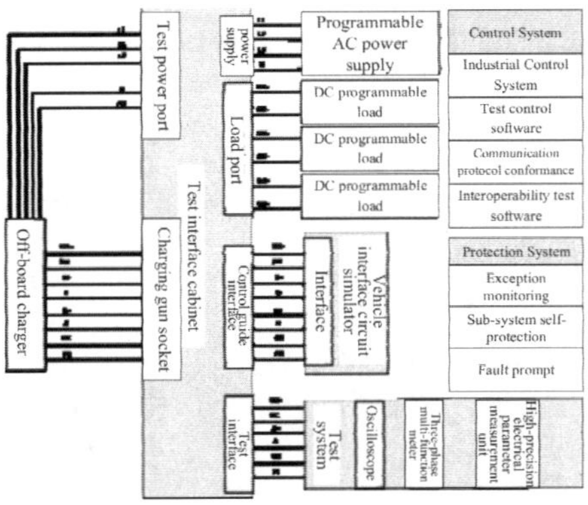

Figure 3. DC charger integrated test system.

The power input of DC charger uses three-phase adjustable power input, and the power docking interface has built-in precision voltage and current sensors to convert the high-voltage, high-current AC signal into a small voltage signal output for DC charger calibration device to collect, and the DC charger calibration device also collects the output DC voltage and DC current of the DC charging station to realize the testing functions of charger efficiency and power factor. Firstly, it is difficult to use electric car as the load in the laboratory due to the site requirements, and secondly, it is convenient to use DC load with BMS simulator to test each calibration point of the charger output, which is not available when electric car is used as the load.

The AC charging pile system consists of AC charging pile calibration device, three-phase adjustable power supply, three-phase standard meter, background calibration software and load box. By controlling the load box load, it can test the human-machine interaction function, metering function, payment transaction function, self-test function, etc. By controlling the AC charging pile calibration device to simulate the electric vehicle control guide function, read the power of the three-phase standard meter in the standby state to test the power consumption of the AC charging pile.

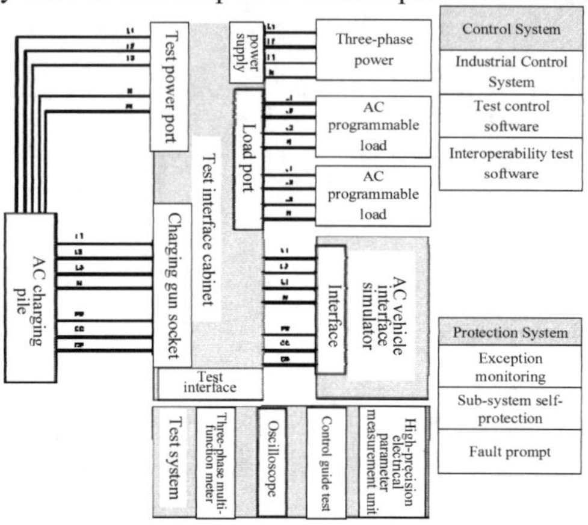

Figure 4. Integrated test system for AC charging piles.

2.2. Hardware Solutions

The hardware component consists of a test platform and a communication concentrator. The test platform is an HP EliteDesk mainframe based on Intel Core multi-core CPU processor with 3.4GHZ, scalable to 3.8GHz, 6M Level 3 cache, 8G DDR4 memory, 2T hard disk, 24" LCD monitor, wireless network card, wired network card, USB, SATA, HDMI, VGA, DisplayPort, Rj45, PCI-E interfaces to meet the demand for strong performance.

2.3. Software Solutions

Software framework is designed using the MVC framework. The model contains the core functionality and data, the view displays the information to the user (consisting of several views), and the controller processes the information entered by the user.

All functions of the software are written using modular design principles to provide maximum flexibility and expandability to suit the needs of different charging pile inspection projects. The software management system mainly contains test control software, communication protocol conformance test software, interoperability test software, communication program module, parameter configuration module, status monitoring module, database, fault monitoring program module, test project program, BMS control program module, report system program and other auxiliary engineering programs. Each program module is designed independently to achieve a good effect of high cohesion and loose coupling.

2.4. Processor Platforms

The processor in the system uses ADSP609, which has two 500MHz DSP cores, which can be turned on simultaneously, and its data processing capability is very high. Bf609 can access 4G address space. In addition to the internal ROM and RAM memory of the BF609 processor, this solution is also designed with DDR2 and NOR FLASH. At the same time, access to both FPGAs and the extended asynchronous bus are mapped uniformly to the BF609 processor's 4G byte address space. The mapped addresses in the scheme are shown in Table 1.

Table 1. Mapping address.

Chip Selection	Data width	capacity (bytes)	starting address	end address	description
D\M\C\0_ \C\S\0\	16	128M	0x0000 0000	0x07FF FFFF	DDR2
S\M\C\0_ \A\M\S\0\	16	16M	0xB000 0000	0xB0FF FFFF	NOR FLASH
S\M\C\0_ \A\M\S\1\	16	384K	0xB400 0000	0xB405 FFFF	IO control of the AD acquisition board
S\M\C\0_ \A\M\S\2\	16	127	0xB800 0000	0xB800 007F	FPGA0
S\M\C\0_ \A\M\S\2\	16	895	0xB800 0080	0xB800 03FF	FPGA_FT3
S\M\C\0_ \A\M\S\3\	16	64M	0xBC00 0000	0xBDFF FFFF	extensions

The system uses DDR2 technology for the memory ram design of the ADSP609 core board. The memory chip uses a 128MB capacity DDR2 chip MT47H64M16, connected to the BF609's DMC interface, which can support up to 256MB capacity of DDR2 or LPDDR memory.

DDR2 reference voltage VREF is generated by two 1K Ohm 1% accuracy resistors to 1.8V divider, clock differential line terminals are connected through a 100 Ohm resistor, and the connection between BF609 and DDR2 chip MT47H64M16 are strung with a 0 Ohm resistor. At the same time, in order to improve the data storage capacity of the system processor motherboard, the program uses a 16MB capacity NOR FLASH chip PC28F128P33, connected to bank0 of the SMC interface of the BF609, which is selected by the chip select signal S\M\C\0_\A\M\S\0\. The SMC0 interface of the BF609 has 25 address lines with a minimum addressing of 2. The reset signal of NOR FLASH is selected by SYS_RSTOUT_N and FLASH_RST_N through 0 ohm resistor "2 select 1" solder, the default solution is that the reset signal is generated by the watchdog chip FLASH_RST_N to NOR FLASH.

3. Experimental Results and Analysis

In the paper, DC metering and AC metering were tested using an integrated AC/DC charging facility testing test system.

3.1. DC Metering

3.1.1. DC Current Metering

Table 2. DC current metering result.

Measurement point (A)	Relative error (%)
5	+0.016
10	+0.010
50	+0.018
100	+0.028
200	+0.018
250	+0.006

3.1.2. DC Voltage Metering

Table 3. DC voltage metering result.

Measurement point (V)	Relative error (%)
100	-0.001
200	+0.002
500	+0.004
1000	+0.002

3.1.3. DC Power Metering

Table 4. DC power metering result.

Measurement points (kW)	Relative error (%)
1 (100V/10A)	+0.009
5 (500V/10A)	+0.014
10 (1000V/10A)	+0.013
10 (100V/100A)	+0.026
50 (500V/100A)	+0.010
100 (1000V/100A)	+0.009
25 (100V/250A)	+0.030
125 (500V/250A)	+0.005
250 (1000V/250A)	+0.007

3.2. AC Metering

3.2.1. AC Current Metering

Table 5. AC current metering result.

current	relative error		
	A-phase	Phase B	C-phase
0.6	+0.005	+0.006	-0.001
1	+0.012	+0.004	+0.008
10	+0.007	+0.005	-0.003
32	+0.003	+0.003	+0.001
63	+0.006	-0.001	-0.002
70	+0.004	+0.001	+0.001

3.2.2. AC Voltage Metering

Table 6. AC voltage metering result.

currents	relative error		
	A-phase	Phase B	C-phase
110	-0.002	-0.008	-0.006
220	-0.003	-0.002	-0.001
264	-0.002	-0.005	-0.004

3.2.3. AC Frequency Metering

Table 7. AC frequency metering result.

currents	current	power factor	frequency	absolute error		
				A-phase	Phase B	C-phase
			47.5	-0.004	-0.004	-0.004
220	12	1.0	50	-0.002	-0.002	-0.002
			52.5	-0.003	-0.003	-0.003

3.2.4. AC Phasor Metering

Table 8. AC phasor metering result.

currents	current	phase (waves)	absolute error		
			A-phase	Phase B	C-phase
		0	+0.006	-0.008	-0.018
220	12	30	+0.012	+0.010	-0.001
		-30	+0.014	+0.019	+0.017

3.2.5. AC Active Power Metering

Table 9. AC active power metering result.

currents	current	power factor	relative error	
			A-phase	compatible
220	0.6	1.0	+0.002	-0.003
	1		+0.009	+0.007
	12		+0.004	+0.003
	32		+0.000	+0.001
	63		+0.004	+0.000
	70		+0.001	+0.001

3.2.6. AC Active Energy Metering

Table 10. AC active energy metering result.

currents	current	power factor	Relative error of the combined phase
220	0.6	1.0	-0.020
	1		+0.004
	12		+0.013
	32		+0.016
	63		+0.010
	70		+0.006

The integrated AC/DC charging facility detection test system is applied to test the DC metering and AC metering respectively, and error analysis is given. The results show that the charging facility metering achieved by using this system is highly accurate and has a wide metering range, which can achieve accurate metering for integrated testing of AC/DC charging facilities.

4. Conclusion

This paper focuses on the comprehensive test system for AC-DC charging facilities, which can simulate and test the operating characteristics and load-bearing capacity of charging piles, and realise the electrical function testing of AC-DC charging piles according to the relevant national and industry technical standards. The experimental results also show that the system is able to complete the testing of DC charging piles and AC charging piles. At the same time, its comprehensive testing of AC and DC charging facilities can achieve a high accuracy and a wide measurement range. The fully automatic testing system for charging facilities studied in this paper is conducive to the development of charging testing facilities in the direction of integration, digitalisation and automation, and can play a positive role in promoting the development of operation and maintenance testing of charging facilities, which has important guiding significance in reducing operation and maintenance costs and improving charging service levels, and is conducive to promoting the benign development of the electric vehicle industry and implementing national long-term development planning and energy-saving and emission reduction policies.

References

[1] Song Peng, Pan Feng, Lin Guoying, et al. Analysis of the influence of ripple on DC energymetering in EV charging [J]. Electrical Measurement & Instrumentation, 2018, 55(1): 46-52.
[2] Hou Huahai. Research on non vehicle DC charger DC electric energy meter verification device for electric vehicle [D]. Suzhou: Jiangsu University Master's Degree Thesis, 2018.
[3] Zeng Yongling. Research and design of DC electricity metering equipment in off-board charger [D]. Nanchang: Nanchang University Master's Degree Thesis, 2011.
[4] Zhou Lei. Research and implementation of portable high-precision DC charging pile measurement and verification system [D]. Chengdu: University of Electronic Science and Technology of China Master's Degree Thesis, 2020.
[5] Zhu Xianwen, Xiao Yong, Tian Yang, et al. Research on DC charging interoperability test method for electric vehicles [J]. Electrical Measurement & Instrumentation, 2020, 57(4): 22-29.

Research on Smart Electricity Meter Based on LoRa Technology

Yang Yang[1], Chuang Zhang[1], Weiwei Yan[2], Qinci Liang[2], Minsheng Xuan[3] and Shuai Guo[4,*]

[1]Heilongjiang Electrical Instrument Engineering Research Center Co., Ltd., China
[2]State Grid Quanzhou Electrdc Power Supply Company, China
[3]Fuzhou Smart Power Technology CO.LTD, China
[4]Heilongjiang Electrical Instrument Engineering Research Center Co., Ltd., China

[*]Corresponding author email: skyard2010@126.com

Abstract. For the current wireless communication methods of smart electricity meter,, the communication cannot simultaneously take into account low power consumption and long distance. In this paper, the smart electricity meter based on LoRa technology is studied, using FM3308 chip as the control core, achieving low-power and long-distance communication through LoRa technology, and the nti-interference design is carried out from both hardware and software aspects. The research content of this paper realizes the coexistence of long-distance transmission and low power consumption of smart electricity meter communication, and maintains the better anti-interference ability of smart electricity meter, which is conducive to the realization of long-distance centralized meter reading behavior and can greatly reduce the arrangement of repeaters, reduces the cost investment while lowering the design requirements and helps to promote the construction of smart grid.
Keyword: LoRa; Smart meters; Long-range communication; Low power consumption.

1. Introduction

As a large population and economic country, China's electricity consumption is very huge, about 6 trillion kilowatt hours last year. At present, the development of smart and strong power grid and ubiquitous Internet of Things is a major macro strategy of the country, and as an indispensable key part of the power grid, the improvement of the degree of intelligence of electricity meters is a manifestation of this goal. Facing the number of meters close to 500 million nationwide, the pressure to solve the reading and inspection of meters is a prominent and the task is heavy. The current smart electricity meters use wired and wireless communication, of which, wired communication methods include RS485 communication and power line carrier (PLC) communication. Wireless communication methods in recent years, rapid development, but these wireless communication methods can not simultaneously take into account the low power consumption and long-distance transmission problems. LoRa technology is an ultra-long-range, low-power IoT solution provided by Semtech with a maximum data transmission speed of up to 300kbps, a maximum link budget of 168dB, a maximum RF transmit power of +20dBm, a receive sensitivity of -146.5dBm, the maximum transmission distance in the suburbs can reach 15km, in the urban area can reach 2-5km.

This paper focuses on the application of LoRa low-power remote communication in smart electricity meter, and designs a smart electricity meter based on LoRa technology according to the existing smart electricity metering scheme. This research plays a positive role in promoting the development of smart electricity meter and accelerating the construction of smart grid.

2. Hardware Design Solutions

2.1. System Design

Based on the common electricity meter and its high performance integrated circuit, it realizes the accurate measurement of energy metering and the automatic control of energy meter, and meets the needs of remote communication and remote measurement and control of electricity meter. MCU adopts FM3308 special chip of Shanghai Fudan Micro Company, and the hardware circuit mainly includes power supply circuit, power metering circuit, communication circuit, display module circuit and relay driving circuit.

Firstly, the voltage and current are sampled and then connected to metering chip RN8209C through the variable ratio back-end, RN8209C provides three ADC instantaneous sampling values,, the sampled data is calculated and stored in the corresponding registers, then RN8209C transmits data with MCU through UART, and MCU communicates with the upper computer through RS232, while controlling the state of the data including LED lights, LCD display. This electricity meter can realize both 485 communication and LoRa wireless communication.

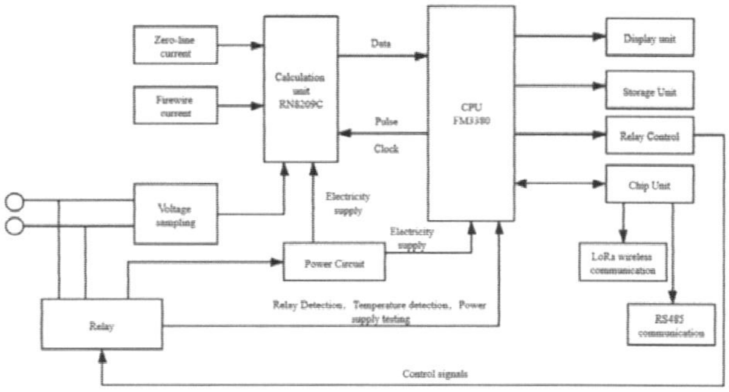

Figure 1. Hardware design framework of the system.

This electricity meter is designed on the basis of the traditional electricity meter and the new wireless communication module, which greatly increases the wireless communication distance.

2.2. Power Supply Circuit Design

The power supply circuit converts the 220V AC power to the power required by each unit circuit and provides power to the main control chip circuit, metering chip circuit, communication circuit, built-in relay control circuit and other circuits.

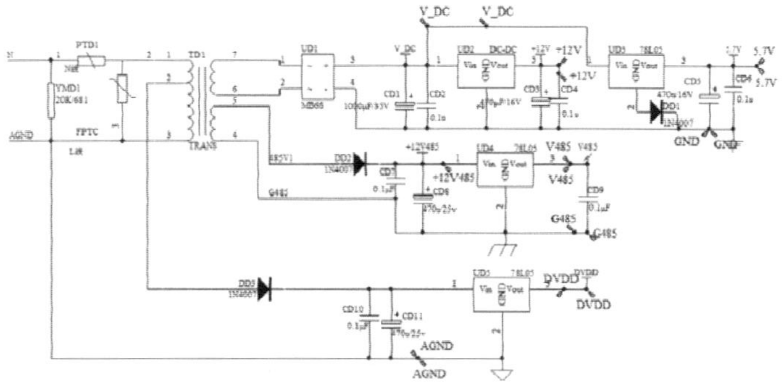

Figure 2. Power supply circuit.

The input is connected in parallel between the L and N lines with a varistor YMD1, which is used to protect the electricity meter. When the surge voltage occurs, the excessive voltage makes the resistance of the varistor decrease, the current flowing through the circuit becomes larger, playing the purpose of shunting, to avoid the voltage is too large; after the surge voltage stops, the resistance of the resistor becomes back to the previous high resistance state.

2.3. Metering Circuit Design

The external unit of the RN8209C metering chip consists of a power supply, SPI, pulse output, and current-voltage analog input circuit. To ensure the EMC performance of the system and to reduce the power supply noise and its ripple impact, decoupling capacitors are connected in parallel to the input of the power supply pins and the PCB wiring process ensures that the analog ground and digital ground are kept in a large area for grounding. The metering module circuit is shown in Figure 3.

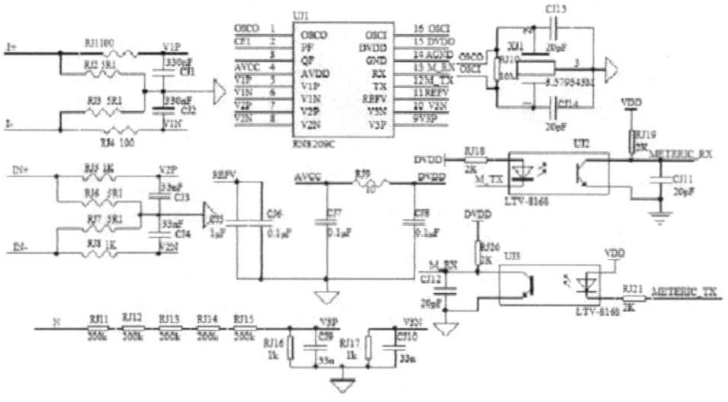

Figure 3. Metering circuit.

2.4. LoRa Communication Module

The LoRa communication module selects Semtech's SX1276 chip, which forms the LoRa wireless transceiver module with the peripheral transceiver circuitry. The module uses a LoRa remote modem, which is suitable for long-distance communication, strong anti-interference ability and low power consumption. Figure 4 shows the analog switch, and UO in the figure is the RF changeover switch. The high and low levels of the CTRL pin and the VDD pin control whether the antenna is connected to a receive or transmit circuit. If VDD is high and CTRL is low, RF2 turns on the RFC, if VDD is low and TRL is high, RF1 turns on the RFC.

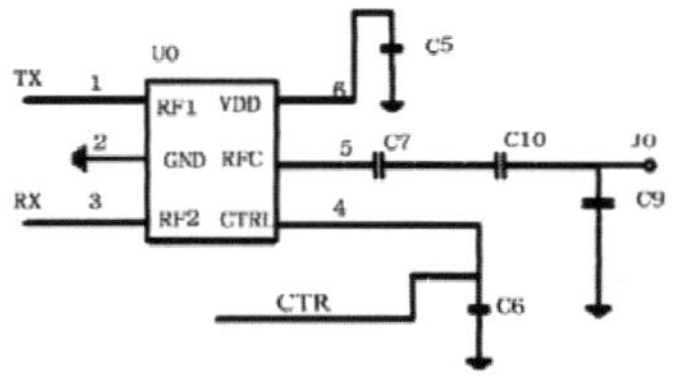

Figure 4. Receive and transmit switch circuit.

Figure 5 shows the SX1276 receiver circuit diagram. L1,L2,C3,C4 form the filter circuit, need to select the appropriate parameters according to the actual communication frequency, U1 is SAW Filter. The SAW Filte is to filter out signals outside of the communication frequency, allowing only signals of a specific frequency to pass through. Without the filtering effect of the SAW Filte, out-of-band signals would interfere with the operation of the SX1276.

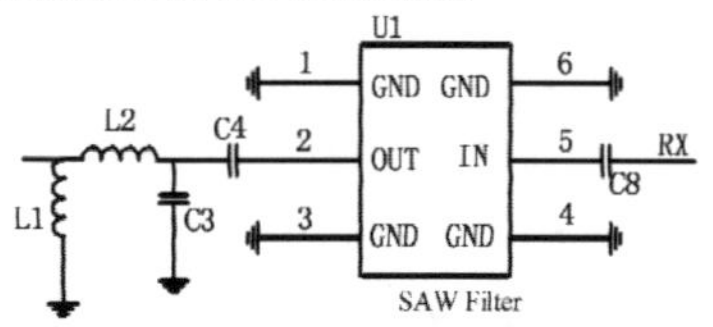

Figure 5. Receiving circuit.

Figure 6 shows the LoRa transmit circuit, the transmit circuit consists of a filter, L3 and C11 form a series resonant circuit, C16,LS and C17,L6 form two parallel resonant circuits, according to the different communication frequencies, choose the appropriate device parameters to achieve the best transmit state.

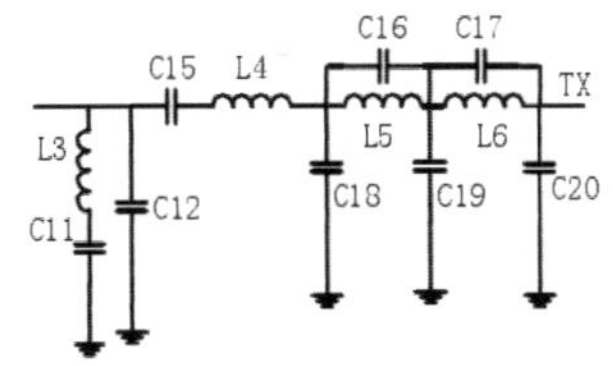

Figure 6. Transmitter circuit

2.5. ESAM Encryption Circuit

The ESAM encrypted security circuit is designed as shown in Figure 7. The ESAM chip communicates with MCU through the 7816 and can be fully powered by MCU due to its own ultra-low power consumption, i.e., powered when needed and powered off when use is complete.

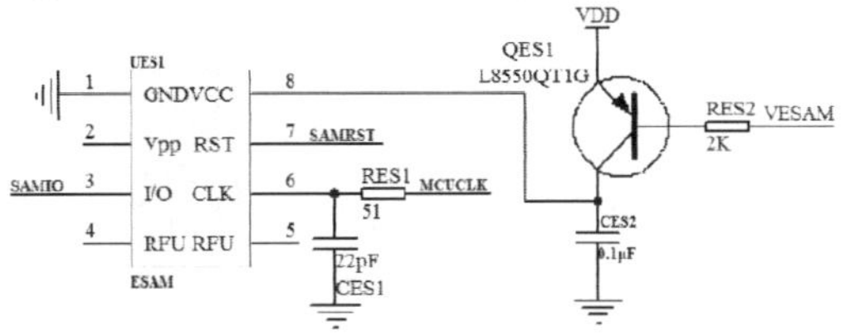

Figure 7. ESAM encryption circuit.

2.6. Software Programmes

The software is developed using embedded C language, with a top-down modular design in terms of the energy meter function hierarchy, identifying the interfaces between the modules. The software as a whole uses a three-layer architecture with a modular design. The data is divided into important data and non-important data according to the importance of the data. The important data, such as power data, is stored in two copies in external EEPROM, and the validity of the data is verified during storage, which greatly increases the reliability of the data; the non-important data, such as record data, is stored in one copy in EEPROM because of the large amount of data and is not very important. This can save EEPROM space and improve the speed of reading and writing data. The whole architecture is based on the idea of single-process multi-tasking state bit processing mechanism.

2.7. Workflow Design

The LoRa wireless module reception mode has two groups of modes: continuous reception and single reception. Figure 8 illustrates the data reception process of LoRa wireless module by continuous reception mode: firstly, the default FIFO address pointer is set to the received base address, and if the module receives the leading code, it receives the data header first, then the data packet, and after receiving the CRC check passes, the data is read and saved. If the CRC checksum does not pass, continue to receive other packets and repeat the above steps. When all packets are received completely, the module will automatically jump to idle mode to reduce the power consumption of the module and end the data reception process of the module.

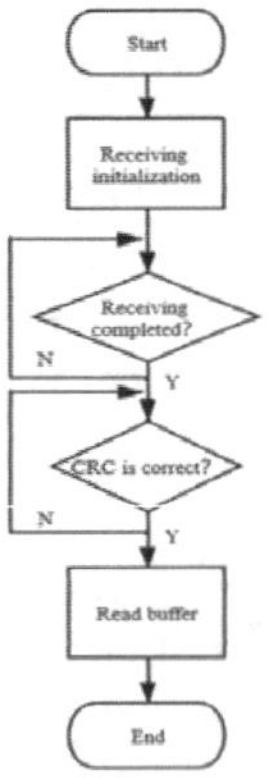

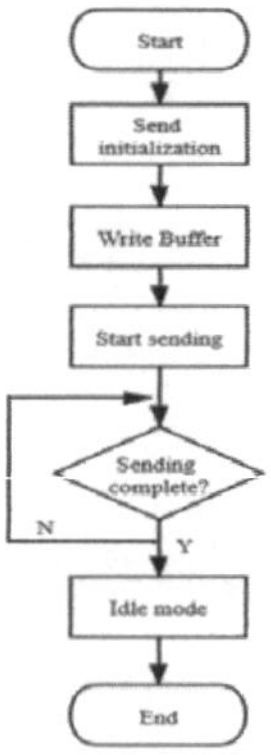

Figure 8. LoRa module data reception flow. **Figure 9.** LoRa module data sending flow.

The LoRa send flow is shown in Figure 9, the send mode starts by initializing the registers, then the module fills the FIFO with data when it is in idle state and writes inside the write buffer. When the data filling is completed, it enters the send state data is sent automatically. When the data is sent, a data send interrupt instruction is generated and then the module enters idle mode again to reduce module power. Data send and receive data is similar, before each fill data must be initialized FIFO pointer and send data base address

2.8. Anti-interference Design

For reducing energy meter errors, this paper focuses on the design of both hardware anti-interference and software anti-interference:

(1) The majority of errors are found to come from interference through test experiments, and interference from the power supply is the main cause. The design of the power supply as a linear power supply and the power supply in modules can greatly reduce the interference of the power supply design part to the accuracy of the energy meter. The power supply is divided into modules, which have the advantages of superior voltage regulation, good stability and low cost.

(2) Circuit design layout to ensure the isolation of strong and weak power, strong printed wiring at least 4-5mm away from weak printed wiring, while digital ground and analog ground should be connected by a point way to improve anti-interference capability. High-voltage alignment should be short, and to keep a distance from the small signal alignment.

(3) The circuit current acquisition process requires the conversion of AC current into equivalent small signal AC current. The power meter designed in this paper uses a manganese copper shunt. Here the same batch and accuracy of 1% chip resistor in series is used to reduce the resistor power consumption and withstand voltage, thus reducing the error.

(4) Watchdog circuit. Use the watchdog circuit in the master chip to set a reset when the program enters a dead loop, reinitialize the program and run the program.

3. Experimental Results and Analysis

The test was performed using a reference voltage of 220V, a reference current of 5A and a reference frequency of 50Hz.

3.1. Power consumption

Table 1. Power consumption test result.

Voltage lines		Current line
active power	apparent power	apparent power
No communication state ≤1.5W ≤3.0W in communication state	≤10VA	≤1VA

3.2. LoRa communication metrics

Table 2. Communication metrics test result.

communication distance	>3000m
transmission rate	18-37.5kbps
power wastage	Emission power consumption: <125mA Receiving power consumption: <16mA Hibernation power consumption: <7uW
Communication success rate	Single read success rate: >98%

3.3. Anti-interference capability

Table 3. Anti-interference capability test result.

Test environment	affected (usually adversely)
electrostatic discharge high frequency electromagnetic field electrofast transient pulse train surge	Not damaged or not materially affected, in accordance with the requirements in GB/T 17215.321-2008

From the experimental results, it can be seen that the LoRa technology-based smart electricity meter designed in this paper has excellent low-power consumption and long-distance communication performance, and has good anti-interference ability, which can avoid interference in the reading process and improve the success rate of meter reading.

4. Conclusion

In this paper, the study of smart electricity meter based on LoRa technology greatly increases the communication distance of electricity meter, which is conducive to the realization of centralized meter reading behavior at a long distance, and at the same time, the increased communication distance can reduce the arrangement of repeaters, which not only reduces the design requirements, but also reduces the laying cost. While ensuring a long transmission distance, the LoRa technology-based smart electricity meter also has the excellent performance of low power consumption, which cannot be satisfied by the traditional wireless communication technology at the same time. In addition to the above advantages, LoRa wireless module terminal nodes take low power design measures, which can greatly improve the battery life cycle, as well as stronger anti-interference and better transmission capability and stability performance. Therefore, LoRa technology-based smart meter technology has far-reaching practical and economic significance, and is the ideal choice for future electricity meter communication.

References

[1] Liu Shiwei 2020 *Design of Meter Reading System for Internet of Things Based on LoRa* (Shan Dong: Shandong University of Science and Technology) p 13
[2] XIAO Siqi 2018 *The Design of Intelligent Meter Reading System Based on LoRa* (Hu Nan: Hunan University) p 4-9
[3] Zhao Taifei 2016 *Design and Implementation of Smart Meter Reading System Based on LoRa (Computer Measurement & Control* vol 29*)* p 298-301
[4] Xin Xiong 2016 *Research and Design of Intelligent Electric Energy Meter Based on LoRa* (Hu Bei: Hubei University of Technology) p 12-15
[5] Torres A, Silva C, Filho H T. An Experimental Study on the Use of LoRa Technology in Vehicle Communication[J]. IEEE Access, 2021, PP(99):1-1.

Flat Currents and Solutions of the $AdS_3 \times S^3$ Superstring

Lixia Zhang[1], Lei Li[1], Zheng Miao[1] and Qi Fan[1]*

[1] Department of Basic Science, Air Force Engineering University, Xi'an, China

*Corresponding author email: 2502171064@qq.com

Abstract. For the covariant κ-symmetric Green-Schwarz superstring action on $AdS_3 \times S^3$ background, we show that Cartan one-forms associated with the flat currents satisfy not only the equations of motion but also the Virasoro constraint. So the flat currents with the spectrum parameter would make for a succession of classical conserved solutions ,which have the similar inexhaustible series of classical conserved quantities.

1. Introduction

The superstring theory on AdS background is important in the study of AdS/CFT correspondence [1], Metsaev and Tesytlin [2]proposed the Green-Schwarz superstring action on $AdS_5 \times S^5$ as a coset space $\frac{SO(4,2)}{SO(4,1)} \times \frac{SO(6)}{SO(5)}$ and the action possesses a local κ-symmetry. For this model, Bena, Polchinski and Roiban[3] found that it has a single-parameter flat current family, which can be available for construct an inexhaustible mass of non-local conserved quantities. This means that it may be precisely integrable. Shortly Hou et al.[4] constructed the flat currents separately in the fixed group base, those two ways are equivalent although their forms are different. Motivated by their work, the study of the integrability and the superstring conserved currents on AdS backgrounds has always been a research hotspot[5-9].

After Metsaev and Tesytlin, Rahmfeld and Rajaraman [10]presented the Green-Schwarz superstring action on $AdS_3 \times S^3$, Chen et al.[11]constructed the infinite flat currents of this model by using the method given by Bena, Polchinski and Roiban[3]. This implies that maybe $AdS_3 \times S^3$ string is integrable. For the ordinary sigma model without the fermionic Wess-Zumino term, the implications of constructing solutions, the solution transformations and the infinite conserved quantities are well known. Although the similar research on the string model has attracted more and more attention, it is still a challenge.

In this paper, for $AdS_3 \times S^3$ string, we show that the Cartan one-forms correspond to the flat currents are not only flat but also satisfy the Viraroso constraint and the equations of motion. Thus a one-parameter family of flat currents related to a whole string of the classical solutions. So a whole new string of solutions probably generated from an existing one. These solutions have the similar endless mass of conserved charges, but not equal to the original.

2. The Green-Schwarz Superstring on $AdS_3 \times S^3$ Background

Firstly, we review the action of Green-Schwarz superstring on $AdS_3 \times S^3$ background.For this case, the super coset manifold is $\frac{SU(1,1|2)_L \times SU(1,1|2)_R}{SO(1,2) \times SO(3)}$,and the bosonic part is $\frac{SO(2,2)}{SO(1,2)} \times \frac{SO(4)}{SO(3)}$. The commutation relations for the generators of the Lie superalgebra $SU(1,1|2)^2$ are

$$
\begin{gathered}
\{Q_I,\overline{Q}_J\} = 2\delta_{IJ}\left(iP_a\gamma^a - P_{a'}\gamma^{a'}\right) + \varepsilon_{IJ}\left(J_{ab}\gamma^{ab} - J_{a'b'}\gamma^{a'b'}\right), \\
[P_a,Q_I] = -\frac{i}{2}\varepsilon_{IJ}\gamma_a Q_J, \qquad [P_{a'},Q_I] = \frac{1}{2}\varepsilon_{IJ}\gamma_{a'}Q_J, \\
[J_{ab},Q_I] = -\frac{1}{2}\gamma_{ab}Q_I, \qquad [J_{a'b'},Q_I] = -\frac{1}{2}\gamma_{a'b'}Q_I, \\
[P_a,\overline{Q}_I] = \frac{i}{2}\varepsilon_{IJ}\overline{Q}_J\gamma_a, \qquad [P_{a'},\overline{Q}_I] = -\frac{1}{2}\varepsilon_{IJ}\overline{Q}_J\gamma_{a'}, \\
[J_{ab},\overline{Q}_I] = \frac{1}{2}\overline{Q}_I\gamma_{ab}, \qquad [J_{a'b'},\overline{Q}_I] = \frac{1}{2}\overline{Q}_I\gamma_{a'b'}, \\
[M_{AB},M_{CD}] = \eta_{BC}M_{AD} + \eta_{AD}M_{BC} - \eta_{AC}M_{BD} - \eta_{BD}M_{AC}, \\
[M_{A'B'},M_{C'D'}] = \delta_{B'C'}M_{A'D'} + \delta_{A'D'}M_{B'C'} - \delta_{A'C'}M_{B'D'} - \delta_{B'D'}M_{AC}. \qquad (1)
\end{gathered}
$$

where $P_a = M_{0a}$, $J_{ab} = M_{ab}$, $P_{a'} = M_{0'a'}$, $J_{a'b'} = M_{a'b'}$, $\eta = (- + +-)$. Note that the indices I, J = 1, 2, a, b = 0, 1, 2, a', b' = 3, 4, 5, $\varepsilon_{12} = -\varepsilon_{21} = 1$, the matrices $\gamma^{a'} = \sigma^{a'-2}$, $\gamma^0 = i\sigma^3$, $\gamma^{1,2} = \sigma^{1,2}$, γ^a is short for $\gamma^a \otimes \mathbf{I}$ as well as primed indices.

The left-invariant Cartan one-forms

$$L^A = dX^M L_M^A, \quad X^M = (x,\theta)$$

are derived from

$$J = G^{-1}dG = L^A T_A \equiv L^a P_a + L^{a'} P_{a'} + \frac{1}{2}L^{ab}J_{ab} + \frac{1}{2}L^{a'b'}J_{a'b'} + \frac{1}{2}(\overline{L}^I Q_I + \overline{Q}^I L_I),$$

where $G = G(x, \theta)$ is a coset representative in $SU(1, 1|2)^2$. it should be noticed that $\overline{L}^{I\alpha\alpha'} = L^{I\beta\beta'}C_{\beta\alpha}C'_{\beta'\alpha'}$ as for $AdS_5 \times S^5$ while here the relation doesn't hold. According to the commutation relations of the generators (1) and the zero-curvature equation $J \wedge J + dJ = 0$, we gain the Maurer-Cartan equations,

$$
\begin{gathered}
dL^a = -\frac{i}{2}\overline{L}^I\gamma^a \wedge L^I - L^b \wedge L^{ba}, \\
dL^{a'} = \frac{1}{2}\overline{L}^I\gamma^{a'} \wedge L^I - L^{b'} \wedge L^{b'a'}, \\
dL^{ab} = -L^{ac} \wedge L^{cb} - L^a \wedge L^b + \varepsilon_{IJ}\overline{L}^I\gamma^{ab}L^J, \\
dL^{a'b'} = -L^{a'c'} \wedge L^{c'b'} + L^{a'} \wedge L^{b'} + \varepsilon_{IJ}\overline{L}^I\gamma^{a'b'}L^J, \\
dL^I = -\frac{i}{2}\varepsilon_{IJ}L^a \wedge \gamma_a L^J + \frac{1}{2}\varepsilon_{IJ}L^{a'} \wedge \gamma_{a'}L^J - \frac{1}{4}L^{ab} \wedge \gamma_{ab}L^I - \frac{1}{4}L^{a'b'} \wedge \gamma_{a'b'}L^I, \\
d\overline{L}^I = \frac{i}{2}\varepsilon_{IJ}\overline{L}^J\gamma_a \wedge L^a - \frac{1}{2}\varepsilon_{IJ}\overline{L}^J\gamma_{a'} \wedge L^{a'} - \frac{1}{4}\overline{L}^I\gamma_{ab} \wedge L^{ab} - \frac{1}{4}\overline{L}^I\gamma_{a'b'} \wedge L^{a'b'}. \qquad (2)
\end{gathered}
$$

The action of the Green-Schwarz IIB superstring on $AdS_3 \times S^3$ space is [11]

$$S = -\frac{1}{2}\int_{\partial M_3} d^2\sigma\sqrt{-g}g^{ij}(L_i^a L_j^a + L_i^{a'} L_j^{a'}) + \frac{i}{4}\int_{M_3} S^{IJ}\left((L^a \wedge \overline{L}^I\gamma^a \wedge L^J + iL^{a'} \wedge \overline{L}^I\gamma^{a'} \wedge L^J) + c.c\right).$$

where $\sqrt{-g} = \sqrt{-\det g_{ij}}$, $g_{ij}g^{jk} = \delta_{ik}$, $i, j = 0$, $S^{IJ} = \mathrm{diag}\{1,-1\}$.

We can get the equations of motion as following due to the variations of the Cartan one-forms

$$\partial_i(\sqrt{-g}g^{ij}L_j^a) + \frac{i}{2}\varepsilon^{ij}S^{IJ}\overline{L}_i^I\gamma^a\overline{L}_j^J + \sqrt{-g}g^{ij}L_i^{ab}L_j^b = 0, \qquad (3)$$

$$\partial_i(\sqrt{-g}g^{ij}L_j^{a'}) + \sqrt{-g}g^{ij}L_i^{a'b'}L_j^{b'} - \frac{1}{2}\varepsilon^{ij}S^{IJ}\overline{L}_i^I\gamma^{a'}\overline{L}_j^J = 0, \qquad (4)$$

$$(L_i^a \gamma^a + i L_i^{a'} \gamma^{a'})(-\varepsilon^{ij} S^{IJ} + \sqrt{-g} g^{ij} \delta^{IJ}) L_j^J = 0, \tag{5}$$

$$\overline{L}_i^I (\gamma^a L_j^a + i\gamma^{a'} L_j^{a'})(\varepsilon^{ij} S^{IJ} + \sqrt{-g} g^{ij} \delta^{IJ}) = 0. \tag{6}$$

and the Virasoro constraint which is from the variation of action over the world sheet metric g_{ij}

$$L_i^{\hat{a}} L_j^{\hat{a}} = \frac{1}{2} g_{ij} g^{kl} (L_k^{\hat{a}} L_l^{\hat{a}}) = \xi' g_{ij}, \tag{7}$$

where $\xi' = \frac{1}{2} g^{kl} (L_k^{\hat{a}} L_l^{\hat{a}})$, $\hat{a} = a, a'$.

3. Flat Connections of the Green-Schwarz Superstring

In this section, we will show that the flat connections with a spectrum parameter satisfy not only the equations of motion but also the Viraroso constraint. Firstly, we introduce the Hodge dual of the Cartan one-form $L^{\hat{a}}$ over the world sheet as following

$$^*L_k^{\hat{a}} = -\varepsilon^{kl} \sqrt{-g} g^{lm} L_m^{\hat{a}} = -\varepsilon^{kl} \gamma^{lm} L_m^{\hat{a}}, \text{ and } \varepsilon^{12} = -\varepsilon^{21} = 1 \tag{8}$$

where $\gamma^{lm} \equiv \sqrt{-g} g^{lm}$, for $\det[\gamma^{lm}] \equiv -1$, it can be proved that

$$^*(^*L^{\hat{a}})_m = L^{\hat{a}}{}_m \tag{9}$$

by the Hodge dual(8), we can get differential form of the equations of motion from (3) to (6)

$$d^*L^a + \frac{i}{2} S^{IJ} \overline{L}^I \gamma^a \wedge L^J + L^{ab} \wedge {}^*L^b = 0 \tag{10}$$

$$d^*L^{a'} - \frac{1}{2} S^{IJ} \overline{L}^I \gamma^{a'} \wedge L^J + L^{a'b'} \wedge {}^*L^{b'} = 0 \tag{11}$$

$$\overline{L}^I \wedge (i\gamma^{a'} {}^*L^{a'} + \gamma^a {}^*L^a) \delta^{IJ} + \overline{L}^I \wedge (i\gamma^{a'} L^{a'} + \gamma^a L^a) S^{IJ} = 0 \tag{12}$$

$$\delta^{IJ} (i {}^*L^{a'} \gamma^{a'} + {}^*L^a \gamma^a) \wedge L^J + S^{IJ} (i L^{a'} \gamma^{a'} + L^a \gamma^a) \wedge L^J = 0 \tag{13}$$

Now we introduce the Cartan one-forms with the spectral parameter σ,

$$\begin{aligned} L^{\hat{a}}(\sigma) &= \frac{1}{2}(-\sigma^{-2} + \sigma^2)^* L^{\hat{a}} + \frac{1}{2}(\sigma^{-2} + \sigma^2) L^{\hat{a}}, \\ L^{ab}(\sigma) &= L^{ab}, L^{a'b'}(\sigma) = L^{a'b'}, \\ L^1(\sigma) &= \sigma L^1, L^2(\sigma) = \sigma^{-1} L^2, \\ \overline{L}^1(\sigma) &= \sigma \overline{L}^1, \overline{L}^2(\sigma) = \sigma^{-1} \overline{L}^2. \end{aligned} \tag{14}$$

It can be proved that $L^a(\sigma)$, $L^{a'}(\sigma)$, $L^{ab}(\sigma)$, $L^{a'b'}(\sigma)$, $L^1(\sigma)$ and $\overline{L}^I(\sigma)$ satisfy all the Maurer-Cartan equations according to the Maurer-Cartan equations (2) and equations from (10) to (13) . A detailed calculation of a similar model was given in Ref [11]. Then we will show the Cartan one-forms (14) satisfy the equations from (10) to (13) and the Virasoro constrait (7) with the original g_{ij}.

Firstly, from the equation(9) , we can get

$$^*L^{\hat{a}}(\sigma) = \frac{1}{2}(\sigma^{-2} + \sigma^2)^* L^{\hat{a}} + \frac{1}{2}(-\sigma^{-2} + \sigma^2) L^{\hat{a}}.$$

Note that $L^{ab} = -L^{ba}$, based on above equation, it follows from the equations of motions and the Maurer-Cartan equations that

$$\begin{aligned} &d^*L^a(\sigma) + L^{ab}(\sigma) \wedge {}^*L^b(\sigma) + \frac{i}{2} S^{IJ} \overline{L}^I(\sigma) \gamma^a \wedge L^J(\sigma) \\ &= \frac{1}{2}(\sigma^2 - \sigma^{-2}) dL^a + \frac{1}{2}(\sigma^2 + \sigma^{-2}) d^*L^a \end{aligned}$$

$$=\frac{1}{2}(\sigma^{-2}+\sigma^{2})d^{*}L^{a}+\frac{1}{2}(\sigma^{2}-\sigma^{-2})\,(-L^{b}\wedge L^{ba}-\frac{i}{2}\overline{L}^{1}\gamma^{a}\wedge L^{1}-\frac{i}{2}\overline{L}^{2}\gamma^{a}\wedge L^{2})$$
$$+\frac{1}{2}(\sigma^{2}+\sigma^{-2})L^{ab}\wedge{}^{*}L^{b}+\frac{1}{2}(\sigma^{2}-\sigma^{-2})L^{ab}\wedge L^{b}+\frac{i}{2}(\sigma^{2}\overline{L}^{1}\gamma^{a}\wedge L^{1}-\sigma^{-2}\overline{L}^{2}\gamma^{a}\wedge L^{2})$$
$$=\frac{1}{2}(\sigma^{2}+\sigma^{-2})(L^{ab}\wedge{}^{*}L^{b}+d^{*}L^{a}+\frac{i}{2}S^{IJ}\overline{L}^{I}\gamma^{a}\wedge L^{J})=0\,. \tag{15}$$

Similarly, we can get

$$L^{a'b'}(\sigma)\wedge{}^{*}L^{b}(\sigma)+d^{*}L^{a'}(\sigma)-\frac{1}{2}S^{IJ}\overline{L}^{I}(\sigma)\gamma^{a'}\wedge L^{J}(\sigma)=0\,.$$

Next, we consider that

$$\overline{L}^{I}(\sigma)\wedge[i\gamma^{a'}{}^{*}L^{a'}(\sigma)+\gamma^{a}{}^{*}L^{a}(\sigma)]\delta^{IJ}+\overline{L}^{I}(\sigma)\wedge[\gamma^{a}L^{a}(\sigma)+i\gamma^{a}L^{a'}(\sigma)]S^{IJ}$$
$$=\overline{L}^{I}(\sigma)\wedge[\frac{1}{2}(\sigma^{2}+\sigma^{-2})\gamma^{a}{}^{*}L^{a}+\frac{1}{2}(\sigma^{2}-\sigma^{-2})\gamma^{a}L^{a}$$
$$+\frac{1}{2}(\sigma^{2}+\sigma^{-2})(i\gamma^{a'})^{*}L^{a'}+\frac{1}{2}(\sigma^{2}-\sigma^{-2})(i\gamma^{a'})L^{a'}]\delta^{IJ}$$
$$+\overline{L}^{I}(\sigma)\wedge[\frac{1}{2}(\sigma^{2}+\sigma^{-2})\gamma^{a}L^{a}+\frac{1}{2}(\sigma^{2}-\sigma^{-2})\gamma^{a}{}^{*}L^{a}$$
$$+\frac{1}{2}(\sigma^{2}+\sigma^{-2})(i\gamma^{a'})L^{a'}+\frac{1}{2}(\sigma^{2}-\sigma^{-2})(i\gamma^{a'})^{*}L^{a'}]S^{IJ}\,.$$

From equation (15), when $I = 1$, we can get

$$RHS=\sigma^{3}[\overline{L}^{1}\wedge(\gamma^{a}{}^{*}L^{a}+i\gamma^{a'}{}^{*}L^{a'})+\overline{L}^{1}\wedge(\gamma^{a}L^{a}+i\gamma^{a'}L^{a'})]=0\,.$$

when $I = 2$, we can get

$$RHS=\sigma^{-3}[\overline{L}^{2}\wedge(\gamma^{a}{}^{*}L^{a}+i\gamma^{a'}{}^{*}L^{a'})-\overline{L}^{2}\wedge(\gamma^{a}L^{a}+i\gamma^{a'}L^{a'})]=0\,.$$

therefore, we can get

$$\overline{L}^{I}(\sigma)\wedge[\gamma^{a}{}^{*}L^{a}(\sigma)+i\gamma^{a'}{}^{*}L^{a'}(\sigma)]\delta^{IJ}+\overline{L}^{I}(\sigma)\wedge[\gamma^{a}L^{a}(\sigma)+i\gamma^{a'}L^{a'}(\sigma)]S^{IJ}=0\,.$$

Similarly, we can prove

$$\delta^{IJ}[i^{*}L^{a'}(\sigma)\gamma^{a'}+{}^{*}L^{a}(\sigma)\gamma^{a})\wedge L^{J}(\sigma)+S^{IJ}(iL^{a'}(\sigma)\gamma^{a'}+L^{a}(\sigma)\gamma^{a}]\wedge L^{J}(\sigma)=0\,.$$

Finally, we should check the Virasoro constraint. Equation (8) Can be expressed as

$${}^{*}\mathbf{L}^{\hat{a}}=-\varepsilon\gamma\mathbf{L}^{\hat{a}}\,,$$

where

$$\varepsilon=\begin{pmatrix}0&1\\-1&0\end{pmatrix},\ \gamma=\begin{pmatrix}\gamma^{11}&\gamma^{12}\\\gamma^{21}&\gamma^{22}\end{pmatrix},\ \text{and}\ \mathbf{L}^{\hat{a}}=\begin{pmatrix}L_{1}^{\hat{a}}\\L_{2}^{\hat{a}}\end{pmatrix}.$$

The Virasoro constraint (7) implies that

$$\sum_{\hat{a}}(\mathbf{L}^{\hat{a}}\mathbf{L}^{\hat{a}^{t}})=\begin{pmatrix}L_{1}^{a}L_{1}^{a}+L_{1}^{a'}L_{1}^{a'}&L_{1}^{a}L_{2}^{a}+L_{1}^{a'}L_{2}^{a'}\\L_{2}^{a}L_{1}^{a}+L_{2}^{a'}L_{1}^{a'}&L_{2}^{a}L_{2}^{a}+L_{2}^{a'}L_{2}^{a'}\end{pmatrix}=\xi'\begin{pmatrix}g_{11}&g_{12}\\g_{21}&g_{22}\end{pmatrix}=\xi\gamma^{-1}\,.$$

Then we can get

$$\mathbf{L}^{\hat{a}}(\sigma)=\frac{1}{2}(\sigma^{2}+\sigma^{-2})\mathbf{L}^{\hat{a}}+\frac{1}{2}(\sigma^{2}-\sigma^{-2})^{*}\mathbf{L}^{\hat{a}}=[\frac{1}{2}(\sigma^{2}+\sigma^{-2})I-\frac{1}{2}(\sigma^{2}-\sigma^{-2})\varepsilon\gamma]\mathbf{L}^{\hat{a}}$$
$$\equiv P(\sigma)\mathbf{L}^{\hat{a}}$$

It can be found that

$$\sum_{\hat{a}}(\mathbf{L}^{\hat{a}}(\sigma)\mathbf{L}^{\hat{a}^{t}}(\sigma))=P(\sigma)\sum_{\hat{a}}(\mathbf{L}^{\hat{a}}\mathbf{L}^{\hat{a}^{t}})M^{t}(\sigma)$$
$$=[\frac{1}{2}(\sigma^{2}+\sigma^{\,2})I-\frac{1}{2}(\sigma^{2}-\sigma^{-2})\varepsilon\gamma]\xi\gamma^{-1}.[\frac{1}{2}(\sigma^{2}+\sigma^{-2})I-\frac{1}{2}(\sigma^{2}-\sigma^{-2})\gamma\varepsilon']$$

$$=\frac{1}{4}(\sigma^2+\sigma^{-2})^2\xi\gamma^{-1}-\frac{1}{4}(\sigma^4-\sigma^{-4})\xi\varepsilon-\frac{1}{4}(\sigma^4-\sigma^{-4})\xi\varepsilon^t+\frac{1}{4}(\sigma^2-\sigma^{-2})^2(\varepsilon\gamma\xi\gamma^{-1}\gamma\varepsilon^t)$$

$$=\frac{1}{4}(\sigma^2+\sigma^{-2})^2\xi\gamma^{-1}-\frac{1}{4}(\sigma^2-\sigma^{-2})^2\xi\gamma^{-1}=\xi\gamma^{-1}=\sum_{\hat{a}}(\mathbf{L}^{\hat{a}}\mathbf{L}^{\hat{a}'}).$$

where we have used $\epsilon\gamma\epsilon=\gamma^{-1}$ and $\epsilon^t=-\epsilon$. Therefore we have

$$L_i^{\hat{a}}(\sigma)L_j^{\hat{a}}(\sigma)=L_i^{\hat{a}}L_j^{\hat{a}}$$

By the equation (7), we can get

$$L_i^{\hat{a}}(\sigma)L_j^{\hat{a}}(\sigma)=\frac{1}{2}g_{ij}g^{kl}(L_k^{\hat{a}}(\sigma)L_l^{\hat{a}}(\sigma))=\xi' g_{ij}$$

Hence, the Cartan one-forms satisfy the equations from (3) to (7).

4. Conclusion

For the Cartan one-forms with the spectral parameter σ satisfying the equations from (3) to (6), the flat currents $J(\sigma)$ can be expressed as

$$J(\sigma)\equiv L^{a'}(\sigma)P_{a'}+L^{a}(\sigma)P_{a}+\frac{1}{2}L^{a'b'}(\sigma)J_{a'b'}+\frac{1}{2}L^{ab}(\sigma)J_{ab}+\frac{1}{2}(\overline{L}^{I}(\sigma)Q_{I}+\overline{Q}^{I}(\sigma)L_{I})$$

which satisfies

$$dJ(\sigma)+J(\sigma)\wedge J(\sigma)=0.$$

Thus, $G(\sigma)^{-1}dG(\sigma)=J(\sigma)$ is integrable leading to an infinite number of nonlocal conserved quantities. In this paper, we have proved that the Cartan one-forms also satisfy the equations of motion and the Virasoro constraint. So with the spectral parameter , the new classical solutions are yielded by the currents $J(\sigma)$

$$G(\lambda\tau,\sigma)=G_0\,\mathrm{P}\exp\int_{\lambda_0\tau_0}^{\lambda\tau}J(\sigma),$$

$$g_{ij}(\lambda\tau,\sigma)=g_{ij}(\lambda\tau).$$

when $\sigma=1$, it resumes to the original solution. So an existing one can generate a set of new classical solutions.

Now with a spectral parameter ρ, the flat currents $J(\sigma,\rho)$ can be constructed from the new solutions $G(\sigma)$ .we find the corresponding flat currents with the parameter $\rho\sigma$ from the new solutions $G(\sigma)$ are just the flat currents of G, They are in the same set. The mass of solutions have the same infinite set of classically conserved quantities. It's argued that when $\sigma=i$, the Maurer-Cartan equations of $J(\sigma)$ is the equations of motion of the system.

References

[1] Itzhaki N and Maldacena J M 1998 Supergravity and the large N limit of theories with sixteen supercharges *Physical Review D* 58(4) 046004

[2] Metsaev RR and Tseytlin AA1998 Type IIB superstring action in $AdS_5\times S^5$ background *Nuclear Physics B* 533(1-3) 109-126

[3] Bena I Polchinski J and Roiban R 2004 Hidden symmetries of the $AdS_5\times S^5$ superstring *Physical Review D* 69(4) 046002

[4] Hou B Y and Peng D T 2004 The Affine Hidden Symmetry and Integrability of Type IIB Superstring in $AdS_5\times S^5$ *arXiv preprint hep-th/0406239*

[5] Kagan D and Young C A 2006 Conformal sigma models on supercoset targets *Nuclear Physics B* 745(3) 109-122

[6] Beisert N and Ahn C 2012 Review of AdS/CFT integrability *Letters in Mathematical Physics* (an overview) 99 (1) 3-32

[7] Wulff L 2014 Superisometries and integrability of superstrings *Journal of High Energy Physics* 1-32

[8] Hoare B and Seibold F 2019 Supergravity backgrounds of the η-deformed $AdS_2 \times S^2 \times T^6$ and $AdS_5 \times S^5$ superstrings *Journal of High Energy Physics* 1-28
[9] Garousi MR 2021 Effective action of type II superstring theories at order α'^3: NS-NS couplings *Journal of High Energy Physics* 1-27
[10] Rahmfeld J and Rajaraman A 1999 Green-Schwarz string action on $AdS_3 \times S^3$ with Ramond-Ramond charge *Physical Review D*60 (6) 064014
[11] Chen B and He Y L 2005 Flat currents of the Green-Schwarz superstrings in $AdS_5 \times S^1$ and $AdS_3 \times S^3$ backgrounds *Physical Review D*71 (8) 086007

A New Generation of Hardware-in-the-loop Simulation Technology Combined with High-performance Computers and Digital Twins

Yehui Shi[1], Yuqiang Cheng[1,*] and Yushan Gao[2]

[1] College of Aerospace Science and Engineering, National University of Defense Technology, Changsha, China
[2] Xi'an Aerospace Propulsion Institute, Xi'an, China

*Corresponding author email: gfkd.sky@foxmail.com

Abstract. Liquid rocket engines are complex equipment with long maintenance cycles and high fault diagnosis costs. Therefore, hardware-in-the-loop simulation combining software and hardware has become the best choice. A hardware-in-the-loop simulation platform framework is proposed based on a V-type development model and related digital twin technology, and the main content of the framework is introduced. Based on the development process of the platform, a certain type of liquid rocket engine is taken as the object to verify the results. The results show that the platform can well simulate the engine changes under different faults, and has versatility and high real-time performance.
Keywords: Hardware-in-the-loop simulation; Digital twin; Liquid rocket engine.

1. Introduction

The engine is the core of the liquid rocket's flight power. Not only is the structure very complex, but also the working conditions are extremely harsh, including high temperature, high pressure, high-frequency oscillation, and so on. In this case, any small anomaly can quickly develop into a very destructive failure, causing the launch to fail and causing huge losses[1]. And because the liquid rocket engine test is extremely expensive, and the special failure test is extremely dangerous, the test data that can be used for failure mode and failure response analysis is extremely limited. Therefore, it is particularly important to conduct hardware-in-the-loop simulation analysis given the long development cycle of liquid rocket engine fault diagnosis and the lack of test data and insufficient fault samples, which lead to the failure to cover all failure modes[2].

Dynamic system simulation is to realize the high-precision, high-fidelity multi-disciplinary coupled numerical simulation of the whole machine, component, or system of the power system in a computer virtual environment[3]. Hardware-in-the-loop simulation combines these numerical simulations with physical models for joint simulation[4]. This type of simulation describes the relatively simple parts of the actual system by mathematical models and converts them into simulation models. For parts with complex structures and uncertain laws, the simulation loop is introduced in the form of physical objects (or physical models)[5]. Wanli Zhao[6] used the rapid prototyping method to build a rocket engine fault diagnosis real-time simulation hardware-in-the-loop platform. In 2020, Wei Xue [7] built a hardware-in-the-loop simulation platform for a liquid rocket engine and verified the improved ARMA algorithm. This not only improves the development efficiency and accuracy of the whole model but also makes the verification process of simulation results very intuitive and greatly shortens the whole development cycle[8]. At the same time, to make the simulation results better presented, this paper combines the digital twin technology to propose a new generation of hardware-in-the-loop simulation technology.

2. Basic Framework and Components of Hardware-in-the-loop Simulation

2.1. Rapid Prototyping

In this way, with the introduction of the related simulation concepts, and based on rapid application development, a more scientific V-type development model has gradually formed. The V-type development model is shown in Figure 1. The first is the functional design of the real-time simulation computer, the functions to be implemented by the real-time simulation computer are determined in detail, then the target program code is generated, and then the realization of the underlying software and hardware of the simulation computer. Finally, after the simulation computer completes the hardware-in-the-loop simulation, it can enter the system integration and test link, and finally, realize the indicators and functions of the initial design[9].

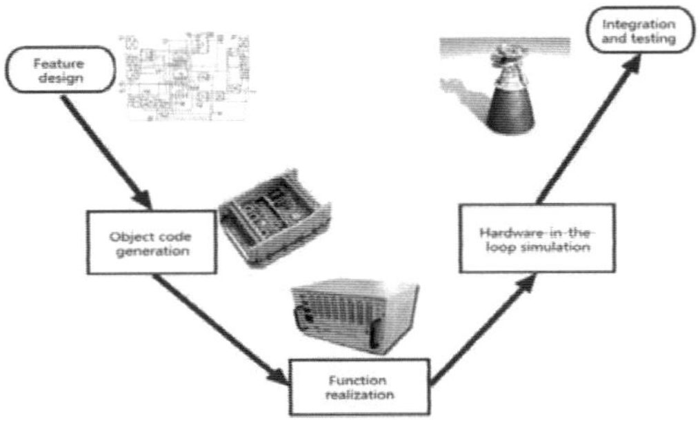

Figure 1. Schematic diagram of V-type development model

2.2. Design a Hardware-in-the-loop Simulation Platform

Based on this development model, a basic hardware-in-the-loop simulation framework is proposed. The framework is mainly composed of five parts: simulation equipment, signal generator, various interface equipment, a test console, and a human-computer interaction system. The functions of each component of the above simulation system are introduced as follows.

Simulation equipment includes real-time simulation computers, simulation control, etc. Its main function is to simulate the working process of a liquid rocket engine and solve problems without real objects. In addition to real-time simulation, offline simulation can also be performed to determine the accuracy of the model. The function of the signal generator is to send out the physical quantity signals required by the engine, such as mass flow, pressure, and so on. The interface equipment mainly includes analog quantity interface, digital quantity interface, and so on. The test console mainly includes the main control computer, whose purpose is to manage and schedule the entire simulation process and realize functions such as operation control and online parameter modification. The human-computer interaction system includes a digital twin system and a database, mainly for data visualization, facilitating researcher model status monitoring, data storage, and post-processing. The hardware platform is shown in Figure 2.

The novel feature of the simulation platform is that a real-time simulation computer structure and a simple digital twin system are proposed. The real-time simulation computer plans to use the NI Veristand real-time simulation platform and Labview software to complete the construction of the engine real-time simulation environment, and the engine simulation computer adopts the upper and lower computer structure. The host computer is a development host, which monitors the status of the real-time emulator through Ethernet, and supports online modification of parameters and storage of real-time data. The lower computer is a real-time simulation computer, which is used to simulate the working state of the engine. And because the above-mentioned simulation is a real-time simulation, the time of each node of the simulation system is required to be consistent, so the upper computer is also responsible for the hardware communication channel matching the engine real-time simulation interface and the data acquisition card. Finally, the output of the real-time engine model is transmitted to the analog output system through the software and hardware communication interface output data and control commands. The upper computer and the lower computer are connected through TPC/IP communication. The structure of the engine simulation computer is shown in Figure 3. The so-called digital twin is to establish a corresponding digital model for the physical entity and simulate the behavior of the physical entity in the real environment through real-time data and historical data transmitted by sensors[10]. Using this feature of digital twins, it acts as a visual computer in the real-time simulation system. At the same time, the digital twin system is mainly divided into the data layer, model layer, control layer, and display layer. The data layer is the key to the digital twin system. It will convert the corresponding transition parameters into digital information through advanced measurement technology, and perform corresponding processing and efficient transmission on it, and at the same time receive control decision instructions. The model layer needs to realize the construction and calculation of simulation models, and it needs to model geometric data and environmental data. The focus of the control layer is to achieve optimization iterations of simulation results and measurement results, and feedback to the data layer and model layer after iteration to improve the accuracy of equipment and simulation models. By creating the display layer of the digital twin system, in addition to engine fault detection and visualization parameters, the application of simulation technology will also be extended to various engine design stages. The system framework is shown in Figure 4.

Figure 2. Hardware schematic

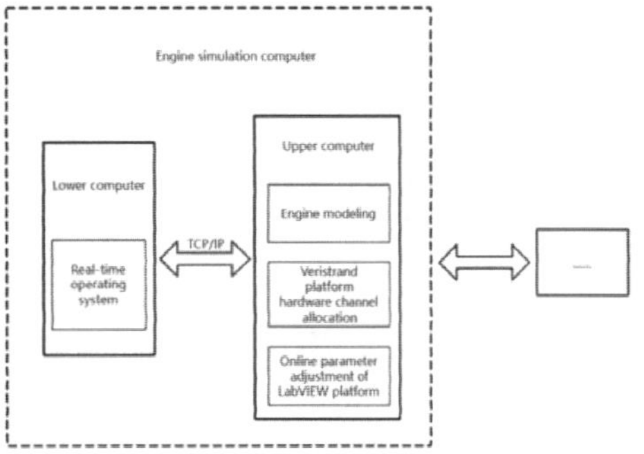

Figure 3. Schematic diagram of engine simulation computer

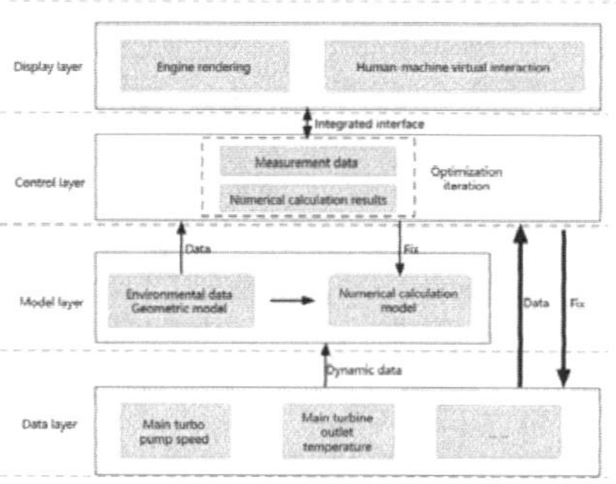

Figure 4. Digital twin system frame diagram

2.3. Hardware-in-the-loop Simulation Development Process

The proposed hardware-in-the-loop simulation platform takes NI Veristand as the core, uses MATLAB software to model, and combines VxWorks simulation to run the closed-loop simulation iterative process[11]. Its main development process is divided into six parts. The main development flow chart is shown in Figure 5.

First, based on the modular modeling and simulation idea, use MATLAB/Simulink to establish the dynamic model of the liquid rocket engine, and profane or offline simulation to ensure the feasibility of the model; then, split and add the model according to the requirements of the general real-time simulation platform NI Veristand Hardware driver modules, etc., to meet the hardware and software real-time communication requirements of the hardware-in-the-loop simulation platform; then, the modified dynamic model is converted into a C/C++ file that can be recognized by a real-time high-performance computer through the SDK encoder and downloaded to the real-time simulation computer. At the same time, set the property parameters of the real-time simulation computer, configure the parameters during monitoring and save the parameter variables to prepare for the subsequent real-time simulation; the next step is to run the simulation system to obtain simulation data under different working conditions. Finally, perform post-processing on the generated real-time simulation data processing, realize real-time storage data upload, format conversion, data playback, and other functions. The development process of the real-time simulation platform is shown in the figure below.

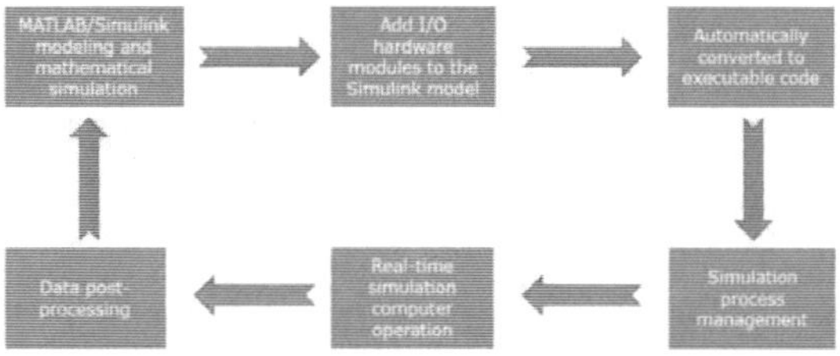

Figure 5. Hardware-in-the-loop simulation platform development process

In this platform, the simulation controller sends instructions to the signal generator, and some of the parameter signals generated by the signal generator are transmitted to the engine simulation computer; the other part is transmitted to the main control computer, and this part of the parameters are fed back to the engine simulation computer in the form of signals. The engine simulation computer calculates various parameters in the process of starting the engine by running the downloaded engine fault model and the historical data and fault data in the database and transmits these parameters to the analog output system to generate a standard voltage signal allowed within the range. After being collected by the data acquisition system, it is output to the simulation integrated control computer for a new round of calculations to form a simulation loop. Finally, the calculation results are transmitted to the digital twin system via Ethernet to realize a real-time display of status detection parameters.

3. A Case

To verify the performance of the architecture, based on the idea of modular modeling, a simulation model of a certain type of liquid rocket engine was established by using Simulink software. The simulation model is shown in Figure 6. Secondly, the changes of flow and pressure under the condition of blockage fault are given.

The results show that the designed real-time simulation platform can not only simulate the changes of the engine under different faults but also can inject faults at any time through online parameter modification. The platform can not only greatly reduce troubleshooting time and improve development efficiency, but also reduce the cost of fault detection and repair. The most important thing is that it can simulate various working conditions of the selected object, simulate complex failure modes and verify failure strategies, and solve the problem. Mode and failure response analysis have limited test data.

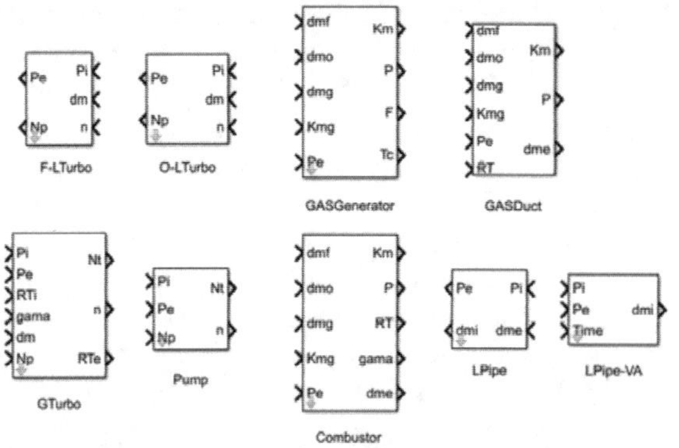

Figure 6. Simulation module library

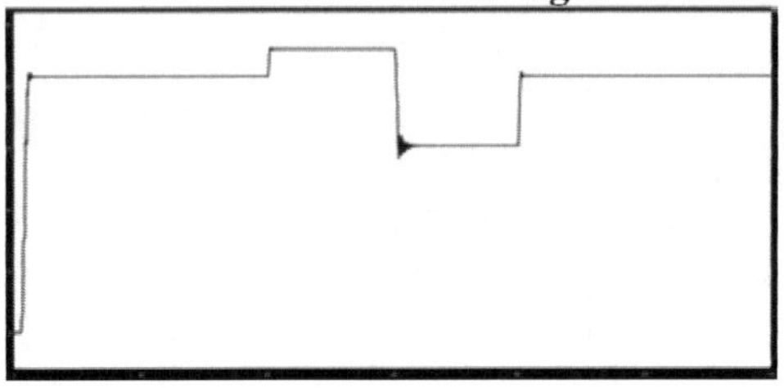

Figure 7. Pressure changes in failure mode

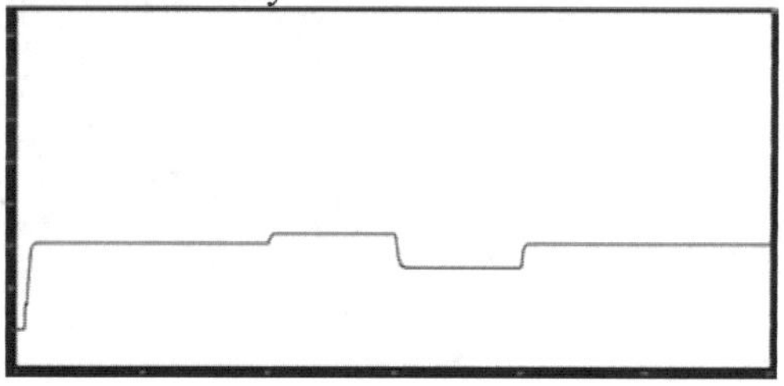

Figure 8. Flow changes in failure mode

4. Advantages and Development Prospects of Hardware-in-the-loop Simulation

A successful hardware-in-the-loop simulation system not only includes a hardware prototype and a real-time two-way interactive simulation model but also must realize the synergy between the prototype and the model. The proposed real-time simulation system not only satisfies this collaborative relationship but also combines digital twin technology to give it unsurpassed advantages.

- Strong versatility.
- Complete complex model calculations within a frame time.
- Good scalability and can be flexibly accessed different hardware and software.
- Good real-time performance.
- More comprehensive forecasting and analysis capabilities.
- Digitization of experience[12].

The real-time simulation technology has the characteristics of a high confidence level and short development cycle and has become an indispensable and important method in the fields of aerospace, weapon systems, and so on. With the continuous development of virtual reality simulation technology[13], environment characteristics simulation technology[14], distributed interactive simulation technology[15], and other technologies, the hardware-in-the-loop simulation platform will adopt a more open architecture and become an intelligent, intelligent, efficient numerical calculation, data processing, and knowledge reserve. Integrated, interactive modeling/simulation software, and hardware systems, with distributed processing capabilities for multi-body physical objects; simulation software will also move towards interoperability, reusability, object-oriented, integration, and intelligence development[16].

5. Conclusion

Hardware-in-the-loop simulation is to connect the selected hardware and other subsystems to form a closed-loop virtual simulation, to realize the whole simulation process. It provides a link between purely mathematical simulations and full physical prototypes, combining the rapidity of the former with the fidelity of the latter. In this paper, a hardware-in-the-loop simulation platform framework and development process with strong versatility combined with digital twin technology is proposed, and each component of the platform is described and explained in detail. Finally, the advantages and development prospects of the platform are explained.

References

[1] Lv, H. , Chen, J. , Wang, J. , Yuan, J. , Liu, Z. . (2021). A supervised framework for recognition of liquid rocket engine health state under steady-state process without fault samples. *IEEE Transactions on Instrumentation and Measurement, PP*(99), 1-1.

[2] Zhang Y. . (1997). Health monitoring of liquid rocket engine-fault analysis and simulation. *Propulsion technology*, 18 (1), 5.

[3] Huang, R. , Li, H. , Liu, J. , Kong, W. . (2019). Status and development trend analysis of aircraft hardware-in-the-loop simulation technology. *Journal of System Simulation* (09), 1763-1774.

[4] Huijgens, L. , Vrijdag, A. , Hopman, H. . (2021).Hardware in the loop experiments with ship propulsion systems in the towing tank: Scale effects, corrections and demonstration. *Ocean Engineering.*

[5] Huang, J. , Ju, J. . (2011). Research status and development trend of hardware in the loop simulation technology. *Ship Electronic Engineering, 31* (7), 4.

[6] Zhao W. , Guo Y. , Yang J. , et al. (2019). Hardware-in-the-loop Simulation Platform for Fault Diagnosis of Rocket Engines[C]. *2019 IEEE 10th International Conference on Mechanical and Aerospace Engineering (ICMAE)*, pp. 268-273.

[7] Xue W. , Zhang Q. , Wu X. . (2019). Research on Real-time Fault Diagnosis Method of Liquid Rocket Engine Based on ARMA Model [J]. *Computer Measurement and Control*, 27(09):4-7+22.

[8] Yan, N. , Zhang, Y. , Liu, C. , Wang, D. (2010). Research on distributed hardware in the loop simulation technology in Matlab environment. *Journal of system simulation* (8), 4.

[9] Li, J. , Tian, G. , Lu, Q. , Chen Q. , Ouyang M. . (2005). Development of vehicle controller for fuel cell hybrid electric bus using v-development mode. *Journal of mechanical engineering, 41* (12), 7.

[10] Ye, Y. , Yang, Q. , Yang, F. , Huo, Y. , Meng, S. . (2020). Digital twin for the structural health management of reusable spacecraft: a case study. *Engineering Fracture Mechanics, 234*(3), 107076.

[11] Zhang, C. , Wang, Y. , Mu, C. , Ma, X. , Li, J. .(2021). Design of an Industrial Robot Polishing Control System. *2021 4th International Conference on Mechatronics, Robotics and Automation (ICMRA)*, pp. 14-18.

[12] Benjamin, Schleich, Nabil, Anwer, Luc, Mathieu, et al. (2017). Shaping the digital twin for design and production engineering. *Cirp Annals Manufacturing Technology.*

[13] Zhang, C. , Liu, B. . (2015). Application Research of virtual reality technology. *Digital technology and application* (11), 1.

[14] Liang, H. , Dong, F. , Wang P. . (2015). Research and future prospects of hardware in the loop simulation technology. *Computer CD software and application, 18* (3), 2.

[15] Wen , LI , Song , Cheng. (2018). Research on Key Technologies of Event-driven Distributed Cooperative Simulation.(eds.)*Proceedings of 2018 International Conference on Computational, Modeling, Simulation and Mathematical Statistics(CMSMS 2018)*(pp.119-125).

[16] Cheng, J. (2007). Development and Prospect of military system modeling and simulation technology. *Command control and simulation, 29* (4), 8.

A Blockchain-based Security Framework for Secure and Resilient Smart Grid

Bin Luo[1,*], Guibin Xu[1], Jinglong Xie[2] and Yufei Wang[1]

[1] Hubei Central China Technology Development of Electric Power C.,Ltd., Hubei, China
[2] Safety Supervision Center, Hubei Electric Power Co., Ltd, Hubei, China

*Corresponding author email: 498478105@qq.com

Abstract. Smart grid integrates the information and communication technologies with the physical grid. It can not only improve security and efficiency of traditional power grids, but also has advantages in promoting environmental protection and sustainable development. However, the existing smart grid systems suffer some security concerns such as privacy leakage and data tampering. In order to address these security issues, we propose a security framework for smart grid using Blockchain. It protects data privacy through cryptographic tools, and it applies Blockchain technologies to prevent data tampering. Compared with the existing smart grid systems, our proposed system enjoys better security and resilience.
Keywords: Smart grid; Blockchain; Safety management.

1. Introduction

Smart grid integrates multi-disciplinary technologies such as computers, communications, networks, and signal processing with the traditional power grids, making them more intelligent thanks to the information technologies, but security is a crucial requirement for smart grid. Otherwise, accidents not only may cause inconvenience to users, but also may cause economic losses and social impacts. Therefore, to prevent possible catastrophic consequences, management entities need to adapt effective preventive measures and safety management policies in order to promote their safety awareness.

1.1. Related Works

With the development of information and communication technologies (ICT), they have shown great potentials to be integrated with the traditional power grids. For example, power grids in the US have established comprehensive information architecture to collect real-time grid data, such as current and voltage, and this helps to improve power management level and energy utilization, as well as meet the safety requirements. In China, people are also paying attention to power safety management. He et al. [1] have introduced a stand-alone standardized management system, which establishes a prototype for power standardization management. But such a stand-alone system cannot meet real-world requirements. An improved system was developed afterwards, which is composed of three parts: power supply, power generation and integration, realizing various functions in power safety [2].
The database of smart grid is vulnerable to various attacks. In 2015, hackers have attacked a substation in Ukraine, which in turn caused about 30 substations to fail, and more than 230,000 users to loss power. Blockchain technologies can provide attractive security characteristics, so this paper considers application of Blockchain with smart grid to promote security and resilience. In the literature, there are some researches on integrating Blockchain with smart grids. For example, the Brooklyn Microgrid

Project [3] can directly sell the electricity generated by users' solar equipment to their neighbours. And these users can directly reach a transaction using Blockchain without the involvement of a power supply company. The electric vehicle charging project [4] uses the smart contract to enable direct transactions between electric vehicles and charging piles, save management costs.
Scholars are also paying great attentions to the application of Blockchain with smart grids. In 2017, Li et al. [5] pointed out that Blockchain has attractive properties for smart grids and they have conducted extensive feasibility analyses on this topic. Feng [6] established a smart grid system based on smart contract, which solved the mutual mistrust among parties among direct transaction of distributed power sources. Gai et al. [7] have considered how to protect the energy trading information using consortium Blockchain in smart grid, and they gave a privacy preserving solution for this problem. Bera et al. [8] have designed a Blockchain assisted access control protocol for IoT enabled smart grid, so that only authenticated entity can access the system. Security survey on implementing Blockchain in smart grid can be found in [9,10]. However, these existing works rarely considered the requirements in operations, such as qualification inspections, operation management, and on-site supervision. This work fills in this gap by introducing a security framework for smart grids using Blockchain, enabling them to meet the desirable security requirements.

1.2. Our Contribution

We propose a Blockchain-based security framework for smart grid. The main contributions can be summarized as follows:
1) Cryptographic building blocks, such as encryption and digital signature, are used to enhance the data privacy and integrity in smart grid. The sensitive data is first encrypted and then digitally signed, so that both data privacy and integrity can be ensured.
2) Blockchain is used to enhance resilience in smart grid. The encrypted data is recorded on the Blockchain. Thanks to its attractive features, such as distributed data storage and tamper resistant, the recorded data is more trustworthy, and any cheating behaviour can be traced to its source.

2. Related Technology

Table 1. Notations.

Notation	Descrption
A_i,B_i,C_i	i-th sensitive information
X_i,Y_i,Z_i	the ciphertext of the i-th sensitive information
PK_i,SK_i	the key pair of the i-th message sender
k_i	temporary session key for the i-th message
$block_i$	the corresponding block responsible for storing the i-th sensitive information
t	timestamp
$a \| b$	splicing element a and element b
pk_i,sk_i	the signing key and de-signing key of the i-th message
S_i	signature of the ciphertext of the i-th job plan information
DP_i	the i-th data packet uploaded to the Blockchain system
Enc	encryption algorithm in hybrid encryption
Sig	signature algorithm in Schnorr
x,y	system public and private key pair

2.1. Blockchain

Blockchain is a shared ledger in a distributed fashion that provides attractive features such as anti-tampering, traceability, and collective maintenance. It has rich applications in many scenarios. The system architecture of Blockchain technology can be divided into three parts: basic module, core module, and interactive module.

The basic module includes two parts. Information infrastructure provides general computing, storage, network and other services. The data layer stores all data in Blockchain, and it also encapsulates hash values, timestamps, public and private keys, etc. Using hash functions and digital signatures, it provides the non-tamperable modification and traceability of the data.

The core modules include network layer, consensus layer, incentive layer and contract layer. The network layer includes related network management, propagation mechanism and user authentication. The consensus layer uses the consensus algorithm, e.g. Proof of Work (PoW) and Proof of Stake (PoS), to maintain and update the Blockchain. The incentive layer includes incentive and distribution mechanisms to encourage nodes to participate in the security verification. The contract layer includes various scripts, codes, smart contracts, and algorithms.

The interaction module includes the application layer, interacting with the actual application. All layers in the Blockchain jointly maintain its stability. The system architecture of the Blockchain is shown in figure 1.

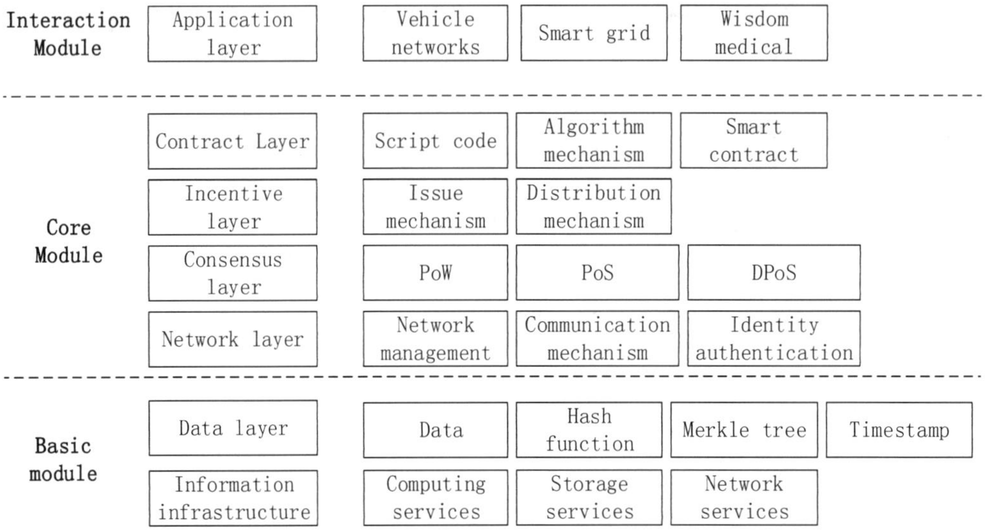

Figure 1. Architecture diagram of the Blockchain system.

Traditional power grids normally use paper materials or electronic files for recording and archiving. It is error prone when filling in data, which is undesirable for effective data analysis. In addition, entities with poor safety awareness may modify the data in violation of disciplines, causing serious consequences. Thanks to the attractive features such as the consensus mechanism, digital signatures, smart contracts, etc., Blockchain provides security properties such as decentralization, no-tampering, and traceability. Therefore, the integration of Blockchain with power grid contributes to efficient and effective smart grids.

2.2. Hybrid Encryption

Encryption schemes can be divided into two categories: symmetric ones and asymmetric ones. Symmetric encryption suffers the key distribution problem. Asymmetric encryption suffers heavy overheads in computation. Therefore, in real-world scenarios, a combination of symmetric and asymmetric encryptions is often used. This work uses a hybrid encryption, combining AES and RSA to protect the private information in the smart grid. The algorithm works as follows:

- **KeyGen:**
  1. One randomly chooses two large primes p,q, and calculates $n = pq$, $\varphi(n) = (p-1)(q-1)$, where $\varphi(n)$ denotes the Euler function.
  2. One chooses an integer e, ensuring that $1 < e < \varphi(n)$ and $\gcd(\varphi(n), e) = 1$.
  3. One calculates d, satisfying $d \cdot e \equiv 1 \bmod \varphi(n)$.

4. The RSA key pair is set to be $(pk,sk)=(\{e,n\},\{d,n\})$.
5. One generates the temporary session key k that is used as the key in AES encryption.

- **Enc:**
  1) The sender uses the session key to encrypt the message as $m: c_1 = E(k,m)$.
  2) The sender uses the public key $\{e,n\}$ of the recipient to encrypt the generated session key as $k: c_2 = k^e \bmod n$.
  3) The sender sends (c_1,c_2) to the recipient.
- **Dec:**
  1) The recipient uses the private key $\{d,n\}$ to decrypt the session key as $k \equiv c_2^d \bmod n$.
  2) The recipient uses the session key to decrypt message as $m = D(k,c_1)$.

2.3. Digital Signature

A digital signature example works as follows:

- **Init:** One chooses two large prime numbers *p,q*, satisfying $q \mid (p-1)$. Then it randomly selects a generator $g \in {}_{p}^{*}$, which satisfies $g^q \equiv 1 \bmod p$. Finally, it randomly selects a private key x, where $1<x<q$, and sets the public key as $y \equiv g^x \bmod p$.
- **Sig:** The sender randomly selects $k: 1<k<q$, and calculates $r \equiv g^k \bmod p$, $e = H(r,m)$, $s \equiv xe + k (\bmod p)$, where H is a secure hash function, $\{e,s\}$ is the user's digital signature for message m.
- **Ver:** After receiving the message m and its signature $\{e,s\}$, the recipient calculates $r' \equiv g^s y^{-e} (\bmod p)$, then verifies whether $H(r',m)=e$. If not, reject the signature.

3. System Model and Security Requirements

3.1. System Model

The outer network includes remote inspection and process supervision. Users can access the system through PC browser or large screen. The inner network focuses on data maintenance and statistical analysis. Users can access the system through PC browser or mobile phone APP. The system model is shown in figure 2.

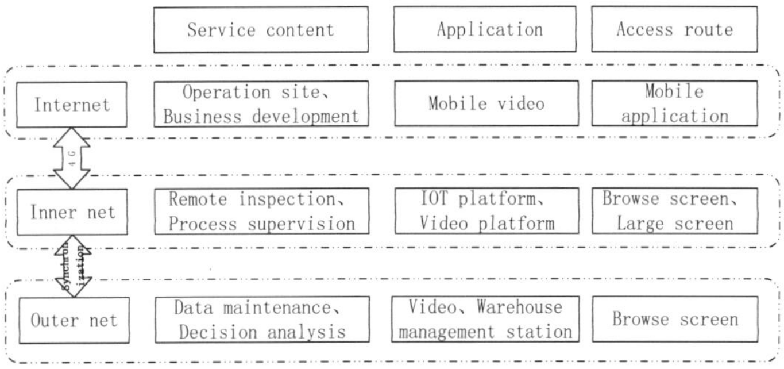

Figure 2. The overall architecture of the system.

3.2. Security Definition

A lot of sensitive data will be generated in the system, including users' identity information and statistical reports. To prevent potential attacks, several security requirements are considered:

- **Privacy**: sensitive information cannot be accessed by unauthorized entities.
- **Integrity**: data cannot be tampered or forged without being detected.
- **Traceability**: data generated in the system can be traced to its source.

4. Our Proposed Security Framework for Smart Grid

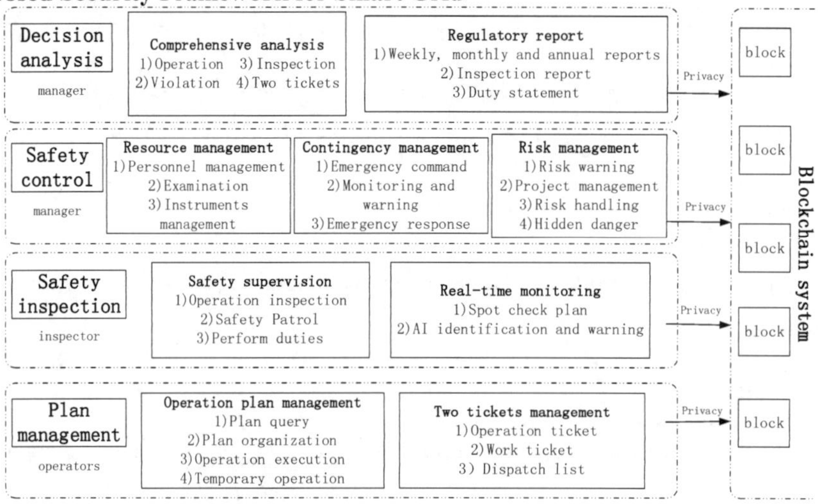

Figure 3. Partition diagram of system functions.

The functional modules are displayed in figure 3, which can be divided into four levels: plan management, safety inspection, safety control, and decision analysis. In each module, information such as job plans, "two-ticket", various reports, etc., will be generated. It is obvious that some information is sensitive. Hence, it is required to be encrypted and then uploaded to the Blockchain for storage. Each block within the Blockchain consists of a block header and a block body. The body stores several records and a binary Merkle tree composed of the hash value of each record. The header generally contains the hash value of the previous block, including random number, root of the Merkle tree, timestamp, etc. These blocks are connected into a chain using the hash values. For example, the data structure of "two-ticket" stored on the Blockchain can be shown as in figure 4.

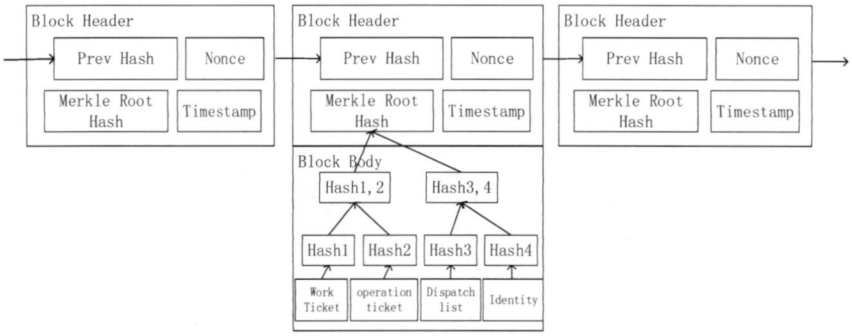

Figure 4. Data structure of the block in the "two-ticket" process.

Since the hash function is one-way and collision-resistant, when one wants to tamper with the record, she has to modify the hash values of the current block as well as all subsequent ones. In addition, the change of the leaf node will lead to the change of the root node. Hence, modification can be identified and traced to its source.

The system considers plan management, personnel management, unit management, and spot management. The plan management includes functions such as operation plan management and risk management. The business diagram of plan management is shown in figure 5.

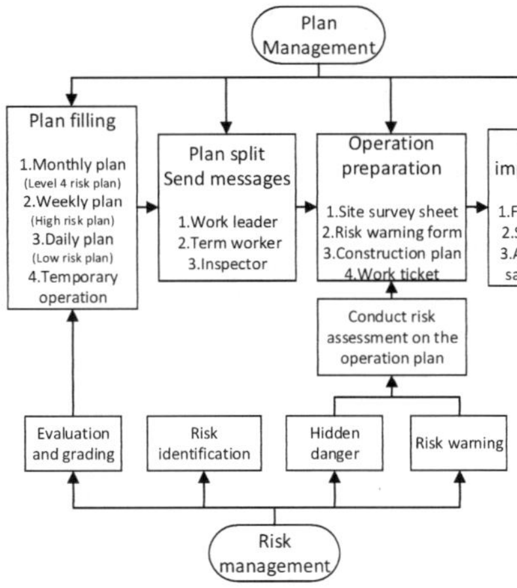

Figure 5. Business graph of plan management.

4.1. Plan Management

The plan management will produce a lot of confidential documents and private information. This information is encrypted and uploaded to the Blockchain system to ensure data security.

The system first obtains the key pairs (PK_i, SK_i) of the i-th sender through the KeyGen algorithm and the temporary session key k_i of the i-th round. The plan details, "two-ticket" data, risk management data and other sensitive information are denoted as $\{A_1, A_2, \dots A_i\}$, the system then encrypts the above sensitive message and the timestamp t through the Enc algorithm to obtain the encrypted ciphertext $(X_{i,1}, X_{i,2})$. As follows, it uses the Init initialization algorithm in the Schnorr signature to obtain key pairs (pk_i, sk_i), and signs the ciphertext by the Sig signature algorithm to obtain S_i. Finally, the system uploads $DP_i = \{(X_{i,1}, X_{i,2}), S_i, t\}$ to the corresponding business node in the Blockchain. Finally, the Blockchain constructs a leaf node in the block body of the $block_i$, and stores it in the Blockchain.

4.2. Unit Management and Personnel Management

The system first analyzes the qualifications and safety files of the units, it then selects safety-conscious and capable units to stay on spot and in charge of the tasks. Those units on the "blacklist" and "negative list", as well as any individual who has violated disciplines or has low safety awareness, are not allowed to enter the spot. The business diagram is shown in figure 6.

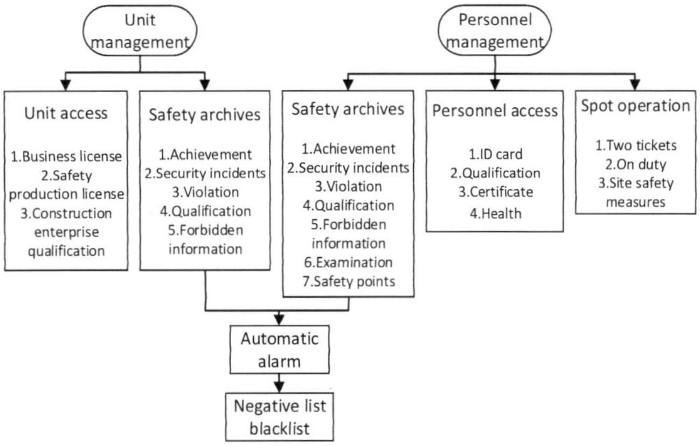

Figure 6. Business graph of unit management and personnel management.

The system establishes a safety file of the units after the qualification review. This information contains various important information of related units. If this information was tampered with, illegal entity may enter the site and cause on-site accidents. Therefore, the information needs to be protected from tampering. Assuming that the file information and personnel information are denoted as $\{B_1, B_2, \ldots B_j\}$. The system first obtains the key pair (pk, sk) and the temporary session key k_i of the round, and then it uses the Enc algorithm to encrypt $\{B_1, B_2, \ldots B_j\}$ and the timestamp t to obtain the ciphertext $(Y_{j,1}, Y_{j,2})$. Finally, the system uploads the ciphertext to the corresponding business node in the Blockchain. It constructs a leaf node in the block body of $block_j$, and stores it in the Blockchain.

4.3. Spot Management

The site is a place where production factors and procedures are gathered. Production factors include units, personnel, materials and other management behaviours, where procedures include planning, organization, and dynamic implementation. Managers and team leaders can use the APP to access the "two-ticket", realizing intelligent management and control of the entire operation processes. Moreover, the "two-ticket" can be associated with weekly and daily plans. Supervisors can use the large-screen visualization system and the external network video monitoring system to carry out the "remote + on-site" inspection, strengthening on-site safety supervision. The business diagram of on-site management is shown in figure 7.

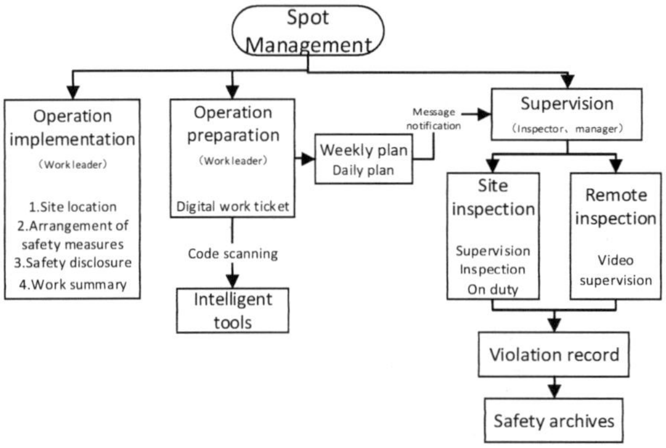

Figure 7. Business graph of spot management.

Once the spot inspection is completed, the inspectors will generate information such as safety inspection reports and duty performance reports. This information is also important to the system and it needs to be handled safely. Assuming that information such as the security audit report and duty performance report is denoted as $\{C_1, C_2 \ldots C_k\}$ .The system first encrypts and protects the above information and the timestamp t through a hybrid encryption algorithm. And then, it uses the Schnorr signature algorithm and the private key of the inspector to sign the ciphertext. Finally, the system uploads $DP_k = \{(Z_{k,1}, Z_{k,2}), S_k, t\}$ to the corresponding business node in the Blockchain. It constructs a leaf node in the block body of the $block_k$ and stores it in the Blockchain.

5. Security Analysis

Theorem 1: This scheme satisfies privacy, i.e. the adversary cannot obtain any information about private data.

Proof: Privacy means that the important sensitive data in the system should be protected, and the adversary cannot obtain any information about the sensitive data. This scheme uses a hybrid encryption algorithm based on the combination of AES symmetric cipher and RSA public key cipher to encrypt data. If the adversary does not have the receiver's private key, she cannot obtain the session key through the RSA decryption algorithm. This implies that the ciphertext cannot be decrypted.

Therefore, based on the security of AES and RSA, this scheme satisfies the privacy of sensitive information.

Theorem 2: The system ensures data integrity, i.e. the data cannot be tampered without being detected.
Proof: After the sensitive information is encrypted, the system uses the Schnorr signature to sign the ciphertext. Then, the system sends the digital signature together with the ciphertext and other relevant information to the recipient. The recipient can calculate a verification value based on the public key of the message sender. Finally, it verifies the message by comparing it with the signature. If they are not equal, the information is invalid. Otherwise, the information is accepted. Therefore, integrity of sensitive messages can be guaranteed in the proposed system.

Theorem 3: The information recorded on the Blockchain in this scheme is traceable.
Proof: The Blockchain records messages in the chronological order, and within the data structure, each block contains the hash value of the previous block. This design allows each block to associate with the nodes before and after it, so that it can be traced back to the starting node. The timestamp in the block is recorded according to the time it was written, thus forming an unchangeable and unforgeable database. Every piece of information on the Blockchain can be traced back to the source through a chain structure, and it can be publicly verified. In addition, the change of each leaf node will also lead to the change of the root node, which can realize fast query and trace the tampered data. Therefore, the information in this solution is traceable after being written to the Blockchain.

6. Conclusion

This paper proposes a Blockchain-based secure framework for smart grids. We use Blockchain to store important data such as "two-ticket" information and operation plan information generated in the smart grid, so that this data cannot be tampered. Even if some misbehaviour happens, it can be traced to its source. Security analyses prove that the system meets the security requirements, such as privacy, integrity, and traceability. The proposed framework contributes to a more secure and resilient smart grid with plan management, personnel management, unit management and spot management. Therefore, it has great potentials in real-world applications. In the future, we plan to further investigate how the framework can be enhance to withstand attacks in the post-quantum era.

References

[1] He Yuan. Design and implementation of BI-based power grid production security management and control analysis platform[J]. Software Guide, (4):90-92, 2014.
[2] D.H.F. Data display and analysis platform based on real-time data model of power grid [J]. Journal of Shanghai University of Electric Power, 12-15, 2017
[3] Mengelkamp, Esther, Gaerttner, et al. Designing microgrid energy markets A case study: The Brooklyn Microgrid [J]. Applied Energy, 2018.
[4] Allison I.RWE and slock.it - Electric cars using ethereum wallets can recharge by induction at traffic lights [J]. International Business Times, 2016
[5] Li Shaolong, Li Shenzhang, Sun Jie. Prospects for the application of Blockchain in power grid reconstruction [J]. Electronic Technology & Software Engineering, (16):219-221, 2017.
[6] Feng Feng. Design and implementation of snart grid system based on biockchain smart contract technology[D]. Shandong University, 2016.
[7] Gai Keke et a. Privacy-preserving energy trading using consortium Blockchain in smart grid [J]. IEEE Trans. On Industrial Informatics, 15(6):3548-3558, 2019.
[8] Bera Basudeb et al. Designing Blockchain-based access control protocol in IoT enabled smart grid system [J]. IEEE Internet of Things Journal, 8(7):5744-5761, 2020.
[9] Mollah M B et al. Blockchain for future smart grid: a comprehensive survey [J]. IEEE Internet of Things Journal 8(1):18-43, 2021.
[10] Zhuang Peng et al. Blockchain for cybersecurity in smart grid: a comprehensive survey [J]. IEEE Trans. On Industrial Informatics, 17(1):3-19, 2021.

An Intelligent Multi-turn Dialogue Classification Method Based on Resampling

Ming Guo, Yunju Zhang*, Qiang Yang and Guangyou Shen

Anshun Power Supply Bureau of Guizhou Power Grid Co. , Ltd. , Guizhou, China

*Corresponding author email: daodan0317@163.com

Abstract. Intelligent multi-turn dialogue classification uses the whole text as input to predict tags. As one of the most popular research topics in dialogue system, intelligent multi-turn dialogue classification has important research significance in academia and industry. In this paper, we focus on removing redundant information from dialogue text. Multi-turn dialogue classification is essentially a text classification problem. We selects Bert as the core model of multi-turn dialogue classification in view of its strong learning and redundancy removing ability. However, Bert limits the length of its input to 512 and the length of dialogues often exceeds 512. A long text segmentation method based on resampling is presented to solve the problem of input length limitation. The resampling mechanism can also further remove the redundant information in multi-turn dialogues. Experimental results show that the proposed method outperforms all the baselines and achieves an F_1 of 53.91%.
Keywords: Convolutional neural network; Multi-turn dialogue classification method; Long text segmentation

1. Introduction

Dialogue system is one of the most meaningful tasks in artificial intelligence. In the early stage of the artificial intelligence development, people have been committed to developing intelligent man-machine dialogue system. In recent years, with the great development of deep learning in the fields of text, voice and image, a dialogue system based on deep learning has emerged. Due to the strong representation ability of deep neural network, the performance of text classification and text generation models has been greatly improved. A variety of technical problems of dialogue system have been solved. Many successful commercial products have been born, such as Siri, Xiaobing and Duer.

According to different application fields, dialogue systems can be divided into task-based and non task-based dialogue systems. Intelligent multi-turn dialogue classification can identify the intention of the whole round of dialogue, evaluate the quality of dialogue, find abnormal events from the dialogue text and realize risk control. It can evaluate whether the dialogue has completed the task in the task-based dialogue. Intelligent multi-turn dialogue classification can be applied in a variety of scenarios. Table 1 shows an example of multi-turn dialogue classification in the telecommunications.. The intention of the dialogue is for "business cancellation". Multi-turn dialogue classification takes the dialogue context as the input to predict the user's intention and understand the semantic information of the whole dialogue. The dialogue classification can identify the type of business that the customer needs to handle in a phone call. As can be seen from the example, intention recognition of multi-turn dialogue requires higher classification ability of the model.

Table 1. An examples of multi-turn dialogue classification in Telecommunications.

dialogue role	dialogue content	dialogue role
customer service	Hello, I'm glad to serve you!	
	Has my 30-yuan flow package been cancelled this month?	customer
customer service	Please wait a minute. There is no 30 traffic package this month.	
	Oh, it's cancelled.	customer
customer service	Yes, it has been cancelled.	
	Oh, OK.	customer
category	business cancellation	

The current research work of the multi-turn dialogue classification tends to directly apply the common text classification model. These models usually deal with the whole conversation text as a long sequence by simply splicing multiple sentences. Then take the sequence into the model—encoding, extracting features and classifying. Researchers have proposed a hybrid model to combine the output characteristics of multiple neural networks, such as bidirectional long-term and short-term memory networks and capsule networks. This method improves the effect of multi round dialogue classification to a certain extent. However, the existing works ignore the self characteristics of dialogue. Compared with the common text classification tasks, multi-turn dialogue text contains more interference information and has poor standardization. Therefore, if it is simply regarded as ordinary text for coding, too much noise will be introduced in the learning process of the model, which will affect the final classification effect of the model. Taking the dialogue in Table 1 as an example, it is one real dialogue of China Mobile customer service dataset. The user is asking the customer service whether the 30 yuan traffic package has been cancelled. However, the example contains a lot of information that is not related to the predicted session category, such as "Hello, I'm glad to serve you!" and "Oh, OK".

Bert is a model released by Google AI team in 2018, which has created the best results in lots of natural language processing fields including text classification. Bert has strong learning ability. Fine tuning on the current task can effectively remove redundant information. Therefore, this paper selects Bert as the core model of multi-turn dialogue classification. However, the Bert model limits the length of its input string to 512, and the length of dialogues often exceeds 512. Therefore, it is necessary to design a method so that the long text of dialogues can be accepted by Bert model. In our work, a long text segmentation mechanism on the basis of resampling is designed to solve the input restriction problem. Meanwhile, based on long text segmentation and resampling, our method can further remove redundancy and learn core information.

2. Related Work

The content of our paper belongs to the task of multi-turn dialogue classification. However, there is less work related to this task. The traditional methods tend to directly deploy the classical neural network model to classify. Among these classification methods, the most classic model is CNN[1]. CNN uses convolution kernel to capture the local semantic features of texts. But it is not suitable for the long-distance dependent scenario of multi-turn dialogue classification. Since then, researchers have put forward their own solutions to the problem of long-distance dependence. Cho et al.[2] proposed an seq2seq coding model based on RNN, which circularly encodes long text. Lai et al.[3] combined the pooling layer in CNN and RNN to utilize the respective advantages of these two models to extract local and global features. Johnson and Zhang[4] proposed word level model to improve the complexity of standard CNN. Then they can obtain the best performance by increasing the network depth without significantly increasing the computing cost. Zhou et al.[5] incorporated the attention mechanism based on RNN to make the model focus on the important parts of the text and avoid the interference of redundant information. Yang et al.[6] further proposed a hierarchical attention frame containing word level and sentence level according to the hierarchical structure of documents to improve the classification effect. Zhou et al.[7] proposed a hybrid model. The model splices the output features of multiple neural networks such as CNN, attention and capsule network[8]. Cao et al.[9] improved TF-IDF algorithm to extract important statements in conversation text and achieved good results.

3. Method

3.1. Pre-processing

In the real dataset, there are redundant information, default value and noise. All data in the proposal of this application are unstructured data. Therefore, data preprocessing is an essential part of the whole classification model. In the data preprocessing step, preprocessing operations such as word segmentation and removal of stop words are performed. In addition, since Chinese text does not use space character to separate words as English text, word segmentation is an indispensable operation in text preprocessing, which converts continuous text into word set. Jieba word segmentation is used to segment the corpus in our work.

3.2. Text Segmentation and Resampling

This step is to segment the long text of the training data set, including the training and test sets. Set the i-th multi-turn text in the data set as C_i. Its length is L_i, which has a label y_i. The i-th existing sample is $d_i = (C_i, y_i)$. After the i-th sample is processed, a sample set D_i will be obtained. For a single sample d_i, the flow of long text segmentation is as follows:

(1) Set the sliding window length as l. Set the moving step as s where $s < l$. Set the sliding window position $k = 0$ and the counter $j = 1$.

(2) If $k + l \geq L_i$, exit the process.

(3) In C_i, intercept the characters from k to $k+l$. set the characters as C_{ij}, and add the sample $d_{ij} = (C_{ij}, y_i)$ to D_i.

(4) Setting $j = j+1$, $k = k+s$; Go back to (2).

After obtaining the result set D_i, each element is input into the classification model. Example of the result of long text segmentation is shown in Figure 1. The short text in D_i with the highest probability is taken as the final resampling text result of C_i.

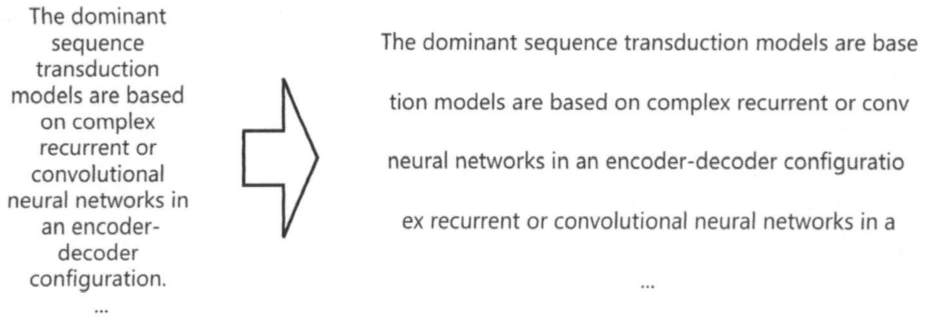

Figure 1. Example of the result of long text segmentation.

It can be seen that this method retains all the semantics of the original text and there are some repeated parts between adjacent short texts. The texts are resampled to obtain the context information between adjacent short texts to a certain extent.

3.3. Classification Model

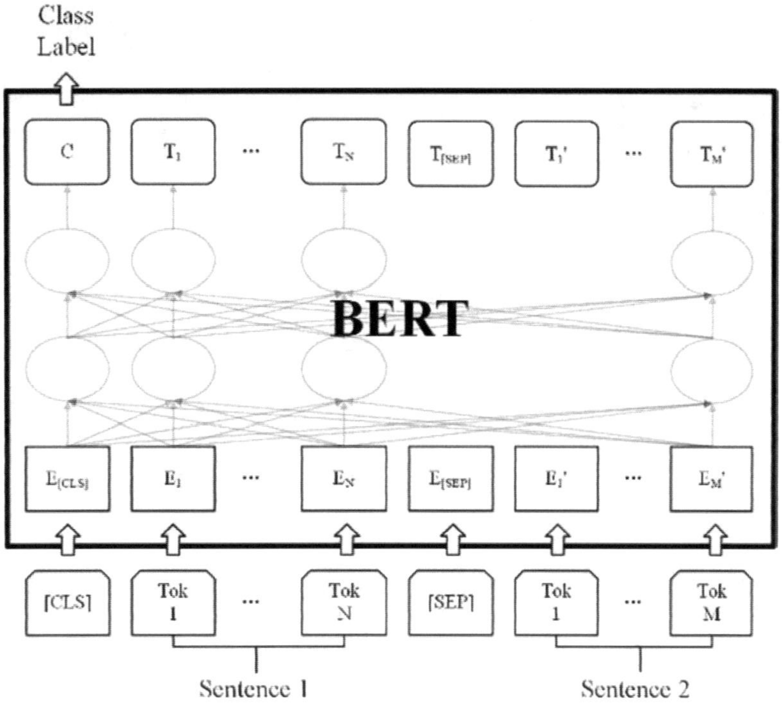

Figure 2. Bert text global feature learning architecture.

In recent years, with the improvement of computing efficiency and the continuous publication of large-scale corpus, more and more pre-training models for general language representation have emerged. The purpose of pre-training is to study the language model on a large-scale unlabelled corpus in advance to obtain general and context sensitive feature representations. The representations can be used to initialize the model and finally fine-tune the parameters in specific downstream tasks to achieve better model effect. The pre-training model can learn a general language representation from large-scale corpus, which is helpful to improve the generalization ability and accelerate the convergence rate. In the existing NLP tasks, the most classic method is to use Bert as the pre-training model to extract the global features of the text, as shown in Figure 2.

In order to support various downstream tasks, Bert designed a general input representation, namely "[CLS], Sentence 1, [SEP], Sentence 2, [SEP]". Sentence 1 and Sentence 2 are text sequences, "[CLS]" is a learnable identifier, which can capture the global information of text input, while "[SEP]" is only the delimiter of Sentence 1 and Sentence 2. After that, Bert inputs it into the bidirectional transformer encoder, and obtains the bidirectional representation of the context through the training of the text from left to right and from right to left. The encode of Transformer is also a multi-layer stacking model. Each layer is composed of sel-attention layer and full connection layer. In each layer, Three matrices consisting of n row vectors is represented as queries $Q \in \mathbb{R}n\times dk$, keys $K \in \mathbb{R}n\times dk$ and values $V \in \mathbb{R}n\times dv$. The attention mechanism calculates the attention weight in the following way.

$$Attention = Softmax(\frac{QK^T}{\sqrt{d_k}})V \tag{1}$$

d_k represents the dimensions of queries and keys and d_v represents the dimensions of values. Then, the input is mapped into h parts. After the attention weight is calculated, the results are spliced to enhance the robustness and generalization.

$$Head_i = Attention(QW_i^Q, KW_i^K, VW_i^V,) \tag{2}$$

$$MHA(Q,K,V) = Concat(Head_1, \ldots, Head_h)W^O \quad (3)$$

$W_i^Q, W_i^K \in R^{d_{model} \times d_k}$, $W_i^V \in R^{d_{model} \times d_v}$, $W^O \in R^{hd_v \times d_{model}}$ are a learnable linear transformation matrix and $d_k = d_v = d_{model} / h$. The output of the multi-head self-attention layer will be transmitted to the full connection layer. Then Bert extracts key information from the input sequence through self alignment. The model of our paper uses the feature vector of fine-tuning Bert as the input of model training.

This paper uses Bert and fully connected network to train the classification model. Our work will fine tune the Bert model with the training set for long text segmentation. And the initialization input of the Bert model is expressed by one hot vectorization.

4. Experiment

Our model is tested on the public China Mobile customer service dialogue dataset. The dataset comes from the CCL2018 China mobile online customer service field user intention classification competition. This task assumes that customer service personnel answer calls from different customers. The purpose is to determine which service is the accurate request of the user's incoming call given the whole conversation history. The conversation dataset is obtained by converting the conversation voice between users and customer service into text through speech recognition technology (ASR). The label of the sample is the predefined business type of China Mobile, such as payment business, opening business and card replacement business.

Our model sets the sliding window length as 100 and moving step as 80. Set the number of stack layers of the Bert as 12. The word vector dimension is 768 and the number of Multi-head Self-attention is 12. The best training parameters of Bert for this task are: dropout=0.1, number of model iterations = 4, amount of training data per batch = 12 and learning rate = 5×10^{-5}.

Table 2. Results of multi-turn dialogue classification

Model	Accuracy	Precision	Recall	F_1
NBC	44.7	33.6	27.8	30.43
SVM	50.9	49.6	40.7	44.71
TextCNN	58.1	55.9	49.6	52.56
Bert	58.7	56.1	50.5	53.15
BERT + resampling	59.9	56.8	51.3	53.91

In order to verify effectiveness and correctness of our method, we selects Naive Bayes(NBC), Support Vector Machine(SVM), TextCNN and Bert as the baseline classifiers for comparison. When Bert is use, the input text is obtained from position 0 of multi-turn dialogue text until the length is conformed to Bert.

The experimental results are shown in Table 2. The results show Bert + resampling has achieved the best results in the four project evaluation indexes, and F1 achieves 53.91%. After Bert integrates the resampling long text segmentation method, the experimental results of the model are further improved. Therefore, the long text segmentation method based on resampling can effectively improve the problem that Bert cannot effectively encode long text and enhance the classification performance of the model

5. Conclusion

This paper proposes multi-turn dialogue classification method based on resampling to obtain the topic semantic information. This paper presents a text segmentation and resampling mechanism in order to solve the problem of Bert input length limitation. Then the sampled text represents the whole conversation and is input into Bert and fully connected network for classification. And the results of experiments show that the F1 of BERT + resampling is 53.91%, which is better than other methods.

Acknowledgments

This work was supported by Science and technology project of China Southern Power Grid Co., Ltd (No. GZKJXM20190503).

References

[1] KIM Y. Convolutional neural networks for sentence classification[C/OL]//Proceedings of the 2014 Conference on Empirical Methods in Natural Language Processing (EMNLP). Doha, Qatar: As sociation for Computational Linguistics, 2014: 1746-1751. https://www.aclweb.org/anthology/ D14-1181. DOI: 10.3115/v1/D14-1181.
[2] CHO K, VAN MERRIËNBOER B, BAHDANAU D, et al. On the properties of neural machine translation: Encoder–decoder approaches[C/OL]//Proceedings of SSST-8, Eighth Workshop on Syntax, Semantics and Structure in Statistical Translation. Doha, Qatar: Association for Computational Linguistics, 2014: 103-111. https://www.aclweb.org/anthology/W14-4012. DOI: 10.3115/v1/W14-4012.
[3] LAI S, XU L, LIU K, et al. Recurrent convolutional neural networks for text classification [C/OL]//BONET B, KOENIG S. Proceedings of the Twenty-Ninth AAAI Conference on Ar tificial Intelligence, January 25-30, 2015, Austin, Texas, USA. AAAI Press, 2015: 2267-2273. http://www.aaai.org/ocs/index.php/AAAI/AAAI15/paper/view/9745.
[4] JOHNSON R, ZHANG T. Deep pyramid convolutional neural networks for text categorization [C/OL]//Proceedings of the 55th Annual Meeting of the Association for Computational Linguis tics (Volume 1: Long Papers). Vancouver, Canada: Association for Computational Linguistics, 2017: 562-570. https://www.aclweb.org/anthology/P17-1052. DOI: 10.18653/v1/P17-1052.
[5] ZHOU P, SHI W, TIAN J, et al. Attention-based bidirectional long short-term memory networks for relation classification[C/OL]//Proceedings of the 54th Annual Meeting of the Association for Computational Linguistics (Volume 2: Short Papers). Berlin, Germany: Association for Computational Linguistics, 2016: 207-212. https://www.aclweb.org/anthology/P16-2034. DOI: 10.18653/v1/P16-2034.
[6] YANG Z, YANG D, DYER C, et al. Hierarchical attention networks for document classification [C/OL]//Proceedings of the 2016 Conference of the North American Chapter of the Association for Computational Linguistics: Human Language Technologies. San Diego, California: Association for Computational Linguistics, 2016: 1480-1489. https://www.aclweb.org/anthology/N16-1174. DOI: 10.18653/v1/N16-1174.
[7] Zhou J Z , Zhu Z K , Zheng-Qiu H E , et al. Hybrid Neural Network Models for Human-machine Dialogue Intention Classification[J/OL]. Journal of Software, 2019.
[8] SABOUR S, FROSST N, HINTON G E. Dynamic routing between capsules[C]//NIPS'17: Proceedings of the 31st International Conference on Neural Information Processing Systems. Red Hook, NY, USA: Curran Associates Inc., 2017: 3859–3869.
[9] CAO B, MA K, LIU Y, et al. Intention classification in multiturn dialogue systems with key sentences mining[J/OL]. Computational Intelligence, 2020. https://onlinelibrary.wiley.com/doi/abs/10.1111/coin.12345.

Face Dataset Augmentation with Generative Adversarial Network

Kelun Cong and Mian Zhou*

School of Computer Science and Engineering, Tianjin University of Technology, Tianjin, China

*Corresponding author email: zhoumian@tjut.edu.cn

Abstract. Face recognition is a widely used scene of artificial intelligence technology. However, some face occlusions cause the face to be unable to be effectively detected in a specific environment. Although many algorithms have been proposed to solve this problem, in essence, a large number of face image data containing occlusion elements is needed to train to improve the detection ability of the algorithm. In recent years, this problem can be effectively solved by using the image generation ability of generative adversarial network. This paper proposes an improved Generative Adversarial Networks (GAN), which improves the effect of occluded face image generation by adding coding module. Through the expansion of data set, the detection accuracy of several classic face detection models for occluded faces is improved by more than 3%. At the moment when the epidemic has not been over, occlusion face data is of great significance to improve the performance of face detection systems in specific public places such as customs security inspection and medical centers.
Keywords: GAN; Neural Style Transfer; Data Augmentation; Face Detection.

1. Introduction

In recent years, deep learning technology has achieved good performance in many fields, which largely relies on promotion of massive data. With the emergence of the new crown pneumonia epidemic in 2020, people need to wear masks to travel in public places according to the epidemic prevention requirements, which poses new challenges to the detection accuracy of face detection tasks in specific places. However, current face detection/recognition systems may not be able complete the task of encountering people with masks. In order to solve the problem, there is a way to recollect masked face images of those people, but this task seems extremely inconvenient and difficult to operate, since it is impossible to call back those people whose faces were collected before. Even in the epidemic situation, collecting real-time face images may expose certain infection risk. In another way, we propose a face dataset augmentation method to extend face images. In fact, we apply unsupervised deep learning approaches to automatically generate more face images which do not exist in the dataset. The expansion and enhancement of image data sets has always been an important research area. However, in the face detection work, the area of the face image is small, which makes most image enhancement methods not applicable. Therefore, expanding more face image data is a more effective method. This is why the generation of adversarial networks is applicable. Generative Adversarial Networks (GAN)[1] is an approach to generative modeling using deep learning methods. The generative model learns the spatial distribution of input data through unsupervised learning mode, and outputs new data according to the distribution. Although GAN successfully dominate generative modeling, directly applying GAN appears some drawbacks such as models collapse, difficulty in training, and poor image generation effect, which causes the generated image examples to be unable to be used for actual data augmented work.

Based on DCGAN (Deep Convolution GAN)[2], this paper proposes a network structure optimized for occluded face image generation tasks. The S-Coder (Source Coder) original domain coding module is added to the network to perform the controllable generation and coverage of occlusion elements. And by modifying the network layer structure, we introduce the Wasserstein distance[3] to optimize network training. On this basis, a consistent training process is used to further increase the stability of network training and improve the quality of image generation. The quality of the image data generated by this method is subjectively and quantitatively compared with the images generated by other GAN approaches. On this basis, the generated occlusion face data set is applied to a variety of face detection methods for training and testing. The experimental results show that the image quality generated by the method proposed in this paper is better and more practical, and the use of the extended face occlusion data set can effectively improve the accuracy of face detection.

2. Principle of Generative Adversarial

A GAN is composed of a generator and a discriminator. The generator uses the random distribution of the input to generate the target image, the discriminator distinguishes the generated images by true or false, and the output is the probability that the generated images are real. So its objective function is

$$\min_{G}\max_{D} V(D,G) = E_{x\sim p_{data}(x)}[log(D(x))] + E_{z\sim p_z(z)}\left[log\left(1 - D(G(z))\right)\right] \quad (1)$$

In training, it firstly optimizes the discriminant network, which we call D-Net. When a real example x is fetched into GAN, we wish the D-Net gives the output $D(x)$ should be as large as possible and eventually close to 1.0, but when a fake example $G(z)$ is fetched into GAN, we wish that the output given by D-Net should be as small as possible and eventually close to 0.0. To unified present the output of the D-Net by a fake example, it is changed to $1 - D(G(z))$, so the combination is the discriminator's optimization formula as follows:

$$\max_{D} V(D,G) = E_{x\sim p_{data}(x)}[log(D(x))] + E_{z\sim p_z(z)}\left[log\left(1 - D(G(z))\right)\right] \quad (2)$$

When optimizing the generation network we call it G-Net, the fake examples need to be closer to the real examples which are labelled as 1.0. Therefore, the larger $D(G(z))$, the better optimization, and then the formula is unified to $1 - D(G(z))$, as shown in the following:

$$\min_{G} V(D,G) = E_{z\sim p_z(z)}\left[log\left(1 - D(G(z))\right)\right] \quad (3)$$

After training, when $D(G(z)) = 0.5$, the fake sample is closer to the real sample, and the balance of the network is achieved.

3. Improved GAN

Deep Convolution Generative Adversarial Network (Deep Convolution GAN, DCGAN) is the combination result between Convolution Neural Networks (CNN) and the idea of Generative Adversarial Networks. In this paper, the residual block structure (Residual Block) is used to replace the original convolution network module, and the final improved network is compared with DCGAN-res. The residual structure is proposed and used by He[4].

3.1. S-Coder Controllable Programming Module

The DCGAN inputs a 100-dimensional noise vector to the deconvolution layer and expands it into a $64 \times 64 \times 3$ feature map. However, since the gradual deconvolution process in the unsupervised training process cannot control the direction which step toward output target, it can only gradually approach the real example according to the distribution in training set. It is impossible to generate multi-class target examples. In order to resolve this problem, the S-coder is added, so that the same network can generate different classes of examples during the supervised training process.

The S-coder is a vector mask that can generate an i-dimensional vector c based on the known labels provided by i classes in data sets, which is the One-Hot encoding. The noise contact operation generated and input are sent to the generating network G-net for the new input. In D-net, samples are sent to a backbone network to extract features, and the features are sent to different network layers respectively, so as to produce the True/False discrimination and the label c. Among them, the image True/False discrimination module D and the image classification module Q share the network parameters of the feature extractor. The network structure is shown in Figure1.

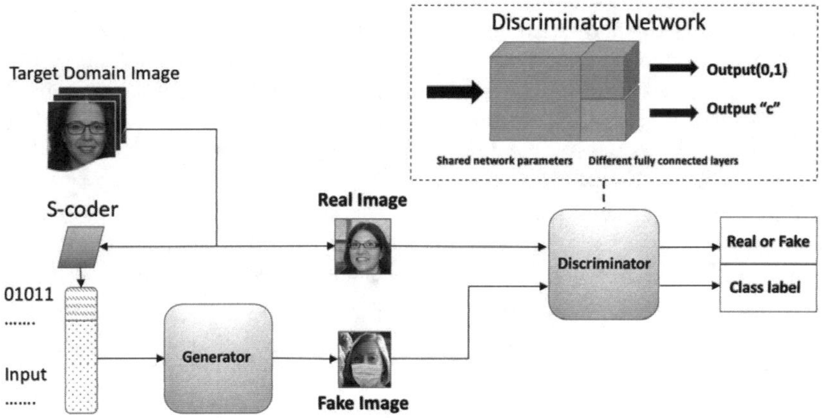

Figure 1. Improved GAN network structure: The Discriminator has two outputs.

The loss of optimizing the discriminator and generator model through the given label encoding is:

$$L_C^D = E_{x,c'}[- \log D (c' \mid x)] \tag{4}$$

$$L_C^G = E_{x,c}\big[- \log D \big(c \mid G(x,c) \big)\big] \tag{5}$$

3.2. Introducing Wasserstein Distance

In traditional GAN training, the model collapses and the generator cannot produce desired images. This problem is well explained in WGAN. The first reason is that the discriminator training is too superior to vanish the gradient of the generator. When the loss of the original discriminator is minimal, the generator loss can be equivalently transformed to minimize the Jensen Shannon divergence(JS divergence) between the real distribution $P(r)$ and the generated image distribution $P(g)$. Since the overlapping area between the distribution of $P(r)$ and $P(g)$ at the initial training stage is 0, or small enough so negligible. As a result, JS divergence always a constant, so that the generator cannot learn effectively. In addition, the discriminator is well optimized, the generator also faces vanishing gradient in some extent. Hence, the result of GAN training is not satisfactory. We use Wasserstein distance in the optimazation of Generators. Wasserstein distance is adopted in learning algorithm when two distributions are in distant far away and with little overlapping. The Wasserstein loss[5] is designed as

$$L_{wl} = E_x[D(x)] - E_{x,c}[D(G(x,c))] - \lambda_{gp} E_{\hat{x}}[(\| \nabla_{\hat{x}} D(\hat{x}) \|_2 - 1)^2] \tag{6}$$

3.3. Consistent Training Process

The Cycle GAN (CycleGAN)[6] proposed by Zhu et al. consists of two pairs of generators and discriminators, the purpose is to effectively transfer the source domain image style to the target domain image style. Its network is two completely symmetrical GANs, which are trained under the constraint of the same period. Use cycle consistency loss to ensure that reconstructed image approaches the distribution of original image to achieve better training stability.

$$\begin{aligned} L_{cyc}(G,F) &= E_{x\sim p_{data}(x)}[\| F(G(x)) - x \|_1] \\ &+ E_{x\sim p_{data}(y)}[\| G(F(y)) - y \|_1] \end{aligned} \tag{7}$$

This paper uses the novel network structure to generate new face images. Through a consistent cycle training process, two identical GAN networks are combined into a symmetric structure, and they are constrained by cycle consistency loss. At the same time, the overall loss of network is as follows. Where L_D represents the loss function of the discriminator. λ_c controls the loss weight of the classifier. L_G represents the generator loss function. λ_{cyc} represents the weight of the consistency loss.

$$\begin{aligned} L_D &= -L_{wl} + \lambda_c L_C^D \\ L_G &= L_{wl} + \lambda_c L_C^G + \lambda_{cyc} L_{cyc} \end{aligned} \tag{8}$$

4. Experimental Results and Analysis

We have conducted two experiments for proving the effectiveness of proposed method. The purpose of this experiment 1 is to check the validity of proposed method by measuring the image quality. In the second, we applied it in face detection scenario to demonstrate the effectiveness in real application.

4.1. Image Quality Experiments

In order to verify the image data generation ability of the proposed algorithm in this paper, first choose to generate augmented sample images on the three-channel image dataset CIFAR-10 to verify the image generation quality and the practicability of the augmented images, thereby verifying the capabilities of algorithm.

4.1.1. Subjective Evaluation. By comparing DCGAN, DCGAN-res (residual structure), WGAN, WGAN-GP, the proposed method in cifar-10 dataset is shown in Figure2. The four category labels are horse, ship, truck, and airplane. We can see that the WGAN-GP and our method present clearer boundary of the main body of object in the images. The outline of the target is clearly displayed.

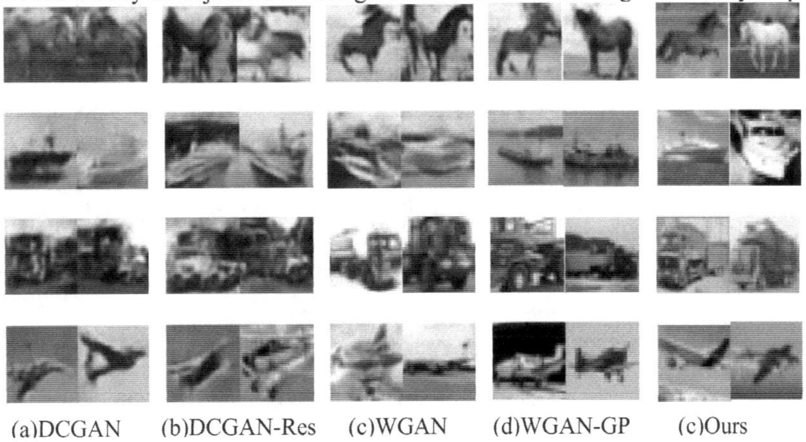

Figure 2. Comparison of cifar10 data set generation in different networks.

4.1.2. Objective Evaluation. In order to better objectively evaluate the quality and diversity of the generated images, some objective metrics are applied to measure generation ability and diversity of different GAN models.

Table 1. IS/FID scores of images generated by different methods.

Dataset	Model	Iterations	IS	FID
Cifar-10	Original	-	11.13	2.06
	DCGAN	300	6.26	13.91
	DCGAN-res	300	8.79	11.24
	WGAN	300	8.87	10.54
	WGAN-GP	300	10.87	7.54
	Our	300	10.36	8.85

We use Inception Score (IS)[7] and Frechet Inception Distance Score (FID)[8] as an objective metric for evaluating the quality of generated images. Because the generator should guarantee the diversity of the generated images. Hence, the larger the IS, the better the image quality. FID is an improved IS metric, but FID calculates the evaluation result obtained by the distribution distance between the generated image and the original image. Therefore, the lower the FID score, the better the quality of the generated image. The experimental results are shown in Table1.It can be seen from the experimental results that the method proposed in this paper on the CIFAR-10 is close to WGAN-GP in image generation quality, and is better than other methods. The experimental results are shown in Table1.It can be seen from the experimental results that the method proposed on the CIFAR-10 is close to WGAN-GP in image generation quality, and is better than other methods.

4.2. Augmentation on Face Data

In this part, we applied our proposed method to a training dataset to generate more occluded face images, then training three standard face detection algorithms with the augmented dataset. Finally, we use another face dataset for evaluating the detection accuracy. Use Adam optimizer for training. The learning rate of the generator network is initially set to 0.001, and the learning rate of the discriminator network is initially set to 0.005.

4.2.1. Face Datasets and Data Reorganizing. The WIDER-FACE dataset[9] is a widely-used dataset for face detection, including 32,203 images and 393,703 annotated faces. The MAFA[10] is a specially proposed occlusion face detection dataset. There are 30,811 images and 35,806 occlusion faces. We use the WIDER FACE dataset as the training set and augmented target. After cleaning and filtering tags such as pose, scale, and face occlusion, we have obtained more than 30,000 cleaned images that meet the augmented requirements, including more than 13,100 original occluded faces, which are used as training images. Use the MAFA as the benchmark dataset, then different face detection algorithms are effectively evaluated before and after augmentation.

4.2.2. Face image generation effect. We use the proposed method to augment the original dataset for adding more occluded faces, i.e., increase mask face occlusion and sunglasses face occlusion, which greatly improve the diversity of faces in the dataset. We have augmented the original widerface dataset into a widerface-mask dataset. After the augmentation, the number of occluded faces in the new dataset will increase by 50%, 100%, 150%, and 200% compared with the original number of occluded faces, and the diversity of face occlusion elements is further improved.

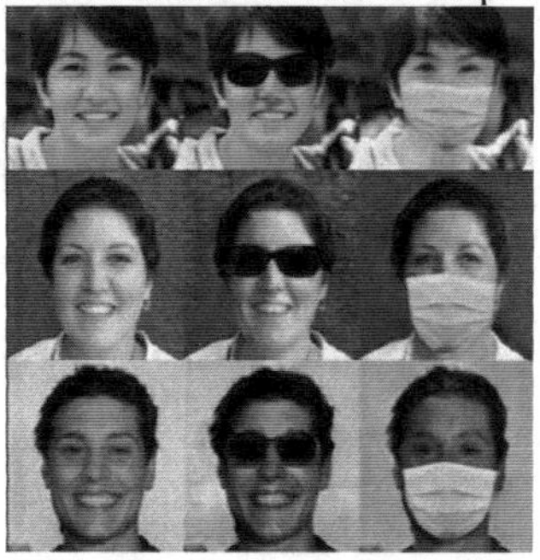

Figure 3. Original face image, shaded by sunglasses, face-occlusion by mask.

The generated face occlusion effect is shown in Figure3, from left to right are the original face image, the sunglasses occlude the face image, and the mask occludes the face image. It indicates that our proposed method can put the right occlusive elements (sunglasses and masks) on the right position in the face images. Even in some images, faces have slightly pose variance, and our method still successfully augments the occlusion elements with appropriate shape.

4.2.3. Comparison Evaluational Results. In this experiment, we select three classic face detection algorithms, namely Single Shot Scale-invariant Face Detector (S3FD)[11], FaceBoxes[12], and

DenseBox[13]. Under the premise that the basic parameters of the three face detection algorithms remain unchanged, we use the WIDERFACE basic dataset(base) and the WIDERFACE-mask augmented dataset(ratios are 50%, 100%, 150% and 200%) for training, then we use the MAFA to evaluate the accuracy of each face detection algorithm. The testing results are shown in Table2.
In this experiment, when only training with the basic data set, three algorithms have achieved the lowest accuracy due to the testing set MAFA has more elements in occlusion. After training with WIDERFACE-mask (50%), the accuracy has made big leap to some extent. When the dataset augmentation ratio increases to 150%, the curve of accuracy is eventually not significantly increasing. It proves that the occluded face image generated by the GAN is effective for improving accuracy.

Table 2. Different algorithms are training and detecting accuracy in the expanded dataset (AP).

Methods	WIDER FACE	WF-MASK (+50%)	WF-MASK (+100%)	WF-MASK (+150%)	WF-MASK (+200%)
S3FD	80.55	83.19	84.19	82.94	85.47
FaceBoxes	76.26	85.25	85.81	86.11	85.29
DenseBox	84.13	91.02	92.02	91.83	91.39

5. Conclusion

Based on the principles of GAN, this paper designs a face image sample expansion algorithm. After a variety of GAN generated image quality comparison experiments, the network in this paper has a good performance in the quality and richness of generated images. In the face dataset expansion comparison experiment, a variety of classic face detection models have improved the detection accuracy of occluded faces to varying degrees. At a time when the epidemic has not completely ended, the application of this technology can provide greater help for special occluded face detection scenarios such as customs security checks and crowded places.

References

[1] Goodfellow, I. J., et al. "Generative Adversarial Networks." *Advances in Neural Information Processing Systems* 3(2014):2672-2680.
[2] Radford, A., L. Metz, and S. Chintala. "Unsupervised Representation Learning with Deep Convolutional Generative Adversarial Networks." *Computer ence* (2015).
[3] Arjovsky, M., S. Chintala, and L. Bottou. "Wasserstein GAN." (2017).
[4] He, K., et al. "Deep Residual Learning for Image Recognition." *IEEE* (2016).
[5] Gulrajani, Ishaan, et al. "Improved Training of Wasserstein GANs." (2017).
[6] Zhu, J. Y., et al. "Unpaired Image-to-Image Translation using Cycle-Consistent Adversarial Networks." *IEEE* (2017).
[7] Barratt, S., and R. Sharma. "A Note on the Inception Score." (2018).
[8] Heusel, M., et al. "GANs Trained by a Two Time-Scale Update Rule Converge to a Local Nash Equilibrium." (2017).
[9] Yang, S., et al. "WIDER FACE: A Face Detection Benchmark." *IEEE Conference on Computer Vision & Pattern Recognition* (2016):5525-5533.
[10] Ge, S., et al. "Detecting Masked Faces in the Wild with LLE-CNNs." *2017 IEEE Conference on Computer Vision and Pattern Recognition (CVPR)* IEEE, 2017.
[11] Zhang, S., et al. "S^3FD: Single Shot Scale-invariant Face Detector." *IEEE Computer Society* (2017).
[12] Zhang, S., et al. "FaceBoxes: A CPU Real-time Face Detector with High Accuracy." (2017).
[13] Huang, L., et al. "DenseBox: Unifying Landmark Localization with End to End Object Detection." *Computer Science* (2015).

Model of Information System for Application of Blockchain Technologies

Evgeniya Ishchukova, Kirill Romanenko* and Vyacheslav Salmanov

Department of Information Security, Southern Federal University, Russia

*Corresponding author email: kirromanenko@sfedu.ru

Abstract. This article describes the current security problems of blockchain networks, identifying the least explored area. The concept of an information system for investigating blockchain incidents is presented. An example of stealing bitcoins from a cryptocurrency wallet was taken as input. This example contains a fairly large number of blocks and transactions, and everything needs to be studied. An overview of the existing vulnerabilities in the Ethereum blockchain platform is presented. A brief description of the basic principles of operation of two largest cryptocurrency platforms - Bitcoin and Ether - is given. Examples of possible incidents that are widespread among users of blockchain systems are considered. The architecture of the designed solutions is described. The first experimental results obtained on the basis of the implementation of the designed algorithms are presented. It is concluded that iterating over a large number of transactions takes a lot of time; in the future, it is planned to apply machine learning methods to analyze such a large data array.
Keywords: Blockchain; Information system; Incident; Transaction; Block; Vulnerability.

1. Introduction

Digital currencies, or cryptocurrencies, are electronic currencies that are generated by networks of computers and are used in place of traditional currencies.

Bitcoin is the most popular cryptocurrency, and at the same time it is one of the first cryptocurrencies for which a certain release limit was originally set.

In 2008, a man known by the pseudonym Satoshi Nakamoto submitted an article [1], which describes the basic principles of the Bitcoin cryptocurrency, including the technology for its generation and management. At the same time, Nakamoto set the maximum number of Bitcoins generated by a distributed network of computers at 21 million. Such a limited supply of Bitcoins ensures the demand for the monetary unit, which, as a result, rises in price. Thus, the price of one Bitcoin changed from $ 1 in 2009 to $ 65,900 in 2021.

Bitcoin is just a convenient way to store and send funds. But the range of actions performed with money is much wider: money can be borrowed, deposited, invested, given in growth, and so on.

The second largest cryptocurrency is Ethereum. The main difference between Ethereum and Bitcoin is the ability to quickly create smart contracts. Smart contract is an alternative to legal contracts. In legal contracts, the third party is the judicial system of the country where the contract is concluded, law institutes are responsible for the execution of the contract.

Smart contract is exactly the same contract, only in digital form. It exists inside the Ethereum system and its execution is guaranteed by a computer program, and in the basis there is a strict mathematical system.

Since 2010, about a dozen cryptocurrency exchanges have been hacked. The losses amount to hundreds of millions of dollars. One example is the hacking of the Bitcoin.org site [2].

The article [3] published examples of vulnerabilities of smart contracts? Written using? Solidity based on the example of attacks on published contracts. It is known that during the exploitation of a vulnerability in a smart contract, which was used by an investor fund, about a third of assets - $ 50 million were stolen [4]. The article [5] shows how minor errors in the Vyper compiler for the Ethereum platform led to the Ethereum creators abandoning the use of Vyper as the official language for writing smart contracts. There are cases when mistakes made in the contract led to large financial losses. So, due to the vulnerability of the Ethereum contract to the callback, the DeFi lending platform lost $ 25 million [6]. And in the DAO project, users could recursively cause the receipt of funds from a smart contract, thereby increasing their capital [7]. The article [8] describes a scenario in which attackers used the hard_fail status and transaction delays to carry out a fake replenishment attack. In addition to errors in smart contracts, when implementing applications, developers often make mistakes in the protocols for generating and distributing keys. For example, attackers managed to obtain the keys of a large number of wallets and some two-factor authorization codes due to an unreliable resource for generating keys [9]. This is just a small part of the possible vulnerabilities of modern blockchain solutions. Today, when the rate of cryptocurrencies, in particular the Bitcoin cryptocurrency, is growing rapidly, the issues of the reliability of blockchain applications are becoming especially relevant.

Such a number of vulnerabilities indicates that at the moment there are no universal methods and techniques using which it would be possible to conduct a security audit for a ready-made solution at the development and testing stage.

Programmers and members of the cryptocurrency community are working hard to identify and fix vulnerabilities in their respective blockchain networks.

Let's consider the main vulnerabilities in the Bitcoin network.

A crypto wallet is a special high-tech program that makes it possible to receive, send and store cryptocurrency. There are crypto wallets that only work with one cryptocurrency. There are also multicurrency crypto wallets.

The wallet is poorly protected from theft. By default, the wallet is not encrypted. For this reason, it becomes an easy target for scammers. However, the latest versions of Bitcoin clients already contain ciphers to protect wallet data, but the user must manually enable the cipher.

The new wallet can be opened with the old password through backups. An old copy of a wallet with an old password can often be easily restored by creating a recovery program (e.g. Apple Time-Machine): restoring an old wallet with a password restores the current wallet and current password. Therefore, frequent password changes are not a guarantee of complete security. In [10], the idea of a Bitcoin wallet key management scheme is considered, this solution, according to the author, is designed to solve the problem with the security of keys storage and access to the wallet.

Sibyl Attack. A hacker can try to fill the network with nodes under his control, and other users will be able to connect only to the blocks created for fraud [12, 13].

Currently, there is a problem of investigating incidents in the field of blockchain technologies and the lack of a solution to this problem is influenced by several reasons. The first reason is the lack of legal regulation in the field of the use of cryptocurrencies and blockchain technologies. The second reason is the complexity of the blockchain technology itself and the lack of a unified incident investigation methodology.

Tracking the history of monetary transactions. The problem of tracking transactions today is acute for law enforcement agencies and companies in the field of information security. There are a number of sites that allow you to display information about the blockchain system in a user-friendly way, for example blockchain.com. Site oxt.me allows you to visualize information from the blockchain system in the form of diagrams and graphs reflecting changes over time for the blockchain. The Maltego service is designed to collect information from various databases of cryptocurrencies, as well as to present them in a graphical format. It allows you to bring together information obtained from open and closed sources, it allows you to visualize aggregated data. The Bitcoin Who's Who service is designed to obtain data on the use of wallets in illegal activities. This service, It "marks" addresses used in fraudulent activities. The service helps to recover stolen Bitcoins, talks about legal services for selling Bitcoins. The Bitcoin Abuse Database service is designed to obtain data on the use of wallets in illegal

activities. It tracks Bitcoin addresses used by ransomware, blackmailers, scammers, etc. The service is open source, located at [14]. Site blockpath.com allows you to visualize the movement of a cryptocurrency in the form of a graph of transitions from one state to another. All the resources considered use a simple display (textual or graphical) of the open data of the blockchain system and do not include elements of an intelligent search engine.
Systems such as Bitcoin contain special mechanisms to obfuscate the transaction trace. Such mechanisms are called Mixers. The problem of using mixers for mixing transactions is considered in [11]. In [11], an algorithm for identifying transactions based on the use of neural networks is proposed. This article is devoted to the design of algorithms for creating an intelligent service for investigating incidents in blockchain systems. The article is organized as follows. Chapter 2 provides a brief description of the basic principles of the considered platforms - Bitcoin and Ether. Chapter 3 discusses examples of possible incidents that are widespread among blockchain users. Chapter 4 describes the architecture of the design solutions. Chapter 5 presents the first experimental results obtained on the basis of the implementation of the designed algorithms. Chapter 6 draws conclusions and discusses future directions of work.

2. Bitcoin and Ethereum

Before moving on to development, we will give a brief description of the blockchain systems under consideration. Bitcoin is an open (public) blockchain network that stores ordered transactions with an indication of the time of their creation. Each node in the Bitcoin network stores the blockchain that it has previously checked. It is believed that when nodes store the same set of blocks, then they are in consensus. The Bitcoin network uses proof of work consensus. A detailed description of how the Bitcoin blockchain works can be found here [18]. All transactions in the Bitcoin system are linked. At first glance, it may seem that coins are simply being transferred from one wallet to another. In fact, the transaction has a strict construction algorithm. Each transaction has inputs and outputs. Inputs are the receipt of money from previously committed transactions, outputs are a waste of money. Money is spent on the basis of previously received transfers from one or more transactions. Therefore, the input of one transaction is the output of the previous transaction. In this case, one transaction can have several outputs at once, that is, the transfer of coins is carried out at once to several different addresses. But the exit from each transaction can be used strictly once. A block for Bitcoin is formed on average once every ten minutes. In this case, the system automatically adjusts the complexity of the search for hash values to fulfill the consensus conditions, depending on the total computing power of the network. Ethereum is the second most popular blockchain platform in the world [19]. An advantageous feature of the Ethereum platform is the ability to apply broader functionality to cryptocurrency. Thanks to the existence of the Ethereum virtual machine and the special programming language Solidity, it is possible to create so-called smart contracts. A smart contract is a special type of transactions that, along with other transactions, are also recorded in the blockchain system. Such transactions contain executable program code that can be run using the Ethereum virtual machine. In fact, a smart contract is a software function that the user of the system launches by passing some parameters to this function. Thus, the Ethereum platform has a more complex architecture compared to Bitcoin and wider functionality. Unlike Bitcoin, Ethereum does not use inputs and outputs to generate a money transfer. Instead, the system stores the current state for each wallet in the system. Ethereum uses Proof of work consensus just like Bitcoin, but the complexity is much lower than Bitcoin. Therefore, blocks are generated on average once every 10 seconds. Due to the fact that Ethereum uses different transactions, the type of these transactions is different and is divided into external (created by users) and internal (created by smart contracts).

3. An Example of an Incident in a Blockchain System

Earlier in one of the popular Telegram channels, a message appeared that on October 15, two masked men, threatening with a pistol, attacked a 24-year-old resident of the city of Tomsk at the entrance of his house. They stole expensive equipment from the guy's apartment, 100 thousand rubles, one thousand dollars and 85 Bitcoins. In total, the attackers "earned" almost 355 million rubles. Having committed the robbery, they tied the victim with tape and fled [15].

The question arises: how to track the transfer? Naturally, the victim knows the address of his wallet and the wallet of the kidnappers. But the further way of moving money is difficult due to the large number of use of mixers and disposable wallets. In our case, we do not even know the addresses of the victim and the kidnapper, since we read this news on the Internet.

There are various information systems that allow us to visualize transactions known to us in advance. An example of visualizing a transaction using the blockpath.com service is shown in Figure 1.

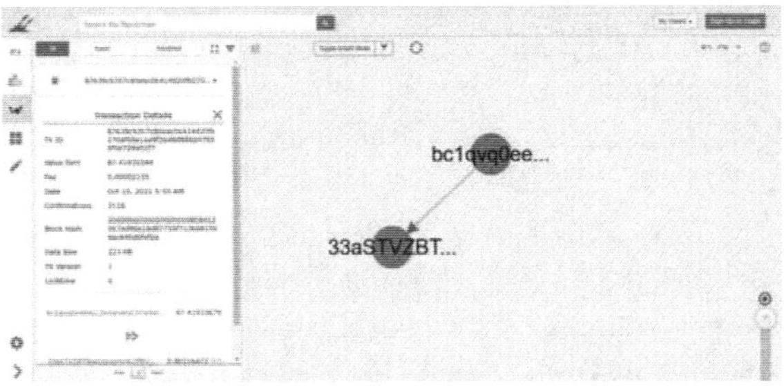

Figure 1. Example of visualizing a transaction.

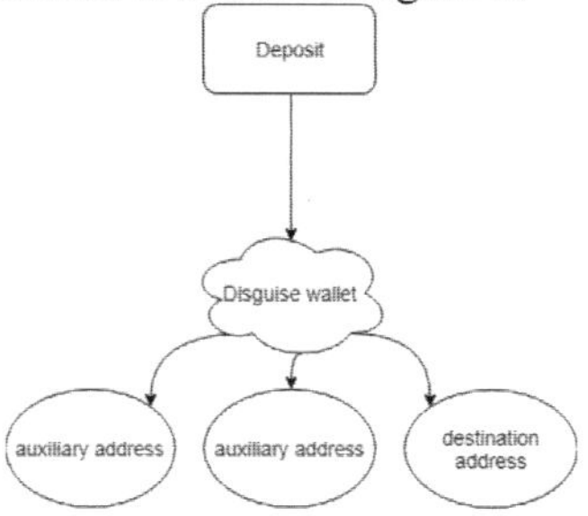

Figure 2. Bitcoin mixing service architecture.

From the advantages of this example, we can note the visual representation of the transfer of funds and the relationship of one wallet with another. The main disadvantage is that we need to know the hash of the transaction in advance, and this does not apply to this example of an incident.

One of the complicating factors in tracking is the mixing of cryptocurrencies. The easiest way to hide the recipient of a stolen cryptocurrency is that bitcoin transactions are performed using a large number of specially created additional wallets. Criminals use bogus (additional) cryptocurrency wallets to hide the illicit source of funds. However, studies have shown that aliases (dummy) bitcoin addresses can be identified by clustering addresses into clusters with identified users, and more and more third-party bitcoin mixing services (creating special dummy cryptocurrency wallets) are emerging to provide additional anonymity. It is reported that at least 4,836 bitcoins stolen from the Binance hack have been laundered using a cryptocurrency mixing service. [16]

The use of bitcoin mixing (concealment) services makes it difficult to track suspicious cash flows because they deliberately hide the relationship between senders and recipients. The architecture of the Bitcoin mixing service is shown in Figure 2.

The principle of operation is that funds are transferred from the victim's wallet to a masking wallet, and then to auxiliary ones (there can be any number of them) and to the attacker's wallet. Thus, it becomes problematic to trace the attacker's wallet.

4. Information System Architecture

Based on the existing problem, it was decided to develop an information system that would track transactions and thereby help in the investigation of incidents in the field of blockchain technologies. Imagine the architecture of the information system in Figure 3.

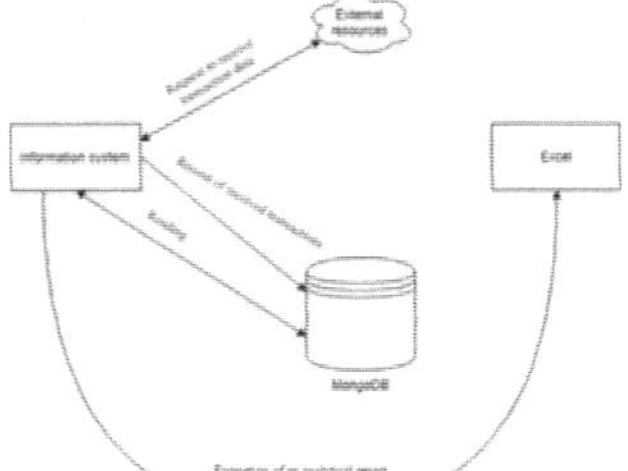

Figure 3. Information system architecture.

The block number is input to the system. The information system returns, based on the request result, transaction data in JSON format, which consists of the transaction hash number, block number, amount and time of the transaction.

Today the use of BPMN (Business Process Model and Notation) as one of the methods for describing business processes is very widespread. BPMN notation is understandable to both business representatives and programs working with business models, it is a standard language that allows you to link business management and the creation of executable algorithms [17]. BPMN is a system of symbols (notation) and their descriptions for modeling business processes. Let's represent a diagram of a business process in BPMN notation (Fig. 4).

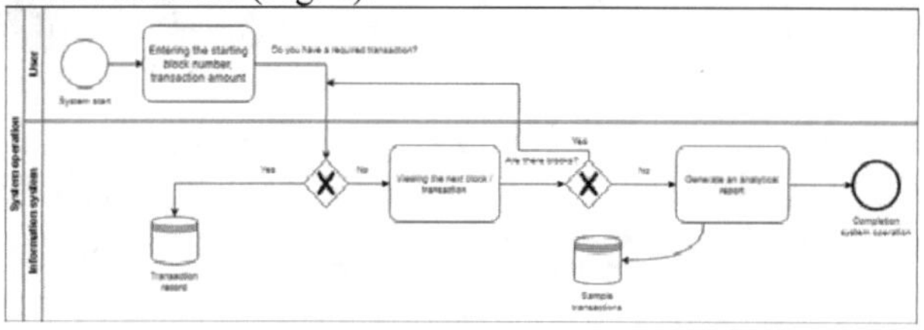

Figure 4. System business process diagram.

The information system makes a request to open sources of data on blockchain transactions, saves them into the database (for convenience and speed, the MongoDB DBMS is used), then, to generate a report, the data is taken from the database at the right time and processed. Data storage in a database is shown in Figure 5.

The block number, transaction hash and transfer amount are entered into the database. In the future, it is planned to increase the amount of data entered into the database.

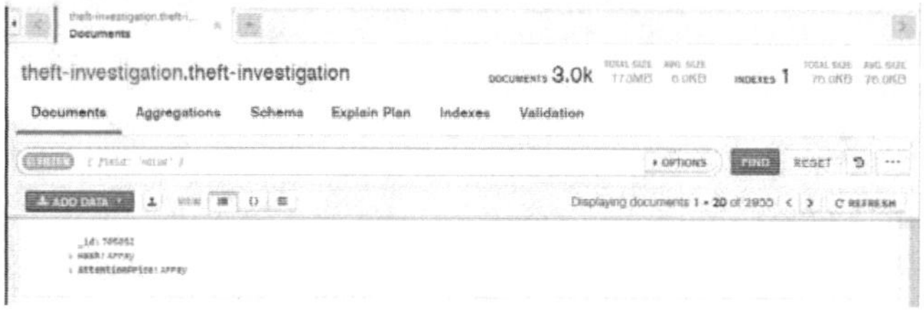

Figure 5. Storing data in a database.

The Node.js framework is used to develop the information system.

5. Experimental Data

Based on the developed algorithms, a software module was developed that allows you to search for transactions for Bitcoin and Ethereum systems according to the specified parameters (transaction amount, transaction time, additional data). So, using this service, we were most likely able to identify the transaction, which was discussed in Chapter 3. The robbers stole Bitcoins in the amount of 85 Bitcoins, leaving 5 Bitcoins to the owner of the cryptocurrency. The transaction we have defined is located at [20]. Experimental data was obtained from an incident related to the theft of 85 Bitcoins. The experiment was carried out using a personal computer. Intel Core i7 processor @ 3.4 GHz RAM: 8 GB DDR3. The calculations were performed in one thread, since Node.js can work in a single thread, it makes sense to parallelize calculations using the golang programming language. Quantitative results of work:

- selected blocks in the range from 705052 to 708006;
- transactions over 85 BTC;
- Number of transactions 3000 pieces;
- Processing time is about 3 hours.

6. Conclusion

This paper describes the modern security problems of blockchain networks, identifies the least explored area of investigation of money laundering in blockchain systems. The concept of an information system for investigating incidents in the field of blockchain technologies was presented. At this stage, a sample of transactions was made for further detailed analysis. At the moment, the system can select transactions for a given initial block and the amount of transfers, it is necessary to

solve the problem of filtering transactions (weed out the addresses of wallets of cryptocurrency exchanges).
In the future, it is planned to speed up the sampling process by parallelizing, and it is also necessary to improve the selection of transactions by filtering out transactions that belonged to exchanges.

References

[1] Satoshi Nakamoto Bitcoin: A Peer-to-Peer Electronic Cash System // https://www.ussc.gov/sites/default/files/pdf/training/annual-national-training-seminar/2018/Emerging_Tech_Bitcoin_Crypto.pdf
[2] Bitcoin.org Hacked, $ 17,000 Stolen From Visitors // https://xakep.ru/2021/09/27/Bitcoinorg-hack/
[3] Aliev I.A. Ethereum blockchain platform smart contracts vulnerabilities. - Scientific notes of young researchers, 2019 // https://cyberleninka.ru/article/n/uyazvimosti-smart-kontraktov-blokcheyn-platformy-Ethereum/viewer
[4] Nathaniel Popper A Hacking of More Than $50 Million Dashes Hopes in the World of Virtual Currency // https://www.nytimes.com/2016/06/18/business/dealbook/hacker-may-have-removed-more-than-50-million-from-experimental-cybercurrency-project.html
[5] Anderey Shevchenko Bugs Found in Compiler for Readable Ethereum Smart Contracts, Team Downplays Concerns // https://cointelegraph.com/news/bugs-found-in-compiler-for-readable-Ethereum-smart-contracts-team-downplays-concerns
[6] Hackers steal $25 million worth of cryptocurrency from Lendf.me platform // https://www.zdnet.com/article/hackers-steal-25-million-worth-of-cryptocurrency-from-uniswap-and-lendf-me/
[7] Smart contract vulnerabilities remain a clear and present danger – https://hackernoon.com/smart-contract-vulnerabilities-remain-a-clear-and-present-danger-59acaf82213f
[8] Details and Solutions for False Top-up Attack (hard_fail status attack) on EOS // https://slowmist.medium.com/details-and-solutions-for-false-top-up-attack-hard-fail-status-attack-on-eos-676b899d27b6
[9] Hack Brief: Hackers Stole $40 Million from Binance Cryptocurrency Exchange // https://www.wired.com/story/hack-binance-cryptocurrency-exchange/
[10] Pal, O., Alam, B., Thakur, V., Singh, S. Key management for blockchain technology (2021) ICT Express, 7 (1), pp. 76-80.
[11] Liu, M., Chen, H., Yan, J. Detecting Roles of Money Laundering in Bitcoin Mixing Transactions: A Goal Modeling and Mining Framework (2021) Frontiers in Physics, 9, article № 665399.
[12] John Maheswaran, Daniel Jackowitz, Ennan Zhai, David Isaac Wolinsky, and Bryan Ford (9 March 2016). Building Privacy-Preserving Cryptographic Credentials from Federated Online Identities (PDF). 6th ACM Conference on Data and Application Security and Privacy (CODASPY) // https://dedis.cs.yale.edu/dissent/papers/cryptobook.pdf
[13] Levine, Brian & Shields, Clay & Margolin, N. (2005). A Survey of Solutions to the Sybil attack. Technical Report of Univ of Massachussets Amherst. 2006–052.
[14] The official repository for BitcoinAbuse.com https://github.com/aarreedd/Bitcoinabuse.com
[15] Tomsk police find out the details of the robbery on Mokrushin // https://riatomsk-ru.turbopages.org/riatomsk.ru/s/article/20211019/razboj-tomsk-mokrushina/
[16] Study: Bitcoins Stolen From Binance Launched Through Chipmixer // https://forklog.com/issledovanie-ukradennye-s-binance-bitkoiny-otmyvalis-cherez-servis-mikshirovaniya-chipmixer/
[17] Object Management Group Business Process Model and Notation // https://www.bpmn.org/
[18] BitcoinDeveloper. Blockchain https://developer.Bitcoin.org/devguide/block_chain.html
[19] Ethereum whitepaper https://Ethereum.org/en/whitepaper/
[20] https://www.blockchain.com/ru/btc/tx/bbc6d052e743e6922a428b39a7892dc025d085495405369300b51e89835dd2d3

Indoor Target Detection and Pose Estimation Based on Point Cloud

Di Zhang[1,2,3], Weimin Zhang[1,2,3,*], Fangxing Li [1,2,3], Ziyuan Guo[1,2,3], Fuyu Nie[1,2,3], Jiahao Jin[1,2,3], Yang Wang[1,2,3], Yongliang Shi[1,2,3] and Qiang Huang[1,2,3]

[1]School of Mechatronical Engineering, Beijing Institute of Technology, Beijing 100811, China
[2]Key Laboratory of Biomimetic Robots and Systems, Beijing Institute of Technology, Ministry of Education, China.
[3]Beijing Advanced Innovation Center for Intelligent Robots and Systems, China.

*Corresponding author email: zhwm@bit.edu.cn

Abstract. The indoor environment faced by service robots is complex, with plenty of objects and mutual occlusion, traditional algorithms cannot detect the target and estimate its pose accurately. We propose a set of target detection and pose estimation algorithm based on point cloud templates to solve the above problems. Through the segmentation and clustering of scenes, the application of the algorithm in complex scenes is realized. Based on the key points to estimate the pose of the target, we achieve the adaptation to mutual occlusion. Besides, we propose an improved RANSAC algorithm, which maintains the accuracy of the original algorithm while greatly improving the speed. Finally, we test the algorithm through the open source model library, simulated and actual scenarios, compared with the SAC-IA algorithm, our algorithm improves the time by 89%, and in the application of complex scenes, the error of the pose estimation is at the millimeter level.
Keywords: Computer Vision; Service Robot; 3D Point Cloud; Scene processing; Target Detection; Pose Estimation.

1. Introduction

At present, the relationship between robots and human beings is becoming increasingly close, and the concept of service robots is gradually emerging. To grasp objects accurately is an indispensable function of the home service robot. In order to realize this function, the correct detection of objects and its posture estimation is one of the key technologies[1-2].

The current mainstream target detection algorithms can be divided into two categories, which are global and local feature-based methods. The former ones such as Global RSD[3], VFH[4], CVFH[5], etc. These methods are faster but have poor adaptability to the occluded environment. The latter methods extract the key points of the scene and the template, calculates the feature descriptors of each key point, and performs target detection and pose estimation according to these descriptors. At present, the mature key point detection methods include Harris 3D[6], LSP[7], etc., descriptor extraction methods include SHOT[8] and ROPS[9], etc. These methods are adaptable to occlusion, but have poor adaptability to complex backgrounds and time-consuming.

We propose a set of target detection and pose estimation algorithm based on template, which will be detailed in Sect. 3. And our contributions are as follows: First, we create a set of algorithm using scene segmentation and target clustering, which eliminates the interference of background and irrelevant clustering so that greatly improved the accuracy, rapidity and adaptability of complex scenes; Second,

We try to use ISS[10] method for indoor object detection and combine it with voxel grid method, which solves the problem of mutual occlusion to a certain extent; Thirdly, We improve RANSAC[11] algorithm and combined it with ICP[12], which improves the speed of the algorithm while ensuring its pose estimation accuracy.

The paper is organized as follows: Sect. I introduces the research background and current situation, as well as our main contributions. And Sect. II describes the method proposed in this article in detail. Experimental results are shown in Sect. III. The conclusions are drawn at the end of the article.

2. Method

This paper presents a set of target detection and pose estimation algorithms based on template point cloud and local features(see Fig 1). Our algorithm can be divided into three stages. The first one is scene pre-processing and object extraction. The second one is key point extraction and descriptor calculation. And the last one is the registration.

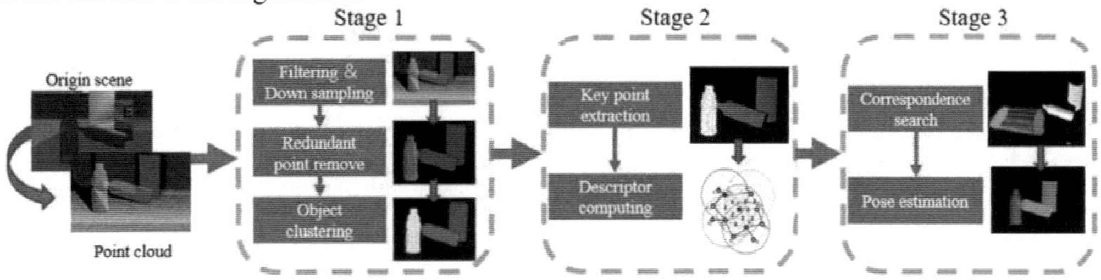

Figure 1. The algorithm flowchart

2.1. Scene Pre-processing and Object Clustering

This part includes point cloud filtering and down-sampling, the removal of the background and the desktop part, the clustering of objects.

The original scene background is complex, but we only care about the scene within the scope of the robot grasp. Thus we filter out points outside the robot's grabbing range.

There are often some points far away from the main point cloud when acquiring the point cloud, we call these points as discrete points or outliers. We statistically analyze the neighborhood of each point, and based on this, remove some points that do not meet the standard.

(1) First, for each point in the input point cloud, query all its neighboring points.

(2) Calculate the distance d_{ij} between each point and its neighbors, where i = {1,2...m}, representing that the point cloud has m points, and j = {1,2...k}, representing that each point has k neighbors.

(3) Assuming that the distance conforms to Gaussian distribution. The distance parameter is modeled using Gaussian distribution, and the calculation method is as follows:

$$\mu = \frac{1}{nk}\sum_{i=1}^{m}\sum_{j=1}^{k} d_{ij} \qquad \sigma = \left(\frac{1}{nk}\sum_{i=1}^{m}\sum_{j=1}^{k}(d_{ij}-\mu)^2\right)^{1/2} \tag{1}$$

(4) For each point, calculate the mean value of the distances between the point and its neighbors;

(5) If the mean value above is outside the 95% confidence interval of Gaussian distribution, the point is considered as a noise point or an outlier, and it is removed. The judgment basis is as follows:

$$\sum_{j=1}^{k} d_{ij} > \mu + 3\sigma \quad or \quad \sum_{j=1}^{k} d_{ij} < \mu - 3\sigma \tag{2}$$

The point scale of the origin scene is large, the appropriate down-sampling of the scene point cloud can shorten the time of subsequent steps. The method is segmenting the point cloud by using small meshes, averaging the points in each mesh.

In the indoor environment faced by the home service robot, the objects are placed on the desktop or other planes, and the plane on which the objects are placed must not be the object grasped by the robot, so before object detection, the large plane in the scene is fitted and removed, so as to simplify the scene.

The specific method is as follows:
(1) Use the minimum data set that can fit the model to estimate its parameters, for the plane, at least three points that are not on the same line are needed.
(2) Use the three points selected, bring the plane equation;
(3) Solve out the plane equation, calculate the distance between each point and the plane, less than a certain threshold, is considered to be the "inner point" in the plane.
(4) Checking the number of the "inner point" meeting the condition in the step (3), returning to the first step if the condition is not met, reselecting the points, and fitting the plane by using the inner points in the step (3) and using a least square method if the condition is met.
(5) Repeating the steps above, and evaluating the model by using the error rate of the local point and the model until an ideal model is fitted or the maximum number of iterations is satisfied.
At the same time, it should also be noted that many daily objects also have planes, such as books, paper boxes and so on. Therefore, before the plane fitting, the number of points included in the plane is judged. The planes with the number of points larger than 1000 are filtered out, and the planes smaller than this value are retained.
After removing the desktop, all the object point clouds are suspended in the air, and the Euclidean distance method is used for further separation, and the process is as follows:
(1) For a certain point P_0 in the space, the kd-Tree search method is used to neighboring point of P_{10}, compare their distance from P_0 with a given threshold, whose distances from P_0 are less than a set threshold are classified into the cluster Q.
(2) Repeat the step (1) for each point in the cluster, if its distance from P_0 is smaller than the threshold value, classifying the neighboring points into the cluster Q, until no new point can be classified into the cluster Q.

2.2. Key Point Extraction and Descriptor Calculation
For the items with large number of points, irregular shape and obvious edges and corners in the scene, using ISS[10] algorithm to extract the key points will get better results.
(1) For each point p_i in a given point cloud, draw a three-dimensional spherical surface with a radius R, and select all points in that spherical region as neighboring points of the point p_i.
(2) For each neighboring point p_j of p_i, the weight between p_j and p_i is calculated as w_{ij}:

$$w_{ij} = \frac{1}{\|p_i - p_j\|}, |p_i - p_j| < r \tag{3}$$

(3) Calculate the weighted covariance matrix of point p_j and each neighbouring point p_i:

$$\mathrm{cov}(p_i) = \sum_{|p_i - p_j| < r} w_{ij}(p_i - p_j)(p_i - p_j)^T / \sum_{|p_i - p_j| < r} w_{ij} \tag{4}$$

(4) Calculate the three eigenvalues $\{k_1, k_2, k_3\}$ of $\mathrm{cov}(p_i)$, and arrange the three eigenvectors from large to small, here we denote the arranged eigenvectors as $\{e_1, e_2, e_3\}$.
(5) Respectively calculating the values of e_2/e_1 and e_3/e_2, and setting th1 and th2 according to the actual situation.

$$\frac{e^2}{e^1} \leq \mathrm{th1}, \ \frac{e^3}{e^2} \leq \mathrm{th2} \tag{5}$$

Then it is considered that p_i is the key point of the given point cloud item.
When the number of key points is less than 5% of the total number of points, the voxel grid is used to extract the key points, the specific process is as follows:
(1) Determine the value of L_x, same method for calculating L_y L_z.

$$L_x = (x_{max} - x_{min}) + \lambda \tag{6}$$

Where x_{max} and x_{min} represent the max and the min value in the x-axis direction respectively. λ is a correction factor to avoid points appearing at the edges of the cube.
(2) Dividing and numbering a cube region, wherein the number of a grid region of any point p_i in the cube is (m_{pi}, n_{pi}, l_{pi}), same method for calculating n_{pi}, l_{pi}.

$$m_{p_i} = \text{ceil}\left(\frac{x_{p_i}-x_{min}}{L}\right) \quad (7)$$

(3) Select the point closest to the center of gravity of the point cloud in the grid as the key point.
After extracting the key points, we use FPFH descriptor to describe the features of key points[13-14]. The specific methods are cited bellow:
(1) For each key point p_s, pick all points in a sphere region as the neighboring point as p_t.
(2) Define the uvw coordinate system, The definition of the uvw coordinate system is as follow:

$$u = n_s, v = (p_t - p_s) \times u, w = u \times v \quad (8)$$

(3) For the point p_t in the neighborhood of the key point p_s, three eigenvalues of each key point are calculated and recorded as the SPFH (Simplified Point Feature Histogram) value of the point:

$$SPFH = \begin{cases} f_1 = \langle v, n_t \rangle \\ f_2 = \langle u, p_t - p_s \rangle / \|p_t - p_s\| \\ f_3 = \text{atan}(\langle w, n_t \rangle, \langle u, n_t \rangle) \end{cases} \quad (9)$$

Where | | pt -ps | | is the euclidean distance between p_t and ps.
(4) For each neighboring point of the key point p, a spherical region is defined, the SPFH value of each neighboring point is calculated in the same way, the SPFH value of p_{ti} is weighted and averaged, and the FPFH value of p_s is calculated as follow:

$$FPFH(p) = SPFH(p) + \frac{1}{k}\sum_{i=1}^{k} \frac{1}{\omega_i} \cdot SPFH(p_{ti}) \quad (10)$$

k represents the number of the neighboring points of p_s, and the weight value ω_i is the distance between p_s and p_{ti}. The three features of each triplet are respectively divided into 11statistical intervals to create a histogram.

2.3. Target Pose Estimation

Anders proposed a speed-optimized in RANSAC algorithms 2014[15], and this paper makes further improvements on the basis of it, performing bidirectional matching in feature point matching, and combining with ICP algorithm.
Similar to the traditional RANSAC algorithms, the ultimate goal of our algorithm is to obtain a transformation matrix T, such that the sum of the squared distances between each point p in the point cloud P and its corresponding point q in the target point cloud Q is minimized after the template point cloud P passes through the transformation matrix T, or less than a predetermined convergence threshold.

$$\hat{\mathbf{T}} = \underset{\mathbf{T}}{\text{argmin}}\varepsilon(\mathbf{T}) = \underset{\mathbf{T}}{\text{argmin}} \sum_{\mathbf{p}\in\mathbf{P}} (\mathbf{Tp} - \mathbf{q})^2 \quad (11)$$

The specific execution process of the algorithm is as follows:
(1) In the template point cloud P, n points (n = 5 in this paper) are randomly searched, respectively denoted by p_1 to p_5. Through the nearest neighbor matching of the feature descriptors, the corresponding points q_1 to q_5 are found in the target point cloud Q.
(2) Aiming at each q_i, re-using a nearest neighbor matching method to find a nearest neighboring point of q_i in the set P, if the point is just p_i, considering that p_i and q_i are a pair of matching pairs and recording as (p_i, q_i), otherwise abandoning the pair of points, re-finding a key point in the set P, and repeating the step (1).
(3) During the translation and rotation, the distance between the points in the same point cloud does not change. Using this feature, we connect the points $p_1 \sim p_5$ randomly selected from the template point cloud P in turn to generate a closed virtual polygon, the side lengths of which are written as d_{pi}, $d_{pi} = \|p_{i+1 \bmod n} - p_i\|$. Their corresponding points $q_1 \sim q_5$, Closed-loop polygons are generated in the same way. Then, calculate the relative difference vector δ:

$$\delta = \left[\frac{|d_{\mathbf{p},1} - d_{\mathbf{q},1}|}{max(d_{\mathbf{p},1}, d_{\mathbf{q},1})} \quad \cdots \quad \frac{|d_{\mathbf{p},n} - d_{\mathbf{q},n}|}{max(d_{\mathbf{p},n}, d_{\mathbf{q},n})}\right] \quad (12)$$

Theoretically, if the matching relationship generated in step (1) is correct, the value of δ Should be 0. In practice, the values of d_{pi} and d_{qi} are not exactly equal because of the influence of noise, sampling and so on, so we usually define a threshold t_{poly} to limit the size of δ, that is:

$$\| \delta \|_{\infty} \leq t_{poly} \tag{13}$$

If δdoesn't meet the condition, directly return to step 1 to re-select random points.
(4) Calculate to obtain a transformation matrix T' by using that match point pairs (p_1, q_1) to (p_5, q_5), and transforming the template point cloud P to obtain the transform point cloud P'.
(5) Search for point pairs with Euclidean distance less than the threshold between P' and Q. If the number of pairs is less than a certain requirement, return to step (1), otherwise, optimize the value of T'.
(6) If the transformation matrix T' meets the condition that mention above, the T' is the final pose transformation matrix, otherwise, the step (1) is returned, and calculation is carried out again.
Compared with the traditional RANSAC algorithm, the proposed algorithm has two additional steps (2) and (3) to ensure the accuracy of the initial matching points. On that in the subsequent step (4), correct relative pose transformation can be obtained more easily, and unnecessary iteration times are reduced, so that the time waste is greatly reduced.
In order to get a more accurate registration result T, based on the matrix T' in the above step, the point cloud registration is carried out. ICP[16] (Iterative Closest Points) is a classical point cloud registration algorithm, unlike the previous methods, ICP does not rely on key points and feature descriptors, and the registration is based on the Euclidean distance in space. The method is as follow:
Assume that the points in the template point cloud P are { p_1, p_2, p_3...p_n } and the points in the target point cloud Q are { q_1, q_2, q_3...q_M } respectively.
(1) Set the initial transformation matrix T', and set the maximum number of iterations t_{max}.
(2) Transforming the template point cloud P by using T', wherein P' = T'P, searching a point q_i with the nearest Euclidean distance from each point p_i in the transformed template point cloud P' in the target point cloud Q, and if the distance between the p_i and the q_i is less than a distance threshold d_{max}, determining that the p_i and the q_i are a pair of matched points, The two points can be used for subsequent registration, otherwise the two points are not considered to be matched, and the two points are eliminated in the matching.
(3) Solving a transformation matrix T_k by using the matching point pairs in the step (2), wherein T_k should satisfy the minimum value of Δ E_k:

$$\Delta E_k = \sum \left| T_k p_i - q_j \right|^2 \tag{14}$$

(4) The template point cloud P is transformed again by using the new transformation matrix T_k, and the steps (2)-(3) are iterated until the convergence condition is satisfied or the T_{max} times are iterated, wherein the convergence condition is as follows:

$$\delta = \frac{|\Delta E_k - \Delta E_{k+1}|}{M} \leq \varepsilon, (\varepsilon > 0) \tag{15}$$

Wherein M is a normalization coefficient, and the output final transformation matrix is an accurate pose transformation matrix of the target relative to the template.

3. Experiment

In this chapter, the accuracy and rapidity of the algorithm in this paper are tested by using 3D model and 2.5D model based on the model library of Stanford University[16], and compared with SAC-IA algorithm[13].The results show that the proposed algorithm can match both 3D point cloud model and 2.5D model with good time efficiency under the condition of ensuring accuracy.
In addition, we also test the effectiveness of the algorithm in simulation and real scenes. The results show that this algorithm can better adapt to the scenes with many indoor objects and high similarity, and its recognition errors are all within mm level. And when things are placed differently or blocked. The algorithm in this paper can still be applied and keep good accuracy.

3.1. Model Library Verification

3.1.1. 3D-3D Pose Estimation. In this section, we use the 3D model of Stanford to test the validity of our algorithm, and compare the results with SAC-IA method, in the accuracy of the time and the pose matrix.

The registration results are shown in Figure 2, where red part is the target, and blue part is the registration result. The higher the coincidence degree is, the more accurate the registration result is.

Figure 2. (a) The registration results obtained by using the SAC-IA method, (b) The registration results obtained by using our method.

The time and error comparison of the two methods are shown in Table 1, where the distance error is the RMSD(root-mean-square error) of xyz, and the angle error is the average of the errors in rpy.

Our method has obvious advantages over SAC-IA method in terms of time efficiency when the accuracy is almost the same. The reason is that firstly ISS method is used to extract key points thus greatly compress point volume, and secondly bidirectional matching and polygon threshold method are used to eliminate wrong matching point pairs in advance.

Table 1. The time and error comparison of the two methods(3D-3D).

Object	SAC-IA Distance error (unit: mm)	Our method Distance error (unit: mm)	SAC-IA Angle error (unit: angle)	Our method Angle error (unit: angle)	SAC-IA Time (unit: s)	Our method Time (unit: s)
Dragon	0.05	0.21	0.02°	0.02°	50.91 s	2.60 s
Bunny	0.02	0.18	0.04°	0.04°	29.86 s	2.48 s
chef	0.33	0.74	0.01°	0.003°	10.20 s	1.93 s

3.1.2. 3D-2.5D Pose Estimation. In practice, the sensor can only scan one side of the object, We call this kind of model scanned by the sensor as 2.5D point cloud model.

Figure 3 shows the registration results of 3D model and 2.5D model. the error and time comparison of the two methods are shown in Table 2.

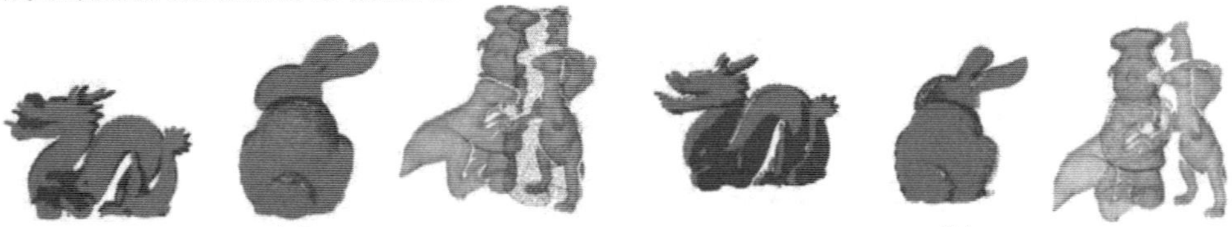

Figure 3. (a) The registration results obtained by using the SAC-IA method, (b) The registration results obtained by using our method.

Table 2. The time and error comparison of the two methods(3D-2.5D).

Object	SAC-IA Distance error (unit: mm)	Our method Distance error (unit: mm)	SAC-IA Angle error (unit: angle)	Our method Angle error (unit: angle)	SAC-IA Time (unit: s)	Our method Time (unit: s)
Dragon	10.27	16.97	4.2°	16.11°	50.91 s	2.60 s
Bunny	33.32	98.48	15.13°	9.58°	29.86 s	2.48 s
chef	107.81	10.43	80.82°	8.87°	10.20 s	1.93 s

When there are many objects and mutual occlusion in the 2.5D point cloud scene, it is difficult to get accurate results by using SAC-IA directly. However, using the method of segmenting and clustering the scene objects, the position and orientation of the target point cloud can be well recognized and matched.

3.2. Simulated Scene Experiment

In the actual scene, it is impossible to accurately obtain the real position of the object, so before the real scene experiment, this paper first uses Gazebo to build the relevant simulation scenes to test the effectiveness of the algorithm.

The image of scene and the test result is shown in Figure 4, Where white part is the original scene, and red part is the result. The error and time are shown in Table 3. We can see that even when there is occlusion exists in the scene, our algorithm still has good adaptability.

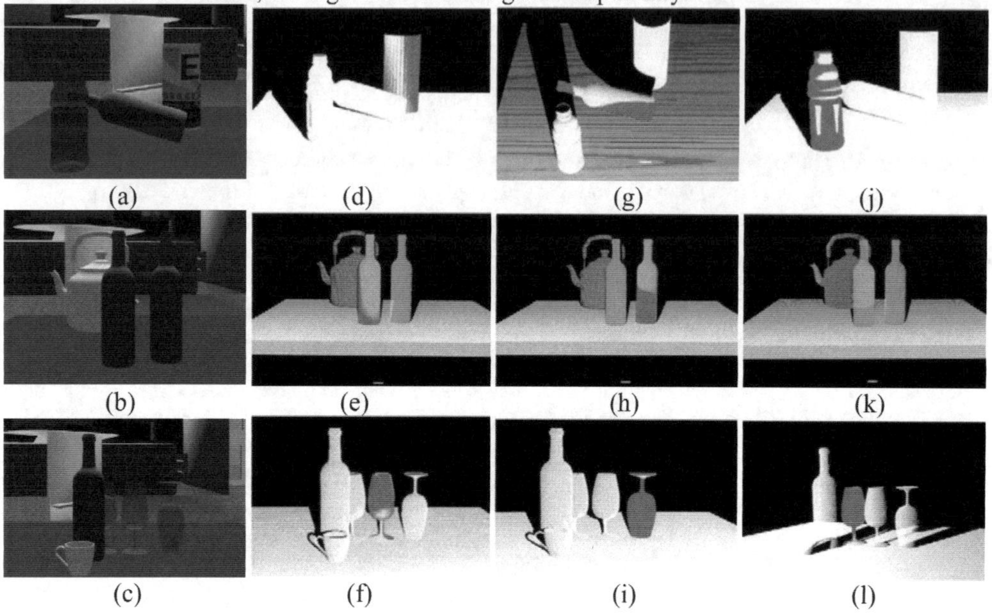

Figure 4. (a)~(c) The origin scene, (d)~(l) The image of the test result.

Table 3. The time and error of simulation scene.

Scene	Object	Distance error (unit: mm)	Angle error (unit: angle)	Time (unit: s)
	Can	0.15	7.00°	0.57
Scene1	Wine	0.36	1.83°	0.48
	Drink	0.29	5.06°	0.42
	Wine(left)	0.22	2.17°	0.46
Scene2	Wine(right)	0.25	7.56°	0.48
	Kettle	0.34	1.72°	0.55
	Glass(middle)	0.34	3.29°	0.37
Scene3	Glass(right)	0.81	0.51°	0.36
	Glass(left)	0.47	6.07°	0.53

3.3. Real Scene Experiment

In the real scene, AprilTag[17] is used to estimate the real pose of objects. The sampling device used in real scenarios is RealSense-D435. Figure 5 shows the origin scenes, Figure 6 shows the results, the error and time are shown in Table 4.

Figure 5. The origin scenes.

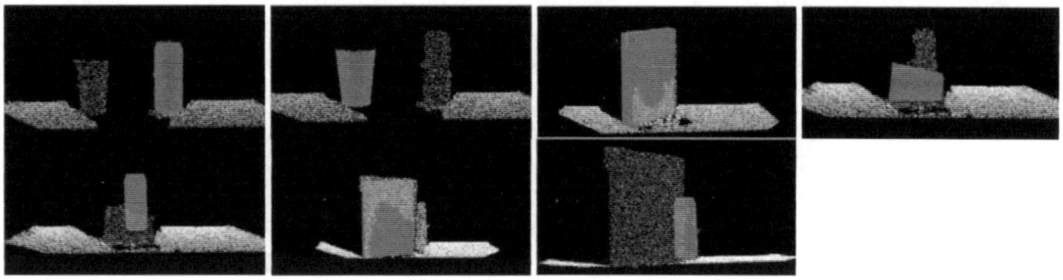

Figure 6. The test results.

Table 4. The time and error comparison of the two methods(3D-2.5D).

Scene	Object	Distance error (unit: mm)	Angle error (unit: angle)	Time (unit: s)
Scene1	Beer	0.20	6.29°	0.34
	Cup	0.45	17.73°	0.34
Scene2	Book	2.47	8.10°	0.95
Scene3	Cup	1.89	12.88°	0.40
	Beer	2.06	15.08°	0.31
Scene4	Book	1.21	5.49°	1.05
	Beer	0.44	13.79°	0.28

4. Conclusion

This paper proposes a target detection and pose estimation algorithm based on point cloud template, and designs its application in indoor environment. Compared with traditional algorithms, our algorithm has better adaptability to complex backgrounds and mutual occlusion of objects. At the same time, our algorithm, compared with the SAC-IA method, has the same accuracy and higher time efficiency. We verify the algorithm through open source model library, simulation scenes and actual scenes, and prove the effectiveness of the algorithm. The experimental results show that the error of our algorithm can be controlled at the mm level in both the simulation scenes and the actual scenes. Meanwhile, the algorithm proposed in this paper is 81% higher in time efficiency than SAC-IA.

The prospect of further research on our algorithm is mainly envisaged in the following aspects: Firstly, how to adapt to a higher occlusion rate is one of the breakthroughs of follow-up research; Secondly, consider further improving the relevant algorithms of rough registration and fine registration to make the algorithm more efficient on the premise of retaining the original accuracy.

Acknowledgments

This work was supported by National Natural Science Foundation of China, Granted No. 61973031.

References

[1] Goldberg, K.. "Editorial: "One Robot is Robotics, Ten Robots is Automation"." IEEE Transactions on Automation Science & Engineering 13.4(2016):1418-1419.

[2] Wang Qi, et al. "Multi-style learning for adaptation of perception intelligence in home service robots." *Pattern Recognition Letters 151*.(2021): doi:10.1016/J.PATREC.2021. 08.026.

[3] Ruiz-Correa, S. , L. G. Shapiro, and M. Meila. "A New Paradigm for Recognizing 3-D Object Shapes from Range Data." IEEE International Conference on Computer Vision IEEE, 2003.

[4] Endres, F. . "Unsupervised discovery of object classes from range data using latent dirichlet allocation." Proceedings of Robotics Science & Systems (2009).
[5] Rusu, R. B. , et al. "Fast 3D recognition and pose using the Viewpoint Feature Histogram." 2010 IEEE/RSJ International Conference on Intelligent Robots and Systems, October 18-22, 2010, Taipei, Taiwan IEEE, 2010.
[6] Mian, A. , M. Bennamoun , and R. Owens . "On the Repeatability and Quality of Keypoints for Local Feature-based 3D Object Retrieval from Cluttered Scenes." International Journal of Computer Vision 89.2-3(2010):348-361.
[7] Hui, C. , and B. Bhanu . "3D free-form object recognition in range images using local surface patches." Pattern Recognition Letters 28. 10(2007):1252-1262.
[8] Salti, S., F. Tombari , and L. D. Stefano. "SHOT: Unique signatures of histograms for surface and texture description." Computer Vision & Image Understanding 125.AUG. (2014): 251-264.
[9] Guo, Yulan , et al. "Rotational Projection Statistics for 3D Local Surface Description and Object Recognition." International Journal of Computer Vision 105.1(2013):63-86.
[10] Yu, Z. . "Intrinsic shape signatures: A shape descriptor for 3D object recognition." IEEE International Conference on Computer Vision Workshops IEEE, 2010.
[11] Fischler, M. A. , and R. C. Bolles . "Random Sample Consensus: A Paradigm for Model Fitting with Applications to Image Analysis and Automated Cartography - ScienceDirect." Readings in Computer Vision (1987):726-740.
[12] Besl, P. J., and H. D. Mckay . "A method for registration of 3-D shapes." IEEE Transactions on Pattern Analysis & Machine Intelligence 14.2(1992):239-256.
[13] Rusu, R. B. , et al. "Aligning Point Cloud Views using Persistent Feature Histograms." 2008 IEEE/RSJ International Conference on Intelligent Robots and Systems, September 22-26, 2008, Acropolis Convention Center, Nice, France IEEE, 2008.
[14] Rusu, Radu Bogdan , N. Blodow , and M. Beetz . "Fast Point Feature Histograms for 3D registration." IEEE International Conference on Robotics & Automation IEEE, 2009.
[15] Buch, A. G. , et al. "Pose estimation using local structure-specific shape and appearance context." International Conference on Robotics and Automation (ICRA) IEEE, 2013.
[16] Gruen, A. , and D. Akca . "Least squares 3D surface and curve matching." ISPRS Journal of Photogrammetry and Remote Sensing 59. 3(2005):151-174.
[17] Wang, J. , and E. Olson . "AprilTag 2: Efficient and robust fiducial detection." IEEE/RSJ International Conference on Intelligent Robots & Systems IEEE, 2016.

A Hand Gesture Recognition Algorithm Based on Multi-scale Hybrid Features

Pu Tu[1,*], Chen Huang[2] and Jinxia Zhu[1]

[1] School of Intelligent Manufacturing, Sichuan University of Arts and Science, Dazhou, China
[2] Department Dazhou Public Security Bureau, Dazhou, China

*Corresponding author email: tupu0803@126.com

Abstract. A human hand gesture recognition algorithm based on multi-scale hybrid features is pro-posed. A large number of gesture data images are organized and annotated by automatic im-age annotation algorithm. Based on dense sampling, the multi-scale Gaussian pyramid is taken into account for enhancing the image details; it can realize the recognition of both large and small target gestures. Meanwhile, according to the aggregation and distribution of gestures, statistical analysis is used to optimize the size and aspect ratio of anchors, which can further improve the accuracy of recognition. The experiment results show that the mean average precisions of this algorithm can be improved to 99.7% and 93.3% for the simple gesture and complex gesture, respectively, which are higher than that obtained by these traditional methods, such as faster region-convolution neural network (Faster R-CNN), You Only Look Once (YOLO), and SSD.
Keywords: Gesture recognition; Multi-scale gaussian pyramid; Faster R-CNN; You Only Look Once (YOLO).

1. Introduction

As a silent visual language, human gestures find practical application in the barrier-free communication for people with disabilities. Therefore, the human gesture recognition methods are very useful in many research areas. Com-pared to traditional machine learning, neural networks stand out in the field of human gesture recognition due to their powerful feature extraction and modeling capabilities. Moreover they have been proven to be effective by experiments [1–3].

Pigou has proposed an algorithm for the effective extraction of image spatial features by convolutional neural network (CNN) in Refs [4]. Zhang Xun et al have proposed an algorithm to improve search efficiency, which used region proposal networks mechanism to extract candidate boxes [5]. In order to form the global association information of sign language, Mokchanovetal et al have taken the 3D convolutional neural net-work model to extract features. This algorithm has a good recognition effect on the multi-modal gesture dataset [6-8]. In order to solve the problems that recurrent neural networks (RNN) training is too difficult and the time-effective information of image sequence cannot be effectively memorized, the Long Short-Time Memory was proposed in Refs [9-10], which was widely used in gesture and sign language modeling. Meanwhile, Shi Yuxin et al have proposed a recognition method based on convolutional neural net-work and random forest algorithm to improve the accuracy of gesture recognition. The recognition rate of this algorithm on the classic MNIST data set is up to 97.9%, which is higher than the traditional convolutional neural network method [11].

Although many gesture recognition methods take the convolutional neural networks (CNN) for basic feature extraction, as the number of convolutional layers increases, more and more gesture details are ig-nored, the accuracy of recognition will decline. Only 0.789% of the total body surface area is found in a single palm of a Chinese adult male [12]. If the details such as finger bending are considered, the propor-tion will decrease sharply. Some researchers have analyzed target recognition algorithms such as R-CNN, Fast R-CNN, Faster R-CNN, YOLO, and SSD, the results show that these methods are mostly suitable for the situation that the scale of the target to be recognized does not change much in the recognition scene. However the recognition accuracy becomes poor for situation with large-scale changes.

To overcome these problems, a multi-scale hybrid features recognition (MHFR) algorithm is pro-posed in this paper, which can enhance image details by multi-scale Gaussian pyramid. Based on dense sampling, the algorithm adopts the strategy of multi-level extraction and hybrid target features, and con-siders the recognition of both large and small targets. According to the aggregation and distribution of the hand gestures in a sign language, the algorithm optimizes the size and aspect ratio of anchors to improve detection accuracy.

2. Recognition Algorithm of Multi-scale Hybrid Features Based on FASTER R-CNN

2.1. Flow Chart of the Algorithm

To solve the problem of low efficiency in manual annotations of the dataset, an automatic image an-notation algorithm is designed to quickly sort out the training data. In addition, a multi-scale Gaussian pyramid procession is added into input images algorithm to enhance the details of gesture images, because the sample size of the images is expanded. Based on the Faster R-CNN feature extraction network, a multi scale feature extraction network is designed to combine the gesture information in different layers for more details on the multi-scale feature maps for prediction and regression bounding boxes and classification inputs. The algorithm adopts a statistical method to set the specifications and aspect ratio of the anchor to further improve accuracy. The feature maps are input to the pooling layer of the region of interest, the repressor, and the classifier for processing. The algorithm flow chart is shown in Figure 1.

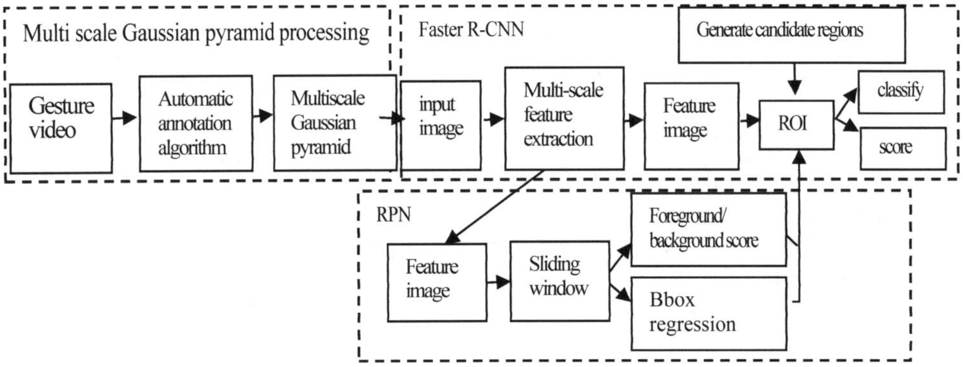

Figure 1. Flow of MSF Faster R-CNN recognition algorith

2.2. Training Data Process

Firstly, we adopt the automatic image annotation algorithm to preprocess the image data before deep learning on the training data. We propose an automatic image annotation algorithm based on correla-tion filtering.Through the spatial reliability map, filter sup-port is adapted to the selected part of a frame for tracking, and an image segmentation method is used to more accurately select the effective tracking target area and achieve high-precision tracking of gesture detail[13]. In this paper, a multi-scale space is established to describe the change of a gesture image's information when its scale changes. The sequence representing the multi-scale space is obtained. The scale space is extracted from the sequence to realize the extraction of image features at different resolutions.

In order to enhance the details of the target image, we first filter the image by Laplace filter, and then we sharp the image details to find the image edge. A multi-scale Gaussian pyramid is constructed for the input image, i.e., down sampling is performed. The image of the current layer of the Gaussian pyramid is generated by Gaussian low-pass filtering on the image of the previous layer, and taking every other row and every other column to generate a down sampled image. The image sequence is obtained by repeated down sampling of the image, with the size of each image layer four-fold smaller than the size of the image in its previous layer. In this paper, in order to enhance the details of gestures, we reduce the size of the upper layer image by a quarter and the resolution by a half respectively to get the lower layer image.

2.3. Multi-scale Feature Extraction

Previous research results show that multi-scale feature fusion can be achieved by connecting feature maps of different scales in ROI region, the mean average precisions of target detection can be improved from 73.9% to 76.4% by the latest PASCAL VOC 2012 [14]. To take into account both large and small targets and generate higher-quality candidate regions, in this paper we adopts a multi-scale feature extraction strategy to improve the Faster-R CNN method based on the VGG16 feature network, and uses batch normalization to enhance the representation power of multi-scale features, while avoiding the problems of the exploding gradient, vanishing gradient, and falling into a local minimum. In batch normalization, each element of the feature is normalized to make the normalized feature have the characteristics of zero mean and unit variance. Inputs with d dimensions, they are normalized in each dimension.

$$\boldsymbol{C}^{(k)} = \frac{x^{(k)} - E[x^{(k)}]}{\sqrt{Var[x^{(k)}]}} \tag{1}$$

Normalization may cause features extracted by the convolutional layer to change, and reduce the representation power of the network. Therefore, a scaling factor and offset parameter are added for each, and are learned together with the network model parameters. The formula is as follow:

$$\boldsymbol{y}^{(k)} = \boldsymbol{\gamma}^{(k)}\boldsymbol{C}^{(k)} + \boldsymbol{\beta}^{(k)} \tag{2}$$

2.4. Design of RPN Network Structures and Setting of Anchor

The most important operation of RPN network is the setting of anchor. We randomly selected 100 single handed and 100 double handed common sign language words from the Lexicon of Common Expressions in Chinese National Sign Language (GF0020-2018) to demonstrate.

Ten participants demonstrate the 200 signs, and the process of demonstration is recorded. The gesture images were recorded at a distance of three meters. The statistics of the areas of the bounding boxes in the hand images (the areas of the annotated regions) were analyzed. The x-axis represents the width of the hand image, and the y-axis represents the height of the hand image. The yellow crosses, solid blue dots, purple crosses, open blue dots, and black squares represent the areas corresponding to the dimensions. The gesture areas show the regional distribution and aggregation characteristics. The gesture areas can be divided into five groups according to the distribution: 102–152, 162–302, 312–502, 512–652 and 702–90×70. The picture must undergo multi-scale Gaussian pyramid processing before the data are trained. The image sizes of the G2 and G3 scale in Figure 2 are reduced to one quarter of the original images. Hence one group, 52–92, is added to the distribution of gesture areas. With these six categories, the anchors can accommo-date the sizes of most gestures. Statistics shows that 95.68% of the aspect ratios of the actual anchor boxes can be classified as 1:2, 1:1, or 2:1. Compared to traditional Faster R-CNN, the sizes and aspect ratios of anchors by the recognition algorithm in this paper is closer to the actual gesture targets.

Experiment

2.5. Experimental Environment

The experimental platform configuration is as follows. The hardware environment is an Intel Core Duo i7-8750H CPU at 2.2 GHz, with 8 GB RAM, an NVIDIA GeForce GTX1050Ti, and a 64-bit operat-ing system. The software environment is Linux Ubuntu 16.04, deep learning framework TensorFlow 1.13.1 GPU, and Python.

2.6. Dataset

Sixteen signs, including "you," "good," "love," "welcome," "like," and "noodles," were selected as examples from the Lexicon of Common Expressions in Chinese National Sign Language. Eight were one-hand signs, and eight were two-hand signs. In the future, the dataset of signs and gestures can be continu-ously expanded to achieve larger-scale recognition. The gestures of Chinese sign language used in this study are shown in Figure 2.

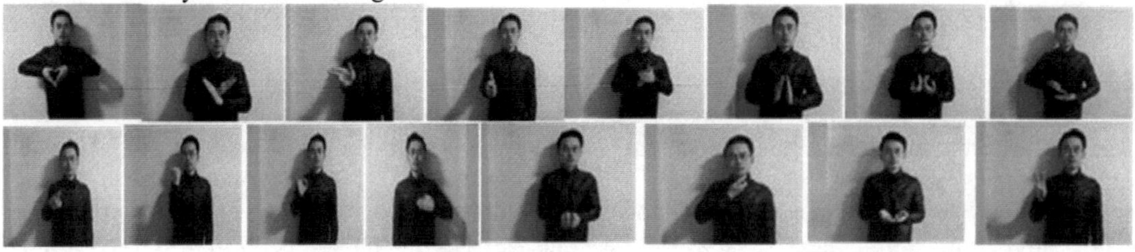

Figure 2. 16 hand gestures in Chinese sign language.

2.7. Scores

In this paper, we take mean average precision (mAP), recall, and accuracy[15] as the scores of model performance. At the same time, to recognize the size of the target, the value of intersection-over-union (IOU) is set as 0.5, so as to comprehensively investigate the accuracy of multi class gesture target recognition.

2.8. Model Training

The VGG-16 network was used as a convolutional network to extract image features in model training. To accelerate training and prevent over fitting, the momentum constant and weight decay are set as 0.90 and 0.0005 correspondingly. RPN training adopts back propagation and stochastic gradient descent training from one node to another node. The anchor parameters were [0.1, 0.2, 0.35, 0.5, 0.65, 1.0] for scales and [0.5, 1.0, 2.0] for aspect ratios. To ensure location accuracy, a non-maximum suppression algorithm was used to twice reduce the redundant boxes. An alternating training strategy was used, taking into ac-count the training for large and small targets. The training cycle was 150,000 iterations. Approaching 90,000 iterations, the recognition precision of the model was consistently at about 99.7%. The recognition precision was 99.74% after completing 150,000 training iterations.

2.9. Experimental Results and Analysis

To determine the effects of multi-scale Gaussian pyramid processing, multi-scale feature extraction, and the anchor size optimization mechanism on the performance of the model, we conducted a comparative experiment on the entire dataset. By comparing combinations of different optimization strategies, the best choices were made. The data of mAP and GPU time before and after the model optimization is shown in Table 1.

Table 1. Network self comparison experiment under different optimization strategies

Methods	mAP%	time/s
FR	97.3	0.324
FR-G	97.9	0.327
FR-M	98.5	0.326
FR-A	98.2	0.343
FR-G-M	98.5	0.328
FR-G-A	99.2	0.345
FR-M-A	99.4	0.345
FR-G-M-A(MSF)	99.7	0.348

Note: FR: Faster R-CNN; suffix-G: Gaussian pyramid; M: multi-scale feature; A: anchor Size optimization.

As shown in Table 1, the mAP of the MSF Faster R-CNN algorithm was improved by a maximum of 2.4% over Faster R-CNN. Four conclusions can be drawn from Table 1:
1) Multi-scale feature extraction is more accurate than the traditional feature method that only uses the final layer, or the shared convolution layer Conv5.
2) Adding the optimization mechanism for anchor size can effectively improve the accuracy of hand gesture recognition.
3) The use of only the multi-scale Gaussian pyramid to process the input image does not significantly im-prove recognition accuracy. The improvement is more noticeable when the multi-scale Gaussian pyramid is implemented together with other optimization strategies.
4) The optimization mechanism of the anchor size contributes most to the computational cost.
From the entire dataset, we extracted 18,150 sign images consisting of three types of complex ges-tures: gestures of small targets, with hand-face contact occlusion, and with motion blur. We take them as the dataset for complex gesture detection.
The comparison results of mAP obtained by he MFF, Fast R-CNN, Faster R-CNN, SSD, YOLO algorithms are shown in Table 2. From the Table 2, we can see that the mAP obtained by MSF algorithm increases by 4.8%, 2.1%, 2.5%, and 9.9% for Fast R-CNN, Faster R-CNN, SSD, and YOLO, respectively. Our model has a higher mAP and a lower missed detection rate.

Table 2. Detection accuracy (%)

Methods	mAP	small	cross	motion
Fast	88.5	89.1	86.3	90.1
Faster	91.2	90.3	90.2	93.1
SSD	90.7	88.3	90.1	93.7
YOLO	83.4	81.5	82.6	86.1
MSF	93.3	93.1	92.4	94.4

Two conclusions can be drawn from Table 2:
1) In complex scenes, the features of small-target gestures are scarcer and more difficult to extract than for large-target gestures. The multiple convolution and pooling operations in Faster R-CNN, Fast R-CNN, YOLO, and SSD cause loss or distortion of the subtle features of small-target gestures. MSF provides denser sampling and finer image features by increasing the image details and multiscale feature fusion methods, thereby effectively extracting features that may be missed due to image scale, and improving the detection accuracy.
2) Adding the optimization mechanism for anchor size can effectively improve the accuracy of hand gesture recognition.

3. Conclusion

Automatic image annotation and multi-scale Gaussian pyramid processing are used to implement dense sampling, and a large number of image data samples are obtained to overcome the problem of ignoring the details of gestures. On this basis, we have proposed an MSF Faster R-CNN network structure and multi-scale feature extraction to enhance the representation power of the network. The aggregation and distribution of hand gestures in a sign language is analyzed and more accurate anchor sizes and aspect ratios are constructed. Then, the experiments with datasets of general and complex gestures are carried out by our algorithm; the results show that this algorithm is very effective for both large and small gesture targets. The accuracy of gesture recognition is significantly improved and the robustness against disturbance is enhanced strongly. However, as a research area with many applications, gesture recognition has room for improvement. For example, in real scenes, the real-time performance must still be improved, and the recognition accuracy in the presence of intersection and occlusion is not so perfect. Therefore, there are the many problems to be researched in the future.

Acknowledgments
This research was supported in part by dazhou science and technology bureau with NO. 19CYR0001 and sichuan provincial public security department with NO. 2021YYCX01.

References

[1] Neverova N,Wolf C,Taylor G W.Multi-scale deep learning for gesture detection and loacalization[C].European Conference on Computer Vision.Springer,2014:475-489.
[2]]Molchanov P,Yang X,Gupta S.Online detection and classification of dynamic hand gestures with recurrent 3d convolutional neural network[C].Proceedings of the IEEE Confernece on Computer Vision arid Pattern Recognition.2016:4207-4215.
[3] Pigou L,Van Herreweghe M,Dambre J.Gesture and sign language recognition with temporal residual networks[C].Procee dings of the IEEE International Conference on Computer Vision.2017:3088-3093.
[4] Pigou L,Dieleman S,Kindermans P-J.Sign language recognition using convolutional neural networks[C].Exiropean Conference on Computer Vision.Springer,2014:572-578.
[5] ZHANG Xun, CHEN Liang, ZHU Xueting, HU Cheng. Gesture Recognition Based on Region Convolutional Neural Network of Faster R-CNN[J]Journal of Donghua University(Natural Science),2019,45(04),559-563.
[6] Molchanov P,Yang X,Gupta S.Online detection and classification of dynamic hand gestures with recurrent 3d convolutional neural network[C].Proceedings of the IEEE Confernece on Computer Vision arid Pattern Recognition.2016:4207-4215.
[7]]Huang J,Zhou W,Li H.Sign language recognition using 3D convolutional neural networks[C].2015 IEEE international conference on multimedia and expo.IEEE,2015:1-6.
[8] LIANG Zhijie. Research on Key Technologies of Sign Language Recognition for Deaf-mutes[D].Wu Han, Central China Normal University,2019.
[9] Venugopalan S,Rohrbach M,Donahue J,Sequence to sequence-video to text[C].Proceedings of the IEEE international conference on computer vision.2015:4534-4542.
[10] Liu T,Zhou W,Li H.Sign language recognition with long short-term memory[C].2016 IEEE International Conference on Image Processing.IEEE,2016:2871-2875.
[11] SHI Yu-xin,DENG Hong-min,GUO Wei-lin. Static Gesture Recognition Based on Hybrid Convolution Neural Network[J]. Computer Science,2019,46(6),165-168.
[12] X. Lu, D. Gao. Palm Area as a Ratio of Body Surface Area in Chinese Adult Male Population[J]. Journal of Forensic Medicine,2015,31(5),359-360.
[13] Lukežič A, Vojř T, Čehovin L, et al. Discriminative Correlation Filter with Channel and Spatial Reliability [C]//CVPR, 2017.
[14] Sean Bell,C. Lawrence Zitnick,Kavita Bala,et al.Inside-Outside Net: Detecting Objects in Context with Skip Pooling and Recurrent Neural Networks[EB/OL].(2015-12-14)[2019-09-21].https://arxiv.org/abs/1512.04143.
[15] REN S,HE K,GIRSHICK R,et al. Faster R–CNN:Towards real-time object detection with region proposal networks[J].IEEE Transactions on Pattern Analysis & Machine Intelligence,2015,39(6):1137-1149.

Research on Mini-EfficientDet Identification Algorithm Based on Transfer Learning

Chao Liu[1] **and Shouying Lin**[2,*]

[1] Graduate student of Fujian Agriculture and Forestry University, mainly engaged in deep learning.
[2] Fujian Agriculture and Forestry University, associate professor, mainly engaged in automatic control. Fujian, China

*Corresponding author email: 1191264001@fafu.edu.cn

Abstract. In this research paper, the Chinese beehive culture in Fujian Province, China is used as the carrier, and the Mini-EfficientDet deep neural network after migration is used to identify common species at the hive door, that is, the identification of Chinese bees, wasps and cockroaches in the form of nymphs. In this paper, we define the modified model as Mini-EfficientDet by compressing the initial EfficientDet model and adding the category imbalance function, which makes it focus more on the recognition and classification of small targets while ensuring the recognition accuracy. Through the test on the MSCOCO2017 data set, it is concluded that when the backbone network adopts EfficientNet B7, it shows strong detection accuracy in detecting targets of various scales, which confirms the role of the category imbalance function proposed in this paper and the efficiency of the improved EfficientDet model. Detection accuracy. The pre-trained model is transferred to the field of beehive species detection through migration learning, that is, after the post-training of the self-collected data set, the detection accuracy of Chinese bee, cockroach, and wasp is 98.66%, 83.71%, and 82.06%. It has made sufficient algorithmic preparations for the later detection and early warning system of beehive species invasion.
Keywords: EfficientDet; Category imbalance function; Mini-EfficientDet; Transfer learning.

1. Introduction

Insects often invade beehives during beehive farming, causing harm to bee colonies, including cockroaches, hornets and ants. In Fujian Province, cockroaches and hornets are the most harmful insects to bee hives. The cockroaches are mainly in their nesting stage, and the hornets are mainly found outside the hives or at the hive doors, attacking and eating the bees coming out.

The common problems in the field of computer image recognition in agricultural farming are mainly in the irregular shape of recognition targets and the large number of targets contained in the images. This paper focuses on the field of Chinese bee hive farming in Fujian Province, China. There are few studies on image recognition in this direction, mainly because the farming field has the characteristics of low investment and high return, which causes the research scholars not too much research. In this paper, according to the environment of Fujian Province and the characteristics of beehive breeding, the problems in this direction are summarized as follows: (1) the recognition target size is small; (2) more recognition targets need to be recognized within the viewpoint[1]; (3) the post-maintenance difficulty of common recognition algorithms is high. In order to solve the above problems, this study adopts the EfficientDet model as the basis[3], which has a high recognition accuracy for multi-scale image recognition, and it also meets the advantages of easy maintenance at a later stage in model

construction, so it can achieve the effect of effective recognition by reducing the model under the premise of no sudden change in recognition accuracy.
Through the improvement of EfficientDet, the improved model pays more attention to the recognition of small targets and prepares the algorithm for the later detection and warning system of beehive species invasion while ensuring the recognition accuracy at the beehive nest gate. In order to improve the above-mentioned shortcomings, this study "slimmed down" the model based on the EfficientDet model by introducing a new category imbalance function, which improved the problem of reduced detection accuracy due to the compression of the model. Later, the pre-trained model was trained by transfer learning with a homemade dataset, and was able to distinguish Chinese wasps, hornets, and cockroaches in worm form efficiently and accurately, thus achieving the expected research goal.

2. Data Set Creation

In this study, both public datasets and homemade datasets are used in the selection of datasets. In order to meet the dataset requirements for later migration learning, the size of the homemade dataset in this study is set to 320×320 pixels, which is equivalent to the size of the public dataset.

2.1. MS COCO Dataset

In this study, after algorithmic innovation, the Microsoft Common Objects in Context 2017 (MS COCO2017) dataset [2] was used for algorithmic accuracy comparison. This dataset contains 91 categories with a total of 123,287 images. Since the MS COCO dataset itself contains images with complex backgrounds, a larger number of targets, and smaller sizes, it is more in line with the research object (Chinese bee) targeted in this study.

2.2. Homemade Datasets

The acquisition of the target data set was accomplished using a camera mounted at the entrance of the hive door. In order to meet the dataset requirements for the later migration learning part, as well as to demonstrate the variable data samples in the daily state as much as possible, this study uses three data enhancement methods on the basis of the original acquisition dataset: (1) by adjusting the brightness and darkness of the acquisition images; (2) by motion blurring the data in the training set according to the principle of point scatter function PSF [3]; (3) by simulating the process of image acquisition The effect of noise generated in the process of image acquisition, this study adds Gaussian white noise to the original image dataset of bees [7]. Multiple species appear in the same image during the natural acquisition process, so this study categorized images containing cockroaches or wasps as cockroaches and wasps in the expansion process, while the illustrations without both were defined as Chinese bees, and finally, as shown in Table 1, the training set was increased from 10,000 to 43,000 by the three methods mentioned above.

Table 1. Data set sum

Quantity/sheet	bee	Cockroach (Wakame)	wasp
Primary	6000	2000	2000
Expanded	24000	10000	9000

3. Structure of the Mini-EfficientDet

In this study, in order to improve the defects such as the model itself is too large, detecting target loss and inaccurate species recognition, the following improvements are made on the basis of the original EfficientDet: (1) the multiple stacking of BiFPN layers of the original model is removed; (2) the category imbalance function is added after the BiFPN layer. This reduces the memory occupation of the initial model on the one hand, and on the other hand, the addition of the category imbalance function focuses more on the recognition of small targets, which is more relevant to the target recognition in this study.

3.1. EfficientNet Features
EfficientNet [4] was proposed by Google in 2019, and the network is known by its own ultra-high efficiency. The features of EfficientNet are (1) to achieve feature extraction, the residual neural network is used to increase the number of layers in the neural network stack; (2) to change the number of layers in the stacked layers for extracting features to further achieve multi-layer feature extraction; (3) to make the network recognition more accurate by adding pictures with different resolutions. Besides its design of this network is completed by deflating the baseline model [5], while adjusting the depth, width and the resolution of the input.

3.2. BiFPN Layers
As shown in Fig. 1 (1) is the most primitive FPN structural model [6], in common convolutional neural networks, the features in the lower layers of the input image have less semantic meaning but highlight the location of the target accurately; the features in the higher layers of the input image are rich in semantic meaning but have vague information about the location of the target. When adopted to the FPN network, the feature maps of the third to seventh layers are extracted and fused, and the size of the feature maps extracted in each layer is 1/2i of the input image.

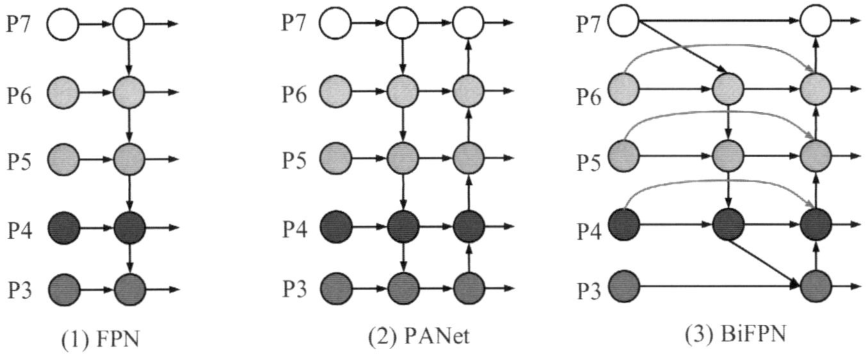

Figure 1. FPN type

In PANet networks [7], bi-directional transmission has been proposed as shown in Figure 1(2). BiFPN is proposed for existing network models.

- Based on PANet, the feature points with only one input i.e., layer 3 and layer 7 are removed.
- Completely connect the feature points of each layer to obtain Figure 1(3); the arcs in the illustration indicate that more features are fused without adding additional computations.

In this paper, when performing pre-feature extraction, the input image is first subjected to different convolutional neural networks for feature extraction, and then stacked with a single BiFPN layer to complete the extraction of deeper semantic information.

3.3. Structure of the Mini-EfficientDet
- Improvement of BiFPN layers

To address the huge computational effort caused by stacking BiFPN layers several times and the accuracy reduction caused by reducing BiFPN layers, this study connects only one BiFPN layer in the original EfficientNet and adds the classification regression loss function after the BiFPN layer to improve the accuracy of recognizing targets. The structure diagram is shown in Figure 2.

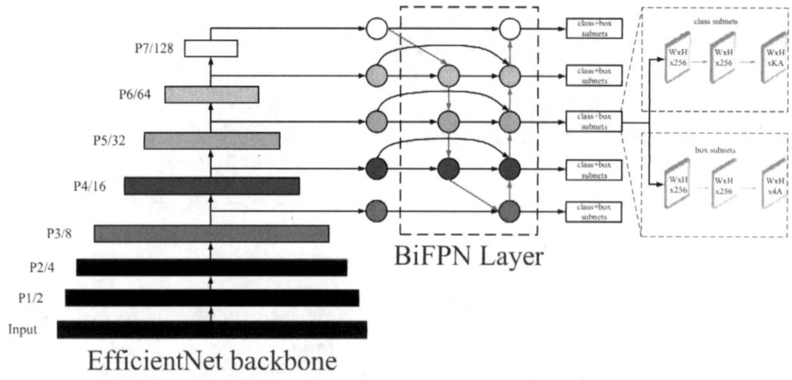

Figure 2. Improved EfficientDet model

- Improvement of Class-Imbalance Loss Function (CIL)

In the improvement process, a new loss function is added, and the standard cross-entropy is used as the basis of the class-imbalance loss function, and the same deflation factor as EfficientNet is introduced ϕ.Taking the binary classification as an example, the category imbalance loss function can be defined as:

$$CIL(p) = -\phi \times CE(p) = \begin{cases} \phi \cdot \log(p) & \text{if } y=1 \\ (1-\phi)\log(1-p) & \text{otherwise} \end{cases} \tag{1}$$

where p is the predicted probability of the label y = 1. As p tends to 0, the category imbalance loss function is significantly larger than the focal loss, which makes the model used in this study focus more on hard-to-classify targets. Comparison with the standard cross-entropy loss function leads to Table 2.

Table 2. Sample probability and weight loss

Sample classification probability (p)	Sample loss weights
0.969	Scaled down 1000 times
0.900	Scaled down 100 times
0.500	Scaled down 2 times
0.100	0.9 times the standard cross-entropy

When $p \approx 0.9$ or $p \approx 0.969$, there is no difference in the adjustment of the loss weights for the category imbalance loss function as well as the focus loss function, when $p \le 0.5$, the loss weight of the category imbalance loss function for the samples is higher than that of the focus loss function, so that it contributes to the differentiation of the positive and negative categories.

To solve the category imbalance problem a common way is to introduce weighting factors $\alpha \in [0,1]$, α is defined as the inverse of the category frequency or the hyperparameter of the cross-test.

$$CE(p)_\alpha = \begin{cases} -\alpha \cdot \log(p) & \text{if } y=1 \\ -(1-\alpha)\log(1-p) & \text{otherwise} \end{cases} \tag{2}$$

$$\alpha = \begin{cases} \alpha & \text{if } y=1 \\ 1-\alpha & \text{otherwise} \end{cases} \tag{3}$$

The above equation is an expansion of the standard cross-entropy loss function. Adding the weighting factor in the experiment $\alpha \in [0,1]$, Compared to CIL(p) without adding weighting factors, the accuracy is improved. Therefore, the expression of the category imbalance loss function is:

$$CIL(p) = -\alpha\phi \times CE(p) = \begin{cases} \alpha\phi \cdot \log(p) & \text{if } y=1 \\ (1-\alpha)(1-\phi)\log(1-p) & \text{otherwise} \end{cases} \quad (4)$$

In this study, the number of Chinese wasps is larger than the number of cockroaches and larger than the number of wasps in the collected data set, and the model established in this data set is easy to recognize Chinese wasps in the training phase, but it is progressively more difficult to recognize other species. The category imbalance loss function adopted in this study makes the overall weight values not consistent, but reduces the weights of the easily classified samples, so that the less easily classified samples get higher weights.

4. Mini-EfficientDet Model Based on Migration Learning

4.1. Transfer Learning

The specific process of transfer learning (TL) is to migrate the weights learned in a neural network to another model to make it a new neural network. The main factors affecting transfer learning are the size of the dataset and the similarity of the new dataset to the dataset used in the pre-training network. In this paper, we use the parameter transfer method to transfer the trained weight parameters from the source model to the target model, so that the target model can extract the effective features faster and perform the recognition task better.

The data involved in this study satisfy the fine-tuning conditions and there is a similar part of the MSCOCO 2017 dataset as the homemade dataset. The MSCOCO 2017 dataset contains more than 90 target categories, i.e., bird, frisbee, mouse, and other small objects. The homemade data contains four categories, i.e., Chinese wasp, cockroach, wasp, and other.

4.2. Transfer Process

The Mini-EfficientDet model after pre-training by MSCOCO 2017 was adapted to the new dataset by fine-tuning the network. Since the weights of the Mini-EfficientDet model were trained based on the MSCOCO 2017 dataset and the homemade dataset resembled some categories in that data, the whole network model was fine-tuned using the homemade dataset.

- Organize the image dataset to ensure that all images are of the same size;
- Modify the model structure by revising the last convolutional layer in the classification sub-network to a fully connected layer with the number of categories as 4, and the last convolutional layer in the regression sub-network to a fully connected layer with the number of outputs as 16;
- The Mini-EfficientDet model trained on the MSCOCO 2017 dataset is used to initialize the backbone network with the migration learning algorithm, and freeze the backbone network, randomly initialize the modified sub-network layers, and train only the task sub-network separately, and the training process is performed by learning a smaller stochastic gradient descent (SGD) and a smaller batchsize, This ensures that the magnitude of updates is kept low;
- Test the accuracy of the migrated model.

5. Experiments and Results Analysis

5.1. Evaluation Criteria

The MSCOCO 2017 dataset was used for the experimental dataset, and for consistency, MSCOCO 2017's own target detection evaluation criteria were used [8]. In the dataset AP is the average of multiple IoUs, specifically using ten different thresholds, i.e., a range of [0.50, 0.95] and a step size of 0.05.

5.2. Experiments and Analysis of Results

This study is based on tensorflow2.1.0 and pytorch1.4.0 frameworks, and the backbone network mainly uses VGG16 [9], ResNet-50 [10], ResNet-101 [11] and EfficientNet B4~7. The size ratio of target detection frames is set to {1:2, 1:1, 2:1}, and the confidence level was set to 0.05. NMS with a threshold of 0.5 was used to detect the highest prediction in each layer. The classification sub-network uses the proposed category imbalance loss function. To prevent the situation of infinite loss and meaningless loss function, the theoretical prediction probability is modified to [0+(1e-5), 1.0-(1e-5)] in this study.

- Ablation experiments

The ablation experiment is mainly used to analyze the existence value of each module in the network. The modules included in the model used in this study are EfficientNet B4~7, BiFPN, and CIL(p). As can be seen in Tables 3 to 6, the addition of CIL(p) in this study improves the overall recognition accuracy, regardless of BiFPN or not. In Table 8, for example, the AP without the addition of CIL(p) is 50.9; without BiFPN, the AP is 52.1, and with the model configured with BiFPN and CIL(p), the AP reaches 53.0, which is 1.04 times higher than without the addition of CIL(p), and 1.01 times higher than without the addition of BiFPN. In contrast, the increase in multiplicity reaches three percent, indicating that the category imbalance loss function can more effectively distinguish hard-to-discriminate targets.

Table 3. Ablation experiment of Efficient B4

Efficient B4	BiFPN	CIL(p)	AP	$AP^{0.5}$	$AP^{0.75}$	AP^{small}	AP^{medium}	AP^{large}
√	√		44.2	55.3	46.1	24.2	44.1	49.3
√		√	46.5	59.7	49.6	26.8	46.6	55.6
√	√	√	48.6	63.2	51.9	28.1	49.8	59.7

Table 4. Ablation experiment of Efficient B5

Efficient B5	BiFPN	CIL(p)	AP	$AP^{0.5}$	$AP^{0.75}$	AP^{small}	AP^{medium}	AP^{large}
√	√		45.9	64.5	49.4	24.9	48.9	56.7
√		√	48.5	66.8	52.1	28.0	50.9	59.1
√	√	√	50.5	68.2	54.3	29.9	53.2	60.8

Table 5. Ablation experiment of Efficient B6

Efficient B6	BiFPN	CIL(p)	AP	$AP^{0.5}$	$AP^{0.75}$	AP^{small}	AP^{medium}	AP^{large}
√	√		46.2	64.0	49.1	30.8	50.5	58.2
√		√	49.3	67.8	53.4	32.1	54.1	60.1
√	√	√	51.1	69.5	55.2	33.4	55.5	62.9

Table 6. Ablation experiment of Efficient B7

Efficient B7	BiFPN	CIL(p)	AP	$AP^{0.5}$	$AP^{0.75}$	AP^{small}	AP^{medium}	AP^{large}
√	√		50.9	67.1	52.1	30.9	51.7	60.8
√		√	52.1	69.2	54.3	33.1	54.6	64.1
√	√	√	53.0	70.2	56.4	34.8	56.6	65.7

- Comparison with common algorithms

For comparing the model of this study with the current algorithms, it can be concluded from Table 7 that it has a higher detection accuracy compared with the common two-stage detection algorithms Faster R-CNN [12] and Mask R-CNN [13]. Compared with the one-stage detection algorithms YOLOv2 [14], YOLOv3 [15], and SSD [16] series, it highlights its own detection accuracy more. In comparison with the RetinaNet [17] model, which combines the advantages of both one- and two-stage models, it presents a higher detection accuracy regardless of the size of the detected target.

Table 7. Comparison of detection accuracy

Methods	Backbone	AP	$AP^{0.5}$	$AP^{0.75}$	AP^{small}	AP^{medium}	AP^{large}
Faster R-CNN	VGG-16	21.7	35.9	26.7	20.5	23.6	25.9
Mask R-CNN	ResNet-101	38.6	60.7	42.0	19.3	43.2	51.2
YOLOv2	DarkNet-19	21.6	44.0	19.2	5.0	22.4	35.5
YOLOv3	DarkNet-19	21.9	47.1	21.5	6.1	25.4	39.1
SSD300	VGG-16	25.1	43.1	25.8	6.6	25.9	41.4
SSD512	VGG-16	28.8	48.5	30.3	10.9	31.8	43.5

RetinaNet	ResNet-101	39.1	59.1	42.3	21.8	42.7	50.2
PANet	ResNet-101	40.2	55.3	44.5	22.5	44.3	51.9
Mini-EfficientDet	EfficientNet-B4	48.6	63.2	51.9	28.1	49.8	59.7
Mini-EfficientDet	EfficientNet-B5	50.5	68.2	54.3	29.9	53.2	60.8
Mini-EfficientDet	EfficientNet-B6	52.1	69.5	55.2	33.4	55.5	62.9
Mini-EfficientDet	EfficientNet-B7	53.0	70.2	56.4	34.8	56.6	65.7

For the algorithm of this study, the detection accuracy of the backbone network is 28.1, 29.9, 33.4, and 34.8 for small-scale target detection when EfficientNet B4~7 is used, showing an overall trend of increasing accuracy; for large-scale target detection, the detection accuracy is 59.4, 60.8, 62.9, and 65.7, showing an overall trend of increasing accuracy though. For medium-scale target detection, the accuracy is 49.8, 53.2, 55.5, 56.6, which is more efficient than small-scale target detection and similar to the accuracy of large target detection, and since the medium-scale target is the same size as the target in this study, it can be obtained that the Mini-EfficientDet algorithm achieves the detection accuracy improvement.

- Detection results after migration learning based

Based on the detection results of the above public dataset, the backbone network of the network model at the time of migration learning intervention in this study adopts EfficientNet-B7. The visualization results obtained after the pre-trained Mini-EfficientDet model is shown in Figure 3 after the training of the homemade dataset. Since the ratio of Chinese wasp, cockroach and hornet in the homemade dataset is 24:10:9, the detected results are included and analyzed to obtain the scatter plot of species detection accuracy in Figure 4, from which we can learn that in the Mini-EfficientDet model, the detection accuracy of Chinese wasp is 0.9866, cockroach is 0.8371 and hornet is 0.8206. In the above analysis, the migrated model achieves the expected requirements for the target recognition in this study.

Figure 3. Visualization result diagram of self-made data set

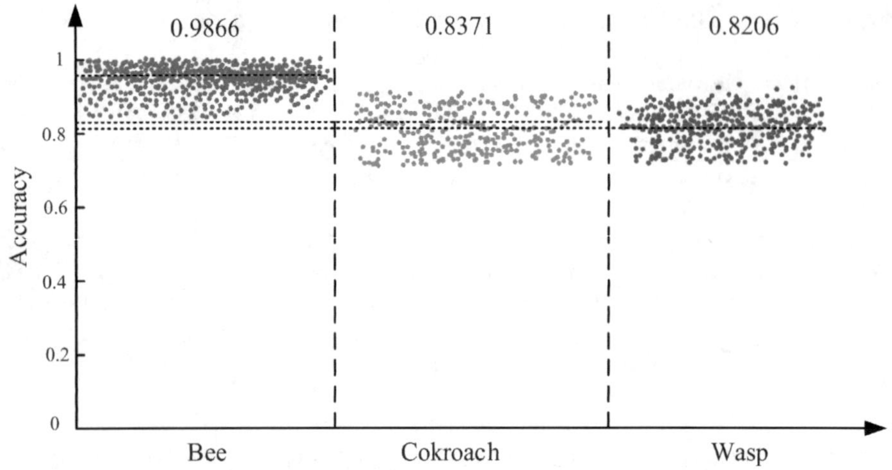

Figure 4. Scatter plot of species detection accuracy

6. Summary

In this study, a Mini-EfficientDet deep neural network is used for image recognition of bees, cockroaches and wasps at the hive door of a Chinese beehive. The initial EfficientDet model was compressed and a class imbalance function was added to make it more focused on the recognition and classification of small targets.

A comparison test based on the public dataset MSCOCO 2017 shows that the Mini-EfficientDet model achieves the desired goal by improving the recognition accuracy for small-sized targets and compressing its own memory size. After adding the pre-trained Mini-EfficientDet model to the homemade dataset for training through transfer learning, the detection accuracy all reached more than 80%, among which the detection accuracy of Chinese bee species reached 98.66%, making the algorithm ready for the later detection and warning system of beehive species invasion.

References

[1] LI Z W, HU A S, WANG X F. Survey of vision based object detection methods[J]. Computer Engineering and Applications, 2020, 56(8): 1-9 .

[2] Tan M, Le QV. 2019. Efficientnet: rethinking model scaling for convolutional neural networks. arXiv preprint.

[3] Tarvainen A, Valpola H. 2017. Mean teachers are better role models: weight-averaged consistency targets improve semi-supervised deep learning results. In: Advances in Neural Information Processing Systems. 1195–1204.

[4] Chen L-C, Papandreou G, Kokkinos I, Murphy K, Yuille AL. 2014. Semantic image segmentation with deep convolutional nets and fully connected CRFS. arXiv preprint.

[5] Deng J, Dong W, Socher R, Li L-J, Li K, Fei-Fei L. 2009. Imagenet: a large-scale hierarchical image database. In: 2009 IEEE Conference on Computer Vision and Pattern Recognition. Piscataway: IEEE, 248–255.

[6] Gatys LA, Ecker AS, Bethge M. 2016. Image style transfer using convolutional neural networks. In: Proceedings of the IEEE Conference on Computer Vision and Pattern Recognition. Piscataway: IEEE, 2414–2423.

[7] Fasihi MS, Mikhael WB. 2016. Overview of current biomedical image segmentation methods. In: 2016 International Conference on Computational Science and Computational Intelligence (CSCI). Piscataway: IEEE, 803–808.

[8] Graves A, Jaitly N. 2014. Towards end-to-end speech recognition with recurrent neural networks. In: International Conference on Machine Learning. 1764–1772.

[9] Ren S., He K., Girshick R., Sun J. Faster R-CNN: Towards Real-Time Object Detection with Region Proposal Networks. IEEE Transactions on Pattern Analysis and Machine Intelligence. 2017;39(6):1137–1149. doi: 10.1109/TPAMI.2016.2577031.

[10] Long J, Shelhamer E, Darrell T. 2015. Fully convolutional networks for semantic segmentation. In: Proceedings of the IEEE Conference on Computer Vision and Pattern Recognition. Piscataway: IEEE, 3431–3440.

[11] Ren S, He K, Girshick R, Sun J. 2015. Faster R-CNN: towards real-time object detection with region proposal networks. In: Advances in Neural Information Processing Systems. 91–99.

[12] HUANG F R, LI Y, GUO S L, et al. Method for detecting surface defects of engine parts based on faster R-CNN[J].Journal of Computer-Aided Design & Computer Graphics,2020, 32(6): 883-893 .

[13] Dorrer M G and Tolmacheva A E 2020 Comparison of the YOLOv3 and Mask R-CNN architectures' efficiency in the smart refrigerator's computer vision J. Phys. Conf. Ser. 1679042022 doi: 10.1088/1742-6596/1679/4/042022

[14] Redmon J., Divvala S., Girshick R., Farhadi A. You Only Look Once: Unified, Real-Time Object Detection. 2016 IEEE Conference on Computer Vision and Pattern Recognition (CVPR) 2016:779–788. doi: 10.1109/CVPR.2016.91.

[15] Redmon, J.; Farhadi, A. Yolov3: An incremental improvement. 2018; arXiv:1804.02767.

[16] W. Liu, D. Anguelov, D. Erhan, C. Szegedy, S. Reed, C.. Y. Fu and A. C. Berg, "SSD: Single Shot MultiBox Detector," In:ECCV, 2016.

[17] T.-Y. Lin, P. Goyal, R. Girshick, K. He, and P. Dollar. Focal loss for dense object detection. In ICCV, 2017.

Classification Prediction Model Based on BP Neural Network

Da Jiang, Changhua Liu* and Zuhao Ou

School of Mathematics and Computer science, Wuhan Polytechnic University, China

*Corresponding author email: liuch@whpu.edu.cn

Abstract. In this paper, a method of plant image processing combined with BP neural network was proposed to predict the disease degree of brassica napus. First, the initial image preprocessing, the roots of brassica napus. were used by Otsu threshold segmentation algorithm is combined with morphological open operation, segment the rape root image with better effect, in order to reduce image size inconsistent on the effects of extracting feature parameters, can be reconstructed images after segmentation, and then obtain the root image multiple characteristic parameters, Through chart correlation analysis to select the feature parameter information, with the characteristics of the rape root disease was positively associated with the degree of disease information to refactor pixel area and the longest diameter, finally establish the BP neural network, two characteristic parameters extracted as input, root disease degree level as the output, training network and prediction and comparison. The experimental results showed that the mean square error of the classification network for brassica olerica was 0.049, and the accuracy rate of the classification results predicted by the model was about 93%. The prediction effect was good and feasible.
Keywords: Image processing; Brassica napus root; BP neural network; MATLAB.

1. Introduction

Many parts of my country's oil production are derived from Brassica napus, and it is also the rape with the largest sowing area in China. It belongs to soil cultivation, it will suffer from the harm of root-edema[1] in the process of growth, thus affecting the production yield and quality of the final oil. The traditional identification method of root-swollen disease is mechanical excavation, which requires removing crops from the soil and cleaning up the soil, but this may damage the root structure of plants and affect the results and accuracy of subsequent research.

In foreign countries, French[2] et al. used visible light imaging technology to collect root images of Arabidopsis thaliana, and obtained root phenotypic parameters such as curvature, growth rate and so on. It lays a foundation for the follow-up root image research. At home, Luo Yanqing[3] and others against root disease kale rape to create new species quality and resistance to the evaluation and prediction, extract the chromosome of brassica napus l. were used, and through the result of hybridization) to determine the root of the relationship between chromosome, obtain rape root disease disease index, accuracy is higher, but the process trival, no feasibility in real life. At the same time, Hang Li et al. [4]used machine learning and image processing technologies to predict the degree of pests and diseases. After a series of image processing operations, they finally used the model of support vector machine to predict diseases, which was highly feasible, but the accuracy was not very high.

In view of the above problems, this paper proposed a classification model of brassica napus l. based on BP neural network. Through Otsu threshold segmentation[5] and morphological processing[6], as well as image reconstruction and correlation analysis, the classification model of BP neural network[7] was established, which could achieve better classification and grading effect of root-swelling disease.

Journal of Physics: Conference Series **2218** (2022) 012040 doi:10.1088/1742-6596/2218/1/012040

2. Image Collection and Preprocessing

2.1. Image Collection
The rape roots with black background cloth were photographed with a high resolution SLR camera. Part of the images taken are shown in Figure 1:

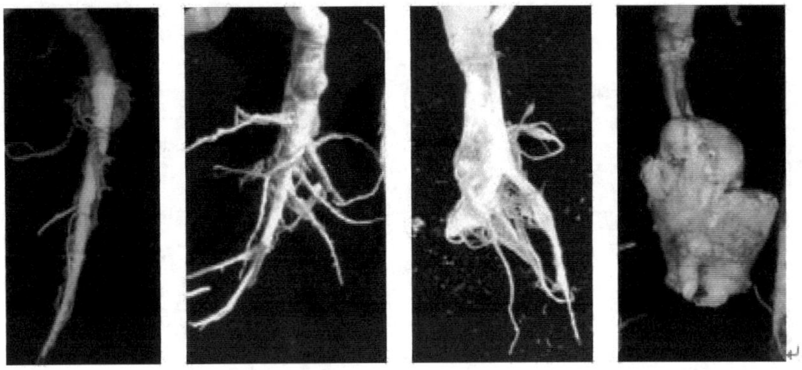

Figure 1. Images of rape roots collected

2.2. Image Preprocessin
In order to obtain the relevant characteristics of brassica napus root, the obtained images will be preprocessed in this paper: First of all to the original image with Otsu threshold segmentation, because the collected pictures to the black background, so the threshold segmentation of rapeseed effect is better, then the segmentation of image morphological open operation, remove the part of the noise in the segmentation process, reduce the noise on the follow-up effect in the process of feature extraction.

2.2.1. Otsu Threshold Segmentation. The between-cluster variance method is also known as Otsu threshold segmentation algorithm, this is a of threshold image segmentation algorithm with self-adaptive properties[8], on the basis of gray image histogram, the use of the characteristics of the gray image, to the best threshold value of the entire image gray value is divided into two parts, namely the target image from background image, make the two parts of variance value maximum, Because the larger the variance value is, the greater the difference between the background image and the target image is, and the target image can be better separated.
In this paper, in order to better separate the background and the target, black cloth is added to the target background during image acquisition, which also improves the efficiency of image segmentation[9] and the segmentation effect. The image after Otsu threshold segmentation is shown in Figure 2:

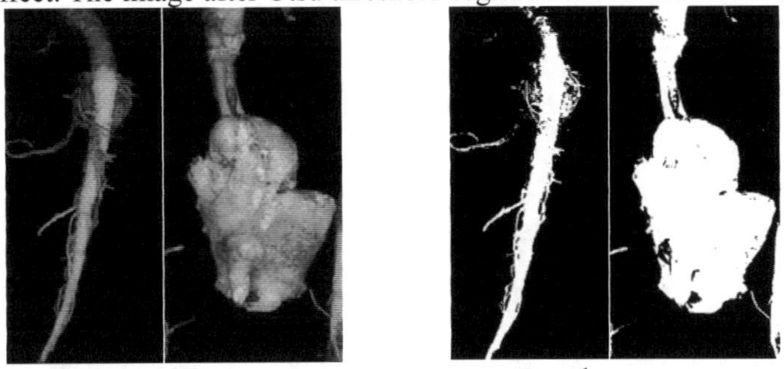

a b

Figure 2. Threshold segmentation of rape root, where **a** is the original rape root image, **b** is the rape root image after threshold segmentation

2.2.2. Morphological open operation. Morphological processing is now used in the field of image processing pattern recognition is a relatively new method, compared with other spatial frequency domain image processing methods, or has some obvious advantages, for example, the image skeleton of morphological processing is more continuous, and easy to parallel processing method. In this paper, in order to reduce the influence of white point noise on subsequent feature extraction, the open operation of morphology is used here.

In this paper, open operation is used to process the segmented images and remove some noise points, so that the images obtained are better, as shown in Figure 3 below:

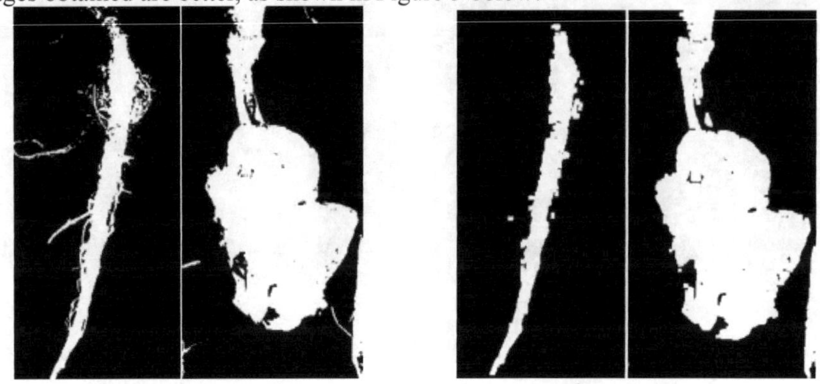

a b

Figure 3. Morphological open operation processing of rape root, where **a** is the rape root image after threshold segmentation, **b** is the rape root image after morphological processing

3. Image Feature Extraction and Analysis

3.1. Image Feature Extraction

Image features can be analyzed based on their color features, texture features, spatial features, and shape features[10]. After the pre-processing of the original image, it was found that the analysis of the shape characteristics of the rape root might be the best choice. At the same time, considering the impact of the inconsistent image sizes in the data set on feature extraction, the size of the image could be reconstructed. Finally, nine characteristic values were selected: Pixel area of root swelling, reconstruction of root swelling area, pixel difference before and after morphology, maximum diameter, minimum diameter, extreme diameter difference, reconstruction of maximum diameter, reconstruction of minimum diameter, reconstruction of extreme diameter difference. The main characteristic parameters are as follows:

(1)Area of root swelling

Calculation of root swelling area: The number of pixel points 1 in the pre-processed binary image was calculated by formula (1) :

$$\text{Area} = \sum_{i=1}^{M}\sum_{j=1}^{N} f(x,y) \quad (1)$$

(2) Reconstruction of root swelling area

In order to reduce the impact of the inconsistent image sizes on the experiment, the reconstructed image size can be uniformly 300*500, so as to calculate the reconstructed root swelling area. The width of the original image is width and the height is height, and the calculation formula is as follows (2) :

$$\text{AreaAvg} = \frac{S\times 300\times 500}{width\times height} \quad (2)$$

(3) Reconstruct the longest diameter

Same as the idea of reconstructing the root swelling area, you can calculate the longest diameter after reconstruction, that is, add the pixels in a row and compare them. You can use a Max built-in function to calculate the longest diameter after reconstruction.

3.2. Image Feature Analysis

This paper uses the method of feature analysis to find the relevant feature[11]parameters, and their characteristic parameters need to be correlated with the disease degree of brassica napus, so it is necessary to analyze the correlation between all characteristic parameters and the disease degree of brassica napus, and the common methods to study the correlation are: Principal component analysis, regression analysis, graph correlation analysis, etc. Here, graph correlation analysis is used to analyze the correlation of features, and the line graph between all characteristic parameters and the degree of root-swelling disease can be drawn to observe whether there is correlation, as shown in Figure 4. It can be found that the reconstructed pixel area and the reconstructed longest diameter are respectively consistent with the degree of root-swelling disease. Therefore, the reconstructed pixel area and the reconstructed longest diameter can be selected as the input values of the subsequent BP neural network, which is more conducive to the training and prediction of network model.

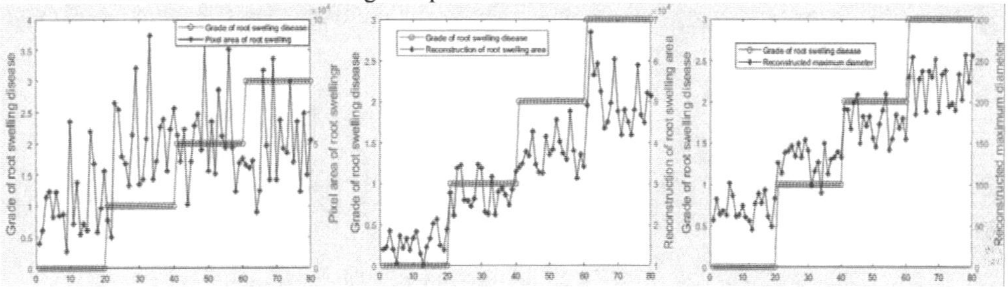

a b c

Figure 4. The characteristic parameter graph was selected, where a was the relationship between the pixel area of the root swelling and the degree of the root swelling disease, **b** was the relationship between the reconstructed area of the root swelling and the degree of the root swelling disease, and **c** was the relationship between the reconstructed longest diameter and the degree of the root swelling disease

4. Classification Model of Rape Root-swelling Disease Degree

4.1. BP Neural Network

BP neural network is also called back propagation neural network. On the basis of artificial neural network, BP neural network adds the process of back propagation, and it is also one of the most basic neural networks. By input layer in the process of the spread of the network, the input data to the hidden layer, and then there is a corresponding weights and bias, and then from the hidden layer, through the whole connected to operational data into the output layer, and get the corresponding loss function, which is known as the objective function, and then through the data transfer from the output layer to input layer, At the same time, the weight and bias of each layer are updated. After several iterations, the objective function value is reduced to the optimal value. The structure diagram of the three-layer BP neural network is shown in Figure 5.

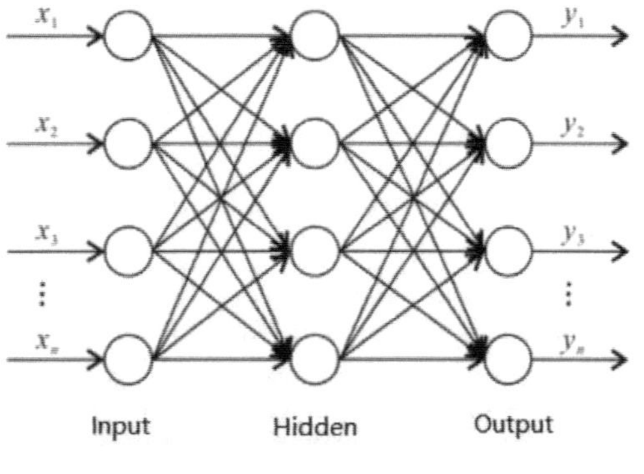

Figure 5. Structure of three-layer BP neural network

The central idea of BP neural network is to gradually optimize the objective function, namely the loss function, by updating weights and bias along the direction of the network. Although the overall structure of neural network is relatively complex, it has high efficiency in the process of data processing. Moreover, nonlinear functions are added to the network to enhance the fitting ability of the network, so it can be applied in more complex occasions.

In this paper, By establishing a BP neural network model and predict and classify the disease degree of the rape root image[12], that is, the disease degree of root swelling can be obtained directly by obtaining the picture of root swelling, so as to save time and effort for manual discrimination of disease degree of root swelling. The neural network structure of one input layer, two hidden layers and one output layer is established. Due to the characteristics of the analysis, the choice is to refactor root area and the longest diameter of two characteristic parameters, so the input layer to the two nodes, and finally need to get the root disease degree level, so the output layer also just a node, at the same time in order to accelerate the convergence rate of network training, appropriate can choose the way to deepen the depth of network model, The given network prediction model has two hidden layers, as shown in Figure 6.

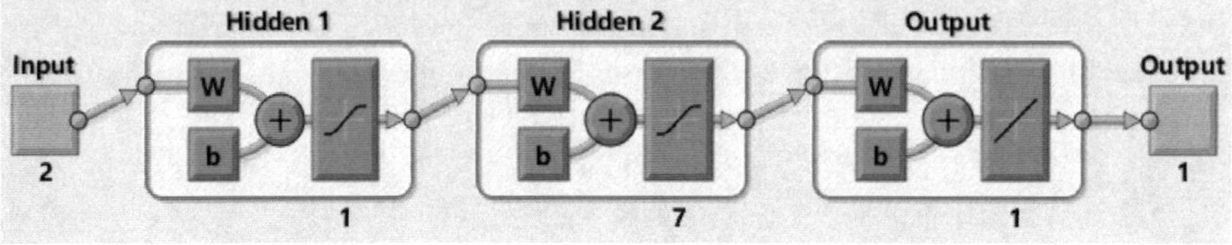

Figure 6. Constructed BP neural network model structure

5. Comparison and Analysis of Experimental Results

5.1. Comparison of Experimental Results

In order to test the accuracy of the network model for grading the disease degree of rapeseed root-swelling, 680 pictures of rapeseed root-swelling with different disease grades could be randomly obtained, because the grades of rapeseed root-swelling were divided into four grades altogether: Normal disease, first-level disease, second-level disease and third-level disease (the classification standard of rapeseed root-swollen disease is derived from the technical specification of resistance Identification of Rapeseed root-swollen disease of Oil Crops Research Institute of Chinese Academy of Agricultural Sciences), so 170 pictures of each grade were given, and 650 of them were randomly selected as the training set of the network. When the number of pictures at each level is not consistent, they are put into the established network model for training. Will be after the completion of the remaining 30 images in hierarchical model has good training level of prediction, and at the same time can be this Zhang Gen 30 images of disease rating prediction results are rounded to an integer, and then compared with artificial recognition level result, calculate the classification model prediction results and known artificial recognition level of the average mean square error (mse) and accurate data, So we can judge the prediction effect of network model. The comparison results of prediction results and artificial recognition grades are shown in Table 1.

Table 1. Comparison results of prediction results and artificial recognition grades

Number	Manual identification level	Predicted results	Results turn the whole	Number	Manual identification level	Predicted results	Results turn the whole
1	3	2.76	3	16	1	1.05	1
2	1	1.15	1	17	2	1.91	2
3	1	1.58	2	18	0	-0.042	0
4	3	3.02	3	19	2	2.17	2
5	3	2.64	3	20	0	0.042	0
6	3	3.04	3	21	0	-0.02	0
7	0	-0.02	0	22	0	-0.04	0
8	3	3.04	3	23	2	1.97	2
9	3	3.06	3	24	0	-0.04	0
10	1	1.01	1	25	3	3.06	3
11	3	2.98	3	26	1	1.47	1
12	0	-0.04	0	27	0	0.06	0
13	3	2.40	2	28	0	-0.02	0
14	2	2.05	2	29	2	2.14	2
15	3	3.06	3	30	0	-0.03	0

5.2. Analysis of Experimental Results

By comparing the disease grades of the above 30 root pictures with the predicted results, it can be seen that the prediction effect of the ESTABLISHED BP neural network model is relatively good. As can be seen from Figure 7, the network has the best performance when it has been iterated for 13 times. Meanwhile, according to Formula (3), it can be obtained that the mean square error of the prediction result and the artificial recognition grade value of 30 root images is 0.049, and the prediction accuracy of the model can reach 93% according to Formula (4). Therefore, it is considered that the BP neural network model has a good prediction ability. In formula (3) and (4), n is the predicted total number of rape root swollen root diseases, and p represents the disease grade predicted by the model, t is the level of artificial recognition, N_t is the number of correct prediction by model, and N_f is the number of wrong prediction by model.

$$\mathrm{MSE} = \frac{\sum_{i=1}^{n}(p-t)^2}{n} \tag{3}$$

$$\mathrm{R} = \frac{N_t}{N_t+N_f} \tag{4}$$

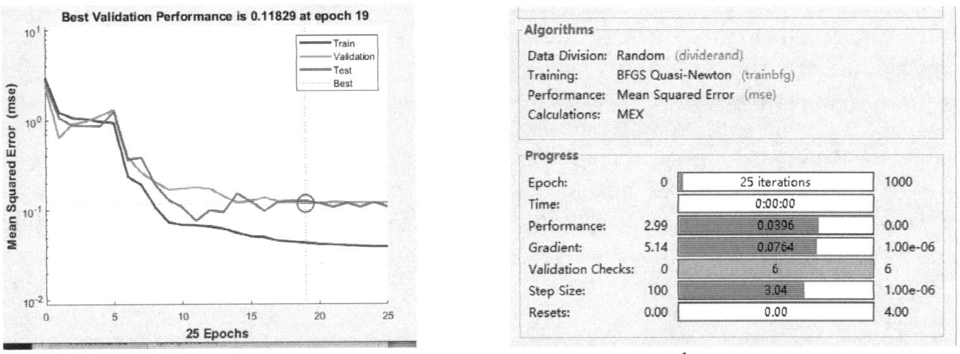

a b

Figure 7. Neural network analysis diagram, where a is the neural network training iterative process diagram, b is the parameter diagram of the neural network

6. Conclusion

This paper established a BP neural network predictive model of rape root disease classification, through threshold segmentation algorithm and morphological open operation for image preprocessing, the roots of the processed root image reconstruction and shape characteristics, combined with the chart analysis method to obtain to refactor pixel area and the longest diameter as characteristic parameters, establish a double hidden layer of BP neural network, Finally, the network model is trained and predicted to judge the prediction effect of the classification network. Through the comparative analysis of the experimental results in this paper, the mean square error of the network predicted value and the real value was 0.049, and the accuracy of the model prediction was 93%, which was in line with the actual demand and feasible, and brought convenience for the classification of the disease degree of brassica napus to a large extent. At the same time, it also laid a foundation for the team to do the application of automatic grading of rape root-swelling disease in the later stage.

Acknowledgments

This study was supported by the Project of Wuhan Polytechnic University (No. 2021XSYL03)

Reference

[1] T. Identification of pathogens and screening of resistant materials for cabbage rooiasis [D]. Northeast Agricultural University,2015. (in Chinese with English abstract)

[2] Sethuraman, V., Taylor, S., Pridmore, T., French, A., Wells, D.. Segmentation and Tracking of Confocal Images of Arabidopsis Thaliana Root Cells Using Automatically-Initialized Network Snakes[P].

[3] Yanqing Luo, Yunyue Wang, Desheng Zhao, Kaiqin Zhao, Picai Zhou, Minglian Fu, Yunsong Dong, Jinfeng Li, Jingqiao Wang. Development and evaluation of new brassica napus l. resistance to root-swelling disease [J]. Southwest China journal of agricultural sciences,2019,32(04):699-705. (in Chinese with English abstract)

[4] Hang Li, Che Jin, Song Peiyuan, Wang Chenyu, Tian Bin. Prediction of insect pests based on machine learning and image processing technology [J]. Journal of southwest university (natural science edition),2020,42(01):134-141.

[5] Mr. Image threshold segmentation technology based on OTSU algorithm [D]. Taiyuan University of Technology,2018.

[6] Li Guoyao, Wang Teng. Building structure, 2020,50(S2):529-533. (in Chinese)

[7] Veryls. Research on image recognition method of cashmere and wool fiber based on BP Neural Network [D]. Hubei University of Technology,2020.

[8] Wang Qiuping, Zhang Zhixiang, Zhu Xufang. A review of image segmentation methods [J]. Journal of information record materials,2019,20(07):12-14.

[9] Liang Lixiu, Ye Junli, Wu Dan, Yang Wanqiang. Research on rice surface root image segmentation algorithm [J]. China agricultural science and technology review,2019,21(01):70-79. (in Chinese with English abstract)

[10] Li Jing, Wu Chenpeng, Liu Muhua, Chen Jinyin, Zheng Jianhong, Zhang Yifan, Wang Wei, Lai Qufang, Xue Long. Spectroscopy and spectral analysis,2020,40(08):2564-2570. (in Chinese)

[11] Research on Random forest Algorithm based on Feature selection and Diversity [D]. Chongqing University of Posts and Telecommunications, 2020.

[12] Liu Jinya. Detection of disease degree of brassica napus based on image processing [D]. Wuhan University of Light Industry, 2020. (in Chinese with English abstract)

An Automatic Marking Method Based on Object Detection Algorithm in Aerial Images

Zuhao Ou, Changhua Liu* **and Da Jiang**

School of Mathematics and Computer science, Wuhan Polytechnic University, China

*Corresponding author email: liuch@whpu.edu.cn

Abstract. Aerial images of brassica napus in field were taken by UAV equipment but often contained unnecessary areas, which caused interferences and inconveniences to the subsequent study of identification of flowering stage by brassica napus images. To solve above problems, YOLO-V4 object detection algorithm based on Convolutional Neural Network (CNN) is adopted to build a model to detect brassica napus distribution area in aerial images, and model is trained iteratively by training data sets and setting initial parameters. As a result, the mean average precision (mAP) of the object detection model reached 96.24%, and the average loss is only 0.5843. In addition, the generalization and robustness ability of model have been evaluated by testing data sets. Conclusion shows the model proposed in this paper can indeed achieve an application of object detection and automatic labeling in aerial brassica napus images with a good performance.

Keywords: Brassica napus; Aerial images; Object detection algorithm; Deep learning; YOLO-V4.

1. Introduction

As a traditional agricultural country, China's supply and demand for food crops are increasing dramatically in recent years. According to the information from National Bureau of Statistics, grain output reached 669.49 million tons in 2020, 5.65 million tons more than the last year, an increase of 0.9 percent, and oil crops output reached 35.85 million tons, an increase of 2.6 percent. As one of the oil crops with the largest planting area in China, the brassica napus is rich in high-quality carotene, vitamin, and a variety of dietary fiber. It is an important oil and food crop with high economic value and nutritional value. Therefore, it is an urgent and significant task to monitor the growth condition and environment of brassica napus to cultivate more excellent rapeseed varieties.

Deep learning technique and all kinds of artificial neural network algorithms have made great contributions to image processing, computer vision and applied mathematics in the rapid development of the past decade. Among them, image object detection technique is the core research focus in this field and provides strong technical support for current industrial automation control technology, intelligent medical equipment, and intelligent agriculture. Therefore, the application of image object detection technique in aerial images of brassica napus highlights obvious practical significance.

To facilitate the real-time tracking of brassica napus growth, experiment adopt image object detection algorithm to build a model in this paper, aiming to locate and label brassica napus area, ignoring some interferences such as breeding greenhouse, canal, and road in aerial images. In this experiment, the labeled aerial image of brassica napus generated by the model can accurately select image coordinate position of brassica napus area, which is helpful to improve the accuracy in the succeeding study of brassica napus flowering classification and identification by brassica napus images.

2. RCNN Object Detection Algorithm

The algorithm model of RCNN[1] stands for "Regions with CNN features" and it was proposed by Girshick et al. in 2013, which is regarded as the pioneering work of image object detection using deep learning method, so that the object detection algorithm proposed later takes it as a reference. Workflow of RCNN algorithm could be divided into the following steps. (1) generate candidate regions[2]; (2) feature extraction[3]; (3) classification judgment[4]; (4) candidate box positioning[5].

But on the other hand, the greatest disadvantage of RCNN is that overlapping candidate regions include same feature part, so all generated candidate regions will cause repeated calculation in the convolution operation, increasing greatly the time of feature extraction. In the section of Experiment and analysis, RCNN and an improved algorithm based on it, Faster-RCNN, are introduced and compared with the object detection algorithm used in this paper.

3. Algorithm Used in This Paper: YOLO-V4

The YOLO[6] (You Only Look Once) series object detection algorithm network launched in 2016 borrowed the advantages of RCNN and made a breakthrough in detection speed. In this study, the object detection model adopts YOLO-V4 version to build, and the network structure and loss function of YOLO-V4 are described in this section.

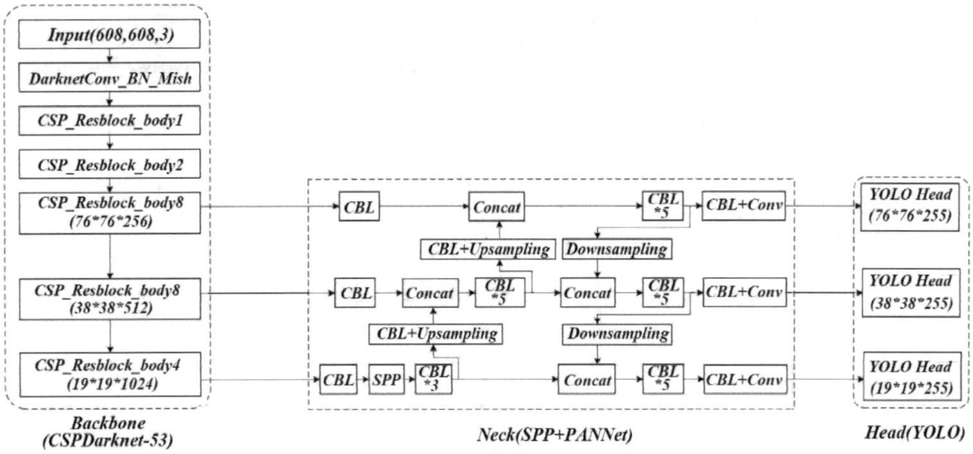

Figure 1. YOLO-V4 network structure.

3.1. Network Structure

The network structure of YOLO-V4 is mainly composed of three parts, as shown in Figure 1, namely backbone network, neck network and head network.

3.1.1. Backbone network. The function of backbone network focuses on feature extraction to input images by CSPDarknet-53. CSPNet[7] (Cross Stage Partial Networks) are used for reference to improve Darknet-53 in the structure of CSPDarknet-53, which can solve problems of gradient disappearance caused by large scale CNN in the process of feature extraction and store the information of gradient change in the feature maps. During the implementation of backbone network, images with a size of 608*608*3 are transmitted into input layer, and the depth of every image is changed to 32 by one CBM module which includes convolution layer, batch normalization, and mish activation function. Next, these changed images will be handled by five CSP modules for downsampling respectively, to extract features from the input images of different sizes. Finally, backbone network could get three feature maps with different sizes of 76*76*256, 38*38*512 and 19*19*1024 and transmitted to neck network.

3.1.2. Neck network. A Feature Pyramid (FPN[8]) structure is formed in neck network by three feature maps output of different sizes from backbone network, which is mainly composed of Spatial Pyramid Pooling (SPP[9]) and Path Aggregation Network (PANet[10]) as shown in Figure 2 and 3. The function of SPP is that feature maps can be directly pooled with a fixed size, which increases the receptive field and can extract important context information from the feature maps without the loss of network processing speed. The workflow of PANet is divided into two parts of sampling and aggregation. Feature maps in upper layer are mostly semantic information of the image, while the lower layer is mostly local information of the image, so feature information between the upper and lower layers can be transferred to each other to obtain richer feature information by multiscale feature fusion operation.

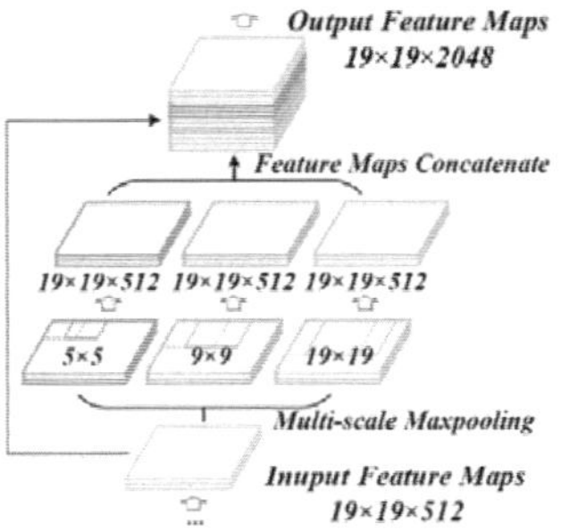

Figure 2. SPP.

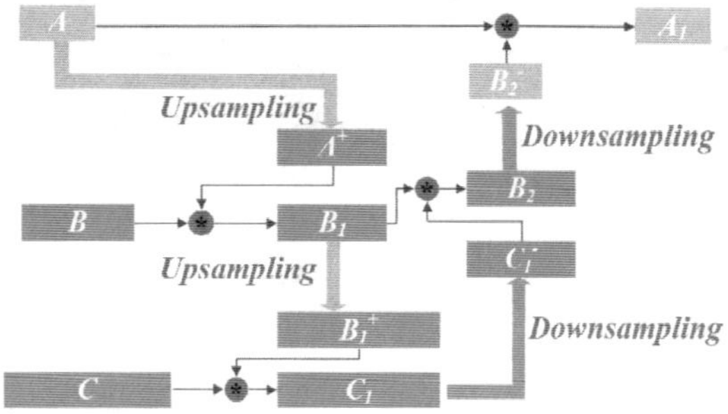

Figure 3. PANet.

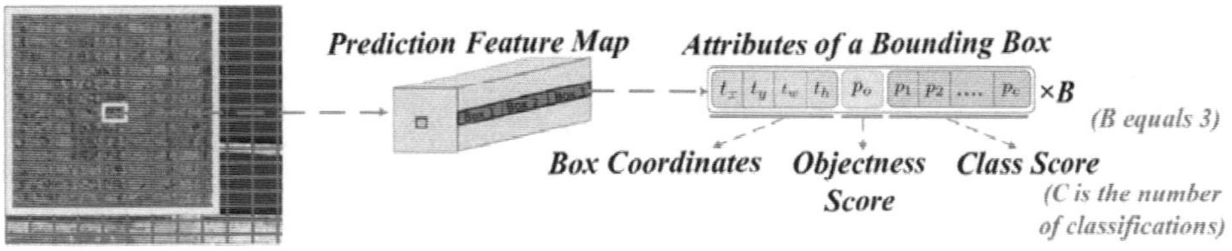

Figure 4. The feature maps divided by grid and detection information in cell grid.

3.1.3. Head network. The last part of network structure, also known as YOLO layer, receives three feature maps of different sizes from neck network, and divides them into grid cells with fixed scale, as shown in Figure 4. Before object detection of feature maps, the depth of each feature map should be adjusted according to actual requirements of study as shown in Formula 1. During object detection process, the center point of ground truth box marked in the image locates in a grid cell, then this grid cell will predict boundary boxes of the object, and each grid cell is set to predict three boundary boxes.

$$Depth\ of feature\ maps\ = B*(5+C) \tag{1}$$

3.2. Loss Function

To make the position and classification result of generated boundary boxes closer to the area of original image, loss function is used in YOLO layer to reduce boundary box loss, object confidence loss and classification loss caused by network training. The overall loss function is shown as Formula 2.

$$Loss = L_{iou} + L_{cls} + L_c \tag{2}$$

3.2.1. Boundary box loss function. The location and size differences of generated boundary box will directly affect the consequence of object detection task, so boundary box loss function is of importance particularly in the overall loss function, as shown below.

$$L_{iou} = \lambda_{iou} \sum_{i=0}^{s^2} \sum_{j=0}^{B} I_{ij}^{obj} L_{ciou} \tag{3}$$

The key point in Formula 3 is the calculation of CIoU loss which considers more geometric measurements based on IoU and DIoU, including the calculation of distance of two center points, overlapping area, and width and height ratio in the two boxes. IoU, DIoU and L_{ciou} are shown as Figure 5 and Formula 4 to 7.

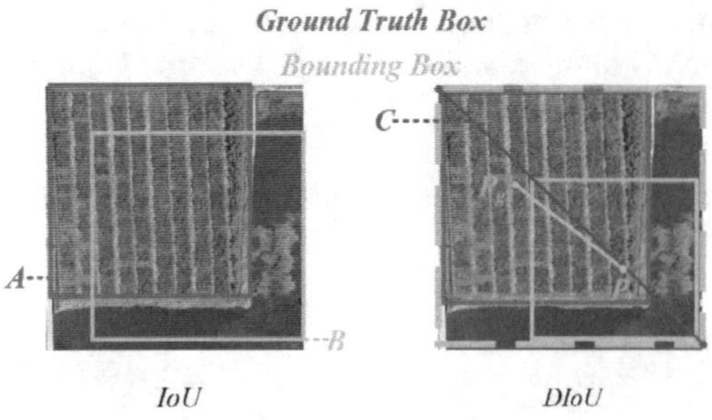

Figure 5. Description of IoU and DIoU.

$$IoU = \frac{A \cap B}{A \cup B} \tag{4}$$

$$DIoU = \frac{Dis\tan ce^2 (P_{gt}, P)}{C} \tag{5}$$

$$L_{ciou} = 1 - IoU + DIoU + \frac{v^2}{(1 - IoU) + v} \tag{6}$$

$$v = \frac{4}{\pi^2}(\arctan\frac{w^{gt}}{h^{gt}} - \arctan\frac{w}{h}) \tag{7}$$

3.2.2. Object confidence loss function. There are two parts in object confidence loss function. The first is the confidence loss of objects detected in the boundary boxes, and the second is the confidence loss of objects not detected in the boundary boxes, as shown in Formula 8.

$$L_{cls} = \lambda_{cls}\sum_{i=0}^{s^2}\sum_{j=0}^{s^2} 1_{ij}^{obj} \lambda_c(c_i - \hat{c})^2 + \lambda_{cls}\sum_{i=0}^{s^2}\sum_{j=0}^{s^2} 1_{ij}^{noobj} \lambda_c(c_i - \hat{c})^2 \tag{8}$$

3.2.3. Classification loss function. In YOLO-V4, multiple independent logistic classifiers are used to obtain the probability of all specific classification labels and calculate classification loss. The classification loss function is shown as Formula 9.

$$L_c = -\sum_{i=0}^{s^2}\sum_{j=0}^{B} 1_{ij}^{obj} \sum_{c \in classes} \lambda_c[\hat{p}_i(c)\log(p_i(c)) + (1 - \hat{p}_i(c)\log(p_i(c))] \tag{9}$$

L_c needs to use binary cross entropy function[11] to calculate the loss for each generated label. In the training stage of network model, classification loss predicted by boundary boxes is exactly the binary cross entropy loss of probability of each category.

4. Experiment and Analysis

4.1. Data Sets Preparation

All brassica napus aerial images data sets used in this study was provided by Oil Crops Research Institute, Chinese Academy of Agricultural Sciences. The aerial images were collected from Yunnan Province, China, and were taken from December 10, 2018, to March 20, 2019, with one week between each shooting process. Researchers took images with UAV equipment above the standardized planting brassica napu field outdoors on sunny, cloud-free days, as shown in Figure 6.

Figure 6. The aerial images of brassica napu field at different locations and time periods.

In this study, aerial images of brassica napu with 5472*3078 of resolution in different time periods and locations are selected as training and testing data sets for the network model. 2500 aerial images training dataset stored in *JPEGImages* folder. Researchers label each training data set image manually by a tool named LabelIng and use rectangular box to mark brassica napu area. Selected areas are regarded as ground truth boxes, and the coordinate information of all ground truth boxes in each image are saved as labeling files which are divided into training samples and validating samples in the ratio of 8:2, stored in *Labels* folder in total 2500. In the testing data sets, 100 original aerial images of brassica napu field stored in *Testfile* folder, which are not labeled and excluded in the training data sets and used to evaluate the generalization ability of network model after completion of training stage.

4.2. Experimental Environment Construction and Parameter Adjustment

The hardware configuration of the experiment is as follows: CPU Intel(R) Core (TM) i7-10870H @ 2.20GHz 8 Cores 16 Threads; Memory Kingston DDR4 RAM @ 3200MHz 16GB; GPU NVIDIA GeForce RTX 2060 @ 6GB. The software configuration is as follows: Operating system Windows 10 professional edition 64- bit DirectX 12; Darknet; python 3.6; OpenCV 3.4.0; CUDA 10.2; cudnn 10.2-windows10-x64-v7.6.5.32; PyCharm 2020.2.1; Visual Studio 2019 community; LabelImg.

In the last step before network training stage, it is the key point to configure network parameters of YOLO-V4 in the Darknet framework. Since the GPU memory is only 6GB, if resolution of image transmitted into model is too large, the problem of GPU memory overflow will be caused. To ensure normal training stage of network model and avoid decrease of model accuracy, the resolution of input images is set as 544*544*3. The depth of feature maps output by neck network should be the same as the number of convolution kernels at this layer, and the number of classifications is 1, so the number of convolution kernels equals 18 according to Formula 1. Model updates all network parameters during the training process when 16 training samples put in, so batch size is set as 16. Until the iterations of training process overstep 4000 times, training stage of model will be finished. The initial learning rate is set as 0.001, but it need change to one tenth of what it was before to ensure better convergence and prediction result closed optimal value of network model, since number of batches reaches 3200 and 3600.

4.3. Model Training and Testing

4.3.1. Standard of model evaluation. The method of confusion matrix[12] shown in Figure 7 can be used to calculate the precision, recall and F1-score[13] of each training stage of model, as shown in Formula 10 to 12.

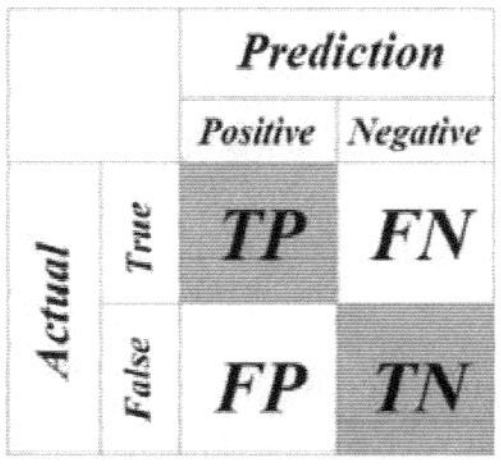

Figure 7. Confusion matrix structure diagram.

$$precision = \frac{TP}{TP + FP} \tag{10}$$

$$recall = \frac{TP}{TP + FN} \tag{11}$$

$$F1\_score = \frac{2 * Precision * Recall}{Precision + Recall} \tag{12}$$

With the number of samples increasingly, the curve of change of precision and recall rate as shown in Figure 8 could indicate that average precision (AP) of the model in a classification refers to the integral area of curve and measures a quality of recognition ability of the model in each classification, while mAP is an evaluation of recognition ability of model in all classifications, summarizing all AP and calculating the average of them. In this study, only a single classification, brassica napu, need to be detected, so the AP of brassica napu calculated is the final mAP of the object detection model.

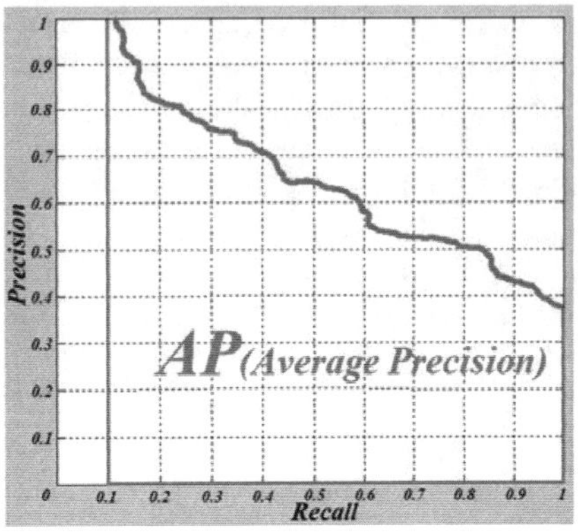

Figure 8. The curve of precision and recall rate on a certain classification during the training process of model.

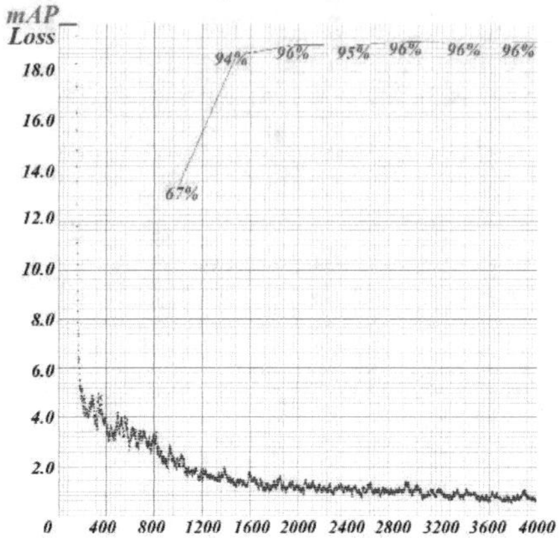

Figure 9. The loss and mAP curves of object detection model in this study.

Table 1. The precision, recall, F1-score, average IoU and mAP of four stages.

Batches	TP	FP	FN	Precision	Recall	F1-score	Average IoU (%)	mAP (%)
1000	904	671	244	0.57	0.79	0.66	40.04	69.85
2000	1092	102	56	0.91	0.95	0.93	70.80	95.45
3000	1093	114	55	0.91	0.95	0.93	74.18	95.50
4000	1090	65	58	0.94	0.95	0.95	83.83	96.24

4.3.2. The result of model. After training stage, the changes of final loss value and mAP of object detection model during the training process are obtained as shown in Figure 9. In the training stage of first 400 batches, the loss value of model varies greatly. With the number of training batches increasingly, the update range of weight parameters of each layer will decrease gradually, and the decline range of loss change curve also slows down, until it turns to converging. When the number of batches equals 1000, model calculates the precision and recall rate in the current training stage and obtains the mAP. During the stages of training, validating samples could find out some problems that the mAP of model does not grow or grows slowly. In the end of training process, mAP of model reaches 96.24%, and loss value of model is only 0.5843. Table 1 displays the precision, recall, F1-score, average IoU and mAP of four stages in the training process.

4.3.3. Model testing. After completed training process of network model, using testing data sets could verify the generalization ability and robustness of object detection model. Figure 10 below shows the testing results of two randomly images from testing data sets.

Figure 10. The output results by the trained network model to the testing image.

Figure 10 displays that the network model executes the operation of object detection on the input aerial image of brassica napus. Red boundary boxes are used to mark brassica napus areas in the original image, and the classification confidence of each boundary box is indicated in two images. It can be seen as Table 2 that the first 20 testing results are shown, including the real number of brassica napus areas, the number of bounding boxes predicted by model and the average confidence of each image.

Table 2. The testing results of the first 20 testing data sets.

Filename	Real	Prediction	Confidence	Filename	Real	Prediction	Confidence
testImg_01	15	15	0.95	testImg_11	14	13	0.81
testImg_02	2	2	0.74	testImg_12	12	12	0.94
testImg_03	14	14	0.92	testImg_13	12	11	0.89
testImg_04	4	4	0.97	testImg_14	8	8	0.98
testImg_05	11	11	0.98	testImg_15	12	12	0.97
testImg_06	14	12	0.82	testImg_16	14	13	0.89
testImg_07	4	4	0.83	testImg_17	12	12	0.95
testImg_08	12	12	0.97	testImg_18	2	2	0.89
testImg_09	9	9	0.87	testImg_19	1	1	0.97
testImg_10	2	2	0.96	testImg_20	6	6	0.95

According to Table 2, unlabeled original aerial images of brassica napus can achieve great labeling results by the trained network model, which can be detected the brassica napus region in each image comprehensively, and the average classification confidence in each image is also satisfied. However, there are also brassica napus areas missed by model in the test data sets. Comparing with the detected brassica napus areas, these areas account for very few proportions in the original image and are located at the edge of images with slightly irregular shapes, leading to nonrecognition or wrong classification.

4.4. Model Performance Comparison

The object detection model concentrating on brassica napus in this paper was trained by 4000 batches in the whole training process with GPU acceleration. To observe the changes of mAP and loss value of model in different iterations, it is necessary to change the number of batches on the previous training data sets and compare the consequences of model mAP and loss value of 4000, 6000 and 8000 iterations. Table 3 shows the performance changes of model after 4000, 6000 and 8000 times of training.

Table 3. The comparison of results under different iterations during training process.

Batches	Training time (s)	Precision	Recall	F1-score	Average IoU (%)	mAP (%)	Loss
4000	19523	0.94	0.95	0.95	83.83	96.24	0.5843
6000	26018	0.95	0.93	0.94	84.61	96.02	0.6021
8000	39216	0.94	0.95	0.94	85.09	96.38	0.5664

As shown from Table 3, with the number of iterations increasingly, the training time of model will be prolonged. However, by observing the evaluation data of model performance, the improvement of performance is quite minimal and the results of mAP and loss values of three models are almost the same in the end of training stage. So, it could be better to consider increasing the number of training data sets for object detection model and adjust the number of batches in the following experiment. To reduce training time, the model with 4000 times iterative training is still selected as the final experimental result.

In the comparison experiment of different object detection algorithms, SRCNN and Faster-RCNN are introduced and compare with YOLO-V4 in before experiment briefly. It is of great concern to find out the differences of performance among three network models by the same training data sets. Table 4 displays the performance comparison of three object detection algorithms.

Table 4. The training results of network models by three object detection algorithms.

The name of object detection algorithm	Batches	Training time (s)	mAP(%)
RCNN	4000	45526	74.61
Faster-RCNN		23411	82.62
YOLO-V4		19523	96.24

According to the description from Table 4, the model obtained by YOLO-V4 algorithm is significantly superior to RCNN and Faster-RCNN in performance of both training time and mAP, when the three object detection algorithms set the same number of batches. RCNN algorithm spent too much time on training process, because the steps in each training stage were relatively tedious and a lot of data was calculated repeatedly, which required a large amount of memory space. Faster-RCNN is an improved algorithm based on RCNN, and the speed of training process of network model is improved obviously. However, due to the differences in network structure between Faster-RCNN and YOLO-V4, the performance of trained model under the same number of batches was still inferior to the determined object detection model in this paper finally.

5. Conclusion

In this paper, an automatic marking method based on object detection algorithm for the area of brassica napus in aerial images is proposed. After training process, the final loss and mAP value of model is satisfied. By adjusting the iterations of network, the most appropriate performance is selected as the final experimental results. Under the same experimental conditions, the object detection model proposed in this paper is significantly more dominant than RCNN and Faster-RCNN. Finally, the abilities of robustness and generalization indicate that model can completely and comprehensively label the brassica napus area in aerial images, achieving successfully the initial experimental requirements and showing strong feasibility. Collecting these labeled aerial images can prepare for the later research about flowering stages recognition and is conducive to improving the accuracy and number of classifications.

Acknowledgments

This study and research were supported by Agricultural Engineering and Information Technology Project (No. 2021XSYL003) of School of Mathematics and Computer science, Wuhan Polytechnic University, China.

References

[1] Girshick R, Donahue J, Darrell T, et al. Rich Feature Hierarchies for Accurate Object Detection and Semantic Segmentation[C]// IEEE Computer Society. IEEE Computer Society, 2013.
[2] Uijlings, J. R. R. van de Sande, K. E. A. Gevers, T. Smeulders, A. W. M. Selective Search for Object Recognition[J]. International Journal of Computer Vision, 2013, 104(2):154-171.
[3] Krizhevsky A, Sutskever I, Hinton G. ImageNet Classification with Deep Convolutional Neural Networks[C]// NIPS. Curran Associates Inc. 2012.
[4] Chen P H, Lin C J, Schlkopf B. A Tutorial on ν-support vector machines[J]. Applied Stochastic Models in Business and Industry, 2005, 21(2):111-136.
[5] Sun C, Vallotton P. Fast linear feature detection using multiple directional non-maximum suppression[J]. Journal of Microscopy, 2009, 234(2):147-157.
[6] Redmon J, Divvala S, Girshick R, et al. You Only Look Once: Unified, Real-Time Object Detection[J]. Computer Vision & Pattern Recognition, 2016.
[7] Wang C Y, Liao H, Wu Y H, et al. CSPNet: A New Backbone that can Enhance Learning

Capability of CNN[C]// 2020 IEEE/CVF Conference on Computer Vision and Pattern Recognition Workshops (CVPRW). IEEE, 2020.
[8] Lin T Y, Dollar P, Girshick R, et al. Feature Pyramid Networks for Object Detection[C]// 2017 IEEE Conference on Computer Vision and Pattern Recognition (CVPR). IEEE Computer Society, 2017.
[9] He K, Zhang X, Ren S, et al. Spatial Pyramid Pooling in Deep Convolutional Networks for Visual Recognition[C]// IEEE Transactions on Pattern Analysis & Machine Intelligence. 2014:1904-16.
[10] Liu S, Qi L, Qin H, et al. Path Aggregation Network for Instance Segmentation[J]. 2018 IEEE/CVF Conference on Computer Vision and Pattern Recognition (CVPR), 2018.
[11] Jennie, Pearce, et al. Evaluating the predictive performance of habitat models developed using logistic regression[J]. Ecological Modelling, 2000.
[12] Kong Ying Hui, beautiful scenery. Research on Classification Method Based on Confusion Matrix and Ensemble Learning [J]. Computer Engineering and Science, 2012, 34(6):111-117.
[13] Wang Yingming, Luo Ying. Research on combination forecasting method based on harmonic mean [J]. Statistical study, 1997, 000(002):66-68.

A Computational Approach of Displacement Prediction in an Engineering Project

Jonathan Chen Tang[1], Liang Peng[2] and Zhenlei Chen[2,*]

[1]Ningbo Hanvos Kent School Ningbo, China
[2]Ningbo University, Ningbo, China

*Corresponding author email: chenzhenlei@nbu.edu.cn

Abstract. In this paper, a fuzzy information granulation method combined with a support vector machine time series prediction model(FIG-SVM) is established for a subway cross passage project of Ningbo Metro Line 3. The FIG-SVM prediction model was used to predict the ground settlement of the subway access during the shield construction process. It was found that the FIG-SVM prediction model can effectively predict the ground settlement trend for the next construction stage based on the historical data of the ground settlement.
Keywords: Cross Passage; Ground settlement; Fuzzy information granulation.

1. Introduction

The world's first mechanical shield method of the Cross Passage has been used in Ningbo Metro Line 3 Yinnan interval. However ground settlement caused by the loss of strata and disturbance of the surrounding soil caused by shield construction are inevitable. If the settlement reaches a certain level, the shield construction may cause damage to the structures and pipelines near the tunnel[1]. Therefore, the studies on the law of the ground deformation and the prediction of the ground settlement caused by the shield construction method are particularly important.

At present, a large number of scholars and experts has conducted in-depth research on predicting ground settlement caused by the shield construction. Some traditional research methods include empirical formula method[2], finite element method[3] and theoretical calculation. However, these methods have certain shortcomings: the empirical formula method is only applicable to some regional environments with simple ground conditions. It is difficult to guarantee the accuracy of its prediction results if it is used to predict some areas with more complicated ground conditions.

Machine learning methods, as an emerging method, have been widely used in various fields because of their powerful nonlinear fitting ability and their ability to make real-time predictions from a large amount of engineering data. The representative methods include neural network[4,5], support vector machine[6], gray theory[7], random forest[8]. Most of the above prediction methods aim at predicting the quantitative value of ground settlement in real time. In order to ensure the safety of shield construction, it is actually necessary to predict the change trend of ground settlement and the spatial change.

The support vector machine method is used to predict settlement by revealing the laws among the influencing factors. It can effectively solve the problem that the model is difficult to determine due to the complex mechanical properties and stresses of the soil[9]. In this paper, a support vector machine model based on information granulation is proposed to predict the trend of the ground settlement caused by the shield construction method of the Cross Passage. It provides support and guidance for the subsequent settlement monitoring of the Cross Passage and the settlement prediction of other Cross Passage construction processes.

2. FIG-SVM Prediction Model

2.1. Fuzzy Information Granulation Method

Information granulation refers to dividing the given information data into multiple sets and representing each set with information granules according to certain characteristics. Fuzzy Information Granulation (FIG) is to represent information grains by fuzzifying the time series with fuzzy set method, which normally includes dividing window and fuzzification. Dividing window means dividing the original time series data into operation windows with appropriate time interval. Fuzzification means taking each window as the object. The particles of fuzzy information are obtained by fuzzifying each window.

At present, the more common fuzzy particles are mainly triangular, trapezoidal, Gaussian, parabolic. In this paper, we use triangular-type particles with the affiliation function [10,11,12,13] as follows.

$$A(x,a,m,b)=\begin{cases}0, & x<a\\ \dfrac{x-a}{m-a}, & a\le x\le m\\ \dfrac{b-x}{b-m}, & m\le x\le b\\ 0, & x>b\end{cases} \tag{1}$$

Where: x is the variable in the domain; a, m, b are the parameters of the affiliation function.

Determine each parameter of the triangular-type fuzzy set separately.

1) Determine the parameter m

For a given set of time series $X=\{x_1,x_2,\ldots,x_N\}$,consider first the single-window problem, and reorder the time series from smallest to largest, and let the sorted set of time series be $X=\{x_1,x_2,\ldots,x_N\}$, then

$$m=X(N/2)$$

2) Determine the lower bound of triangular fuzzy set supporting a

$$Q(a)=\frac{\sum_{x_k\le m}A(x_k)}{m-a} \tag{2}$$

3) Determine the upper bound of triangular fuzzy set supporting b

$$Q(b)=\frac{\sum_{x_k\ge m}A(x_k)}{b-m} \tag{3}$$

After completing the fuzzy granulation of a single window, the original sample data can be windowed using any time series. Then the above fuzzy granulation method can be repeated.

2.2. Support Vector Machine Regression Algorithm

The support vector machine regression algorithm is based on the classification of the support vector machine algorithm. It shows stronger generalization performance than traditional learning methods and better performance in dealing with existing nonlinear problems with small sample size results [14].

For a given data set $\{x_i,y_i\}(i=1,2,\cdots,l)$,where, l is the number of samples, x_i is the input vector, and y_i is the corresponding output target value. The objective of the support vector regression algorithm is to find the regression function $f(x)$ such that the deviation of $f(x_i)$ from y_i is as small as possible. The nonlinear regression function is expressed as

$$f(x)=\sum_{i=1}^{l}\langle\omega x\rangle+b \tag{4}$$

where: ω is the weight vector; b is the bias. Therefore, the essence of prediction model training is the process of solving the parameter matrices ω, b with the objective of minimizing the bias.

2.3. Executive Software Programming

The parameters to be optimized in the SVM model mainly include the kernel function parameter and the penalty factor. In this paper, we mainly use particle swarm optimization (PSO), a global stochastic search algorithm based on population intelligence [15], to initialize the particle swarm with random position velocity before searching for the optimal solution. In the process of comparing the fitness with the individual historical best position and the global best position, each particle constantly updates its velocity and position. When the maximum iteration number is reached or the requirements are satisfied, the algorithm stops outputting the solution for support vector machine model training. Figure 1 shows the parameter optimization process.

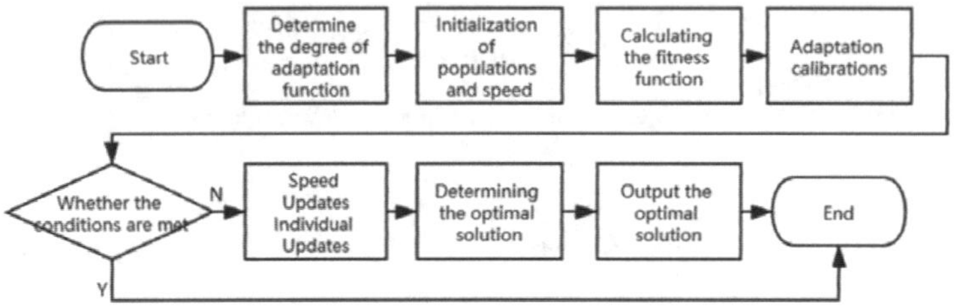

Figure 1. Parameter optimization process

2.4. Judging Criteria

The mean square error and squared correlation coefficients are used to analyze the accuracy of a, m, and b in the training models of the support vector machine.
The mean square error is expressed as

$$E_{ms} = \frac{1}{n}\sum_{i=1}^{n}(Y_i - y_i)^2 \tag{5}$$

where Y_i is the predicted value and y_i is the test value.

The squared correlation coefficient R is expressed as

$$R = \frac{n\sum_{i=1}^{n} Y_i y_i - \sum_{i=1}^{n} Y_i \sum_{i=1}^{n} y_i}{\sqrt{\left[n\sum_{i=1}^{n} Y_i^2 - \left(\sum_{i=1}^{n} Y_i\right)^2\right]\left[n\sum_{i=1}^{n} y_i^2 - \left(\sum_{i=1}^{n} y_i\right)^2\right]}} \tag{6}$$

Assuming that the actual range of variation of the original data is represented by fuzzy particles is $[a,m,b]$, and the range of predicted variation is $[a',m',b']$, the range prediction error E is expressed as:

$$E = \left[\frac{|a-a'|}{a}, \frac{|m-m'|}{m}, \frac{|b-b'|}{b}\right] \tag{7}$$

3. Project Example Analysis

3.1. Project Overview

Ningbo Metro Line 3 liaison channel project is located in the north side of Yinzhou Park East Gate. The liaison channel is 17 m long and the burial depth of the tunnel top is 12.3 m. The project mainly adopts the shield method construction technology. The Yinan interval Cross Passage was started on December 29, 2017, and successfully passed through on January 15, 2018. It is the first shield method of the Cross Passage passed through in the world. Therefore, it has good reference significance and research value.

3.2. Predictive Modeling

To ensure the safety of the construction process of the liaison channel, a total of 43 monitoring points were set for the surrounding ground settlement during the construction of the liaison channel. A total of 51 sets of data were collected. In this paper, 35 sets of ground settlement data were used and 9 monitoring points as shown in Figure 2 were selected as the dependent variable. As an example, Figure 3 shows the interpolated data curves of the changes of the ground settlement at monitoring points D5-5.

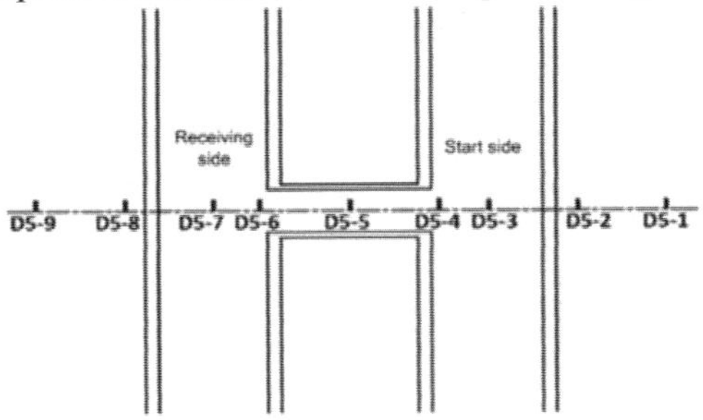

Figure 2. Layout of measuring points

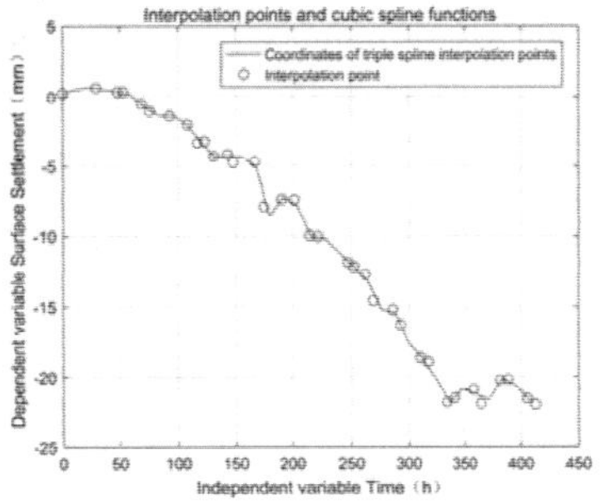

Figure 3. Interpolation data curve of the ground settlement at monitoring points D5-5

The above interpolated data set is divided into two parts, sample data and validation data, as shown in Table 1. The sample data includes first 30 periods of data used as training samples in the FIG-SVM model. The last 5 periods are used as prediction validation data to test the accuracy of the model.

Table 1. Data set

	Sample Data				Validation Data				
periods	1	2	…	30	31	32	33	34	35
Settling /mm	0.24	0.62	…	-20.92	-21.22	-21.63	-20.14	-20.59	-21.79

Fuzzy information granulation of the original data uses triangular fuzzy granulation model. The modeling domain is divided into 6 windows with 5 periods of time as a window. Fuzzy Information granulation of the modeling domain is obtained by Equation (1). Three fuzzy information grains, named Low, R and Up, from each window are obtained corresponding to the three parameters a, m and b in equation (1). The fuzzy information particles of each window are composed of Low, R, and Up in the granulated data. For a single fuzzy particle, Low, R and Up characterize the minimum, average and maximum values of the original data variation, respectively. The final granulation results are shown in Figure 4.

The SVM models were established with Low, R and Up in the granulated data as training sets, respectively. The regression prediction was performed on the granulated data with the objective of minimizing the K-CV validation error. The particle swarm optimization algorithm was used for parameter optimization. The optimization search process is shown in Figure 5 while the optimized the kernel function parameter g and the penalty factor C are shown in Table 2.

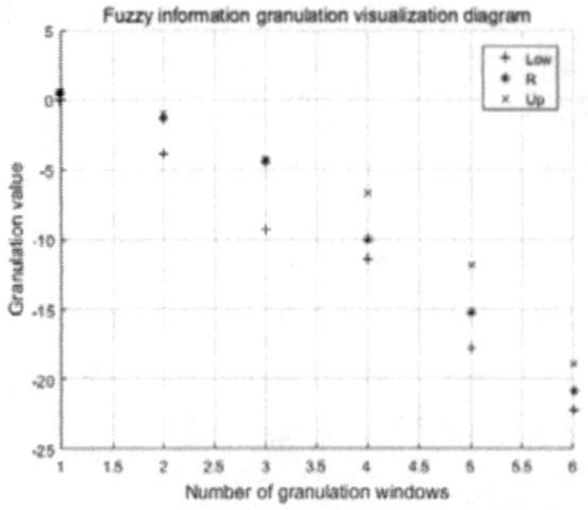

Figure 4. Fuzzy granulation visualization graph

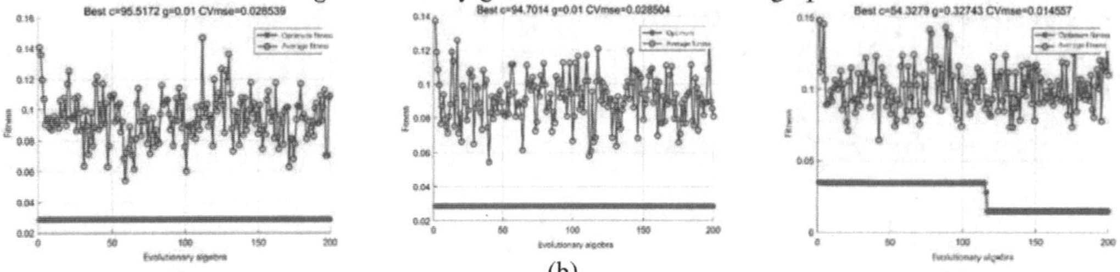

(a) (b) (c)

Figure 5. The optimization process of low (a), R (b), up (c) parameters

Table 2. The best parameters C, g of low, R and up

	C	g
Low	95.5172	0.01
R	94.7014	0.01
Up	54.3279	0.32743

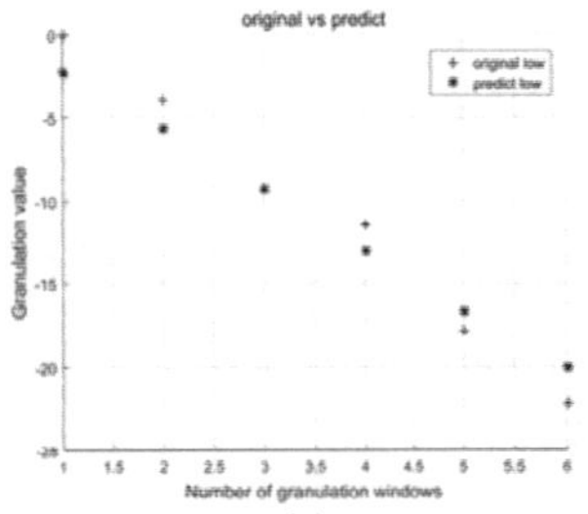

(a)

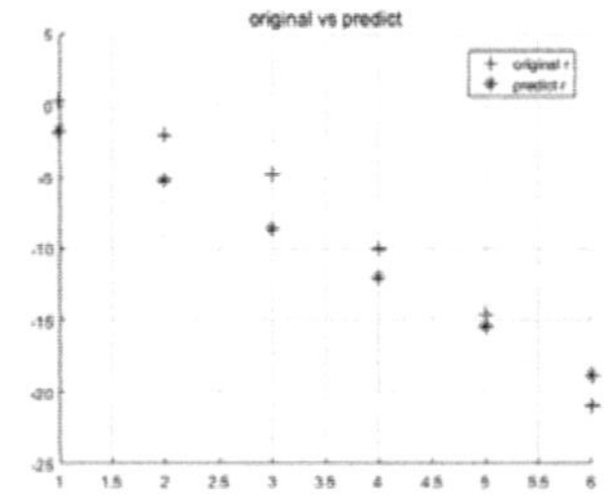

(b)

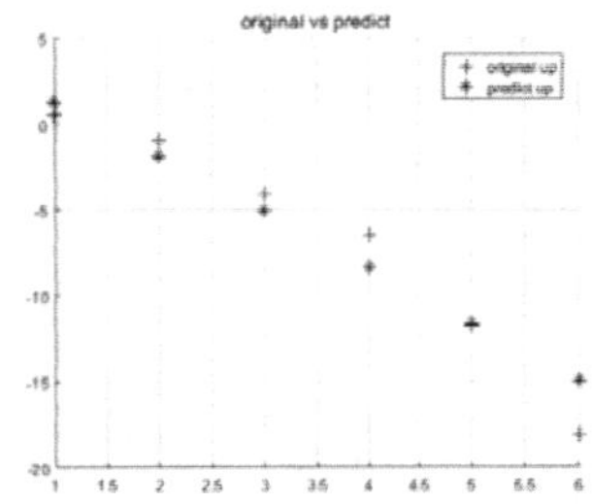

(c)

Figure 6. Low (a), R (b), up (c) regression prediction results

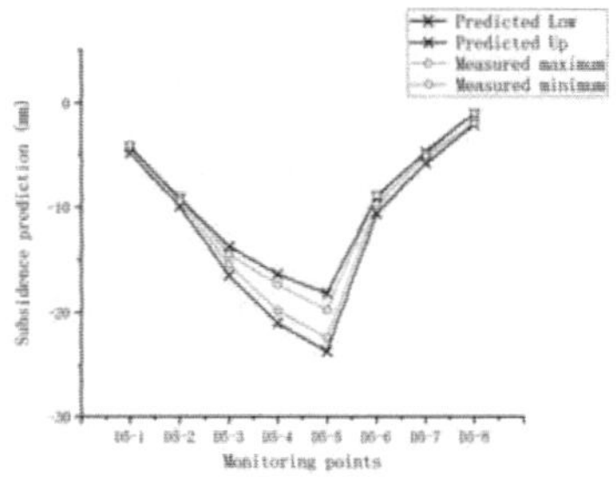

Figure 7. Comparison of predicted and measured values at each monitoring point

Table 3. The best parameters *C*, *g* of Low, R and up

parameters	MSE	R2
Low	0.00615958	0.991892
R	0.0143232	0.970767
Up	0.00758726	0.947813

The SVM is then trained using the above parameters *C* and *g*. Figure 6 is the regression prediction result. The model was evaluated using the mean square error (MSE) and squared correlation coefficient R^2. The result is shown in Table 3 which shows that the SVM regression prediction model with Low, R, and Up parameters has very good R^2 value. The minimum error of the mean square is 0.00615958 while the corresponding R^2 coefficient is 0.991892, which is good enough for the training samples.

3.3. Prediction Effect Verification

As mentioned above, the trained prediction model is used to predict the data of the next five periods as shown in Table 1. To better verify the prediction effect, fuzzy particles are also used to describe the change interval of the measured data in the 31-35 periods. The prediction effect is verified by comparing with the predicted change interval. The actual change interval is described by fuzzy particles after the fuzzy granulation of the measured data information is completed by using equation (1) According to Table 4, it is found that the measured data of 31-35 periods are within the predicted range. The predicted error of ground settlement changed in the above 5 periods is within 10%, which can be used to judge the safety of settlement according to the threshold value. Therefore, it is concluded that the time series prediction model of the FIG-SVM can effectively predict the ground settlement trend based on historical data, and it is suitable for real-time prediction during the construction of the liaison channel.

At the monitoring points D5-1 to D5-4 and D5-5 to D5-8 of the Cross Passage, the time series prediction model of the FIG-SVM was also used to predict the ground settlement. The predicted results are compared with the actual test values of the points, as shown in Figure 7. The relative error of the prediction range is shown in Table 5. The predicted error increases from D5-1 to D5-5 and then decreases from D5-6 to D5-8. The predicted error reaches the maximum at the center axis of the Cross Passage. This is caused by the large fluctuation range of the changes the ground settlement at the central axis. Generally, the predicted range of ground settlement changed at each monitoring point basically meets those during the actual shield construction in the Cross Passage. This further illustrates the effectiveness and possibility of applying the FIG-SVM time series prediction model to similar shield construction projects in the liaison channel.

Table 4. Prediction of variation interval for ground settlement during shield construction of connecting passage

Monitoring point D5-5 surface subsidence prediction 31-35 period data /mm: [-21.22, -21.63, -20.14, -20.59, -21.79]	
Predicted change interval /mm	[-23.65, -22.74, -18.18]
Actual change interval /mm	[-22.35, -21.57, -19.76]
Relative error of prediction E/%	[5.82, 5.42, 8.0]

Table 5. Prediction error of each measuring point

Monitoring points	Relative error of prediction E/%
D5-1	[1.42, 1.74, 1.26]
D5-2	[1.32, 1.12, 1.22]
D5-3	[4.22, 3.45, 5.62]
D5-4	[5.35, 5.63, 7.77]
D5-5	[5.82, 5.42, 8.00]
D5-6	[2.32, 2.54, 2.63]
D5-7	[2.10, 2.22, 2.58]
D5-8	[1.53, 2.53, 3.72]

4. Conclusion

This paper takes the liaison channel project of Ningbo Metro Line 3 as the research background. Various research methods such as field monitoring, prediction are integrated to study the deformation law of the ground surface during the shield construction of the liaison channel. Therefore, the change trend of ground settlement is predicted during the construction process. The main conclusions are as follows.
The FIG-SVM model is a support vector machine time series prediction model based on the fuzzy granulation method. It is a convenient method and has high prediction accuracy. Based on historical data, effective prediction of the change trend of ground settlement is developed in the next phase. The present method is suitable for real-time prediction during the construction of the Cross Passage.

References

[1] Dias D,Kastner R. Movements caused by the excavation of tunnels using face pressurized shields-analysis of monitoring and numerical modeling results[J]. Engineering Geology,2013,152(1) : 17-25
[2] Li Luobin, Gong Xiaonan, Gan Xiaolu, Cheng Kang, Hou Yongmao. Prediction of maximum ground settlement induced by Shield Tunnel Based on recurrent neural network [J]. Chinese Journal of civil engineering, 2020,53 (S1): 13-19
[3] Zujiang Luo,Zhao Li,Jinzhong Tan,Qingshan Ma,Ying Hu. Three-dimensional fluid–soil full coupling numerical simulation of ground settlement caused by shield tunnelling[J]. European Journal of Environmental and Civil Engineering,2020,24(8).
[4] Li Zongliang, Yin Ge. Prediction of ground settlement caused by shield construction using double neural network [J]. Journal of underground space and engineering. 2014 (01)
[5] Ocak I,Seker S E. Calculation of ground settlements caused by EPBM tunneling using artificial neural network,SVM, and Gaussian processes[J]. Environmental Earth Sciences, 2013,70(3) : 1263-12
[6] Song Haiping. Comparative study on grey system theory prediction model of surface subsidence in mining area [J]. Natural resources of North China, 2020 (04): 26-28
[7] Pin Zhang,Ren-Peng Chen,Huai-Na Wu. Real-time analysis and regulation of EPB shield steering using Random Forest[J]. Automation in Construction,2019,106.
[8] Shi Xiangfeng, Wang Lifen, Shenyang, Liu June, Guo Zhanglin. Study on time series prediction of ground settlement in foundation pit construction based on GA-SVM [J]. Construction technology, 2017,46 (08): 16-19 + 31
[9] Chen Benyang, Zhang Chenggang, Ni Ming, et al. Load forecasting of power grid based on fuzzy information granulation support vector machine [J]. Electronic design engineering, 2018, 3 (6): 56-59
[10] Wang Xiaochuan, Shi Feng, Yu Lei, et al. Analysis of 43 cases of MATLAB neural network [M]. Beijing: Beijing University of Aeronautics and Astronautics Press, 2011
[11] Feng Shilin, Zhang Zhongbo, Du Zhong. Study on fault frequency prediction of landing gear system of TONGHANG aircraft [J]. Modular machine tool and automatic processing technology, 2018 (06): 51-55
[12] Li Yanjun, Zhang Jian, Cao Yuyuan, Zhang Lina. Aero engine performance trend prediction based on fuzzy information granulation and optimized SVM [J]. Acta aero dynamics, 2017,32 (12): 3022-3030
[13] Liu Bin, Bi Xiaoxiong, Dang Junpeng, Liu Qi, Wang Jun, Shu Zhengyu. Prediction of substation battery degradation trend based on support vector regression [J]. Acta Sinica Sinica, 2020,18 (06): 207-214
[14] Tang Xiaohong, Hu Junfeng, Xiong Guoliang, Zhang long. Adaptive nonlocal mean and its application in bearing fault detection [J]. Vibration. Test and diagnosis, 2019,39 (01): 61-67 + 221

Pattern Recognition and Analysis: Neural Network using Weighted Cross Entropy

Bingting Guo*, Shuixuan Chen, Zhaobin Hong and Guoqi Xu

School of Mechanical & Automotive Engineering, Xiamen University of Technology, Fujian, China

*Corresponding author email: 785247431@qq.com

Abstract. In order to identify arrhythmia more simply and efficiently, one-dimensional convolutional neural network is used to extract electrocardiogram signal features and a neural network model based on cardiac beat is established, which can identify eight major ECG arrhythmia types. The cross entropy loss function is weighted according with the data distribution to solve the data imbalance problem. Experiments verify the effectiveness of the weighted cross-entropy loss function in ECG recognition, which improves the accuracy of all kinds to more than 99.5% and the sensitivity of all kinds to more than 90%. Compared with other models, this model can improve the performance of ECG arrhythmia recognition.
Keywords: Arrhythmia; Pattern recognition; Convolutional neural network

1. Introduction

In recent years, with the development of semiconductor technology, microelectronic technology and integrated circuit technology, ECG acquisition chip has been widely used in the market, and has changed from large-scale to small-scale, non-intelligent to intelligent. The identification method of atrial fibrillation often needs long-term monitoring. Designing the identification algorithm through deep learning can greatly reduce the workload of medical staff, it can become a quick and instant medical identification method. The heart rhythm recognition algorithm on the portable device can remind the wearer's physical condition in real time and develop a healthy living habit.

Mathews et al. [1] put forward the restricted Boltzmann machine and depth confidence network model, which are used to classify single-lead electrocardiogram with low sampling rate and simple features. Sannino G et al.[2] proposed an automatic classification method for ECG signals of DNN, which has good accuracy, sensitivity and specificity. Yildirim[3] proposed a new model of wavelet sequence for bidirectional LSTM network. JUN T J et al. [4] trained neural network by converting electrocardiogram into grayscale image. ATTIA Z I et al. [5] calculated the confidence interval for each index, thus verifying its effectiveness.

In real images, distortion and noise should be considered. The conversion of ECG signals into images not only increases the training parameters, but also greatly slows down the training speed of the network. In the application scenarios of ECG acquisition equipment such as electronic watches and hand rings, the embedded algorithm can be conveniently used to identify ECG signals. Compared with using image as data and using two-dimensional convolution network, the data processing, model and loss function design in this paper make the algorithm more concise and efficient.

2. Data and Methods

Using the MIT-BIH Arrhythmia Database, the data set consisted of 48 labeled two-lead ECG data, each lasting approximately 30 minutes, with ECG signals sampled at 360hz and the P-QRS-T waveform for each cycle was labeled. The upper lead data were used during training, and the cardiac beats were intercepted from each cycle as training samples.

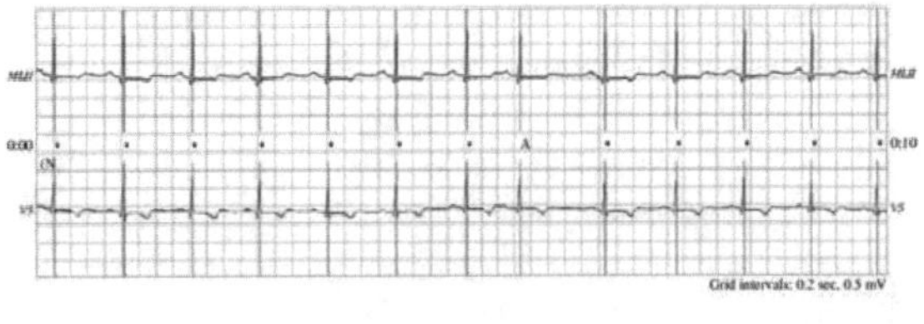

Figure 1. Diagram of ECG in the first ten seconds.

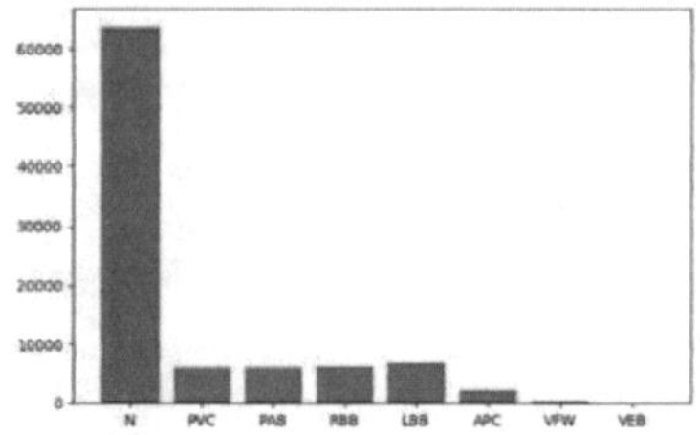

Figure 2. Data distribution.

Approximately 110,000 cardiac beat can be intercepted from that data set. The common ECG recognition research can integrate the first 15 ECG types in the MIT-BIH data set into five categories for recognition according to the AAMI standard. In order to refine the recognition, 8 most common and extensively researched segmentation types are selected from the data set for recognition. They were Normal Heart Rate (NOR), PVC, PAB, RBB, LBB, APC, VFW and VEB. Other types were usually ignored in the classification of arrhythmia. The above 8 rhythm types were numbered sequentially from 0 to 7, followed by one-hot encoding as labels for the training and verification sets. As shown in Figure 2, the numbers of eight types of ECGs were counted. There were too many samples of normal heart beat and the smallest sample of ventricular escape beat, indicating a serious data imbalance.

The time information and R peak location were recorded in the MIT-BIH data set. After the data was read into an array format from the data set, a program was written based on the R peak location for cardiac capture and merging. In this paper, the values of 0.45s before and 0.45s after R peak are intercepted as a beat, that is, 162 sampling points are taken in front of R peak and 162 sampling points are taken after R peak. If the intercepted ECG cardiogram data are converted into pictures as the input of the neural network, the data enhancement needs to be used to reduce the over-fitting, the input data needs to be balanced by adding pictures to achieve higher sensitivity and specificity, and the characteristics of real images need to be considered. Due to the small data set, considering the actual application scenarios, and the existence of data imbalance, we directly uses the ECG signal as the input, and weights the loss function to improve the classification accuracy of each class of the algorithm. See Figure 3 for an example image of each type of cardiac beat.

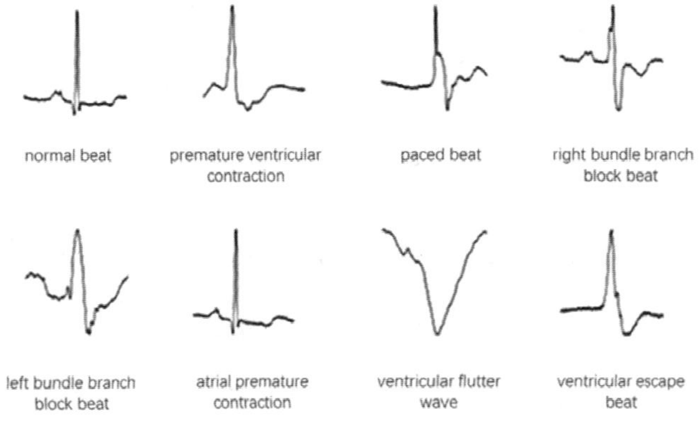

Figure 3. Types of electrocardiogram.

85% of the approximately 110,000 ECG beat signals were used as the training set and 15% as the verification set, and no additional test set was divided. The training set is trained on the designed model and a variety of evaluation indexes are used to evaluate the model performance. Finally, the

Bootstrap sampling is used on the verification set and the confidence interval of AUC is calculated, which reflects the recognition ability of the model on real-world data.

3. Deep Learning Model

3.1. Model Initialization

The model was initialized by Xavier[6]. Random initialization will lead to slow convergence and easy to fall into local minimum. The effect is also poor. Hence, regular initialization is generated, i.e., Xavier initialization. Xavier initialization works for most activation functions, and has gradually become the default initialization method for many frameworks due to its stable performance. Xavier initialization weights conform to the uniform distribution of Equation (1), where n_j and n_{j+1} are the number of neurons in the input and output, respectively.

$$W \sim U\left[-\frac{\sqrt{6}}{\sqrt{n_j+n_{j+1}}}, \frac{\sqrt{6}}{\sqrt{n_j+n_{j+1}}}\right] \tag{1}$$

As shown in Equation (2), with the following learning rate decay strategy, the initial learning rate was set to 0.001, and training was performed for 60 epochs.

$$LearningRate = 0.001 \times 0.8^{\frac{epoch}{5}} \tag{2}$$

The model uses the ELU activation function. Equation (3) is the activation function of ELU.

$$ELU(x) = \begin{cases} x, & x \gg 0 \\ \gamma(e^x - 1), & x < 0 \end{cases} \tag{3}$$

In the process of back propagation using the standard cross-entropy loss function, the loss of more samples will have a greater impact on the weight of the model, relatively ignoring the number of categories, which will ultimately affect the sensitivity and specificity of the model. Because of the imbalance of the data, the loss functions are weighted to balance the directional propagation so that all types of loss functions have the same degree of impact on the model in the back propagation. As shown in Equations (4)-(6), the loss function is the cross-entropy loss function with a certain weight added to balance the directional propagation. $los(i)pos$ is expressed as a positive loss for each type of actual label. $los(i)neg$ is expressed as a negative loss for each type of actual label. $weigh(i)pos$ is the number of negative labels for each class divided by the total number. $weigh(i)neg$ is the number of positive labels for each class divided by the total number. The weight is used in the loss function, which effectively eliminates the influence of different number of sample types on the directional propagation.

$$loss(i)_{pos} = -weight(i)_{pos} \times y(i) \times log(\hat{y}(i)) \tag{4}$$

$$loss(i)_{neg} = -weight(i)_{neg} \times (1 - y(i)) \times \log(1 - \hat{y}(i)) \tag{5}$$

$$loss(i) = \frac{1}{n}(loss(i)_{pos} + loss(i)_{neg}) \tag{6}$$

3.2. Model Structure

The model is a one-dimensional convolutional neural network structure. The output layer of this model performs well with either softmax or sigmoid activation functions, depending on the actual classification task. Googlenet uses 1X1 convolution to reduce the computational complexity and increase the non-linearity and network depth. This model improves the depth of the convolution by adding the convolution with the convolution kernel of 1 after every two convolution layers. At the same time, one-dimensional convolution itself can automatically extract features, which can omit filtering methods such as wavelet denoising.

In the model, a module is stacked four times, and each module has three times of convolution. The convolution kernel of the first module is 16, 16, and 1, respectively, and the number of channels is 64. The convolution kernel of the second module is 8, 8, and 1, respectively, and the number of channels is 64. The third module convolution kernel is 4, 4, and 1, with 128 channels. The fourth module convolution kernel is 2, 2, and 1, with 256 channels. All convolution steps were 1. After that, the global average pooling layer (GAP) and the full connection layer (FC) of 2048 neurons were connected. Finally, sigmoid identified eight classes. Through experiments, the model uses Dropout[7], with a probability of 0.5, and Dropout does not perform better after the global average pooling layer than before the sigmoid output layer.
Local response normalization was used in Alexnet, and Vggnet thought that this method could not improve the performance of the model. However, Googlenet used local response normalization in the same period. Since then, local response normalization has not been widely used, so the effect may be unstable. Therefore, this model uses a more stable BatchNormalization[8](BN), and the position of BN is usually before the activation function and after the convolution layer. However, in the experiment of identifying heart rhythm, it is found that BN is better after the activation function of ELU. Therefore, BN is used before the last ELU of each module activates the function. Adam optimization algorithm is used in the model to speed up the training of neural networks. The structure of the model is shown in Figure 4.

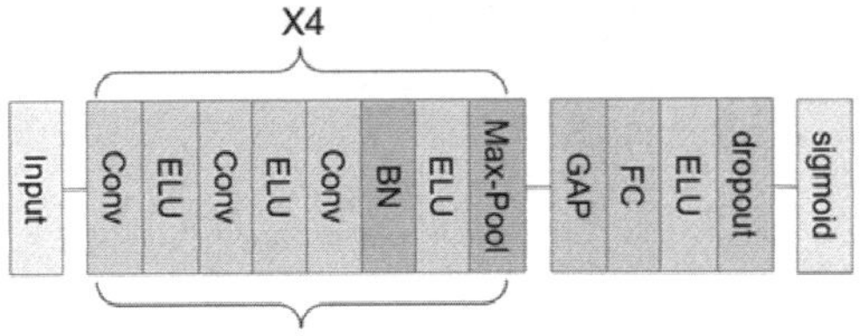

Figure 4. Model structure.

4. Experimental Results

4.1. Confusion Matrix and PRC Curve

After weighting the loss function, the accuracy of all classes is above 99.5%, the sensitivity of APC class has been obviously improved, and other indexes have also been improved to some extent. Experiments show that the weighted loss function has an improved effect regardless of whether the output layer is softmax or sigmoid activation function. The confusion matrix visually shows how many categories are correctly classified. As shown in Figure 5, the 80 samples for APC misclassification were reduced to 33, and the recognition number of other classes has also been improved.

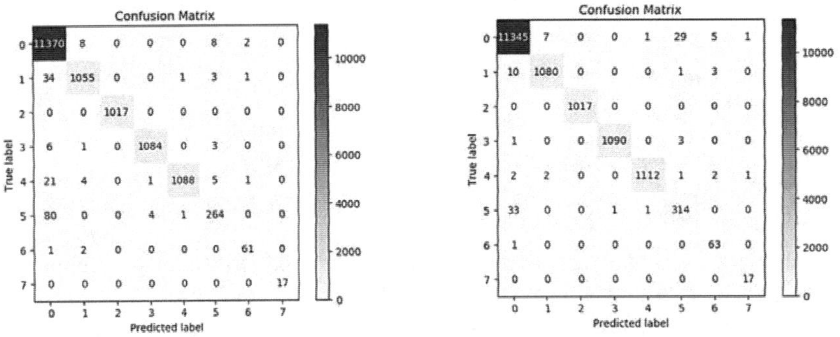

Figure 5. Confusion matrix comparison.

When data are imbalanced, ROC tends to overestimate the quality of the model, PRC is more appropriate. The precision-recall curve (PRC curve) shows the trade-off between precision and recall at different thresholds. If that PRC curve of one classify is completely enclosed by the PRC curve of another classify, the performance of the latter can be asserted to be superior to the former, and if the curve are crossed and cannot be directly judged, the comparison can be made according to the size of the area under the curves. As shown in Figure 6, after weighting the loss function, the PRC curve is closer to the upper right corner, improving the precision and recall of the APC class.

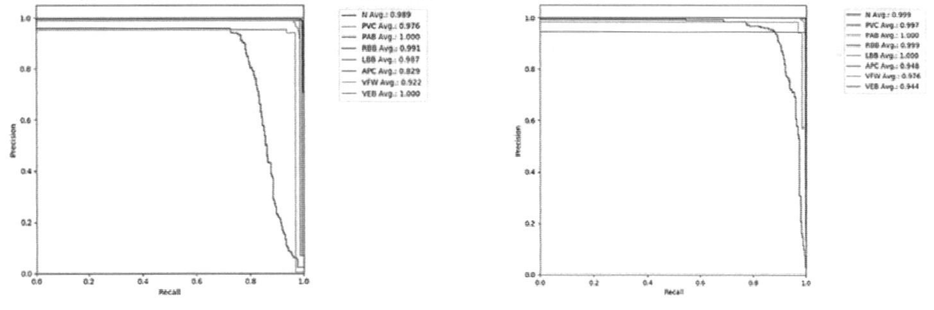

Figure 6. PRC curve comparison.

4.2. AUC Confidence Interval

Bootstrap can be used to create randomness in data. To ensure a representative sample size for each class, 1000 samples were randomly stratified from the validation set. The first category of metrics is calculated first, with actual (0 or 1) and predicted (0 to 1) values for each sample. The AUC score for the first class was calculated based on the actual and predicted values for these 1000 samples. Samples were then stratified and randomly sampled again and counted for 100 cycles, yielding 100 AUC scores. Finally, the mean, maximum and minimum quantiles of AUC are calculated. Similarly for other categories, the confidence intervals of AUC of all categories were obtained. As shown in Table 1, the weighted cross entropy increases the mean AUC and decreases the confidence interval, which indicates a better estimation result for the actual data.

Table 1. Average AUC (CI 5%-95%).

	Without weighted cross entropy	Use weighted cross entropy
N	0.99(0.98-0.99)	1.00(1.00-1.00)
PVC	0.99(0.98-1.00)	1.00(0.99-1.00)
PAB	1.00(1.00-1.00)	1.00(1.00-1.00)
RBB	1.00(0.99-1.00)	1.00(1.00-1.00)
LBB	0.99(0.98-1.00)	1.00(1.00-1.00)
APC	0.97(0.93-1.00)	1.00(0.99-1.00)
VFW	0.99(0.83-1.00)	1.00(1.00-1.00)
VEB	1.00(1.00-1.00)	1.00(1.00-1.00)

4.3. Model Comparison

The model in this paper is implemented by TensorFlow framework. Compared with the classical Alexnet, Vggnet and the traditional machine learning model, the average AUC and other values of the proposed model are significantly improved.

Table 2. Model comparison.

Method	AUC	Accuracy	Sensitivity	Specificity	PPV
Alexnet	0.986	98.85	97.08	99.62	98.59
Vggnet	0.984	98.63	96.93	99.37	97.86
SVM	96.76	99.7	93.94	99.55	98.75
Decision Trees	93.05	99.31	86.75	99.34	88.2
Random Forest	94.89	99.66	90.28	99.53	97.38
AdaBoost	73.44	92.79	52.95	93.99	39.46
Proposed	**0.999**	**99.83**	**98.34**	**99.8**	93.23

5. Conclusion

In this paper, a one-dimensional convolutional neural network model is designed to identify arrhythmia types. Compared with the classical neural network model and the traditional machine learning model, this model has achieved excellent performance in identifying eight ECG types. On the structure of this model, the influence of weighted cross-entropy loss function on classification performance is compared, which shows that the use of weighted cross entropy can improve the ability of the model to train the imbalanced data, so that the model performance is better.

In order to further improve the model performance, Bayesian hyperparametric optimization[9] can be used to obtain a more scientific value for the hyperparametric indicators such as probability of Dropout and learning rate. Electrocardiogram signal recognition based on deep learning can be used as an auxiliary medical diagnosis method, which is helpful to integrate artificial intelligence related technologies with medical treatment and has great application value.

References

[1] Mathews Sherin M and Kambhamettu Chandra and Barner 2 2 Kenneth E. A novel application of deep learning for single-lead ECG classification.[J]. Computers in biology and medicine, 2018, 99: 53-62.

[2] G. Sannino and G. De Pietro. A deep learning approach for ECG-based heartbeat classification for arrhythmia detection[J]. Future Generation Computer Systems, 2018, 86 : 446-455.

[3] Özal Yildirim. A novel wavelet sequence based on deep bidirectional LSTM network model for ECG signal classification[J]. Computers in Biology and Medicine, 2018, 96 : 189-202. KANG

[4] JUN T J, NGUYEN H M, D, et al.. ECG arrhythmia classification using a 2-D convolutional neural network[J]. arXiv:1804.06812 [cs], 2018.

[5] ATTIA Z I, NOSEWORTHY P A, LOPEZ-JIMENEZ F, et al.. An artificial intelligence-enabled ECG algorithm for the identification of patients with atrial fibrillation during sinus rhythm: a retrospective analysis of outcome prediction[J]. The Lancet, 2019, 394(10201): 861–867.

[6] GLOROT X, BENGIO Y. Understanding the difficulty of training deep feedforward neural networks[J]. : 9.

[7] HINTON G E, SRIVASTAVA N, KRIZHEVSKY A, et al.. Improving neural networks by preventing co-adaptation of feature detectors[J]. arXiv:1207.0580 [cs], 2012.

[8] IOFFE S, SZEGEDY C. Batch Normalization: Accelerating Deep Network Training by Reducing Internal Covariate Shift[J]. arXiv:1502.03167 [cs], 2015.

[9] SNOEK J, LAROCHELLE H, ADAMS R P. Practical Bayesian Optimization of Machine Learning Algorithms[J]. arXiv:1206.2944 [cs, stat], 2012.

Visual Pattern Recognition: Identity Authentication Technology Based on Eye Movement Patterns

Jingjing Zhao[1,2] and Xiaoming Wang[1,*]

[1]Xi'an International Studies University, Xi'an, China
[2]Organization Department of Shaanxi Provincial CPC Committee, Xi'an, China

*Corresponding author email:wxm@xisu.edu.cn

Abstract. Existing identity authentication techniques are primarily based on passwords or physical characteristics and are therefore prone to theft and forgery. This can lead to potential security risks for identity recognition. In this paper, we propose a new identity authentication technique based on reading eye movements by introducing eye tracking technology. In this approach, a deep neural network based on multi-input architecture is used to construct a computational model. The model uses the text sequence, the fixation point sequence, and the linguistic feature as inputs, and output the recognition by a neural network. The simulation results showed a recognition accuracy of 89.9%. The eye tracking technique developed in this study may be used as a new and efficient identity authentication method. The data set used in the experiment and the associated codes have been released on GitHub.
Keywords: Identity authentication; Deep neural networks; Eye movement tracking; Eye movement model.

1. Introduction

Identity authentication are divided into two types: physical and behavioral characteristics. Physical characteristics include features as the face, finger vein, fingerprint, retina, iris, and DNA. Behavioral characteristics include features such as an individual's signature, speech, walking gait, and other characteristics [1]. The scope of this research belongs to the second category.

This article's theoretical basis is the difference in eye movement patterns among people in the process of reading words. During the reading process, the reader may feel that the eyes moved smoothly from the left to the right, but in fact, the trajectory of the reader's eye movement consisted of a series of fixations and discrete saccades [2]. This is because that the fovea only affords an equivalent viewing angle of 5 degrees [3]. Therefore, a reader must constantly change the fixation point in a series of saccades, so that the contents of the next image fall in the foveal area of the retina in order to facilitate their reading and understanding the information of the text. Eye tracking is considered an important method for studying reading behavior [4]. This method helps researchers using eye trackers to determine the position and number of times a subject is interested in reading a certain word, and the sequence of eye movements. **Figure 1** shows the eye movement trajectory (eye movement pattern) when an adult reader reads.

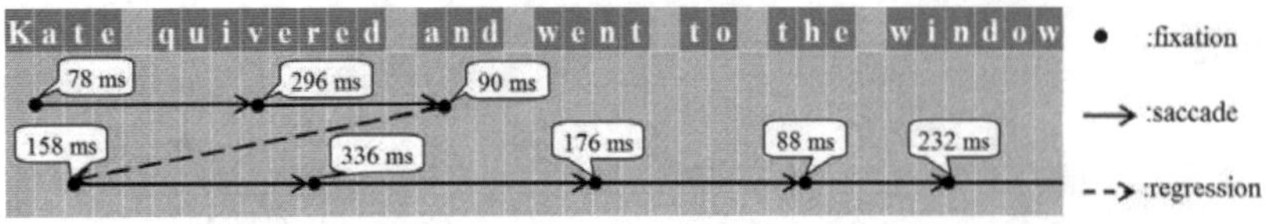

Figure 1. Eye movement trajectory of an adult reader.

Eye movement patterns are significantly different among individuals [5]. The model describing the eye movement pattern is called the eye movement model. Eye movement models are often studied in the field of cognitive psychology [6]. There are also studies of eye movement patterns based on computational models[7][8]. These computational eye movement models have achieved a high degree of prediction accuracy. However, it is not clear whether a reading eye movement model may be used to characterize behavior features for identity recognition, which is the main focus of the current study.

The contributions of this paper are as follows:

(1) This paper proposes a method of identity recognition based on eye movements, which is easy to apply in engineering practice.

(2) The proposed method uses fewer eye movement features to achieve similar accuracy to other methods.

(3) The proposed method makes full use of low-level visual features and high-level language features, which is more in line with the reality of eye movement.

2. Computational Eye Movement Model Based on a Deep Neural Network

The computational model proposed in this paper takes the text sequence, the text linguistic feature sequence and the actual fixation sequence as input, and outputs the recognition result by measuring the similarity distance between the predicted fixation sequence and the actual fixation sequence. Figure 2 shows the architecture of the model.

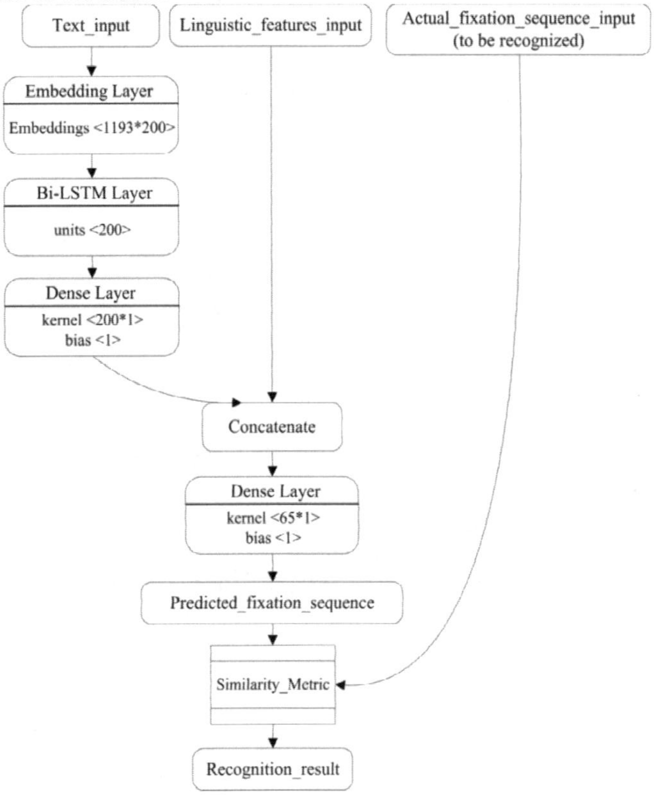

Figure 2. Architecture of a computational eye movement model based on a deep neural network.

3. Experiment

To evaluate the performance of the developed model, we used experimental data to train and test the model. This section provides details on the experimental data set and division, feature selection, data preprocessing, network parameter setting, and parameter optimization. In order to facilitate a wide range of applications, the data set used in the experiment and the codes used for model training and model testing were released on GitHub.

3.1. Data Set and Data Preprocessing

The experiment used data from the 1st–10th test subjects of the Provo Corpus [9], and each of them read 55 sentences. The experiment selects the corpus containing only the text of the 1–45 sentences for model training, and the corpus containing only the text of the 46–55 sentences is used for model testing; the two corpuses do not overlap with each other.The word embedding layer realizes the process of text vectorization. The First 5 lines of the vectorized textare shown in **Table 1**.

Table 1. First 5 lines of the vectorized text

Text_ID	Text	Word_Length	IA_SKIP	Participant_ID
1	[b'are', b'now', b'rumblings', b'that', b'Appl...	[3, 3, 9, 4, 5, 5, 4, 6, 3, 5, 5, 5, 6, 3, 7, ...	[0, 0, 0, 1, 0, 0, 0, 0, 1, 0, 0, 0, 0, 1, 0, ...	0
2	[b'days', b'later,',b'the', b'British', b'ast...	[4, 5, 3, 7, 10, 7, 0, 3, 9, 3, 11, 7, 5, 7, 3...	[0, 0, 1, 0, 0, 0, 0, 0, 0, 0, 0, 0, 0, 0, 0, ...	0
3	[b'agree', b'that', b"California's", b'"three'...	[5, 4, 12, 5, 7, 3, 6, 3, 3, 4, 2, 1, 9, 8, 3,...	[0, 1, 0, 0, 0, 1, 0, 0, 0, 0, 0, 1, 0, 0, 0, ...	0
4	[b'was', b'in', b'a', b'bad', b'temper,',b'fo...	[3, 2, 1, 3, 6, 3, 3, 3, 6, 3, 9, 8, 2, 3, 7, ...	[1, 0, 1, 0, 1, 0, 1, 0, 1, 0, 0, 0, 1, 0, 0, ...	0
5	[b'Darling', b'quivered', b'and', b'went', b't...	[7, 8, 3, 4, 2, 3, 6, 2, 3, 8, 8, 3, 6, 3, 3, ...	[0, 0, 1, 0, 0, 1, 0, 1, 1, 0, 0, 1, 0, 0, 1, ...	0

3.2. Assessment Method and Baseline

After the model was constructed based on a deep neural network, the model was evaluated based on classification accuracy. The baseline is the average value 69.66% for the ratio of the number of fixation words of 10 subjects to the total number of words read.

3.3. VerificationData Setting

By setting the parameter validation_split to 0.2, the program automatically divides the data of each subject into 80% training data and 20% verification data before the training begins. Since all the data have 450 sentences, it is divided into $450 \times 0.8 = 360$ sentences as training data, and $450 \times 0.2 = 90$ sentences as verification data.

3.4. Training Process

Figure 3 shows the learning curve of the model. In **Figure 3**(a), the accuracy curve of the training set is blue (dark) and the accuracy (abbreviated as'acc') curve on the validation set is yellow (light). The curves show that, for both the training set and the verification set, the accuracy of the model becomes higher and higher, and that the verification accuracy is higher than the training accuracy, indicating that the model does not have an over fitting problem.

In **Figure 3**(b), the loss curve on the training set is blue (dark) and the loss curve on the verification set is yellow (light). The results show that, for both the training set and the verification set, the loss of the model gets lower with iterations/learning. In the middle and late stages of the training, the training loss is smaller than the verification error, which is expected.

Since the data type processed by the proposed method is text data and its linguistic features, the problem of computational efficiency can be ignored. Taking the Provo database used in the experiment as an example, the training time of each batch is less than 1 second.

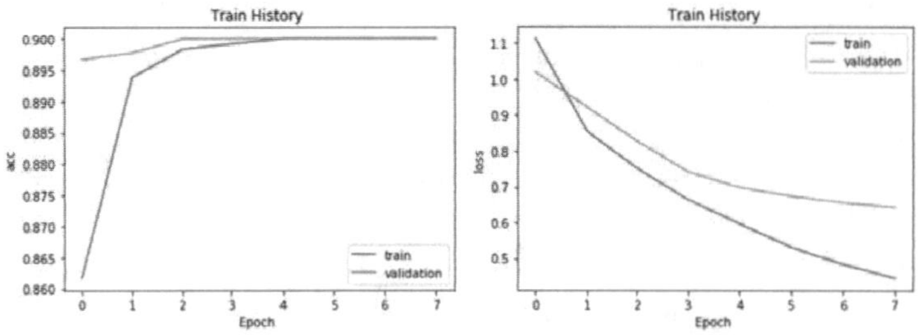

(a) Accuracy of the training set and verification set; (b) Training loss of the training set and verification set.

Figure 3. Learning curve of the model

3.5. Results and Discussion

Using the Keras evaluation method, the accuracy of the model on the test set was 89.9%, which was much higher than the baseline. We believe that there are two reasons for the model's high accuracy. First, it uses a one-dimensional convolution operation on the text sequence to objectively increase the number of training samples, which is favorable for the model to learn the relationship between features and class. Second, the text-based linguistic feature sequence used in this work is a sequence of word lengths, which is not only easy to obtain, but also isthe low level visual features.

As far as we know, the recent state-of-the-art method for identity authentication using eye movements is the method proposed by Harezlak[7] in 2021, and its accuracy rate is 91.2%. Although our method only achieves an accuracy close to that of Harezlak's method, our advantage is that it uses fewer eye movement features. In the process of manual feature extraction, we only use the low-level feature part of the language, and the high-level feature part of the language is handed over to the recurrent neural network for processing.

4. Conclusion

By employing eye tracking technology, we proposed a new technology to realize identity authentication. A computational model was constructed based on a multi-input architecture. The identities of test subjects were predicted using the computational eye movement model. The proposed model is attractive for engineering applications. The simulation results show that the recognition accuracy of the model is 89.9% on the test data set. The experimental results further demonstrate the superiority of the proposed model for Identity authentication.

The limitation of the method proposed in this paper is that we omitted the regressions during reading to facilitate the processing. Further research on this issue is needed in the future, and we must find a method to account for the regressions during reading.

References

[1] Qiu J. H., Feng J., Guo W.: Biometric Recognition: The Revolution of Identity Authentication. Tsinghua University Press, Beijing (2016).

[2] Meng H.X.,Bai X. J., Yan G. L.: The effect of word boundary information on the saccade strategy upon reading the spatially ambiguous words.Journal of Psychological Science38(4), 770–776 (2015).

[3] Fishman G A., Birch D G., Holder G. E.:Electrophysiologic testing in disorders of the retina, optic nerve, and visual pathway. 2nd edn. The Foundation of the American Academy of Ophthalmology, San Francisco (2001).

[4] Rayner K.: Eye movements in reading and information processing: 20 years of research. Psychological Bulletin, 124(3), 372–422 (1998).

[5] Reichle E. D.: Computational Models of Reading: A Primer. Language and Linguistics Compass, 9(7):271-284 (2015).

[6] Clifton J. C., Ferreira F., Henderson J. M.: Eye movements in reading and information processing: Keith Rayner's 40 year legacy. Journal of Memory and Language, 86, 1–19 (2016).

[7] Harezlak, K., Blasiak, M., &Kasprowski, P.: Biometric Identification Based on Eye Movement Dynamic Features. Sensors, 21(18), 6020(2021).

[8] Chowdhary, C.L., Reddy, G.T., Parameshachari, B.D.: Computer Vision and Recognition Systems: Research Innovations and Trends (1st ed.). Apple Academic Press(2022).

[9] Luke S. G., Christianson K.: The Provo Corpus: A large eye-tracking corpus with predictabilitynorms.Behavior Research Methods, 50(2), 826–833(2018).

The Influence of Control Parameters on Precision of Welding Seam Tracking in Manually Control Master-slave Robot Remote Welding System

Hongtang Chen[1], Haijun Ma[2], Hongli Jiang[3], Shenjin Lv[1], Yong Li[1] and Heshan Liu[4,*]

[1]School of Mechanical Engineering, University of Jinan, Ji Nan, China
[2]ShanDong Administration for Market Regulation, Ji Nan, China
[3]BinZhou Polytechnic, Bin Zhou, China
[4]School of Mechanical Engineering, Shan Dong University, Ji Nan, China

*Corresponding author email: liuheshan@sdu.edu.cn

Abstract. The welding seam tracking precision is one of the key factors to ensure welding qua ntity. However, when the operator manually manipulate the master robot to control the slave ro bot tracking welding seam used the master- slave robot remote welding system, seam tracking precision does not meet the requirement of the welding quantity, the influence of control param eters on precision of welding seam tracking is unkown. In this article, a master-slave robot rem ote welding system is set up which includes a master robot with 6 degrees of freedom, a compu ter control system, and robot HP3J carrying a welding torch and laser sensor system. The slave robot tracking welding seam was controlled by computer control system and the human operat or via master manipulator. The influence of control parameters (cross deviation, arc length and welding speed) on precision of welding seam tracking was analysed. The experiment results de monstrate the more control parameters of operator, the lower tracking precision, so in order to guarantee a good welding seam tracking effect, the operator should control less parameters.
Keywords: Seam tracking; Error analysis; Master-slave robot.

1. Introduction

It is necessary to set up a robot remote welding system to perform these tasks, such as maintenance in pipes repairing underground or undersea, construction of space stations, and nuclear power plants, operators and robot are the 2 main performers of robot remote welding tasks. Up to now, current welding robot can execute pre-defined actions generated from off-line programming or taught by robot programmers, but when facing unpredicted disturbances, robot can not perform well, so in order to obtain high quality welding seam, sensors such as vision sensors [1-3], arc sensors [4,5], ultrasonic sensors ,thermal sensors [6,7] and audio sensors [8,9] have been applied to track the welding seam. However, these sensors could not be applied widely due to their respective weaknesses.

Compared with robot, skilled welders could perform well, so human controlled the torch to track welding seam is a viable option for remote robot welding, since robots do not yet have the capability to autonomously perform the remote welding tasks.

The precision of tracking welding seam is one of the key factors to ensure quantity for robot remote welding system, many parameters such as cross deviation, arc length, torch pose, and welding velocity should to be simultaneously controlled in real time, and the influence of control parameters on precision of welding seam tracking is unkown. So in this article, a mater-salve remote robot system is presented, and the influence of control parameters on precision of welding seam tracking is analyzed and discussed.

2. The Master-slave Robot Remote Welding System

2.1. The Master-Slave Robot Remote Welding System

The robot system is shown in Figure.1, welding robot and humans could complete robot remote welding tasks, and the system consists of a navigator space mouse, a laser sensing device, a slave motoman HP3J robot, CCD camera and shutter glasses. The master manipulator navigator space mouse has 6 degree of freedom(DOF), i.e., the operator can move the space mouse right or left and backward or forward and up or down and revolute itself. Coordinate transformation from the space mouse to the welding torch was used to improve the performance, and the corresponding relation of master robot space mouse and slave robot coordinate system is shown in Figure 2. If space mouse is manipulated in the direction of X, Y or Z, welding torch of slave robot moves in the same direction of X, Y and Z, respectively.

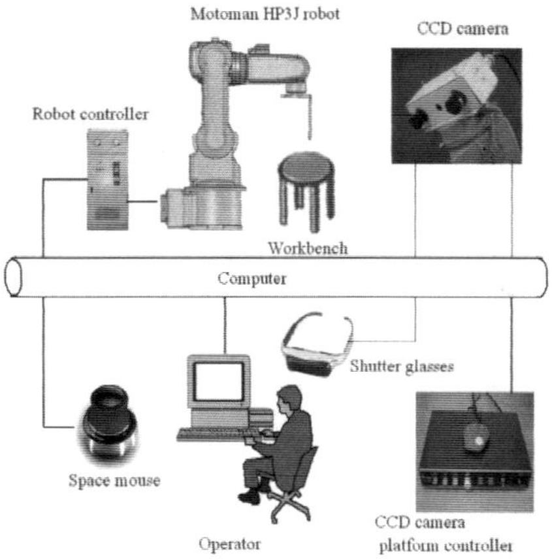

Figure 1. The master-slave robot remote welding system

The slave robot joint information is obtained by the application programming interface(API)---RtlGetJoint, then the joint information is sent to the industrial computer via an I/O card. The I/O card is used for communication between the industrial computer and the slave robot.

The flow chart of master robot controls slave robot is shown in Figure 3.

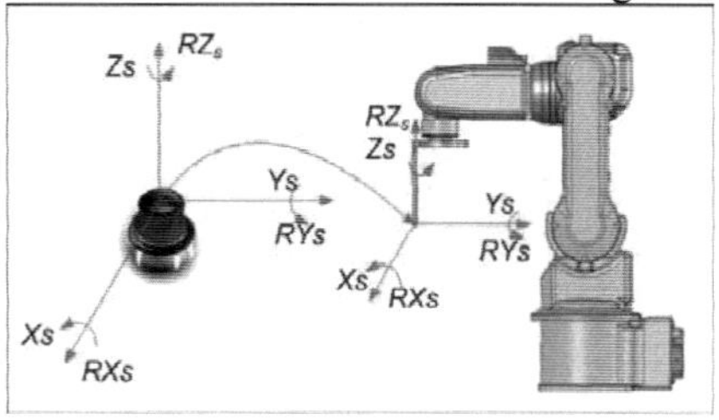

Figure 2. The mapping relationship of spacemouse to robot

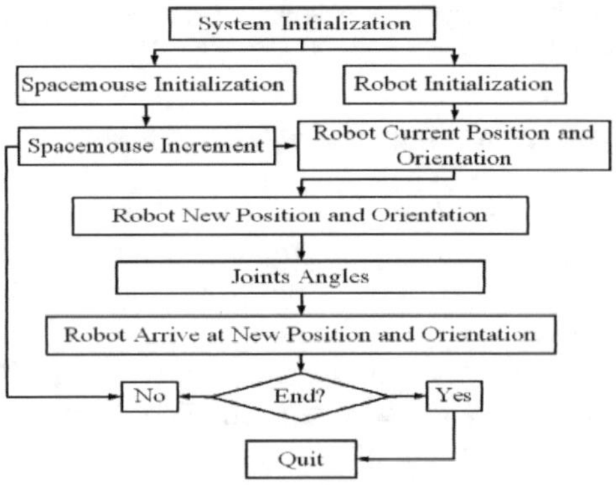

Figure 3. Flow chart of system

The welding quality was affected by the following six parameters: arc length, cross deviation, vertical angle and cross angle, direction angle, and the welding speed.

For simplicity the straight welding seam was used as the tracking target, so the cross deviation, arc length, and the welding speed have more great influence on the weld quality. The sensing devices are two micro-epsilon laser sensors, which are used to obtain the values of the arc length and cross deviation, the welding speed is obtained by the industrial computer. In Figure 4 the laser sensor 1 is used to obtain the values of cross deviation, the laser sensor 2 is used to obtain the values of arc length.

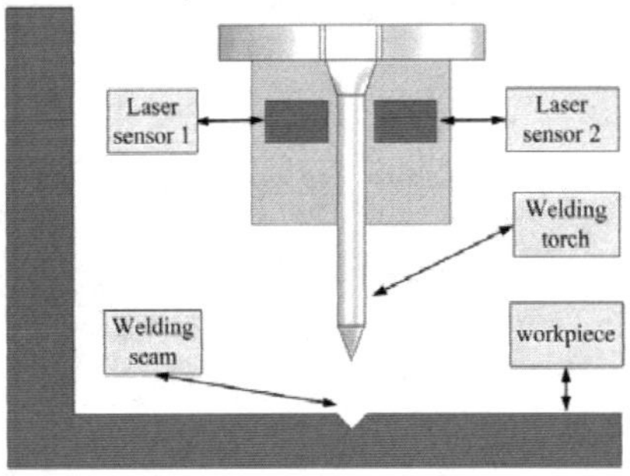

Figure 4. The laser sensor system and the welding torch

3. Experiment

For simplicity straight welding seam is used as the tracking target. "Welding torch" controlled by operator follows the welding seam. For straight welding seam, welding process only requires a constant torch orientation, so in this study only the cross deviation, arc length, and the welding speed were recorded.

The slave robot position signal was the tracking signal, the values of cross deviation and arc length were accessed by the laser sensing system, the welding speed was obtained by the industrial computer. In order to analyze the influence of control parameters on the seam tracking precision, four comparative experiments were done to compare cross deviation:

(1) Only control cross deviation while welding speed and arc length are constant;

(2) Control cross deviation and arc length while welding speed is constant;

(3) Control cross deviation and welding speed while arc length is constant;

(4) Control cross deviation, arc length and welding speed simultaneously

3.1. Only Control Cross Deviation While Welding Speed and Arc Length are Constant

In this mode, arc length and welding speed are simultaneously controlled by computer, which are set to constants, operator only control cross deviation, Figure 5 shows the cross deviation when the track speed of welding torch along the straight welding seam is 5mm/s. It is quite obvious that the cross deviation varies between -0.419mm and 0.481 mm.

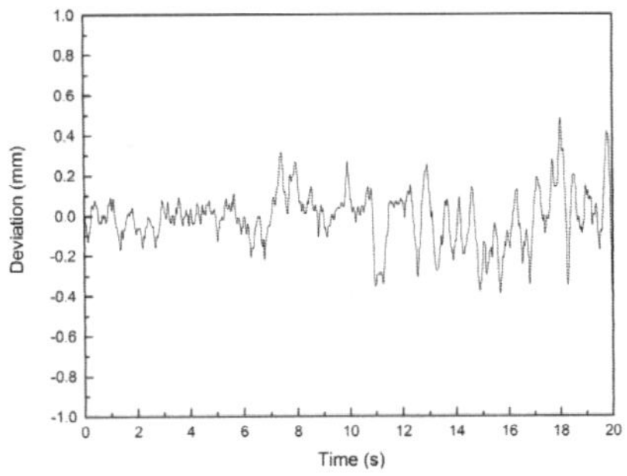

Figure 5. The cross deviation under welding velocity is 5mm/s and the arc length and welding speed are constant

3.2. Control Cross Deviation and Welding Speed While Arc Length is Constant

In this mode, arc length is controlled by computer, which is set to constant, the operator control the cross deviation and welding speed simultaneously. On one hand, the operator should control the cross deviation as small as possible, on the other hand, the operator should control the welding speed as a constant. The Figure 6 shows cross deviation when the welding speed is 5mm/s. It is quite obvious that the cross deviation is -0.437mm--- 0.499mm.

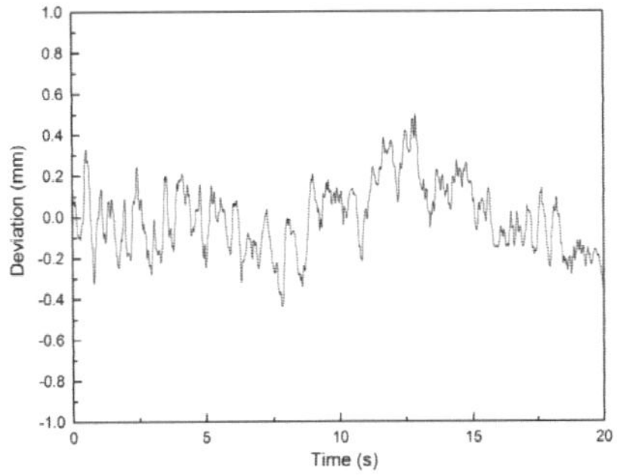

Figure 6. The cross deviation under welding speed is 5mm/s while the arc length is constant

3.3. Control Cross Deviation and Arc Length While Welding Speed is Constant

In this mode, welding speed is controlled by computer, which is set to constant, the operator control the cross deviation and arc length simultaneously. On one hand, the operator should control the cross deviation as small as possible, on the other hand, the operator should control the arc length as a constant. The Figure 7 shows cross deviation when the welding speed is 5mm/s, it is quite obvious that the cross deviation is -0.717mm --- 0.709mm.

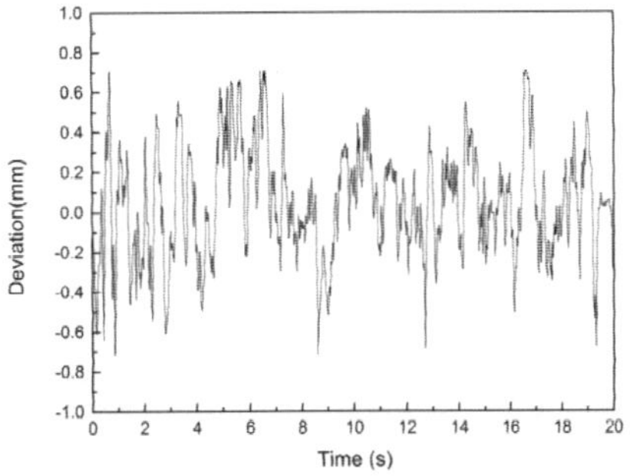

Figure 7. Control the cross deviation and arc length while the welding speed is constant

3.4. Control Cross Deviation, Arc Length and Welding Speed Simultaneously

In this mode, operator control 3 parameters, cross deviation, arc length and welding speed, Figure 8 shows cross deviation of welding speed is 5mm/s, the cross deviation is -0.993mm --- 0.838mm, the cross deviation is the biggest and the cross deviation has the biggest fluctuation, because the operator need to control cross deviation, arc length and welding speed simultaneously, three control parameters distract operator's attention, so there is widely fluctuation in the tracking process.

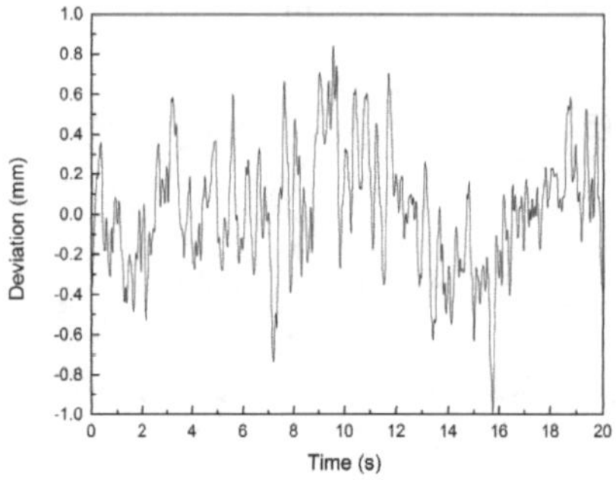

Figure 8. Control the cross deviation, arc length and welding speed simultaneously

4. Result Analysis

Under four different modes, it is found that only control the cross deviation, the cross deviation is -0.419mm --- 0.481 mm, control the cross deviation and welding speed, the cross deviation is -0.437mm --- 0.499mm, control the cross deviation and arc length, the cross deviation is -0.717mm --- 0.709mm, control the cross deviation, arc length and welding speed simultaneously, the cross deviation is -0.993mm --- 0.838mm.

As shown in Figure 5, only control the cross deviation, at first the operator can track the welding seam well, the cross deviation varies between -0.223mm and 0.198mm, but after 7s, the values and fluctuation of cross deviation get bigger.

As shown in Figure 6 and Figure 5, the fluctuation range of cross deviation is about the same, but there is bigger fluctuation in Figure 6.

Through a comprehensive comparison of Figure 5, Figure 6, Figure 7 and Figure 8, increasing the number of parameters increases the difficulty of tracking the welding seam, the more control parameters of operator, the lower tracking precision.

5. Conclusion

(1) A master-slave robot remote welding system is built, which includes a navigator space mouse, a laser sensing device, a slave motoman HP3J robot, CCD camera and shutter glasses.

(2) From the experimental results, it has been shown that, the control ability of the operator is reduced with the number of control parameters increases. When the control parameter is cross deviation, while the welding speed and arc length are constant, the cross deviation is -0.419mm --- 0.481 when the welding speed is 5 mm/s, when control parameters are the cross deviation and welding speed, while the arc length is constant, the cross deviation is -0.437mm --- 0.499mm when the welding speed is 5mm/s, which could only meet the basic requirement of the remote welding tasks, when control parameters are the cross deviation and arc length, the cross deviation is -0.717mm --- 0.709mm and the cross deviation fluctuates wildly, when control parameters are the cross deviation, arc length and welding speed, the cross deviation is -0.993mm --- 0.838mm, It is obvious that the more control parameters the worse tracking accuracy, so in order to obtain better weld quality the operator should control less parameters, while the computer should control more parameters.

Acknowledgments

This work was financially supported in part by Key R & D project of Shandong Province under Grant No. 2019JZZY010452.

References

[1] Lü FL, Chen HB, Fan CJ. A novel control algorithm for weld pool control, Industrial Robot, 2010, 37(1):89-96.
[2] Ma X J, Zhang YM. Gas metal arc weld pool surface imaging: modeling and processing. Welding Journal, 2011, 90(5):85-94.
[3] Liu YK, Zhang YM. Model-based predictive control of weld penetration in gas tungsten arc welding, Control Systems Technology, 2014, 22(3): 955-966.
[4] Li XR, Shao Z, Zhang YM, et al.. Monitoring and control of penetration in GTAW and pipe welding, Welding Journal, 2013, 92(6):190-196.
[5] Wang ZJ, Zhang YM, Wu L. Adaptive interval model control of weld pool surface in pulsed gas metal arc welding, Automatica, 2012, 48(1):233-238.
[6] Wikle HC, Kottilingam S, Zee RH, Chin BA. Infrared sensing techniques for penetration depth control of the submerged arc welding process, Journal of Materials Processing Technology, 2001, 113(1): 228 -233 .
[7] Chandrasekhar N, Vasudevan M, Bhaduri AK, Jayakumar T. Intelligent modeling for estimating weld bead width and depth of penetration from infrared thermal images of the weld pool, Journal of the Brazilian Society of Mechanical Sciences and Engineering, 2015, 26(1):59-71.
[8] Lv N, Xu YL, Li SC, Yu XW. Automated control of welding penetration based on audio sensing technology, Journal of Materials Processing Technology, 2017, 250:81-98.
[9] Lv N, Zhong JY, et al.. Real-time control of welding penetration during robotic GTAW dynamical process by audio sensing of arc length, International Journal of Advanced Manufacturing Technology, 2014, 74(1-4):235-249.

Research on an Improved Sliding Extracting Algorithm of Composite Deviation for Wind Turbine Gear

Bai Zhang[1,2], **Haofang Liu**[1] **and Feng Gao**[1,*]

[1] State Key Laboratory of Power Grid Security and Energy Conservation (China Electric Power Research Institute), Beijing, China
[2] School of Electrical and Information Engineering, North Minzu University, Yinchuan, China

*Corresponding author email: gao_feng@nmu.edu.cn

Abstract. One tooth radial composite deviation and one tooth tangential composite deviation are important parameters of wind turbine gears whose operation life cycle can be influenced. An improved sliding extracting algorithm of composite deviation is proposed. It just only calculates those parts that need to be recalculated, avoids a large number of invalid searching calculations, and greatly improves the efficiency. The experimental results show that the improved sliding extracting algorithm of composite deviation has the same results as the traditional extracting algorithm, and the efficiency of the extracting algorithm is improved nearly 100 times. The improved sliding extracting algorithm of total composite deviation is proposed in this paper, and theoretically the algorithm efficiency is one time than traditional algorithm. The algorithms proposed in this paper can improve the efficiency of radial composite deviation and tangential composite deviation, and can be applied to gear single-flank rolling tester and gear double-flank rolling tester, especially in wind turbine gear field.
Keywords: Wind turbine gear; One tooth radial composite deviation; One tooth tangential composite deviation; Improved sliding extracting algorithm; Algorithm efficiency.

1. Introduction

With the rapid development of manufacturing industry, the quality requirements of industrial products are getting higher and higher. Gears as the most important transmission components, and higher and higher transmission efficiency and smoothness are required in the manufacturing industry. Wind power generation system has higher and higher requirements for the motion stability of gearbox, and the important index of stability evaluation for gear transmission is the gear comprehensive deviation [1-3]. Compared with single deviation measurement instrument, gear comprehensive deviation measurement efficiency is higher, especially gear single-sided meshing measurement is more close to the using status of gears, so it is widely used in rapid gears measurement field. Because the single-side meshing measurement process of gears is close to the using status of gears, and it can directly reflect the smoothness of the meshing process of gears. So it is widely used in the deviation measurement of gears field. Gear deviation data processing algorithm is a core part of gear measurement instrument software. So it is an important research content for gear researchers to research an efficient and stable deviation extracting algorithm. In this paper, the measurement principle of gear composite deviation is declared, the traditional extracting algorithms of gear composite deviation are analyzed, the new improved extracting algorithms are mainly designed and tested by simulation data. The experimental results are show that the efficiency of improved extracting algorithms of composite deviation are significantly higher than the traditional error extraction algorithms.

Journal of Physics: Conference Series **2218** (2022) 012046 doi:10.1088/1742-6596/2218/1/012046

2. Measurement Principle of Gear Composite Deviation

The gear composite deviation includes gear radial composite deviation and gear tangential composite deviation[4-6]. Because the deviation extraction algorithm of gear radial comprehensive deviation is similar to that of gear tangential comprehensive deviation, the deviation extraction algorithm researched in this paper is applicable to the deviation extraction of gear radial comprehensive deviation and gear tangential comprehensive deviation at the same time, and does not specifically refer to radial comprehensive deviation or tangential comprehensive deviation.

2.1. Measurement Principle of Gear Radial Composite Deviation

The gear total radial composite deviation is defined as the difference between the maximum and minimum center distance when the left and right tooth surfaces of product gear are in contact with the measurement gear at the same time in the radial (double-flank) composite deviation measurement instrument in Chinese national standard 10095.2-2008. The one tooth radial composite deviation is defined as the max radial composite deviation value corresponding to one tooth range (360°/z) when the product gear is rotated in a whole circle. The variation curve of the center distance of the product gear in whole cycle can be obtained by double-flank composite deviation tester. By calculating the maximum difference of the center distance from the curve, the total radial composite deviation can be obtained. The maximum difference of the center distance within the corresponding range of the tooth distance is the one tooth radial composite deviation. The definition of gear radical composite deviation is shown in figure 1.

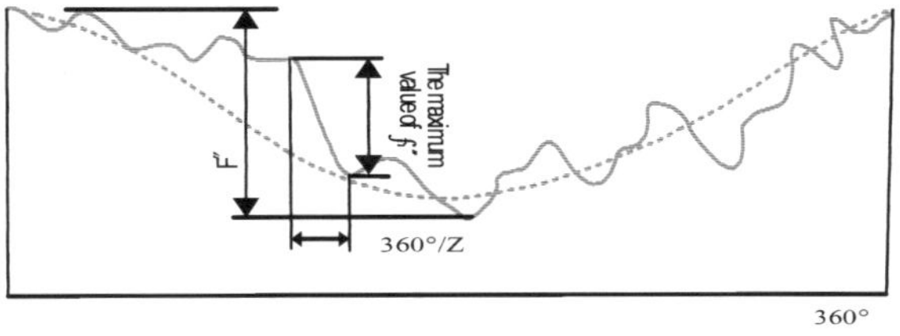

Figure 1. Definition of radial composite deviation of gear

2.2. Measuring Principle of Gear Tangential Composite Deviation

Single-flank testing of gear is a typical way to measure the transmission deviation of gears. The difference between the actual and theoretical rotation angles of the measured gears is defined as transmission deviation [4-7]. The measured gear is driven in single-flank meshing status with backlash under the nominal center distance by master gear (master gear can be replaced by standard worm, standard rack and other measuring elements). Deviation compensation method [8-9] and deviation measurement methodology [10-11] are proposed. The transmission deviation curve can be obtained from difference between the actual and theoretical rotation angles of the measured gears in the whole cycle, and the value of transmission deviation is calculate in reference diameter of measured gear. The difference between the maximum value and the minimum value on the transmission deviation curve is defined as the tangential composite total deviation . In the range of one tooth on the transmission deviation curve, the maximum distance between the vertex of the curve and the bottom of the curve is defined as one tooth tangential composite deviation. The definition of gear tangential composite deviation is shown in figure 2.

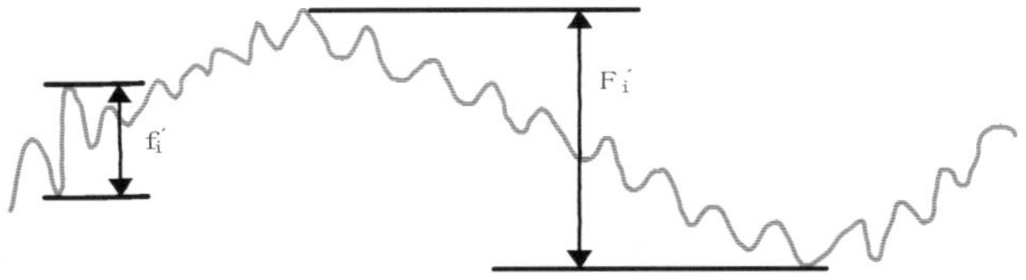

Figure 2. Deviation definition of gear tangential composite deviation curve

3. Extracting Algorithm of Gear Tangential Composite Deviation

3.1. Traditional Sliding Extracting Algorithm of One Tooth Composite Deviation

The traditional algorithm for extracting one tooth radial composite deviation and one tooth tangential composite deviation is similar to slide a ruler with fixed width on the composite deviation curve in turn, and then the maximum distance of whole composite deviation curve with the fixed width can be obtained. This algorithm is named as the traditional sliding deviation extraction algorithm of one tooth composite deviation in this paper. The flow chart of traditional sliding deviation extraction algorithm is shown in Figure 3. The number N and n is set measuring points and the number of tooth, so that corresponding measurement point number of one tooth to N/n can be calculated, the radial composite deviation measurement values are stored in an array. In the process of algorithm implementation, traditional nested loops are used, outer loops range is from 0 to , and inner loops range is from 0 to. A half-ruler sliding deviation extraction method was proposed by author in another paper, but this method is still in a nested loop structure, and the principle of the algorithm has not changed essentially [12].

3.2. Improved Sliding Extracting Algorithm of One Tooth Composite Deviation

The traditional sliding deviation extraction algorithm is deeply researched, and an efficient improved sliding deviation extraction algorithm is proposed in this paper, which significantly improves the computational efficiency. It is named as improved sliding deviation extraction algorithm. Instead of sliding measurement rulers one by one, according to the characteristics of the measurement data of the composite deviation curve, the algorithm decides whether need to compute measurement data of current ruler. It just only calculates those parts that need to be recalculated, avoids a large number of invalid searching calculations, and greatly improves the efficiency of sliding deviation extraction algorithm. The principle of the algorithm is shown in Figure 4. In the sliding process of the measuring ruler, Firstly, it is need to determine whether the sliding point which is shift-out point or shift-in point in the new ruler is greater than the maximum or minimum value, and if so, update the maximum or minimum value. Secondly, it is judged whether the sliding point is equal to the maximum or minimum value of the last ruler, and if so, it is necessary to recalculate the maximum or minimum value. The improved sliding error extraction algorithm eliminates a large number of invalid loop processes, and greatly improves algorithm efficiency.

3.3. Traditional Extracting Algorithm of Total Composite Deviation

The traditional algorithm for extracting total radial composite deviation and total tangential composite deviation is by searching maximum and minimum in the composite deviation curve with the fixed width, all data will participate in the searching algorithm, and then the maximum distance of whole composite deviation curve can be obtained. This algorithm is named as the traditional deviation extraction algorithm of total composite deviation in this paper. Because all data participate in the searching algorithm, so the algorithm efficiency can improve by reducing data in searching algorithm.

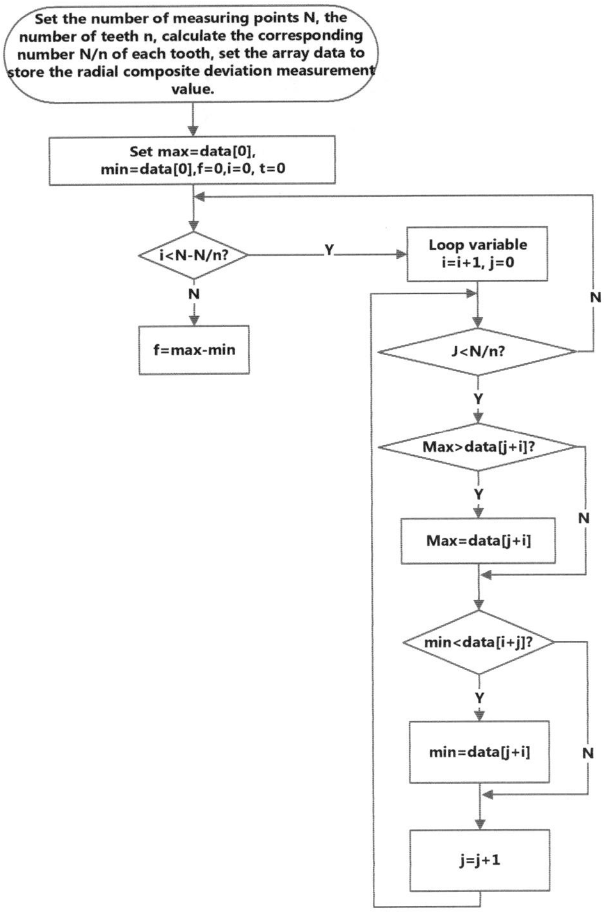

Figure 3. Flow chart of traditional sliding algorithm

3.4. Improved Extracting Algorithm of Total Composite Deviation

The traditional deviation extraction algorithm of total composite deviation is that all data are involved in the comparison, in order to increase algorithm efficiency, an improved deviation extraction algorithm is proposed in this paper, which significantly improves the computational efficiency. According to the characteristics of the measurement data of the composite deviation curve, firstly the algorithm decides to select odd measurement points or even measurement points to searching local maximum and minimum. When the local maximum and minimum of selected odd measurement points or even measurement points are obtained, then searching global maximum and minimum in the interval near the local maximum and minimum. The interval can be set up the measurement point number of three teeth, because of local extremum is close to global extremum, and it is impossible that global extremum span adjacent teeth of local extremum. It just only calculates half measurement data to search local extremum, and then searches global extremum in the small interval. It is just need to search global maximum in the small interval of local maximum, search global minimum in the small interval of local minimum. The improved deviation extraction algorithm of total composite deviation eliminates half number of measurement points, and improves one time algorithm efficiency.

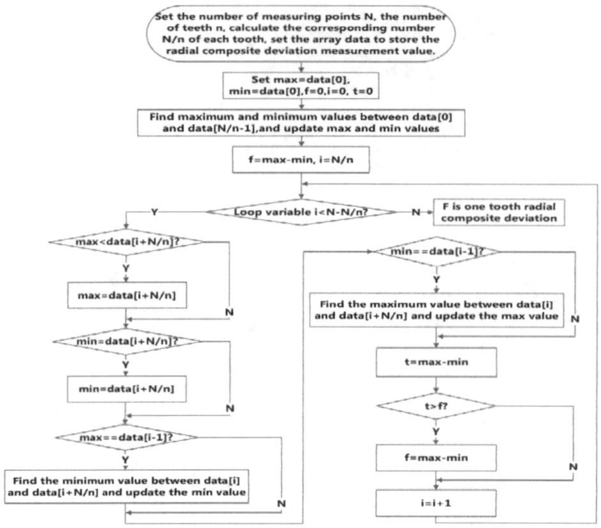

Figure 4. Flow chart of improved sliding algorithm

4. Algorithm Experiment

In order to verify the algorithm efficiency, a series of random numbers were generated and stored in an array to simulate the composite deviation measurements in the process of gear double-flank meshing or single-flank meshing. If the tooth number of gear is 20 and the measuring points of gear composite deviation curve were 10000, the measuring ruler width of each tooth is 500. The deviation extraction algorithms of the same simulation measurement data were carried out on a computer with CPU M520 and 4G memory in this paper. The results of deviation extraction algorithms of 1000 times were identical. The algorithm calculation times are shown in Table 1.

Table 1. Deviation extraction algorithm time-consuming results

The number of experiments	The time-consuming of traditional sliding deviation extraction algorithm (s)	The time-consuming of improved sliding deviation extraction algorithm (s)
1	17.431	0.105
2	23.534	0.107
3	15.744	0.99
4	15.642	0.111
5	18.770	0.120
6	16.160	0.95
7	16.991	0.107
8	23.227	0.133
9	26.149	0.167
10	16.474	0.106

5. Conclusion

The deviation extraction algorithm of radial and tangential composite deviations of one tooth of wind turbine gears is researched in this paper. Aiming at the problem of low efficiency of traditional sliding deviation extraction algorithm, an improved sliding deviation extraction algorithm is proposed in this paper. The efficient calculation of radial composite deviation and tangential composite deviation of one tooth are realized. It can not only ensure the correctness of the deviation calculation results, but also greatly improve the efficiency. The experimental results show that the efficiency of the improved sliding deviation extraction algorithm is more than 100 times than the efficiency of the traditional sliding deviation extraction algorithm. The improved sliding method proposed in this paper can be applied not only in the processing of one tooth radial composite deviation and one tooth tangential composite deviation, but also in the analysis and calculation of other similar data characteristics. The improved deviation extraction algorithm of total composite deviation is proposed in this paper, and theoretically the algorithm efficiency is one time than traditional algorithm. It has good application value and research significance.

Due to the limitation of experimental conditions, the data used in this paper are simulation data by random function, which can not be verified by the measurement data of actual gear single-flank rolling tester and gear double-flank rolling tester. The simulation data is a periodic sinusoidal signal (frequency is 1) superimposed with small attitude sinusoidal functions (frequency is tooth number), and then the random error is superimposed on the simulation data. The overall regular pattern of the simulation data is coincident to the regular pattern of the gear comprehensive error curve. After the relevant experimental instrument are developed later, the algorithm verification with the actual measurement data will carry out to further verify the efficiency and reliability of the algorithm. It should be added that there is little research on extraction algorithm of gear comprehensive deviation. At present, there is no research paper with higher efficiency than the improved algorithms in this paper.

Acknowledgments

Supported by Open Fund of State Key Laboratory of Power Grid Security and Energy Conservation (China Electric Power Research Institute) (No.FXB51201901066).

References

[1] Genshen LIU, Huaiju LIU, Caichao ZHU, et al. Design optimization of a wind turbine gear transmission based on fatigue reliability sensitivity[J]. Frontiers of Mechanical Engineering, 2021, 16(1):61-79.

[2] Yan Y. Load characteristic analysis and fatigue reliability prediction of wind turbine gear transmission system[J]. International Journal of Fatigue, 2020, 130(Jan.):105259.1-105259.9.

[3] Abdulwahed Salem, Ahmed Abu-Siada, Syed Islam. Improved condition monitoring technique for wind turbine gearbox and shaft stress detection[J]. IET Science, Measurement & Technology, 2017, 11(4):431-437.

[4] J Tang, Yang B, Shi Z. Influence on center distance by measurement force in double-flank gear rolling test[J]. Measurement, 2020, 168(1665):108321.

[5] Zhaoyao SHI, Xiaoning LU, Changhe CHEN, et al. Development of single-flank rolling tester for face gears[J]. Chinese Journal of Scientific Instrument, 2013, 34(12):2715-2721.

[6] Xie H K , Fu Y , Feng G . Differential Roll-Scanning Technology for Accuracy Measurement of Fine-Pitch Gears[J]. Advanced Materials Research, 2012, 472-475:3083-3088.

[7] Yin P, Han F, Wang J, et al. Influence of module on measurement uncertainty of gear tooth profile deviation on gear measuring center[J]. Measurement, 2021(2):109688.

[8] Hu Lin, Frank Keller, Martin Stein. Influence and compensation of CMM geometric errors on 3D gear measurements[J]. *Measurement*, 2020, 151.

[9] Zhi-ren Han, Wen-xin Liang, Zhen Jia, Ke Wang. Research on digital measurement error and rectification algorithm for mechanical parts[J]. *International Journal of System Assurance Engineering and Management*, 2017. 8(4): 747-761.

[10] Zhou Qi, Huang Han, Fan Qingrong, et al. Estimation of the instantaneous rotational frequency of gear transmission with large speed variations using short-time angular resampling and ridge-enhancing techniques[J]. *Measurement*, 2021, 172.

[11] Zhou Lihua, Fang Suping, Liu Yongsheng, et al. An alignment angle error compensation method of spiral bevel gear tooth surface measurement based on tooth surface matching[J]. *Measurement Science and Technology*, 2021, 32(10).

[12] Zhang Bai, Tan Wei, Pan Yuanqing, Yunhongyuan. Development of software for comprehensive error measurement of small modulus worm pairs [J]. *Mechanical transmission*, 2011, 35 (01): 62-65.

Design and Performance Analysis of Toothless Slip Type Magnetic Rotary Oil Well Pipe Tong

Jie Zheng[1,2,3,5,*], Cheng Bi[1], Heng Lu[1], Zhenzhen Li[2,3], Xu Yang[1], Yushan Yang[1], Yuan Lu[1], Yihua Dou[2,3] and Yarong Zhang[4]

[1]Xi'an Special Equipment Inspection Institute, Xi'an, Shaanxi 710065， P.R. China
[2]School of Mechanical Engineering, Xi'an Shiyou University, Xi'an, Shaanxi 710065, P.R. China
[3]Xi'an Key Laboratory of Wellbore Integrity Evaluation, Xi'an, Shaanxi 710055, P.R China
[4]School of Science, Xi'an University of Architecture and Technology, Xi'an, Shaanxi 710055, P.R China
[5]School of Power and Energy, Northwestern Polytechnical University, Xi'an, Shaanxi 710072, P.R China

*Corresponding author email: zhjoil@163.com

Abstract. Although the current hydraulic power pipe tong can improve efficiency and reduce strength, there are still some deficiencies. Among them, the inevitable defects are a low degree of automation, high labor intensity, complex mechanical mechanisms, and high maintenance costs. In order to improve its performance, the principle of permanent magnet motor is applied to the processing of shackle of oil pipe tong, and a new type of magnetic rotary oil pipe tong is proposed. The design eliminates the complex planetary gear train transmission mechanism and hydraulic circuit in the traditional tightening device, simplifies the mechanism of the tightening device, reduces the size of the device, saves the manufacturing cost, improves the automation degree of the power tong and reduces the labor intensity of the workers. The triangular chuck is suitable for clamping pipe strings with different diameters. In addition, due to the use of toothless slips, the damage of the tightening device to the sleeve is reduced, the sleeve is protected and the service life of the sleeve is prolonged. Therefore, this new type of magnetic rotary tubing tong has a good application prospect in the up and down operation.
Keywords: Magnetic rotation; Oil well pipe tong; Permanent magnet motor; Slip; Triangle chuck.

1. Introduction

At present, China's oilfield development has entered the middle and late stage, and the difficulty of oil extraction is getting bigger and bigger. With the severe changes in the downhole environment, oil fields have successively replaced gas wells produced by ordinary tubing in corroded areas with 13Cr anti-corrosion tubing. As the oil pipe made of 13Cr has higher requirements on torque and performance of hydraulic tongs. In the past, the hydraulic tongs did some damage to the tubing body in the process of making up, and the damaged part of the gas well in the production process may become the weak link of corrosion, and the anti-corrosion effect of the pipe string may be affected to some extent[1-4]. This requires us to change the previous process and avoid the damage of tubing as much as possible. Although the wide application of traditional hydraulic power tongs improves the efficiency and reduces the strength, it also brings a new problem[5-7] of accelerating the scratch of the outer wall of the oil pipe.

At present, the jaws of power tongs used in oil wells are easy to slip and bite the outer diameter of tubing[8-9]. Every time the common pipe tongs disassemble and assemble the pipe during operation, the surface galvanized layer at both ends of the pipe is completely destroyed, and it will be gnawed with deep bite marks. Sometimes, because the pipe tongs slip on the pipe wall, a layer of iron chips will be cut off or peeled off, and the pipe wall at the clamped place will be crushed. Because of the small contact surface, large radial force, and weak friction strength, the common hydraulic pipe tongs completely rely on the biting force of the tongs to complete the make-up process, which will cause the tubing deformation. When the torque is insufficient or the plier's teeth are damaged, it is easy to cause slippage and peeling. Ordinary hydraulic tongs are suitable for making up carbon steel tubing. Hydraulic tongs have high torque, and ordinary carbon steel oil pipes have low torque requirements, so hydraulic tongs will not cause too much damage to the oil pipe body. However, the corrosion-resistant alloy casing is soft, which will cause some damage to the tubing body when it is screwed up with ordinary hydraulic tongs. When the tubing is lowered into the well, especially the upper tubing receives a huge tensile force, and the damaged part becomes the weak point of corrosion, which easily causes the tubing to break off after long-term production. At present, most of the power tongs used in oil fields control the hydraulic reversing valve manually to control the start and stop of the power tongs. When working, the power tongs are usually suspended on a movable suspension beam through a chain, and the workers move to the casing, and the alignment of casing threads is adjusted by the workers' vision. The main tong and the back tong are often separated, which is inconvenient to use. Even if the back tong is equipped, the casing thread centering[10-12]is realized by the fixed mechanical connection between the main tong and the back tong.

To sum up, the current hydraulic power casing tongs have the following shortcomings: low degree of automation and high labor intensity; The replacement of forceps and the dental plate is time-consuming and laborious; The mechanical structure is complex, and the manufacturing and maintenance costs are high. When using, workers are too close to the construction site, and the safety guarantee is low; The power tongs use hydraulic power, which is bulky, inconvenient to move, complex and harsh on-site environment, and high maintenance cost of hydraulic equipment. Therefore, this paper relies on the advantages of traditional hydraulic power tongs, boldly innovates, subverts the structure of traditional pipe tongs, and combines the working principle of permanent magnet synchronous motor to provide a new type of magnetic rotating oil well pipe tongs to meet the working requirements of the shackle on pipe strings.

2. Magnetic Field 2D Finite Element Analysis

The 2D finite element model of the magnetic field of magnetic rotating oil well pipe tongs lies in XY two-dimensional plane of coordinate system. therefore, when analyzing the 2D finite element model, only the magnetic induction intensity $\boldsymbol{B}$ of the magnetic field of magnetic driving mechanism in XY plane needs to be analyzed, without considering the component of Z-axis, that is, only B_X and B_Y are studied. When analyzing the boundary conditions and results of the model, we only need to analyze its variation with X-axis and Y-axis, which has nothing to do with the Z-axis. In this paper, the 2D finite element method is used to analyze the magnetic field of magnetic rotating oil well pipe tongs, which mainly studies the influence of inner and outer rotor radius, inner and outer rotor thickness, air gap thickness, and magnetic torque. therefore, we can ignore the edge effect of the magnetic driving mechanism and assume that the magnetic driving mechanism is infinitely long, thus simplifying the model and reducing the amount of calculation, which is more convenient than a 3D model.

2.1. Magnetic Field Analysis of 2D Finite Element Model

The magnetic field can be regarded as a static magnetic field when analyzing the magnetic rotating oil well pipe tongs.

In the static magnetic field problem, $\frac{\partial \boldsymbol{E}}{\partial t} = 0$, have to $\frac{\partial^2 \boldsymbol{A}}{\partial t^2} = 0$, available

$$\nabla^2 \boldsymbol{A} = -\mu \boldsymbol{J} \tag{1}$$

In 2D finite element model, because only the component of magnetic induction intensity $\boldsymbol{B}$ in the XY plane is considered, the magnetic vector only has a Z component, so it can be $\boldsymbol{A}_Z$ Substitute into the above formula, that is

$$\nabla^2 \boldsymbol{A}_Z = -\mu \boldsymbol{J}_Z \tag{2}$$

After applying boundary conditions and solving them, the following formula can be used

$$\boldsymbol{B} = \frac{\partial A_Z}{\partial y} \boldsymbol{i}_x - \frac{\partial A_Z}{\partial x} \boldsymbol{i}_y \tag{3}$$

To obtain the magnetic induction intensity $\boldsymbol{B}$.

There is no original power supply for the spacer conductor in the magnetic rotating oil well pipe tongs, but only the velocity effect relative to the magnetic field. The eddy current density caused by the movement can be calculated by the following formula

$$\boldsymbol{J} = \boldsymbol{J}_e = \sigma[-\nabla \phi + \boldsymbol{V} \times (\nabla \times \boldsymbol{A})] \tag{4}$$

The above equation is the basic formula for analyzing the magnetic field in the magnetic rotating oil well pipe tongs, but it is also applicable to the steady magnetic field, alternating magnetic field, and magnetic field with velocity effect of moving conductor.

2.2. Fundamental Assumption

The following two assumptions must be made when conducting 2D finite element research on magnetic rotating oil well pipe tongs:

(1) The magnetic field distribution formed by the inner and outer magnetic rotors in the magnetic rotating oil well pipe tongs is actually a three-dimensional problem, and the magnetic torque is mainly transmitted through the air gap formed between the inner and outer magnetic rotors. It is assumed that the axial length of the inner and outer magnetic rotors is much larger than the size of the air gap formed by the inner and outer magnetic rotors. Only by this assumption can the end effect of the mechanism be ignored, and a three-dimensional distributed magnetic field problem can be transformed into a two-dimensional magnetic field for analysis.

(2) The yokes fixing the inner and outer rotors are both magnetizers, and the size is large enough, so there will be no magnetic saturation at the yokes. From the point of view of practical problems, the yokes are the boundary of the magnetic field, so it is assumed that there is no magnetic field outside the yokes fixing the outer and inner magnetic rotors.

2.3. 2D Modeling, Setting Up Physical Environment

The two-dimensional structure of magnetic rotating oil well pipe tongs is made of air, yoke, stator, air layer, air layer, permanent magnet, yoke, and air from inside to outside.

Permanent magnets have an 8-pole structure, that is, there are 8 permanent magnets on the inner and outer rotors. The materials of permanent magnets in magnetic rotating oil well pipe tongs are as follows Sm_2Co_{17}, the inner and outer radii of the inner rotor are 19 mm and 27 mm with a thickness of 8 mm; The inner and outer radii of the outer rotor are respectively 30.5 mm and 38.5 mm, with a thickness of 8 mm. Yokes are located outside the outer rotor and inside the inner rotor, which is made of silicon steel and has a thickness of 3 mm.

In this paper, the node method is used to analyze the static magnetic field, so the PLANE53 element is selected.

Define material properties for air, yoke, and permanent magnet respectively. The permanent magnet is samarium cobalt with relative permeability $\mu_r = 1.05$, magnetic induction coercivity $H_c = 750000 A/m$; With chromium-nickel alloy, relative permeability $\mu_r = 3$, resistivity is $1.62e-6\Omega \cdot m$.

The distribution of the magnetic rotors is that the magnetic poles are arranged at intervals, so radial magnetization is the most ideal magnetization direction. In this paper, 16 local coordinate systems with the same magnetization direction are established for 16 permanent magnets in the inner and outer rotors by using local coordinate systems.

2.4. Mesh Generation and Boundary Conditions

2.4.1. Mesh generation of 2D model. The modeling of magnetic rotary oil well pipe tongs adopts the PLANE53 element, which is an eight-node quadrilateral element structure with intermediate nodes, so the model is of second order. The mesh generation of the 2D model is quadrilateral mesh generation. It can be seen that the divided grid elements have different areas because they are not uniform elements, which can reduce the amount of analysis and be closer to the actual field strength distribution of materials.

2.4.2. Imposing boundary conditions. (1) the parallel condition of magnetic lines of force is imposed on the inner and outer boundary lines, that is, each node at the boundary $A_z = 0$;

(2) In ANSYS, permanent magnets are defined as components, and force marks are applied to them. Actually, only the magnetic interaction exists on the surface of the magnet, so only the magnetic interaction between the inner side of the permanent magnet on the outer rotor and the outer side of the permanent magnet on the inner rotor needs to be analyzed in the model analysis. According to the above, it can be inferred that in this model, the force arms of the inner and outer magnetic rotors are respectively 19 mm and 30.5 mm. After calculating the force and the force arm, the total torque on the inner and outer rotors can be obtained by the integration of the circumference. In the 2D model, the torque value per unit length of the total torque is N·m/m.

(3) Select all entities to solve the static magnetic field.

2.5. Post Processing Analysis of 2D Finite Element Model(Analysis of Magnetic Induction Intensity)

The results of the magnetic induction intensity $\boldsymbol{B}$ vector analyzed by ANSYS are shown in Fig. 1, and the circumferential distribution curve of air gap magnetic field intensity $\boldsymbol{B}$ at the outer diameter of the inner magnetic pole is shown in Fig. 2.

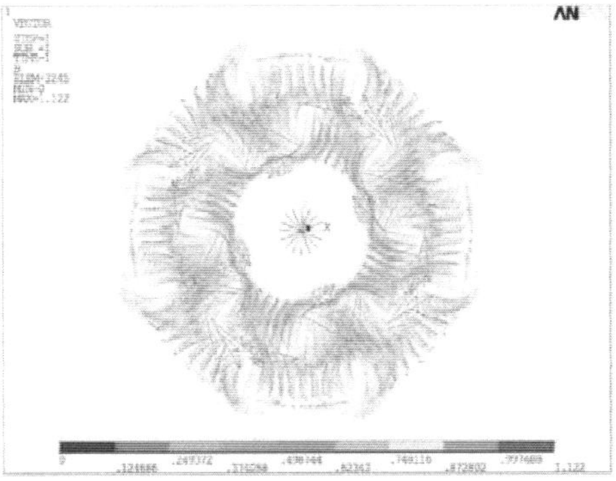

Figure 1. Vector Distribution of Magnetic Induction Intensity

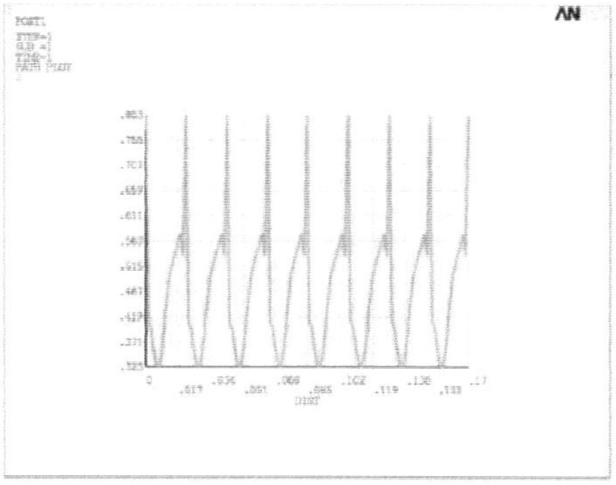

Figure 2. Circumferential Distribution Curve of Magnetic Field Intensity of Inner and Outer Diameter

The magnetic induction intensity distribution diagram shown in Fig. 2 is basically the same as the theoretical assumption. Because there is no passivation treatment at the junction of two permanent magnets, the magnetic induction intensity is large, as shown by the yellow line and the red line in the figure, and the maximum value is 1.122 T. It can be seen from Fig. 3 that the magnetic induction intensity at the inner and outer magnetic diameters is regularly distributed along the circumferential direction, and the maximum value is 0.803 T.

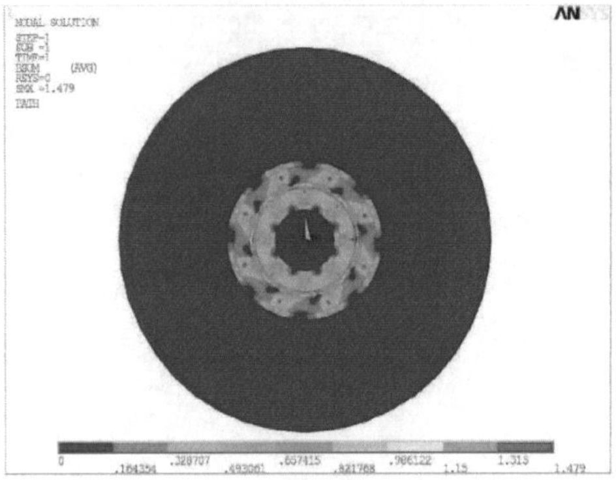

Figure 3. Absolute Value Distribution of Magnetic Induction

3. Control Theory

3.1. Development of AC Permanent Magnet Synchronous Servo System

In the field of AC servo drive, the commonly used AC servo motors mainly include induction asynchronous motors and permanent magnet synchronous motors (PMSM). Induction asynchronous motor is easy to manufacture, low in price and does not need special maintenance. After adopting vector control, the performance of the system is greatly improved, but the system is still relatively complex. Compared with an induction motor, a permanent magnet synchronous motor does not need excitation current, which can significantly improve power factor, reduce stator current and stator resistance loss, and has no rotor resistance loss in steady-state operation.

3.2. Simulation Model and Result Analysis

The vector control system simulation model of permanent magnet synchronous motor is established in Matlab/Simulink/SimpowerSystem environment, as shown in Fig.4.

Among them, the main parts are as follows: the given part, including speed and load torque, is realized by Constant and Step modules; The speed and current adjustment part are realized by PI module. Coordinate transformation part, CLARKE, PARK, and I_Park modules; Permanent magnet synchronous machine module. Module SVPWM realizes voltage space vector modulation; Universal Bridge, a voltage source inverter, is used in the inverter section.

The actual parameters of the selected permanent magnet synchronous motor are DC bus voltage V_{DC}=310 V, carrier period T=10 kHz, given speed n_{ref}=1000 r/min, and given speed as step signal. After debugging, PI parameters of speed loop P=6, I=3, PI parameters of current loop P=10, I=8; The simulation time is 200 ms, and the simulation waveforms are shown in Fig.5, which are respectively three-phase current response waveform, speed response waveform, and torque response waveform.

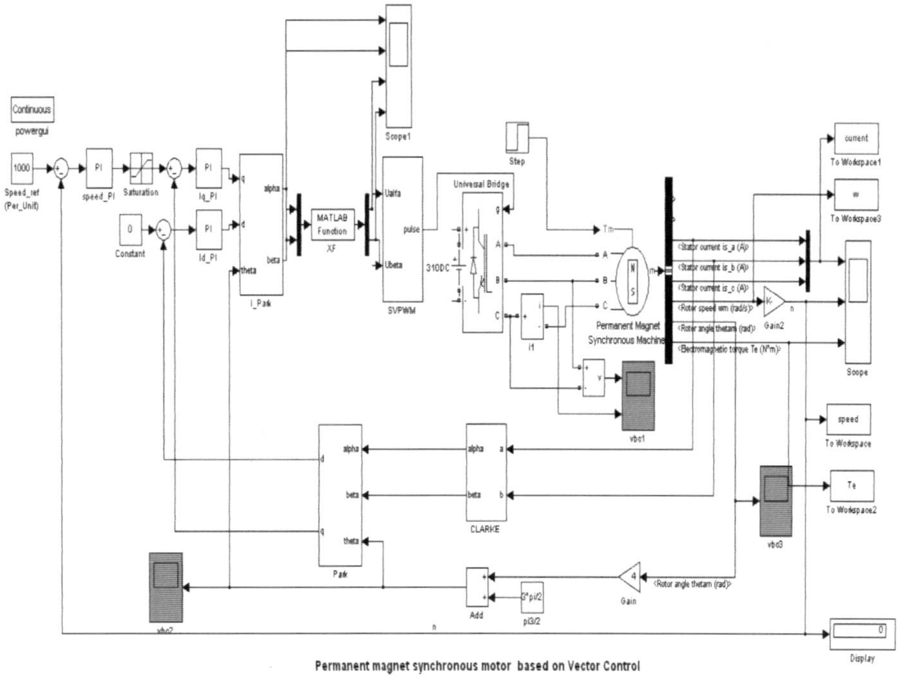

Figure 4. Simulation Model of PMSM Vector Control System

Among them, the main parts are as follows: the given part, including speed and load torque, is realized by Constant and Step modules; The speed and current adjustment part are realized by PI module. Coordinate transformation part, CLARKE, PARK, and I_Park modules; Permanent magnet synchronous machine module. Module SVPWM realizes voltage space vector modulation; Universal Bridge, a voltage source inverter, is used in the inverter section.

The actual parameters of the selected permanent magnet synchronous motor are DC bus voltage V_{DC}=310 V, carrier period T=10 kHz, given speed n_{ref}=1000 r/min, and given speed as step signal. After debugging, PI parameters of speed loop P=6, I=3, PI parameters of current loop P=10, I=8; The simulation time is 200 ms, and the simulation waveforms are shown in Fig.5, which are respectively three-phase current response waveform, speed response waveform, and torque response waveform.

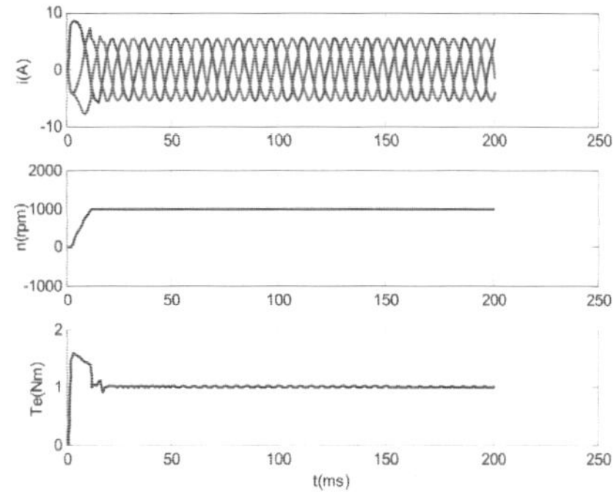

Figure 5. PMSM Three-Phase Current Response Waveform, Speed Response Waveform, Torque Response Waveform

It can be seen from the simulation waveform that the system has a fast response and small overshoot when the speed step is given. There is no static difference in the speed during steady-state operation, and the electromagnetic torque during startup is large, which has good static and dynamic characteristics.

4. Conclusion

The magnetic rotary oil well pipe tongs adopt a new structure of tongs teeth, use brand-new tongs teeth materials, and completely fit the surface of the pipe string so that the force acting on the pipe string is more uniform. It fundamentally changes the power transmission mode, makes the whole structure simpler, and greatly improves the mechanical efficiency of the power tongs. The volume of the power tongs is reduced, the automation degree of the power tongs is improved, and the labor intensity of workers is reduced. The designed new plier's teeth can effectively prevent the oil casing from being corroded because of reducing tooth marks and protecting the surface of the oil casing from being scratched and have a good application prospect in saving pipe resources.

(1) During the screwing process of the magnetic rotating oil well pipe tongs, the complicated planetary gear train transmission mechanism and hydraulic circuit in the conventional screwing device are omitted, the mechanism of the screwing device is simplified, the volume of the screwing device is reduced, the manufacturing cost is saved, the transmission principle of the upper shackle work is fundamentally changed, and new toothless tongs are designed to replace the conventional power tongs, thus reducing the damage of the screwing device to the casing, protecting the casing and prolonging the casing life.

(2) The clamp teeth of the electromagnetic power machine use the motor to rotate and control the triangle chuck to hold the pipe string, which is a one-to-one correspondence with the clamp teeth, tooth seat, and pipe diameter of the conventional power pliers, so it is suitable for pipe strings with different diameters. An end cover with a conductive groove is used as a control circuit for a triangular chuck motor, so that winding of the control circuit of a stepping motor in the process of screwing up and screwing off is avoided.

(3) The electromagnetic rotation principle is adopted to design, which breaks the limitation of the traditional hydraulic tongs, and can rotate forward, backward, or simultaneously, thus expanding the application range and improving the working efficiency.

(4) A simple closed loop is adopted to centrally control the whole screwing system, which realizes the automation of the screwing process and reduces the labor intensity of workers and the incidence of accidents. In addition, the number of screwing turns is accurately controlled, and the system has fast response speed, good stability, and high shackle quality.

Acknowledgement

This project was supported by the National Natural Science Foundation of China (No. 52004215, 12101482, 51674199).

References

[1] Fletcher S. Increased US workover-rig activity forecast [J]. Oil and Gas Journal, 2001, 99(5):45-46.

[2] Cassee, Udo, Kara, Danny T, Rock, Devin P, McLaughlin, Sean M. True hybrid operations combining coiled-tubing drilling and conventional rig workover techniques and practices [J]. SPE Drilling & Completion, 2006, 21(04):248-253.

[3] Pietras B G, Carlsson A, Schulze-Beckinhausen J E. WRENCHING TONG [J]. 2011.

[4] Hansen M D, Abrahamsen E. Improving safety performance through rig mechanization [M]. 2001.

[5] Henson, Jonathan. Safety on a drilling rig: Is it safety culture?[J]. Dissertations& Theses - Gradworks, 2013.

[6] Mansour H G, Ahmad M M, Ahmad H. A practical method for evaluating operational performance of workover activites in Sarir oilfield[C]// International Conference on Flexible Automation & Intelligent Manufacturing. 2012.

[7] Ribeiro G M, Desaulniers G, Desrosiers J. A branch-price-and-cut algorithm for the workover rig routing problem [J]. Computers & Operations Research, 2012, 39(12):3305-3315.

[8] Dayan de Castro Bissoli, Gisele de Lorena Diniz Chaves, Glaydston Mattos Ribeiro. Drivers to the workover rig problem [J]. Journal of Petroleum Science and Engineering, 2016, 139.

[9] Desrosiers D J. A branch-price-and-cut algorithm for the workover rig routing problem [J]. Computers & Operations Research, 2012.

[10] Fu H D, Yang W, Wang S G. Finite element analysis and optimal design for the derrick of workover rig [J]. Advanced Materials Research, 2013, 900-904.
[11] Gardner G B, Snisarenko P P. Mobility and speed in a workover/ completion rig for the arctic [C]// SPE/IADC Drilling Conference. 1989.
[12] Safar E, Al-Khatrash M, Al-Hazeem F. Rigless pre-workover diagnostic of a dual string completion [C]// SPE Russian Petroleum Technology Conference. 2018.

Automatic Detection of Avionics Electrical Connector Pins

Xiaolin Zhang[1,*], Hanqing Wu[1], Yanxiang Wu[2] and Lei Cao[2]

[1]School of Instrumentation of Science, Harbin Institute of Technology, Heilongjiang, China
[2]Capital Spaceflight Mechinery Company, Fengtai District, Beijing, China

*Corresponding author email: zhangxiaolin@hit.edu.cn

Abstract. Avionics electronic connectors have a wide use and play a significant role in instrumentation and aerospace system. To substitute manual detection, we propose an automatic in-station connector pin detection system based on computer vision, machine vision, deep learning, and industrial robots. This system provides a high-reliable, high-accuracy, traceable, quantifiable, and real-time detection for connector pins.
Keywords: Pin detection; CNN; Character recognition; Template matching.

1. Introduction

Spacecraft is a complex and sophisticated system, which is consisted of various circuit modules. Avionics electronic connectors, providing fast connection or disconnection for the ends of wires and cables, are vital interconnection components[1] to ensure the accurate transmission of power and signals. Being widely used in aerospace systems[2], any connector failure may cause safety problems and substantial economic losses of the whole system[3].

To prevent unreliable connections caused by deflection, bending, and damage of connector pins, it is necessary to detect in advance. Now, it is usually done manually. Though easy, manual detection highly depends on environment, work experience, and human physical condition. Thus, it leads to high cost, low reliability and efficiency, and inability to form a unified industry standard[4]. While, the development of online inspection technology has provided faster and more accurate parts inspection. It is foreseeable that automatic factories will be a major trend[5]. Following it, we propose an automatic in-station pin detection system integrating computer vision, machine vision, deep learning, and industrial robots.

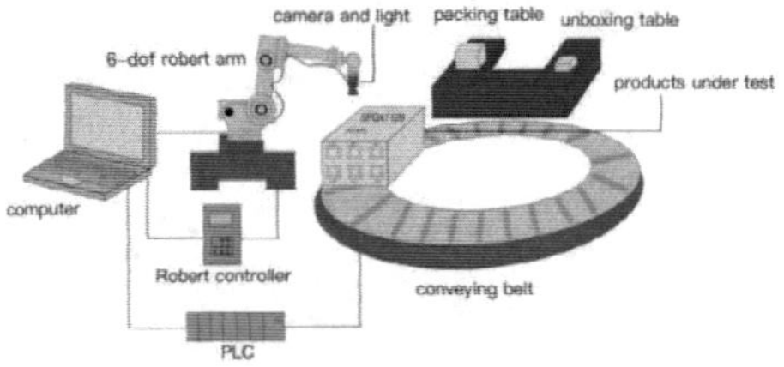

Figure 1. Schematic diagram of an automatic in-station pin detection system

2. System Introduction

This system includes functional modules above, and specific steps are described as follows:

1. The computer communicates with the belt through PLC and controls the start, stop, and speed of the belt to transmit the tested products to the testing area.

2. Guide the robotic arm equipped with an imaging system (including an industrial camera and a ring light source) to photograph products' upper surface for extracting information on it. Apply text recognition algorithm to acquire product model and then match it with template datasets.
3. Robot arm's joint motion is under the control of computer through industrial Ethernet. Based on the acquired product model, the arm is adjusted to specific areas for photographing each connector. Finally, judge pins and find out the defective pins via extracting feature points, registrating point sets, and comparing with template.
4. After completing the test, the conveyor belt transports the tested products into the sorting area and classifies them on the basis of judgment results. Then the robot arm is reset and waits for the next avionics electronic connector.

software		hardware
1.File management subsystem	5.Algorithm for judging pins' qualification	Image acquisition system
2.Image acquisition algorithm	6.Algorithm for 6-dof robot arm	6-dof robot arm
3.product model extraction algorithm	7.Algorithm for conveyor belt	conveyor belt
4.product model extraction algorithm	8.GUI: Graphical User Interface	main computer

Figure 2. System structure

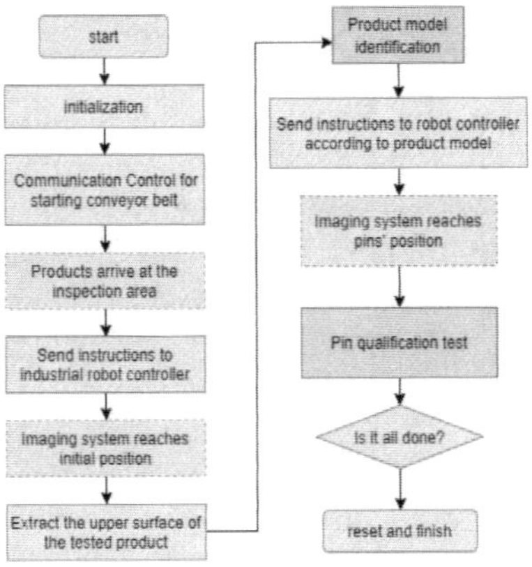

Figure 3. the system flowchart

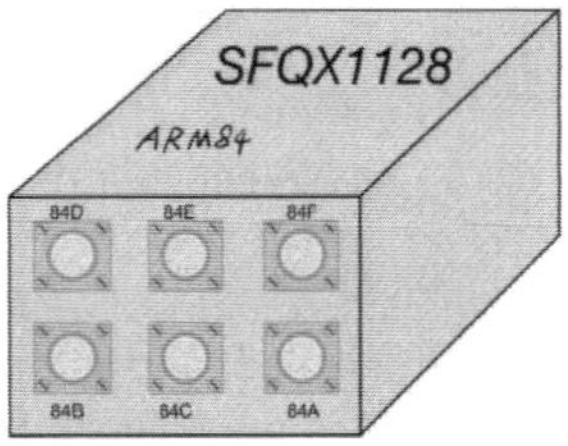

Figure 4. A connector box

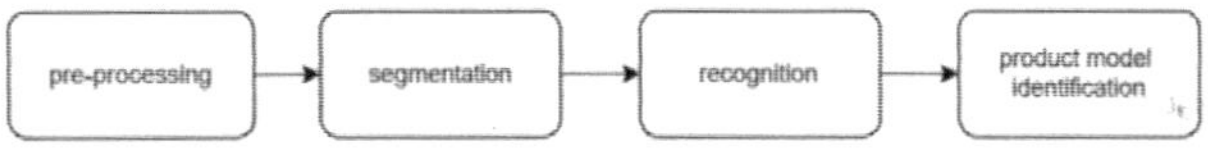

Figure 5. Flowchart of product model identification

3. Recognition of Product Model

Avionics electrical connectors have various types and complicated standards[8], which brings a huge workload and reduces detection efficiency. However, the connector boxes are affixed or engraved with organized serial numbers representing their product model (as shown in figure 4, its product model is SFQX1128 ARM84). If we extract its product model and then match it with preset templates, it will extremely reduce the workload for detecting connector pins. Therefore, we propose a recognition method (shown in figure 5) according to the shared product model characteristics of connectors: numbers and letters, handwriting and printing, imprinting and post-it notes on the box surface are mixed, black characters on a blue background, white on black and black on white are mixed.

3.1. Image Pre-processing

Considering obtained images mainly contain Gaussian noise, they are first subjected to mean filtering. Then the obvious color difference between character areas and the background can be well expressed through binarization. Since the equipped industrial camera and ring light source could provide a stable lighting condition, we select the OSTU method[6]with better speed. Considering the information posted on boxes in the form of a sticky note, we use find contours function to extract edge information and then only retain the character area.

3.2. Character Segmentation

In the application scenario, the product model is clearly written in standard (no overlap). Thus, a projection segmentation algorithm[7] is selected to calculate the number of black pixels in vertical and horizontal directions successively, and plot the corresponding histogram. The point where the trough value is zero is the interval between two characters. When characters are overlapped accidentally (presented as long blocks without zero points), they could be segmented according to pre-counted characters' length. After cutting, it is filled with padding and normalized to the same size.

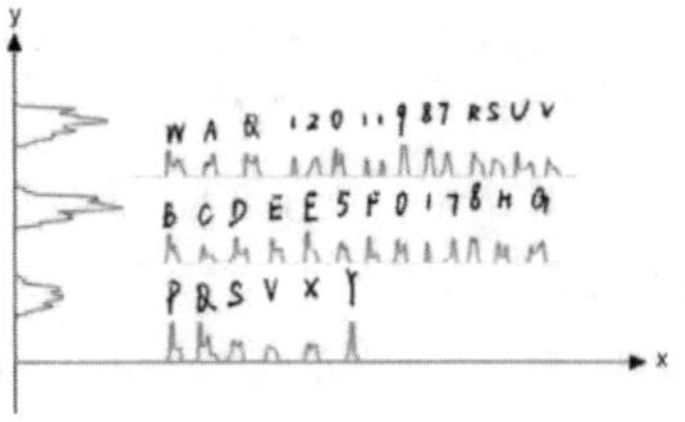

Figure 6. Projection segmentation algorithm

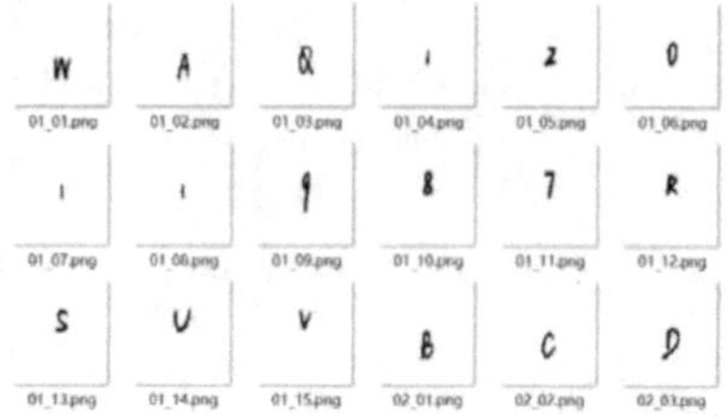

Figure 7. Normalized effect chart

3.3. Character Recognition

The hand-crafted feature extraction technique[8][9],such as random forests and support vector machines[10], not only requires high staffing but also lacks stability and universality. The developing deep learning field has been researching in automatically learning higher-level feature representations[11][12]. Such a method can quickly generate many valuable features that are difficult to create manually[13][14], enhancing charactor classifying performance. The convolutional neural network (CNN) can not only input two-dimensional images directly without changing its topological structure but also avoid the complex process of feature extraction and data reconstruction in the traditional recognition algorithms. To meet the needs of extracting high-level features and adapting to industrial characters with different fonts, our test recognition module bases on the CNN architecture in figure 8.

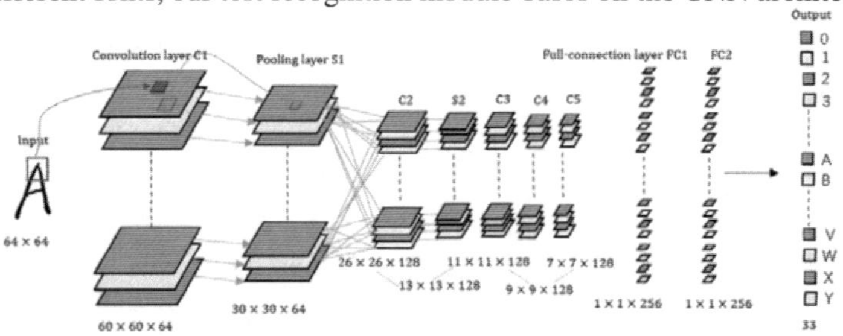

Figure 8. CNN architecture with convolutions, pooling and fully connected layers for softmax activation function

Since the letters O, I, and Z are highly similar in morphology to 0, 1, and 2 and number's using frequency is higher than letters in the encoding process, O, I and Z are discarded. Thus, the training database has 33 categories, and each contains 98 images (48 from handwritten character library EnglishHnd, 24 printed and 24 personal writing patterns). We trained it for 2000 epochs and the results are shown in figure 9.

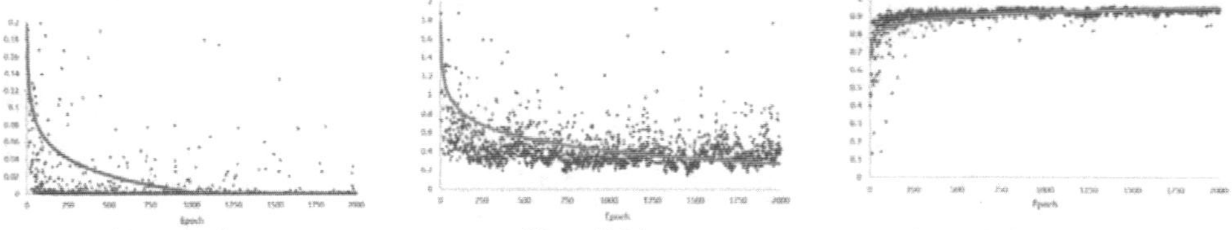

(a) train loss (b) valid loss (c) valid accuracy

Figure 9. Train results

3.4. Product Model Identification and Testing Results

Since we added some different fonts of characters into the test set to test the method's adaptability to various fonts, the recognizing accuracy of the printed characters is littler lower than that of printed characters.

Table 1. Testing results.

Testing set	Number	Letter	Handwritten character	Printed character
Recognizing accuracy	98.18%	96.39%	98.43%	97.50%

4. Detection of Pins

According to the matched template, the imaging system is adjusted to specific positions for photographing. After segmenting each electrical connector, pin detection is performed.

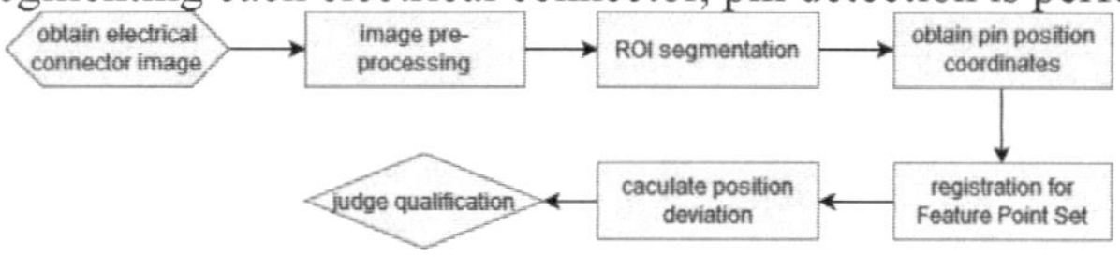

Figure 10. Processing flow of pin detection

4.1. Image Pre-processing

In order to highlight image features and facilitate subsequent processing, the pre-processing process includes filtering, binarization, dilation, and corrosion.

- A Gaussian filter $G(x) = \frac{1}{2\pi\sigma^2} e^{-\frac{(x-\mu)^2+(y-\gamma)^2}{2\sigma^2}}$ is required for image denoising.
- The Otsu method is used to segment the adaptive threshold
- Images after dilation and corrosion will have a smooth boundary which is helpful in extracting the edge contour

Figure 11. grayscale image (left) and image after pre-processing (right)

4.2. ROI Segmentation

Extract the outer contour pixels of the plug interface with findcontours function in OpenCV and establish the ROI mask. Apply the mask to the binarized image and its effect is shown below, accurately filtering out all the background and only retaining the interface area.

Figure 12. ROI mask (left) and segmentation results (right)

4.3. Obtain Pin Position Coordinates

Canny edge detection[15] is applied to extract edge contours, as shown in figure 13. Under the light source's illumination, a circular spot will form at the center of the pin. Hough circle transformation [16]was used to extract circles from ROI regions and their centers are regarded as pin's position. Constrain the Hough circle extraction process with the minimum distance between the center of each circle, the maximum and minimum radius of the circle.

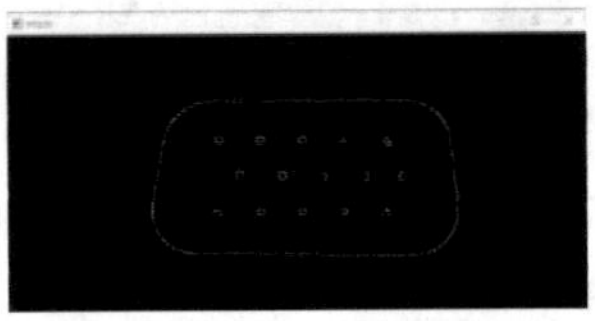

Figure 13. Edge contour extraction

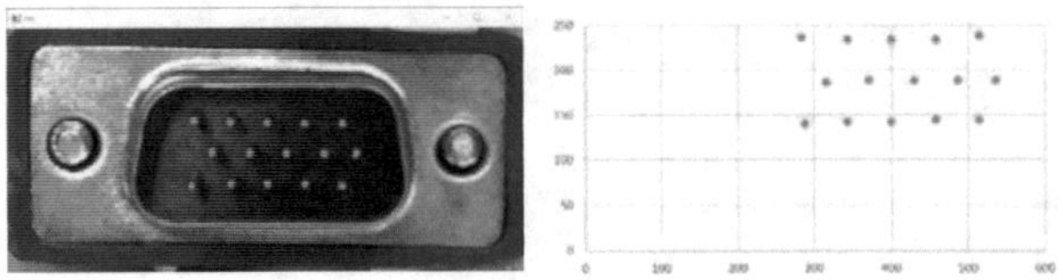

Figure 14. Pin position extraction effect

4.4. Registration for Feature Point Set

There are detailed series standard[17] of FDB1504-M0DS300K6KC, as shown below.

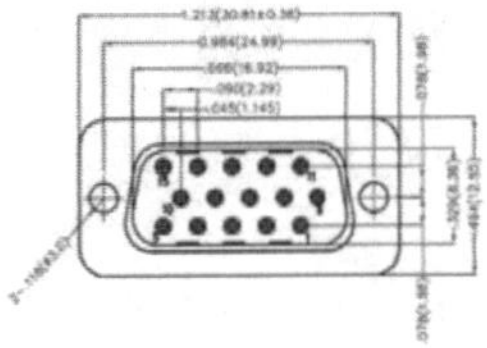

Figure 15. VGA pin position regulation

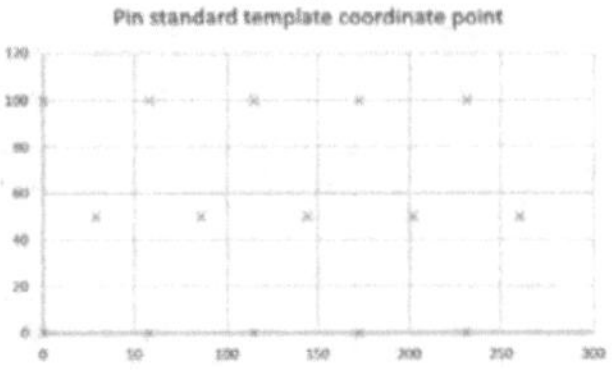

Figure 16. Standard template

After using the calibrated pixel equivalent as the weight calculation, the obtained pin coordinates and the template coordinates still have the relationship between translation and rotation changes. In order to calculate pins' position deviation, it is necessary to convert the coordinates in the image coordinate system to the world coordinate system. The conversion relationship of the coordinates can be obtained by the following algorithm.

a. **Normalization of the rotation center:** Because the pins are generally arranged in concentric circles, the least square method is used to fit the measured point set to a circle and take it as the rotation center. Then normalize the measured point set and move the rotation center to the origin for facilitating subsequent processing.

b. **Rotation Angle iteration:** The normalized set (x_i, y_i)has a rotation difference with the standard point set(u_i, v_i). We define their matching loss as: $L = \sum_{i=0}^{n}\sqrt{\frac{(x_i-u_i)^2+(y_i-v_i)^2}{n}}$. L is iteratively calculated by changing rotation angle and take that belongs to the minimum L as the result. To simplify calculation, a variable step search is used: set the step as 10°, iterate it in the interval$[0°, 360°]$and confirm the rotation angle(k); change step to 1°and iterate it in$[k - step, k + step]$, and so on. When the step is less than 0.01°, output it.

c. **Point set matching:** According to the rotation center and rotation Angle, match the tested pinpoint with the standard one. Take two times the average distance between the two sets as the judgment threshold, it is regulated that the pins exceeding it are unqualified.

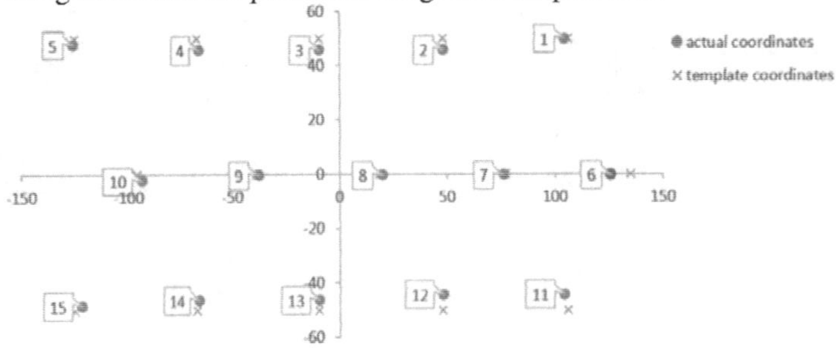

Figure 17. Effect of pin qualification judgment

After normalizing the rotation center of extracted actual pins' coordinates, we moved the two sets' center to the same origin. Then we iterate the rotation angle and calculated the matching loss between the two sets. According to the minimum matching loss, we finished the matching and began to calculate the average distance between the two sets. Taking twice the distance as the judgement, we judge that the pin 6 is unqualified.

5. Conclusion

Electrical connectors, playing a significant role in signal and power transmission, are widely used in aerospace system engineering. Based on this background, this paper designed a pin automatic detection system and described core software in details. The design of hardware structure and experiment results will be described in a later article.

References

[1] Fenwei Yang. Common fault inspection of aerospace electrical connector assembly[J]. ELECTROM ECHANICAL components,2005:40-52
[2] Xiaoyuan Li. Current situation and failure analysis of electrical connector standard[A]. China Standardization Association,2017:5.
[3] Yongying Du. Reliability Study on Terminal Contact Parts of Electrical Connector of JF Series[J]. Design and research,2015:11-15.
[4] Langming Zhou. Study on the Method of Vision Measurement and Reconstruction for mini(small) objects[D].WuHan University,2013.
[5] Songping Liu.Automated Non-destructive Testing Techniques and Its Applications[J]. Aviation manufacturing technology,2009(04):26-31.
[6] Lijing Tong, Comparison and an Analysis of Several Document Image Binarization Algorithms[J].J.North China Univ. Of. Tech.2011,23(01):25-33.
[7] Lingfeng Ran.An Algorithm of Characters Segmentation based on Vertical Projection for License Plate[J]. Communications Technology,2012,45(04):89-91+98
[8] Yokobayashi M , Wakahara T . Binarization and Recognition of Degraded Characters Using a Maximum Separability Axis in Color Space and GAT Correlation[C]// International Conference on Pattern Recognition. IEEE Computer Society, 2006.
[9] Weinman J J. Typographical Features for Scene Text Recognition[C] //Proceedings of the 20th IEEE International Conference on Pattern Recognition, Istanbul, Turkey, 2010.
[10] Machine Vision[J]. Metallurgical Industry Automation, 2013, 37(4): 12-15.
[11] Bengio Y. Learning Deep Architectures for AI[J]. Foundations & Trends in Machine Learning, 2009, 2(1):1-127.
[12] Song H A, Lee S Y. Hierarchical Representation Using NMF[C]// International Conference on Neural Information Processing. 2013.
[13] Lee H, Grosse R, Ranganath R, et al. Convolutional Deep Belief Networks for Scalable Unsupervised Learning of Hierarchical Representations[62].
[14] Bengio Y, Courville A, Vincent P. Representation Learning: A Review and NewPerspectives[J]. 2012.
[15] Canny J. A Computational Approach to Edge Detection[J]. IEEE Transactions on Pattern Analysis and Machine Intelligence, 1986, PAMI-8(6):679-698.
[16] D.H. Ballard.Generalizing the Hough transform to detect arbitrary shapes [J]. Pattern Recognition, 1981, 13(2):111-122.
[17] D-SUB Connectors 15 Circuits 180d Vertical Male[S].TXGA

A New Optimization Method for Pole-arc Parameters of Variable-gap Two-phase Switched Reluctance Motor

Ruihao Wang[1], Junling Huo[1], Xiaowei Pan[1], Qihui Wang[1] and Hongxing Wu[2,*]

[1] Department of Technical Engineering, Jinan Keya Election and Technology., LTD, No.1 Zidong Avenue, Tianqiao District, Jinan, China.
[2] School of Electrical Engineering and Automation, Harbin Institute of Technology, 92 West Direct Street, Nangang District, Harbin, China.

*Corresponding author email: whx0422@sina.com

Abstract. As a switched reluctance motor with simple structure, a two-phase switched reluctance motor (SRM) has the advantages of the above. In contrast, the complexity of the motor structure and the manufacturing cost are low, and only one-way operation is required. The control strategy is simple. Based on the analysis of the self-starting mechanism of the two-phase switched reluctance motor with variable air-gap rotor structure, the structure scheme of the variable air-gap rotor is determined by the finite element method. Two-dimensional model of two-phase switched reluctance motor with vertical air gap structure. To minimize the electromagnetic torque ripple, the structure parameters of the motor such as air gap, stator-rotor pole arc and winding turns are optimized. To solve the finite element optimization simulation time process, a numerical batch interpolation method is developed to optimize the non-uniform pole arc of the rotor to improve the finite element simulation speed. Finally, the fixed-rotor pole arc and winding turns of the motor are optimized. The motor parameters are further optimized to make the static performance of the motor more stable.

1. Introduction

As a new speed control system, switched reluctance motor has advantages of both DC and AC speed control systems. Comparing other AC and DC motors, the advantages are a simple and reliable power circuit, can work independently, high system stability, etc. Two-phase switched reluctance motors are a kind of less phase switched reluctance motors, which has typical double salient structure, simple motor structure and convenient control strategy, high reliability, and tolerance, has been concerned in various fields.

The structure and working principle of switched reluctance motors have a larger difference from traditional motors. In the reference [1], the unique double salient structure leads to the periodic change of the reluctance of the air-gap magnetic field. The motor's magnetic circuit is highly nonlinear, which makes the power supply current and voltage waveforms of each phase winding more complex. Therefore, the design method of switched reluctance motors is different from other motors. In reference to [2], the design method of switched reluctance motors, based on a genetic algorithm, is proposed. This design method is advantageous in solving nonlinear problems, and the switched reluctance motors are optimized based on this algorithm.

Document [3] optimizes the design of switched reluctance motor using genetic algorithm, establishes the optimization objective function and constraint inequality, obtains the expected optimization results in the process of motor instance optimization, and proves that the improved hybrid genetic algorithm has certain advantages in global accuracy and convergence speed, which is helpful for the optimal design of switched reluctance motor. In reference [4], the optimal switch reluctance motor design uses the two-dimensional finite element FEM numerical analysis method. Two optimal design methods are introduced: the minimum torque fluctuation design and the maximum average torque design methods. The performance of the designed switched reluctance motor is compared. Document [5] introduces the main methods to optimize the performance of switched reluctance motors. In the motor optimisation process, some variables that have a significant influence on the structure of the motor are defined as input variables, and other motor parameters must be optimized according to these input variables. In the literature [6], the pole arc parameters of the stator and the rotator have been taken as critical geometric parameters. The optimal design results of these parameters are obtained by finite element simulation. Reference [7] EFM and fuzzy optimization algorithm are used to optimize the motor's output torque, to improve the output waveform of the torque so that the switched reluctance motor can produce a more significant average torque. Reference [8] introduces a new type of switched reluctance motor with a 4/5 structure. To verify the performance of the motor, a simulation model of the switched reluctance motor is established in FLUX software. Finite element analysis of switched reluctance motor (SRM) is carried out by ANSOFT software, and the distribution of the electromagnetic field of the motor is observed. The electromagnetic field analysis of the motor is carried out [9] [10].

Based on the previous calculation of the structure parameters of the motor by magnetic circuit method, this paper establishes the simulation model of the traditional two-phase switched reluctance motor in FLUX software according to the main structural parameters of the motor. The working mechanism of self-starting of two-phase switched reluctance motor with variable air-gap rotator structure is analysed, and the structure scheme of the variable air-gap rotator is determined. A two-dimensional model of a two-phase switched reluctance motor with variable air gap structure was established by FLUX finite element method. Based on the minimum electromagnetic torque ripple, the structure parameters of the motor such as air gap structure, stator-rotor pole arc, winding turns were optimized. A method based on digital batch interpolation to optimize the non-uniform polar arcs of rotors is studied, which improves the speed of finite element simulation.

2. Mechanism Analysis of Two-dimensional

Compared with other numerical calculation methods, the two-dimensional finite analysis method is more accurate and faster. It can solve many complex problems, such as nonlinear problems, variable field calculation and complex boundary problems. The basic idea of the two-dimensional finite element analysis method is to use a series of small elements to divide the solution domain of the problem and then connect these small elements with nodes. The physical quantity inside the small cell can be calculated by the physical quantity of the small cell node through the selected functional relationship difference. The small element structure is usually relatively simple. It is easy to establish the equations between nodes from the balanced relationship and energy relationship. Then all the small element equations are combined to form an overall algebraic equation system, which can be solved through the boundary conditions, so the physical values on the nodes can be solved more accurately. Therefore, the two-dimensional finite element analysis method is a more effective method to obtain the magnetization curve of the motor.

According to the structural characteristics of a two-phase switched reluctance motor, the following assumptions are made: the current density on all wires is evenly distributed; Ignoring the external magnetic field of two-phase switched reluctance motor. The external magnetic field of the motor stator is zero; The permeability of the iron core is isotropic, that is, it has a single value B-H curve; Assuming that the iron core is infinitely long, the magnetic field is analysed in two dimensions.

For the two-dimensional plane field, let the current density J and the magnetic vector potential a be parallel to each other and parallel to the *Z-axis*, that is, $A_z = a, A_x = A_y, J_z = J_y = 0$; At the same time, the components of magnetic induction intensity B and magnetic field intensity H in the Z-axis direction are zero. The differential equation of the magnetic field is

$$\frac{\partial}{\partial x}\left(v\frac{\partial A}{\partial y}\right)+\frac{\partial}{\partial y}\left(v\frac{\partial A}{\partial y}\right)=-J \tag{1}$$

Where A is the value of the appropriate magnetic potential, the magnetic vector potential component in the Z-axis direction; J is the component of current density in the Z-axis direction. Suppose there is ferromagnetic material in the electromagnetic field because magnetic field intensity H and magnetic induction intensity B are nonlinear. In that case, the magnetoresistance is related to the coordinate system and magnetic induction intensity B.
For two-dimension magnetic field,

$$B=\nabla\times A=\frac{\partial A}{\partial y}i-\frac{\partial A}{\partial x}j \tag{2}$$

In the equation, i and j are the unit vector of X-axis and Y-axis. Therefore,

$$B=\sqrt{B_x^2+B_y^2}=\sqrt{\left(\frac{\partial A}{\partial y}\right)^2+\left(\frac{\partial A}{\partial x}\right)^2} \tag{3}$$

In practical electromagnetic calculation, for the differential equation of two-dimensional plane field, there are two kinds of boundary conditions to determine the unique solution, namely the first kind of boundary conditions and the second kind of boundary conditions

1) The first boundary conditions

The distribution of vector magnetic potential A on the boundary is known, which is,

$$A(x,y)=A_G(x,y) \tag{4}$$

Under special circumstances

$$A=0 \tag{5}$$

2) The second boundary conditions

The boundary is the symmetry plane of vector magnetic potential a, and the change rate of magnetic vector potential along the average direction outside the boundary is zero, i.e.

$$\frac{\partial A}{\partial n}=0 \tag{6}$$

3. Finite Element Modelling and Optimization Method of Variable Air Gap Two-phase Switched Reluctance Motor

The stator structure of switched reluctance motor remains unchanged, and the rotor adopts uniform variable air gap structure. The structure diagram of two-phase switched reluctance motor is shown in Figure 1. The technical indexes of the two-phase switched reluctance motor are shown in Table 1.

Table 1. Technical specifications of two-phase switched reluctance motor

Parameters	Values	Parameters	Values
Motor Structure	4/2	Motor Phases	2
Rated Power	1.5kW	Rated Voltage	250V(DC)
Rated Speed	2500 r/min	Motor Efficiency	84%

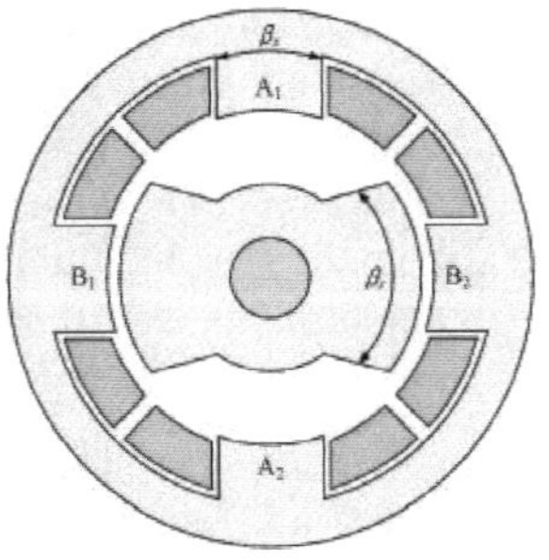

Figure 1. Structure diagram of two-phase switched reluctance motor

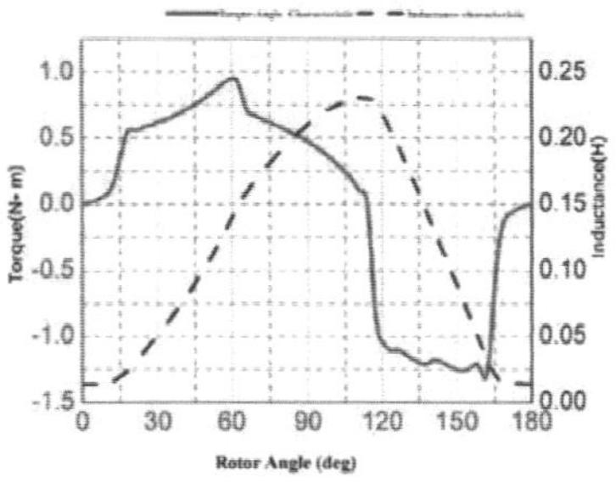

Figure 2. Curve of Moment Angle and Inductance Characteristics for Uniform Variable Gap

The rotor pole arc is bounded by the central axis of the rotor pole. The left side is the uniform variable air gap pole arc, and the right is the uniform air gap pole arc. Only the position points at both ends are set at the variable air gap side, and the points at both ends transition evenly with a smooth arc. The minimum air gap is 0.30mm, and the maximum air gap is 0.38mm. On the uniform air gap side, the air gap width is 0.30mm. The simulation model of the motor is established in the flux software, and the torque angle characteristic curve and inductance characteristic curve of the motor is obtained through simulation, as shown in Figure 2.

As shown in Figure 2, when the rotor angle is from 0 ° position angle to 110 ° position angle, the inductance shows an upward trend and generates positive torque. From 110 ° position angle to 180 ° position angle, the inductance gradually decreases and generates negative torque; In the positive torque area, the rotor position angle increases from 0 degrees to 60 degrees, the torque reaches the maximum value of 2.78n · m at 60 degrees. Then the torque gradually decreases to 0.

3.1. Structure Analysis of Non-uniform Variable Gap Rotor

The pole arc of the rotator is also bounded by the central axis of the rotator pole, with a non-uniform variable gap pole arc on the left and uniform gap pole arc on the right. Unlike uniform variable gap structure, non-uniform variable gap structure sets 25 locations at equal distances on the variable gap side to fine-tune the local gap width. The air gap variation is no longer uniform, with the maximum gap 0.40mm and the minimum gap 0.31mm. On the uniform gap structure side, air gap width is 0.30mm. Local air gap width can be adjusted by fine-tuning different locations on the variable air gap side. The motor stator structure is unchanged, and the rotor structure is a non-uniform variable air gap structure, as shown in Figure 3.

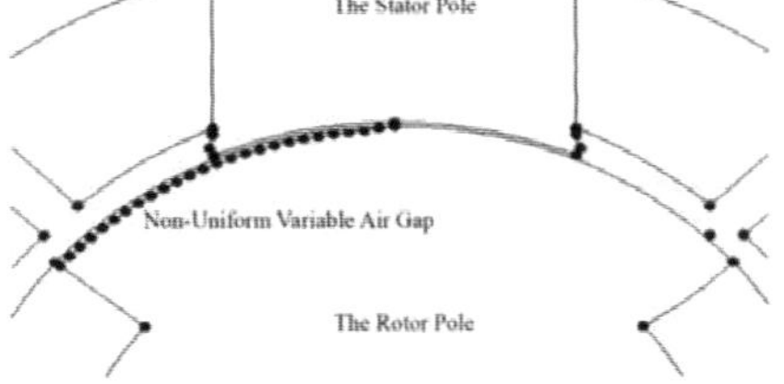

Figure 3. Structure diagram of non-uniform variable air gap rotor

The optimum design method of two-phase switched reluctance motor with non-uniform variable air-gap rotor structure is as follows:

- Firstly, the 25 position points on the variable air gap side are arranged into position sequences, numbered A1 to A25 from the central axis of the rotor pole. The air gap structure of the motor can be adjusted by adjusting the value of the air gap width corresponding to each position from A1 to A25.
- Then, by optimizing the gap width values corresponding to each location from A1 to A25 in turn, the air gap structure of the motor is precisely optimized to obtain a stable and smooth

output of the torque, which minimizes the torque pulsation. Finally, the structure optimization model of the asymmetric variable-gap rotors of the motor is obtained.

3.2. Optimization of Air Gap Parameters Based on Digital Batch Interpolation

Because the traditional finite element optimization simulation takes a long time to improve the efficiency and reduce the simulation time, this paper presents a digital batch interpolation method to optimize the non-uniform polar arc of the rotator to improve the finite element simulation speed. The optimization design process based on the digital batch interpolation method is as follows:

Firstly, a small number of location points are set at the uneven variable gap side of the motor rotator, and the gap width at the first set of location points is adjusted by finite element simulation to optimize the motor's output torque to reduce the torque pulsation. Then, the second set of location points are inserted between the two locations with large torque pulsation, and the gap width at the second set of location points is optimized by simulation, in turn. Finally, through the batch interpolation optimization of 25 locations, the final optimization results of the motor structure with non-uniform variable air-gap rotors are obtained.

By changing the gap width value corresponding to position A1 in a particular range, the moment angle characteristics of the motor under rated conditions are simulated and analyzed, and the moment angle characteristics curve of the motor is obtained, as shown in Figures 4 and 5.

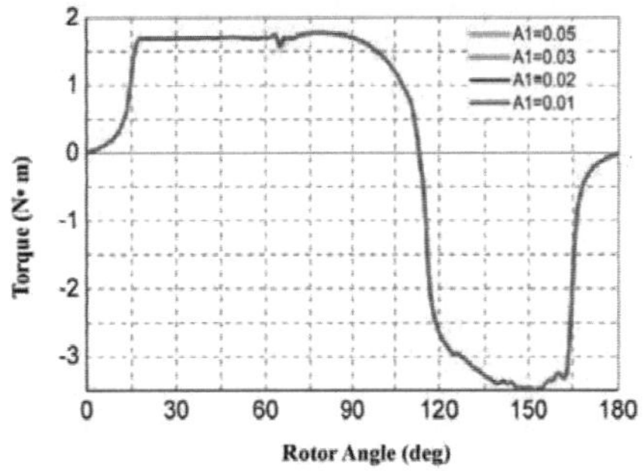

Figure 4. Optimization of air gap width at non-uniform variable air gap position

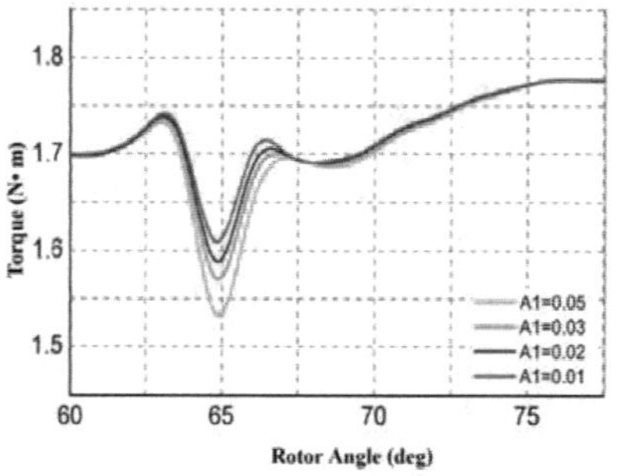

Figure 5. Enlarged drawing of 65-degree torque angle characteristic curve of rotor

As shown in Fig. 4 and 5, the change of air gap width at position A1 only affects the torque waveform near the 65-degree position angle of the rotor. With the increase of A1, the torque pulsation near the 65-degree angle increases gradually. When A1=0.01, the maximum torque of the rotor at the 65-degree position angle is 1.61N_m. Therefore, to reduce the torque pulsation and obtain the ideal output torque waveform, select position A1. The result of optimizing the corresponding gap width is 0.01 (mm). By optimizing the gap width value at each location by the above method, the final optimizing result of non-uniform variable gap rotator structure of the motor is obtained, which is shown in Table 2.

Table 2. Structural optimization results of non-uniform variable air gap rotor

Position Point	A1	A2	A3	A4	A5	A6
Air Gap Increment (mm)	0.010	0.138	0.320	0.320	0.350	0.360
Position Point	A7	A8	A9	A10	A11	A12
Air Gap Increment (mm)	0.365	0.370	0.372	0.379	0.380	0.385
Position Point	A13	A14	A15	A16	A17	A18
Air Gap Increment (mm)	0.385	0.390	0.390	0.390	0.395	0.395

Position Point	A19	A20	A21	A22	A23	A24	A25
Air Gap Increment (mm)	0.400	0.400	0.400	0.400	0.400	0.400	0.400

Based on the results shown in Table 2, the fitting curves of the gap increment corresponding to the different locations are made. As shown in Figure 7, the curve of the gap increment along the variable gap side of the motor rotator with the position angle is obtained. The fitting function is as follows:

$$g(\theta) = 0.00314 + 0.10585\theta - 0.01198\theta^2 + 0.000067\theta^3 \tag{7}$$

Where theta is the position angle, that is, the angle between the calculated position point and the axis line and the central axis of the rotor pole, and g(theta) is the gap increment at the position angle theta.

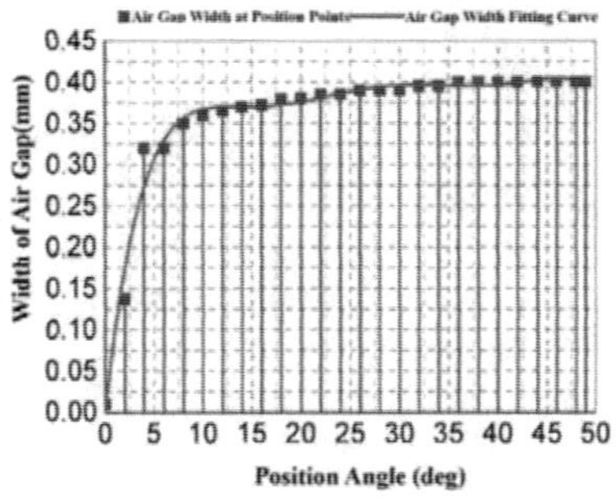

Figure 6. Non uniform variable air gap width and fitting curve

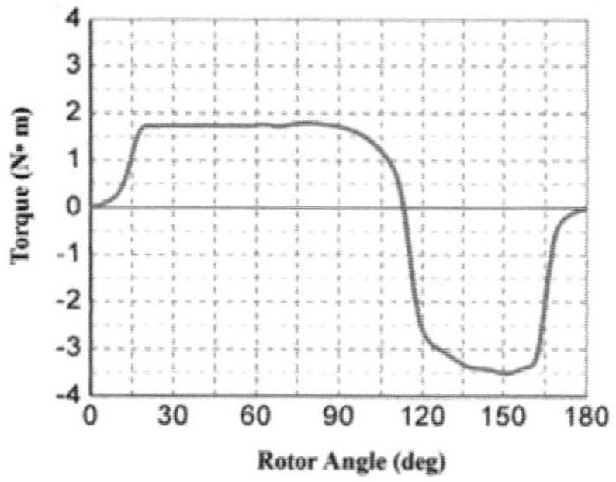

Figure 7. Torque angle characteristic curve at rated current after optimization

Based on the gap width at different locations shown in Table 2, the final optimum model of the motor can be established, and the winding current is set to 5.5A. The optimum moment-angle characteristic curve of the motor can be obtained. As shown in Figure 7, the analysis shows that the torque characteristic of the motor has been improved to some extent.

4. Optimization of Motor Structure Parameters

The following further optimizes the motor structure parameters, such as the stator-rotor pole arc width and winding turns, to improve the static characteristics of the motor and reduce the torque pulsation. The static characteristics of the motor are analysed by FLUX software, and the above structure parameters of the motor are adjusted appropriately to make the structure design of the motor more reasonable and the working performance more stable.

4.1. Polar Arc Width Optimization of Stator

Set the winding current to 5.5A and the motor operates in two phases. The effect of changing the stator pole arc width on the motor output torque is studied when the width of the pole arc of the rotator is fixed. Based on the analysis of the motor torque characteristics, the effects of changing the stator pole arc width on the motor torque characteristics are studied by simulation respectively when the width of the pole arc of the rotator is 98 degrees and 100 degrees. The simulation results are as follows, figures 8 and 9 show the calculation of maximum T_{max}, minimum T_{min}, average T_{av} and torque pulsation K_{ripple}. The results are shown in Table 3.

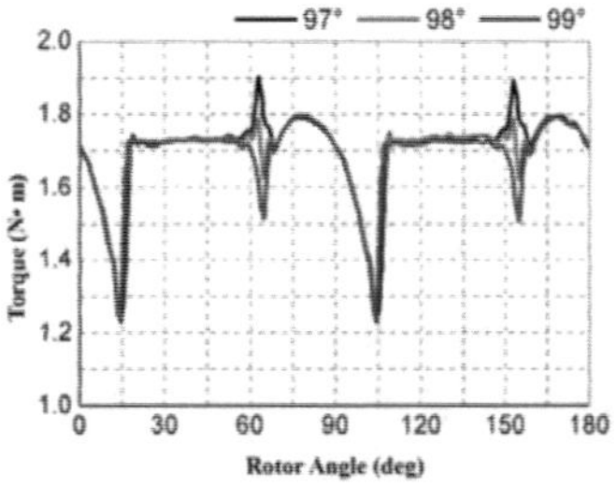

Figure 8. Torque characteristics of 98 ° rotor pole arc

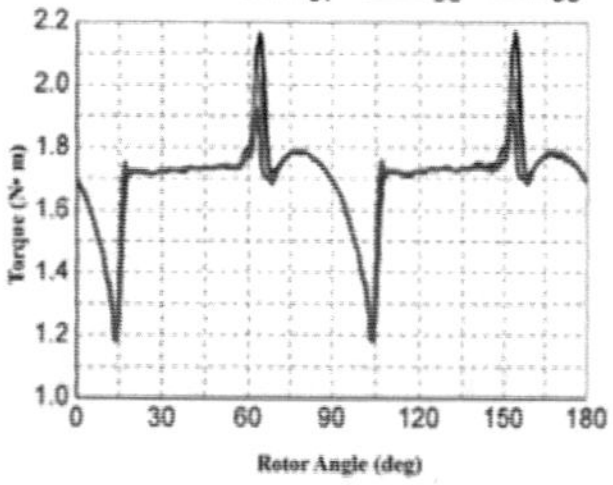

Figure 9. Rotor pole arc 100 ° torque characteristics

Table 3. Influence of stator pole arc on torque characteristics

Rotor Pole Arc (°)	98			100		
Stator Pole Arc (°)	45	46	47	45	46	47
T_{max}(N·m)	1.801	1.794	2.032	1.785	1.773	1.768
T_{min}(N·m)	1.256	1.235	1.181	1.061	1.229	1.183
T_{av}(N·m)	1.676	1.698	1.719	1.647	1.653	1.662
K_{ripple}(N·m)	0.326	0.329	0.495	0.440	0.329	0.352

- When the pole arc width of the rotor is 98 degrees, it can be seen from Fig. 8 and Table 3 that changing the pole arc width of the stator will not change the general trend of the torque characteristic curve, but will affect the torque pulsation of the motor, which is reflected in the sudden decrease or surge of the torque at the 63-degree position angle of the rotor. When the width of the stator pole arc is 45 degrees, the torque at position angle 63 degrees produces a sudden drop, which affects the minimum torque T_{min} of the motor and increases the torque pulsation. When the width of the stator pole arc is 47 degrees, the torque at position 63 degrees of the rotor position increases sharply, which affects the maximum torque T of the motor. Max, increases the torque pulsation; when the stator pole arc width is 46 degrees, the motor output torque is stable, the torque pulsation is small, and the motor performance is stable.
- When the width of the pole arc of the rotator is 100 degrees, its torque characteristics are shown in Figure 9. Similar to 98 degrees of the pole arc width of the rotator, changing the width of the stator pole arc only affects the change of the motor's torque at the position angle of 63 degrees of the rotator, and the change trend is the same. When the sub pole arc width is 47 degrees, the motor output torque is stable, the torque pulsation is small, and the motor performance is stable.

4.2. Rotor Polar Arc Width Optimization

Setting the winding current to 5.5A and the motor operates in two phases, the simulation method is similar to that when the width of stator pole arc is studied. Based on the analysis of motor torque characteristics, the influence of changing the width of stator pole arc on motor torque characteristics is studied by simulation respectively when the width of stator pole arc is 46 degrees and 47 degrees. The simulation results are shown in Figs. 11 and 12, and the maximum value of torque T_{max} is calculated., minimum T_{min}, average T_{av}, and torque pulsation K_{ripple}. The results are shown in Table 4.

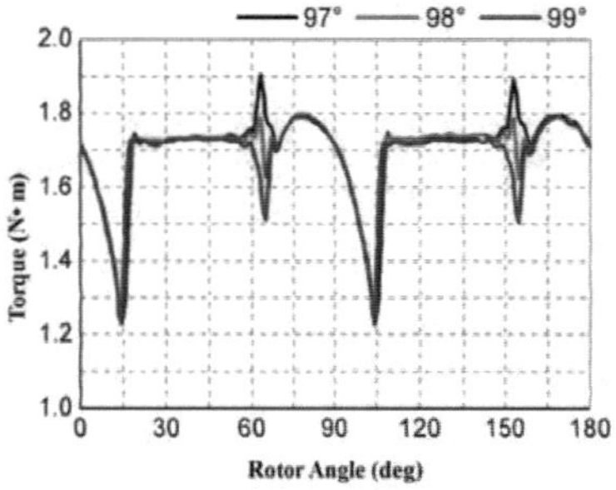

Figure 10. Torque characteristics at 46 ° stator pole arc

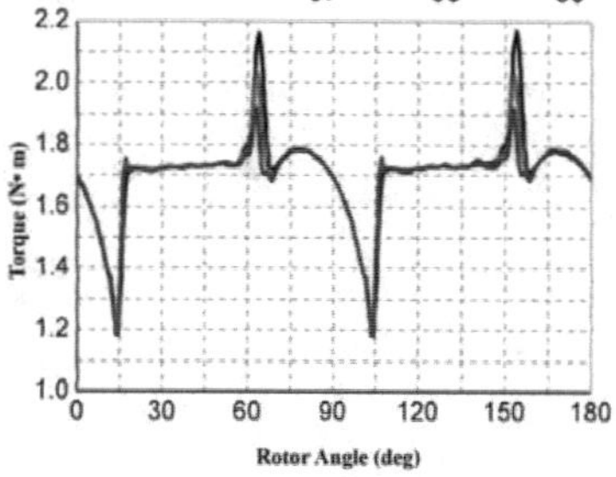

Figure 11. Torque characteristics at 47 ° stator pole arc

Table 4. Influence of rotor pole arc on torque characteristics

Stator Pole Arc (°)	46			47		
Rotor Pole Arc (°)	97	98	99	97	98	99
T_{max}(N·m)	1.903	1.794	1.797	2.118	2.031	1.921
T_{min}(N·m)	1.228	1.235	1.239	1.184	1.182	1.185
T_{av}(N·m)	1.702	1.698	1.687	1.710	1.705	1.697
K_{ripple}(N·m)	0.397	0.329	0.331	0.546	0.498	0.434

When the width of the stator pole arc is 46 degrees, it can be seen from Fig. 10 and Table 4 that changing the width of the rotor pole arc will not change the general trend of the torque characteristic curve, but will affect the motor's torque pulsation, which is reflected in the sudden decrease or surge of the torque at the 63 degree position angle of the rotor. Near the 63 degrees of the rotor position angle, the torque decreases with the increase of the width of the rotor pole arc, and the decrease amplitude is more obvious. When the pole arc width of the rotor is 97 degrees, the torque at 63-degree of the rotor position angle increases sharply, which affects the maximum torque T_{max} of the motor and increases the torque pulsation. When the pole arc width of the rotor is 99 degrees, the torque at 63 degrees of the rotor position angle decreases sharply and the torque pulsation increases. When the pole arc width of the rotor is 98 degrees, the output torque of the motor is stable, the torque pulsation is small, and the working performance of the motor is stable.

When the width of the stator pole arc is 47 degrees, the torque characteristics are shown in Fig.11. Similar to 46 degrees of the stator pole arc width, changing the width of the stator pole arc only affects the change of the motor's torque at the 63-degree position angle of the rotator, and the change trend is the same. When the width of the rotator pole arc is 97 degrees, the torque at the 63-degree position of the rotator increases sharply, and the increase is 46 degrees greater than that at the width of the stator pole arc. Large, affects the maximum torque T_{max} of the motor, increases the torque pulsation; When the pole arc width of the rotor is 98 degrees, 99 degrees, the torque at 63 degrees of the rotor position angle also increases sharply, affects the maximum torque T_{max} of the motor, and increases the torque pulsation.

4.3. Effect of Winding Turns on Motor Output Torque

Set the winding current to 5.5A and the motor operates in two phases to study the effect of winding turns on the output torque of the motor. Based on the analysis of the motor's torque characteristics, the changes of the motor's torque characteristics when the winding turns change are studied by simulation. The simulation results are shown in Fig. 12. The maximum and minimum values of Tmax, the minimum values of Tmin, the average value of T_{av} and the torque pulsation K_{ripple} are calculated. The results are shown in Table 5.

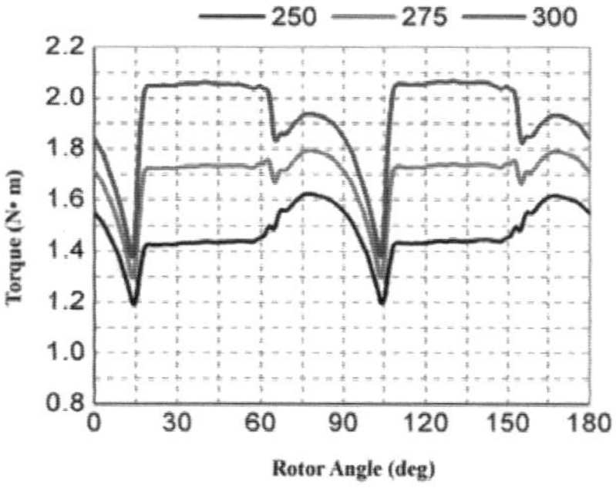

Figure 12. Influence of winding turns on torque

From Figure 12 and Table 5, it is known that the motor torque characteristic curve moves as the number of winding turns increases in a certain range. When the angle of the rotor position is fixed, the torque increases with the number of winding turns, and the maximum and minimum torque values and average torque increase with the increase of winding turns. However, due to the constraints of other structural parameters of the motor, the number of winding turns is limited to a certain range, and the motor transmission output Torque will also be limited.

Table 5. Influence of winding turns on torque characteristics

Winding turns	250	275	300
T_{max}(N·m)	1.625	1.794	2.073
T_{min}(N·m)	1.152	1.235	1.300
T_{av}(N·m)	1.469	1.698	1.928
K_{ripple}(N·m)	0.322	0.329	0.401

5. Conclusion

Based on the traditional two-phase switched reluctance motor simulation model established in this paper, a numerical batch interpolation method is studied to optimize the non-uniform pole arc of the rotor. The analysis shows that the method optimizes the non-uniform pole arc gap parameters of two-phase switched reluctance motor and improves the finite element simulation speed by adjusting the optimization accuracy of the motor. The gap structure of the motor and the optimization model of the variable air gap two-phase switched reluctance motor are established. Finally, the variable air gap rotator structure is selected as the optimal structure of the two-phase switched reluctance motor, and the working characteristics of the motor optimization model are analysed. The motor structure parameters, such as the pole arc width of the stator and the number of windings turns, are optimized to make the performance of the motor more stable and reduce the torque pulse.

Acknowledgment
In this paper, the research was sponsored by Natural Science Foundation of China (Project No. 516770385) and China Postdoctoral Science Foundation Funded Project.

Reference

[1] C Choi, K Park. Fuzzy Design of a Switched Reluctance Motor Based on the Torque Profile Ptimization[C]. IEEE International Magnetics Conference, 2000, 36(5):256-256.
[2] DH Lee, JW Ahn. Performance of High-Speed 4/2 Switched Reluctance Motor [J]. Journal of Electrical Engineering & Technology, 2011, 6(5): 640-646.
[3] ES Afjei, HA Toliyat. A Novel Multilayer Switched Reluctance Motor[C]. IEEE Power Engineering Review, 2002, 17(2): 217-221.
[4] Hongji Zhu, Fei Zhou, Suichun Zhai. Finite element analysis of high speed switched reluctance motor based on Ansoft[J]. Journal of Hunan University of Technology. 2013, 27(4): 66-70
[5] MM Nezamabadi, E Afjei, MR Naemi. Design and 3D-FEM Analysis of a Rotary Inear Switched Reluctance Motor[J]. International Symposium on Power Electronics, 2016:430-434.
[6] MR Feyzi, SRM Aghdam, Y Ebrahimi .A Comprehensive Review on the Performance Improvement in Switched Reluctance Motor Design[C]. Canadian Conference on Electrical, 2013:348-353.
[7] M Tanujaya, DH Lee ,YJ An. Design and Analysis of a Novel 4/5 Two-Phase Switched Reluctance Motor[C]. International Conference on Electrical Machines, 2013:1-6.
[8] S Brisset, P Brochet. Optimization of Switched Reluctance Motors Using Deterministic Methods with Static and Dynamic Finite Element Simulations[J]. IEEE Transactions on Magnetics, 1998 ,34(5):2853-2856.
[9] Shoujun Song, Lefei Ge, Hucheng Liu. Design of switched reluctance motor and multi-objective optimization method[J]. Journal of Electrotechnics. 2014, 9(5): 197-204
[10] Youn-Hyun Kim, Jae-Hak Choi, Sung-In Jung. Optimal Design of Switched Reluctance Motor Using Two-Demensional Finate Element Method. Journal of Applied Physics, 2002, 91(10):6967-6969.
[11] Zhiwei Xu, Yongjian Li. Genetic algorithm optimization design and electromagnetic field analysis of switched reluctance motor. Journal of Hunan Institute of Engineering, 2013, 23(2): 9-11

Metal Defect Detection Based on Yolov5

Kun Wang, Zixuan Teng and Tengyue Zou*

Fujian Agriculture and Forestry University, Fuzhou, China

*Corresponding author email: ztybaby@qq.com

Abstract. Metal surface defect detection has been a challenge in the industrial field. The current metal surface defect algorithms target only at a few types of defects and fail to perform well on defects with different scales. In this paper, a large number of metal surface defects are studied based on GC10-DET data set. An improved yolov5 detection network is designed targeting defects of various scales, especially of small-scaled objects, using a specific data enhancement method to regularize and an effective loss function to address data imbalance caused by small-scaled object defects. Finally, the comparative experiment on GC10-DET data set proves the major improvements on accuracy superiority of the proposed method.
Keywords: Surface defect detection; Object detection; Small object detection.

1. Introduction

For national construction, metal materials are important materials that are widely used, diverse and indispensable.Metals are essential and important materials for national construction, which are widely used and varied. Factors such as production machines and environment often lead to the inevitable production of defective products in the production process. In order to prevent economic losses and casualties after being supplied to other industries, The detection of metal surface defects on the production line is the top priority[1]. Therefore, Therefore, improving the detection effect of metal surface defects is an enduring problem in industrial production.

In addition to manual visual inspection, the popular method of defect detection is traditional image processing technology [2,3]. Manual inspection may be inefficient and the detection effect is unstable. Even though the traditional image processing method can alleviate the cost and efficiency problems of manual visual inspection, the difficulty of parameter setting and poor applicability still cannot meet the needs of the industry [4]. After nearly ten years of development, deep learning has gradually changed the task of machine learning, the recognition and detection breadth can be comparable to humans [5].

In this work, we come up with a new defect detection network based on yolov5. during networks training , the data augmentation method of matting, copying and pasting is adopted, and the new loss function Asymmetric loss is adopted to small defects detection. finally, it is verified in GC10-DET.

In a nutshell:

- We dish a detection network based on yolov5.
- Specific data enhancement method and new loss function are used for GC10-DET metal surface defect detection task.
- Extensive experiments on GC10-DET illustrate the importance of each component. In addition, compared with proved SSD network, Faster-RCNN, EfficientDet, CenterNet, yolox and SaccadeNet. Experimental proved that we can get better results on GPU.

2. Related Works

Lv, X et al. put forward a improved SSD detector, which can be deployed on the server. L. Qiu et al. come up with an efficient algorithm for pixel surface defect based on deep learning[6]. A novel adversarial network based on unsupervised anomalous feature editing (AFEAN) is proposed to accurately check various texture defects, especially using paired input composed of defect-free images and artificially defective images for training by H Yang et al.[7]. A new type of defect detection framework is proposed, which is based only on positive sample training [8]. Now that we have entered the industry 5.0 era, production automation is gradually becoming more popular. The advancement of intelligent manufacturing is inseparable from the support of enabling technology. Artificial intelligence to which computer vision belongs has played an important role in this field [9].

The complexity of threshold setting makes traditional image processing technology gradually eliminated. In deep learning, the accuracy of existing works, especially for object detection, needs to be improved. The detection of large object is gradually becoming perfect, but for small object detection, There are still many things we need to do.

From an optimization point of view, we provide an alternative solution for a wide range of metal detection tasks[10]. In addition, the improved yolov5 has achieved gratifying performance.

3. Methods

3.1. Yolov5 Model

The basic idea of the Yolo series is continued by yolov5, and has been further improved on yolov5[11]. Although there is no revolutionary change in object detection, Yolov5 still combines speed and accuracy very well, Favored by the industry. As shown in Figure. 1,the general detection pipeline can be seen as consisting of the following three parts: backbone, encoder, and decoder. The structure of Yolov5 algorithm mainly includes: backbone feature extraction network, strengthen feature extraction network, and use the obtained features to make predictions.

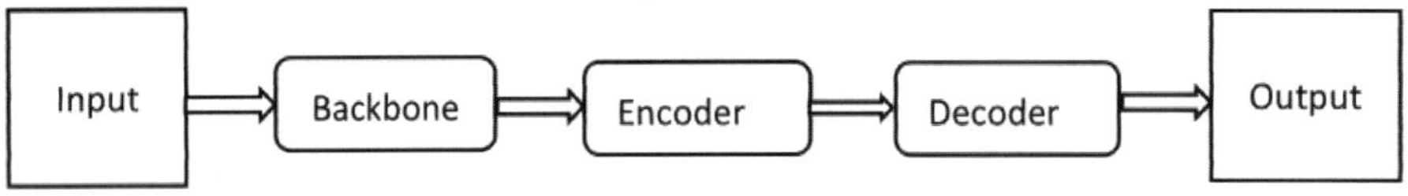

Figure 1. The general detection pipeline

3.2. Improved Network Structure Based on Yolov5

The proposed method uses yolov5 as the baseline, and we propose a metal surface defect detection algorithm. Firstly, it is an image preprocessing process. Adjusted to 608x608, the image is input to the backbone for subsequent normalization operations. When training the network, Yolov5 uses Mosaic data enhancement operations, and proposes a method of adaptive anchor frame.In this case, the speed of fitting and the accuracy of training have been significantly improved. For the backbone feature extraction network, The structure of CSPDarknet53 is retained, in addition to using a brand-new Focus structure in the base network. Multiple feature maps are generated after the image goes through the backbone feature extraction network, and then they are further input into the neck network, where features are repeatedly extracted and enhanced to become more robust. SPP[12] module and FPN[13]+PAN module has also been applied to yolov5. The result of object detection is output in the head. Not Smooth L1 Loss function, GIOU_Loss is chosen to focus on accuracy.

For the one-stage detection network, it refers to the architecture that uses a single feed forward CNN to directly predict the class probability and boundary box offset from the complete image in the overall setting, and encapsulates all calculations. To put it simply, the picture is actually divided into several grids, and each grid point predicts a corresponding area on the original image. As long as the object falls in this area, the object is determined by this grid point. As we all know, the ability of image recognition with greatly high accuracy can be achieved through a very deep network. The deeper the network layer, the problems of gradient disappearance and explosion are extraordinary obvious, and the network even degenerates. Darknet53 uses a residual structure to improve the performance of the

Journal of Physics: Conference Series **2218** (2022) 012050 doi:10.1088/1742-6596/2218/1/012050

model. On this basis the CSPnet structure is used. At this time, the Darknet53 was changed to CSPDarknet53.

With the deepening of the network, the deeper feelings are bigger, stronger, robust to changeful and insensitive to details, which is a big goal. However, the lower layer has small receptive field, is more sensitive to details, and is suitable for small targets. Therefore, if the target object is small, it needs low-level detailed information, and it is likely to be lost with the deepening of the network. Obviously,It is difficult to detect small-scaled objectss. In the mainstream detectors of target detection, such as Cascade R-CNN, Mask R-CNNand FCOS, feature pyramids are almost everywhere. The most common application of feature pyramid is feature pyramid network (FPN), the FPN can merge high-level and low-level features, so that the entire network can achieve better performance, and can detect different levels of object scales [14]. Therefore, in view of the large scale span of metal surface defects and the lack of available features for small targets, based on three effective feature layers with shape of 76x76, 38x38 and 19x19 extracted from CSPDarknet53, the main part of yolov5,Add a feature layer with a shape of 152x152 and input it into neck. Like other feature layers, after repeated feature extraction, it is output to Head for prediction, thus optimizing target detection.

3.3. Data Augmentation Methods

Among the existing benchmark data sets ImageNet, MS COCO and Open Images, MS COCO has increasingly become the most commonly used benchmark data set, but according to statistics, even in MS COCO, the small scaled object only account for 31.62%. So far, The result of deep learning is positively correlated with massive data. In the same way, the improvement of the performance of small-scaled object also depends on the number of small-scaled object. Different scenes have different definitions of small-scaled object, and a unified standard has not yet been formed. The existing definition methods of small goals are consist of two parts, the definition based on relative scale and the definition based on absolute scale [15].

In GC10-DET data set, the ratio of the width and height of the bounding box of the foreign object to the width and height of the image is far less than 0.1, which is a small-scaled object in relative size. therefore, in view of the problem that the number of small-scaled objects in the same image is small and the resulting data is unbalanced [16], as shown in Figure. 2, we cut out the small-scaled object image from the Mask divided in the picture. Rotating and slightly scaling it and then using copy and paste, the number of small scaled object in the same picture increases, and more positive samples are matched in the method of Anchor strategy. As shown in Figure. 3, we demonstrate the small target defect after zooming in.

5

Figure 2. Data augmentation

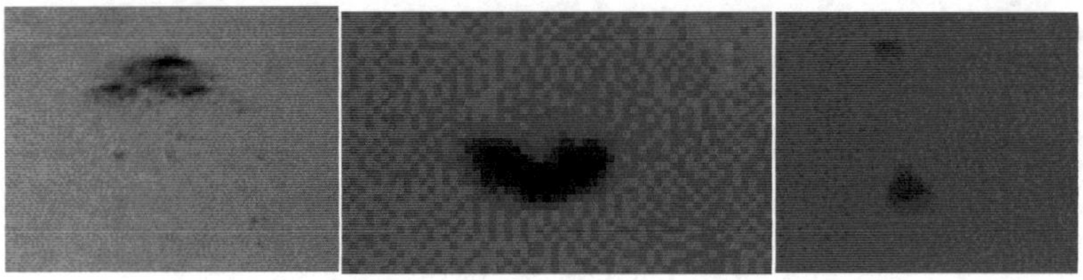

Figure 3. Small scaled metal defect

3.4. Loss Function

So as to weigh the degree of fit between the predicted value and the actual value, the loss function is proposed, that is, the operation of the algorithm.

In GC10-DET data set, although the number of small target categories is equal to that of other categories, its contribution characteristics are still far from those of medium and large targets, so it is urgent to improve this imbalance. Therefore, in the calculation of loss, we artificially increase the weight of loss of a small number of classes, so as to improve the contribution of this class to the loss, thus achieving the purpose of data balance.

There are two main types of target deep learning architecture, one-stage and two-stage. Because the positioning and category prediction of one-stage have the same network output, although the speed is faster than that of two-stage, there is a big difference in accuracy. In this paper, it is considered that it is caused by class imbalance in the training process. In the two-stage algorithm, most of the backgrounds have been filtered out in the first stage. Narrow the target to a certain extent. In the first stage, there are a lot of backgrounds without targets. In order to alleviate this situation, Focal Loss [17] is proposed, which can improve the inherent defects of one-stage by adjusting the loss function of the model to control the contribution of different types of samples to model training. Tal Ridnik et al. proposed an improvement on focal loss, They decoupled the Gamma parameter and introduced the penalty parameter for Easy negative Sample [18], which made further improvement. Therefore, we apply the ASL function in the classification loss.

Two categories of cross entropy loss:

$$CEL = -y * \log(p) - (1 - y) * \log(1 - p) \tag{1}$$

The focal loss of is:

$$\mathrm{FL}(\mathrm{p_t}) = -\alpha_t (1 - \mathrm{p_t})^{\gamma} \log(\mathrm{p_t}) \tag{2}$$

When doing multi-classification, softmax is generally used as the final activation function, and the output has multiple values, corresponding to the probability value of each classification, and the sum is 1. The formula for calculating the cross entropy loss of multi-classification tasks with sofmax as the activation function is:

$$\mathrm{c} = \mathrm{CEL} = -\sum_{0}^{C-1} \mathrm{y_i} * \log(\mathrm{p_i}) = -\log(\mathrm{p_c}) \tag{3}$$

The focal loss of multiple classifications is:

$$\mathrm{FL_{softmax}} = -\alpha_c (1 - \mathrm{p_c})^{\gamma} \log(\mathrm{p_c}) \tag{4}$$

The Asymmetric Loss (ASL)[19]:

$$\mathrm{ASL} = \begin{cases} \mathrm{L_+} = (1 - \mathrm{p})^{\gamma} + \log(\mathrm{p}) \\ \mathrm{L_-} = (\mathrm{p_m})^{\gamma} - \log(1 - \mathrm{p_m}) \end{cases} \tag{5}$$

4. Experiments

4.1. Ablation Experiment

All our experiments were conducted on GC10-DET data set. GC10-DET data set is a surface defect data set collected in actual production line. It contains 10 kinds of surface defects. Part of the detect results are shown in the Figure 4.

We have proved the effectiveness of each part through ablation experiments, as shown in Table 1, which can prove that the accuracy of our proposed methods for surface defect detection has been improved, especially the accuracy of small-scaled objects.

1, 2, and 3 respectively represent the three methods of Yolov5, Improved Yolov5+ data enhancement and Improved Yolov5+ data enhancement+Asymmetric loss in Table 2.

After traditional rotation, scaling, cropping and other data enhancement processing, and the cutout for small scaled object defects, the original image is rotated and slightly zoomed, which increases the actual number of small scaled object defects, balances the number of various samples, and makes the accuracy of object types has been improved to a certain extent. Similarly, the addition of the loss function is also very beneficial to the detection of small scaled object. It shows that the AP in these 10 categories have been improved in the 6 categories (the font has been bolded). It can be seen that the method we proposed is very effective for improving the accuracy of surface defect detection, especially the accuracy of small scaled object.

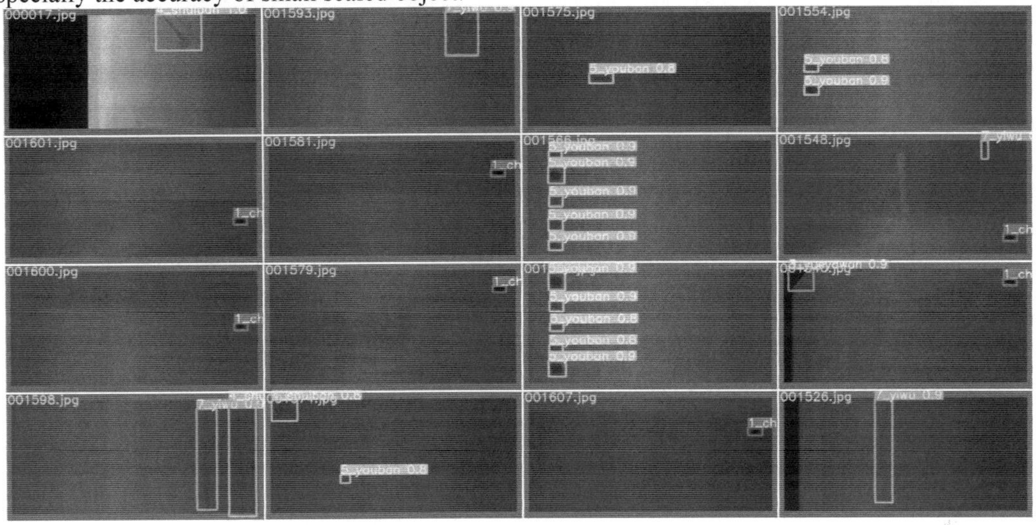

Figure 4. Part of the detect results

Table 1. The data results comparison of 3 schemes.

scheme	map
Yolov5	0.702
Improved Yolov5+ Data augmentation	0.722
Improved Yolov5+ Data augmentation +Asymmetric loss	0.741

Table 2. The data results in GC10-DET

scheme	Pu	Wl	Cg	Ws	Os	Ss	In	Rp	Cr	Wf
1	0.953	0.877	0.96	0.76	0.627	0.687	0.367	0.539	0.427	0.822
2	0.961	0.868	0.957	0.78	0.639	0.657	0.406	0.72	0.421	0.81
3	0.957	0.85	0.953	**0.79**	**0.654**	0.653	**0.43**	**0.741**	**0.542**	**0.831**

4.2. Comparative Experiment

In Table 3, we compare the dataset method with several classic reference networks, a new network proposed by GC10-DET data set provider,and recent state of the art, which shows the effectiveness of our network. After adding a new effective feature layer and fusing with effective feature layers of other scales, although the accuracy of larger-scale defects has been slightly reduced, the accuracy of small-scale defects has been greatly improved. Overall, the map has also been obtained. effective improvement. In addition to the good performance on the map, it is mainly in the small target category Os, and the accuracy performance on Ss, especially In, is higher than other methods.

Table 3. Comparative experiment results in GC10-DET

Detecter	Pu	Wl	Cg	Ws	Os	Ss	In	Rp	Cr	Wf	map
FasterRcnn	0.899	0.554	0.87	0.599	0.653	0.579	0.194	0.364	0.736	0.818	0.627
CenterNet	0.94	0.31	0.88	0.77	0.36	0.33	0.03	0.01	0.07	0.49	0.418
EfficientDet	0.95	0.03	0.93	0.58	0.20	0.13	0.08	0.07	0.01	0.13	0.31
Modified-SSD	0.90	0.885	0.848	0.558	0.622	0.650	0.256	0.364	0.521	0.919	0.651
SaccadeNet	0.962	0.47	0.925	0.764	0.52	0.471	0.39	0.68	0.34	0.54	0.606
Yolox	0.966	0.84	0.932	0.825	0.651	0.62	0.40	0.747	0.63	0.51	0.712
way	0.957	0.85	0.953	0.79	0.654	0.653	0.43	0.741	0.542	0.431	0.741

5. Conclusion

In this paper proposes, an improved metal surface defect detection algorithm based on yolov5, which adopts a specific data enhancement method and a new loss function for the characteristics of small-scale objects, which greatly improves the accuracy of small-scale objects in the detection of metal defects. The GC10-DET data set performs well. In addition, although the anchor mechanism-based detector in the target detection algorithm is still the mainstream, many anchorfree detectors have emerged in recent years, and different fields of deep learning learn from each other, achieving exciting results, especially recently In the past two years, many detectors have been added to the tansformer layer. We know that because anchors are often preset, which restricts the detection effect of small targets to a certain extent, in the next step, we will try to use Anchor free detectors such as DETR to test the accuracy of small-scale objects in metal defects. Whether there is room for further improvement, try to make the accuracy of small-scale defects comparable to that of large-scale defects.

Acknowledgments

I would like to thank Mr. Tengyue Zou for the financial assistance provided in the experiment, as well as the key guidance and problem solving for this paper. Finally, thanks all the people who have helped this research.

References

[1] Kim, S.; Kim, W.; Noh, Y.K.; Park, F.C. Transfer Learning for Automated Optical Inspection. In Proceedings of the 2017 International Joint Conference on Neural Networks (IJCNN), Anchorage, Alaska, 14–19 May 2017; pp. 2517–2524.

[2] Z. Li, J. Zhang, T. Zhuang and Q. Wang, "Metal surface defect detection based on MATLAB," 2018 IEEE 3rd Advanced Information Technology, Electronic and Automation Control Conference (IAEAC), 2018, pp. 2365-2371, doi: 10.1109/IAEAC.2018.8577540.

[3] M. Senthikumar, V. Palanisamy and J. Jaya, "Metal surface defect detection using iterative thresholding technique," Second International Conference on Current Trends In Engineering and Technology - ICCTET 2014, 2014, pp. 561-564, doi: 10.1109/ICCTET.2014.6966360.

[4] Lv X, Duan F, Jiang J-j, Fu X, Gan L. Deep Metallic Surface Defect Detection: The New Benchmark and Detection Network. Sensors. 2020; 20(6):1562. https: //doi.org/10. 3390/s2006 1562

[5] Liu, L., Ouyang, W., Wang, X. et al. Deep Learning for Generic Object Detection: A Survey. Int J Comput Vis 128, 261–318 (2020). https://doi.org/10.1007/s11263-019-01247-4

[6] L. Qiu, X. Wu and Z. Yu, "A High-Efficiency Fully Convolutional Networks for Pixel-Wise Surface Defect Detection," in IEEE Access, vol. 7, pp. 15884-15893, 2019, doi:10. 1109/ACCE SS.2019.2894420

[7] H. Yang, Q. Zhou, K. Song and Z. Yin, "An Anomaly Feature-Editing-Based Adversarial Network for Texture Defect Visual Inspection," in IEEE Transactions on Industrial Informatics, vol. 17, no. 3, pp. 2220-2230, March 2021, doi: 10.1109/TII.2020.3015765.

[8] Zhao Z., Li B., Dong R., Zhao P. (2018) A Surface Defect Detection Method Based on Positive Samples. In: Geng X., Kang BH. (eds) PRICAI 2018: Trends in Artificial Intelligence. PRICAI 2018. Lecture Notes in Computer Science, vol 11013. Springer, Cham. https://doi.org/10. 1007/ 978-3-319-97310-4_54
[9] Praveen Kumar Reddy Maddikunta, Quoc-Viet Pham, Prabadevi B, N Deepa, Kapal Dev, Thippa Reddy Gadekallu, Rukhsana Ruby, Madhusanka Liyanage,Industry 5.0: A survey on enabling technologies and potential applications,Journal of Industrial Information Integration,2021,100257,ISSN 2452-414X,https:/+/doi.org/10.1016/j.jii.2021.100257.
[10] Xu, Yiming , K. Zhang , and L. Wang . "metal surface defect detection using modified yolo.".
[11] Bochkovskiy, Alexey et al. "YOLOv4: Optimal Speed and Accuracy of Object Detection." ArXiv abs/2004.10934 (2020): n. pag.
[12] He K., Zhang X., Ren S., Sun J. (2014) Spatial Pyramid Pooling in Deep Convolutional Networks for Visual Recognition. In: Fleet D., Pajdla T., Schiele B., Tuytelaars T. (eds) Computer Vision – ECCV 2014. ECCV 2014. Lecture Notes in Computer Science, vol 8691. Springer, Cham. https://doi.org/10.1007/978-3-319-10578-9_23
[13] T. Lin, et al., "Feature Pyramid Networks for Object Detection," in 2017 IEEE Conference on Computer Vision and Pattern Recognition (CVPR), Honolulu, HI, USA, 2017 pp. 936-944. doi: 10.1109/CVPR.2017.106
[14] Chen, Q., Wang, Y., Yang, T., Zhang, X., Cheng, J., & Sun, J. (2021). You only look one-level feature. In Proceedings of the IEEE/CVF Conference on Computer Vision and Pattern Recognition (pp. 13039-13048).
[15] Gao Xinbo,Mo Mengjingcheng,Wang Haitao,et al.Recent Advances in Small Object Detection[J].Journal of Data Acquisition and Processing,2021,36(03):391-417.
[16] OKSUZ K,CAM B C,KALKAN S,et al.Imbalance problems in object detection: A review[J].IEEE Transactions on Pattern Analysis and Machine Intelligence, 2020. DOI: 10. 110 9/TPAMI.2020.2981890.
[17] AGARWAL S,TERRAIL J O D,JURIE F.Recent advances in object detection in the age of deep convolution alneural networks [EB/OL] .(2018 09 10) [2019 08 20]. https://arxiv.org/a bs/1809.03193.
[18] Kisantal, M., Wojna, Z., Murawski, J., Naruniec, J., & Cho, K. (2019). Augmentation for small object detection. arXiv preprint arXiv:1902.07296.
[19] Ben-Baruch, Emanuel, et al. "Asymmetric loss for multi-label classification." arXiv preprint arXiv:2009. 14119 (2020).

Practical Application for Reactor Installation to Reduce Harmonics Resonance of Steel Factory

Hashim Hasabelrasul[1], Zhenjiang Cai[1,*], Mohaned Kamal[2], Lei Sun[1] and Xuesong Suo[1]

[1]College of Mechanical and Electrical Engineering, Hebei Agricultural University, No. 289 Lingyusi Street, Baoding 071001, China
[2]Power up for Electrical Engineering CO., Khartoum, Sudan

*Corresponding author email: Czj65@hebau.edu.cn

Abstract. Harmonicis one of the problems in the power system. It needs more focus due to the number of electrical devices that generate a considerable amount of harmonics. In this paper, a reactor was designed, installation and implementation for a steel factory. The installation objective is to solve harmonic resonance and reduce billing penalties for reactive power KVA by the electric power utility—technical analysis applied to the factory main panel. The ETAP software is used to discover the problem and evaluate the solution results. The power flow and harmonics analysis of an 11 kV system is modeled and implemented. Finally, the proper reactor was calculated and installation as practical to solve the harmonic and economic issues.
Keywords: Harmonic order; ETAP; Reactor; Tuned filter.

1. Introduction

Operating steel plants at power factor low leads to several problems, including system lower voltage used in the plant buses, increased operating costs, and additional costs from the electrical energy company in the form of penalties [1,2]. However, harmonic sources come from plant equipment. The impacts of harmonics differ based on the load class. Any non-linear load generates a harmonic current, which passes from a load inside the main source system. Warped network system voltage is created when the voltage is harmonically combined with the 50 or 60 Hz voltages [3-5]. The IEEE standard 519 that regulates allowed a warped harmonic at customer favor portals, has been adopted by electric utilities to solve this problem. Many more standards exist that specified limitations for such harmonic penetration and safeguard both companies and utilities [6].

Harmonic studies have a great concern in power quality research. Many studies have been discussed in contemporary literature for this harmonic problem. Article [7] described the sizing of the design process of centralized units in a steel manufacturing harmonics environment. The authors emphasized the importance of thorough measurement campaigns as the only way in an industrial environment rich in harmonics and proper deformation pre-analysis aimed at a design approach to avoid probable failures. In terms of the major characteristics of an existing steel facility, particularly some essential production tools directly related to the generation and contributing, the harmonics to the system are studied in [8]. According to the paper's author, a harmonic diagnostic can give critical information for increasing the quality of electricity utilized in manufacturing operations. In terms of harmonic simulation and analysis research, researchers offered a more profound knowledge of the harmonic impacts on the power system and a technique for solving harmonic analysis difficulties [9-13]. A novel configuration technology of power factor adjustment and harmonic reduction in the electrical system was proposed in reference [14]. The authors concluded the technology might not need highly complex computer work algorithms or microprocessors. According to experimental and simulation data, the novel technology eliminated many harmonic currents, transients, and flickers. The article [15] discussed the many uses, problems, and trends for classical harmonic filters for harmonic distortion reduction and power quality improvements.

There is 11KV overloading and screaming to operate the steel factory. Therefore, the effect of this problem appeared in many aspects of factory operation and created a direct economic impact. The main purpose of this study was to install a practical reactor filter by analyzing the harmonic and calculating parameters for a suitable reactor. This work is structured as: Section 2 discusses the practical case study in the ABOGASM steel factory. In section 3, we presented a detuned filter design. Results are given in section 4. Lastly, the study concludes in section 5.

2. A Practical Case Study in the ABOGASM Steel Factory

The research presented in this article was carried out at the ABOGASM steel factory and located in Sudan- Khartoum city, specializing in steel iron production.

2.1. The Steel Factory Scheme Description

ABOGASM factory system consists of one bus with bus-coupler and is fed from utility plants at 33 kV through two transformer 10MVA, 33/11kV delta/star. The transformers built up to feed different loads included driver load and two induction furnaces with special trans delta/star/delta. The system is supplemented with two capacitor banks (CAP) 4500KVAR for each bus. Part of the single line scheme of the ABOGASM electric system is presented in Figure 1.

The factory information equipment: Two induction furnaces with special trans delta/star/delta and a lot of LV drivers for motor controlling, CAP specification: U_n=6432V, C_n=19.683, 250 KVAR double star 3 stage, stage power: 6cell/1500 KVAR panel 4500KVAR, configuration: Single Incoming feeder 33K, Tow transformer 2*10MVA 33KV delta/star Z: 7%, 2*11KV bus with bus-coupler, CAP bank 4500KVAR for each bus, and actual VAR: 1250KVAR for each bus, 2500KVAR total plant consumption.

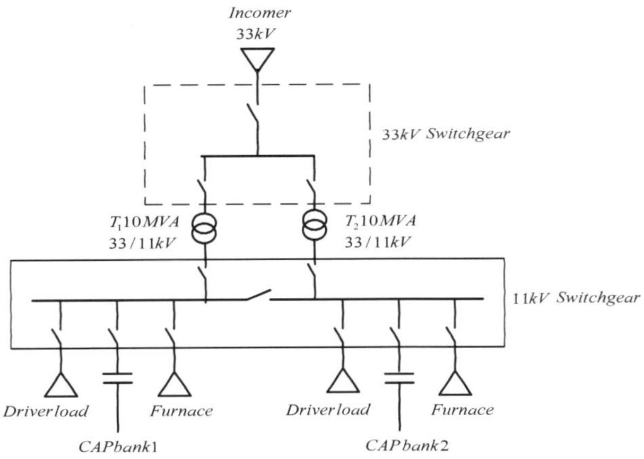

Figure 1. Steel factory partial single line diagram.

2.2. Measurement for Main Panel Harmonics

The measurement is done for the main factory panel. To collect the most general data from the main system to study the factory harmonics, we employed a Fluke 434ii PQ analyzer. The team members have conducted the measurements. The total harmonic distortion (THD) was measured for order values. The measurement of harmonics shown resonance case:

Amplification of order 11,13th to high value 53% of the fund, and amplification of 5,7th order to 35% at another period.

2.3. Technical Analysis

To determine the harmonics in voltage and current technical analysis was carried out. In this case, a load flow and harmonics analysis are performed to a 33/11 kV main panel using ETAP software. The load flow analysis for induction and a rolling mill with both capacitor banks is shown in Figure 2.

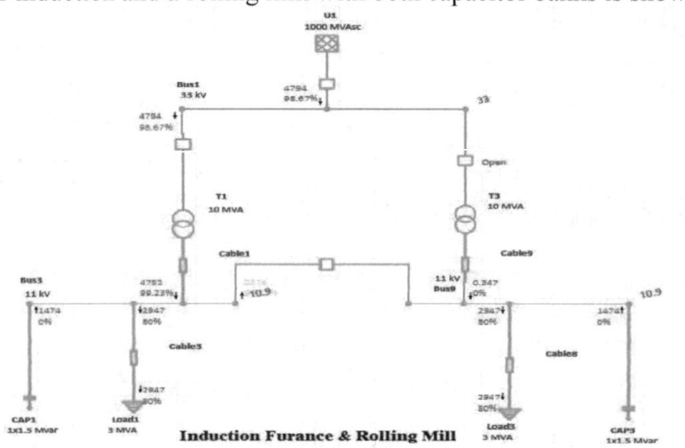

Figure 2. Load flow analysis: with both capacitor banks are on to reach 0.98 power factor.

The harmonics analysis without capacitor THD of current was 16%. Capacitors were purchased and installed for the factory without filters. To find out the effect of the harmonic system with capacitors, the harmonic analyzed in the presence of capacitors as in Figure 3. Then it became clear harmonic increased as seen in the figure the THD of the current from 16% to 40%.

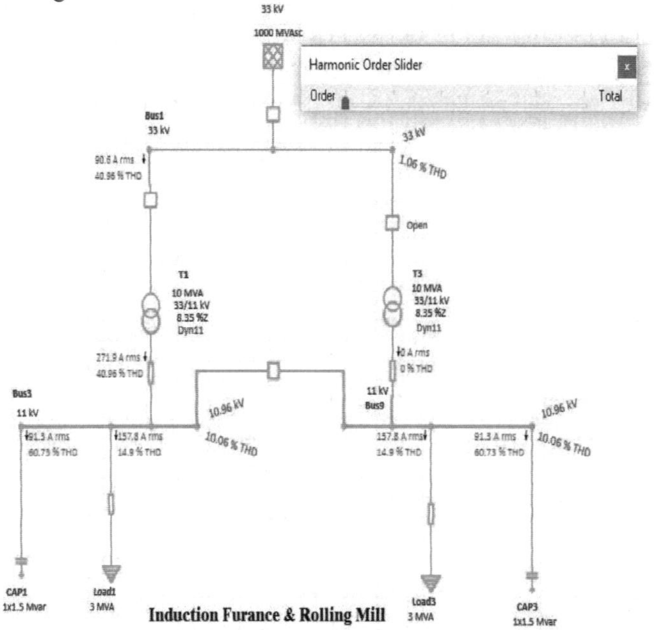

Figure 3. Harmonics analysis with capacitor shows THD of current 40%.

3. Detuned Filter Design

A detuned harmonic filter consists of a tuning reactor. The fixed capacitor banks are connected to it in series. For design reasons, data on load specifics, power factor, reactive power, voltage, and current THD were obtained. This data was obtained from a power analyzer and technical analysis. We started by calculating the detuning factor, which defines the series resonance frequency between the capacitor and reactor is necessary for the blocking and filtering effects. The detuning factor p is given by

$$p = \frac{X_L}{X_C} \times 100 \tag{1}$$

Capacitor voltage V_C is given by

$$V_C = \frac{V_N}{\left[1 - \frac{p}{100\%}\right]} \tag{2}$$

The series resonance frequency f_r calculated as

$$f_r = f_N \times \sqrt{\frac{100\%}{p}} \tag{3}$$

Where V_N is the rated system voltage, f_N is the rated system frequency, X_L and X_C are the reactance's of reactor and capacitor, respectively.
Reactor calculated rated factory values presented in Table 1.

Table 1. Reactor design calculated value.

Unit	Value
P	5.7%
V_N	10.5kV
V_C	11kV
C per phase	32.2 F
X_c	80Ω
X_L	4.3Ω
L	14 mH

4. The Results

The reactor value is calculated according to the factory main panel requirement to maintain the power factor, voltage, and current. We point out that the reactor was designed inside the factory in Sudan by the authors. Then after that, the design was sent with one of the authors for manufacturing in China. After installing the reactor in series with the capacitors then repeated the technical analysis by ETAP V19 to prove sizing and harmonics analysis. Load flow analysis with combination of capacitor bank and reactor to reach 0.99 power factor shows in Figure 4. Figure 5 shows the harmonics analysis with reactor 5.7%. It shows THD of current 10% (Filtration) to avoid harmonics resonance, amplification, and filtration.

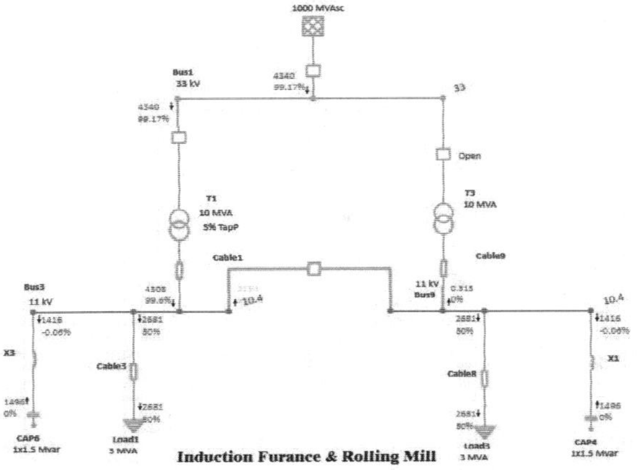

Figure 4. Load flow analysis with a combination of capacitor bank and reactor.

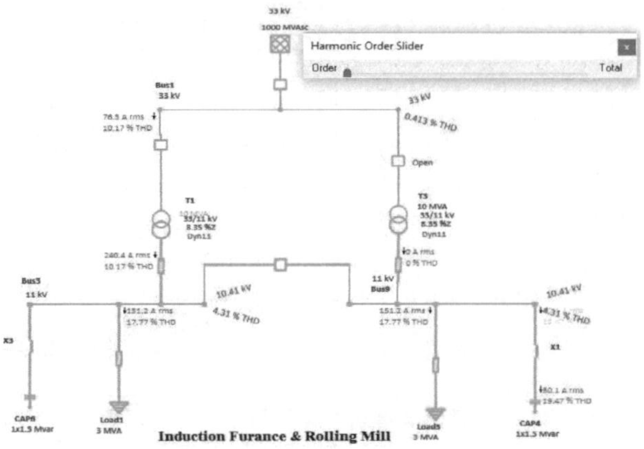

Figure 5. Harmonics analysis with a reactor.

Harmonics analysis resonance at 5th order filtration is shown in Figure 6.

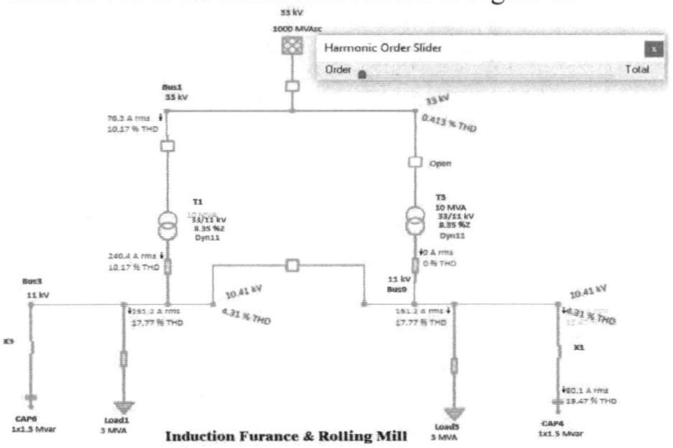

Figure 6. Harmonics analysis 5th order filtration with a reactor.

The reactor acceptance test before installation and the reactor connection installation is shown in Figure7.

(a) (b)

Figure 7. Typical reactor installation for the steel factory: (a) Acceptance test before installation; (b) Reactor connection installation.

The electrical measurements load current was taken with fluke 434ii PQ analyzer. The results are presented in Figure 8. Table 2 presented the commissioning, measurement, and testing.

From the practical result, we can clearly understand the harmonic effect over the capacitor banks in the electrical factory system. The system without capacitor banks generates harmonic of THDi-16.5% & THDv-7 %. When the same procedure is energized with a capacitor bank without a series reactor, the capacitor banks amplify the 5th, 7th, 11th, and 13th order of current and voltage, which outcomes in a rise of THDi value from 16.5% to 40%. Then leads to an increase in THDv from 7% to 10 %, which is very dangerous to the factory system and the capacitors connected to the system. Now, look at system configuration when the capacitor is associated with a detuned reactor tuned at 189 Hz. We can notice that the harmonic current decreases from 40% to 10%. Also, the harmonic voltage decreases from 10% to 4 %—the results in minimizing the effect of resonance over the higher-order harmonics.

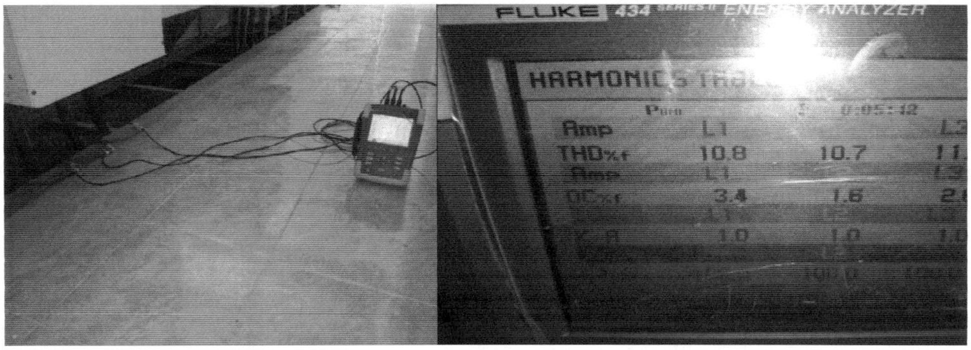

(a) (b)

Figure 8. Typical reactor installation for the steel factory: (a) fluke 434ii PQ analyzer connected with the main panel; (b) THD of load current.

Table 2. The commissioning, measurement, and testing.

		Measurement at 11kV-Bus, without CAP		
THDorder	5	7	11	13
% f value	12.85	5.5	7.5	4
THD_i	16.5%			
THD_v	7%			
		Measurement at 11kV-Bus, with CAP In		
THDorder	5	7	11	13
% f value	37.5	14.3	4.7	1.6
THD_i	40%			
THD_v	10%			
		Measurement at 11kV-Bus, with reactor		
THDorder	5	7	11	13
% f value	5.4	4	6	3.6
THD_i	10%			
THD_v	4%			

As for the economic aspect of the factory consumption bill, the reactive power KVAR consumption is 3000KVAR per hour. Electricity board penalty: 1KVAR cost 0.002 USD for P.F less than 0.95. The P.F penalty cost: 6 USD per hour – for a day 144 USD, and one month, 4,320 USD.

5. Conclusions

By practical measurements and monitoring the order harmonics of the main panel of the steel factory, we found overload harmonics and screaming that led to a problem of factory operation. Technical analysis has been conducted for the main panel and non-linear load. ETAP load flow is applied to determine the voltage drop at different points of the system through each branch and feeder. Reactor designed and installed practically to reduce harmonics for the steel factory were mostly for the 5th, and 7th, 11th, and 13th order. The power factors have increased to 0.95 as a result of the installation of the specified reactor. The reactor installation also helped end the bills compulsory by the electric utility from 4,320 USD per month to zero USD. The authors strongly recommend frequent system assessments, especially when new equipment is introduced to the plant, and measurements and analysis of planned reactors with harmonic and other benefits. It will also be essential to do power quality tests to double-check the order of the harmonic discovered through technical analysis.

Acknowledgments

We are grateful for the Postdoctoral program of Hebei Agricultural University. And thanks to Power up for Electrical Engineering Company for participating practically in this work. This research was funded by the Postdoctoral program of Hebei Agricultural University, Science and Technology Support Program of Baoding (18ZG011), and "Key R&D project of Hebei Province: (19227401D, 19227206D. The University Science and Technology Research project of Hebei Province (QN2020444).

References

[1] Mekhamer S F.; Abdelaziz A Y.; Ismael S M. Design Practices in Harmonic Analysis StudiesApplied to Industrial Electrical Power Systems.*Engineering, Technology and AppliedScience Research*2013, 3(4), pp. 467–472.

[2] Arrillaga; D. A. Bradley; P. S. Bodger. Power System Harmonics. John Wiley & Sons, New York, 1985.

[3] G. T. Heydt. Electric Power Quality. Stars in Circle Publications, West LaFayette, IN, 1991.

[4] R. C. Dugan. Simulation of Arc Furnace power systems.*IEEE Trans. on Industry Application*1980, 16(6), pp. 813–818.

[5] M. F. McGranaghan; R. C. Dugan; H. W. Beaty. Electrical Power Systems Quality. New York: McGraw-Hill, 1996.

[6] IEEE Recommended Practice and Requirements for Harmonic Control in Electric Power Systems, IEEE Std. 519-2014 (Revision of IEEE Std 519-1992), June 2014.

[7] Baggini A; Bua F; Buratti F; et al. PFC units sizing in steel factory harmonics environment: a case study. International Conference on Electrical Power Quality & Utilisation, Barcelona, Spain, 9-11 Oct, 2007.

[8] Ferreira; D. M.; P. R. S. Jota; C. P. Souza. Harmonics in electrical industrial systems: A case study in a steel manufacturing facility. 11th International Conference on Electrical Power Quality and Utilisation. IEEE, Lisbon, Portugal, 17-19 Oct., 2011.

[9] Wu G.; Zhong X. Harmonic analysis and simulation research of 10kV series reactor failure. International Conference & Exhibition on Electricity Distribution, Stockholm, 10-13 June, 2013.

[10] Yan Y H; Chen C S; Moo C S; et al. Harmonic analysis for industrial customers.*Industry Applications IEEE Transactions*1994, 30(2), pp. 462–468.

[11] Damjanović; Ivana; et al. Harmonic Performance Analysis of Static Var Compensator Connected to the Power Transmission Network. Second SEERC Conference: energy transition and innovations in electricity sector, Kyiv, Ukraine, 12-13 June, 2018.

[12] Abreu; J. P. G.; H. Arango; F. N. Belchior. Harmonics: A proposal for new terms. 13th International Conference on Harmonics and Quality of Power, Wollongong, NSW, Australia, 28

Sept.-1 Oct., 2008.
[13] Sooriyagoda D; Kumara S; Navaratne S; et al. Harmonic Analysis of New Breaker Switched Capacitor Bank Installations: A case study. IEEE 14th Conference on Industrial and Information Systems, Peradeniya, Sri Lanka, 18-20 Dec., 2019.
[14] KasikciI . A new method for power factor correction and harmonic elimination in power systems. Ninth International Conference on Harmonics and Quality of Power. Proceedings (Cat. No.00EX441), Orlando, FL, USA, 1-4 Oct,2000.
[15] Tischer H.; Pfeifer T. Hybrid filter for dynamic harmonics filtering and reduction of commutation notches - a case study. 17th International Conference on Harmonics and Quality of Power (ICHQP), Belo Horizonte, Brazil, 16-19 Oct., 2016.

Application of Similarity Theory and Dimensional Analysis for Helical Gear Fatigue Model Test

Chengli Zhu*

Department of Industrial Engineering, Tsinghua University, Beijing, China

*Corresponding author email:zhcl@tsinghua.edu.cn

Abstract. Gear fatigue test is an important way to verify the reliability of gears, but the test period is too long and costly, especially for the high-speed and heavy-load helical gears, the working process of them are difficult to simulate in the laboratory. For reducing the cost of gear fatigue test, we can use the scaled similar model to replace full-size gear in the fatigue test, process and the results can be applied to the full-size gear according to the similarity principle. In this paper, the similarity principles of helical gears are established with dimensional analysis method. A scale-up and a scale-down similarity gear models are designed for a helical gear pair, and the fatigue damages of them are analysed under eight-stage load spectrum. The results show that the scaled similarity gear models and the full-size gears have nearly the same fatigue damages. This study can bring us an economic and effective fatigue test method for the helical gears.

Keywords: Helical gear; Similarity; Dimensional analysis; Fatigue life; Model test.

1. Introduction

Helical gears are widely used in heavy-load mechanical transmissions due to their advantages of larger load capacity, less impact, stable transmission, lower vibration and noise [1]. However, helical gears are easily failure under dynamic loads. Many studies show that fatigue damage is the main cause of helical gear failures [2-4]. In order to ensure the developed helical gears have required reliability and safety, we must do the experimental fatigue test on bench test systems [5]. However, gear fatigue test is a long and expensive process, especially for the large and heavy helical gears, their long-time fatigue tests consume a lot of materials and energy, and their complicated working processes are very difficult to simulate in the laboratory [6].

In this paper, dimensional analysis method was used to establish the similarity principles in geometry, bending and contact stress, cycle number and material for helical gears based on similarity theory [7, 8]. Then, the similarity fatigue test method of helical gear transmission was proposed. After that, the similarity helical gear models were designed and their contact fatigue life and bending fatigue life were calculated, which proved the effectiveness of the proposed method. This provides us an economical and time-saving helical gear fatigue test method.

2. Fatigue Similarity Principle of Helical Gears

The fatigue life of helical gear is determined by the load magnitude, load cycles, materials, processing technology, lubrication conditions and working environment [9]. Based on the analysing of these factors we deduced the fatigue similar principles for helical gears.

2.1. Geometric Similarity Analysis for Helical Gears

According to the similarity theory about geometry [8], If two helical gears have the same ratio of length dimension, they can be considered as similar in geometry. The standard geometric parameters of helical gear refer to the normal parameters. So, the geometric similarity requirements of helical gears are: they must have the same helical angles and the same ratios of length parameters, such as reference diameter, face width, etc.

According to the knowledge of gear geometry, if the corresponding wheels and pinions have equal helical angle, working pressure angle, tooth width and number, the helical gear pairs are similar in geometry and they have equal transvers contact and overlap contact ratios.

2.2. Stress Similarity Analysis for Helical Gears

The fatigue life of helical gear depends on the following governing parameters: tooth contact stress, root bending stress, material performance and the number of stress cycles[10].If two helical gears have the same material performance and the equal values of contact stress, bending stress, and cycle numbers, they should have the same fatigue life[11].

The contact line of two helical gears is an oblique line, and the contact stress calculation formula of them is [12]:

$$\sigma_H = \sqrt{\frac{2KT}{\phi_d d^3 \varepsilon_a} \cdot \frac{u \pm 1}{u}} \cdot Z_H Z_E \tag{1}$$

Where ϕ_d is tooth width coefficient: $\phi_d = b/d$, Z_E is the gear material elastic coefficient, Z_H is the area coefficient (pitch point).

The important variables affecting helical gear contact stressare: transmitted torque *T*, tooth number ratio μ, helical angle β, working pressure angle α_t, pitch circle diameter *d*, contact ratio ε_a, material Poisson's ratio υ, material elastic module *E*.

Helical gear's contact stress can be written as:

$$\sigma_H = f(T, d, b, u, \varepsilon_\alpha, \alpha_t, \beta, v, E) \tag{2}$$

Where, ε_a, α_t, β, υ are dimensionless. *T*, *d*, *b*, *u*, and *E*, can be expressed as the exponentials of the fundamental quantity dimensions: length L, mass M), time T. The variables affecting the contact stress of helical gears and the dimensions of them are list in Table 1.

Table 1. Dimensions of the contact stress variables

	[σ_H]	[*T*]	[*d*]	[*b*]	[μ]	[ε_a]	[α_t]	[υ]	[*E*]
[L]	-1	2	1	1	0	0	0	0	-1
[M]	1	2	0	0	0	0	0	0	1
[T]	-2	-2	0	0	0	0	0	0	-2

As we can see from Table 1, only 2 dimensional independent variables. There are 7 dimensionless Pi terms (9-2 = 7). We can select *E*,*d* as the fundamental dimensional independent variables.

The Pi terms can be expressed as follows:

$$\pi_1 = \frac{\sigma_H}{Td^{-3}}, \quad \pi_2 = \frac{E}{Td^{-3}}, \quad \pi_3 = \frac{b}{d}, \quad \pi_4 = u$$
$$\pi_5 = \varepsilon_a, \quad \pi_6 = \alpha_t, \quad \pi_7 = \beta, \quad \pi_8 = \upsilon \tag{3}$$

Let's assume that we have two helical gears, they have the same material, and they are similar in geometry. We can obtain: $\pi_3 = L_r = \text{const}$, $\pi_4 = 1 = const$, $\pi_5 = const$, $\pi_6 = const$, $\pi_7 = const$, $\pi_8 = const$, Moreover, π_1 and π_2 are also should be constant according to the similarity theorem [7].

$$\pi_1 = \frac{\sigma_H}{Td^{-3}} = const \tag{4}$$

$$\pi_2 = \frac{E}{Td^{-3}} = const \tag{5}$$

According to equation (4) and (5), let the geometric similar helical gears have equal contact stress, the torqueof them should satisfy the relationship:

$$\frac{T_m}{T_p} = \frac{d_m}{d_p} = L_r^3 \tag{6}$$

In equation (6), T_m and T_p are the torqueof the similar gears: the model and prototype gears.
From the above analysis, we can know that for the geometric similar helical gears, if their material have equal elastic modulus and Poisson's ratio, and the ratio of their torque is equal to the cube of their geometric similarity ratio, they must have equal contact stress and root bending stress.
The number of stress cycles N can be written:

$$N = 60njL_n \tag{7}$$

In equation (7), n is the gear speed, its unit is r/min. j is the engagement times, it has no unit. L_n is the working time of the gear, its unit is hour.
For geometric similar helical gears, they have the same value of 'j'. Moreover, if the product of their speed and working time are equal they must have equal contact stress cycles and bending stress cycles. That means they have the same N values.

2.3. The Fatigue Model Test Principle of Helical Gear

Through the analysis of geometric similarity, stress similarity and stress cycle number of helical gears, it can be found that if they are similar in geometric, and they have the same tooth contact stress, bending stress, and number of stress circles, they should have the same fatigue damage. And they have the same fatigue life.
In order to maintain the similarity of helical gear fatigue life as much as possible, the other factors which can affect the helical gear fatigue life, such as accuracy of machining, accuracy of assembly, technology for heat treatment, linear speed and the runtime environment should be consistent as much as possible.

3. Fatigue Damages Analysis for Similar Helical Gears

For verifying the effectiveness of the helical gear fatigue model test principle proposed in this paper, a helical gear pair is designed, and their eight-stage load spectrum are given. Its scale-up and scale-down similarity gear models are designed by changing the gear size. According to the fatigue similarity requirements of helical gears, the load spectrum of the full-scale gear can be converted into the load spectrum of similar gears, and the fatigue damages of them can be calculated respectively. The variables and valuesof three similar helical gear pairs are listed in Table 2.
The eight-stage standard load spectrum (binomial distribution, P=5/6) is defined for the full-scale helical gear pair. The load spectrum of the similar scale-up and scale-down gear pairs can be calculated by the following equation:

$$T_m = T_p L_r^3 \tag{8}$$

The contact fatigue damages and the root bending fatigue damages of all the similar helical gears under the various eight-stage loads are calculated during the design life period and the whole fatigue life period. As for the hardened helical gears with good lubrication, the root bending fatigue life is much longer than its contact fatigue life. In the design life period and the whole fatigue life period, the bending fatigue damage of the tooth is relatively very little and can be ignored.

Table 2. The parameters of the similar helical gear models

Parameters		GP1 (scale-down)		GP2 (full-scale)		GP3 (scale-up)	
		Pinion	Wheel	Pinion	Wheel	Pinion	Wheel
Tooth number	z	17	40	17	40	17	40
Face width	b (mm)	10	10	20	20	25	25
Normal module	m_n (mm)	1.5	1.5	3.25	3.25	4	4
Pressure angle	α_n (°)	20	20	20	20	20	20
Helical angle	β (°)	20	20	20	20	20	20
Precision grade	Q	6	6	6	6	6	6
Modification coefficient	x^*	0	0	0	0	0	0
Centre distance	a (mm)	47.56		103.05		126.84	

The contact fatigue damages of the pinions and the wheels of the similar helical gear pairs under eight-level loads in the design life are shown in Figure 1 and Figure 2. The contact fatigue damages of the pinions and wheels in the whole fatigue life period are shown in Figure 3 and Figure4.

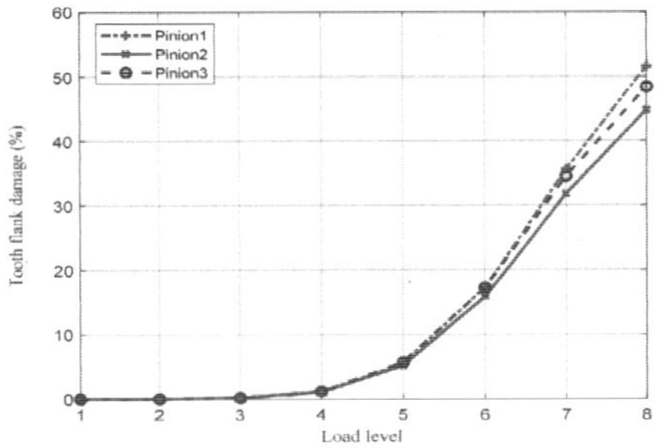

Figure 1. Contact fatigue damages of the similar pinions in the design life period.

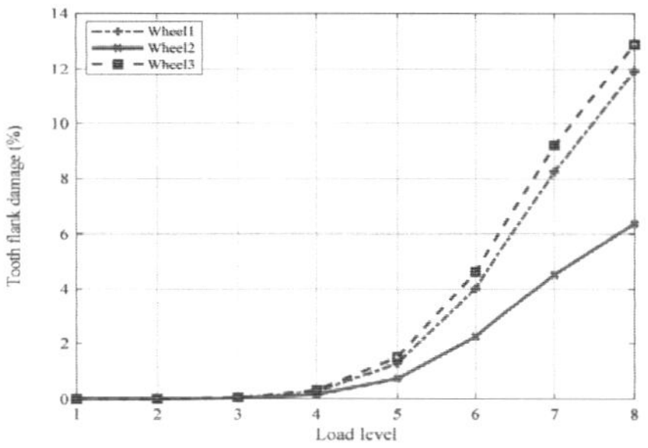

Figure 2. Contact fatigue damages of the similar wheels in the design life period.

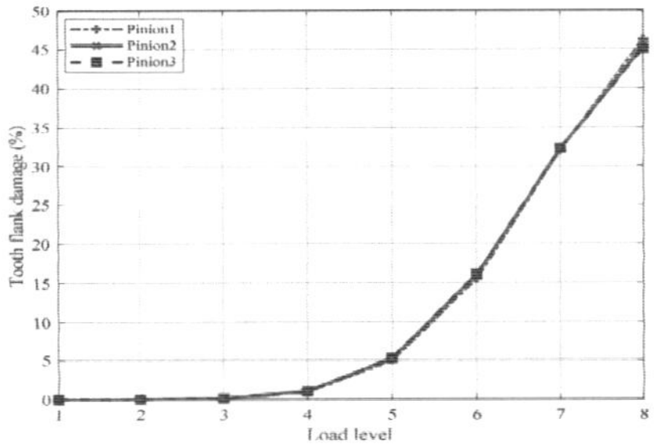

Figure 3. Contact fatigue damages of the similar pinions in the whole fatigue life period.

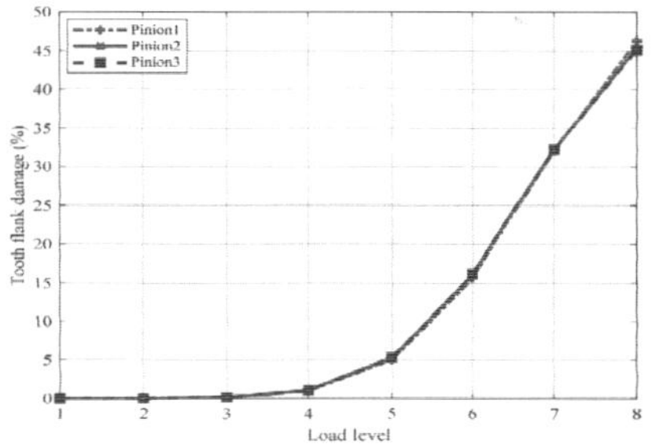

Figure 4. Contact fatigue damages of the similar wheels in the whole fatigue life period.

It can be seen from Figure 1 to 4 that the contact fatigue damages of the pinion gears and the wheel gears of the similar helical gear pairs under eight-stage load spectrum are very close during the design life period and the whole fatigue life period. From Figure 1 and Figure 3, it can be know that the contact damages of the similar pinion gears have good consistence. While, from Figure 2 and Figure 4, we can see that the contact damages of the wheel gears have a little poor consistence with their geometric size scaling up and scaling down.

4. Conclusion

The conclusions are:

- If helical gears have equal helical angle and equal proportional ratios of module, tooth width, reference diameter, and parameters of length dimension. They can be considered as similar in geometric.
- Geometric similar helical gears must have equal contact stress and root bending stress,when their torque ratio is equal to the cube of their geometric similarity ratio, and materials of them have equal Poisson's ratio and elastic modulus.
- The full-size helical gear and its fatigue similar models have close fatigue lives and contact fatigue damages.

Acknowledgments
This work was supported by Tsinghua University Laboratory Innovation fund.

References
[1] Jia X P, Wang W Z, et al. 2014 A contact fatigue model of helical gear under elastohydrodynamic lubrication *Tribology* (vol 34) pp 8-14
[2] Pantazopoulos G A. 2015 Bending fatigue failure of a helical pinion bevel gear *Journal of Failure Analysis and Prevention* (vol 15) pp 219-226
[3] Cao W, Ren S, et al. 2020 Microstress cycle and contact fatigue of spiral bevel gears by rolling-sliding of asperity contact *Friction* (8) pp 1083-1101
[4] Wang H, Zhou C, et al. 2019 An adhesive wear model for helical gears in line-contact mixed elastohydrodynamic lubrication *Wear* (426) pp 896-909
[5] Hoeprich M R. 2001 Analysis of micropitting of prototype surface fatigue test gears *Tribotest* (val 7) pp 333-347
[6] Wei P, Zhou H, et al. 2019 Modeling of contact fatigue damage behavior of a wind turbine carburized gear considering its mechanical properties and microstructure gradients *International Journal of Mechanical Sciences* (vol 156) pp 283-296
[7] Cheng Y and Cheng C. 2004 Scaling, Dimensional Analysis, and Indentation Measurements *Materials Science & Engineering R* (vol 44) pp 91-149
[8] Levin M 2016 Dimensional Analysis *How to Scale-Up a Wet Granulation End Point Scientifically* (Boston: Academic Press) Chapter 2 pp 3-17
[9] Zhao G, Dong H, et al. 2016 Whole Fatigue Life of Helical Gear Covering Crack Initiation Stage and Propagation Process *Tribology* (vol 36) pp 643-9
[10] Zhang J , Fang Z , et al. 2007 The modified pitch cone design of the hypoid gear: Manufacture, stress analysis and experimental tests *Mechanism and Machine Theory* (vol 42) pp 147-158
[11] Patil S S , Karuppanan S and Atanasovska I. 2016 Experimental measurement of strain and stress state at the contacting helical gear pairs *Measurement* (vol 82) pp 313-322
[12] Sánchez M B, Pedrero J I and Pleguezuelos M. 2013 Contact stress calculation of high transverse contact ratio spur and helical gear teeth *Mechanism and Machine Theory* (vol 64) pp 93-110

Permanent Magnet Synchronous Motor Control Method Based on Fuzzy PI and Sliding Mode Observer

Linlin Sha[1,*], Yi Song[1] and Zijin Li[1,2]

[1]School of Electrical Engineering, Shanghai DianJi University. Shanghai, 201306, China
[2] Minge New-Type Motor and Electronic Control Research Institute. Foshan City, Guangdong Province, 528599, China

*Corresponding author email: 206001010108@st.sdju.edu.cn

Abstract. In order to solve the many shortcomings of the current servo system using mechanical position sensors, this article combines the traditional fuzzy PI control and sliding mode observer, and proposes a control method of PMSM based on fuzzy PI and sliding mode observer. Simulink is used for simulation analysis, and the results are compared. It can be seen from the simulation results that the designed control method can efficiently control the permanent magnet synchronous speed control system, so that the system has excellent starting characteristics and dynamic stability.
Keywords: PMSM; Fuzzy PI; Sliding mode observer.

1. Introduction
The traditional PI controller has a good control effect when the system is simple and linear, but for permanent magnet synchronous motors (PMSM) with complex mathematical models, strong coupling, and nonlinearity, the control effect is often poor. The fuzzy PI controller is mainly based on the experience of the engineer, which adjusts the deviation amount and deviation change rate online, and has the characteristics of high control accuracy [1]. Literature [2] uses fuzzy PI control method for the PMSM vector control system. At this time, the system adjustment time is shorter than that of ordinary PID controllers.
The vector control method of PMSM is to install position sensors to collect and process data, and use sensors to measure the position and speed of the rotor to achieve closed-loop control of the system. Because the hardware circuit of the sensor is very complicated, the cost is high, and its application in production and life is limited, so sensorless control is currently a major control hotspot [3]. Simple design and strong robustness are the advantages of the sliding mode observer method, but due to the switching function idea, it will be accompanied by jitter in the control, that is, chattering will occur.
This paper combines fuzzy PI control and sliding mode observer system, and proposes a PMSM control method based on fuzzy PI and sliding mode observer. Aiming at the traditional PI control method, the PI link is introduced into the fuzzy control to improve the control performance of the motor, reduce its chattering phenomenon. Finally, the stability, rapid response and robustness of the proposed method are verified by simulation experiments.

2. Mathematical Modelling of PMSM
For the later design of the velocity loop, select the appropriate coordinate transformation to reduce the order and decouple the system. In order to control the model simply, it's usually to use the basic

mathematical model of d and q axis that ignores the iron loss, taking the surface mount PMSM motor as an example [4], as shown in the following formula:

$$\begin{cases} u_d = Ri_d + L_s \frac{di_d}{dt} - p_n\omega_m L_s i_q \\ u_q = Ri_q + L_s \frac{di_q}{dt} + p_n\omega_m L_s i_q + p_n\omega_m \psi_f \\ J\frac{d\omega_m}{dt} = \frac{3}{2}p_n\psi_f i_q - T_L \end{cases} \tag{1}$$

In formula (1), u_d and $u_q$are the voltages on the d and q axes; i_d and i_q are the currents on the d and q axes; the inductance on the d and q axes is L_s; $\psi_f$is the flux linkage between the permanent magnet and the stator; R is the stator winding resistance; $p_n$is the number of electromagnetic pole pairs. For surface-mounted PMSM, usually the rotor field-oriented control method with i_d=0 can be used to obtain a better control effect, as shown in the following formula:

$$\begin{cases} \frac{di_q}{dt} = \frac{1}{L_s}(-Ri_q - p_n\omega_m\psi_f + u_q) \\ \frac{d\omega_m}{dt} = \frac{1}{J}(-T_L + \frac{3p_n\psi_f}{2}i_q) \end{cases} \tag{2}$$

3. Controller Design

3.1. The Mathematical Model of the Sliding Mode Observer

Define the switching surface function as

$$s(x) = \hat{i}_s - i_s \tag{3}$$

In the formula: $\hat{i}_s$——Estimated current value, $\hat{i}_s = [\hat{i}_\alpha \quad \hat{i}_\beta]^{\mathrm{T}}$;
i_s——Actual current value, $i_s = [i_\alpha \quad i_\beta]^{\mathrm{T}}$.
Select the switching surface as

$$s(x) = \hat{i}_s - i_s = 0 \tag{4}$$

The Sigmoid function is selected as the switching surface control function, and the expression is

$$F(s) = \frac{2}{1+e^{(-as)}} - 1 \tag{5}$$

In the formula: a———Positive constant, which is an adjustable parameter.
The sliding mode observer is constructed as

$$\begin{cases} \frac{d\hat{i}_\alpha}{dt} = \frac{R_s}{L_s}\hat{i}_\alpha + \frac{1}{L_s}u_\alpha - \frac{1}{L_s}kF(\bar{i}_\alpha) \\ \frac{d\hat{i}_\beta}{dt} = \frac{R_s}{L_s}\hat{i}_\beta + \frac{1}{L_s}u_\beta - \frac{1}{L_s}kF(\bar{i}_\beta) \end{cases} \tag{6}$$

In the formula: $\bar{i}_\alpha$, $\bar{i}_\beta$——Current observation error;
k——Sliding mode gain.
The estimated value of back-EMF is

$$\begin{cases} \hat{e}_\alpha = kF(\bar{i}_\alpha) = \frac{2k}{1+e^{(-a\bar{i}_\alpha)}} - k \\ \hat{e}_\beta = kF(\bar{i}_\beta) = \frac{2k}{1+e^{(-a\bar{i}_\beta)}} - k \end{cases} \tag{7}$$

From this, the estimated angle and estimated speed can be obtained as

$$\begin{cases} \hat{\theta} = \arctan\left(-\frac{\hat{e}_\alpha}{\hat{e}_\beta}\right) \\ \hat{\omega} = \frac{d\hat{\theta}}{dt} \end{cases} \tag{8}$$

3.2. Stability Analysis

Subtracting formula (6) and PMSM from the mathematical model formula in the two-phase stationary coordinate system can be obtained

$$\begin{cases} \frac{d\bar{i}_\alpha}{dt} = -\frac{R_s}{L_s}\bar{i}_\alpha + \frac{1}{L_s}e_\alpha - \frac{k}{L_s}F(\bar{i}_\alpha) \\ \frac{d\bar{i}_\beta}{dt} = -\frac{R_s}{L_s}\bar{i}_\beta + \frac{1}{L_s}e_\beta - \frac{k}{L_s}F(\bar{i}_\beta) \end{cases} \tag{9}$$

The Lyapunov function constructed is as follows:

$$V = \frac{1}{2}i_\alpha{}^2 + \frac{1}{2}i_\beta{}^2 \tag{10}$$

It can be seen that the functional formula is always greater than zero, and the derivative of it and combining with formula (9) can be obtained

$$\dot{V} = \bar{i}_\alpha\dot{\bar{i}}_\alpha + \bar{i}_\beta\dot{\bar{i}}_\beta = -\frac{R_s}{L_s}\left(\bar{i}^2{}_\alpha + \bar{i}^2{}_\beta\right) + \frac{1}{L_s}\{\bar{i}_\alpha[e_\alpha - kF(\bar{i}_\alpha)] + \bar{i}_\beta[e_\beta - kF(\bar{i}_\beta)]\} \le 0 \tag{11}$$

The conditions for the observer to be stable are

$$k > max(|e_\alpha|, |e_\beta|) \tag{12}$$

3.3. Fuzzy PI Controller Design

Fuzzy control is largely related to expert experience and has strong robustness. The fuzzy logic theory is introduced into the traditional PID control theory, and the fuzzy PI controller is designed, in which the input precision variable deviation signal e(k) and its rate of change $e_c(k)$ are fuzzy and fuzzy After reasoning and defuzzification, the precise control quantity, proportional coefficient K_P and integral coefficient K_I, are output. This paper uses this characteristic of the fuzzy controller to implement PID parameter correction for the original permanent magnet synchronous motor PI control system to achieve a certain adaptive ability.

The fuzzy subsets of parameters e, $e_c(k)$, K_P, and K_I are all {NB, NM, NS, ZO, PS, PM, PB}, and the domain is {-6, 6}, as shown in Figure 1.

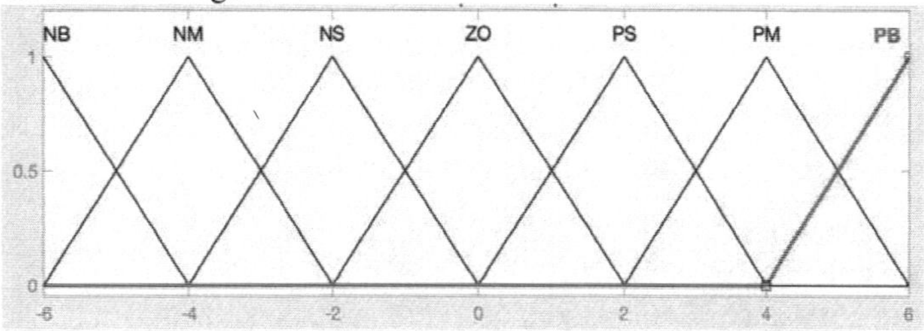

Figure 1. Membership functions of e, $e_c(k)$, KP, KI

The control rules need to be designed and the design process of the fuzzy control rule is mainly determined according to the expert experience and the characteristics of the controlled object. Table 1 and Table 2 list the fuzzy control rule table in this paper.

Table 1. K_P fuzzy rules table

$e_c(k)$ / e	NB	NM	NS	ZO	PS	PM	PB
NB	PB	PB	PM	PM	PM	PS	ZO
NM	PB	PB	PM	PS	PS	ZO	NS
NS	PM	PM	PS	ZO	ZO	NS	NS
ZO	PM	PS	ZO	ZO	ZO	NS	NM
PS	PS	PS	ZO	ZO	NS	NM	NM
PM	PS	ZO	NS	NS	NM	NM	NB
PB	ZO	NS	NM	NM	NB	NB	NB

Table 2. K_I fuzzy rules table

$e_c(k)$ / e	NB	NM	NS	ZO	PS	PM	PB
NB	NB	NB	NM	NM	NM	NS	ZO
NM	NB	NM	NM	NM	NS	ZO	PS
NS	NM	NM	NS	ZO	ZO	PS	PS
ZO	NM	NM	ZO	ZO	ZO	PM	PM
PS	NM	NS	ZO	ZO	PS	PM	PM
PM	NS	ZO	PS	PS	PM	PB	PB
PB	ZO	PS	PM	PM	PM	PB	PB

4. Simulation Results and Analysis

According to the designed sliding mode observer and fuzzy PI controller, a PMSM speed sensorless control system with fuzzy PI and sliding mode observer is established. In the Simulink environment, the entire system model was built to verify and analyze the effectiveness and stability of the system. The specific system flow chart simulation diagram is shown in Figure 2.

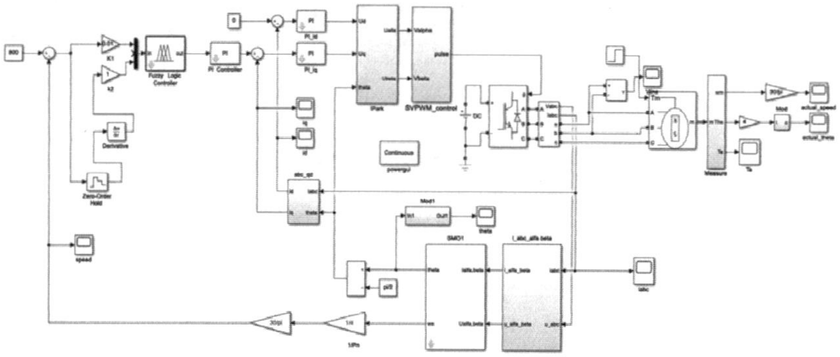

Figure 2. Flow chart of the speed sensorless control system of fuzzy PI and sliding mode observer PMSM

The PMSM parameters used in this chapter can be seen in Table 3 below:

Table 3. Parameters of PMSM

parameter	Parameter value
Permanent magnet flux /Wb	0.175
Wire resistance /Ω	2.875
Line inductance /mH	8.5
Moment of inertia /($g \cdot m^2$)	0.003
Number of pole pairs	4

Set the motor's given speed to 800r/min, the simulation time to be 0.1 s, and the initial load torque to be 0N. At the same time, give the motor a load of 5 N at 0.05s to verify the anti-disturbance ability of the motor speed. Simulation results as shown in Figure 3 and Figure 4.

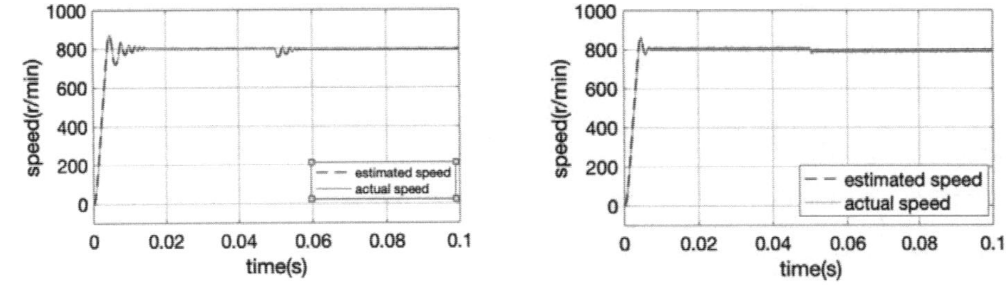

(a) Speed Simulation of Traditional SMO (b) Speed Simulation of Fuzzy PI and SMO

Figure 3. Comparison of rotational speed simulation between traditional SMO and fuzzy PI and SMO

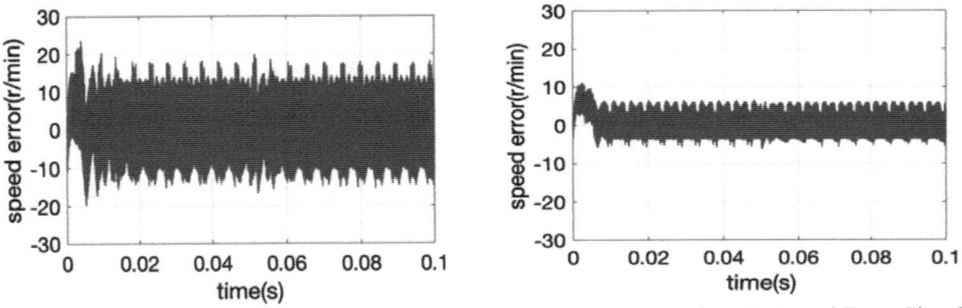

(a) Simulation of Speed Error of Traditional SMO (b) Fuzzy PI and SMO Speed Error Simulation

Figure 4. Comparison of speed error simulation of traditional SMO and fuzzy PI and SMO

Figure 3 and Figure 4 are the speed comparison and the speed error comparison between the traditional sliding mode observer and the control method based on sliding mode observer and fuzzy PI. The fuzzy PI and sliding mode observer systems have faster response time and smaller speed fluctuations. After the 0.05s sudden increase in complexity, the recovery time after interference is shorter, the stability is faster, and the robustness is better. It has better control effect and better performance than traditional sliding mode observer.

5. Conclusion

Due to the poor anti-disturbance ability of existing sliding mode control, this paper combines fuzzy PI control and sliding mode observer system to propose a PMSM control strategy based on fuzzy PI and sliding mode observer. The controller is designed and adopted Simulink conducts simulation analysis and compares the results. The simulation experiment results confirm the feasibility and effectiveness of the method posed in this article. Compared with existing sliding mode control, the control strategy posed in this article has shorter system revise time and better property. The system overshoot of the control strategy in this paper needs to be improved. In the future Industry 5.0 era, the control strategy proposed in this article can be used to control permanent magnet synchronous motors.

References

[1] Psonis T, Nikolakopoulos P G. Electromagnetic Analysis of Fuzzy Controlled Active Magnetic Bearings [C]. MATEC Web of Conferences,2018: 04011.

[2] YE S. Fuzzy sliding mode observer with dual SOGI-FLL in sensorless control of PMSM drives[J]. ISA Transactions,2019,85: 161.

[3] LI Neng. Modeling and Simulation of permanent magnet synchronous motor vector control system[J]. Mechani□cal engineer, 2017, (2):47-48.

[4] WAN Dechao, SUN Shumin, YIN Wenquan. Control and application of flywheel energy storage system based on MRAS[J]. Smart Power, 2019, 47(6): 101-106.

Waveform Correlation Analysis Method for Mechanical Characteristics of Circuit Breaker in Intelligent GIS Equipment

Zhongqing Sang, Xiaodong Gu*, Yongxin Wang, Weiyun Jiang and Wenjuan Cheng

Xiamen University of Technology, 361024, Fujian, China

*Corresponding author email: 2022031073@s.xmut.edu.cn

Abstract. Gas-insulated electrical appliances (GIS) have the characteristics of small footprint and low cost. Therefore, GIS is widely used in intelligent substations. However, once the GIS fails, there will be shortcomings of long maintenance cycle, which seriously affects the normal life and production activities. Based on this, on the basis of summarizing and analyzing the GIS intelligent technology, this paper studies the correlation analysis method of the mechanical characteristic waveform of the intelligent GIS equipment circuit breaker from the aspects of equipment intelligence, device development and software development,corresponding theoretical guidance is provided for the practical application of intelligent GIS equipment circuit breakers.
Keywords: Intelligent GIS equipment; Breaker; Mechanical properties; SF6.

1. Introduction

In recent years, there have been more and more studies on the mechanical characteristics of intelligent GIS circuit breakers. Xiao Tianwei and others discussed in detail the analysis method and realization principle of the correlation of the waveforms of the mechanical characteristics of the intelligent GIS equipment circuit breakers, and analyzed the waveforms of the curves formed at different degrees of closing, and set the reference evaluation criteria of the mechanical state under the angle of deviation(Xiao et al, 2013). Zheng boqian and others pointed out that it is an important part of checking the operating mechanical characteristics of circuit breakers that the results of mechanical inspection can be carried out regularly under the condition of power failure. Therefore, a series of experiments were carried out in accordance with the standardized process, and the experimental results were effectively analyzed in order to discover the hidden dangers and defects of the circuit breaker, and put forward suggestions for improving the safe and stable operation of the related hidden dangers(Zheng et al, 2019). Liu Hongzheng and others introduced the GIS high-frequency partial discharge, micro-water measurement, lightning protection online monitoring system in terms of structure, application, function, etc., based on the selection principle of new electronic transformers, summarized the possible installation methods, and finally proposed related designs requirements and development direction(Liu et al, 2014). Yue Guohui pointed out that as a new type of equipment, GIS equipment is widely used in power system. Therefore, this paper introduces the basic process of GIS equipment installation, and discusses the method of GIS loop resistance measurement and resistance maximization(Yue, 2013). Gu Xianhua and others analyzed the actual situation of the current 500kv and tank double-break opening and closing times, and pointed out that the current 500kv does not fully reflect the specific working conditions of the circuit breaker, so it is necessary to propose multiple circuit breaker opening and closing times for specific work(Gu et al, 2010). Si Wei and others studied the fault accidents of 220kv GIS disconnectors, and proposed that when GIS disconnectors use spring devices, they have poor stability. Therefore, it should be replaced with an electric device(Si and Si, 2018).

With the proposal and continuous application of the concept of energy internet, domestic investment in infrastructure for smart grids and UHV transmission and transformation projects has been increased, so higher-level requirements can be put forward for safe operation. This is mainly because the substation is an important node for electric power to perform electric energy reception, flow direction control, voltage change, power flow and control, voltage adjustment and other actions, and can effectively connect the power plant and the user. In recent years, with the continuous development of electronic information technology, substations have developed from traditional power stations to intelligent substations with advanced application functions such as automatic real-time control, intelligent adjustment, online analysis and decision-making, and collaborative interaction. Intelligent power equipment is the main body of the smart grid, and the intelligent gas-insulated closed combined electrical appliances have the advantages of high reliability and little external influence, and it was widely used in the design of smart grids. Specifically, the intelligent GIS encloses the isolating switch, transformer, connector and circuit breaker in a metal shell connected to the ground for specific work and has high safety. After years of development, GIS has been called maintenance-free equipment. However, looking at the specific operations of GIS in the actual work of the global power system, it is found that in recent years, safety accidents often occur during the operation of GIS. The failure rate of GIS is far greater than the regulations of the International Electrotechnical Commission and the International Large Grid Conference, this time it is every 100 intervals / 0.1 times per year. In addition, because of the encapsulation characteristics of GIS, once a failure occurs, there will be a longer maintenance cycle, which will cause a longer power outage. Therefore, in view of the problems in the current application process of GIS online detection in smart substations, this paper conducts an in-depth study on the smart GIS online detection system, and further designs the system to detect its operating status based on the smart GIS technology. The purpose is to provide a solid operation for the smart substation basis.

2. Intelligent GIS Mechanical Parameter Acquisition and Waveform Principle

GIS intelligence refers to the fusion of power electronic technology, sensor technology and electrical control technology, and the application of GIS primary and secondary departments to monitor the GIS operating status in real time. In the actual operation process, the GIS operation status information is timely transmitted to the unit with control, protection, and measurement functions, and the intelligent operation is fully realized, which greatly improves the safety and reliability of the GIS operation process.

2.1. Parameter Analysis

The GIS equipment is filled with SF_6 gas, SF_6 gas has the characteristics of odorless, odorless, non-flammable, and it is an excellent insulator material. At room temperature, SF_6 has relatively stable properties and is very suitable for use in the industrial field. The specific performance in the inspection is that once the insulation defects occur inside the combined electrical, the arc will generate thousands of degrees of high temperature to decompose the SF_6 gas (see Figure 1), producing SOF_2, HF and other gases. Therefore, the use of SF_6 gas to detect the internal pressure can effectively analyze the cause of GIS internal faults.

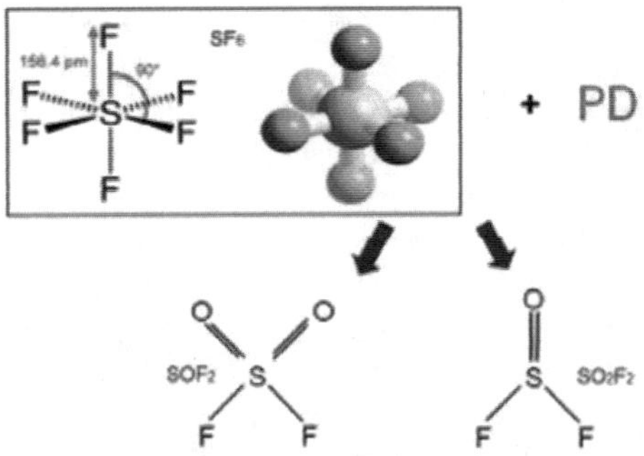

Figure 1. SF6 Decomposition.

In view of the failure of the gas insulator assembly, the smart gas insulator assembly mainly includes temperature state parameters such as SF6 gas and partial discharge contacts. The specific parameter settings are shown in Table 1.

Table 1. Formatting sections, subsections and subsubsections.

Monitoring object	State parameter
Circuit breaker air chamber	Opening and closing time
	Opening and closing speed
	Working state of energy storage motor
	temperature

2.2. Skills Requirement

When the intelligent GIS on-line monitoring system for gas insulated combined electrical appliances achieves the monitoring objectives, it is necessary to follow the corresponding technical requirements, as shown in Figure 2. In Figure 2, there are eight technical requirements: network system architecture, information sharing platform, device status visualization, monitoring target panorama, information display integration, information data, communication protocol standardization, and monitoring function odularization.

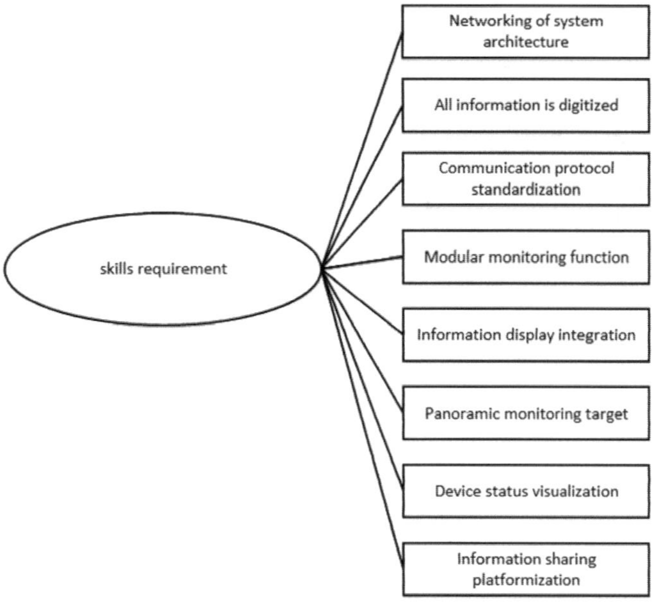

Figure 2. Technical Requirements.

In Figure 2, the network architecture of the system mainly requires the segmented control of the system structure within the station, which mainly includes the station control layer, process layer and bay layer, etc., and performs corresponding data transmission and control according to the substation communication network and system standards. Digitalization in the station mainly requires that the body and accessories of the station's digitized equipment must have the data required for intelligent control. The standardization of the protocol in the communication process mainly means that the data communication of the entire site must strictly follow the internal network communication standard of the substation. The detection function modularization mainly means that the detection function can be flexibly configured according to the actual situation, and the corresponding communication protocol is followed between different modules to realize timely insertion and use. The integration of information display mainly means that users can obtain relevant information through the information platform on the station, get the most detailed fault diagnosis information, and maximize the system availability. Panoramic detection is mainly for the detection of key equipment in substations in all aspects. For example, carry out a full range of tests on switches, surge arresters, transformers, etc., to ensure that all instruments in the station are visualized. The device status visualization mainly refers to the analysis of the detection data and the data obtained by the communication exchange by the intelligent device, so as to realize the visualization processing of the data. Information platformization mainly means that the data information in the station should be shared to meet the needs of centralized detection, and the station control layer should use the integrated information platform to connect to the data.

2.3. Waveform Principle

The characteristic value of the intelligent switch mainly includes the current characteristic value and the time characteristic value in the current waveform of the opening and closing coil. Current characteristic value and time characteristic value in current waveform of energy storage motor. The characteristic value and time characteristic value of the opening (closing) displacement of the time displacement waveform during the opening (closing) process. Therefore, important information for mechanical fault diagnosis can be extracted from the current waveform of the DC electromagnetic coil.

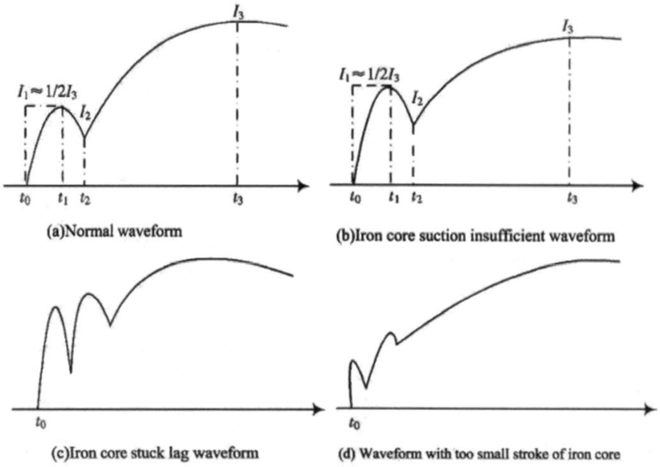

Figure 3. The Typical Waveform Of A Circuit Breaker Operating Coil Current

In Figure 3,several typical waveforms of circuit breaker operating coil current are shown.Figure 3(a) is the current waveform of the coil in normal operation, the time t0-t1 is the triggering stage of the iron core, and the time t1-t2 is the time when the iron core starts to accelerate under the action of electromagnetic force, overcoming the resistance of gravity and spring , until the core reaches the contacts.As shown in Figure 3(b) and Figure 3(c), the suction force of the iron core is insufficient or the resistance is too large, and whether the iron core is stuck or the mechanical load of energy release changes. The time from t2 to t3 is the process of driving the circuit breaker contacts to open and close through the transmission system. The iron core stops moving at t2, and the contact starts moving after t2. t3 is the cut-off time of the auxiliary contact of the circuit breaker, so t0~t3 or t2~t3 can reflect the state of the operation of the circuit breaker operating system.Therefore, by analyzing the current waveform of the coil and the characteristic quantities of t1, t2, t3, I1, I2:, I3, etc., parameters such as the starting time of the iron core, the time of movement, and the time of the coil energization can be analyzed and calculated, so as to obtain the iron core According to the parameter range of the circuit breaker itself, compare and judge whether the operating mechanism has iron core empty stroke, spring stuck and other faults, so as to monitor the operating status of the operating mechanism. Forewarning of failure precursors.

3. System Design

Based on the analysis of the waveform-related theory of the mechanical characteristics of the intelligent GIS circuit breaker, a corresponding system is constructed to evaluate it online. The GIS detection system can be divided into two parts, namely hardware and software systems. Among them, the hardware system is an important part of the GIS detection system, mainly composed of the sensor type and detection IED. Therefore, it is necessary to analyze the mechanical characteristics of GIS circuit breakers during system design and introduce various detection IED (Liang and Lin, 2019). In addition, during the system design, the article also puts forward the information integration platform model construction scheme and overall architecture, and conducts a more detailed analysis.

3.1. Hardware System

The mechanical characteristics of the circuit breaker require that the contact stroke be detected during the detection process, and the main goal is to calculate the opening and closing time and synchronization. Therefore, it is required that the mechanical property sensor of the gas insulated combined electrical appliance belongs to the displacement sensor. The specific execution process is that when the circuit breaker is opened and closed, the rotation of the main shaft drives the contacts to separate (Li et al, 2016). During operation, there is a correlation between the rotation angle of the main shaft and the linear displacement of the contact, so the displacement angle of the contact can be further calculated by measuring the displacement angle of the main shaft (Zhang et al, 2015). In addition, the contact is at a high potential level, and the main shaft is at a low potential level. There is a certain difference between the potential levels of the two. Therefore, there is a certain distance between the two and the high voltage part of the circuit breaker, so the high potential isolation can be considered differently during operation.

As can be seen from the previous section, SF6 gas has a very good effect in the insulated combination electrical appliances. After the SF6 gas reaches a certain level of water content, if a partial discharge occurs, the arc in the station reaches a high temperature of thousands of degrees will cause SF6 to undergo a chemical reaction, decomposing into S atomic gas and F atomic gas, which will chemically corrode the equipment and affect it electrical equipment is operating normally (Liu et al, 2013). Therefore, in the system design, it is necessary to detect the SF6 micro water content in time to ensure the safe operation of electrical equipment.

3.2. Monitoring IED Development

Circuit breaker mechanical characteristic detection IED can effectively analyze waveform related methods, and meet the waveform state detection between various types of circuit breakers from 100KV to 1000KV. Specifically, the intelligent characteristics of GIS switches determine the specific operating conditions of circuit breakers through a variety of mechanical characteristics and online detection of coil current and breaking current between opening and closing. When testing IED, it is necessary to use SF6 gas to detect IED. In actual testing, the density of SF6 gas will have a profound impact on the insulation and power-off effects of SF6 power equipment. In gas insulated combined electrical appliances, if the SF6 gas micro-water content reaches a certain limit, the SF6 gas will undergo chemical action at high temperature of the arc, which will affect the normal operation of the equipment. In the actual detection process, the SF6 gas detection IED can effectively meet the detection of SF6 gas density and micro water content in GIS, and observe the changes in micro water content in GIS in time. SF6 gas density and micro water detection sensor, high-precision pressure sensor and temperature sensor, convert the measured data into digital signal through analog to digital, input the corresponding processing cell for compensation operation and processing, and finally pass the data during the processing through RS485 . First feedback to the SF6 gas density micro water detection IED to realize the detection of micro water content in SF6 gas.

In the actual testing process, the SF6 gas density micro water testing IED testing supervision has in many cases exceeded the national standards and more accurate data can be obtained. Therefore, in the actual detection process, electrical equipment such as circuit breakers and intelligent GIS can be tested for a long time, and the content of micro water in SF6 gas in related equipment can be recorded. When the relevant indicators in SF6 gas change, the corresponding trend chart can be drawn according to the actual changes. When the relevant index breaks through the warning line, the detector will automatically trigger the alarm device to generate an alarm function. When the corresponding indicator seriously exceeds the warning line in a dangerous state, the detector will automatically trigger the alarm device and activate the locking device to prevent the circuit breaker from continuing to work after the alarm occurs, and is designed to ensure the safe operation .

After the SF6 gas detection IED is completed, the IED partial discharge needs to be detected. Specifically, for the detection of IED partial discharge, the detection of partial discharge signals is carried out by the UHF method. When the partial discharge detection IED realizes the sampling of partial discharge signals at various points through a certain period, the sampling period can be adjusted according to the actual needs of the site. Among them, the shortest sampling period can be adjusted to 1min. IED partial discharge detection has four synchronous high-speed sampling ports, so four UHF sensors can be configured at the same time, which can effectively reduce noise interference. In the actual detection process, the GIS partial discharge detection IED mainly analyzes and detects the characteristic parameters such as the phase of the maximum discharge volume, the average discharge volume and the number of discharges in the partial discharge. In addition, the IED partial discharge detection can also use solid-state hard disk storage technology to achieve timely storage of test data, which can effectively save historical data from 24 hours to more than one year. When needed, we can query through the network and use historical data, we can query the original data and result data.

4. System Design

4.1. Platform Construction

In order to effectively monitor intelligent GIS online monitoring and diagnosis, it is necessary to build an information integration platform model for research. In the modeling process, a unified modeling language can be used to establish the association between classes and classes, and to complete panoramic information modeling. There are three issues to pay attention to when expanding CIM. First, whether new packages need to be created; second, which types need to be extended; Third, which existing type properties and attribute values should be extended. The specific steps are as follows:
(1)Determine the main object of modeling and determine whether there is a model package related to this equipment in the standard.
(2)Determine whether the waiting extension type exists in the new model package. If it exists, consider expanding the attributes. If it does not exist, consider creating a corresponding attribute package.
(3)Describe the relationship between the type to be expanded and the formed type, mainly including the relationship between aggregation and inheritance.
(4)Determine whether the data type of the newly created type attribute and the extended type attribute already exists. If it exists, you need to use the relevant data type; if it does not exist, you need to create a new data type.
There are three principles that need to be followed in the specification. First, the multi-attribute class is decomposed into multiple single attribute classes; second, redundant data is eliminated; Thirdly, try to select the nouns in the examples as the decomposition class names, and define the key nouns of multiple categories.

4.2. Overall Architecture

In the online detection system of intelligent substation combined appliances, the station control layer data can comprehensively analyze the system software operating platform,and the embedded real-time multitask Linux operating system and corresponding PI background data are used in the analysis. The constructed system is mainly based on object-oriented technology and modular design concept, encapsulating functions in reusable modules, and each module can be connected to each other through a standard interface, which can be configured according to demand in practical applications. It can reflect the characteristics of strong scalability and high reusability of modules, and carry out practical research. In the specific development of system software, we need to build a detailed framework based on J2EE architecture model, integrating browser and server, client and server. The construction model can support cross-platform operation, provide corresponding data interaction and shared collaborative application environment, and support distributed maintenance and management of data.
The system station control layer comprehensive analysis software is mainly embedded into the information integration platform through modular form, real-time detection of field equipment operation parameters, and diagnosis according to the obtained data. At the same time, it can also provide the corresponding system login, historical data, spectrum data management and other functions to meet the functions of the platform.

4.3. Comprehensive Diagnostic Evaluation

The intelligent GIS circuit breaker mechanical characteristic evaluation method is mainly to accurately evaluate the state of the circuit breaker by detecting the contact formation and operation characteristics of the circuit breaker. Two methods of GIS equipment SF6 gas leakage fault diagnosis threshold comparison method and relative leakage rate detection are often used. Specifically, the threshold comparison method mainly detects whether the SF6 gas pressure is less than a preset value. If the measured gas pressure is less than the predicted value, it is considered that SF6 gas has a leakage phenomenon; if the change rate is greater than the preset value, it means that SF6 gas has the possibility of leakage. Therefore, at this time, the SF6 gas detection IED needs to have a self-assessment function, and the self-assessment results need to be sent to the system service through the process network by setting the message format of the failure part, failure mode, and equipment status according to user needs. And when the assessment risk increases by 2%, the message will increase once.

When conducting comprehensive diagnosis and evaluation, it is necessary to measure the leakage current of the arrester in time, and then realize the dynamic monitoring of the arrester. The specific process is to encapsulate the data according to the IEC61580 protocol through the data and equipment status, realize the fiber optic Ethernet to transmit the data to the information integration platform, and realize the dynamic monitoring of the insulation status.

5. Conclusion

Through the selection of GIS circuit breaker mechanical characteristics, SF6 gas and arrester current detection sensors, the corresponding detection sensors are designed, which can effectively reduce the sampling efficiency and realize the collection of local electrical signals. By showing the selection and development process of GIS intelligent circuit breaker mechanical characteristics, SF6 gas, partial discharge, lightning arrester current and detection IED, it is possible to effectively detect the GIS circuit breaker mechanical characteristic waveform correlation. At the same time, the GIS intelligent circuit breaker based on the information integration platform can be used for online detection, diagnosis and evaluation software, which can realize the comprehensive diagnosis of GIS equipment and effectively improve the accuracy of the diagnosis results. In addition, the construction system can accurately find the corresponding equipment type, name and obtain GIS detection information to ensure the normal operation of the equipment.

References

[1] Xiao T.W, Li G, Qin Y, et al. (2013). Waveform Correlation Analysis Method for Mechanical Characteristics of Intelligent GIS Circuit Breakers. China Science and Technology, 12(23), 158-158

[2] Zheng Bo.Q, Jiang B.L, Zhang X.D, et al. (2019). Key Points of High-voltage Circuit Breaker Mechanical Characteristics Test. Electrical Switch, 57(2), 10-12

[3] Liu H.Z, Qu W.T, Ping H.T. (2014). Research and Application of GI Intelligent Design. Shandong Electric Power Technology, 41(5), 1-5

[4] Yue G.H. (2013). Several Problems in the Installation and Debugging of GIS Equipment. China Science and Technology Investment, 12(24), 235-235

[5] Gu X.H, Cao Y, Li K. (2010). 500 kV GIS and Tank Circuit Breaker Opening and Closing Time Test Results. Shandong Electric Power Technology, 17(2), 18-20, 25

[6] Si W, Si H.X.(2018).220kV GIS Research and Improvement on Mechanical Characteristics of 220 kV GIS Disconnector. High Voltage Electrical Appliances, 54(3), 243-250

[7] Liang J.F, Lin M.W. (2019). Development of a Special Bracket for the Mechanical Characteristic Sensor of the JK-550 Rapid Ground Knife in Xikai. Electrical Technology,40 (8), 47-48

[8] Li X.G, Wu X.T, Zhu H.B, et al. (2016). Research on Contacting Fault of GIS Contact Based on Vibration Signal. High Voltage Electrical Appliances,52(10), 165-169,175

[9] Zhang X, Tao W, Zhao Y.M, et al. (2015). Statistics and Analysis of Circuit Breaker Mechanical Characteristic Data in GIS above 110kV. High-voltage Electrical Appliances, 49(9), 135-139

[10] Liu F.Y, Su X.P, Wang Y.B. (2013). Application of Angular Displacement Sensors in the Testing of Mechanical Characteristics of Intelligent Circuit Breakers. Machine Tool and Hydraulics, 41(12), 23-27

Design and Analysis of a New Cooling Structure for Interior Permanent Magnet Synchronous Motor with Built-in Fan

Jin Wang, Lihui Chen, Zhanyang Yu, Zicheng Wang and Yan Li*

National Engineering Research Center for Rare Earth Permanent Magnet Machines, Shenyang University of Technology, China.

*Corresponding author email: eeliyan@126.com

Abstract. Aiming at the overheating problem of interior permanent magnet synchronous motor (IPMSM) for compressors due to the limitation of working environment, a new type of IPMSM with built-in fan based on the axial ventilation cooling structure is proposed. Firstly, an axial fan is installed inside the rotor without affecting the electromagnetic performance of motor, and the air velocity distribution and temperature distribution characteristics inside motor are analysed by computational fluid dynamics (CFD). Secondly, the influence of the position of the inlet and outlet on the heat flow distribution is analyzed. Finally, for the problem that the heat dissipation effect of the ventilation slots on casing is not obvious, the ventilation slots are removed and the air outlets are installed on front cover. And the results of heat flow characteristics show that this cooling structure is optimal.
Keywords: IPMSM; CFD; Cooling structure; Flow-heat coupling.

1. Introduction
In order to solve the overheating problem of driving motor, its cooling structure is also the focus of the designer. According to the different cooling medium, the cooling method can be mainly divided into air cooling and liquid cooling. The air cooling structure is simple, mainly including natural cooling and forced air cooling[1]. Liquid cooling methods mainly include water cooling and oil cooling. Compared with air cooling, the cooling effect is better, but the cost is increased. A cooling structure of wafters on the rotor of the axial flux PMSM to improve the heat dissipation capacity of the motor was proposed in [2]. The influence of different ventilation cooling methods on the heat dissipation of the rotor was analysed in [3]. In [4], the cooling capacity of the radial forced-ventilation permanent magnet synchronous wind generator was improved by comparing the heat performance of different air cooling channels and channel widths. The thermal analysis of a air-cooled 8-poles 6-slots switched reluctance motor(SRM) with an internal windings cooling structure was researched in [5]. In order to reduce the peak temperature of stator, the model of motor was parameterized, and the analysis showed that setting the vent at the best position can reduce peak temperature of the stator[6]. The air-cooling effect of high-speed motor in modern high-frequency traction drive was analysed in [7].

2. Physical Model of a New Structure of IPMSM with Built-in Fan
A new cooling structure of IPMSM mainly includes built-in fan on the rotor, air inlets on the rear end cover, and air outlets on the front end of casing, as shown in Figure 1.

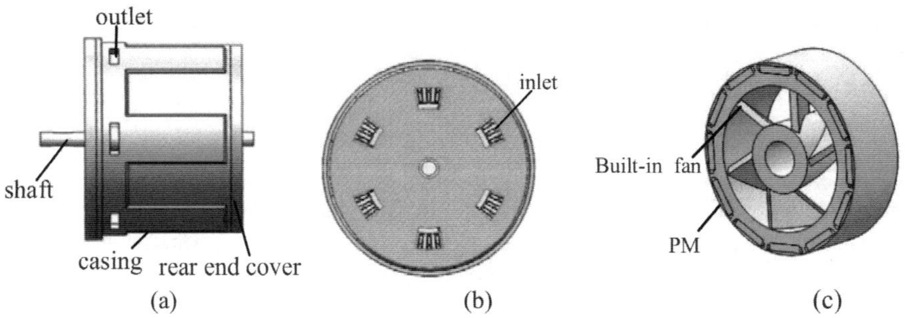

Figure 1. Physical model of cooling structure. (a) Front view. (b) Rear view. (c) Rotor structure.

2.1. Loss Calculation

2.1.1 Copper loss calculation. Copper loss mainly refers to the loss caused by the current flowing through the stator winding. The calculation formula is

$$P_{cu} = mI^2R \tag{1}$$

Where, I is the phase current, and R is phase winding resistance under rated conditions.

2.1.2 Iron loss calculation. The iron loss is the power loss caused by alternating or pulsating magnetic field in the iron core. According to the classic iron loss theory, the iron loss calculation is mainly classified into hysteresis loss and eddy current loss. The iron loss calculation formula is as follows.

$$\begin{cases} P_{Fe} = P_h + P_e \\ P_h = k_h f B_m^{\alpha} \\ P_e = k_e f^2 B_m^2 \end{cases} \tag{2}$$

Where, P_h is the hysteresis loss, P_e is the eddy current loss, K_h and K_e are related calculation coefficients, B_m is the magnetic density amplitude, f is the power frequency and α is the empirical coefficient.

2.1.3 PM eddy current loss calculation. The change of the magnetic field density will generate eddy currents and result in eddy current loss in the PMs. The calculation formula is as follows:

$$P_{eddy} = \int_V E \cdot J dV = \int_V \frac{J^2}{\sigma} dV = \int_V \rho J^2 dV \tag{3}$$

Where, E is the electric field strength, J is the eddy current density, σ is the conductivity of PM, ρ is the resistivity of PM, and when it is isotropic, $\rho = 1/\sigma$. And V is the volume of PM.

2.2. Basic Assumptions and Boundary Conditions

Certain complex-shaped structures increase the difficulty of forming high-quality grids, and have almost no effect on the simulation results. Therefore, the motor model can be appropriately simplified and the following assumptions can be made:

- The flowing air is mainly caused by rotating rotor and fan. The flow velocity is much smaller than the sound speed, and its Mach number is very small. So air in the motor can be regarded as an incompressible fluid.
- All materials do not change with temperature so the thermal conductivity of material is constant.
- The copper wires are equivalent to copper block by the principle of area equivalence, and the skin effect of winding is ignored.

The pressure-based solver and the RNG k-ε turbulence model are chosen, the boundary conditions used are as follows:

- The air inlet is the pressure inlet condition.
- The air outlet is the pressure outlet condition.

- The fluid in the rotating area adopts the MRF model, and the given rated speed is 3000r/min.

3. Distribution of Flow Heat Characteristics Inside Motor

3.1. The Distribution of Internal Air Velocity

As shown in Figure 2, the speed on both sides of fan is high, and the maximum speed is 21.14m/s. The air in air-gap has an axial flow velocity, and the direction of air velocity in the ventilation slot of casing is the same as that in air-gap. It can be obtained that after the air enters motor from air inlet, it flows to the front end-cavity under rotating pressure of fan.

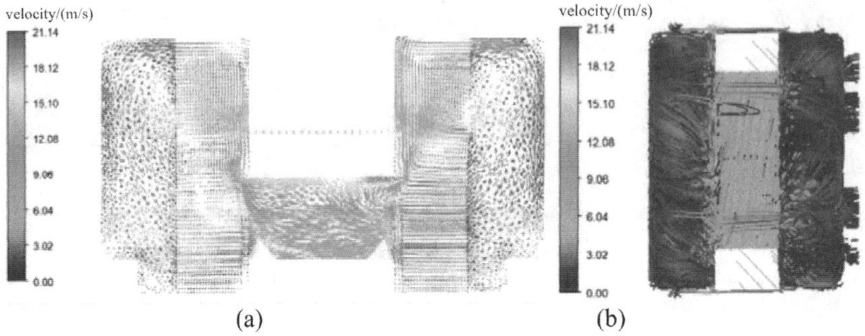

Figure 2. The velocity distribution of internal air. (a)Velocity vector. (b) Flow trail.

3.2. Results of Temperature Field

As shown in Figure 3, the winding temperature reaches 112.21℃ in the entire motor domain. The temperature at the air inlet and outlet is close to outer temperature due to direct convective heat exchange with external environment. The temperature distribution along radial direction of the motor is shown in Figure 4. It can be obtained that the upper winding temperature is highest, the change gradient at air gap is very large, and rotor temperature is low because it is directly affected by the flowing air under fan rotating.

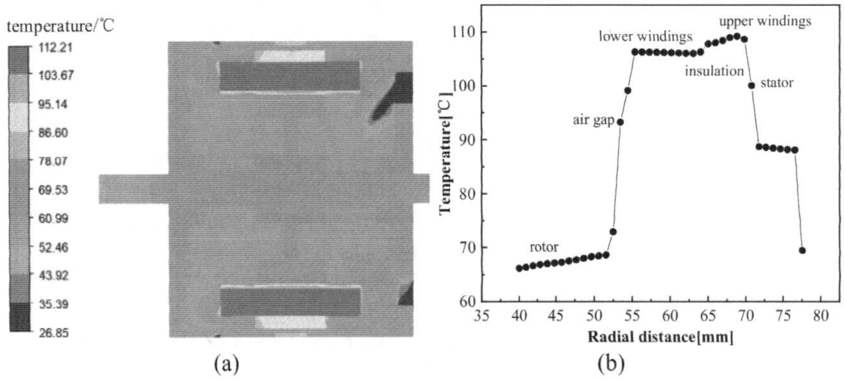

Figure 3. The distribution of heat source temperature. (a) The overall temperature. (b) Radial temperature curve.

The temperature of rotor core, built-in fan and PMs are shown in Figure 5. The temperature at the front end is lower than at the back end because the air velocity increases after the fan rotating, that is, the front end has a better heat dissipation effect.

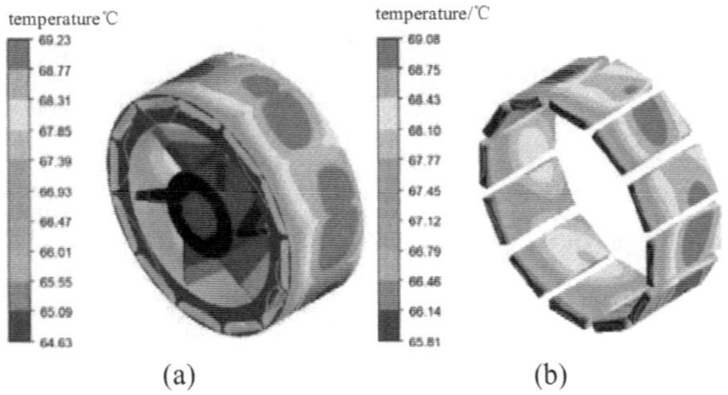

Figure 4. The temperature of PMs and rotor. (a) The temperature of rotor. (b) The temperature of PMs.

4. Improvement of Cooling System

4.1. The Inlet and Outlet Positions Are on the Same Side

The air inlet is changed to the center of rear end cover, and the air outlet is changed to the same side. It can be obtained from the Figure 5 that after the air flows in from the air inlets, it passes through the air-gap and reaches the front end cover, and then returns to the air outlets along the ventilation slots of casing.

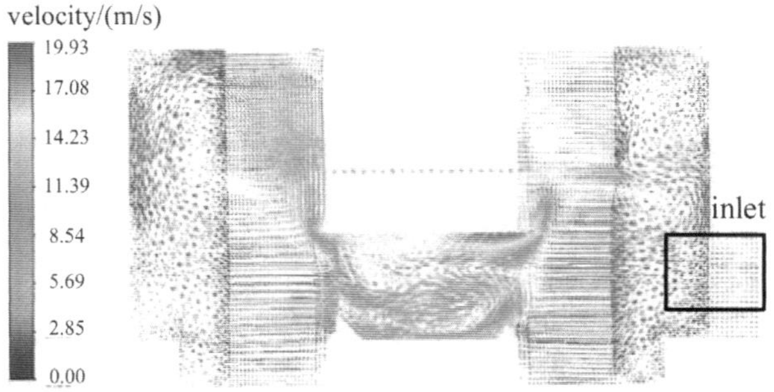

Figure 5. The distribution of air velocity(air inlet and outlet on same side).

It can be obtained from the Figure 6 that the highest winding temperature is 110.33°C, which is 1.88°C lower than before improvement, and the lowest temperature is concentrated near the air inlets and outlets. The results shows that the cooling effect has been improved.

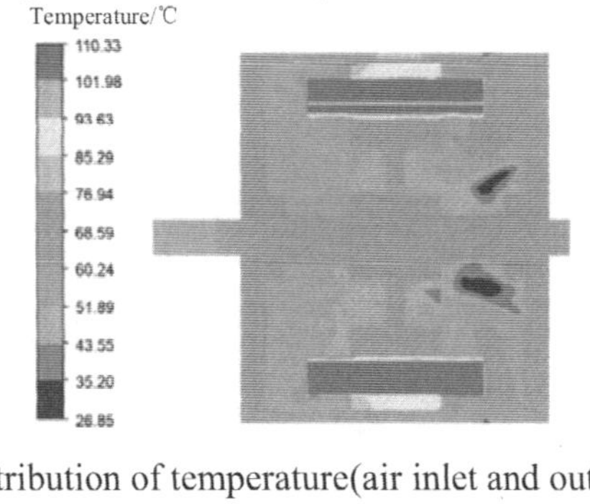

Figure 6. The distribution of temperature(air inlet and outlet on same sides).

4.2. The Air Inlets and Outlets Are Located on the Front and Rear Covers, and the Ventilation Slots of the Casing are Removed

Aim at the overheating problem of stator core above-mentioned, the air inlets and outlets are respectively installed at the same center of the end covers, and the ventilation slots on casing are removed.

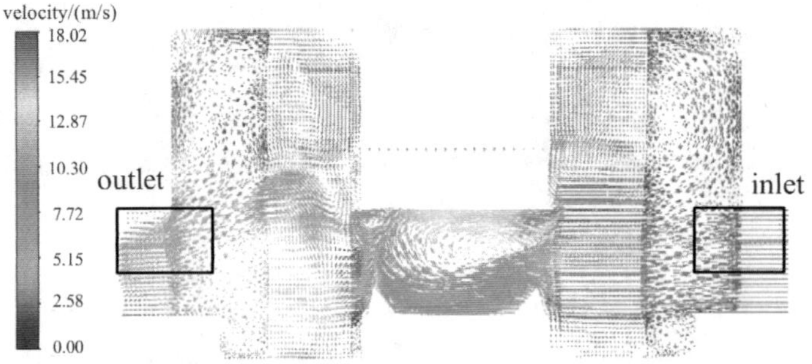

Figure 7. The distribution of air velocity(air inlet and outlet on different side).

The temperature distribution is shown in Figure 8. It can be obtained that the peak temperature is effectively reduced, and the stator temperature is similar to casing. This is due to the stronger heat conduction compared to the previous heat convection. Meanwhile the range of temperature close to external environment is larger, so this cooling structure has a better heat dissipation effect.

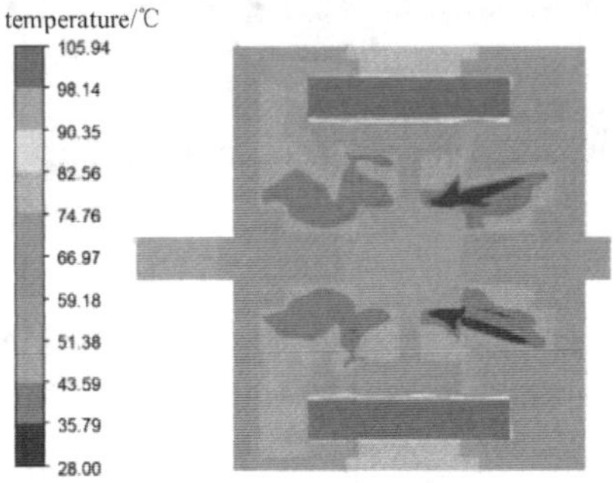

Figure 8. The distribution of temperature(air inlet and outlet on different side).

5. Conclusion

A new cooling structure of IPMSM with built-in fan is designed and analysed by combining CFD method. When the air outlets position are installed at the center of end cover, the flow resistance of flow channel is the smallest, and the direction of flow velocity vector of inlet and outlet is basically parallel. And the final improved cooling structure has the best heat dissipation efficiency, the maximum vaule of whole machine temperature has been reduced by 6.27℃, and the temperature of the stator has also been decreased. So the optimal structure of cooling structure with built-in fan is that the air inlet and outlet are at the same centerline of the end caps on both sides, which provides a reference for the engineering project.

6. Future work

We gratelfully acknowledge National Engineering Research Center for Rare Earth Permanent Magnet Machines for the hardware support provided in the research project. In the future, a prototype will be manufactured and the cooling effect will be tested, which will certify that the cooling structure has good heat dissipation characteristics.

References

[1] Bin, X., GuoBiao, G., Lin, R., & JinShui, G. (2014). Studies on the structure of radial ventilation channel to improve the cooling capacity of large turbo generator stator. In *2014 17th International Conference on Electrical Machines and Systems* (pp. 354-357). Hang zhou,China.

[2] Fawzal, A. S., Cirstea, R. M., & Blundell, M. (2016). The fan design impact on the rotor cooling of axial flux permanent magnet machines. In *2016 XXII International Conference on Electrical Machines* (pp. 2725-2731). Lausanne, Switzerland.

[3] Huang, Z., Fang, J., Liu, X., & Han, B. (2015). Loss calculation and thermal analysis of rotors supported by active magnetic bearings for high-speed permanent-magnet electrical machines. *IEEE Transactions on Industrial Electronics*, 63(4), 2027-2035.

[4] Fan, X., Qu, R., Li, J., Li, D., Zhang, B., & Wang, C. (2017). Ventilation and thermal improvement of radial forced air-cooled FSCW permanent magnet synchronous wind generators. *IEEE Transactions on Industry Applications*, 53(4), 3447-3456.

[5] Rahman, N. A., Bostanci, E., & Fahimi, B. (2016). Thermal analysis of switched reluctance motor with direct in-winding cooling system. In *2016 IEEE Conference on Electromagnetic Field Computation* (pp. 1-1). Miami, FL.

[6] Bersch, K., Connor, P. H., Eastwick, C. N., Galea, M., & Rolston, R. (2017). CFD optimisation of the thermal design for a vented electrical machine. In *2017 IEEE Workshop on Electrical Machines Design, Control and Diagnosis* (pp. 39-44). Nottingham,UK.

[7] Jaeger, M., Ruf, A., Hameyer, K., & Grosse-von Tongeln, T. (2018). Thermal analysis of an electrical traction motor with an air cooled rotor. In *2018 IEEE Transportation Electrification Conference and Expo* (pp. 467-470). Long Beach, USA.

Sliding Mode Control of Permanent Magnet Synchronous Motor Based on Disturbance Observer

Zehang Song[1,*], Linlin Sha[1] and Xiaodong Yao[1,2]

[1]School of Electrical Engineering, Shanghai DianJi University. Shanghai 201306, China
[2]Minge New-Type Motor and Electronic Control Research Institute. Foshan City, Guangdong Province, 528599, China

*Corresponding author email: 206001010310@st.sdju.edu.cn

Abstract. A sliding mode control method of surface mount permannent magnet synchronous motor (SPMSM) based on disturbance observer was proposed to solve the problem of internal parameter disturbance caused by temperature change during PMSM operation. Firstly, a disturbance observer based on parametric disturbance is established. Secondly, a new double power approach law is used to design the speed ring controller, and the stability of the controller and observer is proved by Lyapunov stability theory. The results turns out that the new method can clearly improve the performance of the motor, weaken chattering, enhance the robustness and anti-interference ability.
Keywords: Disturbance observer; Sliding mode control; New approach law; Vector control.

1. Introduction

In recent years, aiming at the high performance control problem of SPMSM, many control methods have been applied in SPMSM control, such as new approach law sliding mode control [1], model reference adaptive control [2], improved super-spiral observer control [3].However, in the actual operating conditions, the motor is a complex system, and the internal parameters of the motor may produce parameter disturbance during operation, so it is difficult to achieve high precision control of the system. Therefore, in view of the problem that the performance of the system is degraded due to parameter disturbance, disturbance observer and adaptive sliding mode control are proposed in literature [4].In literature [5], a new approach law is proposed, and a new method is proposed to modify the approach law. Literature [6] proposed that the use of sliding mode observer in brushless DC motor control method increases the possibility. However, too many parameters are designed, which makes it more difficult to adjust parameters. Therefore, a new sliding mode control method based on sliding mode disturbance observer is proposed in this artical. The proposed method is based on the motor vector control model, combined with the improved new approach law function to set the sliding mode speed controller, and the sliding mode disturbance observer (SMDO) is used to observe the disturbance value of parameters and perform feedback compensation, so as to obtain better robustness.

2. Mathematical Model of Permanent Magnet Synchronous Motor

Generally, identify the SPMSM runs under ideal conditions, the mathematical model under synchronous coordinates is

$$\begin{cases} u_d = \mathrm{R}i_d + \dot{\psi_d} - \omega_e\psi_q \\ u_q = \mathrm{R}i_q + \dot{\psi_q} + \omega_e\psi_d \end{cases} \tag{1}$$

$$\begin{cases} \dot{\omega}_m = \frac{1}{J}(T_e - T_L - B\omega_m) \\ T_e = \frac{3}{2}P_n\psi_f i_q \end{cases} \tag{2}$$

$$\begin{cases} \psi_\mathrm{d} = \mathrm{L}i_\mathrm{d} + \psi_\mathrm{f} \\ \psi_\mathrm{q} = \mathrm{L}i_\mathrm{q} \end{cases} \tag{3}$$

However, in the actual operation of the motor, the internal parameters of the motor, the inductance and other internal parameters of the motor will change under the influence of temperature and other factors. Therefore, considering the influence of uncertain disturbance on the motor. Equation (2) can be written as

$$\dot{\omega}_m^{\mathrm{ref}} = \frac{3P_n(\psi_f+\Delta\psi_f)}{2(J+\Delta J)}i_q - \frac{T_L+\Delta T_L}{J+\Delta J} - \frac{B+\Delta B}{J+\Delta J}\omega_m \tag{4}$$

$$= \dot{\omega}_m - \mathrm{d}_1 i_q - \left(\frac{1}{J} - \mathrm{d}_2\right)T_L + \mathrm{d}_2 \cdot \Delta\, T_L + \mathrm{d}_3\omega_m$$

Where $\Delta\psi_f, \Delta T_L, \Delta B$ are the perturbations of parameter changes, and d_1, d_2, $\mathrm{d}_3$are defined as uncertain perturbations for convenience

$$\mathrm{D} = -\mathrm{d}_1 i_q - \left(\frac{1}{J} - \mathrm{d}_2\right)T_L + \mathrm{d}_2 \cdot \Delta\, T_L + \mathrm{d}_3\omega_m \tag{5}$$

In order to estimate the uncertain disturbance D, the sliding mode disturbance observer is designed

$$\dot{\hat{\omega}}_m = -\frac{B}{J}\hat{\omega}_m + \frac{T_e}{J} - \frac{T_L}{J} + \hat{D} \tag{6}$$

Assume that the error equation of the observer can be obtained if Equation (6) is subtracted from Equation (2) as required by the estimated speed and disturbance of $\hat{\omega}_\mathrm{m}$ bearing m and $\hat{\mathrm{D}}$ bearing respectively

$$\dot{\hat{\omega}}_m - \dot{\omega}_m = \dot{e}_\omega = -\frac{B}{J}e_\omega + \hat{D} \tag{7}$$

Where $\mathrm{e}_\omega$is the velocity observation error, $\dot{\mathrm{s}}_\omega = \dot{\mathrm{e}}_\omega$ is obtained by selecting the power reaching law and the sliding mode surfaces

$$\dot{S}_\omega = \dot{e}_\omega = -\mathrm{k}|\mathrm{s}|^{\mathrm{a}}\mathrm{sgn}(\mathrm{s}) \qquad 0 < \mathrm{a} < 1,\ \mathrm{k} > 0 \tag{8}$$

Assume that Equation (8) is substituted into Equation (7) to obtain the estimated disturbance $\hat{\mathrm{D}}$ bearing

$$\hat{D} = -\mathrm{k}|\mathrm{s}|^{\mathrm{a}}\mathrm{sgn}(\mathrm{s}) + -\frac{B}{J}e_\omega \tag{9}$$

3. Speed Sliding Mode Controller Design

According to Formula (2), the speed ring control law is designed, and select speed error as the controller state variable

$$\begin{cases} x_1 = \omega_m - \omega_{ref} \\ x_2 = \dot{\omega}_m \\ \dot{x}_2 = -\frac{B}{J}x_2 + \frac{3P_n\psi_f}{2J}\dot{i_q} - \frac{\dot{T}_L}{J} \end{cases} \tag{10}$$

Because of the uncertainty perturbation, the state equation (10) is written as

$$\begin{cases} \dot{x}_1 = x_2 \\ \dot{x}_2 = -\frac{B}{J}x_2 + \frac{3P_n\psi_f}{2J}\dot{i_q} + \widehat{D} \end{cases} \tag{11}$$

Select the linear sliding surface

$$\mathrm{s} = \mathrm{c}x_1 + x_2 \tag{12}$$

The approach law of traditional sliding mode control is shown in Formula (13),

$$\dot{\mathrm{s}} = -\mathrm{k}_1\mathrm{sgn}(\mathrm{s}) - \mathrm{k}_2 s \qquad k_1,\ k_2 > 0 \tag{13}$$

A new double power approach law is used

$$\dot{\mathrm{s}} = -\mathrm{k}_1|\mathrm{s}|^{a_1}\mathrm{sgn}(\mathrm{s}) - k_2|\mathrm{s}|^{a_2}\mathrm{sgn}(\mathrm{s}) - k_3\begin{cases} s & |s| \le \mathrm{q} \\ \mathrm{sgn}(\mathrm{s}) & |s| > \mathrm{q} \end{cases} \tag{14}$$

Where $a_1 > 1, 1 > a_2 > 0, k_1 > 0, k_2 > 0, \mathrm{q} > 0$, substitute Equations (12) and (14) into Equation (11), the current is controlled

$$i_q^{ref} = \frac{2J}{3P_n\psi_f}\int -\mathrm{k}_1|\mathrm{s}|^{a_1}\mathrm{sgn}(\mathrm{s}) - k_2|\mathrm{s}|^{a_2}\mathrm{sgn}(\mathrm{s}) - k_3\begin{cases} S \\ \mathrm{sgn}(\mathrm{s}) \end{cases} - \left(c - \frac{B}{J}\right)\dot{\omega}_m\, \mathrm{dt} \tag{15}$$

4. Stability Analysis

In order to verify the stability of the designed sliding mode controller, the Lyapunov function is selected Lyapunov function $\mathrm{V} = 0.5S^2$, When the sliding mode arrival condition is satisfied, the system is stable.

$$\dot{V} = S\dot{S} \le 0 \tag{16}$$

Substitute Equation (14) into Equation (16) to obtain

$$\dot{V} = S\dot{S} = \mathrm{s}\left(-\mathrm{k}_1|\mathrm{s}|^{a_1}\mathrm{sgn}(\mathrm{s}) - k_2|\mathrm{s}|^{a_2}\mathrm{sgn}(\mathrm{s}) - k_3\begin{cases} s & |s| \le \mathrm{q} \\ \mathrm{sgn}(\mathrm{s}) & |s| > \mathrm{q} \end{cases}\right) \tag{17}$$

Where $\mathrm{s} < \mathrm{q} < 0$ or $0 < \mathrm{s} < \mathrm{q}$:

$$\dot{V} = -\mathrm{k}_1|\mathrm{s}|^{a_1}|\mathrm{s}| - k_2|\mathrm{s}|^{a_2}|\mathrm{s}| - k_3 s^2 < 0 \tag{18}$$

Where $\mathrm{q} < \mathrm{s} < 0$ or $0 < \mathrm{q} < \mathrm{s}$:

$$\dot{V} = -\mathrm{k}_1|\mathrm{s}|^{a_1}|\mathrm{s}| - k_2|\mathrm{s}|^{a_2}|\mathrm{s}| - k_3 s \cdot \mathrm{sgn}(\mathrm{s}) < 0 \tag{19}$$

For eq. (17) and (18), $\dot{V}$=0 if and only if S =0.To sum up, formula (16) is always less than or equal to 0, so the new approach law is designed to meet the asymptotic stability of sliding mode.

5. Simulation and Experiment Comparison

The motor simulation parameters are as follows: stator resistance R=2.875Ω, winding inductance $L_q = L_d = 8.5\mathrm{mH}$, rotor flux $\psi_f = 0.175\mathrm{Wb}$, pole pairs $\mathrm{P}_n = 4$, moment of inertia $\mathrm{J} = 0.003\mathrm{kg}\cdot m^2$.Viscous friction coefficient $\mathrm{B} = 0.008\mathrm{N}\cdot\mathrm{m}\cdot\mathrm{s}$,DC voltage=311V. The control system block diagram of the motor is shown in Figure 1.The initial motor speed was set at 1000r/min, the simulation time was set at 0.4s, and the load torque was 0N*m. At 0.2s, the speed suddenly increased to 1200r/min. The simulation results in Figure.2 and Figure.3. Speed of the motor was set at 1000r/min, the simulation time was 0.4 s, and the initial load torque was 0N·m. At the same time, a load of 10 N·m was given to the motor at 0.2s to verify the antijamming capability of the motor speed. The simulation results in Figure.4 and Figure.5.

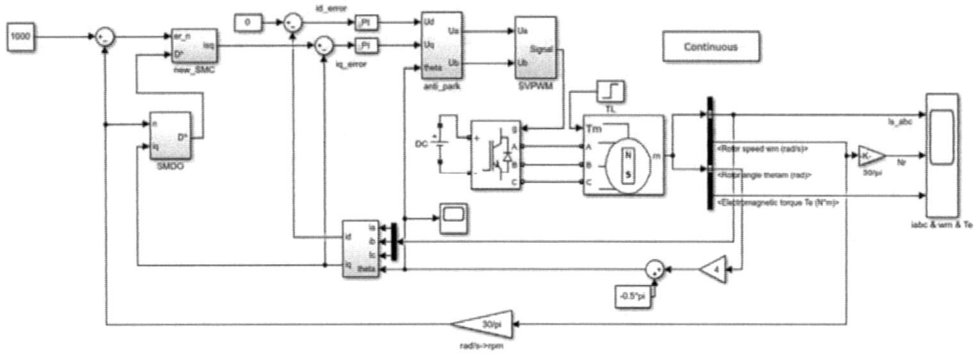

Figure 1. Block diagram of a new approach law sliding mode control system

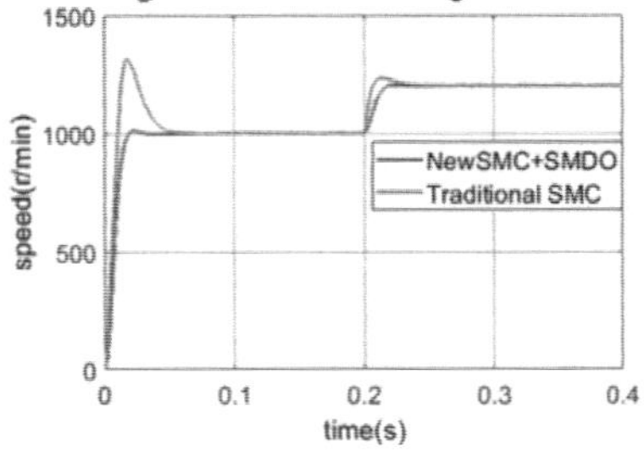

Figure 2. Speed comparison diagram of variable speed under no load

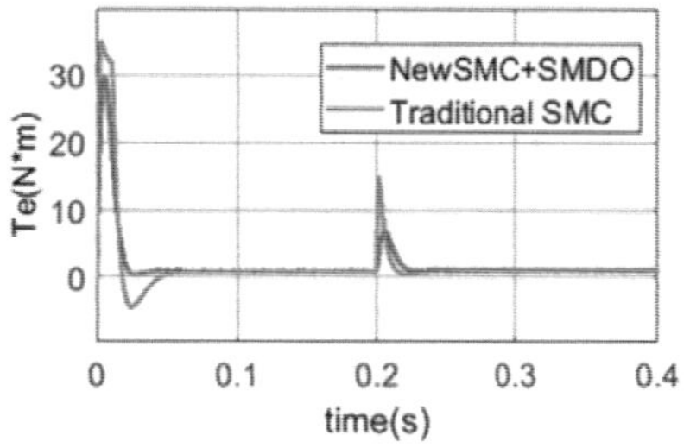

Figure 3. Electromagnetic torque comparison diagram of variable speed under no-load condition

Figure 2 and Figure 3 show the simulation diagram of speed and electromagnetic torque of the proposed new approaching law sliding mode disturbance observer applying the mutation speed under no-load condition. As can be seen from Figure.2, when t = 0.2s, the motor speed setting changes from 1000r/min to 1200r/min. When the fluctuation range reaches 300r/min, the recovery time is 0.23s, while the new SMC+SMDO proposed in this paper has a fluctuation range of 50r/min when the speed changes, and the recovery time is 0.22s. According to Figure.3, when t=0.2s, the fluctuation range of conventional SMC electromagnetic torque is 15N·m, while the fluctuation range of electromagnetic torque under the control of NewSMC+SMDO is only 7N·m.

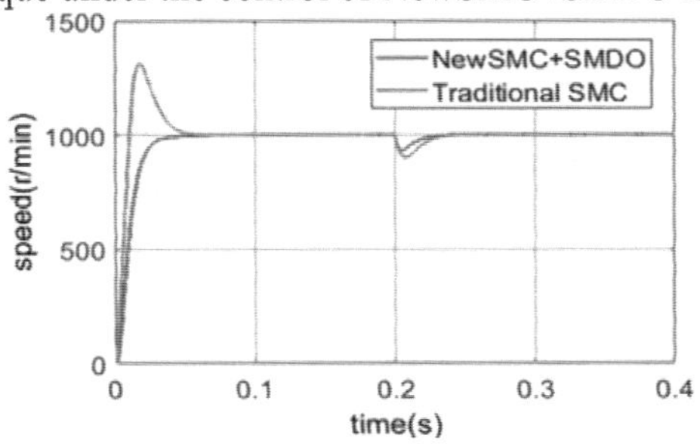

Figure 4. Comparison diagram of variable load speed under constant speed

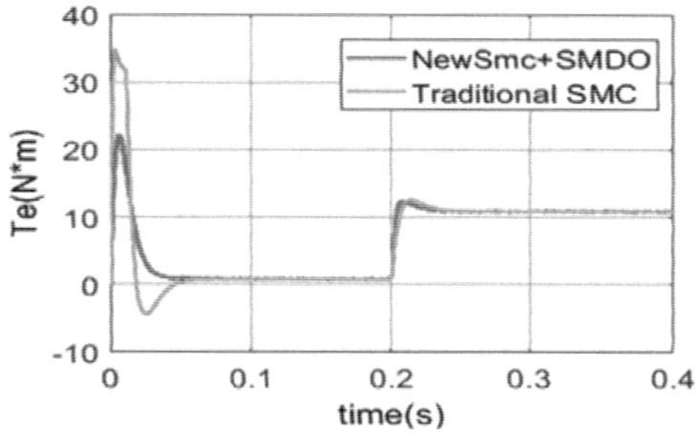

Figure 5. Comparison diagram of variable load electromagnetic torque under constant speed

Figure 4 and Figure 5 show the simulation diagram of speed and electromagnetic torque of the proposed new approach law sliding mode disturbance observer under variable load torque. In Figure.4 that when t = 0.2s and the load is increased by 10N·m, under NewSMC+SMDO control method, the speed disturbance reaches 50r/min and returns to steady state at 0.23s, while that of traditional SMC reaches 70r/min and returns to steady state at 0.25s. Figure.5 shows the change of electromagnetic torque. It is obvious that under the control of NewSMC+SMDO, the no-load electromagnetic torque fluctuation is relatively stable, and the load increases by 10N·m at 0.2s, and the fluctuation range is 2N·m. However, under the control of traditional SMC, the no-load electromagnetic torque fluctuation range is obvious, and under the influence of 10N·m load. It fluctuates by 3N·m. According to four simulation results, it is clearly that the proposed NewSMC+SMDO has good control effect and improves the control precision of the motor.

6. Conclusion

A sliding mode control strategy for SPMSM based on sliding mode disturbance observer is proposed in this artical. When using the sliding mode disturbance observer to estimate the motor running process, the internal parameter resistance and inductance of the motor will be disturbed due to the temperature change. A new double power approach law sliding mode control is designed to compensate the disturbance generated by the motor feed forward, and the stability conditions are determined by using Lyapunov stability criterion. And it has smaller estimation error than traditional disturbance observer. Finally, compared with the traditional sliding mode approach law, the superiority and convergence of the proposed method are verified from the image. For the future improvement, on the one hand, the double power approach law can be further modified and its convergence time can be specified. On the other hand, the disturbance observer can be further improved to reduce its observation error.

References

[1] Taochang Li. Nonlinear integral sliding mode control method based on new reaching law. Control engineering of China[J],2019,26(11):2031-2035.
[2] Kaltenbacher Barbara, Ngoc Nguyen Tram Thi. A model reference adaptive system approach for nonlinear online parameter identification[J]. Inverse Problems,2021,37(5):
[3] Jingli Miao, Yuhang Zhang, Wangyu Qin. Sensorless control strategy of permanent magnet synchronous motor based on superhelix algorithm. [J]Modern electronic technique,2020,43(17):162-165+170.
[4] Yongting Deng, Jianli Wang, Hongwen Li, Jing Liu, Dapeng Tian, Adaptive sliding mode current control with sliding mode disturbance observer for PMSM drives, ISA Transactions, Volume 88,2019,Pages 113-126,ISSN 0019-0578
[5] Gu Longrui, CEN Jianjun, Zhang Qinjin, Guo Haohao. Journal of Shanghai Institute of Electrical Engineering,2017, V. 20; No.123(01):22-26+62.
[6] HE Jun. Application of Sliding Mode Observer in Position Sensorless Direct Torque Control of Brushless DC Motor [J]. Journal of Shanghai Institute of Electrical Engineering,2018, V. 21; No.131(03):6-12.

Torque Transmission Law of Wind Turbine Drive System under the Excitation of Uneven Wind Load

Baiyun Qian[1], Kuncheng Zhou[2], Xiaoxia Gao[2] and Wei Liu[3,*]

[1,3] State Grid Xinjiang Electric Power Co., Ltd. Electric Power Research Institute, No. 200, Hengda Street, Xinshi District, Urumqi, Xinjiang, 830001
[2] Department of Power Engineering, North China Electric Power University, Baoding 071003, Hebei, China

*Corresponding author email: linxgreat@126.com

Abstract. A random wind speed time history model based on Weibull distribution is constructed. According to the known mean and variance of wind speed, the time fitting curve of random wind speed is obtained through MATLAB simulation method. According to the wind speed range of the actual operation of the wind turbine, the obtained wind speed curve is transformed, and then according to the obtained relationship between the wind speed of the wind turbine and the input torque, the input time-varying torque of the actual operation of the wind turbine is obtained. Establish the rigid shaft model of the wind turbine gearbox, calculate the gearbox transmission ratios at all levels, and then obtain the torque output curves of the gearbox planetary stage, medium-speed stage, and high-speed stage under random wind speed input to study wind power the torque transmission law of the machine drive system. The ability to better analyse the transmission structure of the gearbox and the torque transmission law of key components has a certain significance for the design and manufacture of the wind turbine transmission chain.
Keywords: Wind power generation; Random wind speed; Transmission system; Rigid axis model; Torque.

1. Introduction

At present, human beings are facing the energy crisis to a considerable extent due to the excessive exploitation of conventional energy resources, and many non-renewable energy resources are facing the situation of exhaustion. Wind energy is a very key component in a variety of new energy. There is a very large amount of wind energy savings in the whole nature. Due to the intense radiation of the sun all the time, the earth can get more than 3.75×10^{16} kW·h of wind energy almost every year. Wind energy will be one of the most critical energy sources in the earth at present and in the future. Through full development and utilization of wind energy, it can provide new ideas for solving the problem of energy shortage that is becoming increasingly serious at present.

China, by contrast, has a complex terrain and vast territory. Wind resources throughout the country are also very rich. However, because of the complexity of the terrain, the distribution of wind resources in China is uneven. Data show that the annual average wind speed in some coastal areas of southern China exceeds 3 m/s, and the effective wind energy density in northern China is even better than that in southern China, exceeding 75 W/s2. However, wind energy density and wind speed are low in central China and parts of southern land. It presents obvious inhomogeneity [1]. As a result, wind energy resources in China are much more difficult to use than those abroad.

But in the face of such a huge amount of wind energy, Chinese scholars are constantly improving wind power technology, and today, China's total installed wind power capacity ranks first in the world.

According to the statistical report released by the World Wind Energy Association (WWEA) in 2019, before 2020, China's accumulated installed capacity reached 209.33 million kW, an increase of 13.7% over the same period; In 2019, China added 20.94 million kw of wind power installations, up 31.37 % from the same period. With the continuous improvement of the single capacity of the wind turbine and the increasing blade size of the wind turbine, the excitation response degree to the uneven wind load in the actual operation of the wind turbine becomes larger, and the structural strength of the whole machine is required to be higher. The main shaft of the fan is an important part connecting the hub of the wind wheel and the gear box, so it is of great significance to study the torque variation rule when it is under uneven excitation, and take this as an important indicator to check the safety of the fan.

In this paper, the simulation of wind speed in a particular area of China is carried out, and the model based on Weibull distribution is planned to simulate the actual wind speed. The rigid shaft model of wind turbine transmission system was established, and the torque transfer law of wind turbine spindle and gear transmission system was taken as the research object to analyze and solve the variation law of torque transfer under uneven wind load excitation.

2. Weibull Random Wind Speed Model

Currently, generally accepted wind speed models mainly include Pearson, Rayleigh and Weibull distribution [2]. Due to relevant statistical studies, the distribution of natural wind speed has a positive skewness distribution with statistical characteristics, so weibull two-parameter curve is considered to be a function that can accurately describe the characteristics of wind speed to a certain extent in relevant studies [3]. In fact, Weibull two-parameter model measures the distribution characteristics of local wind sources through the statistical frequency distribution of wind speed. The most commonly used statistical value of wind speed frequency is wind frequency, which reflects the probability of the occurrence of a certain wind speed in a certain period of time in the stroke field.

The probability density function of Weibull distribution is:

$$f(v)=\frac{K}{C}\left(\frac{v}{C}\right)^{K-1}e^{-\left(\frac{v}{C}\right)^{K}} \tag{1}$$

In the formula, K —shape parameter;

C —position parameter;

V —wind speed, m/s.

By integrating formula (1), the Weibull model distribution function of local wind speed can be obtained as follows:

$$F(v)=\int_{0}^{+\infty} f(v)dv=\int_{0}^{+\infty}\frac{K}{C}\left(\frac{v}{C}\right)^{(K-1)}e^{-\left(\frac{v}{C}\right)^{K}}dv=1-e^{-\left(\frac{v}{C}\right)^{K}} \tag{2}$$

Using Weibull function distribution to simulate random wind speed, first of all, two fixed parameters must be determined. There are many methods to determine Weibull distribution parameters, the common ones are average wind speed method, maximum wind speed method, standard deviation and average wind speed $\overline{v}$, cumulative distribution function fitting, etc. [4]. However, research shows [5] that the average wind speed and the wind speed standard deviation S_v are used to estimate the shape and position parameters of the Weibull distribution, and the results have the best fitting effect.

When the average wind speed $\overline{v}$ and the wind speed standard deviation S_v are known, the sample mean μ and standard deviation σ can be approximated by formulas (3) and (4).

Use the average wind speed $\overline{v}$ to estimate the mean value μ of the sample, and use the sample variance S_v to estimate the standard deviation σ.

$$\mu=\overline{v}=\frac{1}{n}\sum_{i=1}^{n}(v_i-\overline{v})^2 \tag{3}$$

$$\sigma = S_v = \frac{1}{n}\sum_{i=1}^{n}(v_i - \bar{v})^2 \tag{4}$$

According to the relationship between the mean and variance of the Weibull distribution:

$$\bar{v} = E_{(v)} = \int_0^{+\infty} vf(v)dv = C\Gamma(1+\frac{1}{K}) \tag{5}$$

$$\left(\frac{\sigma}{\bar{v}}\right)^2 = [\frac{\Gamma\left(1+\frac{2}{K}\right)}{\Gamma^2\left(1+\frac{1}{K}\right)}]-1 \tag{6}$$

The empirical formula of Gamma function Γ(a) in formulas (5) and (6) is:

$$\Gamma\left(1+\frac{1}{K}\right) = \left(0.568+\frac{0.434}{K}\right)^{\frac{1}{K}} \tag{7}$$

From formula (5) to formula (7), we can get:

$$K = \left(\frac{\sigma}{\mu}\right)^{-1.086} \tag{8}$$

$$C = \frac{\mu}{\Gamma\left(1+\frac{1}{K}\right)} \tag{9}$$

After calculating the shape parameters and Weibull distribution position parameters using the above formula, use any number of the Weibull distribution random wind speed function (wblrnd) in MATLAB to generate a series of discrete random wind speed data, and then configure the corresponding wind speed for each data point one by one. At sampling time points, the two are fitted to each other, and a random wind speed time history chart is drawn.

When the average wind speed of a given wind field is 12 m/s and the variance of wind speed is 9 $(m/s)^2$, the shape parameter calculated according to the above equations (8) and (9) is 4.506477; the position parameter is 13.140088; it is simulated by software 150 data points are randomly generated, and the wind speed time history curve obtained by one-to-one correspondence with the time points is shown in Figure 1.

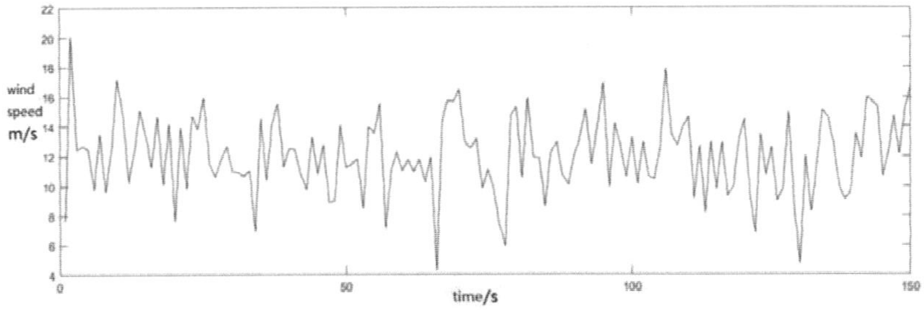

Figure 1. Random wind speed time history curve based on Weibull distribution.

3. Dynamic Model of Wind Turbine Drive System

In the discussion of this article, the most widely used and best overall effect of large-scale wind turbine gear transmission form [6] is adopted, and the form of one-stage planetary gear transmission plus two-stage fixed-axis gear transmission is adopted. The schematic diagram of the gear box structure is shown in Figure 2. Among them, v represents the impeller, c represents the planet carrier, R represents the internal gear, pi represents the planet gear, 4, 5, 6, 7 represent the two-stage fixed shaft helical gear, T_{in} is the torque input, and T_{out} is the output torque.

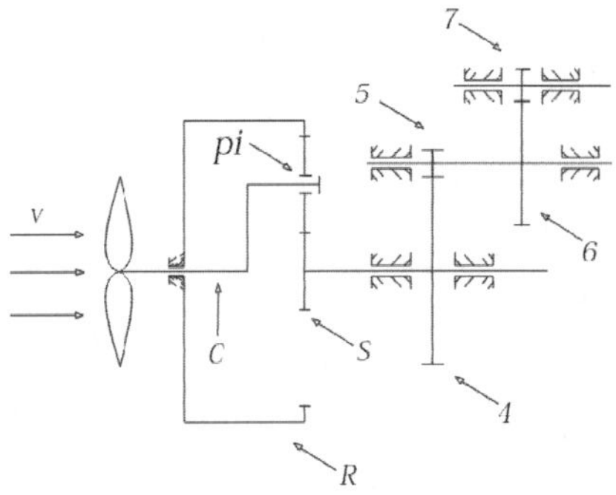

Figure 2. Schematic diagram of wind turbine gear transmission system.

Among them, the single-stage planetary gear transmission adopts the 2Z-X(A) type planetary transmission. This kind of planetary transmission gear train has relatively high transmission efficiency, strong carrying capacity, simple structure and good manufacturability. It adopts the basic form of planetary frame input, solar wheel output, fixed gear ring. Three same-sized planetary gears are used to share the concentrated load. The medium and high-speed helical gearboxes are driven by fixed-axis helical gears.

3.1. Transmission Ratio of Transmission System

The transmission mode of one-stage planetary gear plus two-stage parallel shaft gear, the transmission ratio of each stage can be determined by superimposing the single-stage transmission ratio.

3.1.1. Planetary gear ratio

In the model, the front end of the gearbox uses a 2Z-X planetary gear transmission. For the 2Z-X type planetary transmission with planet carrier, sun gear, ring gear, S, R, C in Fig. 3-1, the angular velocity relation of the planetary gear transmission can be obtained as [7] formula (10) :

$$\begin{cases} w_S = i_{SR}^{c} w_R + i_{SC}^{R} w_C \\ w_R = i_{RS}^{c} w_S + i_{RC}^{S} w_C \\ w_C = i_{CS}^{R} w_S + i_{RC}^{S} w_R \end{cases} \tag{10}$$

The kinematic equation of planetary gear transmission is:

$$\begin{cases} w_S + p w_R \ -(1+p) w_C = 0 \\ w_R + \dfrac{1}{p} w_S - (\dfrac{1+p}{p}) w_C = 0 \\ w_C - \dfrac{1}{1+p} w_S - \dfrac{p}{1+p} w_R = 0 \end{cases} \tag{11}$$

In the formula, p is the characteristic parameter of 2Z-X (A) planetary gear transmission system,

$p = \frac{Z_R}{Z_S}$.

When the input of components is known, equation (11) can be used to obtain the transmission ratio of the planetary gear transmission.
Since the inner ring gear of the planetary gear train is fixed, w_R=0,

$$w_S = (1 + p)w_C \tag{12}$$

Substituting formula (12) into the above formula (11), when the torque is input by the arm C, the transmission ratio can be obtained as:

$$i_{CS}^R = \frac{w_C}{w_S} = \frac{1}{1 + p} \tag{13}$$

3.1.2. Medium-speed gear ratio

The medium-speed stage of the gearbox uses 4 and 5 two helical gears to form a medium-speed fixed-axis gear system. According to the calculation formula of the transmission ratio of the fixed shaft gear, the transmission ratio of the medium-speed stage is:

$$i_{45} = \frac{w_4}{w_5} = -\frac{Z_5}{Z_4} \tag{14}$$

3.1.3. High-speed transmission ratio

The high-speed stage of the gearbox uses 6 and 7 two helical gears to form a high-speed fixed-axis gear system. According to the calculation formula of the transmission ratio of the fixed shaft gear, the transmission ratio of the high-speed stage is:

$$i_{67} = \frac{w_6}{w_7} = -\frac{Z_7}{Z_6} \tag{15}$$

3.2. The Relationship between Wind Speed and Torque

The input load of the fan drive system is the random wind speed time-varying load obtained in the MATLAB software according to the Weibull distribution above. When the wind wheel is in a balanced working state, and there is no unstable inflow around the wind wheel, the Cp (λ, β) power coefficient characteristic can be used to calculate the input power of the wind turbine wheel. For the variable speed control type wind turbine, since the pitch angle β is constant, the aerodynamic characteristics of the wind wheel can be converted into the characteristics of Cp(λ). From this, the mechanical efficiency of the stall control wind turbine can be obtained [8]:

$$P_M = \frac{1}{2}\rho V^3 \ \pi R^2 C_p(\lambda) \tag{16}$$

In the formula, P_M——input power of wind power gearbox, W;
ρ——air density, kg/m^3;
V——wind speed, m/s;
R——impeller radius, m;
C_p——wind energy utilization coefficient.

When the wind speed is less than the cut-in and start-up wind speed of the wind turbine, the wind turbine rotation speed is zero, and the unit does not produce electric energy, and it is feathered to protect the unit from being damaged by strong wind; when the wind speed is greater than the cut-out wind speed of the wind turbine, the wind turbine shuts down. When the machine stops, the pitch angle is changed to 90°. When the wind speed is higher than the cut-in start wind speed but lower than the rated operating wind speed, the wind turbine obtains the maximum C_P and maximum output power P_{max} by changing the tip speed ratio and the pitch angle. At this time, the following relationship exists between the input power and speed of the gearbox [9]:

$$p_{\max} = K_w \omega^3 \tag{17}$$

In the formula, K_ω is the coefficient related to the structure; ω is the speed of the wind wheel, as well as the input speed of the gearbox, rad/s

Formula (16) shows that the maximum input power is the cube relationship of wind speed, and formula (17) shows that the maximum input power is also the cube relationship of the wind wheel speed. Based on the above two conclusions, it can be concluded that the optimal speed of the wind wheel is directly proportional to the wind speed.

Similarly, when the wind speed is higher than the starting wind speed but lower than the rated wind speed, the wind speed of the wind power gearbox can also be determined by the following formula (18):

$$P = \omega T_{in} \tag{18}$$

From formula (17) and (18), can get:

$$T_{in} = K_w \omega^2 \tag{19}$$

Equation (19) shows that when the wind speed is higher than the cut-in wind speed and lower than the rated wind speed for normal operation, the input torque of the gearbox is the square of the speed, and since the speed of the wind wheel is proportional to the wind speed at this time. The input torque of the hour gearbox is also the square of the wind speed.

When the wind speed is higher than the cut-in start wind speed and lower than the rated operating wind speed, the relationship between the input torque of the gearbox and the wind speed is shown in formula (20) [6]:

$$T_{in} = \frac{T_{rate}}{v^2{}_{rate}} \cdot v^2 \tag{20}$$

When the operating wind speed of the wind turbine is higher than the cut-out wind speed or lower than the cut-in wind speed, the wind turbine is in a shutdown state and the gearbox input torque is zero; when the operating wind speed of the wind turbine is higher than the rated wind speed of the unit and lower than the cut-out wind speed of the wind turbine, The input torque of the gearbox of the wind turbine is controlled to the rated torque.

The relationship between the input torque of the wind turbine gearbox and the wind speed is shown in Figure 3 below, V_{out} in is the cut-in wind speed, V_{rate} is the rated wind speed, and $V_{out\ off}$ is the cut-out wind speed.

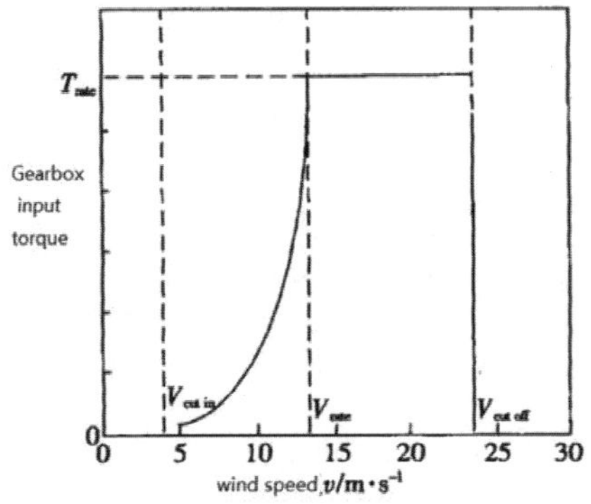

Figure 3. The relationship curve between the input torque of the wind turbine gearbox and the wind speed.

Suppose the parameters are as follows: when the wind turbine cut-in wind speed is 4 m/s, the rated wind speed is 12 m/s, the cut-out wind speed is 20 m/s, the rated power is 2 MW, and the rated speed is 14.8 r/min, the formula (17), (18), (19), the rated torque of the wind turbine can be obtained as 1.29×106 N·M.

According to Equation (20), when the wind speed is greater than the cut wind speed and less than the rated wind speed, the input torque of the fan is

$$T_{in} = \frac{1.29\times10^6}{12^2}\cdot v^2 \tag{21}$$

In the formula, v is wind speed, m/s.

When the wind speed is greater than the rated wind speed and the wind speed is less than the cut-out wind speed, the input speed and input torque of the wind turbine at this time are both the rated speed and rated torque of the wind turbine.

When the wind speed is lower than the cut-in wind speed or higher than the cut-out wind speed, the wind turbine is in a shutdown state at this time, and the input speed and input torque are both zero.

From the above, the input torque and wind speed of the gearbox have the following relationship:

$$T_{in} = \begin{cases} 0 & v < 4 \cup v > 20 \\ \frac{1.29\times10^6}{12^2}\cdot v^2 & 4 \le v < 12 \\ 1.29\times10^6 & 12 \le v < 20 \end{cases} \tag{22}$$

The random wind speed data generated by the simulation in MATLAB is converted by the above formula, and the input torque of the wind turbine gear box under the action of random wind is obtained as shown in Figure 4 below:

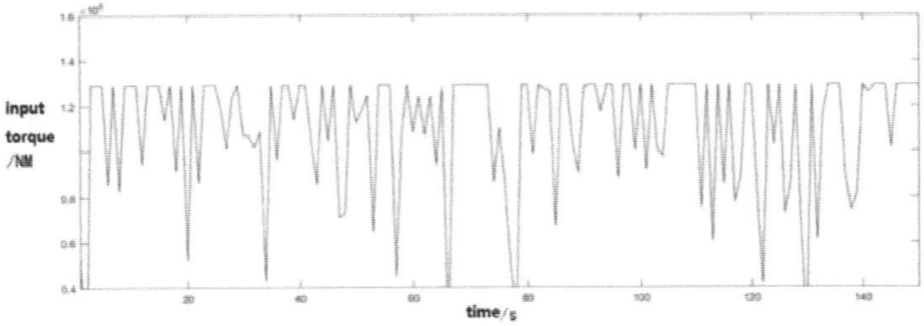

Figure 4. Wind turbine gearbox input torque curve.

3.3. Gearbox Rigid Shaft Model

For the rigid shaft model of the wind turbine system, the assumption made in advance is that the torsional rigidity of the transmission system of the gearbox is very large, and its dynamic characteristics are not considered. Therefore, the rigid shaft model of the gear transmission system of the wind turbine is established. As shown in Figure 5.

The rigid shaft model theory believes that [10]: The main shaft of the fan, namely the low-speed shaft, gearbox, and high-speed shaft are rigid. There is no more than one degree of freedom between the rotor and the generator. The ratio of the gearbox transmission ratio changes according to the fixed value.

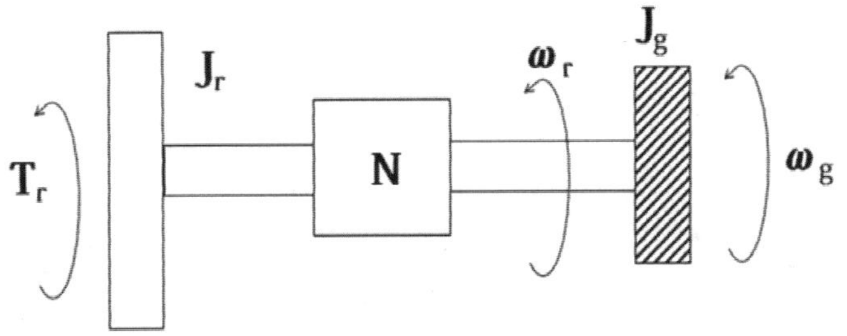

Figure 5. Schematic diagram of torque transmission of rigid shaft drive system of wind turbine.

The overall model of the rigid transmission chain regards the gearbox as a single-stage coupled speed increaser with a transmission ratio of i and an efficiency of η. In this case, the input torque of the generator shaft relative to the wind wheel is reduced by i times and the speed Increase i times. In the model assumptions shown in the figure above, the dynamic model of the wind energy conversion system under the high-speed axis or the low-speed axis can be expressed by the following equations (23) and (24) [11].

$$J_H \frac{d\Omega_e}{dt} = \frac{\eta}{i} M(\Omega, v) - M_e(\Omega_e, c) \tag{23}$$

$$J_L \frac{d\Omega_r}{dt} = M(\Omega, v) - \frac{i}{\eta} M_e(\Omega_e, c) \tag{24}$$

In the formula, Ω——wind wheel rotation angular velocity, rad/s;

Ω_e——generator rotor mechanical angular velocity, rad/s;

V——wind speed, m/s;

M (Ω, v) ——aerodynamic torque, N·M ;

M_e (Ω_e, c) ——Electromagnetic torque, generally using c as the load variable as a parameter;

J_H, J_L——Equivalent moment of inertia value of high-speed shaft and low-speed shaft, kg·m^2;

In the research of this section, consider the rigid transmission gearbox transmission. The input end is the low-speed gearbox input shaft, and the output end is the generator shaft. It is considered that the input and output ends of the transmission shaft are massless shafts, and the moment of inertia of the gearbox is not considered. Compared with the moment of inertia of the generator rotor, the moment of inertia of the gearbox is much smaller than, about 1/30[11]. In engineering calculations, the calculation of the dynamic characteristics of the entire transmission system is basically not affected. Therefore, it is reasonable to establish the rigid shaft model of the gearbox.

4. Torque Transmission of the Transmission System

In the rigid shaft model of the gearbox, the output torque of the wind wheel-main shaft is the input torque of the wind turbine gearbox. The output torque of the gearbox is the input torque of the generator. According to the structural characteristics of the wind turbine gearbox, the wind turbine gearbox is divided into planetary, medium-speed, and high-speed. Calculate the torque transmission changes of each stage separately. The relationship between the torque output of the transmission system at all levels and the torque input of the gearbox is shown in the following formula (25):

$$\begin{cases} T_{out}^{Planetary} = T_{in} \cdot i_{CS}^{R} \\ T_{out}^{Medium} = T_{in} \cdot (i_{CS}^{R} \cdot i_{45}) \\ T_{out}^{High} = T_{in}(i_{CS}^{R} \cdot i_{45} \cdot i_{67}) \end{cases} \tag{25}$$

The basic parameters of the wind turbine gearbox are shown in Table 1 below.

Table 1. Geometric parameters of 2Z-X (A) gear transmission system of wind turbine.

Planetary	Number of sun gear teeth Z_S	Number of planetary level Z_P	Number of teeth of internal gear Z_R	Modulus m
	27	44	117	13
Medium	Number of driving gear teeth Z_4	Number of teeth of driven gear Z_5	Normal modulus m_n	Helix angle $\beta/(°)$
	104	23	10	10.500
High	Number of driving gear teeth Z_6	Number of teeth of driven gear Z_7	Normal modulus m_n	Helix angle $\beta/(°)$
	98	25	6.5	12.500

Bringing the gearbox parameters into the above equations (13), (14), (15), the gearbox planetary gear ratio i_{CS}^{R} of the wind turbine is 0.1875, and the transmission ratio i45 of the medium-speed stage is 0.2216. The transmission ratio i67 of the stage is 0.2551.
Put the above transmission ratio into formula (25), and calculate the input torque of the gearbox under random wind speed, the output torque of each level of the wind turbine gear transmission system can be obtained, and the graph is shown in Figure (6) below.
It can be seen from the figure that the torque transmission of the wind turbine gearbox is basically the same at all levels. The uneven wind load of a wind turbine is a decisive factor for its change. And the size of the torque decreases step by step, the planetary star is the largest, the medium-speed class is the second, and the high-speed class is the smallest, gradually decreasing to about one-fifth of the previous class. When checking the torsional rigidity of the shaft and the meshing strength of the gears, priority should be given to the planetary level.

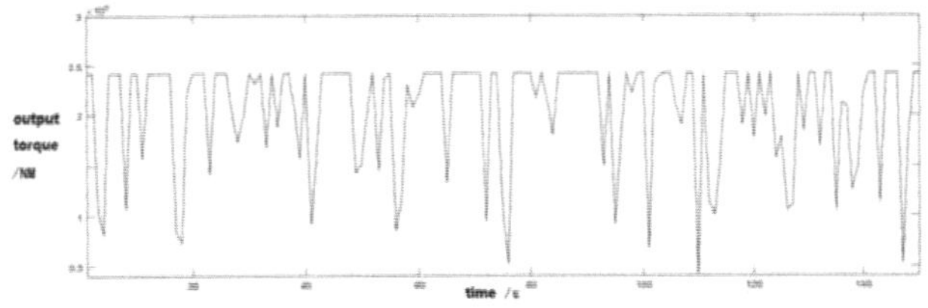

a. planetary level

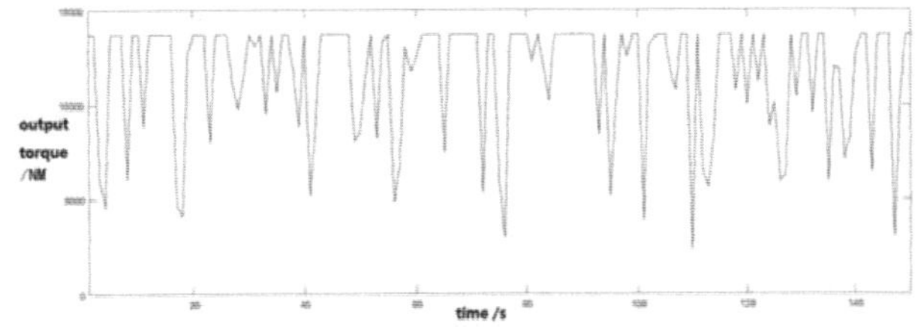

b. medium level

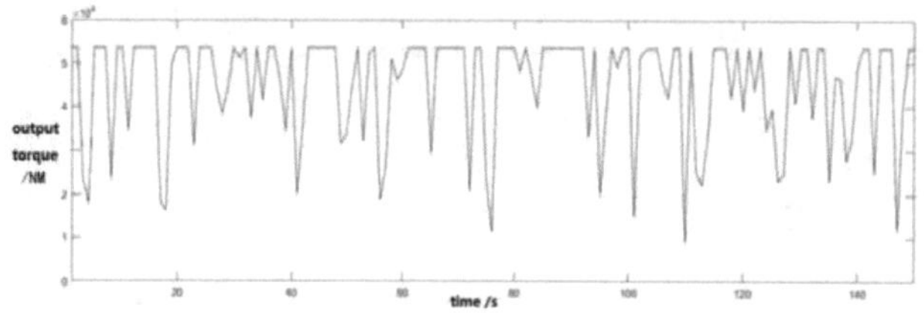

c. high level

Figure 6. Torque curve of wind turbine gearbox.

5. Conclusions and Prospects

(1) For the torque transmission of the wind turbine gearbox, the fluctuation law at all levels is basically the same. The uneven wind load of a wind turbine is a decisive factor for its change. The rigid shaft model can intuitively reflect the relationship between the output and input of the wind turbine, that is, the influence of the external wind speed on the system.

(2) The torque at all levels of the wind turbine gearbox decreases step by step, with the planetary star being the largest, the medium-speed class being the second, and the high-speed class being the smallest, gradually decreasing to about one-fifth of the previous level. When designing the gear transmission ratio of the wind turbine, it should be as close as possible to this value. At this time, the gearbox has a higher load-bearing capacity. When checking the torsional stiffness of the shaft and the meshing strength of the gears, the planetary star should be given priority.

(3) For the same gearbox mechanical model, under the simulated random wind speed, the torque transmission changes with the wind speed basically the same.

In this paper, the rigid shaft model of the wind turbine is established, and the torque transmission law of the wind turbine under these two models is analyzed. Regardless of the dynamic characteristics of the wind turbine gearbox, the torque change of the wind turbine gearbox at all levels of transmission has an obvious positive correlation with the wind speed change, and the gearbox torque transmission change is closely related to the transmission ratio.

This article can be used as the prerequisite basis for the follow-up study of the wind turbine transmission system, and the rigid shaft model established will be fully applied in the future research, so as to establish the overall dynamic analysis of the wind turbine.

References

[1] FENG et al. "The temporal and spatial distribution characteristics of wind energy resources in China from 1960 to 2016." Letters of Atmospheric and Oceanic Sciences: English Edition 2(2020): 136-145.

[2] Li Hui, Sun Hongbin, and Zhang Fang. "Summary of Research on Wind Speed Distribution Model of Wind Farm." New Technology of Electrical Engineering and Energy 33.8(2014):5

[3] Cai Hao. "Comparison of wind speed Weibull distribution parameter estimation methods." Statistics and Decision 16(2016): 3.

[4] He Yulin et al. "The application of the revised average wind speed and maximum wind speed method in the estimation of wind speed distribution." Power System Technology 3(2010): 4.

[5] Xu Weimin, Zhang Xinglin, and Kong Xinhong. "Comparison of Three Kinds of Wind Speed Weibull Distribution Parameters." Jiangxi Electric Power 01 (2007): 4-6.

[6] Datong Qin, Chao Li, and Feixiong Tang. "Study on the Dynamic Characteristics of Variable Speed Wind Turbine Transmission System under Variable Wind Load." Acta Energia Sinica 34.2(2013): 10.

[7] Liu Zhongming, Duan Shoumin, and Wang Changlu. "Development and Prospects of Wind Power Gearbox Design and Manufacturing Technology." Mechanical Transmission 6(2006): 7.

[8] Rui Xiaoming, Liu Yibing, and Ma Zhiyong. Wind Turbine Design. Mechanical Industry Press, 2010.

[9] Akhmatov, V.. Induction generator for wind turbines. Beijing. China Electric Power Press, 2009.

[10] Novak, P., et al. "Modeling and control of variable-speed wind-turbine drive-system dynamics." IEEE Control Systems Magazine 15.4(1995):28-38.
[11] Yao Xingjia. Theory and Design of Wind Turbine Generators. China Machinery Industry Press, 2013.

Permanent Magnet Synchronous Motor Control Based on New Sliding Mode Observer

Yi Song[1,*], Zehang Song[1] and Xiaodong Yao[1,2]

[1]School of Electrical Engineering, Shanghai DianJi University. Shanghai, 201306, China
[2]Minge New-Type Motor and Electronic Control Research Institute. Foshan City, Guangdong Province, 528599, China

*Corresponding author email: 206001010221@st.sdju.edu.cn

Abstract. Aiming at the problems of large chattering and slow response time existing in the pre-existing sliding mode observer permanent magnet synchronous motor control, this paper proposes a new sliding mode observer control way. The model of the permanent magnet synchronous motor was established, and design a new function to take the place of the pre-existing transforming function. And use Lyapunov stability criterion to analyse the system stability. At the same time, Use Matlab/Simulink simulation software to carry out simulation experiments, build simulation experiment models, and obtain the control result waveforms of the integrated sliding mode observer control system and the newly designed sliding mode observer control system. Compared to pre-existing sliding mode observers, it can be known that, constrasted with the pre-existing sliding film observer, better performance and stronger stability are possessed by the new sliding mode observer proposed in this artical.
Keywords: Sliding mode observer; PMSM; Lyapunov stability criterion.

1. Introduction
Permanent Magnet Synchronous Motor (PMSM) is diffusely used in various industries [1]. In order to obtain better control result, experts and scholars suggested some sliding mode control methods [2,3]. , The system is prone to jitter when moving along the sliding mode surface toward the equilibrium point. Literature [4] designs a non-singular fast terminal sliding mode controller, and uses a interference observer to compensate the system model, which effectively shortens the PMSM system response time, improves robustness, and enable the control effect to avoid the influence from the system parameters and structure. Literature [5] proposed a new approaching law in the speed loop, and used fuzzy adaptive method to adjust the parameters, this design can reduce chattering.

2. Mathematical Model of PMSM
The mathematical equation of PMSM is generally on the d-q axis coordinate system. This conclusion is mainly on account of the theory of coordinate transformation.

$$\begin{cases} u_d = R_s i_d + \frac{d\psi_d}{dt} - \omega_e \psi_q \\ u_q = R_s i_q + \frac{d\psi_q}{dt} + \omega_e \psi_d \end{cases} \tag{1}$$

Among them, the stator flux equation is:

$$\begin{cases} \psi_d = R_s i_d + \frac{d\psi_d}{dt} - \omega_e \psi_q \\ \psi_q = L_q i_q \end{cases} \tag{2}$$

Among them, u_d and u_q express the stator voltage component; i_d and i_q express the stator current component; ψ_d and ψ_q express the stator flux component; ω_e express the electrical angular velocity of the sub; L_d and L_q express the stator inductance , Rs express the phase resistance. All parameters in the above formula are parameters on the d-q axis coordinate system.

The variables in the PMSM mathematical model are all disassembled, which greatly simplifies the complexity of the PMSM mathematical model, and can realize individual control of the torque. It can be seen that the equation of electromagnetic torque at this time can be voiced as:

$$T_e = \frac{3}{2} p_n (\psi_d i_q - \psi_q i_d) = \frac{3}{2} p_n [\psi_f i_q + (L_d - L_q) i_d i_q] \tag{3}$$

To obtain the mathematical equation of the PMSM on the stationary axis coordinate system, it is only necessary to transform the equation on the d-q axis coordinate system to the α-β axis coordinate system using inverse Park. Therefore, the voltage equation in the α-β coordinate system can be expressed as:

$$\begin{cases} u_\alpha = R_s i_\alpha + \frac{d\psi_\alpha}{dt} \\ u_\beta = R_s i_\beta + \frac{d\psi_\beta}{dt} \end{cases} \tag{4}$$

Among them, u_α, u_β express the stator voltage components; i_α, i_β express the stator current components; ψ_α, ψ_β express the stator flux components. All parameters in the above formula are parameters in the d-q coordinate system.

Among them, the flux linkage component model is:

$$\begin{cases} \psi_\alpha = L_s i_\alpha + \psi_f \cos\theta \\ \psi_\beta = L_s i_\beta + \psi_f \sin\theta \end{cases} \tag{5}$$

At the same time, the current state equation of PMSM can be obtained from formula (4) and (5) as:

$$\begin{cases} \frac{di_\alpha}{dt} = -\frac{R_s}{L_s} i_\alpha + \frac{1}{L_s} u_\alpha - \frac{1}{L_s}\frac{d\psi_\alpha}{dt} \\ \frac{di_\beta}{dt} = -\frac{R_s}{L_s} i_\beta + \frac{1}{L_s} u_\beta - \frac{1}{L_s}\frac{d\psi_\beta}{dt} \end{cases} \tag{6}$$

The back-EMF equation is:

$$\begin{cases} e_\alpha = -\frac{d\psi_\alpha}{dt} = -\psi_f \omega_e \sin\theta \\ e_\beta = \frac{d\psi_\beta}{dt} = \psi_f \omega_e \cos\theta \end{cases} \tag{7}$$

Among them, e_α and e_β on behalf of the components of the back electromotive force on the α and β axes. The mathematical expression of PMSM under the α-β coordinates is:

$$\begin{cases} \frac{di_\alpha}{dt} = -\frac{R_s}{L_s} i_\alpha + \frac{1}{L_s} u_\alpha - \frac{1}{L_s}\frac{d\psi_\alpha}{dt} \\ \frac{di_\beta}{dt} = -\frac{R_s}{L_s} i_\beta + \frac{1}{L_s} u_\beta - \frac{1}{L_s}\frac{d\psi_\beta}{dt} \\ e_\alpha = -\psi_f \omega_e \sin\theta \\ e_\beta = \psi_f \omega_e \cos\theta \end{cases} \tag{8}$$

3. Design of a New Sliding Mode Observer

The estimated stator current is

$$\hat{i}_s = [\hat{i}_\alpha \quad \hat{i}_\beta]^T \tag{9}$$

The measured value of the stator current is

$$i_s = \begin{bmatrix} i_\alpha & i_\beta \end{bmatrix}^T \tag{10}$$

The sliding surface is used:

$$S(x) = \hat{i}_s - i_s = \begin{bmatrix} \hat{i}_\alpha - i_\alpha \\ \hat{i}_\beta - i_\beta \end{bmatrix} \tag{11}$$

The pre-existing sliding mode observer's mathematical expression is:

$$\begin{cases} \frac{d\hat{i}_\alpha}{dt} = -\frac{R_s}{L_s}\hat{i}_\alpha + \frac{1}{L_s}\hat{u}_\alpha - \frac{K}{L_s}\mathrm{sgn}(\hat{i}_\alpha - i_\alpha) \\ \frac{d\hat{i}_\beta}{dt} = -\frac{R_s}{L_s}\hat{i}_\beta + \frac{1}{L_s}\hat{u}_\beta - \frac{K}{L_s}\mathrm{sgn}(\hat{i}_\beta - i_\beta) \end{cases} \tag{12}$$

Design a new piecewise function as

$$Y(x) = \begin{cases} 1, & x \geq a \\ \frac{x^3}{a^3}, & 0 \leq x < a \\ -\frac{x^3}{a^3}, & -a < x < 0 \\ -1, & x \leq -a \end{cases} \tag{13}$$

Among them, α is the thickness of the boundary layer. The value of α determines the degree of chattering of the sigmoid function.

The function line is visualized as a process in which the system state points run across the sliding surface. Compared with the pre-existing symbolic function, the piecewise exponential function can be approximately parallel to the X axis when passing through the interface zero point and approach the horizontal axis asymptotically. Far away from the origin, that is, after passing through the sliding surface, it is guided according to the exponential function movement trajectory, resulting in a slower deviation from the sliding surface of the system. This speed allows the system state point to be adjusted in time and return to the sliding surface in time.

According to equations (8), (12), (13), the designed sliding mode observer's mathematical expression is:

$$\begin{cases} \frac{d\hat{i}_\alpha}{dt} = -\frac{R_s}{L_s}\hat{i}_\alpha + \frac{1}{L_s}\hat{u}_\alpha - \frac{K}{L_s}y(\hat{i}_\alpha - i_\alpha) \\ \frac{d\hat{i}_\beta}{dt} = -\frac{R_s}{L_s}\hat{i}_\beta + \frac{1}{L_s}\hat{u}_\beta - \frac{K}{L_s}y(\hat{i}_\beta - i_\beta) \end{cases} \tag{14}$$

4. Stability Analysis

This chapter uses the Lyapunov stability theory to analyse the stability of the designed new sliding mode observer.

Subtract formula (14) and formula (8) can be obtained:

$$\dot{s} = -\frac{R_s}{L_s}s + \frac{1}{L_s}[e_s - Ky(x)] \tag{15}$$

Among them, s is the sliding surface; es is the back electromotive force of the motor.

Then the Lyapunov stability equation can be established as

$$\dot{V} = s^T\dot{s} = -\frac{R_s}{L_s}s^T s + \frac{1}{L_s}s^T[e_s - Ky(x)] \tag{16}$$

Among them, V is the Lyapunov function; V˙ is the derivative of the Lyapunov function.

To meet the stability conditions required by the Lyapunov function, need

$$V = s^T\dot{s} < 0 \tag{17}$$

From equations (16) and (17), we can get:

$$-\frac{R_s}{L_s}s^T s < 0 \tag{18}$$

$$\frac{1}{L_s}s^T[e_s - Ky(x)] < 0 \tag{19}$$

Since $-\frac{R_s}{L_s}$ is always less than 0, only the equation (19) needs to be satisfied to make the system reach the stability condition, namely

$$\max(|e_\alpha|, |e_\beta|) < 0 \tag{20}$$

5. Simulation Experiment

The following table shows the PMSM's parameters in this article:

Table 1. PMSM parameters used in the simulation

parameter	Parameter value
Permanent magnet flux /Wb	0.0385
Wire resistance /Ω	3.2
Line inductance /mH	10.15
Moment of inertia /(g·m^2)	0.01814
Number of pole pairs	4

The built Simulink simulation model is shown in the figure below

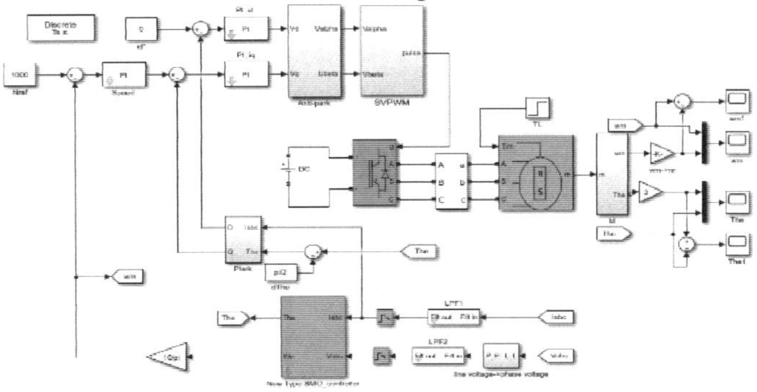

Figure 1. Simulink simulation diagram

The motor speed is set to 1000r/min, the initial value of load torque is 0N/m, and the value of load torque is increased from 0 to 10N/m in 0.05s.Figure 2 shows the rotor speed estimation errors obtained by the pre-existing sliding mode observer and the new sliding mode observer under the above conditions.It can get that the estimated error is significantly smaller compared to the actual value the speed error is smaller, and the speed fluctuation after stability is smaller.

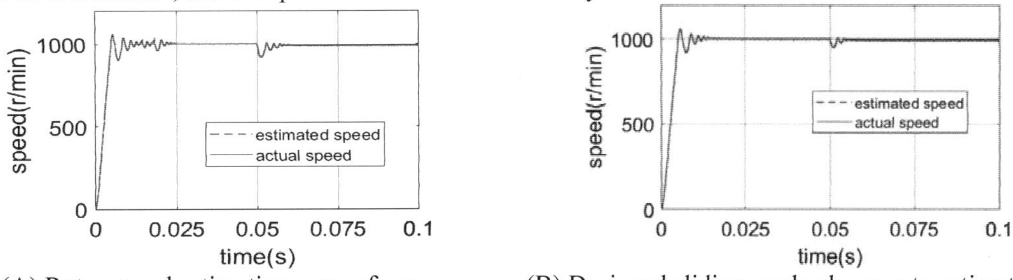

(A) Rotor speed estimation error of pre-existing sliding mode observer

(B) Designed sliding mode observer to estimate rotor speed

Figure 2. Comparison diagram of rotor speed

Figure 3 shows the error comparison of the estimated rotor speed estimated value and the practical value of the pre-existing sliding mode observer and the new type sliding mode observer. It can get that the new sliding film observer has a smaller rotor speed estimation error, and has a shorter recovery time after being disturbed, and has better robustness.

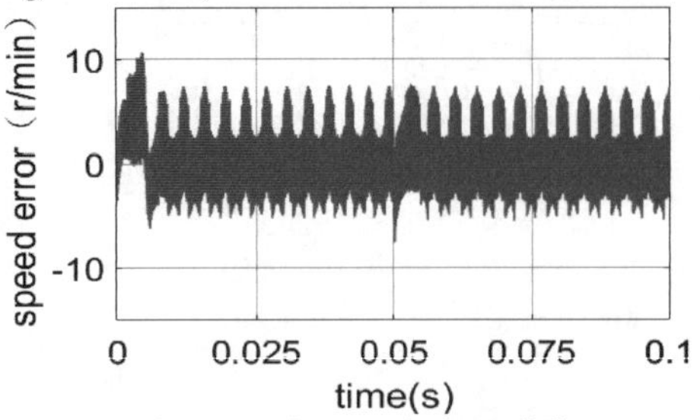

(A) Rotor speed error of pre-existing sliding mode observer

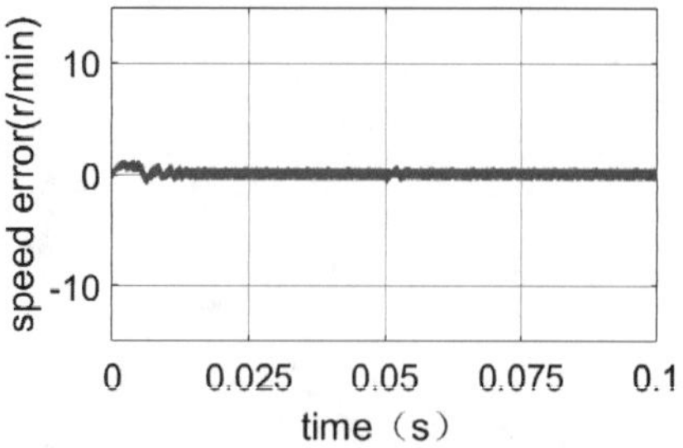

(B) Rotor speed error of new sliding mode observer

Figure 3. Rotor speed error

6. Conclusion

Through simulation analysis and experimental results, it can get that the chattering phenomenon is serious when the pre-existing sliding mode observer estimates the motor rotor's speed, while the new sliding mode observer can effectively suppress the chattering while improving the dynamic performance of the observer. Vibrate. Therefore, for the PMSM control system in operation, when it is at a higher speed, the slip film observer designed in this paper has a higher accuracy for the detection of the rotor position. Therefore, the motor control The reliability is greatly improved. The simulation experiment proves that the control accuracy and response speed are greatly improved due to the introduction of the control method in this article. But this method also has certain shortcomings. Because the sliding mode requires sufficient back-EMF signals and cannot achieve precise control at low speeds, it is necessary to study high-precision control schemes in the low-speed range to achieve the practical conditions in more occasions.

References

[1] Cui J R, Gao J F, Zhang B, et al. Robust Control of Synchronous Motor Based on Sliding Mode Variable Structure[J]. Electric Machines and Control, 2016, 20(5): 84-89.
[2] Cai R, Zheng R, Liu M, et al, Robust Control of PMSM Using Geometric Model Reduction and μ-Synthesis[J]. IEEE Transactions on Industrial Electronics, 2018, 65(1): 498-509.
[3] Luo D R, He R Z, Huang S D, et al. Dynamic Sliding Mode Control for Single Stator Dual Rotor Disc Counter-rotating Permanent Magnet Synchronous Motor[J]. Transactions of China Electrotechnical Society, 2019, 34(9): 1806-1814.
[4] Wang Y M, Niu Z M, Mai Y F, et al. Double Closed-loop Smooth Non-singular Terminal Sliding Mode Control of PMSM Based on Anti-windup Method[J]. Electric Machines and Control, 2020, 24(3): 138-146.
[5] Miao J L, Zheng D W, Zhou C X. Direct Torque Control of SPMSM Based on Hybrid Sliding Mode Controller and Arctangent Observer[J]. Control and Decision, 2019, 34(9): 1831-1839.

Research on Microgrid Connected and Islanded Switching Control Based on Switched System

Liying Sun* **and Yang Liu**

School of Electrical Engineering, Liaoning University of Technology, Jinzhou, China

*Corresponding author email: sunliying@lnut.edu.cn

Abstract. This paper focuses on the switching stability control of microgrid in the process of grid-connected and island operation. Firstly, the grid-connected operation model and island operation model are established, and they are regarded as two subsystems of switched system; then the cross backstepping method is used to design the switching controllers of the two subsystems, and the common Lyapunov function is constructed at the same time. Finally, matlab simulation verifies the effectiveness of the designed controller and switching law.
Keywords: Microgrid; Inverter; Switched Systems; Cross-backstepping.

1. Introduction

In the recent two decades, microgrid has high application value and many countries have carried out research because of the increasingly serious energy shortage and environmental pollution [1-2]. Microgrid includes three categories: AC, DC and AC/DC hybrid microgrid [3], which can operate not only in grid-connected mode, but also in island mode.

DG based on power electronic inverter generally adopts PQ, VF, Droop or VSG control. The research on the switching of microgrid mostly focuses on the four control methods. Switching between these traditional voltage and current controls can cause severe transient shocks and oscillations. In addition because the main circuit is power electronic devices, using the traditional synchronous generator grid-connected mode, the inverter will have a great current impact at the moment of grid-connected. Smooth switching technology restricts the further development of microgrid[4]. When the microgrid runs in the grid-connected state, the large grid can provide frequency and voltage support, inverter is generally controlled by PQ [5].At present, it is a great significance to study how to realize smooth switching between grid-connected and isolated operation modes of microgrid.The author in [6] designed a linear robust controller based on Lyapunov stability for grid-connected and island modes. However, microgrid is nonlinear in practical application and contains uncertain factors and external interference [7]. The stability of the switched system can be proved by the combination of the nonlinear switched system and the microgrid switching. At present, the research of smooth switching control for this method is relatively few.

In this paper, the nonlinear switching system theory is applied to the grid-connected and island switching. Firstly, the nonlinear switching system model is established. Then, the nonlinear switched controller and common Lyapunov function of the two subsystems are designed by using the cross backstepping method. Finally, using matlab simulation proves the effectiveness of the switched controller and switching law designed in this paper.

2. Switching Control and Stability Analysis

2.1. System Description

Several assumptions are made to establish the mathematical model of inverter: the grid voltage is a stable three-phase pure chord function; Constant DC voltage source is used as the input voltage of system instead of DG in microgrid [8]. So the grid-connected subsystem 1 topological structure is given in figure 1. The island subsystem 2 topological structure is given in figure 2.

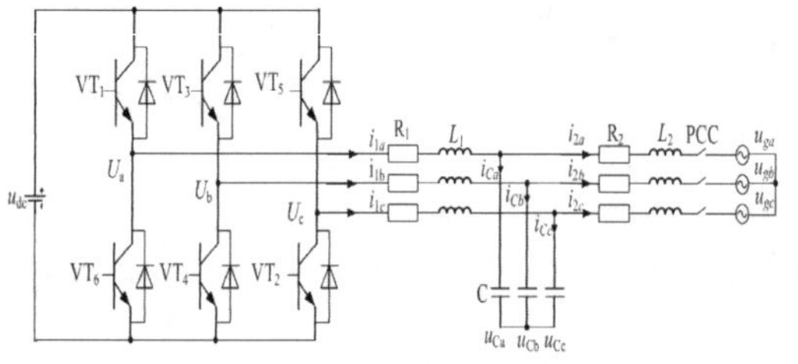

Figure 1. Grid-connected topological structure.

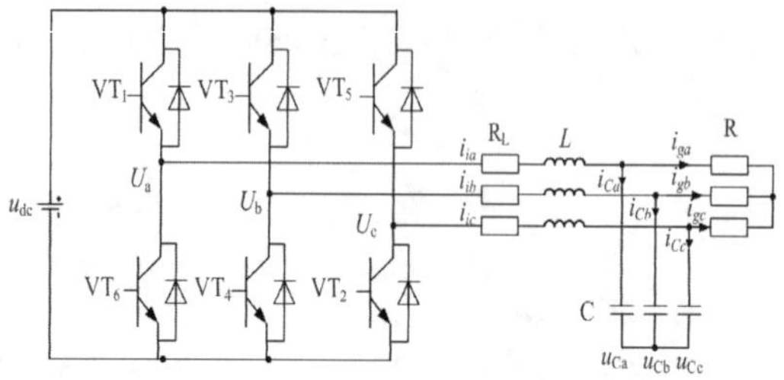

Figure 2. Islanding topological structure.

In figure 1, L_1 represents the inverter-side inductor, C is the filter capacitor, L_2 represents the grid-side inductor, these passive components formed a three-order LCL filter together; The parasitic resistances of L_1 and L_2 are defined as R_1 and R_2 respectively; u_{dc} represents the input voltage on the DC side of the inverter; $VT_1 \sim VT_6$ are power electronic switching devices IGBT; $U_a \sim U_c$ are inverter output voltages; $i_{1a} \sim i_{1c}$ are inverter side currents; $i_{2a} \sim i_{2c}$ are grid side currents; $i_{Ca} \sim i_{Cc}$ are the currents flowing into C; $u_{Ca} \sim u_{Cc}$ are the voltage across C; $u_{ga} \sim u_{gc}$ are three-phase vectors AC voltage on the grid side. In figure 2, L represents the inverter-side inductor. C is the filter capacitor, they formed LC filter together; R_L is parasitic resistances; $i_{ia} \sim i_{ic}$ are inverter side currents; R is the load; $i_{ga} \sim i_{gc}$ are the currents flowing into the load. (Other parameters have the same physical meaning as grid connected). Based on KVL grid-connected subsystem model can be described by Equation (1).

$$\begin{cases} \frac{di_{1\alpha}}{dt} = \frac{1}{L_1}u_{r\alpha} - \frac{1}{L_1}u_{C\alpha} - \frac{R_1}{L_1}i_{1\alpha} \\ \frac{di_{1\beta}}{dt} = \frac{1}{L_1}u_{r\beta} - \frac{1}{L_1}u_{C\beta} - \frac{R_1}{L_1}i_{1\beta} \\ \frac{di_{2\alpha}}{dt} = \frac{1}{L_2}u_{C\alpha} - \frac{1}{L_2}u_{g\alpha} - \frac{R_2}{L_2}i_{2\alpha} \\ \frac{di_{2\beta}}{dt} = \frac{1}{L_2}u_{C\beta} - \frac{1}{L_2}u_{g\beta} - \frac{R_2}{L_2}i_{2\beta} \\ \frac{du_{C\alpha}}{dt} = \frac{1}{C}i_{1\alpha} - \frac{1}{C}i_{2\alpha} \\ \frac{du_{C\beta}}{dt} = \frac{1}{C}i_{1\beta} - \frac{1}{C}i_{2\beta} \end{cases} \tag{1}$$

Similarly, based on Clark transformation and KVL island subsystem model can be described by Equation (2)

$$\begin{cases} \frac{di_{i\alpha}}{dt} = \frac{1}{L}u_{\alpha} - \frac{1}{L}u_{C\alpha} - \frac{R_1}{L}i_{i\alpha} \\ \frac{di_{i\beta}}{dt} = \frac{1}{L}u_{\beta} - \frac{1}{L}u_{C\beta} - \frac{R_1}{L}i_{i\beta} \\ \frac{du_{C\alpha}}{dt} = \frac{1}{C}i_{i\alpha} - \frac{1}{C}i_{g\alpha} \\ \frac{du_{C\beta}}{dt} = \frac{1}{C}i_{i\beta} - \frac{1}{C}i_{g\beta} \end{cases} \tag{2}$$

In this paper, grid-connected operation mode and isolated island operation mode are regarded as subsystem 1 and subsystem 2 respectively. According to Equation (1), we can properly select $x_1 = i_{2\alpha} - i_{2\alpha0}$, $x_2 = i_{2\beta} - i_{2\beta0}$, $x_3 = u_{c\alpha} - u_{c\alpha0}$, $x_4 = u_{c\beta} - u_{c\beta0}$, $x_5 = i_{1\alpha} - i_{1\alpha0}$, $x_6 = i_{1\beta} - i_{1\beta0}$ as the states variables. Selecting $u_1 = u_{r\alpha}$, $u_2 = u_{r\beta}$ as the control input of the grid-connected subsystem. And $u_{r\alpha} = u_{dc}K_\alpha$, $u_{r\beta} = u_{dc}K_\beta$ is obtained by switching period average method. In order to design common Lyapunov function extend Equation (2), we can select states: $x_1 = 0$, $x_2 = 0$, $x_3 = u_{c\alpha} - u_{c\alpha0}$, $x_4 = u_{c\beta} - u_{c\beta0}$, $x_5 = i_{1\alpha} - i_{1\alpha0}$, $x_6 = i_{1\beta} - i_{1\beta0}$ and $u_1 = u_\alpha$ $u_2 = u_\beta$ as the control input of the islanding subsystem, where $u_\alpha = u_{dc}K_\alpha$, $u_\beta = u_{dc}K_\beta$. Therefore, the two subsystem models of the switching system can be described by Equation (3) and Equation (4) respectively.

$$\begin{cases} \dot{x}_1 = \frac{1}{L_2}(x_3 + u_{C\alpha0}) - \frac{R_2}{L_2}(x_1 + i_{2\alpha0}) - \frac{1}{L_2}u_{g\alpha} \\ \dot{x}_2 = \frac{1}{L_2}(x_4 + u_{C\beta0}) - \frac{R_2}{L_2}(x_2 + i_{2\beta0}) - \frac{1}{L_2}u_{g\beta} \\ \dot{x}_3 = \frac{1}{C}(x_5 + i_{1\alpha0}) - \frac{1}{C}(x_1 + i_{2\alpha0}) \\ \dot{x}_4 = \frac{1}{C}(x_6 + i_{1\beta0}) - \frac{1}{C}(x_2 + i_{2\beta0}) \\ \dot{x}_5 = \frac{1}{L_1}u_1 - \frac{1}{L_1}(x_3 + u_{C\alpha0}) - \frac{R_1}{L_1}(x_5 + i_{1\alpha0}) \\ \dot{x}_6 = \frac{1}{L_1}u_2 - \frac{1}{L_1}(x_4 + u_{C\beta0}) - \frac{R_1}{L_1}(x_6 + i_{1\beta0}) \end{cases} \tag{3}$$

$$\begin{cases} \dot{x}_1 = 0 \\ \dot{x}_2 = 0 \\ \dot{x}_3 = \frac{1}{C}(x_5 + i_{l\alpha0}) - \frac{1}{C}i_{g\alpha} \\ \dot{x}_4 = \frac{1}{C}(x_6 + i_{1\beta0}) - \frac{1}{C}i_{g\beta} \\ \dot{x}_5 = \frac{1}{L}u_1 - \frac{1}{L}(x_3 + u_{C\alpha0}) - \frac{R_L}{L}(x_5 + i_{1\alpha0}) \\ \dot{x}_6 = \frac{1}{L}u_2 - \frac{1}{L}(x_4 + u_{C\beta0}) - \frac{R_L}{L}(x_6 + i_{1\beta0}) \end{cases} \tag{4}$$

2.2. Controller Design

Gong et al. proposes a wider class of nonlinear multi-input systems (its cross strict feedback system) and a cross backstepping method is proposed for the first time [9-11]. This paper applies this method to design controllers for grid-connected operation mode systems. In order to better explain the design process of controller, the flow chart is introduced.

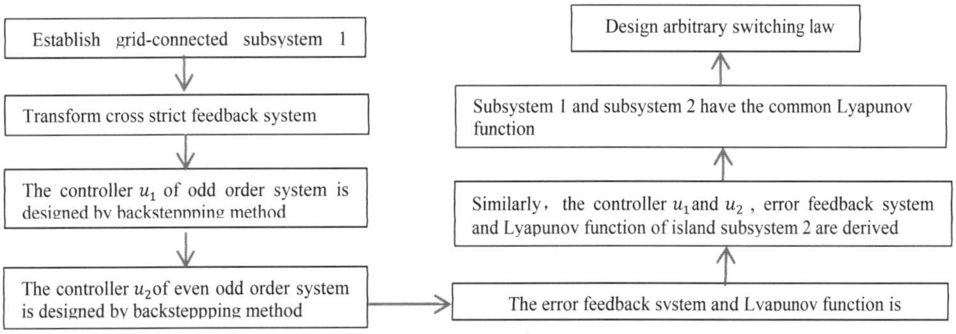

Figure 3. The design flowchart.

Firstly, the Equation (3) is divided odd order Equation (5) and even-order Equation (6).

$$\begin{cases} \dot{x}_1 = \frac{1}{L_2}(x_3 + u_{C\alpha 0}) - \frac{R_2}{L_2}(x_1 + i_{2\alpha 0}) - \frac{1}{L_2}u_{g\alpha} \\ \dot{x}_3 = \frac{1}{C}(x_5 + i_{1\alpha 0}) - \frac{1}{C}(x_1 + i_{2\alpha 0}) \\ \dot{x}_5 = \frac{1}{L_1}u_1 - \frac{1}{L_1}(x_3 + u_{C\alpha 0}) - \frac{R_1}{L_1}(x_5 + i_{1\alpha 0}) \end{cases} \tag{5}$$

$$\begin{cases} \dot{x}_2 = \frac{1}{L_2}(x_4 + u_{C\beta 0}) - \frac{R_2}{L_2}(x_2 + i_{2\beta 0}) - \frac{1}{L_2}u_{g\beta} \\ \dot{x}_4 = \frac{1}{C}(x_6 + i_{1\beta 0}) - \frac{1}{C}(x_2 + i_{2\beta 0}) \\ \dot{x}_6 = \frac{1}{L_1}u_2 - \frac{1}{L_1}(x_4 + u_{C\beta 0}) - \frac{R_1}{L_1}(x_6 + i_{1\beta 0}) \end{cases} \tag{6}$$

The design of the grid-connected subsystem controller equation (3) is described in detail. Firstly, in the light of the odd order Equation (5):
Step 1: Define the error as $z_1 = x_1$, its dynamics are as follows:

$$\dot{z}_1 = \dot{x}_1 = \frac{1}{L_2}(x_3 + u_{C\alpha 0}) - \frac{R_2}{L_2}(x_1 + i_{2\alpha 0}) - \frac{1}{L_2}u_{g\alpha} \tag{7}$$

Let x_3 be regarded as virtual control, and choose the virtual stabilization function as follows:

$$x_{3d} = -u_{c\alpha 0} + L_2\left[\frac{R_2}{L_2}(x_1 + i_{2\alpha 0}) + \frac{1}{L_2}u_{g\alpha} - c_1 z_1\right] \tag{8}$$

Where $c_1 > 0$ is constant. Define the error $z_3 = x_3 - x_{3d}$, so z_1 dynamics are as $\dot{z}_1 = \frac{1}{L_2}z_3 - c_1 z_1$. Let the Lyapunov function of the Equation (5) 1th as $V_1 = \frac{1}{2}z_1^2$. So the time derivative of V_1 along the trajectory of the Equation (5) satisfies $\dot{V}_1 = z_1\dot{z}_1 = -c_1 z_1^2 + \frac{1}{L_2}z_1 z_3$.
Step 2: Let the Lyapunov function of the Equation (5) 2th as $V_3 = V_1 + \frac{1}{2}z_3^2$. Such that the time derivative of V_3 along the trajectory of the system (5) satisfies

$$\dot{V}_3 = \dot{V}_1 + z_3\dot{z}_3 = \dot{V}_1 + z_3\left(\left(\frac{1}{C}(x_5 + i_{1\alpha 0}) - \frac{1}{C}(x_1 + i_{2\alpha 0}\right)\right) \tag{9}$$

Let x_5 be regarded as virtual control, and define the error as $z_5 = x_5 - x_{5d}$, choose the virtual stabilization function as follows:

$$x_{5d} = -i_{1\alpha 0} + C\left[-\frac{1}{L_1}z_1 + \frac{1}{C}(x_1 + i_{2\alpha 0}) + az_1 + bz_3 - c_3 z_3\right] \tag{10}$$

Where $c_3 > 0$ is constant and $a = -\frac{c_1 L_2 R_2}{L_1} - L_2 c_1^2$, $b = \frac{R_2}{L_1} - c_1$. So that $\dot{V}_3 = -c_1 z_1^2 - c_3 z_3^2 + \frac{1}{C}z_3 z_5$.
Step 3: Let the Lyapunov function of the Equation (5) 3th as $V_5 = V_3 + \frac{1}{2}z_5^2$. The time derivative of V_5 along the trajectory of the system (5) satisfies as follows:

$$\dot{V}_5 = -c_1 z_1^2 - c_3 z_3^2 + z_5\left(\frac{1}{C}z_3 - \frac{R_1}{L_1}(x_5 + i_{1\alpha 0}) - \frac{1}{L_1}(x_3 + u_{c\alpha 0}) + \frac{1}{L_1}u_1 - \dot{x}_{5d}\right) \tag{11}$$

Define the error is $z_5 = x_5 - x_{5d}$, so that

$$\dot{z}_5 = \dot{x}_5 - \dot{x}_{5d} = \frac{1}{L_1}u_1 - \frac{1}{L_1}(x_3 + u_{C\alpha 0}) - \frac{R_1}{L_1}(x_5 + i_{1\alpha 0}) - \dot{x}_{5d} \tag{12}$$

Select the feedback control law as follows

$$u_1 = L_1\left[\frac{R_1}{L_1}(x_5 + i_{1\alpha 0}) + \frac{1}{L_1}(x_3 + i_{1\alpha 0}) + mz_1 + nz_3 + pz_5 - \frac{1}{C}z_3 - c_5 z_5\right] \tag{13}$$

Where $c_5 > 0$ is design constant and $m = b - c_5$, $n = \frac{Cc_1}{L_2} - c_1 - ac_1C$, $p = \frac{C}{L_2^2} + \frac{1}{L_2} + \frac{aC}{L_2} - bc_5C + c_5^2C$. So that

$$\dot{V}_5 = -c_1z_1^2 - c_3z_3^2 + z_5\left(\frac{1}{C}z_3 - \frac{R_1}{L_1}(x_5 + i_{1\alpha 0}) - \frac{1}{L_1}(x_3 + u_{c\alpha 0}) + \frac{1}{L_1}u_1 - \dot{x}_{5d}\right) = -c_1z_1^2 - c_3z_3^2 - c_5z_5^2 \tag{14}$$

With the same method, the controller of Equation (6) and the time derivative of the Lyapunov function can be obtained:

$$u_2 = L_2\left[\frac{R_1}{L_1}(x_6 + i_{1\beta 0}) + \frac{1}{L_1}(x_4 + u_{c\beta 0}) + jz_1 + kz_3 + lz_5 - \frac{1}{C}z_4 - c_6 z_6\right] \tag{15}$$

$c_2 > 0$, $c_4 > 0$, $c_6 > 0$ are constants. $l = h - c_4$, $j = \frac{Cc_2}{L_2} - c_2 - gc_2C$, $p = -\frac{C}{L_2^2} + \frac{1}{L_2} + \frac{gc}{L_2} - hc_4C + c_4^2C$.

$$\dot{V}_6 = -c_2z_2^2 - c_4z_4^2 + z_6\left(\frac{1}{C}z_4 - \frac{R_1}{L_1}(x_6 + i_{1\beta 0}) - \frac{1}{L_1}(x_4 + u_{c\beta 0}) + \frac{1}{L_1}u_2 - \dot{x}_{6d}\right) = -c_2z_2^2 - c_4z_4^2 - c_6z_6^2 \tag{16}$$

With the controllers of u_1, u_2, the closed-loop error system equation of the whole system is as follows:

$$\begin{cases} \dot{z}_1 = -c_1z_1 + \frac{1}{L_2}z_3 \\ \dot{z}_2 = -c_2z_2 + \frac{1}{L_2}z_4 \\ \dot{z}_3 = -c_3z_3 - \frac{1}{C}z_5 \\ \dot{z}_4 = -c_4z_4 - \frac{1}{C}z_6 \\ \dot{z}_5 = -c_5z_5 \\ \dot{z}_6 = -c_6z_6 \end{cases} \tag{17}$$

Finally, combining odd-order Equation (5) and even-order Equation (6), the global Lyapunov function can be obtained as follows:

$$V = \frac{1}{2}z_1^2 + \frac{1}{2}z_2^2 + \frac{1}{2}z_3^2 + \frac{1}{2}z_4^2 + \frac{1}{2}z_5^2 + \frac{1}{2}z_6^2 \tag{18}$$

The method of applying cross backstepping method to design the controller for the island model is the same as the principle of grid connection. Because of the limited space, I will not repeat it here and give the result directly. The controller and the closed-loop error system equation are as follows:

$$\begin{cases} u_1 = L\left[-\frac{1}{C}z_5 + \frac{1}{L}(x_3 + u_{c\alpha 0}) + \frac{R_L}{L}(x_5 + i_{\alpha 0}) + Cc_3^2z_3 - c_3z_5 - c_5z_5\right] \\ u_2 = L\left[-\frac{1}{C}z_4 + \frac{1}{L}(x_4 + u_{c\beta 0}) + \frac{R_L}{L}(x_6 + i_{\beta 0}) + Cc_4^2z_4 - c_4z_6 - c_6z_6\right] \end{cases} \tag{19}$$

$$\begin{cases} \dot{z}_1 = 0 \\ \dot{z}_2 = 0 \\ \dot{z}_3 = -c_3z_3 + \frac{1}{C}z_5 \\ \dot{z}_4 = -c_4z_4 + \frac{1}{C}z_6 \\ \dot{z}_5 = -c_5z_5 \\ \dot{z}_6 = -c_6z_6 + \frac{1}{C}z_6 \end{cases} \tag{20}$$

3. Stability Analysis

If all subsystems have a common Lyapunov function, the switch system is asymptotically stable under any switching signal [12].According to analysis Equation (3) and Equation (4) of the switching system have a common Lyapunov function (18). Therefore, the equilibrium point of the system is asymptotically stable. Because all subsystems have a common Lyapunov function, the switched system is asymptotically stable under the action of arbitrary switching law.

4. Switching Law Design

Since there is the common Lyapunov function in the switched system, any switching law can be designed [13]. In this section, the island detection technology is not considered, and the state information of the system is known. Therefore, regard whether the voltage and frequency of the power grid are within the normal operation range as the switching law. When the power grid is in normal condition, it runs in case 1, $u_{\sigma(t)}=U_1$. Otherwise, it runs in case 2, $u_{\sigma(t)}=U_2$.Where $\sigma(t)=[0,\infty)\rightarrow\{1,2\}$ represents switching signal, $\sigma(t)=1$ represents subsystem 1 and $\sigma(t)=2$ operation subsystem 2. Design the switching law as follows:

$$\sigma(t)=\begin{cases}1, & \text{case } 1, u_{\sigma(t)}=U_1 \\ 2, & \text{case } 2, u_{\sigma(t)}=U_2\end{cases} \tag{21}$$

$$U_1=\begin{cases}u_1=L_1\left[\frac{R_1}{L_1}\left(x_5+i_{1\alpha 0}\right)+\frac{1}{L_1}\left(x_3+i_{1\alpha 0}\right)+mz_1+nz_3+pz_5-\frac{1}{C}z_3-c_5z_5\right] \\ u_2=L_2\left[\frac{R_1}{L_1}\left(x_6+i_{1\beta 0}\right)+\frac{1}{L_1}\left(x_4+u_{c\beta 0}\right)+jz_1+kz_3+lz_5-\frac{1}{C}z_4-c_6z_6\right]\end{cases} \tag{22}$$

$$U_2=\begin{cases}u_1=L\left[-\frac{1}{C}z_5+\frac{1}{L}\left(x_3+u_{c\alpha 0}\right)+\frac{R_L}{L}\left(x_5+i_{i\alpha 0}\right)+Cc_3^2z_3-c_3z_5-c_5z_5\right] \\ u_2=L\left[-\frac{1}{C}z_4+\frac{1}{L}\left(x_4+u_{c\beta 0}\right)+\frac{R_L}{L}\left(x_6+i_{i\beta 0}\right)+Cc_4^2z_4-c_4z_6-c_6z_6\right]\end{cases} \tag{23}$$

5. Simulation Analysis

In this section, the switching technology from island subsystems to grid-connected subsystem is more difficult. So take the island to grid-connected switching as an example to simulate in Matlab/Simulink, in which the controller uses S function module. Initial condition for the grid-connected subsystem 1 and island subsystem 2 are as follows:

$$(x_1,x_2,x_3,x_4,x_5,x_6)=(1,-2,1,-2,1,-2)\,,\ (x_1,x_2,x_3,x_4,x_5,x_6)=(0,0,1,-2,1,-2)$$

Figure 4 is the response curve of grid-connected operation state variables, and figure 5 is the response curve of island operation state variables. It can be seen that the state variables can be stabilized at the equilibrium point when the two subsystems operate independently.

Taking island switch grid-connected as an example (see Fig.6 ~ Fig.11). Figure 6 and Figure 7 are the simulation of the voltage and the current wave of island mode switching to grid-connected mode. The voltage is kept at 311V during the switching from island mode to grid-connected mode at 0.1s. During switching the output current of the inverter is smoothly switched from 5A to 10A. Figure 8 is a simulation of any phase voltage and the corresponding input current wave in the process of switching from island to grid-connected. Taking A phase as an example, the voltage is constant and the input current is switched from 0A to 10A without abrupt change at 0.1s. Figure 9 is island switches to grid-connected any phase voltage and the corresponding output current wave of the inverter. Where the voltage is constant and the output current of the inverter is switched from 5A to 10A, there is no abrupt change in the smooth stability at the switching time. In Figure 10, although there is a fluctuation at the time of 0.1s switching, it can be seen the fluctuation can be recovered within a few seconds, and the fluctuation error is always within the standard range. In Figure 11, the active power is switched from 500W to 1500W at 0.1s. The switching time has some abrupt changes, but the stability is restored within 1s, the design of the visible controller can effectively reduce the time of mutation recovery and can keep the stability of the system by combining the rules of the switching law.

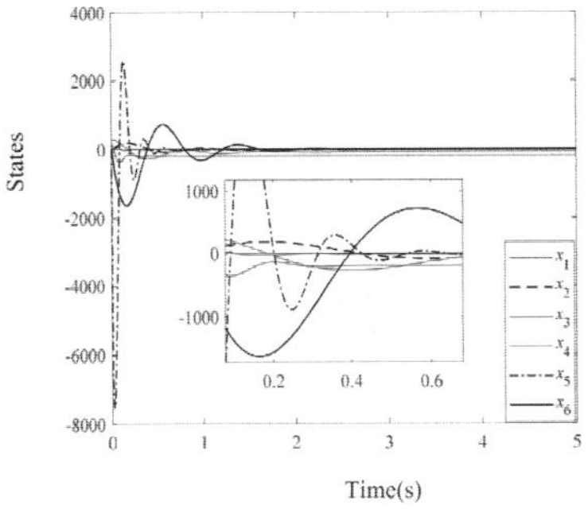

Figure 4. State response curve of grid-connected system.

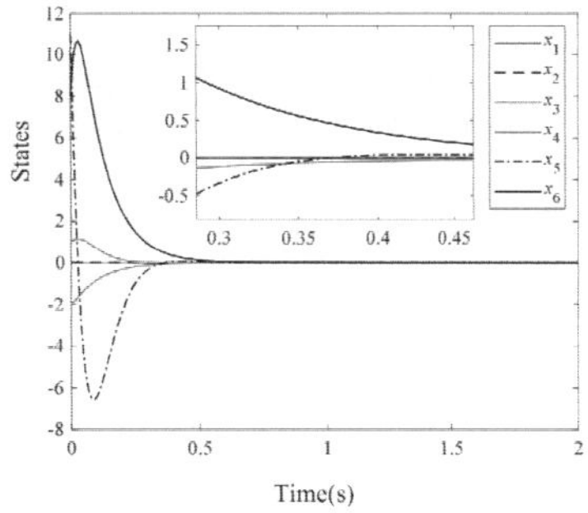

Figure 5. State response curve of island system.

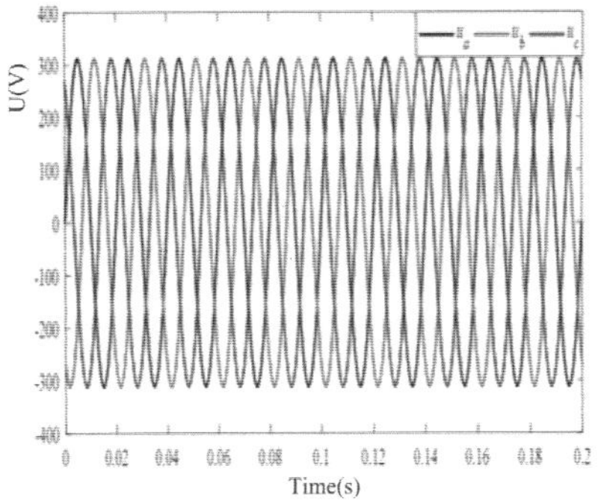

Figure 6. The simulation diagram of voltage switching from islanding to grid side.

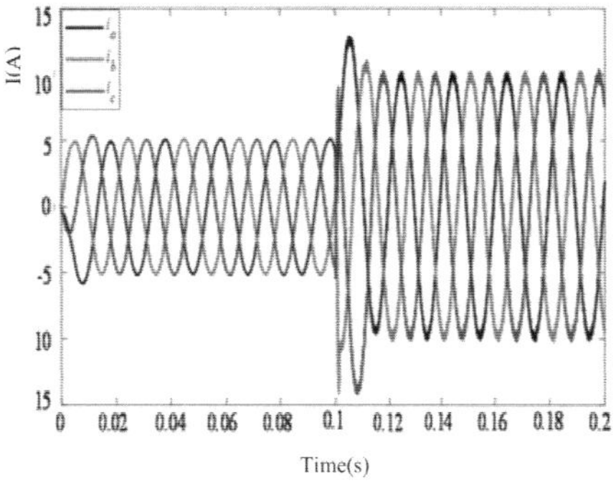

Figure 7. The simulation diagram of current switching from islanding to grid side.

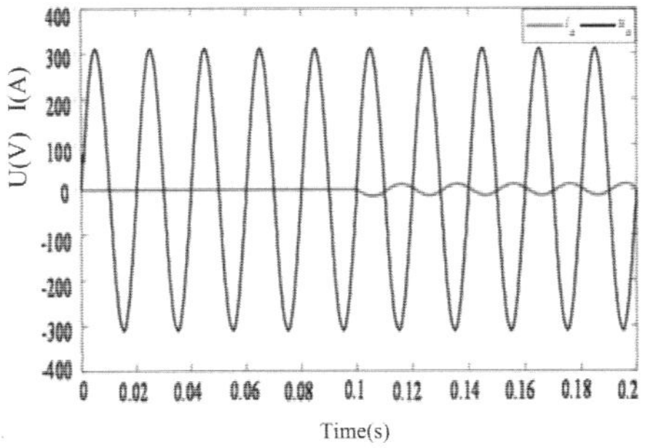

Figure 8. The diagram of voltage and current switching from island to grid connected at PCC.

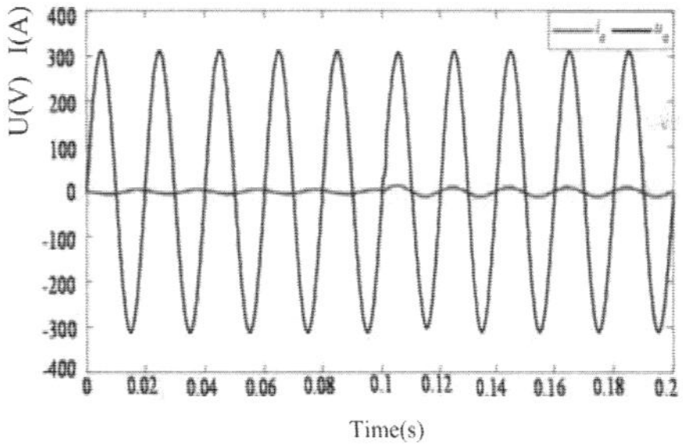

Figure 9. The voltage and current switching for island switch grid connected at inverter.

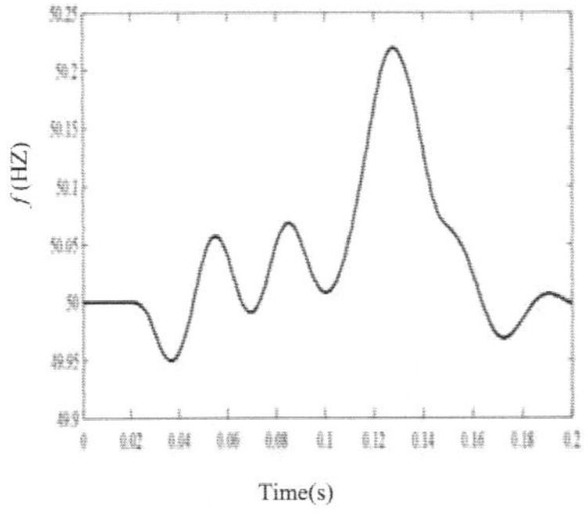

Figure 10. The picture of frequency curve.

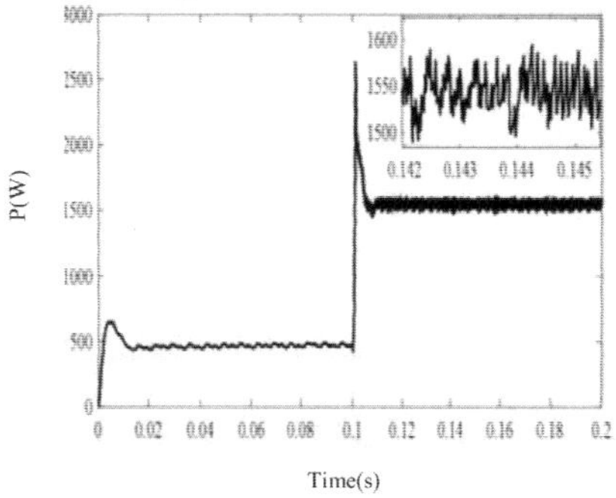

Figure 11. The picture of power curve.

6. Conclusion

In this paper, the theory of switched system is applied to microgrid, and the grid-connected operation mode and island operation mode are regarded as two subsystems of switched system. Designing switch controllers by using the cross backstepping algorithm, meanwhile the common Lyapunov function is constructed. It is shown that arbitrary switching law can ensure the stability of the switched system. In addition the method can effectively suppress the system oscillation caused by the sudden change of voltage and current in the switching process, realize the smooth switching of grid-connected and island, improve the power quality and ensure the normal operation of the important loads. However, the method presented in this paper is only applicable to nonlinear systems that can be transformed into cross strict feedback system.

Acknowledgments

This work was supported by the Key Project of Liaoning Provincial China Department of Education under Grants JZL201915401.I would like to express my heartfelt thanks.

References

[1] Blaabjerg F, Teodorescu R, Liserre M and Timbus A V 2006 Overview of control and grid synchronization for distributed power generation systems *(IEEE Transactions Industrial Electronics* vol. 53*)* no. 5, pp.1398-1409

[2] Espina E, Llanos J, Burgos-Mellado C, Cardenas-Dobson R and Mart´ınez-G M 2020 Distributed control strategies for microgrids : overview *(IEEE Access*, vol. 8, no. 4*)* pp. 193412-193448

[3] Zheng Z, Hui L and Ran L 2016 Review of AC microgrid inverter control srategy *(Power System Automation* vol 40*)*, no 9, pp. 142-151

[4] Majumder R and Ghosh A. 2008 Control of parallel converters for load sharing with seamless transfer between grid connected and islanded modes (*IEEE Power and Energy Society General Meeting - Conversion and Delivery of Electrical Energy in the 21st Century*)

[5] Y. Y. Xu, H. E. Nian, Y. M. Wang and X. P. SunD 2020 Impedance modelling and stability analysis of VSG controlled grid-connected converters with cascaded inner control loop *(Energies* vol 13*)* no. 19, pp. 5114-5135

[6] Che L, Shahidehpour M, Alabdulwahab A.and Turki Y A 2015 Hierarchical coordination of a community microgrid with AC and DC microgrids *(IEEE Transactions on Smart Grid* vol 6*)* no. 6, pp. 3042-3051

[7] X. L. Li, Y. C. Wen, C. Chen and Q. W. Xu 2021 Adaptive rsesilient secondary control for microgrids with communication faults *(IEEE Transactions on Cybernetics)*

[8] C Xie, D Liu, K Li, J X Zou, K L Zou and Josep M G 2021 Passivitybased design of repetitive controller for LCL-type grid-connected inverters suitable for microgrid applications *(IEEE Transactions on Power Electronics* vol 36*)* no. 2, pp. 2420-2431

[9] Q Gong and Y P Tian 2000 A constructive designmethod of nonlinear crossed strict feedback system *(Acta Automatica Sinica*, vol 26*)* no. 4, pp. 447-453

[10] D. J. Li, S. M. Lu and L. Liu 2021 Adaptive NN cross backstepping control for nonlinear systems with partial time-varying state constraints and its applications to hyper-chaotic systems *(IEEE Transactions on Systems,Man, and Cybernetics: Systems* vol 51*)* no. 5, pp. 1-12

[11] D. Sun, X. Y. Shao and G. Feng 2007 A model-free cross-coupled control for position synchronization of multi-axis motions: Theory and Experiments *(IEEE Transaction on Control System Technology* vol 15*)*, no. 2, pp. 306-314

[12] Liberzon D and Morse A S 1999 Basic problems in stability and design of switched systems *(IEEE Control Systems Magazine*, vol. 19*)* no. 5, pp. 59-70

[13] Hajiahmadi M, Schutter B D and Hellendoorn H 2016 Design of stabilizing switching laws for mixed switched affine systems *(IEEE Transactions on Automatic Control*, vol. 61*)* no. 6, pp. 1676 - 1681

Inspection of Synchronous Belt in Lifting Appliances

Hu Chen[1*], Peng Hu[1], Pengzhao Zhang[1], Qin Lv[1] and Qin Zhang[2]

[1] Shanghai Institute of Special Equipment Inspection and Technical Research, 915 Jinshajiang Rd. Putuo District, Shanghai, China, 200062
[2] Huzhou Vocational & Technical College, Huzhou, Zhejiang, China

*Corresponding author email: bahatiger@qq.com

Abstract. Introduced the inspection requirements of the main transmission components in the hoisting machinery in the current special equipment regulation; Checked the technical requirements related to safety in the relevant standards of the synchronous belt. Based on the physical structure, failure mode and regulatory standard requirements of the synchronous belt, summarized the inspection method of the synchronous belt.
Keywords: Synchronous belt; Crane; Inspection; Failure.

1. The Current Regulations for Lifting Appliances Drivetrain Parts Inspection Requirements

In the currently produced lifting appliances, power transfer devices or components in addition to the gearbox mainly have reels, wire rope (for roll drive), walking gears, racks (for gear rack drive), sprockets, chains (for sprocket drive), wheels, rails (wheel friction drive), in addition to the balance valve, hydraulic lock, hydraulic cylinder and other hydraulic system in the hydraulic drive in the car-specific lift is also more common.

In China, the safety technical specifications mainly included "TSG Q7015-2016: lifting appliances periodical inspection regulation", "TSG Q7016-2016: lifting appliances installation, modernization & general overhaul supervision inspection regulation" and "TSG Q7002-2019: regulation for type test of lifting appliances". There are clear requirements for the inspection of the above-mentioned transmission components. Take the "TSG Q7015-2016: lifting appliances periodical inspection regulation" for example, in the appendix, it clarifies the inspection of the main components, including the wire rope (The fixing requirements of the rope end of the wire rope, the number of safety laps on the drum and the specific requirements of the rope end fixing device on the drum), hydraulic system, sprockets and chains and gears and racks, etc.

In TSG Q7002-2019: regulation for type test of lifting appliances, for drive components, the following provisions are made:

(1) Wire rope should be fixed, properly wear around and neatly arranged. When the lifting device is at the lowest point of the working position, the drum winding wire rope, in addition to the number of rings fixed rope tail, the number of safety laps retained on the drum should not be less than 2 laps (tower crane, simple lift should not be less than 3 laps). When the lifting device is at the highest point of the working position, there is at least 1 full circle of winding rope margin on the drum, when the pressure plate is fixed, there are at least 2 press plates (no less than 3 pieces of electric reeds) on the wire rope fixing end on the drum, and the chain and sprocket should engage correctly, without stuck and impact The safety factor of wire rope and chain of mechanical parking equipment shall be as follows: no one is less than 5,quasi-unmanned is not less than 7, and someone is not less than 12.

(2) The hook shall be set up in accordance with the design requirements of the safety device against accidental decoupling of heavy objects.

(3) The sheave shall have a device or structure to prevent the wire rope from coming out of the rope groove. The gap with the sheave body at the side plate of the sheave cover and the curved roof plate shall not exceed 0.5 times the nominal diameter of the wire rope.

2. The Application of Belt Transmission in Mechanical Parking Equipment

According to the "Special Equipment Catalog" issued by the State Council of China in 2014, mechanical parking equipment is a category of crane equipment in the catalogue. As the national living standard improves, the number of family cars in 100 households has increased from 16.9 in 2013 to 35.3 in 2019[1]. The manufacturing industry for mechanical parking equipment is also booming, and new materials and technologies are emerging. The lifting mechanism of mechanical parking equipment is usually driven by traditional wire rope forced drive or traction, but also by hydraulic drive and chain sprocket drive.

Recently, due to the development of belt transmission technology and the improvement of the level of technology, synchronous belt drive is also gradually applied to some mechanical equipment. According to the principle of belt drive, it can be divided into frictional type and meshing type. Frictional type belt overload will slip, there is a certain overload protection effect when we use frictional type belt drive, but the transmission ratio is not accurate. Meshing type belt can ensure synchronous transmission, synchronous profile belt belongs to the meshing belt transmission. synchronous cog belt drive technology was born in the United States in 1964[2]. This transmission technology achieves synchronous operation of the transmission power by synchronizing the meshing of the gears with the wheels and the toothing of the synchronous profiles. According to the profile, it can be divided into trapezoidal tooth synchronization belt and arc tooth synchronization belt, arc tooth synchronization belt working wing root stress concentration is smaller than trapezoidal tooth synchronization band. Compared with the technology of wire rope reel roll drive, chain sprocket drive, synchronous tooth belt by meshing drive, carrying layer to ensure that the tooth pitch is unchanged, the transmission ratio is accurate, the shaft pressure is small, with the advantages of light equipment, small size, and high precision. In addition, because the base material of the synchronous tooth belt is rubber or polyurethane and other polymer materials, oil resistance, wear resistance is better, no lubrication is required, easy maintenance. Using the synchronous cog belt for the transmission system, the transmission mechanism of mechanical parking equipment is becoming more and more compact, the transmission accuracy is also higher, the synchronous profile belt is also small in weight, more conducive to the speed of the lifting mechanism precise control. Therefore, in recent years, some manufacturers of mechanical parking equipment have replaced the traditional wire rope roll drive and chain sprocket transmission in the crane industry with synchronous cog belt transmission.

Although used as a drive component, the synchronous belt has many advantages for the application of wire ropes in mechanical parking equipment. However, as the main components of lifting appliances, we lack some regulations about the synchronous tooth belt for safety requirements. When Inspector perform inspection tasks of a synchronous belt transmission used mechanical parking equipment, it cannot be inspected according to the requirements of the wire rope or the requirements of the sprocket chain. Apparently, if we ignore the safety conditions of the synchrotron belt, we do not meet the requirements of our work to ensure the safety of special equipment.

3. The Main Form of Failure in Synchronous Cog Belt

Figure 1 is a structural diagram of a synchronous cag belt (single-sided arc tooth), which consists mainly of a strong layer, a substrate material, and a scrim. The substrate material is composed of polymer materials, which are processed into various shapes suitable for various operating conditions, providing friction, it has anti-wear and corrosion characteristics. Scrim is a fiber material covering the outer surface of the profile of the substrate material, which can increase the coefficient of friction between the synchronous profile belt and the synchronous belt gear and increase the wear resistance of the substrate material:

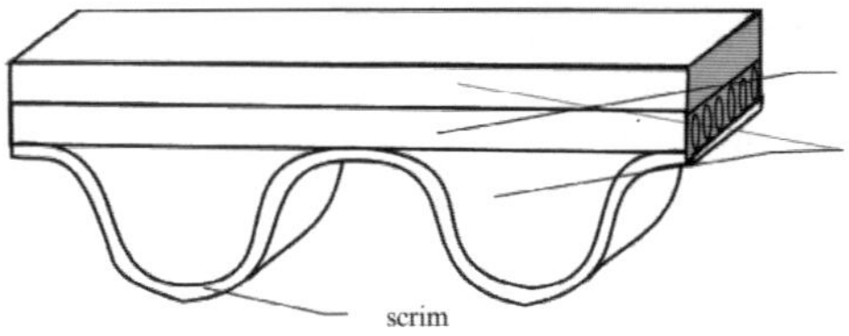

Figure 1. The synchronous cog belt structure chart (Single-sided arc tooth).

Figure 2 is the object picture of a kind of synchronous belt.

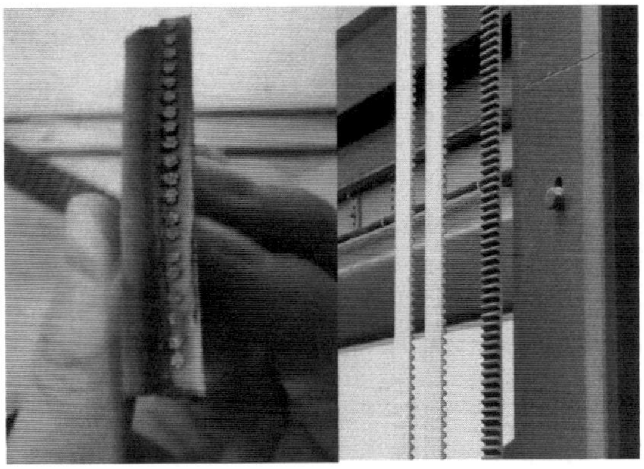

Figure 2. The object picture of a kind of synchronous belt.

As a major component of special equipment, the synchronous belt belongs to the consumables, and the inspection agencies should focus on the failure effect of fatigue and wear on the synchronous belt. Research in references[3][4] shows that, there are two forms of fatigue damage in the synchronous belt, one is the internal damage of the substrate material, and the other is the synchronous cog belt tooth root crack. The synchronous cog belt is a composite structure consisting of substrate material, scrim, strong layer, etc. According the study of fatigue fracture mechanism of the synchronous belt, Suyu Li, Yunwen Shen [5] found that, the weak link of fatigue fracture in the synchronous cog belt is the substrate material rather than the strong layer, so the defect in the use of the wire rope core in the strong layer of the synchronous cog belt should not be the main concern of the synchronous belt. Inspectors should pay more attention to the crack and wear defect of the synchronous belt substrate layer and the tooth root.
According the experiment, Cong Wu, Shujun Li [6] found that, under the influence of large tension, the failure type of the belt is mainly the whole belt pulled off, tooth break and flank wear, under the influence of small tension, the main failure is mostly broken belt and tooth fracture, and the tooth wear is not the main failure form. High tension can cause the synchronous band tooth part to cut or break, and excessive tension can increase wear, it is worth noting that the tension is too high synchronous belt flank usually has the teeth mark of the synchronous belt wheel; Excessive relaxation will result in synchronous belt failure and transmission efficiency decreases, moreover, the more the relative motion of the wheel and the synchronous belt increased, the more the friction and tooth impact load increases. it will inevitably cause the synchronous belt tooth wear to intensify and tooth root cracks. When the belt works in the Excessive low tension for a long time, it will also appear obvious slip marks[7]. Therefore, the tension of the synchronous cog belt should also be controlled within a reasonable range. In designing, when we select the synchronous cog belt, the tighter side tension value is according to the formula 1:

$$F_1 = 1250\frac{P_d}{v} \tag{1}$$

The loose side tension value is according to the formula 2:

$$F_2 = 250\frac{P_d}{v} \tag{2}$$

In formula 1 and 2, P_d is the design power, v is the speed of the belt.

4. Technical Requirements for Synchronous Cog Belts in the Current National Standard

In the current special equipment safety technical regulations, the requirements for major components are mostly reference to the technical standards of the corresponding components. For instance, the safety requirements for wire ropes are mainly referred to GB/T 3811 "Design rules for cranes", GB/T 6067.1 "Safety rules for lifting appliances – part 1: General", and GB/T 20118 "Steel wire rope for general purposes" etc. So, we can reference the technical standards for the synchronous cog belts. In GB/T 13487-2017 "Synchronous belts for general drive", the physical properties of the synchronous belts for general transmission are as Table 1(Take the single-sided arc tooth synchronization band as an example)[8].

Table 1. Physical properties of the single-sided tooth synchronous belts.

item	tensile strength N/mm ≥	reference elongation		jacket fabrics adhesive strength N/mm ≥	steel-cord adhesive strength N ≥	teeth shear strength N/mm ≥
		force	elongation % ≤			
3M	90	70	4.0	—	—	—
5M	160	130		6	400	50
8M	300	240		10	700	60
14M	400	320		12	1200	80
20M	520	410		15	1600	100

Appearance quality aspect, the standard provides: The teeth with the synchronous belts should be full, the tooth surface must not have folding, breakage and other phenomena; The synchronous straps after cutting should be flat and free of burrs; At the same time, if there is one of the following circumstances, it is a non-conforming product:

(1) More than 3 consecutive pores or bubbles were found.
(2) A pore or bubble with a diameter of more than 0.8 mm.
(3) There are pores on both sides of a tooth.

At the same time, the standard also provides that the appearance of the synchronous belts inspection should be carried out article by article inspection, the main size can be spot checked; The same model with the same batch of the same material synchronous belts should be carried out not less than 1 time a month of physical properties testing, the test has non-conforming items, should be in the same batch of synchronous belts to take another double sample for non-conforming project testing. If the re-inspection project still has a result is not qualified, then the batch of products as unqualified.

In GB/T 30172-2013 "Synchronous belt drives--Technical specification for installation", there are also provisions for inspection items before the synchronous belt drivetrain is operational: After 10 minutes of commissioning of the drivetrain, check the items listed in Table 2.

Table 2. Synchronous belts inspection items (GB/T 30172).

No.	the content of the item	resolution measures
1	the synchronous belt appears to climb up the wheeled edge	Adjust the parallel conditions of the wheel
2	resonance or loud noise occurs in the synchronous belt	Adjust the mounting tension of the belt to check the design and machining parts
3	tooth jump occurs	Increase tension. If the tooth-jumping phenomenon still occurs after tensioning, it may be that the transmission system failure or the load on the synchronization belt exceeds the rated load, and a redesign should be considered
4	The tension force of the belt is abnormally decreased	Check the sliding base for looseness. If there are still anomalies, the belt should be re-tightened

The standard inspection project is mainly to avoid the installation of the synchronous belt caused by the failure of the synchronous belt, to avoid running bias, resonance, too loose and other problems.

5. Conclusion

For synchronous belts, we believe that the failure modes, laws, regulations, and technical standards are very important. Our inspection work must be based on this. So when the special equipment inspectors check the synchronous belts, should focus on the following points:

(1) The selected synchronous belt should match the synchronous wheel, and when repairing, the different models of synchronous belt and synchronous wheel shall not be replaced without the confirmation of the design and installation unit.

(2) Investigate the synchronous belt's quality documents and design calculations, for verifying whether the chosen belts' physical properties meet the demand of the designed strength.

(3) Check the inspection record in the equipment maintenance records of the synchronous belts, ensure the record is integrity.

(4) Visual inspection the belt, Appearance should not appear more than 3 consecutive pores or bubbles and the diameter of the pores or bubbles should not be greater than 0.8mm, the teeth should not appear on both sides of the pores.

(5) Synchronous cag belts teeth should be full, tooth surface should be no folding, breakage, and other phenomena.

(6) There should be no obvious cracks, pits, breakages, and other defects on the synchronous cog belts substrate material.

(7) The synchronous belts should not show obvious tension or looseness, such as checking the sync tooth strap for wheeled tooth indentations or visible slip marks.

(8) During the operation of the equipment, the synchronous belts should not appear run-off phenomenon, should not appear the phenomenon of tooth jumping, the base of the synchronous belts should not be loose.

(9) Since the safety factor of the wire rope and chain used in mechanical parking equipment is not less than 5, the safety factor of the synchronous belt should not be less than 5.

(10) The safety factor of the synchronous band connector must not be less than the safety factor of the synchronous belt.

Acknowledgements

Thanks to the Huzhou Natural Science Foundation (2019YZ09) for financial support.

References

[1] 2020 China Statistical Yearbook [J]. Theory and practice of statistics,2021(01):2.
[2] Yuping Yang, Cao Qinglin, Shen Shide. The research status and development of synchronous cog belts [J]. Journal of Nantong Institute of Technology,2000,16(1):12-16.
[3] Hiroshi Iizuka etc. Qi Bing (Interpret). Research about the fatigue failure of the synchronous belts, rubber belt industry, 1993(6)
[4] Hiroshi Iizuka etc. Qi Bing (Interpret). Research about the tooth crack mechanism of the synchronous belts, rubber belt industry,1995(7)
[5] Suyou Li, Shen Yunwen. The fatigue failure mechanism and life evaluation of the synchronous belts [J]. Mechanical drive,1993,17(3):28-31.
[6] Cong Wu, Li Shujun. Research about the synchronous belts' transmission failure and fatigue life. [J]. Mechanical Science and Technology,1996,15(3):415-417.
[7] Kuiliang Han. Research about the synchronous belts' failure mode [J]. Development and innovation of the mechanical and electrical products,2012,25(5):84-86.
[8] The State Administration of Quality Supervision, Inspection and Quarantine of the People's Republic of China, China's National Standardization Management Committee. Synchronous belts for general drive: GB/T 13487-2017[S]. 2017.

Control Strategy of Rare Earth Permanent Magnet Synchronous Motor

Xichen Liu*

Jiangsu chentai Construction Engineering Co., Ltd Jiangsu, China

*Corresponding author email: Xichenliu@163.com

Abstract: Rare earth permanent magnet materials and magnetic composites help permanent magnet synchronous motor well developed. And it also has the advantages of small rotor inertia, high efficiency, high power density and high reliability. It has a good application prospect in high-performance servo system. This paper discusses the control strategy of permanent magnet synchronous motor.
Keywords: Permanent magnet synchronous motor; Control strategy; Rare earth.

1. Introduction

Recently, new material technology was developed very well; permanent magnet motor is more and more active in various industrial fields. Permanent magnet synchronous motor is a new type of motor, and its development time is not long. This motor has the advantages of small rotor inertia, high efficiency, high reliability and high power density. It has a good application prospect in high-performance servo system. For example, permanent magnet synchronous motor is used in CNC machine tools, machining centers and other equipment and devices. Because the permanent magnet synchronous motor does not need excitation winding, it can reduce the motor volume, reduce the weight, improve the efficiency and reduce the motor calorific value. Therefore, it can improve the efficiency and power factor and effectively save electric energy.

2. Types and Basic Structure of Permanent Magnet Synchronous Motor

According to the rotor structure, permanent magnet synchronous motor can be divided into three basic forms: convex type, and embedded type. The first two are also collectively referred to as external structures. The convex type has the advantages of simple structure, low price and many applications. Due to the small diameter of motor rotor and the small inertia of motor, it is often used in servo system. The embedded structure can increase the flux leakage linkage, resulting in the increase of pole angle and the decrease of torque. Compared with the built-in motor, its structure is more complex and the price is higher, but this structure has high air gap flux density and can produce greater torque than the convex motor, so it is more suitable for use under the condition of large torque.

Permanent magnet synchronous motor is composed of rotor, end cover, stator and other components. The obviously difference between rotor structure and asynchronous motor is that there are high-quality permanent magnet poles on the rotor. According to the different positions of permanent magnets on the rotor, it is usually called surface rotor structure and built-in rotor structure.

3. Characteristics of Permanent Magnet Synchronous Motor

Compared with common asynchronous motor, permanent magnet synchronous motor has the following characteristics.

3.1. Low Loss and Low Temperature Rise

Permanent magnet provide the magnetic field, compared with magnetic field generated by excitation current, the excitation loss will be avoid; and the temperature rise of the motor was low. Under the same load, the temperature rise can be reduced by more than 20K.

3.2. High Power Factor and Good Energy Saving Performance

Permanent magnet synchronous motor's power factor is high and independent of the motor stage. Under full load condition, the power factor is close to 1. In this way, compared with asynchronous motor, tit has efficiency. With high power factor, the power supply (transformer) capacity of motor can be reduced theoretically, and the specifications of supporting switchgear and cables can be reduced at the same time.

3.3. High Efficiency

Compared with asynchronous motor, the efficiency value of permanent magnet synchronous motor is much higher under light load. Its efficient operation range is wide, and the efficiency is greater than 90% in the range of 25% ~ 120%. The rated efficiency of permanent magnet synchronous motor can meet the level 1 energy efficiency requirements of the current national standard, which is its biggest advantage over asynchronous motor in energy saving. In addition, permanent magnet synchronous motor can still maintain high efficiency in light load area.

3.4. Other Advantages

It also has the advantages such as high starting torque, short starting time and high overload capacity. And it has convenient control, constant speed, it depends on frequency. The stability and reliability was good.

4. Control Strategy

Compared to DC motor, the control system of permanent magnet synchronous motor is more complex. The control strategies of permanent magnet synchronous motor mainly include external synchronization control strategy, self synchronization control strategy and flux weakening control.

4.1. External Synchronization Control Strategy

External synchronous control means that the permanent magnet synchronous motor uses an independent variable frequency power supply, and the speed of the synchronous motor is related to the power frequency, also known as other controlled variable frequency speed regulation.

4.2. Self Synchronization Control Strategy

Self synchronous speed regulation control is divided into direct torque control, automatic frequency conversion speed regulation and vector control. Vector control, Siemens Engineer balschke first proposed vector control theory in 1971. That is, taking the rotor flux rotation space vector as the reference coordinate, the stator current is decomposed into two orthogonal components. One is in the same direction as the flux, representing the stator current excitation component, and the other is orthogonal to the flux direction, representing the stator current torque component. They are controlled respectively. It is characterized by simple control structure and easy realization of control software, so it is widely used in various speed regulation systems. However, the vector control method needs complex coordinate transformation and accurate observation of rotor flux. Moreover, it depends greatly on the parameters of the motor, and the control effect is limited.

The basic idea of direct torque control is to realize direct torque control by controlling stator flux. This control method does not need the equivalence and transformation of AC motor and DC motor, and simplifies the coordinate transformation; the stator magnetic field orientation is adopted to realize the observation and control of motor flux and torque, directly. The estimation of stator flux only involves stator resistance, reduces the dependence on motor parameters, and overcomes the shortcomings of vector control to a great extent. This technology is developed after vector control. It was generally

applied in induction motor at first, and then gradually applied to permanent magnet synchronous motor.
Automatic frequency conversion control is similar to the control method of changing armature voltage of DC motor. It can be divided into two strategies: power supply type and voltage type.

4.3. Flux Weakening Control
Using flux weakening control, it can output constant torque in the condition of low speed and constant power at high speed, and has a wide speed regulation range. The basic idea of flux weakening control of the motor is to weaken the magnetic field of the motor by using the direct axis armature reaction, so as to achieve the effect of equivalent magnetic field weakening. In addition, different parameters of permanent magnet synchronous motor will cause different change trajectories of voltage and current phases of the motor. Based on this, researchers have improved the motor structure to improve the flux weakening control ability of permanent magnet synchronous motor.

4.4. Other Advanced Control Modes of Permanent Magnet Motor
In addition to the common control methods of permanent magnet motor introduced above, many advanced control strategies and theories are gradually applied. Such as：Adaptive control: variable structure control and intelligent control. At present, adaptive control is mainly used in parameter identification and speed and position identification in permanent magnet synchronous motor control to reduce cost and improve system reliability. Sliding mode variable structure control method is also used in the field of controlling system of the permanent magnet synchronous motor. In addition, in order to deal with the complexity and uncertainty of permanent magnet motor control, functional control systems such as artificial neural network are also gradually applied, so that the control quality and stability of the system have been significantly improved.

5. Focus of Future Research
Permanent magnet synchronous motor has been greatly developed. In the future, with the progress of material science and technology, the volume of permanent magnet motor will be further reduced. In terms of motor control strategy, whatever the control strategy is, there are still some problems. We need to comprehensively use all this control strategies. The research on the penetration and combination of various control strategies is the development direction of motor control technology. Various strategies work together to improve the performance of AC speed regulation system. Compound control strategy will be the main trend of future development.

6. Conclusions
The main ideas of the control strategy of rare earth permanent magnet synchronous motor were introduced. The types and basic structure and the characteristics of permanent magnet synchronous motor were described, and the control strategy was also analyzed, and the future research direction was pointed out.

Reference
[1] Leng Zaixing. Flux weakening speed control of permanent magnet synchronous motor [D]. Huazhong University of science and technology, 2006
[2] Zhang Yu, Zhang Xing. Improvement of direct torque control system of permanent magnet synchronous motor [J]. *Journal of Hefei University of technology*, 2007
[3] Xu Junfeng, Feng Jianghua, Xu Jianping. Overview of permanent magnet synchronous motor control strategy [J]. *Locomotive electric drive*, 2005 (03): 7-11 + 76
[4] Liu X, Yu H, Gerada D, et al. Comparison of rare-earth and hybrid-magnet mover configurations for a permanent magnet synchronous linear motor[J]. IET Electric Power Applications, 2021, 15(3).

Statistical Method of Low Efficiency Wind Turbine Generating Wind Power Curve Based on Operating Data

Zhannan Chen[1], Jiechang Wang[1], Jianfei Guo[1], Yan Dong[1], Jinglong Yu[1], Runan Zhou[1] and Huanxiu Xiao[2,*]

[1] New Energy Branch of Huaneng Jilin Power Generation Co. LTD, Changchun, Jilin, 130000

[2] School of Energy and Power Engineering, Huazhong University of Science & Technology, 430074 Wuhan, China

*Corresponding author email: 18810986221@163.com

Abstract. Aiming at the low efficiency of wind turbines in service, this paper proposes a statistical method of low efficiency turbines based on wind power curve generated from operating data. According to the data distribution characteristics of the wind-power data scatter diagram, a data preprocessing model was established based on the quartile method and quadratic clustering method, which can effectively mark the data of shutdown, invalidity, undergeneration, overgeneration and power limit, and draw the wind power curve which can accurately represent the fan's performance. On this basis, a mining method for low efficiency units based on the difference rate of lost electric quantity and power consistency coefficient is established, which can make statistics of high loss electric quantity units, units deviating from the designed power curve and units with degraded performance regularly, and evaluate the generation improvement before and after maintenance. Through the actual operation and maintenance data of all 814 wind turbines of a certain company, the effectiveness of the method is verified.

Keywords: Low efficiency unit; Data preprocessing; Wind power curve; Evaluation of power generation improvement.

1. Introduction

With the continuous development of social economy, the shortage of traditional energy has become a key problem restricting the development of social economy. As a clean renewable energy, wind power has the advantages of huge reserves, wide distribution and low development and utilization cost. It is the most promising energy situation among clean renewable energy sources. With the rapid growth of the number of operating units, due to factors such as inaccurate micro-site selection, unreasonable equipment selection, deterioration of fan performance, creep defects of equipment and so on, part of the wind turbine generating capacity is low after it is put into operation, and the operating data and benefits of the wind turbine are far less than the feasibility evaluation value and design value.

In order to find inefficient wind turbines, power generation, capacity coefficient, Time Based Availability (TBA) and other indicators are usually used to carry out benchsetting error detection, locate inefficient turbines, and carry out targeted operation and maintenance work. The above indicators and assessment methods are the summary of experience in the industry and play an important guiding role in the operation and maintenance management of wind farms. However, due to the unstable and uncertain changes of wind speed and direction, it is difficult for wind turbines to maintain a stable wind energy conversion state, which also directly leads to the uncertainty of generating state. When these two indexes locate inefficient wind turbines, there are many problems such as unclear directivity, unknown responsible party and difficult quantitative measurement. For example, for the same 2000 hours of electricity generation, the year of strong wind and the year of light wind are different. Therefore, when the energy generation and CF are used as assessment indicators, people often simply refer to the average wind speed of the statistical cycle to get a rough comparison scale, but it covers the real wind energy and power generation potential. Similarly, TBA conceals the potential of wind turbines. According to the statistics of the Global Wind Energy Council and the Energy Bureau, with the continuous improvement of the reliability of wind turbine equipment, the TBA of most wind farms at home and abroad has reached more than 97%, but there are still great differences in power generation level between different projects. The average annual power generation hours of typical wind farms in Europe and the United States reach 2500 hours, while the average level of wind farms in China is only about 1700-1900 hours. The nearly identical TBA level cannot explain the generation gap of 600-800 hours. The reason lies in that the loss of shutdown for 1 hour in high wind is often more than the loss of shutdown for 10 hours in low wind. The measurement of pure time dimension cannot measure the energy loss caused by shutdown, so it cannot accurately locate the inefficient group effectively.

The Supervisory Control and Data Acquisition (SCADA) system installed in the wind farm collects the wind farm power curve based on the wind turbine operating data (hereinafter referred to as the operating wind power curve). According to the specific machine position, the 10-minute data in the normal power generation period is selected, and the high randomness and fluctuation of wind speed under the actual operation condition of the fan is considered, as well as the response of the unit control system when the wind speed changes, which can more accurately reflect the power generation performance of the unit under the actual environmental conditions. Through actual output power wind turbine and wind speed of wind - power curve, according to the engine room 10 minutes wind speed to match the actual power curve for the month, calculation of this abnormal operation period should be output can locate the abnormal running time of power loss, can be more accurate positioning problem, quantify the each link in the process of production lead to the loss of power. At the same time, by comparing and analyzing the monthly power curve, the deviation from the design power curve and the analysis of performance deterioration can be effectively located. Therefore, this paper generates the wind power curve based on the actual operation data of the fan, and on this basis, puts forward a set of methods that can effectively locate the inefficient unit.

2. Wind Power Curve

2.1. The Current Situation

There are a lot of abnormal data in the operation data collected by SCADA system. There are three main reasons for abnormal data. First, the bad operating environment of the fan is easy to cause the sensor to fail, and the planned shutdown of the wind turbine due to maintenance failure may also cause the measurement data to deviate from the normal value. Second is to abandon the wind power rationing is to accept the local power grid caused by inadequate ability to abandon the wind power rationing will lead to the unit output power at a lower level, even if the wind speed exceeds the rated wind speed, the output power will be lower than the rated power, which will lead to a large number of abnormal data in the original data, and the abnormal data will directly affect the operation accuracy based on the wind power curve modeling. Therefore, in order to improve the data quality of the operation data, it is necessary to clean the operation data collected by SCADA system to improve the quality of the operation data. At present, the pre-processing of wind speed and power scatter data can be divided into three categories. The first is the cleaning method based on mathematical statistics (such as density, distance average, median, variance, etc.), but the method of distribution density anomaly data recognition effect is limited, such as the literature [1] by using the optimal interclass variance cleaning algorithm, detection unit the abnormal state of the electricity generating performance, reduce the rigid requirements for detection tools and data dimension, However, abnormal data accounts for a large proportion in wind power data, so this method will remove a large number of normal data. The second type is the method based on data mining, that is, in a large number of data, use unsupervised learning methods to identify data with abnormal characteristics for data analysis and processing. This kind of method does not need to establish a model in advance and has good anti-interference. For example, in reference [2], the improved K-means method was used to clean data for the missing and abnormal characteristics of historical data to ensure the validity of data. Third class method based on the data distribution, assume that the wind - power is normal distribution, the probability density of error is located on the outside of the wind - power curve data for abnormal data, such as the literature [3] is proposed based on "3 - sigma law" methods to detect outliers, but the wind - power probability density are often multiple peak, accuracy is not high.

2.2. Data Preprocessing

In view of the above mentioned data quality problems that lead to abnormal wind power curve, this paper proposes a data state marking method for wind power scatter graph combining the characteristics of mathematical statistics and data mining methods. The method includes marking data as the following states:

- Invalid data: Filter noise data that is obviously out of limit
- Downtime: mark according to wind speed, pitch Angle and active power
- Underpass: abnormal data based on quartile method
- Overshoot: Abnormal data based on quartile method
- Power limiting: Abnormal data based on k-means clustering algorithm
- Normal: Normal can be used to evaluate the wind power curve data points

2.2.1. Downtime data marking. For the data point of wind speed is greater than cut in wind speed, power is less than power set threshold (default value 5kW) is judged as stop; For data points with pitch Angle is greater than 85°, wind speed is less than set threshold of wind speed (default value 6m/s) is judged as stop.

2.2.2. Invalid data marking. Wind turbine SCADA data contain both normal operation data and other abnormal noise data. In this paper, the 10min average value of SCADA wind speed and output power was first used to mark the following data categories as invalid data by comparing the wind turbine operating conditions and logs:

- Data whose output power is 0 or negative during shutdown or startup;
- Data points where the active power significantly exceeds the rated power;
- The wind speed is significantly higher than the cut wind speed or the wind speed is less than 0;
- Fault or abnormal operation data recorded in the operation log;
- Data point of measurement error caused by sensor failure;
- Data points of fan load limiting operation recorded in operation log;

2.2.3. Under-transmission and over-transmission. In descriptive statistics, quartiles refer to three data points that divide a sorted data sample into four parts, and each part contains 25% of the data volume of the entire data sample. There are many calculation methods for quartiles. For an ascending sample X[x1,x2,...,xn], the quartile calculation method used in this article is as follows. Data outside the inner limit [Fl,Fu] are all Is an outlier.

Take a fan with a wind speed of 25m/s and rated power of 2000kW as an example, the identification process of abnormal data is as follows.

1) The horizontal quartile method was used to eliminate the anomalous data points with horizontal distribution.

Firstly, the wind speed power data are sorted according to the power from small to large, and the power of [0,2000]kW is divided into equal intervals with 25kW as the interval unit. The quartile method was adopted for the wind speed data set within each power range, and the data points outside the inner limit [Fl,Fu] of the wind speed data set were eliminated. Among them, the wind speed Wind speed is greater than Fu, identified as underdevelopment.

2) Longitudinal quartile method was used to eliminate the scattered abnormal data points on the left.

The data were sorted according to the wind speed from small to large, and the wind speed of [0,25]m/s was divided into equal intervals with 0.5m/s as the interval unit. The quartile method was adopted for the power data set corresponding to each wind speed interval, and the data points whose power data set was larger than the upper limit Fu were eliminated. If the power is greater than Fu, it is identified as oversending.

The reason why the lower limit value Fl is not considered is that the abnormal data is located below the data area, and the data volume of the abnormal power rationing data cluster in a wind speed interval may be equal to the normal data volume or even more than the normal data volume. In this case , The quartile method instead treats normal data as abnormal data and eliminates it, causing a large amount of normal data to be deleted by mistake.

2.2.4. Power limit data mark. Take a wind turbine with a cut-off wind speed of 25m/s and a rated power of 2000kW as an example. The identification process of power limit is as follows.

The wind speed power data are sorted according to the wind speed from small to large, and the wind power of [0,25]m/s is divided into equal intervals with 0.5m/s as the interval unit. The corresponding power data sets in each wind speed range are analyzed by clustering.

Because the number of abnormal data clusters in the data of each typhoon generator set is different, and the number of abnormal data clusters of each typhoon generator set cannot be known in advance when the algorithm is implemented, that is, the value of input parameter K of the clustering algorithm of each typhoon generator set cannot be determined in advance. When the clustering algorithm is adopted in this system, k is taken as a large value so that K can cover the abnormal data clusters of each typhoon generator unit. In the previous process, abnormal data on the left and top of the scatter graph had been removed, so the normal data expected to be retained in the end must have been gathered in the uppermost and leftmost parts of the scatter graph. Therefore, the clustering results with larger mean values should be retained as much as possible, so the following quadratic clustering idea is adopted:

① For each wind speed range, it is assumed that the clustering result obtained by direct clustering method is $\{C_1,C_2...,C_k\}$, and calculate the clustering center of each cluster as $\{w_1,w_2... ,w_k\}$;

② Find the $\{w_1, w_2,... ,w_k\}$ is the largest clustering center w_{max}, and its corresponding clustering is C_{max};

③ Calculate and w_{max} $\{w_1, w_2,... ,w_k\}$, the distance of each cluster center except w_i is $d_i = w_{max} - w_i$;

④ Compare the size of each d_i with the preset threshold T(such as 0.2× rated power). If $d_i < T$, merge C_i and C_{max} into a cluster; otherwise, eliminate C_i;

⑤ The final retained cluster is the expected normal data within the wind speed range, and the remaining data clusters are marked as power limit data.

2.3. Construction of Wind Power Curve

The paper uses the "bin method" provided by the IEC61400-12-1 standard to construct the actual wind power curve of the wind turbine. In this paper, the wind speed under the modified standard air density is taken as reference, and the interval is divided according to the integral multiple of the wind speed of 0.5m/s, and each interval is divided continuously according to the integral multiple of the wind speed of 0.5m/s. The interval is closed at the front and opened at the back. Each wind speed interval divided can be expressed as $[v_{ni} - 0.25, v_{ni} + 0.25]$, where v_{ni} is the central wind speed between the regions. For example, the wind speed point 3.5m/s represents the wind speed interval [3.25,3.75]. Then for each wind speed area, the wind speed marked as normal and the corresponding power are counted as the average value, and the statistical result is the measured power curve.

3. Low Efficiency Unit Statistical Method

3.1. Statistics of High Loss Electric Generating Units

Based on the operating power curve, the mining index based on the difference rate of electric quantity loss of the fan is established. After 0 o 'clock every Monday, the system automatically excavates the high-loss units of the whole company last week according to the latest excavation conditions. The difference rate of lost electricity is defined as follows:

$$Y_t = \frac{E_t / P_{rated}}{Days} \tag{1}$$

$$Y_{loss} = \frac{E_{loss} / P_{rated}}{Days} \tag{2}$$

$$D = \frac{Y_{loss}}{Y_t} \tag{3}$$

In this expression:

D: Difference rate of lost electric quantity
Y_t: Average theoretical generating hours per day
Y_{loss}: Average number of hours lost per day
E_t: Total theoretical generating capacity
E_{loss}: Total lost electricity
P_{rated}: Rated power
$Days$: Days in a statistical period

Based on the theoretical power generation and lost power curve of the fan, the wind speed measured by the fan wind meter is used for calculation. When the fan has no communication or fault shutdown without wind speed signal, the average wind speed of the whole field is used for calculation.
The excavation process of high loss electric generating unit is as follows:
(1) Excavate the generating unit with high loss once a week;
(2) The fans whose average daily loss hours are in the top 20% and the difference rate is greater than 5% are marked as high loss units;
(3) The fans whose average daily loss hours are in the top 40% and the difference rate is greater than 5% are marked as intermediate loss units, but the fans whose average daily loss hours are in the top 40% and the difference rate is greater than 5% are not treated;
(4) The excavation process requires the removal of fans that have been marked as long shut down units.

3.2. Deviation from the Design Curve Unit Statistics

In this paper, the consistency coefficient F is used to evaluate whether the operating power curve of wind turbines deviates from the design power curve. The consistency coefficient f of each fan is defined as follows. Take the 10-minute data of the fan within the statistical period and calculate f for the data marked as normal by the following formula:

$$f = \left(1 - \frac{\sum_{1}^{n} = 1 \frac{P_d(v_i) - p_i}{P_d(v_i)}}{n} \right) \times 100\% \tag{4}$$

In this expression:

i is the 10-minute sampling point;
n is the number of valid data points within a time range;
$P_d(v_i)$ is the power of the designed power curve corresponding to the actual wind speed at the sampling point;
p_i is the actual power of the sampling point.

The design mining strategy is as follows: calculate the consistency coefficient f of all the fans in the whole company, take the consistency coefficient f of all fans in the same type as the set, calculate the inner limit of the set $[F_l, F_u]$according to the quarterle method, and mark the fans whose consistency coefficient is less than F_l as units deviating from the design curve. Similarly, the excavation process requires the removal of fans that have been marked as long shut down units.

3.3. Performance Deterioration Unit Statistics
In this paper, consistency coefficient is also used to evaluate degraded wind turbines. The mining strategy is set as follows: The consistency coefficient f of each fan in the whole company for 12 consecutive weeks is calculated as $\{F_1, \cdots, F_{12}\}$, and the least square method is used to perform linear fitting for $\{F_1, \cdots, F_{12}\}$, and the slope of the fitting is used as the performance deterioration coefficient. The performance degradation coefficient of each fan in the same type was taken as the set, the inner limit $[A_l, A_u]$ of the set was calculated according to the quartile method, and the fan with a consistency coefficient less than the consistency coefficient was marked as the performance degradation unit. Similarly, the excavation process requires the removal of fans that have been marked as long shut down units.

3.4. Evaluation of Power Generation Enhancement
For the low efficiency unit after treatment, the increase of generating capacity is evaluated by the operation power curve before and after the treatment time point. Evaluation methods are as follows:

① Calculate the operating power curve P_a before treatment and the operating power curve P_b after treatment;

② Take the wind speed and actual power experienced by the fan during the evaluation period, and write it as $\{v_i, p_i\}$

③ The following formula is used to calculate energy generation improvement:

$$E_\Delta = \sum_1^n =_1 \left(p_i - P_b\left(v_i\right)\right) t_i \quad (5)$$

④ The proportion of energy generation increase is calculated by the following formula:

$$\delta = \left(\frac{E_\Delta}{\sum_{i=1}^n P_b\left(v_i\right) t_i} \right) \times 100\% \quad (6)$$

⑤ Evaluate the time period during which fan faults and fan downtime are to be eliminated.

In the above expression:
$P_b(v_i)$ is the power of wind speed v_i corresponding to the operating power curve before processing.

4. The Example Analysis
According to the previous research results, all 814 fans of a company were statistically analyzed to verify the statistics of low-efficiency fans.

4.1. Data Preprocessing
Take fan no. 33 in a wind field as an example, as shown in Figure 1 below. Take the 10-minute data of the fan in a year and mark the data status according to the running status given by the fan SCADA, as shown in Figure 1_A), where 82.60% are normal data points. After drawing the scatter diagram of wind speed and power, it can be found that SCADA marked part of the power limit data (FIG. 1_A) and the undergeneration data (FIG. 1_A) as normal power generation data, while some normal power generation data (FIG. 1_A) were marked as shutdown data. When these abnormal data are included in the data set used to draw the wind power curve, the drawn wind power curve will greatly deviate from the actual fan operation performance. After data is preprocessed by the above research method, as shown in FIG. 1_B), it can be seen that only 55.08% of the data is marked as normal data, and the data cleaning intensity is about 66.7%. The power limit data and undergeneration data in FIG. 1_B have

been accurately marked (areas ② and ③ in FIG. 1_B). In this way, the operating power curve reflecting the actual operating performance of the wind turbine can be drawn.

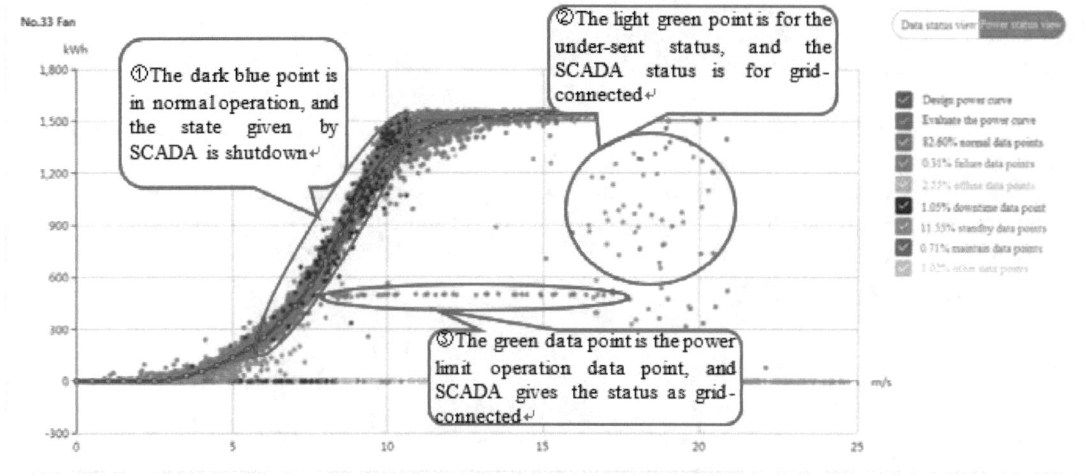

a) Data state before data preprocessing

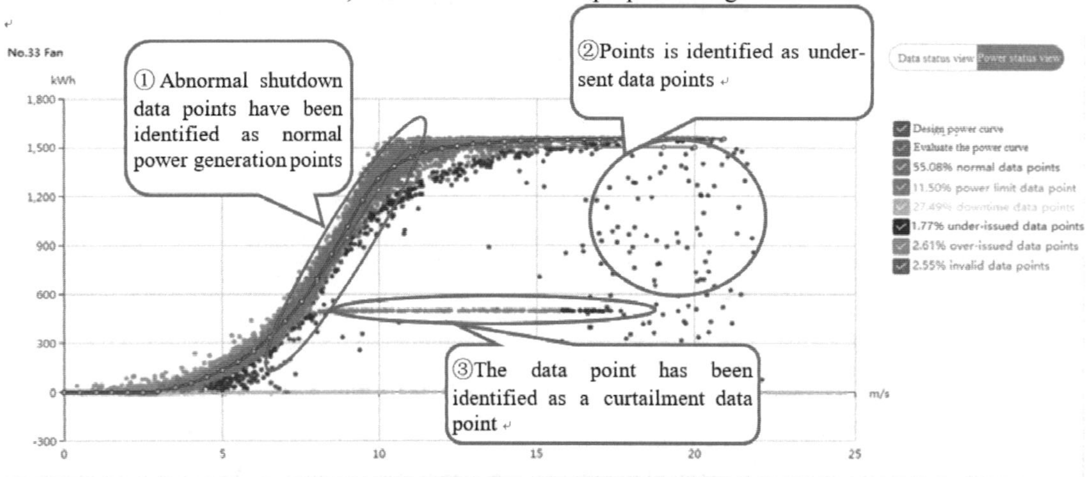

b) The state of data after data pretreatment

Figure 1. Data status before and after data pretreatment.

4.2. Mining Efficiency Statistics of Inefficient Units

According to the mining strategy designed in this paper, a total of 287 low-efficiency units will be excavated from January 1, 2021 to September 30, 2021, as shown in Table 1.

Table 1. Results statistics of mining low-efficiency units.

No.	Category	Mining were	Processing were
1	High loss generating unit	108	96
2	Deviation from design curve unit	103	47
3	Performance degradation unit	76	59
Total		287	202

After delivering the wind field, the total number of processing is 202, and the statistical accuracy is 70.3%. Before and after the treatment, the accumulative power generation was increased by 5.319 million kWh, with an increase ratio of 4.81.

For example, D30 unit (Yunda WD88-1500A) in a wind farm was marked as low efficiency unit in June 2021, and maintenance personnel boarded the aircraft for processing on June 13. The wind power curve before and after treatment is shown in the figure below. It can be obviously found that the power curve of low and medium wind speed section has been improved after treatment. FIG. 3 shows the lost generating capacity of the fan in 4 weeks before treatment (compared with the operating power curve after maintenance) and the increase in generating capacity in 12 weeks after maintenance (compared with the operating power curve before maintenance). The statistical results show that from June 14, 2021 to September 30, 2021, the actual generating capacity of the fan is 866,300 kWh, and the total generating capacity is 82,700 kWh, and the generating capacity increase percentage is 9.55%.

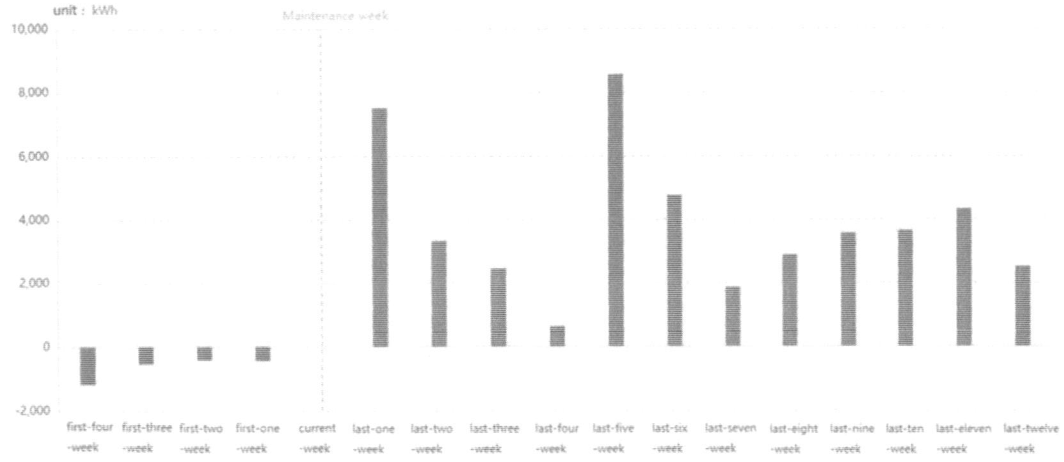

Figure 2. Comparison of wind power curves of D30 units in a wind field before and after treatment.

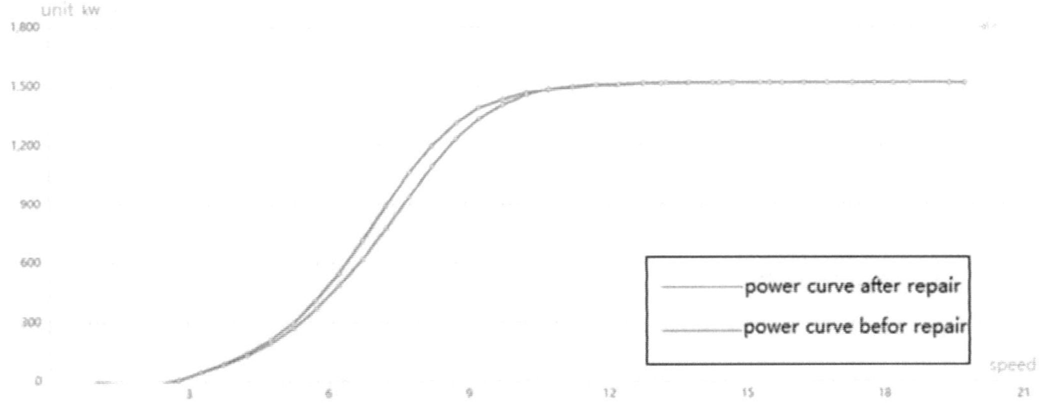

Figure 3. Evaluation of power generation improvement of D30 unit in a wind field.

5. Conclusion

Aiming at the low efficiency of wind turbines in service, based on the historical operation data of wind turbine SCADA system, a set of data preprocessing method was proposed to mark and eliminate the original data from abnormal state data such as shutdown, invalid, overgeneration, undergeneration and power limit, so as to generate the operating power curve of wind turbines. Then, the difference rate of lost electric quantity is used to excavate the unit with high loss electric quantity, and the consistency coefficient of power curve is used to excavate the unit with deviation from the design curve. Further, the performance degradation coefficient is defined on the basis of consistency coefficient to excavate the unit with degraded performance. Based on the actual operation and maintenance data of all 814 fans in a company, the statistical method of low efficiency units proposed by the local government is verified. The results show that the method can accurately locate low efficiency units, so as to guide operation and maintenance personnel to maintain low efficiency units, and significantly improve the power generation of fans.

References

[1] Lou Jianlou, XU Jia, LU Heng, QU Chaoyang, LI Shaowu, LIU Ruihua; Wind Turbine Data Cleaning Algorithm based on Power Curve [J].;power system automation,2016,40(10)
[2] Huang Qiujuan; Research on Wind Turbine Power Curve Anomaly Identification Based on Data Drive [J]; Journal of Anhui Agricultural Sciences; 2006-01 Shenyang University of Technology; 2019.05
[3] Zhao Yongning, YE Lin, Zhu Qianwen; Analysis of wind Abandoning Anomaly Data Cluster in Wind Power Plant [J]; Journal of Anhui Agricultural Sciences; 2006-01 Automation of Electric Power Systems,2014(21)

Selecting Sensitive Features Using CDET for Bearing Fault Diagnosis

Haifeng Se[1], Lei Zhou[2], Shiming Xu[3] and Song Kai[1,*]

[1] School of Electrical Engineering and Automation, Harbin Institute of Technology, Harbin, China
[2] Beijing Institute of Aerospace Test Technology, Beijing, China
[3] School of Instrumentation Science and Engineering, Harbin Institute of Technology, Harbin, China

*Corresponding author email: kaisong@hit.edu.cn

Abstract. The rolling bearing fault detection technology solution can be used to maintain the machinery in advance and avoid downtime and safety accidents. However, the current studies primarily focus on utilizing complex algorithms to improve the accuracy of fault diagnosis, while neglecting the influence of extracted features on diagnosis results, resulting in redundancy and conflict among features. To address such an issue, a feature selection method based on compensation distance evaluation technique (CDET) is proposed to optimize the feature set extracted from acoustic emission (AE) signals. Firstly, multiple failure data of the bearing are modeled by the fault simulation experiment platform. Then, the proposed feature selection method is used to evaluate the sensitive features. Finally, hidden markov model-support vector machine (HMM-SVM) series model is used to predict works. The experimental results show that the proposed method selects 7 sensitive features from the feature set composed of 13 features and achieve a high accuracy (99.3%).
Keywords: Bearing fault diagnosis; AE; Feature selection; CDET; HMM; SVM.

1. Introduction

Acoustic emission (AE) technology is a dynamic non-destructive examination, and the AE signals excited in the early defect development stage of bearings can be easily detected [1-3]. It has the advantages of short response time, wide frequency band range, convenient sensor installation and high signal-to-noise ratio [4]. Therefore, AE technology has been widely applied in the research field of fault diagnosis based on rolling bearings in recent years [5,6]. Žvokelj *et al.* [3] designed an early fault detection method of bearings based on multi-scale independent component analysis and ensemble empirical mode decomposition algorithm (EEMD), which adaptively decompositions signals into different time scales. Yang *et al.* [7] proposed a low-speed bearing AE fault diagnosis method combining complementary EEMD (CEEMD) and support vector machine (SVM). Manjurul *et al.* [8] extracted fault features from the mixed domain of AE signals, and used the optimized least squares hyper-sphere multi-class SVM (LSHS-MCSVM) classifier for fault diagnosis.

Although the above methods use different complex machine learning algorithms to improve the classification accuracy, they ignore importance of extracted features, resulting in redundancy and conflict among features, which ultimately leads to unsatisfactory performance of the algorithm. This paper proposes a method for selecting sensitive features to improve the accuracy of bearing fault diagnosis algorithm. Compensation distance evaluation technique (CDET) is used to screen out sensitive features that are sufficient to characterize the bearing state from the feature set. Finally, the extracted sensitive features are input to hidden markov model-support vector machine (HMM-SVM) series model for bearing fault diagnosis.

2. Feature Selection Method

2.1. Wavelet Threshold Denoising

The useful signal and noise can be separated in wavelet domain by wavelet threshold method. The main process of denoising by wavelet threshold method is to decompose the AE signals and get the wavelet coefficients. The denoised signal can be obtained by threshold processing and wavelet reconstruction.

2.2. VMD

The VMD algorithm is an adaptive signal processing method, which has been applied in many fields [9-11]. It defines the intrinsic mode functions (IMFs) is defined:

$$u_k(t) = A_k(t)\cos(\phi_k(t)) \tag{1}$$

where $A_k(t)$ stands for the instantaneous amplitude and $\phi_k(t)$ denotes phase.

$f(t)$ can be decomposed into K components $u(t)$. The Hilbert transformation is performed on $u(t)$ to obtain its analytical signal, which is then mixed with the estimated center frequency ω_k. The constraint is that the sum of $u(t)$ is equal to signal $f(t)$, so the variational problem is:

$$\begin{cases} \min\limits_{\{u_k\},\{\omega_k\}} \left\{ \sum\limits_k \left\| \partial_t \left[\left(\sigma(t) + \frac{j}{\pi t} \right) * u_k(t) \right] e^{-j\omega_k t} \right\|_2^2 \right\} \\ s.t. \sum\limits_k u_k(t) = f(t) \end{cases} \tag{2}$$

where $\{u_k\}$ is the set of IMF components; $\{\omega_k\}$ is the set of frequency centers; $\sigma(t)$ is the impact function.

The Lagrange multiplier operator $\lambda(t)$ are introduced to transform the problem into the unconstrained variational problem and obtain augmented Lagrange function. The equation (3) is solved by alternate direction multiplier method (ADMM), and its optimal solution is obtained.

$$\begin{aligned} L(\{u_k\},\{\omega_k\},\lambda) = \alpha \left\{ \sum_k \left\| \partial_t \left[\left(\sigma(t) + \frac{j}{\pi t} \right) * u_k(t) \right] e^{-j\omega_k t} \right\|_2^2 \right\} \\ + \left\| f(t) - \sum_k u_k(t) \right\|_2^2 + \left\langle \lambda(t), f(t) - \sum_k u_k(t) \right\rangle \end{aligned} \tag{3}$$

where α represents the quadratic penalty factor.

2.3. Establishment of Feature Set

The feature set consists of four parts, including singular value (*SV*), permutation entropy (*PE*), energy (*E*), barycenter frequency (*FC*).

SV: The singular values of IMFs matrix are different under different bearing states.

PE: the difference degree of envelope can be quantitatively described by permutation entropy. It can not only reflect the uncertainty of non-stationary and nonlinear signals, but also has strong anti-noise ability and simple calculation.

E: the energy distribution of IMFs in different frequency bands is also different, that is, the failure state is intrinsically related to the energy distribution.

FC: the *FC* can represent the frequency center distribution of signals. Therefore, the *FC* of AE signal envelope can be used as a feature of fault diagnosis.

2.4. *Features Selection Method Based on CDET*

This paper uses CDET technology to screen the above features. It is assumed that the average intra-class distance of the *k*-th feature in the feature vector in *M* classes is D_k, and the inter-class distance is $D_{k'}$. According to the ratio of $D_{k'}$ to D_k, the distance evaluation factor is:

$$\beta_k = \gamma_k \frac{D_{k'}}{D_k} \tag{4}$$

where β_k can comprehensively evaluate the separability of the *k*-th feature to *M* classes. For the normalized treatment of β_k, the evaluation index of compensation distance is:

$$\overline{\beta}_k = \frac{\beta_k}{\max(\beta_k)} \tag{5}$$

If the threshold $\phi(\phi \in [0,1])$ is set, the features that meet the distance evaluation index $\overline{\beta}_k \geq \phi$ are regarded as sensitive features.

The specific process of feature selection method: firstly, the original AE signal is denoised by wavelet threshold. Then, on the one hand, the *FC* of envelope in the signal is calculated, on the other hand, high dimensional IMFs matrix is obtained by VMD, and *SV*, *E* and *PE* are calculated respectively. Finally, CDET is used to evaluate the feature set composed of the above four features so as to select sensitive features, which are used in the subsequent fault diagnosis model (see figure 1).

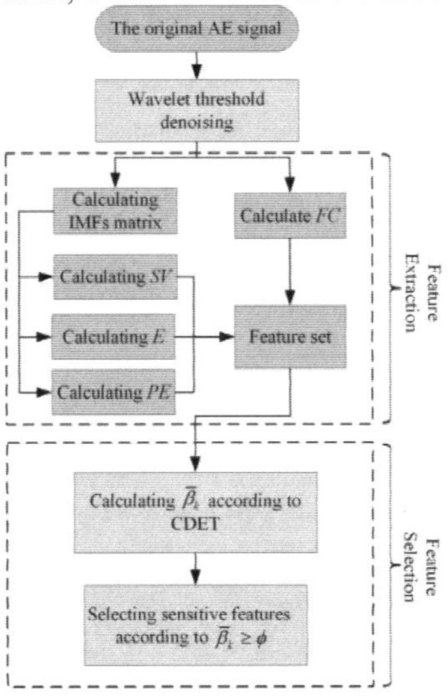

Figure 1. Feature selection method.

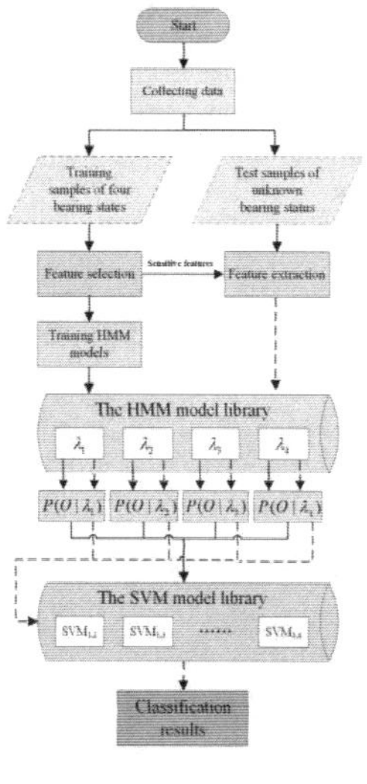

Figure 2. Fault diagnosis flow chart of HMM-SVM series model.

3. The Fault Diagnosis Model

HMM is suitable for processing continuous dynamic signals. SVM is good at processing classification and reflects the differences between categories to a greater extent. Therefore, combining the advantages of the two algorithms, HMM-SVM series model is used to diagnose bearing faults.

The working process of HMM-SVM series model is as follows: sensitive features extracted from training samples are fed into HMM model, and the probability values output by training samples under each HMM model are calculated, which are then fed into SVM model as new features for training. Finally, the HMM-SVM series model is applied to diagnose the unknown signals, as shown in figure 2.

4. Experiments and Results

4.1. The Experimental Data

In this paper, the simulation experiment of rolling bearing fault was carried out by using the fault simulation experiment platform. Four kinds of single point faults, including normal state, inner ring fault, outer ring fault and rolling element fault, were simulated on the bearing of fault simulation experiment platform. The AE sensor (SR150M) was selected to collect the above four bearing status signals. The SR150M sensor was adsorbed near the bearing seat. The frequency range of the SR150M sensor is 60~400 kHz and the resonant frequency is 150 kHz, the frequency of the acquisition card is set at 200 kHz. Therefore, the bearing status signal collected by the SR150M sensor basically covers the AE frequency range of metal materials, and the signal contains sufficient bearing fault information, which is conducive to the subsequent feature extraction and fault diagnosis operations. Finally, 48 samples were collected for each state, including 12 samples for training and 36 samples for testing. Therefore, the total number of training samples is 48, and the total number of test samples is 144. The data length of each sample was 10000 points (50 ms). Each sample is divided into 10 frames on average, thus the data length of each frame is 1000 points, and the length of feature vector sequence of each sample is 10. The frame processing of AE signals is shown in figure 3.

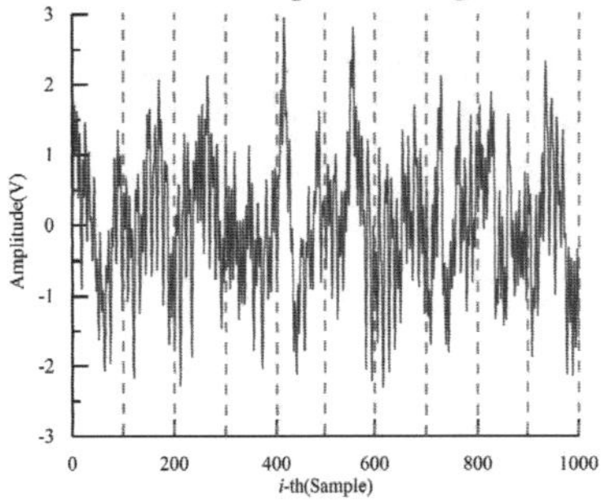

Figure 3. Frame processing schematic diagram of AE signal.

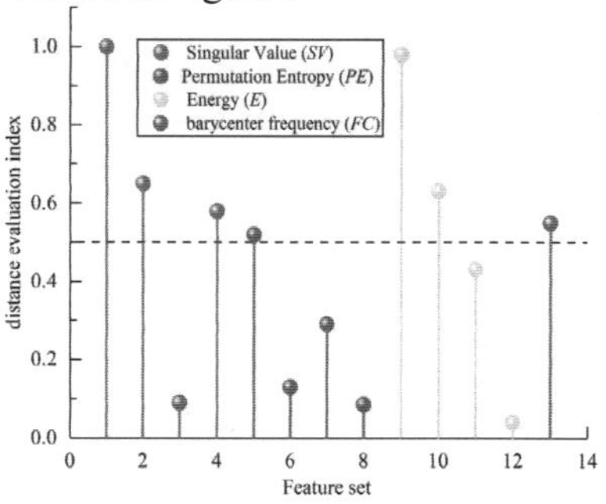

Figure 4. Feature evaluation result.

4.2. Feature Extraction and Evaluation

Firstly, VMD is used to process each frame signal after denoising and the 4×1000 IMFs matrix is output (K = 4). Secondly, four singular value features are obtained by singular value decomposition of the matrix, that is, $SV = \{SV_1, SV_2, SV_3, SV_4\}$. Thirdly, the energy value, permutation entropy feature and the barycenter frequency are calculated successively. Thus, each frame signal is composed of a feature vector of 13-dimensional, that is, $V = \{SV_1, SV_2, SV_3,SV_3, E_1, E_2, E_3, E_4, PE_1, PE_2, PE_3, PE_4, FC\}$, and a sample consists of a 10×13 feature matrix.

In order to screen out the features with the best classification effect, the CDET is used to evaluate and analyze the feature set V, thus obtaining the evaluation factors of 13 features, as shown in figure 4. When $\bar{\beta}_k \geq 0.5$ (ϕ =0.5), the corresponding features are selected as sensitive features. Therefore, the sensitive feature set consists of 7 sensitive features (*i.e.* SV_1, SV_2, ,SV_4, E_1, PE_1, PE_2 and FC) and is used as the input of the subsequent HMM-SVM series model. Meanwhile, the training sample is composed of 480×7 feature matrix, as shown in figure 5.

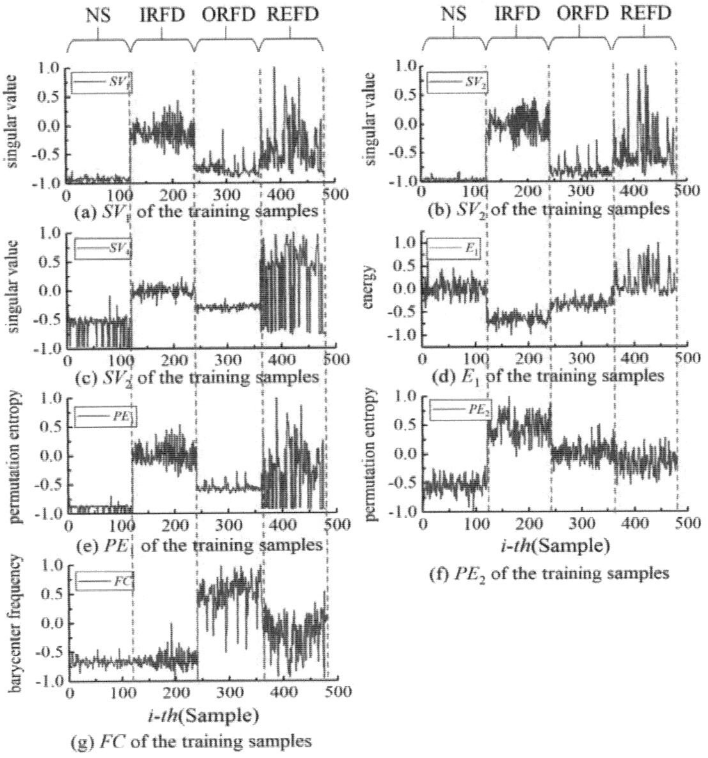

Figure 5. Sensitive feature set of training samples. Where the "NS" represents normal state data, the "IRFD" represents inner ring fault data, the "ORFD" represents outer ring fault data, the "ORFD" represents rolling element fault.

4.3. Fault Diagnosis Result

In the training process of HMM-SVM series model, the sensitive features are first input to the HMM model. Then the values output by HMM model are input to SVM model for training.The test samples are input into the HMM-SVM model, and the results are shown in table 1. The HMM-SVM series model has a very high recognition accuracy (99.3%). In addition, a single HMM model and a single SVM model are tested according to the same steps mentioned above, and the results are 89.6% and 88.2%, respectively. Therefore the HMM-SVM series model has superior performance.

Table 1. Fault diagnosis results of HMM-SVM series model. Where "NS" represents normal state, "IRF" represents inner ring fault state, "ORF" represents outer ring fault state, "OEF" represents rolling element fault state, "*Acc*" represents the accuracy of the results, "*Aver_Acc*" represents average accuracy.

		Predicted Results				*Acc*	*Ave_Acc*
		NS	IRF	ORF	REF		
Actual Status	NS	36	0	0	0	100%	99.3%
	IRF	0	36	0	0	100%	
	ORF	0	0	36	0	100%	
	REF	0	0	1	35	97.22%	

$$Acc = \frac{N_{pre}^{k}}{N_{act}^{k}} \tag{6}$$

$$Aver\_Acc = \sum_{i} \frac{N_{pre}^{i}}{N_{act}^{i}} \quad (7)$$

where N_{pre}^{k} indicates the number of class k bearing status data that is correctly classified, N_{act}^{k} refers to the number of class k bearing status data, k = "NS", "IRF", "ORF", "OEF".

5. Conclusion

The feature evaluation method based on CDET is studied, the sensitive feature which is beneficial to classification is obtained by multi-dimensional feature evaluation, which effectively solves the problems of redundancy and conflict among multi-dimensional features. In addition, the proposed HMM-SVM series fault diagnosis models combine the dynamic analysis capability of HMM with the small sample classification capability of SVM. And single HMM and single SVM model are compared. The results show that the accuracy of HMM-SVM hybrid model is 99.3%, which is higher than the other two models.

Acknowledgments

The work was supported by Technical Foundation Project of Quality and Reliability (Grant No. JSZL2018203C012). The authors would like to thank the colleagues of the project, as well as friends that have supported this work.

References

[1] Glowacz A, Glowacz W, Glowacz Z, et al. Early fault diagnosis of bearing and stator faults of the single-phase induction motor using acoustic signals[J]. Measurement, 2018,113:1-9.

[2] He P J, Liu G Y, She T T. Research on monitoring of aircraft structural components based on blind separation of single channel AE signal [J]. Chinese Journal of Scientific Instrument, 2018,39(8): 34-42.

[3] Žvokelj M, Zupan S, Prebil I. EEMD-based multiscale ICA method for slewing bearing fault detection and diagnosis[J]. Journal of Sound and Vibration, 2016,370:394-423.

[4] Hemmati F, Orfali W, Gadala M S. Roller bearing acoustic signature extraction by wavelet packet transform, applications in fault detection and size estimation[J]. Applied Acoustics, 2016,104:101-118.

[5] Zhang Y X, Cheng H X, Song S J. Fault diagnosis of rolling bearing based on improved RBF neural network [J]. Industrial Instrumentation & Automation, 2018(6): 31-34.

[6] Elforjani M, Shanbr S. Prognosis of Bearing Acoustic Emission Signals Using Supervised Machine Learning[J]. IEEE Transactions on Industrial Electronics, 2018,65(7):5864-5871.

[7] Yang J, Zhang P L, Liu Z T, et al. Acoustic emission diagnosis of low-speed bearing faults based on CEEMD energy entropy and SVM[J]. Nondestructive Testing, 2017,39(09):1-6.

[8] Manjurul Islam M M, Kim J. Reliable multiple combined fault diagnosis of bearings using heterogeneous feature models and multiclass support vector Machines[J]. Reliability Engineering & System Safety, 2019,184:55-66.

[9] Dragomiretskiy K, Zosso D. Variational Mode Decomposition[J]. IEEE Transactions on Signal Processing, 2014,62; 62(3; 3):531-544.

[10] Huang Y, Lin J, Liu Z, et al. A modified scale-space guiding variational mode decomposition for high-speed railway bearing fault diagnosis[J]. Journal of Sound and Vibration, 2019,444:216-234.

[11] Jiang J, Liu W. VMD and HMM based rolling bearing fault diagnosis. 2018 IEEE International Instrumentation and Measurement Technology Conference, pp:1-6.

Experimental Study on Electric Harvesting of Combine Harvester

Zhengtao Yang*, Haili Wang, Han Sun, Pengfei Wang and Qixin Cao

School of Mechanical and Engineering, Shanghai Jiao Tong University, China

*Corresponding author email: yzt0803@sjtu.edu.cn

Abstract. The electrification and intelligent transformation of harvester to make it have independent operation ability is an important embodiment of the development trend of agricultural automation and intelligence. Based on a full feed combine provided by relative company, this paper makes an electric transformation for its reaping and threshing part, including mechanical design, electric circuit construction, brushless DC(Direct Current) motor installation and test. Based on the finite automatic state mechanism, the control system of the electric operation part is built. Its sequential start and stop functions, reel speed following function and blockage detection and discharge function are tested in no-load experiment and farmland harvesting experiment. Through the continuous wheat harvesting test for 30 minutes, the power characteristics when harvesting wheat are measured, and the stability of its work is verified.
Keywords. Combine; CAN bus; Electrification; Finite state machine.

1. Introduction

At present, the power source of combine harvester used in China is mainly internal combustion engine such as diesel engine, which drives the working parts of reaping and threshing part through mechanical transmission such as belt. This transmission mode has fixed transmission ratio, which cannot be flexibly adjusted, and is easy to cause engine power wasting. In the actual harvesting process, the driver needs to rely on experience to adjust the traveling speed and header height of the harvester according to the current crop conditions, so as to achieve the ideal comprehensive efficiency[1][2].

The c-series combine of German Fendt company uses CAN(Controller Area Network) bus to build a communication network to ensure the high-speed communication between the sensor and the control computer, so as to detect the state of the harvester. Sensors are installed in each link of harvesting, threshing and cleaning. Sensors are also set at the top of the barn to detect the grain yield in the barn. At the same time, the cab is equipped with a touch screen to select the crop type and display the rotation speed of all transmission shafts, so as to display the current load and whether blockage may occur[3].

The full feed 4l-z-2.5 (pro688q) combine harvester of Kubota company in Japan has detection elements for parameters such as vehicle speed, feeding amount and stubble height[4]. The 4lbz-145b combine harvester of Jingguan company of Japan realizes the key press or remote control of the moving functions such as forward and turning, as well as the functions such as feeding depth and granary unloading[5]. Seher kadirova et al. Put forward an electrification modification scheme of combine header, using DC brushless motor and microcontroller to design the control system of header [6]. Ildar badretdinov et al. Established a three-dimensional model of the air screen in the threshing device of the combine, analyzed the flow field, and gave the method to determine the structural parameters and process parameters of the screen[7].

Li YaoMing's team[8] in China designed and installed a grain loss flow sensor composed of piezoelectric materials at the straw outlet of the harvester to detect the entrainment loss during the removal of straw debris through the piezoelectric effect caused by the impact of grains on the sensor thin plate [8]. On this basis, the speed of the fan and the running speed of the vibrating cleaning screen can be adjusted to reduce the loss.

The future development trends for the combine harvester electrification are unmanned and robotic. Industrial automation technology must be applied to agricultural machinery. Its electrification process will be optimized and the energy consumption will be lower. E-drive technique has obvious advantage in the new agricultural machinery[9][10][11].

Based on a full feed combine provided by a relative company, this paper lays a foundation for the realization of adaptive control of header and threshing part. Based on the electric platform, the output power of each working shaft is automatically adjusted with the change of vehicle speed and cutting width, which effectively improves the work efficiency and reduces the energy consumption. In view of the problem that the transmission channel between the reaping and the threshing part is easy to be blocked in the harvesting process, this paper adopts the sequential control based on the finite state machine, and monitors the speed and current of the motor in the transmission channel, which effectively reduces the possibility of blocking.

2. Electric Modification for Harvesting Unit

The harvesting unit of full-feed combine harvester is composed of two parts: reaping unit and threshing unit.

2.1. Objective of Electric Modification

The threshing part of the harvester has a total of 7 power components, including threshing rotor, swing shaft, fan and conveying shaft. The original harvester uses a diesel engine to drive each shaft to work synchronously. The transmission mode is belt and chain, and the transmission ratio is fixed.

The goal of electrification is to drive each power shaft independently by motor, so that the speed and power of each shaft can be adjusted freely. The electric modification process of each shaft is similar. In order to save space, the front rotor is introduced as an example below.

2.2. Mechanical Design and Verification

The front rotor of a harvester described in this paper is located between the feeder house and the threshing rotor, which plays the role of conveying grain.

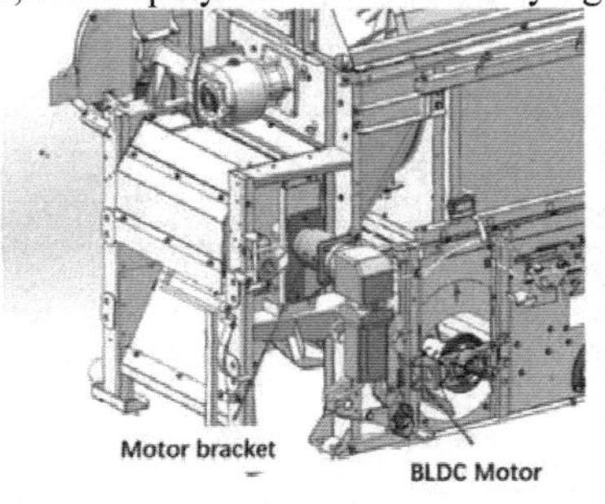

(a)

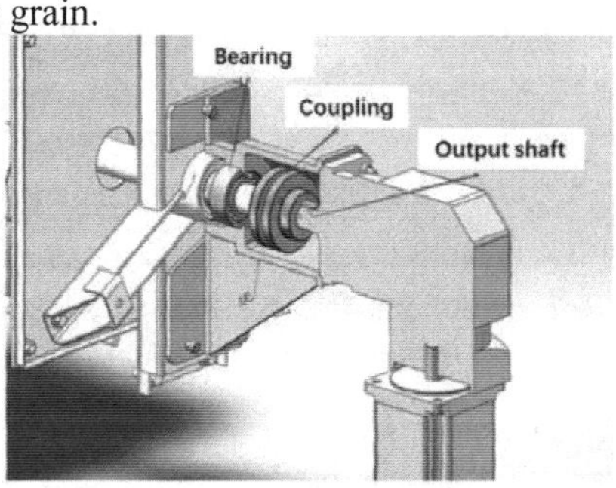

(b)

Figure 1. (a) Installation layout of front rotor. (b) Section views of front rotor transmission method.

With the change of vehicle speed and cutting height, the real-time feeding amount of grain is changing. The rotating speed of the front rotor needs to meet the demand of feeding amount to avoid blockage and avoid power waste as much as possible.

The rated speed and rated power of each power shaft are measured by the enterprise through sensors. The rated power of the front drum is 1.5KW, and the rated speed and torque are 600 rpm and 25 Nm respectively. The mechanical design process is as follows: 1) select the reducer and brushless DC motor that meet the requirements; 2) Design suitable motor fixing bracket; 3) Finite element analysis and practical test

The front rotor of the combine and other input shafts are the design mode of adaptive belt drive. In the process of electrification transformation, it needs to be adapted with the output shaft of the reducer to transmit torque. In this design, the installation and positioning accuracy of the support is improved by using the positioning bearing, and the single diaphragm coupling is used to ensure sufficient coaxiality tolerance.

2.3. Electric Drive Method

During the harvesting operation of the combine, the working load fluctuates greatly and the working condition is bad. Therefore, the selected motor and motor controller shall have strong adaptability and have locked rotor and overcurrent protection mechanism to prevent the controller from burning out due to grain blockage during operation.

This paper adopts the common brushless DC motor. The communication mode of the matched motor controller is RS485, which can read and control the motor speed in real time through 485 bus, and the speed control error is less than 50rpm.

At the same speed, the working current of the motor reflects the output load. In the harvesting process, the load of each power shaft is measured and monitored in real time, and the current comprehensive working state of the harvester can be obtained, which is convenient to adjust the traveling speed and feeding amount of the harvester. In this paper, the current loop based on Hall effect is used to detect the input current of motor controller, and the analogue-CAN module is used to send the current information to the industrial computer in the form of CAN frame message.

3. The Control Framework of Electric Harvesting Unit

3.1. RS485 Bus and CAN Bus Framework

The RS485 bus is used to control and read the speed of each motor. The CAN bus is set as the current value transferring and the reading of velocity of harvester, which is shown in Figure 2.

The control platform is an industrial computer, the model is nuvo-5095gc, and its CPU is Intel Core i7. The power supply is 24V DC. The operating system is Ubuntu 18.04, which is a stable operating system based on Linux. Based on the Ubuntu operating system, we choose ROS (robot operating system) as our control core implementation platform. And the ROS version is ROS melodic.

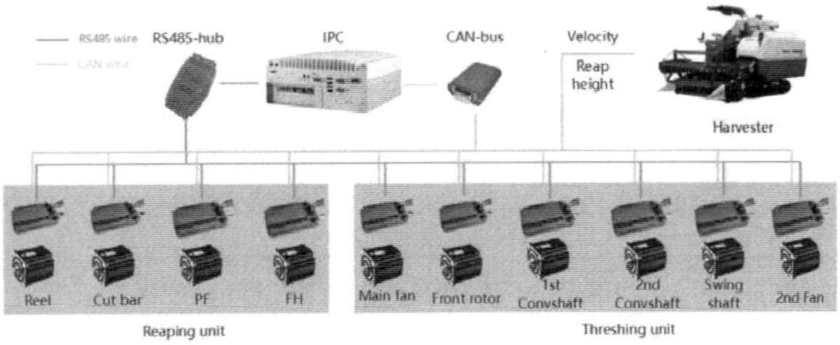

Figure 2. Framework of electrification control system.

3.2. The Control Method of Multi Motors Based on Work Load

The load during the operation of the harvester is mainly related to the current grain feeding amount q (kg / s), and the calculation formula of the feeding amount can be expressed as:

$$q = H\rho_w v \tag{1}$$

And in the equation: H – cutting width(m), ρ_w -- the average density (kg/m^2), v – forward velocity of harvester (m/s).

In the harvesting process, the four links of cutting, threshing, cleaning and grain storage of the combine harvester process the fed grain in sequence, which is similar to the production line. When the processing speed of the link located at the downstream of the assembly line slows down due to excessive threshing load and too small spacing between the threshing drum and the sieve plate, and at the same time, the cutting part at the upstream still runs at high speed and continues to feed grain, it is easy to be blocked. On the other hand, the moment of inertia of each working shaft is different, and the response time required for the driver to control the motor to accelerate or decelerate to the specified speed is different. When the vehicle speed changes, the order of each axis reaching the specified speed cannot be synchronized.

This paper mainly focuses on the speed synchronous control of each motor, so it mainly takes the motor speed and actual speed as the control target. Taking the reaping unit as an example, according to the order of rice contact during harvesting, it is reel (reel), cut bar (CB), auger shaft (PF) and feeder house (FH). If the feeding amount per unit time increases when the vehicle speed increases, the speed of each power shaft shall be increased proportionally to avoid blockage. Due to the different acceleration time of each shaft, if the rear conveyor chain does not accelerate to the working speed, the front reel has accelerated to a higher speed and transmitted a large amount of grain, it is easy to cause blockage. Therefore, this paper sets up a finite automatic state machine to control the acceleration and deceleration transition of each working motor.

As Figure 3 shows, during acceleration, the conveyor chain and auger shaft at the rear accelerate to a stable state and then feed back to the client, and then the reel and cutter at the front begin to accelerate to the target state. This method can effectively avoid the inconsistent acceleration and deceleration time caused by different axle loads and the resulting blockage problem. When the harvester decelerates or stops, the work flow is consistent with the acceleration stage, and the motor sequence needs to be reversed.

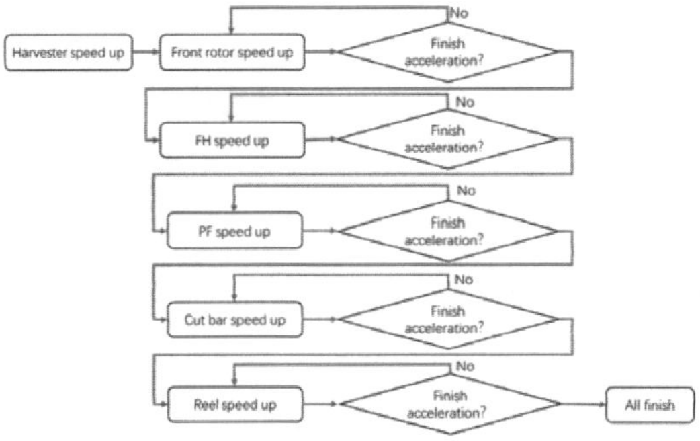

Figure 3. Workflow chart of acceleration process of reaping unit.

4. Experiments and Analysis

4.1. Sequential Start Experiment

Figure 4 shows the acceleration sequence of all motors from the grain flow sequence.

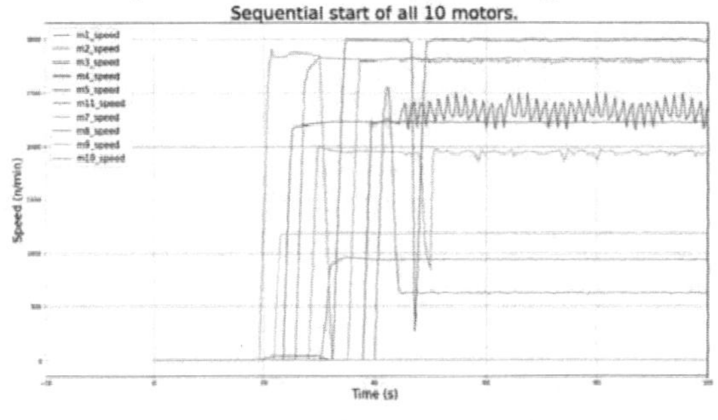

Figure 4. Speed diagram of sequential start of all 10 motors.

There is fluctuation of motor speeds, such as m1(Feeder House) and m4(Cut bar). The reason of m1 speed fluctuation is that the speed of m1 is easily affected by feed rate. And the reason of m4 speed fluctuation is that the eccentric mechanism transforming the rotation to translation, which causes the output torque and speed change periodically.

4.2. The Speed Follow Control of Reel

During the harvesting operation of the combine, as long as the reel plays the role of righting and feeding crops, the speed of the reel needs to be adjusted in combination with the vehicle speed, the current grain density and whether there is lodging. When the reel rotates, the teeth on the reel not only move in a straight line with the progress of the harvester, but also rotate around the shaft. Therefore, the motion track relative to the ground and crops is a cycloidal track.

As Figure 5 shows, the coordinate origin o is set at the reel shaft, the X axis is the forward direction of the harvester, the Y axis is perpendicular to the ground and downward, and the rotation direction of the reel is anti-clockwise.

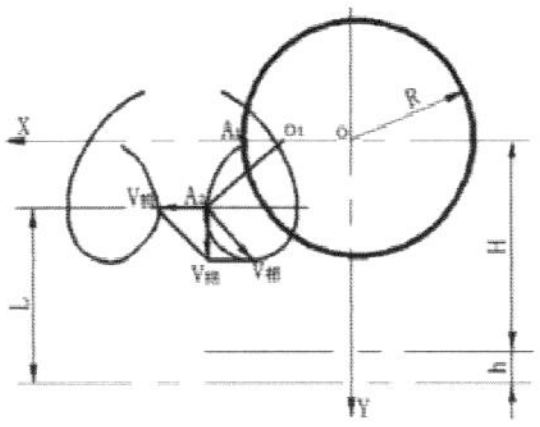

Figure 5. Schematic drawing of the reel motion[13].

Establish the speed and height control model of reel according to Figure 5:

$$\begin{cases} X = v_h + R \cdot \cos wt \\ Y = H - R \cdot \sin wt + h \end{cases} \tag{2}$$

In equation (2), *R*-the radius of reel, ω-the angle velocity of reel, v_h --the forward velocity of harvester, *H*-the installation height of reel shaft; *h*-the stubble height; *L*-height of grain. In order to make the horizontal relative partial velocity of the reel teeth in contact with the crop 0 and avoid impact, the expression of the reel shaft height is obtained as listed:

$$H = L + \frac{R}{\lambda} - h \tag{3}$$

In equation (3), λ is the ratio between the peripheral speed of reel and the forward speed of harvester. For wheat harvesting, λ is between 1.2 and 1.7. For rice harvesting, λ is between 1.3 and 2.0. During the harvesting process, the reel is rotating as an initial speed even the harvester is static. Hence the relation between the speed of reel and harvester is:

$$w = \max\{k \cdot v_h + w_{bias}, 3000\}, 0 \le v_h \le 1.2m/s \tag{4}$$

In this experiment, the setting is $k = 1160(rpm/m \cdot s^{-1})$, $w_{bias} = 800rpm$.

Hence, both the target speed and the actual speed of reel motor during harvesting could be obtained. Then, the accuracy of motor speed control could be calculated, and the result is as Table 1 shows. The control accuracy verifies the performance of motor control framework.

Table 1. Reel motor speed control accuracy (Target & Actual).

Experiment	Max speed difference(rev/min)	Average speed difference(rev/min)	Average speed difference (%)
Test A	329.1	18.8	1.24
Test B	335.5	21.7	1.73

5. Conclusion

(1) The electric operation system of combine harvester based on 48V DC brushless motor and RS485 / CAN bus system can stably and continuously complete the harvesting task when harvesting rice and wheat. The system operation is monitored and adjusted automatically by collecting motors' speed and current in real time.

(2) The finite automatic state machine control system based on the client server communication mechanism in ROS system can control the acceleration and deceleration of each motor in combination with the vehicle speed, feeding amount, current speed and target speed of each motor. Effectively avoid the blocking phenomenon caused by the inconsistent acceleration and deceleration time of each working shaft.

(3) The electric operation system described in this paper has verified the stability of its control algorithm logic, including reel speed following and multi motor acceleration and deceleration control. The experimental results show that the construction scheme and control system of the electric platform are practical and popularized, which can provide a basis for the design, development and promotion of electric harvester in the later stage.

References

[1] Qingquan Wang, Bo Yang. Overall design of grain combine harvester [J]. Heilongjiang Science and technology information,2017(11):87.

[2] Lianghao Zhu. Research on jam fault monitoring system of combine harvester [J]. Agriculture and technology, 2020, 40(02):71:73.

[3] Qingfen Wu. Fendt company C series new combine harvester[J]. Agriculture engineering, 2013, 3(02):107-111

[4] Dongming Cai, Zhiyin Xu, Yangqun Hu, Hetao Lei. Introduction to the operation of Kubota full feed combine harvester [J]. Agriculture and technology, 2017, 37(23):42:44

[5] Zhigang Xue, Hua Wang, Jingguan 4lbz-145b combine harvester [J]. New Agriculture, 2011(10):51-52

[6] Kadirova, Seher & Nenov, Teodor. (2017). Microcontroller system for control of the harvester's header propulsion with BLDC motor. 149-154. 10.1109/SIITME.2017.8259879.

[7] Ildar Badretdinov,Salavat Mudarisov,Ramil Lukmanov,Valery Permyakov, Radik Ibragimov, Ruslan Nasyrov. Mathematical modeling and research of the work of the grain combine harvester cleaning system[J]. Computers and Electronics in Agriculture,2019,165

[8] Yaoming Li. Design and analysis of grain combine harvester[M]. Beijing: China Machine Press, 2014:164-165

[9] Pramod, A.S. and T.V. Jithinmon. Development of mobile dual PR arm agricultural robot. In 2nd International Conference on New Frontiers in Engineering, Science and Technology, NFEST 2019, February 18, 2019 - February 22, 2019. 2019. Kurukshetra, Haryana, India: Institute of Physics Publishing.

[10] Panchenko, A.V., et al., The automated testing facility based on machine vision for optimizing grain quality control technology. Ninth International Conference on Machine Vision. Vol. 10341. 2017: SPIE.

[11] Sembiring, A., A. Budiman, and Y.D. Lestari. Design and Control of Agricultural Robot for Tomato Plants Treatment and Harvesting. in International Conference on Information and Communication Technology 2017, IconICT 2017, August 25, 2017 - August 26, 2017. 2017. Medan, Sumatera Utara, Indonesia: Institute of Physics Publishing

[12] Wang P, Tian M, Wang H , et al. Electrical Modification and Experimental Study of Combine Harvester Reaping Unit[J]. IOP Conference Series Materials Science and Engineering, 2020, 790:012168.

[13] Baofa Li, Agriculture Mechanics[M]. Beijing, Agriculture Press, 2009:408.

Modeling Simulation and Experiment of Wide-range Current Transformer Based on New Nanocrystalline Alloy Material

Bing Lu[1], Xiwen Chen[1], Zhongda Xu[2], Gaoning Nie[1] and Xiaojun Tang[2,*]

[1] China Electric Power Research Institute, Wuhan, China
[2] School of Electrical Engineering, Xi'an Jiaotong University, Xi'an, China

*Corresponding author email: xiaojun_tang@xjtu.edu.cn

Abstract. In the traditional electromagnetic current transformer, when the operating current is too large or too small, the measurement error will fluctuate greatly, resulting in a sharp deterioration of the measurement accuracy. In order to solve this problem, the article first established a simulation model, measured the characteristic curve of the new nanocrystalline alloy material, and imported it into the simulation model. Compare the wide-range error conditions of the current transformer; then, use the new nanocrystalline alloy material as the new wide-range current transformer core, adjust the size of the wide-range current transformer core, and increase the diameter of the secondary winding for simulation ; Finally, according to the simulation results, a prototype of a new wide-range current transformer is prepared and error tests are carried out. The test results of the experiment show that the primary side of the new wide-range current transformer is within the range of 0.1%-200% of the rated current. The limit of the ratio difference and the angle difference reaches ±0.2% and ±10', which meets the measurement requirements of class 0.2 current transformers.
Keywords: Current Transformer; Wide Range; New type nanocrystalline alloy; Iron Core.

1. Introduction

As one of the indispensable equipment in the power system, the current transformer converts the large current signal on the primary side into a small current signal according to a certain ratio for secondary measurement, which plays a role in signal conversion and safety isolation[1]. The traditional electromagnetic current transformer can only guarantee the accuracy of the measurement under the range of 1%-120% of the rated current of the primary side current. When the current to be measured exceeds this range, the electromagnetic current transformer's the accuracy of the measurement drops rapidly. In the power grid, there are periodic fluctuations in the power load, and there are a large number of users with a larger range of power load fluctuations. After stopping high-power equipment, their actual load current is usually smaller than the rated current value of the configured metering current transformer. On the other hand, when using a transformer lower than its actual load, the primary current of the current transformer can even reach 200% of the nominal rated capacity. However, the traditional electromagnetic current transformer does not It is not fully applicable to all electric energy metering requirements. Therefore, it is of great significance to solve the problem of electric energy metering loss caused by the range limitation of traditional metering current transformers.

Regarding the improvement of the measurement accuracy of the current transformer for measurement and the improvement of the measurement range of the upper and lower limits, it is mainly achieved through active compensation and changing the core material of the transformer [2]-[3]. Increasing the upper limit of current transformer measurement is mainly to study the saturation of the current transformer core, including digital compensation, air gap compensation and other methods [4]-[16].

Literature [5]-[9] studied the method of identifying the saturation of the core of the current transformer and the method of locating the saturation area of the secondary side current. Literature [11] proposes a dual-core current transformer that combines an air gap current transformer and an ordinary current transformer, which improves the anti-saturation capability of the current transformer, but the overall structure is more complicated. Increasing the lower limit of current transformer measurement is mainly achieved by changing the core material of the current transformer and performing active compensation. The article firstly tests current error of traditional 0.2-level current transformer in the range of 1%-120% of the rated current on the primary side, and then measures the traditional 0.2-level current transformer in a wider range, that is, 0.1 of the primary side rated current. The error conditions in the range of %-200% are tested and measured. Secondly, the current transformer is modeled and simulated, and the characteristic curve of the new nanocrystalline alloy material is measured and imported into the simulation model. The wide-range error conditions of the current transformer with iron core and traditional nanocrystalline alloy iron core are compared. Finally, the new nanocrystalline alloy material is used as the iron core of the wide-range current transformer, and the core size and secondary side of the wide-range current transformer are changed. The winding wire diameter is simulated, the appropriate core size is selected according to the simulation results, and the current error test of the wide-range current transformer of the new nano-alloy material is compared with the simulation results. The results show that the current on the primary side is 0.1% of the rated current In the range of -200%, the ratio and angular difference limits of the new wide-range current transformer reach the same value, which meets the measurement requirements of class 0.2 current transformers.

2. Current Transformer Wide-range Error Test and Simulation Model Building

2.1. Current Transformer Wide Range Error Test

The current transformer is generally composed of three parts, which are the primary coil, the secondary coil and the iron core. For an ideal current transformer, the iron core does not consume active power and does not require reactive power for excitation. At this time, the apparent power of the primary coil and the secondary coil are equal [17]. However, in practice, both the iron core and the coil have losses, causing the conversion error of the current transformer, which mainly includes capacitive error and magnetic error. Among them, the error caused by the capacitive leakage current between the primary coil, the secondary coil itself and the two is called the capacitive error [18]. The value of the capacitive error is very small and can be ignored under normal circumstances. . The errors caused by magnetic leakage, excitation current and air gap magnetic potential are called magnetic errors [19].

Traditional current transformer error test, the measurement range is 1%-120% of the rated current, which can be completed under the traditional current transformer error test platform. We can see the current transformer error test platform in Figure 1.

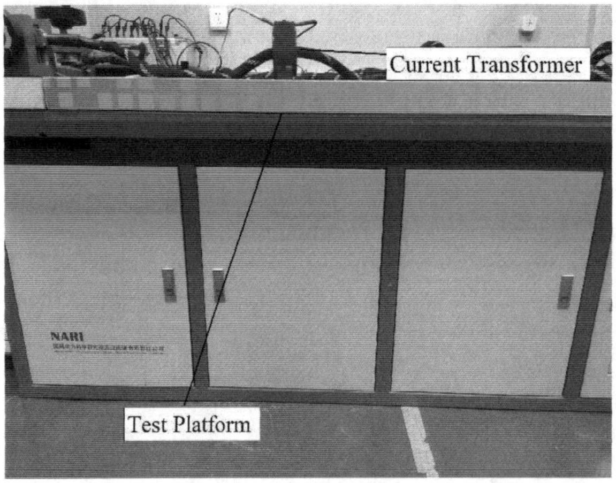

Figure 1. Current transformer error test platform.

Under this test platform, the traditional current transformers with rated currents of 200A, 600A, 1500A on the primary side were tested for error under the condition of 5VA on the secondary side. The test results are shown in figure 2.

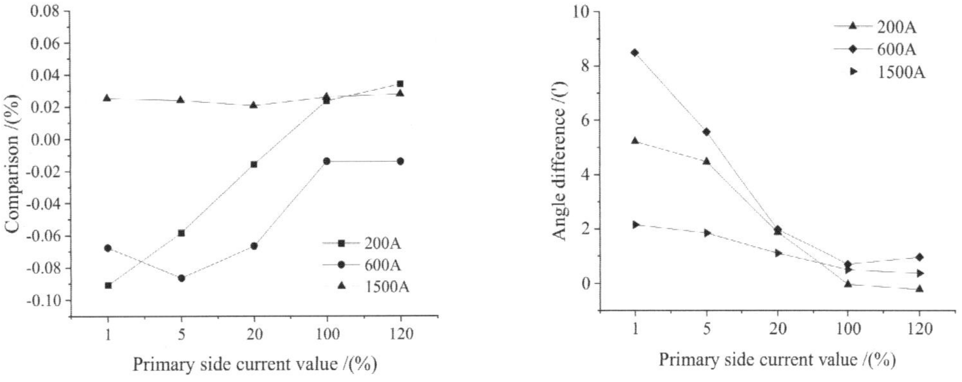

(a) (b)

Figure 2. Error test results of traditional current transformers.

It can be seen from the test results that traditional current transformers meet the measurement requirements of class 0.2 current transformers within the range of 1%-120% of the rated current on the primary side. Since fractional turns compensation is applied to the traditional current transformer under test, the partial ratio difference measurement result is positive. To further test the error conditions of traditional current transformers in a wider range, the required instruments include current transformers, wide-range standard current transformers, wide-range current transformer load boxes and transformer calibrators, which are compatible with the wide-range standard current Use transformers together, as shown in figure 3.

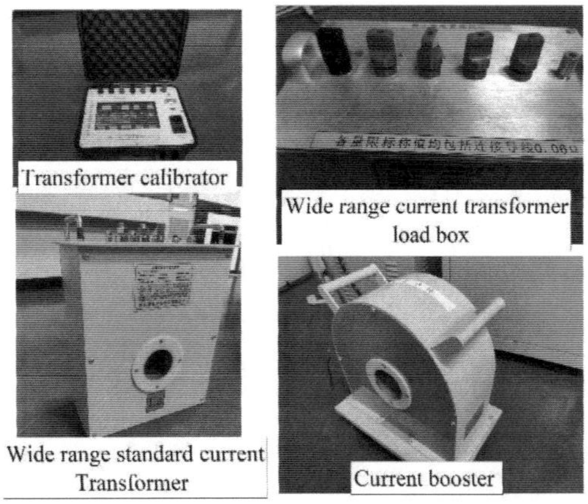

Figure 3. Current transformer wide-range error test platform.

Here, a traditional current transformer with a rated current of 300A is tested for the error of the rated current on the primary side. We can see the test results in figure 4.

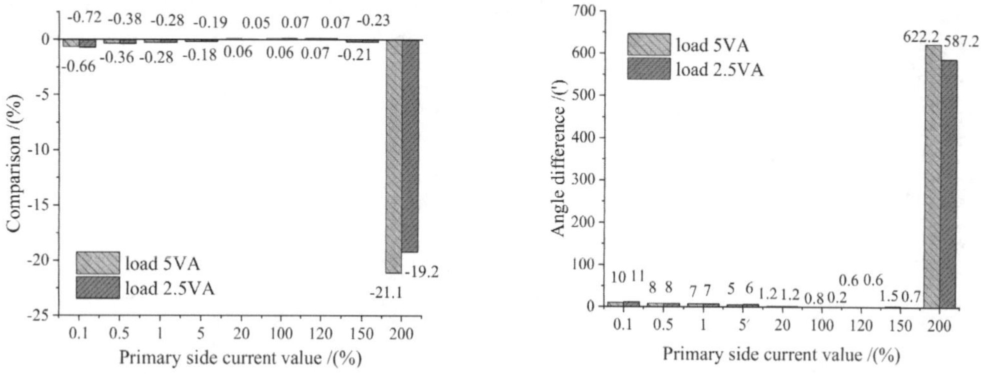

(a) (b)

Figure 4. Wide-range error test results of traditional current transformers.

We can observe from the test results that when the primary current of the current transformer under test is less than 1% of the rated current, the ratio difference does not meet the 0.2 level measurement requirements. When the primary side current is 200% of the rated current, the ratio difference and angle The difference can even reach -20% and 600' respectively. Therefore, in order to further broaden the range of the current transformer on the premise of ensuring the measurement accuracy, it is necessary to change the core material and adjust the core size.

2.2. Construction of Simulation Model of Current Transformer

In order to analyze the influence of the core material on the upper and lower limit errors of the current transformer measurement, and to further adjust the core size of the current transformer, broaden the range of the current transformer, a simulation model of the current transformer is built in SIMULINK as shown in figure 5.

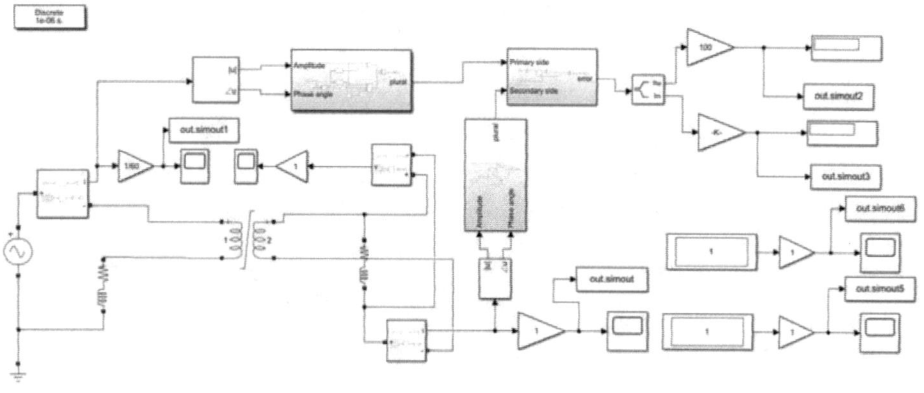

Figure 5. Simulation model of current transformer.

Among them, the amplitudes and phase angles of the fundamental components of the primary and secondary currents decomposed by the Fourier module are converted into complex numbers through the calculation of the Subsystem module, and finally the error of the current transformer is obtained by formula (1).

$$\tilde{\varepsilon} = -\frac{K_{in}\dot{I}_2 - \dot{I}_1}{\dot{I}_1} \times 100\% = -\frac{\dot{I}_0}{\dot{I}_1} \times 100\% = f + \mathrm{j}\delta \quad (1)$$

In the formula, Kin is the current transformation ratio of the transformer, f is the ratio difference, which is the angle difference. When the initial value of the vector angle of the secondary current leads the primary current, the phase difference is positive, otherwise it is negative.

The relevant parameters of the simulation model mainly include primary and secondary winding resistance, leakage reactance, secondary side load, primary side current, and the magnetization characteristic curve of the core material of the current transformer.

The core material of the tested current transformer can be set to silicon steel, permalloy, traditional nanocrystalline alloy material, and new nanocrystalline alloy material. The primary side rated current is 300A, the secondary side rated current is 5A, and the secondary side load is 5VA, calculate the primary side impedance, secondary side impedance and load of the current transformer and import them into the Saturable Transformer and Series RLC Branch modules respectively. Set the primary AC Voltage Source module to a sine wave with a voltage of 10kV and a frequency of 50Hz, and Set the resistance value of the Series RLC Branch module on the primary side to 100/3Ω, and finally import the scattered point value of the magnetization characteristic curve of the measured material into the Saturation characteristic column of the Saturable Transformer module, and then the error of the current mutual inductance can be simulated.

3. Comparison and Selection of Iron Core Materials for Wide-range Current Transformers

3.1. Comparison of Magnetization Characteristics of Current Transformer Core Materials

The iron core of the traditional current transformer is generally made of cold-rolled silicon steel sheet. The core material of the current transformer with higher precision is permalloy or nanocrystalline alloy, but it is still difficult to complete the wide range of the current transformer. The measurement requirements are that the new nanocrystalline alloy material changes the material ratio of the traditional nanocrystalline alloy material in the preparation, and adopts different operation methods in the steps of smelting, annealing, dipping and spraying, and the initial magnetic core of the final produced iron core Both the permeability and the saturation magnetic flux density have been greatly improved, realizing that the iron core of the same size has a higher initial magnetic permeability and magnetic flux density. The comparison of the magnetization characteristic curves of different core materials is shown in figure 6.

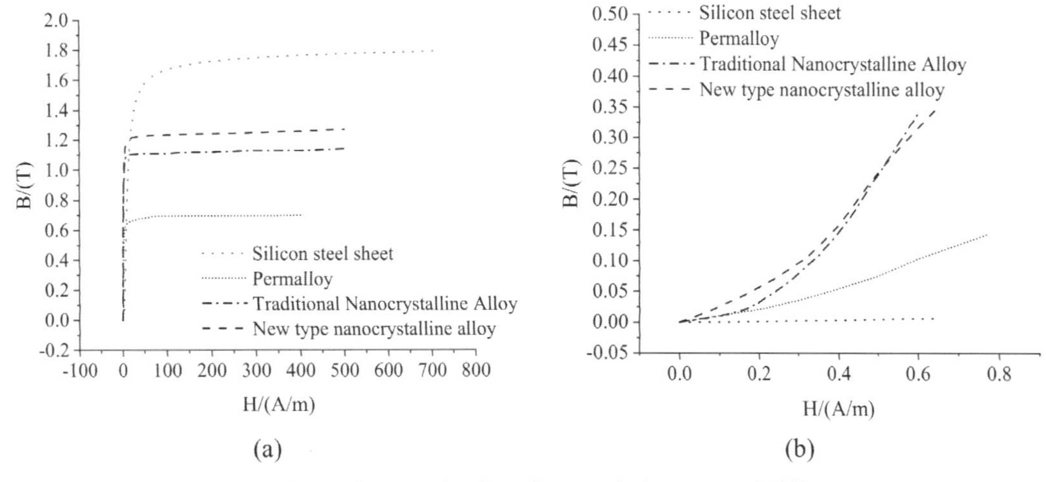

Figure 6. Comparison of magnetization characteristic curves of different materials.

We can observe from figure 6 that the saturation magnetic flux density of the silicon steel sheet is the largest, which can meet the upper limit of the measurement of a wide-range current transformer. However, compared with the other three materials, the initial permeability of the silicon steel sheet is too small to meet the requirements. The accuracy of the lower limit of measurement is required, and the initial permeability of the new nanocrystalline alloy material is greater than that of the other three materials, which meets the requirements of the lower limit of the measurement of a wide-range current transformer, and the saturation magnetic flux density of the new nanocrystalline alloy material is greater than that of the traditional nanocrystalline alloy. The alloy material is much larger than the permalloy material.

3.2. Simulation Analysis of Different Core Materials of Wide-range Current Transformer

Bring the scatter plots of the magnetization characteristic curves of different core materials into the simulation model. The core size, secondary winding, secondary load and core insulation parameters are consistent with those in Section 2.1. On the other hand, the stack of silicon steel sheets is set the flake coefficient is 0.95, the lamination coefficient of permalloy material is 0.9, the lamination coefficient of traditional nanocrystalline alloy and new nanocrystalline alloy material is 0.7, and other conditions remain unchanged.

The simulation results of wide-range error of current transformers with different core materials are shown in figure 7.

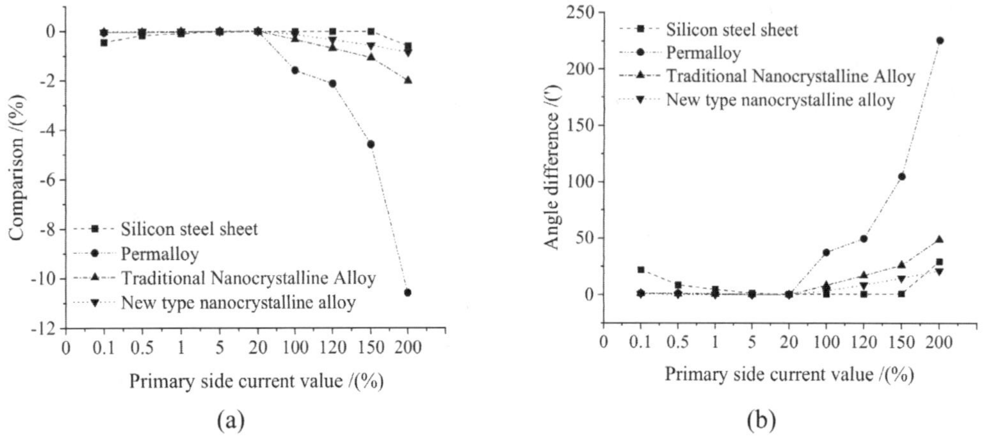

Figure 7. Simulation results of wide-range error of current transformers with different core materials.

We can observe from figure 7 that in terms of measurement error, the measurement error of the silicon steel sheet iron core current transformer is lower than that of the other three materials., but the accuracy of the lower limit of measurement is much lower than that of the other three materials, so it cannot be used as a wide-range current transformer. The new type of nano-alloy material core current transformer has extremely high measurement accuracy at the lower limit of measurement, but its upper limit of measurement still cannot meet the accuracy requirements of class 0.2 current transformers, so the core size can be changed to increase Large winding wire diameter.

In summary, a new type of nanocrystalline alloy material is selected as the material for preparing wide-range current transformers.

4. Simulation Analysis and Test of a New Wide-range Current Transformer

4.1. Simulation Analysis of a New Wide-range Current Transformer

Adjust the iron core size of the current transformer, that is, increase the cross-sectional area of the toroidal iron core, reduce the average magnetic circuit length of the iron core, and increase the winding wire diameter. The final selection of the iron core size is 80mm outside diameter and 60mm inside diameter, the height is 25mm, the diameter of the secondary winding wire is set to 1.90mm, and the magnetic permeability remains unchanged. Because the compensation value is too large when full-turn compensation is used, the method of fractional-turn compensation is considered, that is, two single wires of the same diameter are used for the second time, and the compensation value is not set in the simulation analysis, so the new wide-range current mutual inductance The simulation result of the device is shown in figure 8.

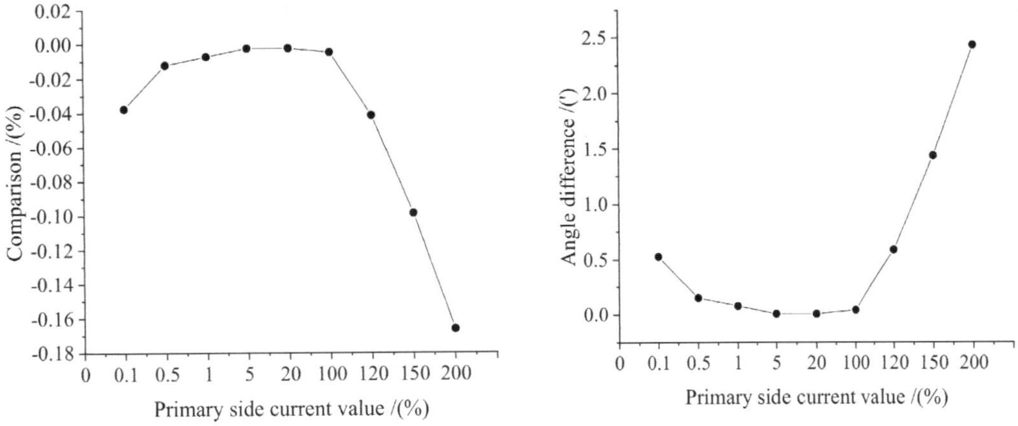

(a) (b)

Figure 8. Simulation results of the new wide-range current transformer.

The simulation results show that using the new nanocrystalline alloy as the core material of the current transformer, adjusting the core size, and increasing the coil wire diameter can effectively broaden the current transformer range while ensuring the accuracy of the measurement.

4.2. Test and Analysis of a New Wide-range Current Transformer

According to the simulation analysis results, the core material, size and winding wire diameter are determined, and a wide-range current transformer prototype is prepared. The prototype uses different fractional turn compensation values and carries out a wide-range error test on different machines. The test results are shown in figure 9 Shown.

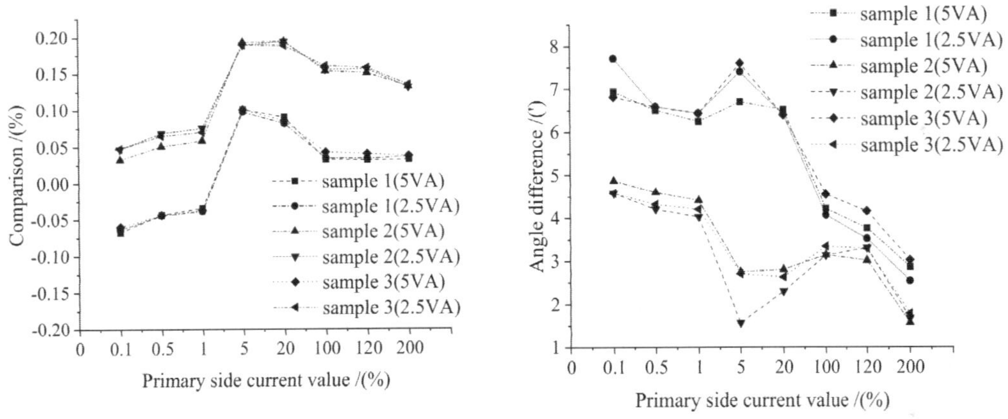

(a) (b)

Figure 9. Test results of a new wide-range current transformer.

Three different new wide-range current transformer prototypes were tested, and the secondary side loads were 5VA and 2.5VA respectively. Under the range of 0.1%-200% of the rated current of the primary side current, three new wide-range current transformers Both the ratio difference and angular difference limit of the current transformer prototype reach the same value, which meets the measurement requirements of the 0.2-level current transformer.

5. Conclusion

The traditional electromagnetic current transformer guarantees the accuracy of the measurement under the range of 1%-120% of the rated current of the primary current. The measurement accuracy of the current transformer begins to decline rapidly when the measured current exceeds this range. The lower limit error of the current transformer measurement is mainly related to the initial permeability of the

iron core material, and the upper limit error of the measurement is mainly caused by the saturation of the iron core. The iron core material with better preparation performance can effectively reduce the error of the lower limit of measurement. Adjusting the core size and increasing the cross-sectional area of the winding can reduce the upper limit error value of the measurement.
Under the goal of widening the range of the current transformer while ensuring the measurement accuracy of the current transformer, this paper adopts a new type of nanocrystalline alloy as the core material of the current transformer. This material realizes that the iron core of the same size has a higher initial Magnetic permeability and magnetic flux density, and optimize the size of the core, increase the winding wire diameter, and finally the new wide-range current transformer prepared achieves its primary current within the range of 0.1%-200% of its rated current. Both the ratio difference and the angular difference limit reached the sum, which met the measurement requirements of the 0.2-level current transformer.

Acknowledgments
Our thanks to all of the people who have contributed to this paper. My paper is Supported by Science and Technology Project of State Grid Corporation of China: Research on Key Technology of Wide Range Current Transformer for Metering (5600-202055169A-0-0-00).

References
[1] Makarand Sudhakar Ballal and Manish Ganesh Wath and Hiralal Murlidhar Suryawanshi. Measurement Current Transformer Error Compensation by ANN Methodology[J]. Journal of The Institution of Engineers (India): Series B: Electrical, Electronics & Telecommunication and Computer Engineering, 2020, 101(3): 261-271.
[2] Bao Hai, Ai Xin, Yang Yihan. Study on a large range measurement current transformer[J]. Automation of Electric Power Systems, 2003(10): 63-66.
[3] Chen Hongjiang, Jiang Zhijun, Jin Shaoping, et al. Realization of a wide-range measurement of current transformer based on Permalloy cores[J]. Experimental Technology and Management, 2016(07): 43-47.
[4] Liu Jie, Pei Jie, Tian Ming, Zhu Xuan. Research on nonlinear correction method of current transformer based on support vector machine[J]. Journal of Electrical Machines and Control, 2020, 24(10): 130-138.
[5] Liu Huanzhang, Wang Xingguo, Zhou Zexin, Guo Yarong, Du Dingxiang. A CT Saturation Identification Method Using Sampled Current Fault Component Data[J]. Power System Technology, 2016, 40(11): 3574-3579. DOI: 10.13335 /j.1000-3673.pst.2016.11.043.
[6] Liu Yiqing, Gao Weicong, Wang Chengyou, Sun Tiande, Gao Houlei. Detection Method for Current Transformer Saturation Based on Characteristic of Differential Current by Half-Cycle Integral Algorithm[J]. Power System Technology, 2016, 40(09): 2889-2896.
[7] Xing Jiawei, Jin Neng, Lin Xiangning, Liu Yao, Li Zhen. A Novel Strategy of Current Transformer Saturation Identification and Valid Data Application Based on Wavelet Packet[J]. Transactions of China Electrotechnical Society, 2019, 34(06): 1170-1179.
[8] Xu Feng, Li Kaicheng, Jin Bo, Wang Ke. Research on Locating saturation of current transformer[J]. Electrical Measurement & Instrumentation, 2016, 53(17): 1-5+17.
[9] Wang Lingyun, Li Kaicheng, Xiao Xianying, Tian Zhaoguang, Meng Qingxu, Cai Delong. Current transformer saturation detection based on Savitzky-Golay filter[J]. Electrical Measurement & Instrumentation, 2018, 55(6): 76-81.
[10] Chen Haibin, Yang Shunan, Chen Liwen, Zhen Haohan, Chen Shengze, Zhang Mengyu. Research on current transformer error testing technology based on field signal simulation technology[J]. Electrical Measurement & Instrumentation, 2021, 58(2): 133-138.
[11] Xiong Kui, Yue Changxi, Li Dengyun, Zhu Kai, Li Zhicheng. Research of theory and error characters of high magnetic permeability ratio two cores current transformer[J]. Electrical Measurement & Instrumentation, 2018, 55(17): 114-119.
[12] Wang Zikai. Research on large error deviation in field test of current transformer[J]. Transformer, 2018, 55(05): 67-68.

[13] Xu Lingjie, Chen Xiao, Liu Yong, Guo Peng, Zhang Weihua, Zhu Chongye. Application of Different Frequency Measurement Method in Field Measurement of Current Transformer[J]. Transformer, 2021, 58(03): 70-75.
[14] Shi Kaimeng, Zong Ming. Power analysis and design method of fast-saturation current transformer[J]. Advanced Technology of Electrical Engineering and Energy, 2020, 39(06): 74-80.
[15] Li Min, Yang Aichao, Wu Yu, Jin Shaoping, Zhu Liang, Li Dongjiang. Research on anti-DC performance of air-gap-core current transformer[J]. Electrical Measurement & Instrumentation, 2019, 56(13): 123-128+135.
[16] Yan Qin, Li Shisong, Ye Yuanyu, Xiong Ru, Zhao Wei. Magnetic model and measurement characterization of current transformers with DC bias current[J/OL]. Electrical Measurement and Instrumentation: 1-6[2021-11-19]. http://kns.cnki.net/kcms/detail/23.1202.th.20201127.1045.012.html.
[17] K Draxler and J Hlavacek and R Styblikova. Determination of instrument current transformer errors in frequency range up to 1 kHz[J]. Journal of Physics: Conference Series, 2018, 1065(5).
[18] Zhang Hao, Zhao Wei, Qu Kaifeng, et al. Research on influence of External Current to Error of Current Transformer[J]. Transformer, 2008(09): 28-32.
[19] Zang Dianhong, Kang Zhongjian, Zhao Youjun. Development on the Current Sensor with Zero Flux based on Active Compensation Method[J]. Power Electronics Technology, 2009(06): 56-57

A Fuzzy Evaluation Method for the Drivability of New Energy Vehicles Based on Similarity

Xincheng Huang*

School of Automotive Engineering, Wuhan University of Technology, Wuhan, Hubei, 430070, China

*Corresponding author email: huangxincheng1027@163.com

Abstract. The purpose of this paper is to propose a fuzzy evaluation method for the drivability of new energy vehicles. This method divides the evaluation processes of the vehicle drivability into two stages. In the first stage, the vehicle drivability test experts describe the importance of each drivability index through linguistic variables, and then the concept of similarity between two triangular fuzzy numbers is used to determine the weights of experts after converting linguistic variables into triangular fuzzy numbers. The weights of indexes are determined according to the combination of subjective and objective weights, in which the objective weights are determined according to the conflict degree between two fuzzy numbers for the rating of importance given by different experts under the same index. In the second stage, for each index, the experienced drivers give the rating of the vehicle drivability, and then calculate the weighted fuzzy value of the drivability indexes according to the indexes weights. The evaluation results are explained by numerical drivability and linguistic drivability. The former is obtained from the total fuzzy drivability through defuzzification, while the latter is determined according to the similarity between the total fuzzy drivability and the fuzzy number for the rating of the drivability. We give a numerical example to evaluate the drivability of three new energy vehicles.
Keywords: Drivability; New energy vehicles; Triangular fuzzy number; Similarity.

1. Introduction

New energy vehicle with low energy consumption and light pollution can improve the problems of energy shortage and environmental pollution, so it has attracted great attention of governments all over the world. With the development of new energy vehicle industry, the competition between the manufacturers becomes more and more intense. Meanwhile, due to the gradual decline of government subsidies, consumers' purchase behavior of new energy vehicles has become more and more rational, and their requirement for the drivability of the whole vehicle has become higher and higher. In this context, the drivability evaluation of new energy vehicles is very important for vehicle manufacturers to understand the drivability of their vehicles.

The research on the drivability of new energy vehicles in the existing literature mainly focuses on the technical breakthrough, while there is little research on the evaluation of the drivability. In the existing literature on the evaluation of the vehicle drivability, He [1] used the objective evaluation system to evaluate the drivability of AMT commercial vehicles. Liu et al. [2] divided the evaluation of vehicle drivability into five aspects: WOT (Wide Open Throttle) response, POT (Partial Open Throttle) response, accelerator pedal feeling, gear shift smoothness and tip in/tip out transient response, and proposed a comprehensive evaluation method of drivability using objective quantitative analysis. Chen et al. [3] objectively tested and evaluated the vehicle drivability with AVL DRIVE system combined with the tested acceleration, current and CAN network signals. In these studies, the evaluation indexes are scientific and reasonable, the evaluation methods are objective, the evaluation objects are general vehicles, and the evaluation values of drivability are deterministic.

However, the existing evaluation indexes are for general vehicles, and a new index system needs to be constructed for the drivability evaluation of new energy vehicles. In addition, in the subjective evaluation, the vehicle drivability test experts or drivers are usually difficult to accurately judge the drivability of the vehicle, resulting in inaccurate drivability scores. Using linguistic variables to express the evaluation values of the drivability can effectively avoid this problem. This paper constructs the drivability evaluation indexes of new energy vehicles that can be easily understood by the drivers, introduces the linguistic variable for the rating of the drivability, and uses the fuzzy evaluation method to evaluate the drivability of new energy vehicles.

2. Preliminaries

The use of numerical values in rating the importance of the drivability indexes or the drivability of vehicles may have certain limitations in dealing with uncertain data, while fuzzy numbers can usually be used to solve this problem of uncertain data. In practical application, the fuzzy models using triangular fuzzy numbers have been proved to be very effective for solving the problem with imprecise information [4].

In this section, we will briefly review some basic definitions and arithmetic operations of triangular fuzzy numbers [4-7].

Definition 1. A fuzzy number $\widetilde{A} = (l, m, u)$ is said to be a triangular fuzzy number if its membership function is given by

$$\mu_{\widetilde{A}}(x) = \begin{cases} \frac{x-l}{m-l}, & l \le x \le m \\ \frac{m-x}{u-m}, & m \le x \le u \\ 0, & \text{otherwise} \end{cases}$$

where l, m, and u are the lower, middle, and upper crisp numerical values of the triangular fuzzy number, respectively.

Definition 2. Let $\widetilde{A} = (l, m, u)$ be a triangular fuzzy number then the defuzzified crisp number is given by $D(\widetilde{A}) = \frac{l+2m+u}{4}$.

Definition 3. Let $\widetilde{A} = (l_a, m_a, u_a)$ and $\widetilde{B} = (l_b, m_b, u_b)$ be two triangular fuzzy numbers then the operations between them are defined as follows:

(i) $\widetilde{A} \oplus \widetilde{B} = (l_a + l_b, m_a + m_b, u_a + u_b)$

(ii) $\widetilde{A} \otimes \widetilde{B} = (l_a \times l_b, m_a \times m_b, u_a \times u_b)$

(iii) $k\widetilde{A} = (kl_a, km_a, ku_a)$

where $l_a, l_b, k \ge 0$

Definition 4. Let $\widetilde{A} = (l_a, m_a, u_a)$ and $\widetilde{B} = (l_b, m_b, u_b)$ be two triangular fuzzy numbers, then the distance between them is defined as $d(\widetilde{A}, \widetilde{B}) = \sqrt{\frac{1}{3}[(l_a-l_b)^2+(m_a-m_b)^2+(u_a-u_b)^2]}$

Definition 5. Let $\widetilde{A} = (l_a, m_a, u_a)$ and $\widetilde{B} = (l_b, m_b, u_b)$ be two triangular fuzzy numbers, then the similarity $S(\widetilde{A}, \widetilde{B})$ and conflict degree between them is defined as $S(\widetilde{A}, \widetilde{B}) = 1 - \frac{d(\widetilde{A},\widetilde{B})}{\|U_{ab}\|}$, $CD(\widetilde{A}, \widetilde{B}) = \frac{d(\widetilde{A},\widetilde{B})}{\|U_{ab}\|}$, respectively, where $\|U_{ab}\| = \max(u_a, u_b) - \min(l_a, l_b)$

The concept of similarity and conflict degree in fuzzy numbers is very important in the field of fuzzy evaluation. This paper uses the two concepts to calculate the weight of experts and the objective weights of the indexes, respectively.

3. Evaluation Indexes and Evaluation Framework of the Drivability

3.1. Evaluation Indexes

Based on the analysis of relevant literature, an index system for evaluating the drivability of new energy vehicles is preliminarily constructed. Through the investigation and interview with the vehicle test experts and experienced drivers, the following eight evaluation indexes most concerned by experts and drivers are finally selected from the index system: Braking performance (C1), Trafficability (C2), Handling characteristics (C3), Steering characteristics (C4), Safety and comfort (C5), Vehicle NVH (Noise, Vibration, Harshness) performance (C6), Battery characteristics (C7), and Intelligent configuration (C8). The first five indexes are the vehicle dynamic indexes. When evaluating the vehicle drivability, we should also consider the importance of the indexes. Thus, in addition to rating the drivability for each index, the experts or drivers should also be required to rate the importance of each index.

3.2. Evaluation Framework

The problem for evaluating the drivability of new energy vehicles could be described as: there are *m* possible alternatives ($A_1, A_2, \cdots, A_m$) among which the vehicle test experts have to evaluate, and each new energy vehicle is represented as an alternative. The vehicle drivability evaluation indexes (C_1, C_2, $\cdots$, C_n) are used to evaluate the alternatives. There are *h* experts ($D_1, D_2, \cdots, D_h$) to rating the importance of the drivability indexes. For each selected new energy vehicle, the owner will rate the drivability of his own vehicle under each index. The evaluation process of the drivability is divided into two stages: calculating the weights of indexes and evaluating the drivability of new energy vehicles.

4. The Proposed Method for Evaluating the Drivability

4.1. Calculation of Indexes Weights

In this subsection, the weights of the indexes are calculated by using the fuzzy evaluation value given by experts when rating the importance of the indexes. The two key points of the method are as follows: (1) the expert weights are calculated by using the similarity of triangular fuzzy numbers, so as to avoid the unreasonable impact of individual evaluation values of an expert on group evaluation results. When rating the importance of indexes, the total similarity between the fuzzy evaluation values given by one expert and that given by other experts is measured to reflect the consistency of individual and group evaluation information. The greater the total similarity, the stronger the consistency of individual and group evaluation, and the greater the corresponding experts weight. (2) The objective weights of the indexes are calculated by using the conflict degree of triangular fuzzy numbers. For each index, the sum of the conflict degree between the fuzzy evaluation values given by any two experts is calculated. The greater the sum corresponding to the index, the better the index can distinguish the differences between alternatives, so the greater the weight of the index.

The main steps for calculating the weights of indexes could be described as followings:

Step 1: Define linguistic variables and corresponding triangular fuzzy numbers, which are used to measure the importance of each index.

Step 2: Construct a matrix used for determining weights of indexes. Let fuzzy evaluation value of the jth index importance given by the ith expert be $\widetilde{w}_{ij}$, then the matrix for determining the weights of indexes can be expressed as follow:

$$\widetilde{w} = \begin{bmatrix} \widetilde{w}_{11} & \widetilde{w}_{12} & \cdots & \widetilde{w}_{1n} \\ \widetilde{w}_{21} & \widetilde{w}_{22} & \cdots & \widetilde{w}_{2n} \\ \vdots & \vdots & \ddots & \vdots \\ \widetilde{w}_{h1} & \widetilde{w}_{h2} & \cdots & \widetilde{w}_{hn} \end{bmatrix}, \tag{1}$$

Step 3: Calculate expert weights. According to Definition 5, the similarity of two triangular fuzzy evaluation values given by experts i and k are calculated as:

$$S(\widetilde{w}_i, \widetilde{w}_k) = \sum_{j=1}^{n} \left[1 - \frac{d(\widetilde{w}_{ij}, \widetilde{w}_{kj})}{\|U_{ik}\|}\right], i, k = 1,2,\cdots,h \tag{2}$$

Then the sum of similarity of triangular fuzzy evaluation values given by expert i and other experts is:

$$S(\tilde{w}_i) = \sum_{k=1}^{h} S(\tilde{w}_i, \tilde{w}_k)\text{-}n, \ i = 1,2,\cdots,h \tag{3}$$

The expert weights are calculated using the following expressions:

$$\alpha_i = \frac{S(\tilde{w}_i)}{\sum_{i=1}^{h} S(\tilde{w}_i)}, \ i = 1,2,\cdots,h \tag{4}$$

Step 4: Calculate the subjective weight of the indexes. First, calculate the aggregated subjective importance weights $\tilde{w}_j^s$ using the following expressions:

$$\tilde{w}_j^s = \sum_{i=1}^{h} \alpha_i \tilde{w}_{ij}, \tag{5}$$

Then defuzzify fuzzy weights $\tilde{w}_j^s$ of each index and derive their crisp values $\overline{w}_j^s$ according to definition 2. Finally, calculate the subjective weight of the indexes using the following formula:

$$w_j^s = \frac{\overline{w}_j^s}{\sum_{j=1}^{n} \overline{w}_j^s}, \tag{6}$$

Step 5: Calculate the objective weights of the indexes. According to definition 5, the total conflict degree between any two triangular fuzzy evaluation values of index j can be calculated as

$$CD_j = \sum_{i<k} \frac{d(\tilde{w}_{ij}, \tilde{w}_{kj})}{\|U_{ik}\|}, \ i,k = 1,2,\cdots m \tag{7}$$

The objective weight of index j is

$$w_j^o = \frac{CD_j}{\sum_{j=1}^{n} CD_j}, \tag{8}$$

Step 6: Calculate the comprehensive weights of the indexes. The following linear weighting method is used to obtain the comprehensive weight of index j.

$$w_j = \beta w_j^s + (1-\beta) w_j^o, \tag{9}$$

Where, parameter $\beta \in [0,1]$

4.2. Evaluation of the Drivability

In this subsection, for each new energy vehicle that needs to evaluate its drivability, the driver gives the drivability rating of the vehicle under each index, and then calculates the weighted fuzzy value of the vehicle drivability according to the index weight, and explains the fuzzy evaluation results of the vehicle drivability. The main steps of the vehicle drivability evaluation could be described as followings:

Step 1: Define linguistic variables and corresponding triangular fuzzy numbers used for the rating of the vehicle drivability by drivers.

Step 2: Construct an evaluation matrix. Let the fuzzy evaluation value for ith vehicle regarding jth index be $\tilde{x}_{ij}$, then evaluation matrix can be expressed as follow:

$$\tilde{x} = \begin{bmatrix} \tilde{x}_{11} & \tilde{x}_{12} & \cdots & \tilde{x}_{1n} \\ \tilde{x}_{21} & \tilde{x}_{22} & \cdots & \tilde{x}_{2n} \\ \vdots & \vdots & \ddots & \vdots \\ \tilde{x}_{m1} & \tilde{x}_{m2} & \cdots & \tilde{x}_{mn} \end{bmatrix} \tag{10}$$

Step 3: Calculate the weighted fuzzy value of the drivability for the ith vehicle, and the calculation formula is

$$\tilde{x}_i = \sum_{j=1}^{n} w_j \tilde{x}_{ij}, \tag{11}$$

Step 4: Explain the vehicle drivability of the ith vehicle. Two methods are used to explain the vehicle drivability: (1) Crisp value evaluation of the vehicle drivability: Defuzzify the weighted fuzzy value of the vehicle drivability and derive their crisp value of the drivability of the ith vehicle according to definition 2. (2) Linguistic evaluation of the vehicle drivability: Calculate the similarity between the weighted fuzzy value and each triangular fuzzy number for the rating of the vehicle drivability, respectively. The linguistic variable corresponding to the triangular fuzzy number with the greatest similarity is selected as the linguistic evaluation of the vehicle drivability.

5. Illustrative Example

In this section, we will present a numerical example to illustrate the application of the proposed method in the drivability evaluation of new energy vehicles. In the following, we first calculate the weights of the indexes, and then evaluate the drivability of the selected new energy vehicles.

5.1. Calculating Weights of Indexes

A new energy vehicle company selects five vehicle drivability test experts to rate the importance of the drivability indexes. The processes for determining indexes weights could be described as followings: (1) Define linguistic variables and corresponding triangular fuzzy numbers, as shown in Table 1. (2) The experts use linguistic variables to rate the importance of each index, and the results are shown in Tables 2. Linguistic values in the Table 2 are converted to fuzzy numbers using Table 1. (3) Calculate expert weights. Firstly, the similarity between the triangular fuzzy evaluation values given by two experts is calculated using Eq. (2), and the similarity matrix is obtained. Then use Eqs. (3) and (4) to calculate the sum of the similarity and expert weights, respectively. These results are shown in Table 3. (4) Calculate the subjective weight of the indexes. First, calculate the aggregated subjective importance weights $\tilde{w}_j^s$ using Eq.(5), we have $\tilde{w}^s$ = [(0.4204, 0.5407, 0.6610), (0.4488, 0.5872, 0.7052), (0.4894, 0.6278, 0.7458), (0.3848, 0.5441, 0.6831), (0.6045, 0.7452 , 0.8452), (0.4148, 0.5735, 0.7142), (0.5355, 0.6535, 0.7535), (0.4999, 0.6616, 0.7825)]. Then defuzzify fuzzy weights $\tilde{w}_j^s$ of each index and derive their crisp values $\overline{w}_j^s$ according to definition 2. Finally, calculate the subjective weight of the indexes using Eq. (6). The results are shown in rows 2 and 3 of Table 4, respectively. (5) Calculate the objective weight of the indexes. Using Eqs. (7) and (8), the total conflict degree and the objective weight of index j can be calculated, and the results are shown in rows 4 and 5 of Table 4, respectively. (6) Calculate the comprehensive weights of the indexes. Given $\beta = 0.7$, using Eq. (9), the comprehensive weight value of index j is obtained, and the results are shown in the last row of Table 4.

Table 1. Linguistic variables and fuzzy numbers for the rating of indexes importance.

Linguistic variables	Abbreviation	Triangular fuzzy numbers
Very important	VI	(0.7,0.9,1.0)
Important	I	(0.6,0.7,0.8)
Medium important	MI	(0.4,0.5,0.6)
Moderate	M	(0.1,0.3,0.5)
Medium unimportant	MU	(0.0,0.2,0.3)
Unimportant	U	(0.0,0.1,0.2)
Very unimportant	VU	(0.0,0.0,0.2)

Tables 2. Rating of the indexes importance given by experts (linguistic values).

	C1	C2	C3	C4	C5	C6	C7	C8
D1	I	MI	VI	I	VI	M	I	I
D2	I	VI	I	VI	I	I	MI	VI
D3	M	MI	MI	MI	VI	M	I	VI
D4	MI	I	I	M	I	I	MI	M
D5	MI	M	M	M	MI	VI	VI	MI

Tables 3. Calculation process of expert weights.

	D1	D2	D3	D4	D5	$S(\tilde{w}_i)$	α_i
D1	8.000	4.569	5.868	3.884	3.672	17.993	0.2032
D2	4.569	8.000	4.168	5.803	3.488	18.028	0.2036
D3	5.868	4.168	8.000	3.988	3.975	17.998	0.2033
D4	3.884	5.803	3.988	8.000	4.852	18.528	0.2093
D5	3.672	3.488	3.975	4.852	8.000	15.987	0.1806

Table 4. The process for calculating the comprehensive weights of the indexes.

	C1	C2	C3	C4	C5	C6	C7	C8
$\overline{w}_j^s$	0.5407	0.5821	0.6227	0.5390	0.7350	0.5690	0.6490	0.6514
w_j^s	0.1106	0.1191	0.1274	0.1103	0.1503	0.1164	0.1328	0.1332
CD_j	4.0305	4.7444	4.7124	4.8433	3.9643	4.4625	4.0983	4.8772
w_j^o	0.1128	0.1328	0.1319	0.1355	0.1109	0.1249	0.1147	0.1365
w_j	0.1113	0.1232	0.1287	0.1178	0.1385	0.1189	0.1273	0.1342

From Table 4, we can get the following observations: (1) Index C5 has the largest subjective weight, followed by index C8, which shows that for the test experts, safety and comfort performance are the main factor affecting the vehicle drivability, followed by intelligent configuration. (2) The objective weight of index C8 is the largest, followed by index C4, which shows that the test experts have the most controversy over the intelligent configuration when rating the importance of the index, followed by steering characteristics. (3) Index C5 has the largest comprehensive weight, followed by index C8, which shows that in general, safety and comfort performance have the greatest impact on the vehicle drivability, followed by intelligent configuration.

5.2. Evaluating the Vehicle Drivability

This subsection uses the proposed method to evaluate the drivability of three new energy vehicles. The main steps could be described as followings: (1) Define linguistic variables and corresponding triangular fuzzy numbers, as shown in Table 5. (2) The vehicle drivers use linguistic variables to assess their own vehicle drivability with respect to each index, and the results are shown in columns 2 to 9 of Table 6. (3) Linguistic values of the vehicle drivability are converted to fuzzy numbers using Table 5. Then, the weighted fuzzy values of the vehicle drivability are calculated according to the Eq. (11), and results are shown in column 10 of Table 6. (4) Defuzzify the weighted fuzzy value of the vehicle drivability and derive their crisp value of the vehicle drivability according to definition 2. The results are shown in the last column of Table 6. It shows that the numerical vehicle drivability of A1, A2 and A3 are 83, 73 and 91, respectively. (5) Calculate the similarity between the weighted fuzzy value and each triangular fuzzy number for the rating of the vehicle drivability, respectively, and the results are shown in Table 7. It shows that the linguistic vehicle drivability of A1, A2 and A3 are High (H), Medium high (MH) and High (H), respectively.

Table 5. Linguistic variables and corresponding fuzzy numbers for the rating of the vehicle drivability.

Linguistic variables	Abbreviation	Triangular fuzzy numbers
Very high	VH	(90, 100, 100)
High	H	(70, 90, 100)
Medium high	MH	(50, 70, 90)
Moderate	M	(30, 50, 70)
Medium low	ML	(10, 30, 50)
Low	L	(0, 10, 30)
Very low	VL	(0, 0, 10)

Table 6. Process of the vehicle drivability calculation.

	C1	C2	C3	C4	C5	C6	C7	C8	$\tilde{x}_i$	$D(\tilde{x}_i)$
A1	VH	H	VH	H	MH	H	M	H	(67, 85, 95)	83
A2	MH	H	M	MH	VH	H	M	MH	(55, 74, 89)	73
A3	VH	VH	MH	VH	H	VH	H	VH	(80, 93, 99)	91

Table 7. The similarity between $\tilde{x}_i$ and the fuzzy number for the rating of the vehicle drivability.

	VH	H	MH	M	ML	L	VL
A1	0.507	0.858	0.706	0.498	0.383	0.273	0.165
A2	0.420	0.683	0.904	0.612	0.457	0.330	0.211
A3	0.650	0.803	0.541	0.396	0.310	0.215	0.114

6. Conclusion

This paper proposes a fuzzy method to evaluate the drivability of new energy vehicles, which is regarded as a two-stage evaluation problem. The first stage is to determine the evaluation model, and the second stage is to evaluate the vehicle drivability using the model. In the method, linguistic variables are used to rate the importance of the indexes and the vehicle drivability for each index, and the expert weights are determined according to the similarity between the fuzzy evaluation values of indexes importance. The weights of the indexes are determined according to the combination of subjective and objective methods, in which the objective weights are determined based on the conflict degree of the fuzzy evaluation value. The evaluation results are expressed by numerical drivability and linguistic drivability. The numerical drivability is obtained from the integrated fuzzy drivability through defuzzification, while the linguistic drivability is determined according to the similarity between the integrated fuzzy drivability and the fuzzy number for the rating of the drivability.

From the numerical example, we can get the following findings: (1) For the vehicle drivability test experts, safety and comfort performance are the main factor affecting the vehicle drivability, followed by intelligent configuration. (2) The experts have the most controversy over the intelligent configuration when rating the importance of the index, followed by steering characteristics. (3) In general, safety and comfort performance have the greatest impact on the vehicle drivability, followed by intelligent configuration. The research results of this paper provide a theoretical and practical basis for vehicle manufacturers to understand the advantages and disadvantages of their vehicle drivability and the subjective preferences of drivers, so as to enhance their competitive advantage over competitors in the market.

In the future, we will expand the second stage of this method, use fuzzy TOPSIS and other methods to rank the mainstream new energy vehicles according to the vehicle drivability, and select the new energy vehicle that the drivers are most satisfied with the drivability.

References

[1] Hu D 2016 Application of Objective Evaluation System in AMT Commercial Vehicle Driving Performance Evaluation. Mechanical Engineering & Automation vol 23(4) pp 204-205

[2] Liu P and Li Y 2017 Objective characterization and simulation analysis of vehicle drivability. Chinese journal of construction machinery vol 15(4) pp 370-376

[3] Chen M, Huang J and Wei X 2021 Application of AVL DRIVE in Vehicle Drivability Development. Automobile Applied Technology vol 28(12) pp 117-119

[4] Krohling R A and Campanharo V C 2011 Fuzzy TOPSIS for group decision-making: A case study for accidents with oil spill in the sea. Expert Systems with Applications vol 38 pp 4190–4197

[5] Kumar A, Kaur J and Singh P 2011 A new method for solving fully fuzzy linear programming problems. Applied Mathematical Modelling vol 35 pp 817–823

[6] Valenzuela O and Pasadas M 2011 A new approach to estimate the interpolation error of fuzzy data using similarity measures of fuzzy numbers. Computers and Mathematics with Applications vol 61 pp1633–1645

[7] Ogundoyin S O and Kamil I A 2020 A Fuzzy-AHP based prioritization of trust indexes in fog computing services. Applied Soft Computing Journal vol 97 pp 1-16

Research on the Friction Pairs in Water Hydraulic Piston Pumps

Yousheng Yang*, Richmond Polley Yankey and Hao Li

College of Engineering, Ocean University of China, 238 Songling Road, Qingdao, China

*Corresponding author email: ysyang@ouc.edu.cn

Abstract. Power control hydraulics (PCH) has one of the most versatile research fields and technologies that has many applications especially in the mechatronic industries. It also comes with drawbacks especially with issues concerning the use of mineral oil as a pressurized fluid in most of it applications. Furthermore, in the mechatronic field, researches in the ocean environment has become more difficult due to mineral oil incompatibility with the ocean water. One of the most accepted solution is to find alternatives to mineral oil as a PCH pressurized fluid which has led to water being the safest candidate. As a result, more researches have been carried out on water as a pressurized fluid in PCH systems. Axial hydraulic piston pump has been one of the best pumps for PCH systems. It was realized in this paper that, more options for material selection for the friction pairs were available and therefore water hydraulics can realize its main goals of providing a safe environment as well as being cost-effective. This paper seeks to review the various material friction pair selection that has been suggested by various researchers in this field and analyze their various methodologies and experimentations.
Keywords: Axial piston pump; Swash plate; PEEK; Water hydraulic; Materials.

1. Introduction

With an increasing concern on the protection of the environment by various stakeholders worldwide, there has been the need to employ practices which are safe to use in the environment. One of the increasing concerns in the field of mechanical engineering is applications where the pressurized fluid in a PCH system is mineral oil. Mineral oil as a pressurized fluid has been in existence since the 20th century. Oil was the preferred fluid in hydraulic systems due to the technological leaps in electrohydraulic controls and the successive development of the electrohydraulic servo valve [2]. Oil took over in hydraulic systems except in areas where the environment was a sensitive issue like food production companies where there is a risk of contamination. Mineral oil for applications of pressurized fluid in PCH systems was superior as a result of lubrication due to its high viscosity, high unlikeliness to cause corrosion, lower risks of freezing at lower temperatures, lower bulk modulus and little to no leakages in hydraulic systems. Hydraulic systems like the hydraulic axial piston pump consist of components which are constantly sliding over each other. Due to this situation, the fluid used in these systems should be able to present a level of lubrication between its components which reduces friction. Mineral oil is known to have one of the best lubrication which encouraged its usage in a lot of industrial machinery and processes. Furthermore, mineral oil has a low viscosity which implies less leakage.

However, in the late 20th century, there has been a search for alternatives to oil as a fluid in hydraulic system. There have been increasing concerns about the use of mineral oil due to its pollution, bio-degradability, fire and other safety reasons. Mineral oil if not disposed of properly may lead to the contamination of the environment especially seepage into underground water reservoirs. Furthermore, there are risks of fire hazards and explosions. The spillage of oil from hydraulic systems leads to harsh hazardous environment conditions [2]. All these are the various reasons that led to the search for an environmentally friendly option. An option was to use bio-degradable fluids. But these fluids may contain chemicals which will still be harmful in case of a spillage. Water as an option for a pressurized fluid and an alternative to mineral oil has been a more promising prospect due to its various environmental friendliness and other characteristics which makes it a viable option. The utilization of water as a pressurized fluid has been in existence for over 200 years [16].

Water is characterized with low viscosity, poor lubrication, has a very corrosive nature, to name but a few. Water is known to have a kinematic viscosity of about 0.7 m^2/s at 40℃ as compared to mineral oil which has a kinematic viscosity of about 32m2/s at the same temperature [3]. The low viscosity implies that the formation of a hydrodynamic film between the contact or friction pairs will be difficult [3]. Low viscosity also leads to poor lubricity which is a well-known characteristic of water. From the general knowledge of lubricity, poor lubricity implies high corrosion and abrasive wear between the friction pairs which were major problems facing the use of water as a pressurized fluid and in the long run, led to the shift to mineral oil as a pressurized fluid in a hydraulic system [21]. Mineral oil although has good lubrication properties, high viscosity and the likes, the negative impact of its usage outweighs the positive. Mineral oil has been known to be environmentally unfriendly due to how difficult and expensive it is to dispose of them properly. One example which may be related to the cost of mineral oil spillage to the environment is that of the deepwater horizon oil spillage that occurred in 2010. It is known in history as one of the largest oil spillages that has ever occurred [26]. Birds, fishes and oceanic living organisms were affected greatly which distorts the normal ecological order. The cost of the spillage was estimated to be about $65 billion [24]. Mineral oils have a high risk of causing fire if not handled properly. All these arguments and most importantly environmental protection led to the search for an alternative to mineral oil for use in hydraulic systems.

This is where water as a pressurized fluid got more attention even though researches were already being done to make it more efficient. Water is the safest, environmentally friendly and fire retardant of all other alternatives which includes biodegradable fluids which to a certain level contain chemicals which may be harmful to the environment. Water is very abundant on earth and can be reused over and over again with little to no treatment. And with the increasing focus on green energy, water (seawater, tap water) are the most viable candidates as a pressurized fluid for hydraulic systems. With water becoming the ideal replacement for mineral oil, there are a lot of challenges that need to be solved in other for it to be more useful as in the case of mineral oil. One of the most important applications of water hydraulics is in the area of marine engineering which is a very sensitive area. The start of 20th century saw an explosion of deep blue sea resource developments for economic growth [17]. With increasing concerns of finding a different means of electricity generation through the utilization of natural resources as well as the increase in the depletion of land resources and pollution of the environment, developed countries are turning their heads towards the ocean. Also, the recognition of how problematic climate change led to the raised interest in marine renewable energy [17]. And in other to prevent the reoccurrence of the incidence in 2010, seawater has been the ideal choice for a pressurized fluid for hydraulic systems in ocean development. Environmental friendliness, economical, ease in disposing of working media, etc., are some of the advantages when water hydraulic transmission is incorporated in the development of ocean machinery [17].

Even though, water seems to be the way forward, it does not make it flawless. Water like any other natural water compared with hydraulic oil, has properties which physically and chemically make it quiet difficult for the application in a hydraulic system. The explanation for some of these properties are as follows:

- Lower viscosity: As stated earlier, water has a kinematic viscosity 0.7 m^2/s at 40℃ as compared to mineral oil which has a kinematic viscosity of about 32m2/s at the same temperature. Lower viscosity can imply low power loss due to less flow friction [2]. Which is a good situation because all the power generated will be utilized for its intended purpose. But lower viscosity also results in leakages both internally and externally when similar allowances applied to hydraulic oil. As a result the components of water hydraulic systems should be adjusted with the suitable clearances.

- Lubrication: Poor lubricity properties of water as result of lower viscosities makes its utilization in hydraulic systems nearly impossible without the right materials for the components. The friction between these components in water hydraulic system made it nearly unbearable for these kind of applications. This delayed the realization of water as a pressurized fluid in hydraulics over a long period of time. Recent advancement in technology has paved way to research into materials that can work in such conditions with little to no detrimental effects on the components.
- Corrosion: Corrosion is a complex electrochemical process which involves metallic materials and their environment interacting with each other and as a result lead to the deterioration of the materials. Corrosion is one of the key factors that made it very difficult for the realization of water (especially seawater) in hydraulic systems. Seawater can be lethal to components in marine equipment due to the presence of dissolved gases such as oxygen, chlorine, etc., [2]. Which implies, the choices of materials for water hydraulics are very important without it can lead to corrosion which is very destructive. Furthermore, seawater is known to be electrically conductive and with the presence of certain impurities can act as an electrolyte which leads to an electrolytic corrosion [3].

Furthermore, water hydraulic systems require the use of smaller clearances than that of the oil hydraulics in other to give enough volumetric efficiency. But the use of smaller clearances directly translate to the high possibility of sliding surfaces having a direct contact [18]. In this paper, a review is done on the various researches that has been done on the materials used in water hydraulics system and various approach and findings are analyzed.

Various kinds of procedures and material combinations were applied in the experimentation processes by various researchers. Material combination such as polymer and a ceramic material, metallic alloys and polymeric materials, etc. The polymeric material can be seen as one of the favorite across board as one of the pairs which suites the applications in PCH. Furthermore, other directions such as applying thin films on surfaces of materials to greatly increase the corrosion resistance as well as improve on other properties have been employed. Water as a pressurized fluid in PCH system greatly known to have smaller clearances can be quite harsh on these films.

In other to determine the appropriate material pair for applications in water hydraulic systems, experimentation and analysis environment should be as close to the natural working environment as much as possible. To achieve this simulation for the natural environment of their specimen, [4] [1] [3] produced artificial seawater according to the GB 17378.4-2007 and ASTMD 1141-98 standard respectively. This may prove very useful in cases where natural seawater is inaccessible. [12] used natural water in their experiment. To further improve the simulation environment, [12] [4] poured a certain volume seawater on the specimen in other to ensure the seawater lubrication state is achieved. The typical experimentation procedure for the various experimentation processes includes the use of the tribotester, ball on disk [1] and pin on disk [4] [20]. The important parameters looked out for are the coefficient of friction and wear rates in these kind of test. One's knowledge on how certain factors like wear rates and coefficient of friction affect the overall performance and efficiency of the friction pairs of the swash plate piston pump is really important [29]. Lower friction coefficient implies less friction therefore these components can freely slide over each other which is more desirable in such setups. Furthermore, lower wear rates is much more desirable in other to save cost of materials. An optical microscope or a scanning electronic microscope is then used for the surface examination which can give the type of wear on the surfaces of the specimen [11] [20]. An important parameter for the evaluation of the efficiency of the friction pair is the Pv (MPa m/s) value where p represents the applied pressure and v represents velocity of the friction pairs Yinshui et al [10].

The various methods and materials that have been used so far have immensely contributed to the growth of water as a pressurized fluid in PCH. The experimentation of using artificial seawater in the analysis has great advantages especially in areas where getting seawater is not an option. But in the sense of creating the ideal natural environment for experimentation might not be realized since there are various factors in the sea environment that are not constant (such as varying temperatures in various levels, varying dissolved oxygen, etc.) and may greatly increase the rate of corrosion process. Therefore, that might be a little hindrance in the part of creating an artificial seawater for the experiment.

2. Wear Analysis

For water hydraulic systems, the wear analysis of components is one of the most important factors to look into when investigating the materials for friction pairs. Due to poor lubricity, lower viscosity and other factors, its quiet significant to study the effects of wear on the material selection process. Hsu et al [33] studied the wear prediction of metals. They realized that, most wear models were condition specific or material specific. Which implies that, there was no general method for wear analysis. The developed wear analysis at the time was not comprehensive enough therefore [33] sought to develop wear prediction mechanisms. Hsu et al [33] further explained that, temperature as well as sliding conditions which deposits some sort of debris on the softer from the harder material may greatly affect the selection of materials. They concluded that, lubricating environment should be factored in the general wear analysis in the general wear prediction analysis since the same lubricating environment will exist no matter the type of material. Wear analysis of the slipper pad and swash plate were researched using oil as the pressurized fluid by [34]. They highlighted the effects of temperature, outlet pressure and rotational speed of the pump. They realized that increments in temperature as well as the outlet pressure increased the wear rate. Higher rotational speed of the pump led to a decrease in wear rate. This presents a great literature in the applications of water hydraulics and the selection of materials based on their resistance to certain conditions presented during their operations.

Yang et al [38] also approached from the point of non-probabilistic reliability theory. The efficiency of convex model's technique in analyzing friction pairs made it the right choice over the interval model. They realized the data from the initial probabilistic reliability theory suggested by [34] had insufficient data for analysis of the friction pairs of water hydraulics motors. The whole analysis required a lot of information in other for the realization of the probabilistic reliability theory. The non-probabilistic reliability was therefore suggested to be the best option due to limited information on the probabilistic reliability theory. Furthermore, they also suggested that, the utmost economical and logical approach in analyzing the friction pairs' wear reliability in water hydraulic systems is the convex model.

3. Material Selection for Friction Pairs

The first point of interaction is the matching schemes where each engineering materials pair would have to be selected based on this scheme. The scheme includes a hard-to-hard configuration, (for example a ceramic sliding over another ceramic material), and a hard to soft configuration (for example ceramic material sliding over a plastic material, etc.).Yinshui et al [10] upon further investigation on these material selection configurations found out that, hard to soft configurations were one of the best material selection schemes with the bottom being the hard material (swash plate) and top being the soft material (slipper pad). This has greatly been translated into most ongoing researches on friction pairs' choices of materials in water hydraulic systems. This configuration does provide the best solution to the first step in the selection of the materials since the main aim is ensuring there is less wear as well as coefficient of friction. This can only be achieved in this configuration when water serves as the pressurized fluid in power control hydraulic systems. Zhang et al [6] investigated the use of seawater as a lubricating medium when CFR polyetheretherketone slide against silicon carbide ceramic material. It was realized that the silicon carbide ceramic material had significant corrosion and wear resistance and that the carbon reinforced polyetheretherketone was not able to cause any damage as it slide on the ceramic material with seawater as the lubrication. Ceramic material are generally known for their high corrosion and wear resistance nature. Their brittleness limit their applications in various industries. Different corrosion resistant alloys which were slide against carbon fiber reinforced polyetheretherketone were studied by [4]. Their main emphasis was on their tribological nature. The medium of lubrication was an artificial seawater. The experiment was done using the pin-on-disk configuration for testing friction. The corrosion resistant alloys include AISI 630 and AISI 316 stainless steel, C-Ti6Al4V QAl9-4, Ti6Al4V and N-Ti6Al4V alloys. It was realized in their experiment that, Ti6Al4V and N-Ti6Al4V alloys exhibited poor tribological properties when sliding against CF-PEEK whiles AISI 630 and AISI 316 stainless steel generally exhibited a better tribological properties when sliding against CF-PEEK. Another research from [10] studied cermets when slide over silicon carbide with their main point of interest being their tribological properties in these conditions when seawater serves as the lubrication medium. The cermets used in this experiment include WC-TiC-Co-Ni, WC-6Ni and WC-6Co. From this work, with sliding speed as the main factor, it was realized that, WC-6Co although had a better wear resistance as compared to WC-6Ni as a result of it being harder, WC-6Ni was the general recommendation by the authors.

Liang et al [12] decided to go with another perspective of the selection of the materials. They investigated the tribological properties of the selected materials based on their surface smoothness.

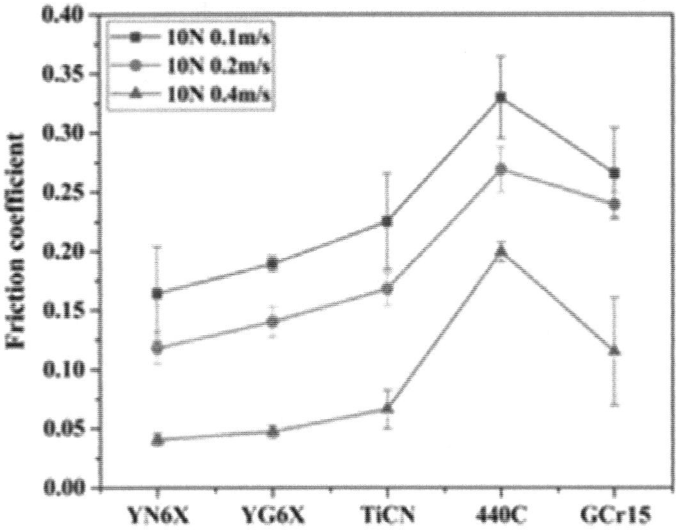

Figure 1. Mean steady state coefficient of friction of different types of steel balls and cermet sliding over SiC at various sliding speeds with artificial seawater as the pressurized fluid [10].

From Figure 1, both WC-6Co (represented with YG6X) and WC-6Ni (represented with YN6X) had the smallest coefficient of friction under various varying speeds. This was as a result of WC-6Ni having nickel which excellently stood a chance against the corrosion properties of seawater. Based on previous research work, it was realized that, the roughness between two parallel sliding surfaces can lead to hydrodynamic lubrication. Therefore Liang et al [12] based on this went ahead to investigate the tribo properties when 316L stainless steel slides against that of carbon reinforced polyetheretherketone and compared it with the same specimen but this time around, the surfaces were smooth. Seawater was the lubrication medium in both instances. It was realized from their experiment that the smooth surfaces had both abrasive and adhesive wear with large friction coefficient which is undesirable. Those on the surfaces with some slight roughness experienced some slight adhesive wear and significantly reduced abrasion wear. The reduction was as a result of the surface roughness serving as a reservoir for both seawater. This created a hydrodynamic effect which effectively reduced the coefficient of friction.

To further investigate the type of surface roughness, Liang et al [12] in their experiment created five different surfaces namely; tri-prism, conical, cylindrical, hemispherical, and elliptical. The tri-prism pits as well as the elliptical demonstrated the excellent abrasion resistance and drag reduction causing a significant decrease in coefficient of friction. In studying the tribological properties, Wang et al [13] investigated from another angle where coupling of ZrO_2, SiC, Si_3N_4 , Al_2O_3 and WC was done with Graphite-like Carbon (GLC) with a chromium element films doping. Graphite-like carbon (GLC) film are known to have high hardness, anticorrosion properties and significantly better wear resistance and when doped with chromium have significantly improved mechanical and tribological properties. From their experiment, it was realized that upon sliding the chromium/graphite like carbon against that of the engineering ceramic materials in an artificial seawater, Si_3N_4 and SiC had lower friction coefficient and wear as compared to WC, Al_2O_3 and ZrO_2. This behavior of these two ceramics has a lot of different school of thought backing it. One of them suggests that there is a possible formation of silicon-oxygen-silicon inorganic network basically known as the condensed silica gel which induces a hydrodynamic lubrication [15].

Guan et al [30] studied the friction pair; slipper- crankshaft based on a novel water hydraulic motor. The material used were a 316L stainless steel ($Cr_{17}Ni_{14}Mo_2$) and polyetheretherketone with 30% carbon fiber composition (PEEK-CF30). To diversify their investigations, the effects of surface coatings on 316L stainless steel was introduced. OVINO-TAC as well as the OVINO-GIC were coated in some of the materials as a form of surface treatment. It was realized from their investigation that the surface treated stainless steel as compared to the untreated pair had a significant decreasing coefficient of friction as friction time increased. This was mainly seen in the surface treated OVINO-TAC 316L stainless steel paired with PEEK-CF30 [9]. In their experiment were able to find out that, the corrosion resistant alloy laser cladding has a potential as a tribo material when it slides against composite materials. Various new engineering materials have the potential to sustain seawater hydraulic systems, as a result various researches should be continued to find an ideal candidate for the various friction pairs in the water axial piston pumps. The wear and friction of polyphenylenesulphide and PEEK were studied by [22] as they slide against steel. Polyphenylenesulphide was found to be suitable in water due its formation of iron sulphide (FeS) which in turn reduces wear as well as coefficient of friction. With the various trends in the selection of the materials, it can be noted that, seawater hydraulics has not gotten to the level oil hydraulics due to some few design problems associated with various water hydraulics. But the high optimism of various researches to find the most appropriate tribopair for these components is highly applaud able.

Yin et al [35] also did a study on the axial piston pump on the basis of the slide bearing. Different material combinations were made in other to choose the appropriate friction pair for the research based on their superiority in terms of wear rate and friction coefficient. The material friction pairs includes a quench-polish-quench (QPQ) hardened 0Cr17Ni4Cu4Nb for increasing the corrosion resistance of the metal and engineering plastics PEEK 450-FC30 and PEEK CA30. PEEK 450-FC30 is an engineering plastic made up of 30% carbon reinforced fiber, graphite and PTFE whiles PEEK CA30 is made up of 30% carbon reinforced fiber. The selected friction pair among other combinations was PEEK 450-FC30 and 0Cr17Ni4Cu4Nb which had excellent friction and wear. They concluded that, quench-polish-quench hardening approach provided a high micro hardness on the surfaces of the metal. This has encouraged the application of surface treatment on metals to enhance their applications in areas such as water hydraulics where wear rate is very important.

As seen from the various research directions in this paper, one can identify PEEK as one of the most versatile material choice and therefore really help step up the use of PCH with the evolution of water as a pressurized fluid in such systems.

4. Conclusions

From this review of the materials used in water hydraulic systems, the broad conclusions that can be drawn about the current state of seawater hydraulic systems includes;

1. Water hydraulic systems are very vital for the next biggest industrial revolution that is going to happen due to increased attention on marine energy resources.
2. With the current status of research done on the materials for applications in power control water hydraulic systems, this review paper has shown the various progress that has been made towards the goal of optimizing the axial piston pumps. Optimizing the key components in axial piston pumps through the selection of the appropriate materials can in a long way make the axial piston pump more efficient. Also, various analysis on the various wear propagation in the components of water hydraulics should be done. In other to tackle the issue of wear, there is the need to understand it and know the detailed mechanisms involved.
3. More materials with the ability of forming a film should be investigated more. This has been seen be more useful in reducing the coefficient of friction as wear in materials used in water hydraulic systems.

Acknowledgements

This research was partly financed by the Key Research and Development Plan of Shandong Province of China (2017GGX30106); the National Natural Science Foundation of China (51475197, 51275495, 51708528); the National College Student Innovation and Entrepreneurship Training Program; the

Science and Technology on Underwater Vehicle Technology Research Fund (SXJQR2017KFJJ04) and the Natural Science Foundation of Shandong Province of China (ZR2018MEE023)

References

[1] Yin, F., Wang, Y. Ji, H. 2021 Impact of Sliding Speed on the Tribological Behaviors of Cermet and Steel Balls Sliding Against SiC Lubricated with Seawater. *J. Tribol* Lett **69** 39. https://doi.org/10.1007/s11249-021-01413-1

[2] Krutz, G.W., Chua, P.S.K. 2004 Water Hydraulics-Theory and Applications Workshop on Water Hydraulics. *Agricultural Equipment Technology Conference AETC'04*, Feb, Louisville, Kentucky, USA

[3] Wang, D., Li, Z. and Zhu, Y. 2003 Lubrication and tribology in seawater hydraulic piston pump. *J. JMSA* **2,** 35–40. DOI: 10.1007/BF02935573

[4] Yin, F., Zhou, Z., Nie, S., Ji, H. Hu, Z. 2019 Tribocorrosion Behavior of Several Corrosion-resistant Alloys Sliding against CF-PEEK: Application for Hydraulic Valve in Seawater *Int. J. Electrochem. Sci.* 14 4643 – 4658. DOI: 10.20964/2019.05.66

[5] Harris, R. M., Edge, K. A. Tilley, D. G. 1996 Predicting the Behavior of Slipper Pads in Swash plate-Type Axial Piston Pumps *J. Journal of Dynamic Systems, Measurement, and Control.* 118(1) 41. DOI:10.1115/1.2801149.

[6] Zhang, J.J., Liu, J.C., Wang, Z.X., Chen, W., Hu, B., Zhang, Y., Liao, H.M. Ma, S.D. 2020 Tribological behavior and lubricating mechanism of Si_3N_4 in artificial seawater. *J. Ceram. Int.* 46, 14361–14368.

[7] Zhang G, Liu Y and Wang Y. 2017 Wear behavior of WC-Ni sliding against graphite under water lubrication. *J. Journal of Materials Science & Technology.* 33(11): 1346-1352.

[8] J.M. Bergada, S. Kumar, J. Watton, Axial piston pumps, new trends and development, Nova Science, USA.

[9] Schenk, A.T. 2014 Predicting lubrication performance between the slipper and swash plate in axial piston hydraulic machines. open Access Dissertations. 359.

[10] Yinshui, L., Wu, D., He, X., Zhuangyun, L. 2009 Materials screening of matching pairs in a water hydraulic piston pump Industrial Lubrication and Tribology. J. Vol. 61 No. 3, pp. 173-178. DOI: 10.1108/00368790910953695

[11] Zhang Z, Nie S, Liao W, Li, L., Yuan S. 2014 Tribological behaviors of carbon fiber reinforced polyetheretherketone sliding against silicon carbide ceramic under seawater lubrication. Proceedings of the Institution of Mechanical Engineers *Part J: Journal of Engineering Tribology*, 228 (12):1421-1432. J. DOI: 10.1177/1350650114541108

[12] Liang, Y., Gao, D., Zhao, J. 2020 Tribological Properties of Friction Pair between 316L Stainless Steel and CF/PEEK with Nonsmooth Surface under Seawater Lubrication, *Tribology Transactions J* DOI: 10.1080/10402004.2020.1734704.

[13] Wang, C., Ye, Y., Guan, X., Hu, J., Wang, Y., Li, J. 2015 An analysis of tribological performance on Cr/GLC film coupling with Si3N4, SiC, WC, Al2O3 and ZrO2 in seawater *Tribology International J* DOI: 10.1016/j.triboint.2015.12.010

[14] Soukissian, T.H., Denaxa, D., Karathanasi, F., Prospathopoulos, A., Sarantakos, K., Iona, A., Georgantas, K., Mavrakos, S. 2017 Marine Renewable Energy in the Mediterranean Sea:Status and Perspectives Energies. 10, 1512. DOI: 10.3390/en10101512

[15] Gates, S.R., Hsu, S.M. 2004 Tribochemistry between water and Si3N4 and SiC: Induction time analysis. J. DOI: 10.1023/B:TRIL.0000044488.84437.68

[16] Trostmann, E., Frolund, B., Olesen B. H., Hilberecht, B. 2001 Tap Water as a Hydraulic Pressure Medium New York: Marcel Dekker.

[17] Hammar, L., Gullstr€om, M., Dahlgren, T.G., Asplund, M.E., Goncalves, I.B., Molander, S. 2017 Introducing ocean energy industries to a busy marine environment. Renew, Sustain, Energy Rev. 74, 178–185. J. DOI: 10.1016/j.rser.2017.01.092

[18] Rokala, M., Calonius, O., Koskinen, K.T., Pietola, M. 2008. Study of lubrication conditions in slipper-swash plate contact in water hydraulic axial piston pump test rig Proceedings of the JFPS International Symposium on Fluid Power. *J. The Japan Fluid Power System Society,* pp. 91–94.

[19] Nie, S.L., Huang, G.H., Li, Y.P. 2006 Tribological study on hydrostatic slipper bearing with annular orifice damper for water hydraulic axial piston motor. *J. Tribology International*, Volume 39, Issue 11, 1342-1354, DOI: 10.1016/j.triboint.2005.10.007

[20] Andersson, P. Water lubricated pin-on-disk tests with ceramics *Wear* **154** 1992 37-47.

[21] Meng, Y., Xu, J. and Jin, Z. 2020 A review of recent advances in tribology *Friction* **8,** 221–300 DOI: 10.1007/s40544-020-0367-2

[22] Yamamoto Y and Takashima T. 2002 Friction and wear of water lubricated PEEK and PPS sliding contacts. J. *Wear*; 253(7): 820–826.

[23] Michael Smith Engineers ltd https://www.michael-smith-engineers.co.uk/resources/useful-info/positive-displacement-pumps.

[24] The maritime executive 2018 https://www.maritime-executive.com/article/bp-s-deepwater-horizon-costs-reach-65-billion

[25] Amutha R. D., Yoshizawa Y., Hyuga, H. 2004 Tribological behavior of ceramic materials Si3N4, SiC and Al2O3 in aqueous medium. .*J. Eur Ceram Soc*; 24(10): 3279–3284.

[26] Pallardy, R. Deepwater Horizon oil spill environmental disaster Gulf of Mexico https://www.britannica.com/event/Deepwater-Horizon-oil-spill.

[27] Xia, W. Z., Zhao, J. W., Wu, H., Zhao, X. M., Zhang, X. M., Xu, J. Z., Jiao, S. H., Wang, X. G., Zhou, C. L., Jiang, Z. Y. 2018 Effects of oil-in-water based nano lubricant containing TiO2 nanoparticles in hot rolling of 304 stainless steel. *J. Mater Process Technol* 262: 149–156.

[28] Chen, W., Wang, Z. X., Liu, X. 2020 Effect of Load on the Friction and Wear Characteristics of Si3N4-Hbn Ceramic Composites Sliding against PEEK in Artificial Seawater Tribology International 141 pp. 105902.

[29] Qiu, B., Zhao, J., Man, J. 2019 Comparative study of materials for friction pairs in a new high water-based hydraulic motor with low speed and high pressure. Industrial Lubrication and Tribology 71(1) 164–172. DOI: 10.1108/ilt-01-2018-0006.

[30] Guan, Z., Wu, D., Cheng, Q., Wang, Z., Tang, M., Liu, Y. 2020 Friction and Wear characteristics of CF/PEEK against 431 stainless steel under high hydrostatic pressure water lubrication. *Materials & Design*. 196, 109057. DOI:10.1016/j.matdes.2020.109057

[31] Meng, F., Gong, X. Kang, X. 2018 Research on Tribological Characteristics of 316L Stainless Steel against PEEKHPV under Water Lubrication Current Trends in Computer Science and Mechanical Automation Vol.2 edited by Shawn X. Wang. Warsaw Poland De Gruyter Open Poland pp. 449-457 DOI: 10.1515/9783110584998-047

[32] Sun, Y.G. 1995 Reliability evaluation in the wear behavior of friction pair in axial piston pump *Constr Mach Equip* 1: 25–28

[33] Yin, F., Nie, S., Zhang, Z., Zhang, X. 2013 Research on the sliding bearing pair of water hydraulic axial piston pump. Proceedings of the Institution of Mechanical Engineers *Part C Journal of Mechanical Engineering Science* 227(9), 2049–2063 DOI:10.1177/0954406212470364

[34] Hsu, S. M., Shen, M. C., Ruff, A. W. 1997 Wear prediction for metals. Tribology International 30(5), 377–383. DOI: 10.1016/s0301-679x (96)00067-9

[35] Yang, L., Nie, S., Zhang, A. 2012 Non-probabilistic wear reliability analysis of swash-plate/slipper of water hydraulic piston motor based on convex model. Proceedings of the Institution of Mechanical Engineers *Part C Journal of Mechanical Engineering Science* 227(3), 609–619. DOI: 10.1177/0954406212463501

Vibration Analysis on the Rack of Electric Mini-tiller

Guojun Huang[1,2], Hongrong Wang[1] and Xincheng Liang[3,*]

[1]State Key Laboratory of Vehicle NVH and Safety Technology/ P. R. China;
[2]Southwest University, College of Artificial Intelligence/ P. R. China
[3]Southwest University, College of Engineering and Technology / P. R. China

*Corresponding author email: dylb1978@swu.edu.cn

Abstract. The conventional internal combustion mini-tiller is a vitally agricultural tool with multiple functions, meanwhile the strong vibration may impose damaging effect on the user's health. Hence, it is inevitable to replace the internal combustion by the electric motor. Although the performance of the electric mini-tiller has been enhanced greatly, the remaining vibration is still distinct and some diseases can be occurred incontrovertibly. The rack is one of the most significant components to connect the different parts for the electric mini-tiller, and the index may be improved through redesigning the structure of rack. In this study the rack has been analysed under different structure by Ansys workbench software, and simulations show that the vibration acceleration from the rack to the human body decreases greatly after the structure had been redesigned, which signifies the handling stability and riding comfort of mini-tiller can be promoted without the extra cost. Although the total vibration acceleration is reduced a lot, the vibration of cutting roller is more distinct than that of before. Therefore, the cutting configuration is needed to be optimized, and vibration transmitting path of cutting roller should be intervened as well in the future, enhancing the handling comfort and energy conservation of electric mini-tiller.
Keywords: Mini-tiller; Rack; Vibration; FEM.

1. Introduction

The mini-tiller is the significant agricultural tools for peasants [1]. In the mountainous regions, the use of large and middle- size mechanical equipment is restricted highly, so such tillers with small size are very popular. In figure 1, it is not hard to see the structural features of mini-tiller: the rack is the main support component and the rest are installed on it. When the mini-tiller is working, the vibration generated in the engine and cutting roller will be transmitted from the rack to the human body. It is common sense that vibration of engine is much bigger than that of electric motor under the same operating condition, so the power system in figure 1(a) is replaced by the electric motor naturally, and the detailed structure is shown in figure 1(b).

(a)

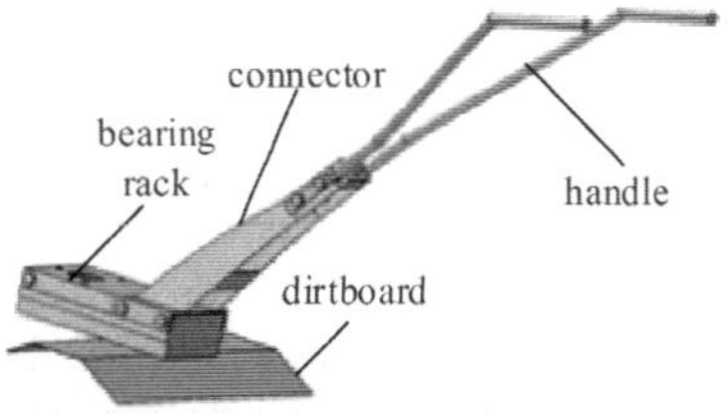

(b)

Figure 1. The mini-tiller and the structure of rack

Although the vibration of electric mini-tiller has been reduced obviously compared with the conventional one, the vibration of cutting roller of electric mini-tiller becomes more distinct. The vibration of cutting roller is the main part for the electric mini-tiller, while it is tough to be diminished through arrangement mode of cutter, influenced by the cost, manufacturing technique and so on. Since the rack is the one and only component to transmit vibration from cutting roller to human body, it is significant to analyse the process of vibration generating and transmitting.
Contrasted with the vehicle, the vibration amplitude of mini-tiller is far big, resulting in diseases such as the organ's disorders and neurological disease [**2**]. Consequently, relative problems regarding the vibration of mini-tiller have been studied. The vibration of conventional mini-tiller and electric mini-tiller has been investigated under different operation gears [**3~6**]. In fact, extra vibration of the mini-till is associated with other elements such as the firmness of soil, the moisture content of the soil, tilling depth and so on, these elements have been researched as well [**7~9**]. In addition, the vibration characteristics of mini-tiller is also studied under non-operational state, and the advice to perfect the handling comfort is recommended [**10**].
Though the performance of the electric mini-tiller has been improved evidently according to the researches in the preceding paragraph, the indexes are still expected to be enhanced. The rack plays a crucial role to transmit the vibration from the source to the human body, unfortunately rare investigations have been implemented to date. Therefore, the purpose of this study is pointing out the unreasonable structure for the electric mini-tiller, and some pieces of advice can be offered to modify the design and obtain the better performance. Briefly, the extra vibration of electric mini-tiller still exerts a negative effect on people's health, then it is necessary to intervene the vibration transmission process.

2. Vibration Model

Just like any a mechanical system, the rack can be viewed as multiple degrees of time-invariant system, and the differential equation of motion is depicted as follow

$$[M]\{\ddot{X}\}+[C]\{\dot{X}\}+[K]\{X\}=\{F\} \tag{1}$$

where M is the mass of system, [kg]; C is the damping of system, [Ns/m]; K is the stiffness of system, [N/s]; F is the exciting force vector of system, [N]; X is the displacement response vector of system, [m].
In addition, equation (1) can be rewritten after Laplace transformation

$$(s^2[M]+s[C]+[K])\{X(s)\}=\{F(s)\} \tag{2}$$

where $X(s)$ is the Laplace transformation of displacement response; $F(s)$ is the Laplace transformation of exciting force. If the symbol s in equation (2) is replaced by $j\omega$, then the equation (2) is expressed as

$$(-\omega^2[M]+j\omega[C]+[K])\{X(\omega)\}=\{F(\omega)\} \tag{3}$$

Considering the variables coupled in equation (3), the modal coordinates $\{X\}=[\phi]\{q\}$ will be introduced, and the equation (3) is signified as

$$(-\omega^2[M]+j\omega[C]+[K])[\phi]\{q\}=\{F\} \tag{4}$$

For most mechanical systems, the orthogonality relation among stiffness matrix [K], mass matrix [M] and vibration mode [ϕ] is satisfied, and thus matrix [K] and [M] is feasible to be diagonalized with matrix [ϕ].The damp matrix [K] is also disposed of using vibration mode [ϕ], especially for a linear vibration system with small damping coefficient. Assuming the inverse matrix of $[\phi]$ is $[\phi]^T$, the equation (4) multiplies $[\phi]^T$ and can be rewritten as

$$(-\omega^2[M]+j\omega[C]+[K])\{q\}=[\phi]^T\{F\} \tag{5}$$

After related transformation, the differential kinematic equation has been decoupled independent equation under the modal coordinate. That is to say the *i*th equation is indicated as

$$(k_i-\omega^2 M_i+j\omega C_i)q_i=\sum_{i=1}^{n}\phi_{ji}f_i\left(j=1,2,3,...,n\right) \tag{6}$$

meanwhile, the response under *l* coordinate can be

$$X_l=\sum_{i=1}^{n}\phi_{li}q_i \tag{7}$$

with helping equations above, the characteristics of mechanical system will be obtained theoretically. However, it is tough to acquire the vibration mode and frequency of the rack of electric mini-tiller, and the reasons are expressed as follow. Firstly, most of components have been welded firmly, so the elastic coefficients and damping factors of rods are challenging to be confirmed. Secondly, the mass matrix in equation (1) is untoward to be gotten accurately as well, for the intricate topological relations among rods. In a word, the mass-elastic-damping matrixes are not easy to be obtained through simplifying the structure of rack of electric mini-tiller, so it is natural to use another method—Finite Element Method (FEM), which is easy to obtain the vibration characteristics but workload can be reduced greatly.

According to the rules of FEM, the rack of mini-tiller has been selected as the research target and disposed as follow in this study. Firstly, the 3D model of rack is established in the CATIA software. Secondly, the model is imported in ANSYS. Lastly, the model is analysed using Mechanical module under free mode. Virtually, the natural frequency and modal shape are determined by the mass distribution and stiffness of the rack structure. Therefore, the details including the tiny chamfer, screw thread, fillet and so on will be neglected when the elements are generated. Such simplifying is of benefits—the calculation accuracy can be guaranteed, and computing workload be decreased as well.

After simplifying the model, the metric is chosen as units. The material is viewed as structure steel, which is of advantages such as isotropy and uniform density distribution. The parameter of material can be indicated in table 1. Considering the parts of mini-tiller are weld by the screw thread, the deformation is hard to be measured, so the contact among different parts can be regarded well. In the Mechanical module the contact is set as bonded mode. Then the grid of main structure is generated automatically, decreasing the workload, while the parts such as the handle whose curvature is big will be modified by hand. In free state it is easier to investigate the structural rationality of mini-tiller, so any load or constraint will be exerted on the rack. Finally, 98613 nodes and 52465 elements have been produced.

Table 1. The material parameters of model

Attribute	Unit	Value
Elasticity E	Pa	2.06e11
Poisson ratio μ	-	0.3
Density ρ	kg/m^3	7.8e3

It is easy to find that some components including the electric motor from figure 1, the battery pack are fastened on the rack. The relationship of vibration frequency and correlated variables is expressed as

$$f=\frac{1}{2\pi}\sqrt{\frac{k}{m}} \tag{8}$$

where *m* is the mass of rack, kg; *k* is the stiffness of rack, N/m.

Apparently, if some components are appended the rack, the model of rack with components is more accurate than the model in the figure 1. Undoubtedly, the disposing process of new model is similar to the one without other components, while the precision of calculating results will be enhanced evidently-151422 nodes and 81416 elements are generated, and this is the reason why the extra time will be consumed. Consequently, relevant outcomes will be described as follow based on the different models.

3. Results Analysis

Helping of the Workbench, the outcomes of model can be obtained easily. The natural vibration frequency from the first order to the eighth order for the mini-tiller will be list subsequently. Correspondingly, the vibration mode of rack will be indicated in the table 2 and figure 2.

Table 2. The $1^{st}\sim8^{th}$ natural frequency of non-rigid body modes of rack without mass

Mode	Frequency/Hz
1	7.1488
2	16.825
3	28.072
4	30.356
5	41.431
6	49.242
7	58.822
8	85.692

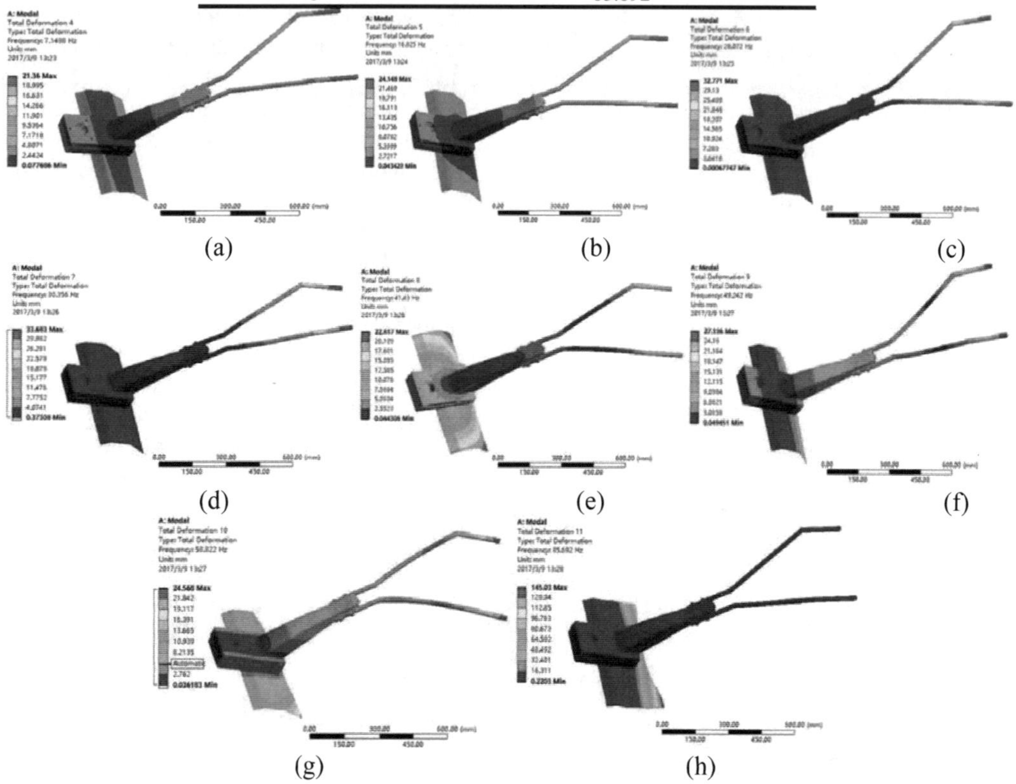

Figure 2. The $1^{st}\sim8^{th}$ modal shape of rack

When the model of rack with extra mass is imported the Workbench, the vibration frequency and corresponding modal shape can be indicated in table 3 and figure 3.

Table 3. The $1^{st}\sim8^{th}$ natural frequency of non-rigid body modes of rack with mass

Mode	Frequency/Hz
1	16.688
2	23.592
3	27.995
4	30.284
5	44.863
6	84.511
7	104.25
8	106.33

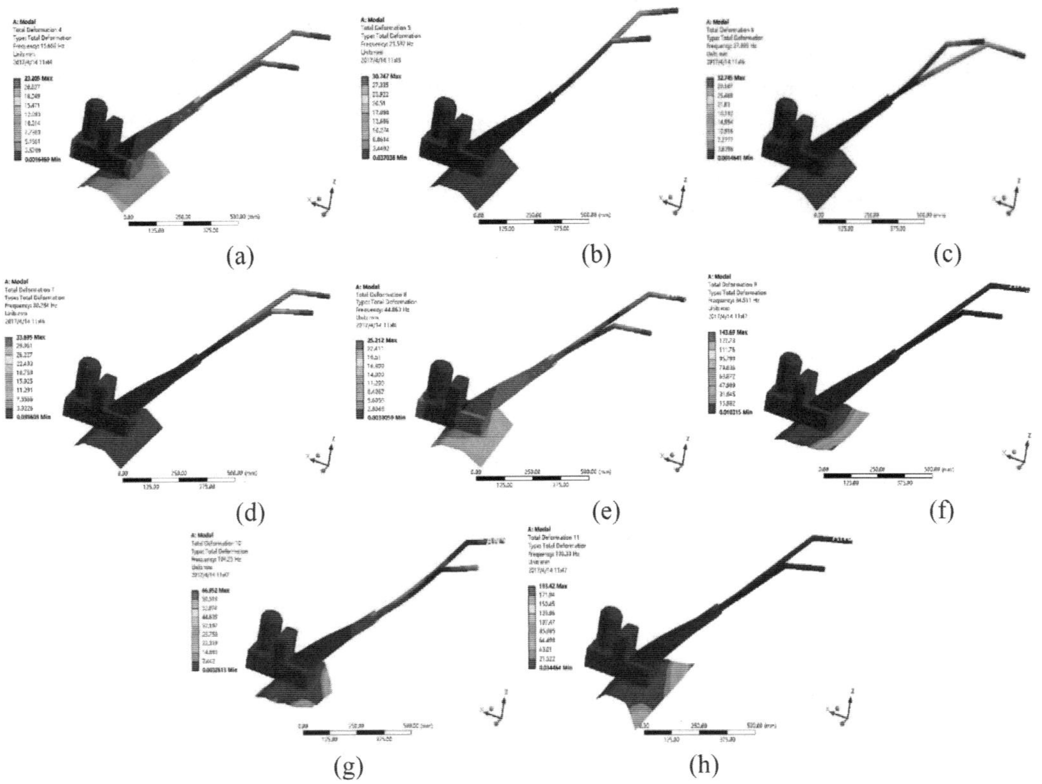

Figure 3. The 1st~8th modal shape of rack with extra mass

In figure 3, the mass of electric motor is 40kg, and the mass of battery pack is 20kg. If more batteries are fixed on the rack, longer operating time can be gained and modal shape of rack will be changed according to the equation (8). In order to research on the vibration of mini-tiller, it is imperative to compare the figure 2 and figure 3.

For the first order vibration mode, it is seen that the vibration of rack in figure 2(a) is around Y-axis, and the maximum displacement is 24.15mm at the end of handle. In addition, hardly any deformation is generated for the bearing frame and dirtboard. For the model with extra mass, the more evidently flexural vibration around Y-axis can be observed in the dirtboard and handle whose maximum displacement is 23.21mm at the end of handle in figure 3(a). For the second order vibration mode, the distinct deformation is about Z-axis in figure 2(b), and the maximum displacement is 24.15mm at the end of handle. Relatively, in the figure 3(b) the large-scale vibration deformation is around Y-axis, and the maximum displacement is 30.75mm at the end of handle. For the third order mode, only evidently flexural vibration has been generated around the X-axis in the figure 2(c), and the maximum offset is 32.77mm at the distal end of handle. While in figure 3(c), the obviously flexural deformation is around X-axis and the maximum displacement is 32.75mm at the end of right handle. For the fourth order vibration mode, the modal shape is similar to the figure 2(d) and figure 3(d). For the fifth order vibration mode, the different deformation of rack is generated in the figure 2(e) and figure 3(e), but the location generated the maximum displacement is diverse. Specifically, in the figure 2(e) 22.62mm offset is produced in the two sides of dirtboard, while the maximum offset is 25.21mm shown at the distal end of handle in the figure 3(e). For the sixth order vibration mode, the evidently flexural vibration is around the Y-axis, and the maximum displacement is 27.20mm at the end of handle in the figure 2(f). Reversely, the vibration deformation of dirtboard in the figure 3(f) is similar to the fifth order mode, and the maximum is 143.69mm signified at the rim of dirtboard. For the seventh order vibration mode, the different deformation is expressed for the parts of rack, and the Z-axis deformation is generated in figure 2(g). Simultaneously, the maximum offset is 24.57mm shown at the end of handle. But in the figure 3(g), the rack is of flexural vibration deformation around Y-axis, and the maximum displacement is

66.95mm behind the left side of dirtboard. For the eighth order vibration mode, in the figure 2(h) only the dirtboard is of 145.03mm offset and other parts of rack keeps still. In the figure 3(h) the vibration deformation is analogical to the figure 2(h), while the displacement is 193.42mm shown in the two sides of dirtboard.

From the figure 2 and figure 3, the evident vibration is indicated at the end of handle, which is easy to transfer to the human body directly. As a result, the extra scale vibration will exert harmful impact on the operator's health, and structure improvement of rack for electric mini-till is urgent.

Considering the evident deformation is shown frequently in the dirtboard and the end of handle, concerned structure of rack should be redesigned. Firstly, the reinforcing rib should be used on the dirtboard. Secondly, anti-roll bar is expected to be appended in the handle. Thirdly, the cross-sectional area of handle will be enlarged appropriately. The redesigned model of rack is indicated in figure 4 based on the above design philosophy.

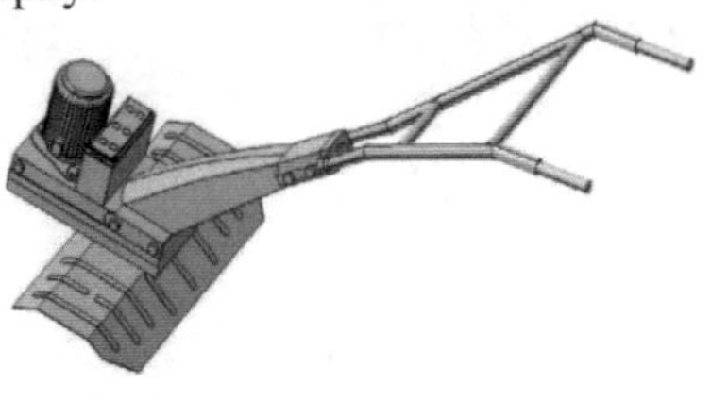

Figure 4. The redesigned model of rack

Similarly, the simulations of the modified model are showed subsequently.

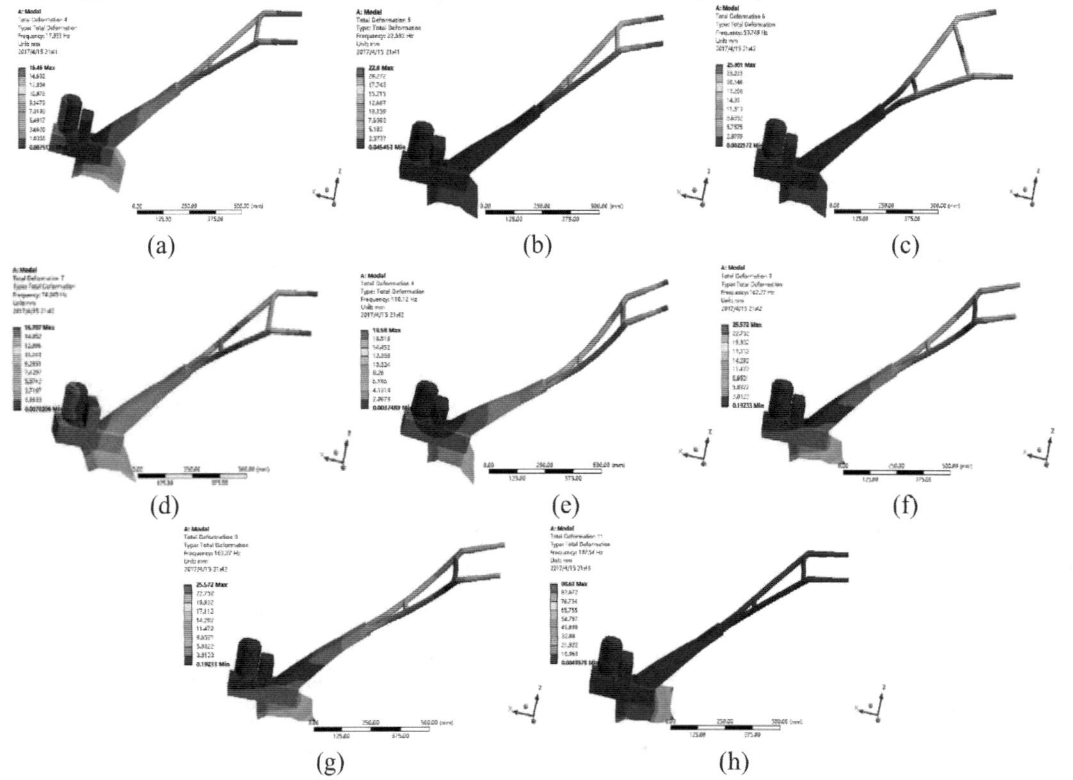

(a) (b) (c) (d) (e) (f) (g) (h)

Figure 5. The 1st~8th modal shape of optimized rack with extra mass

Compared with figure 3, the vibration deformation and maximum displacements are all decreased evidently from the first order to the eighth order modal shape. Especially for the sixth order modal shape in figure 5(f), the maximum displacement can be reduced 118.12mm, which demonstrates that the performance of rack has been enhanced distinctly.

4. Conclusions

Compared with the traditional mini-tiller, the performances including handling stability and riding comfort of the electric mini-tiller have been improved greatly, while the reduced vibration can still exert a negative impact on the driver's body, leading to the symptoms such as multiple neuritis, elevated blood pressure and changes in pulse pressure. The other influencing factors, for example, the mass of the battery, will be researched in the future. The conclusion will be list as follow

(1) The structure of rack is vital to the performance of electric mini-tiller, and the similar suspension should be used for the cutting roller to link the rack, reducing the vibration transmitting.

(2) The mass of electric motor and battery, the rotate speed of electric motor may impose the damaging influence on the property of the electric mini-tiller, and relevant research should be studied in the future.

Acknowledgements

This paper was financially supported by the Open Fund for the State Key Laboratory of Vehicle NVH and Safety Technology (NVHSKL-202006)

References

[1] Chen J., Chen C., Chen H. Three new challenges micro tillers face in southwest China and study of countermeasures. *Journal of Agricultural Mechanization Research*, 2014, 10:245-248.

[2] Ragni L. Vibration and Noise of Small Implements for Soil Tillage. *Journal of Agriculture Engineer Research*, 1999,74:403-409.

[3] Liang X. C., Chen J., Wang Z. Research on the vibration of mini-tiller. *INMATEH Agriculture Engineering*, 2018, 56(3):17-24.

[4] Liang X. C., Zhao J. D. Research on the vibration of mini-tiller. *INMATEH Agriculture Engineering*, 2019, 59(3):237-244.

[5] Niu P., Chen J., Zhao J. D., et al. Analysis and evaluation of vibration characteristics of a new type of electric mini-tiller based on vibration test [J], *International Journal of Agricultural and Biological Engineering*, 2019,12(5):106-110.

[6] Niu P., Yang M. J., Chen J. et al. Structural optimization of a handheld tiller handrail by vibration modal analysis. *INMATEH Agriculture Engineering*, 2017, 52(2):91-98.

[7] Zhang Y. H., Yang L., Niu P., et al. Study on the scoop angle characteristics of a handheld tiller's rotary blade, *INMATEH Agriculture Engineering*,2016, 49 (2):5-12.

[8] Wang Z., Chen J., Wang S. M., et al. Simulation and experimental study on the vibration response of the pre-stressed mini-tiller handle, *Journal of Agricultural Mechanization Research*,2018, 40(4):195-199.

[9] Deboli R., Calvo A., Preti C. Whole-body vibration: Measurement of horizontal and vertical transmissibility of an agricultural tractor seat. *International Journal of Industrial Ergonomics*, 2017, 58: 69-78.

[10] Xu H. B., Chen Y. J., Liu Y., et al. Vibration testing and response analysis of power tiller in non-operational state. *Journal of Chinese Agricultural Mechanization*,2016, 37(11):1-5.

Dynamic Analysis of Rudder Transmission Mechanism Considering Clearance Joint and Friction

Tao Zhang[1,2], Geng Liu[1,*], Shangjun Ma[1]

[1] Shaanxi Engineering Laboratory for Transmissions and Controls, Northwestern Polytechnical University, Xi'an, China
[2] CALTR&D Center, China Academy of Launch Vehicle Technology, Beijing, China

*Corresponding author email: npuliug@nwpu.edu.cn

Abstract. Based on the theory of multi-body dynamics, this paper proposes typical impact model and friction model to research the dynamic characteristics of the transmission mechanism with clearance joints in ADAMS softwear environment; On this basis, considering the hinge gap and friction factors under the rigid condition of the mechanism, a multi-body model is developed. Dynamic simulation analysis has studied the influence of five hinge gaps and three friction coefficients on the speed and output displacement of the rudder transmission mechanism. The research results lay an important foundation for the further development of the rigid-flexible coupling and thermal-mechanical coupling dynamic characteristics of the rudder transmission mechanism base.
Keywords: Dynamic analysis; Rudder Transmission Mechanism; Clearance Joint.

1. Introduction

The rudder transmission mechanism is the key servo system to ensure the flight quality of the aircraft and the precise control of the flight attitude. The pros and cons of its motion characteristics are affected by many factors [1]. Among them, the spatial layout of the rudder surface transmission mechanism, that is, the size of the space position, is one of the key factors restricting the improvement of its transmission performance. Existing studies mostly focus on considering the stiffness, strength, or weight reduction of members, and optimize the design of member cross-sectional dimensions, without fully considering the effects of multivariable parameters such as rod length conditions, speed conditions, acceleration conditions, and displacement conditions [2-5]. When high precision and high-speed operation are required, the gap between the kinematic pairs and the size and shape of the gap need to be considered for the kinematics and dynamics of the steering gear. Friction and wear are inevitable during the movement. Consider the analysis of the life of components after friction and wear.

2. Construction of the Impact Force Model

The transmission mechanism of the electromechanical actuation subsystem is mainly composed of a frame, a steering gear, a connecting rod, and a rudder shaft. The main types of its motion pairs are sliding pairs and rotating pairs, and there must be gaps and frictions between the motion pairs. Both of these factors will affect the motion characteristics of the electromechanical actuation subsystem transmission mechanism. Since the sliding pair is mainly the movement between the steering gear's actuating rod and the steering gear body, this paper mainly considers the clearance and friction of the rotating pair at the joint of each link hinge. The multi-body dynamics simulation software Adams is used to analyze the clearance and friction of the electromechanical system. The influence of the motion characteristics of the transmission mechanism of the actuation subsystem.

There are two methods for calculating contact force in ADAMS, one is the compensation method; the other is the impact function method. Among them, the impact function method is to calculate the contact force between two elements by the Impact function. The contact collision model summarizes the collision process as "free movement-contact deformation". It establishes a constitutive relationship describing the force and contact deformation in the collision process by taking into account the elasticity and damping of the contact surface of the collision body [6]. In ADAMS, a nonlinear equivalent spring-damping method is used to simulate the contact process, and the model is mathematically easy to handle [7]. The normal contact force F can be expressed as:

$$F = kg^e + c\frac{dg}{dt} \tag{1}$$

Where: k is the contact stiffness of $Hertz$; e is the index of 1; g is the normal penetration distance of the contact point; c is the damping factor.

The tangential contact force F_t adopts the Coulomb friction law and can be expressed as:

$$F_t = \mu(|V_t|)sign(V_t)F \tag{2}$$

Where: V_t is the tangential speed; μ is the friction coefficient.

The definition of friction has a great influence on the result of contact force. Defining friction can make the contact force curve smoother and closer to reality. The friction force in the contact force is the product of the positive contact pressure and the friction coefficient. The friction in ADAMS contact force is a nonlinear friction model, which is the mutual conversion between dynamic friction and static friction according to the relative slip speed of the two contacting objects.

The compliance of the object can be approximated by associating k with the material and geometric parameters of the object; in doing so, in order to incorporate the general material constitutive relationship of the contacting object and facilitate time integration, Adams Solver (c++) uses nonlinearity The displacement-related viscous damping term expands the previous expression. Then the general form of the impact force function is:

$$F_n = Kg^e + STEP(g,0,0,d_{max},c_{max})\frac{dg}{dt} \tag{3}$$

Where, k is spring stiffness, e is shape index, $d_{\max}$ is the Maximum permissible penetration depth, and $c_{\max}$ is the maximum damping value applied when the maximum penetration depth is reached.

Adams Solver (c++) uses a relatively simple velocity-based contact friction model. Specifying the friction behavior is optional. Fig. 1 below shows the variation of the friction coefficient with the slip speed, which can be written as

$$F_T = -\mu(v_T)F_N sign(v_T) \tag{4}$$

where $\mu(v_T)$ is the dynamic friction coefficient, which is written as

$$\mu(V_T) = \begin{cases} \mu_d sign(v_T) & |v_T| > v_d \\ -\left\{\mu_d + (\mu_s - \mu_d)\left(\frac{|v_T| - v_s}{v_d - v_s}\right)^2\left[3 - 2\left(\frac{|v_T| - v_s}{v_d - v_s}\right)\right]\right\}sign(v_T) & v_s \le |v_T| \le v_d \\ 2\mu_d\left(3\left(\frac{v_T + v_d}{2v_d}\right)^2 - 2\left(\frac{v_T + v_d}{2v_d}\right)^3 - 0.5\right) & |v_T| < v_s \end{cases} \tag{5}$$

where, v_s and v_d represent critical velocities of static friction and dynamic friction, μ_s and μ_d are coefficients of static friction and dynamic friction, respectively.

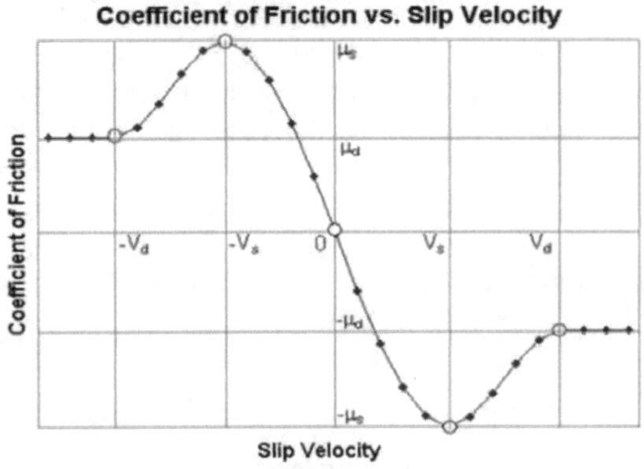

Figure 1. Sliding speed diagram.

3. Three-dimensional Model of Rudder Transmission Mechanism

3.1. Organization Diagram

The three-dimensional rudder surface transmission mechanism model is shown in Fig. 2.

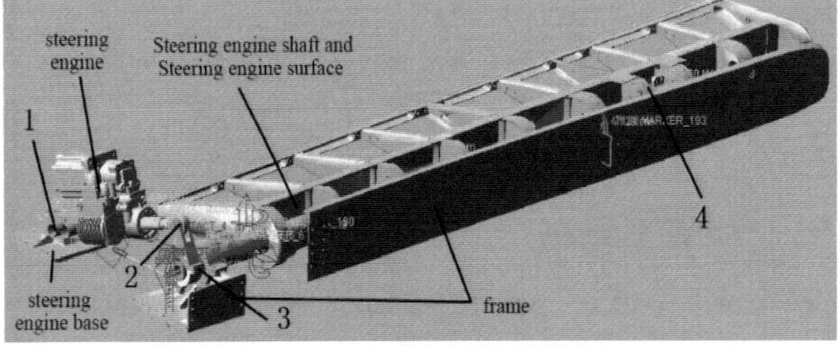

Figure 2. Three-dimensional solid analysis model.

As shown in Fig.2, point 1 is the connection between the rear end of the steering gear and the base of the steering gear, point 2 is the connection between the front end of the steering gear and the rocker arm, and point 3 is the connection between the rudder shaft and the bearing near the steering gear, as well as the connection between the rudder shaft and the bearing away from the steering gear.

3.2. Definition of Driving Parameters

In order to facilitate the analysis of the influence of different clearances and friction coefficients on the movement characteristics of the transmission mechanism, set the design variables DV_1 to drive the diameter of 1 pin, DV_2 to drive the diameter of the pin at 2 locations, DV_3 to drive the diameter of the rudder shaft at 3 locations, and DV_4 to drive 4 locations. The diameter of the rudder shaft, DV_5 drive friction coefficient. The process of defining design variables is: Build-Design Variable-New.

3.3. Definition of Constraints

The fixed pair is defined between the steering gear base and the frame, the frame defines the fixed pair, and the moving pair is defined between the steering gear rod and the servo body. Definition of the moving pair: First, establish two mark points on the actuating rod along the length of the rod, and set the position of the moving pair on one of the mark points and point to the other mark point. After the definition of the moving pair is completed, ensure that the Z axis of the two mark points of the moving pair are along the direction of the actuating rod.

3.4. Definition of Contact

The contact definition dialog box is shown in Fig. 3.

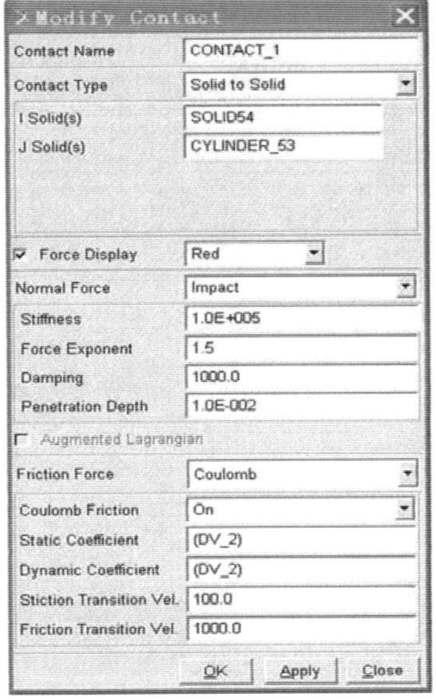

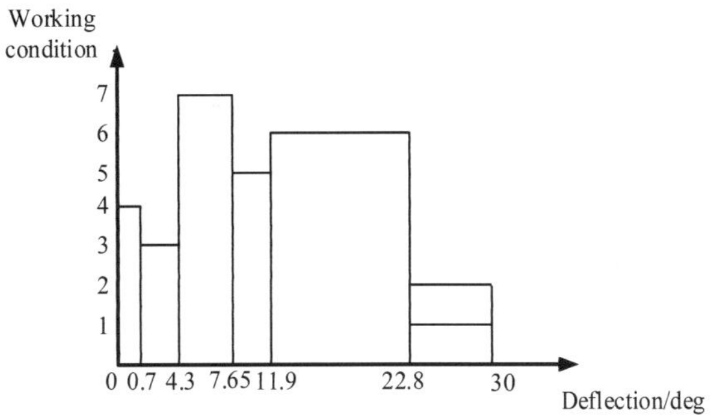

Figure 3. Contact definition dialog box. **Figure 4.** Rudder surface loading method.

The friction coefficient is the value given in the technical requirements, where the friction coefficient is driven by the design variable DV_2. In order to facilitate the analysis of the influence of the friction coefficient on the motion characteristics, the dynamic and static friction coefficients are set to the same value.

In this model analysis, eight contacts are defined, namely: steering gear tail bearing and pin, steering gear tail bearing and steering gear seat mounting lugs, steering gear front end bearing and pin, steering gear front end bearing and rocker arm Install lugs, two bearings and rudder shaft, two bearings and corresponding limit sleeves.

3.5. Definition of Drive and Load

Add displacement drive on the moving pair between the steering gear actuator and the steering gear body. According to the given steering gear motion parameters, the driving function is the deflection angle of the corresponding rudder surface is 0~30°.

Due to the fluctuation of the load on the rudder surface, different load conditions are presented. In this simulation analysis, seven operating conditions are set. When loading, it can be loaded in the neighborhood corresponding to the seven working conditions. The loading mark point is fixed on the rudder surface, and the mark point rotates with the rudder surface. The direction of the coordinate axis is consistent with the given technical requirements, and the specific loading method is shown in Fig. 4.

4. The Effect of Clearance on the Movement Characteristics

When analyzing the influence of the clearance on the motion characteristics, the friction coefficient (DV_5) is taken as 0.08, and the maximum clearance of each fit is taken as the standard clearance, and then the clearance of 0.8, 1.2, 1.4 and 1.6 times of the clearance on the motion characteristics of the transmission mechanism is analyzed respectively. Influence. Take the calculated reference clearance as the maximum clearance δ of the fit. According to the given fit tolerance, the values of each design variable can be obtained as shown in Tab.1.

Table 1. Value combinations of design variables.

	δ	0.8δ	1.2δ	1.4δ	1.6δ
DV_1	19.987	19.9896	19.9844	19.9818	19.9792
DV_2	19.987	19.9896	19.9844	19.9818	19.9792
DV_3	50	50	50	50	50
DV_4	24.957	24.9656	24.9484	24.9398	24.9312

(1) Simulate the system under ideal conditions (no gap and no collision between rigid bodies) and 1.2δ respectively, and the velocity curves and output displacement curves in the two cases can be obtained as shown in Fig. 5 and respectively. As shown in Fig. 6.

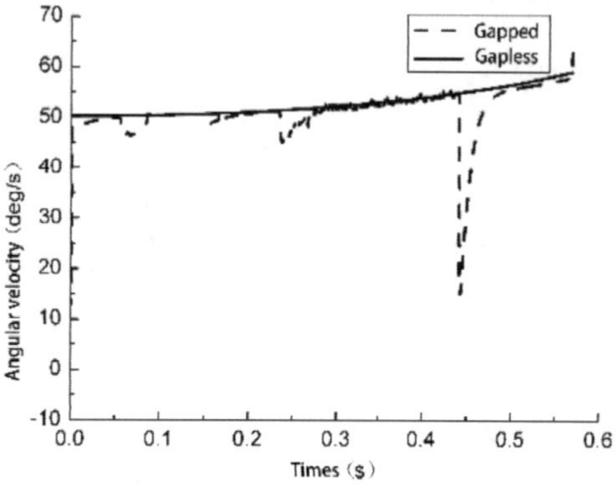

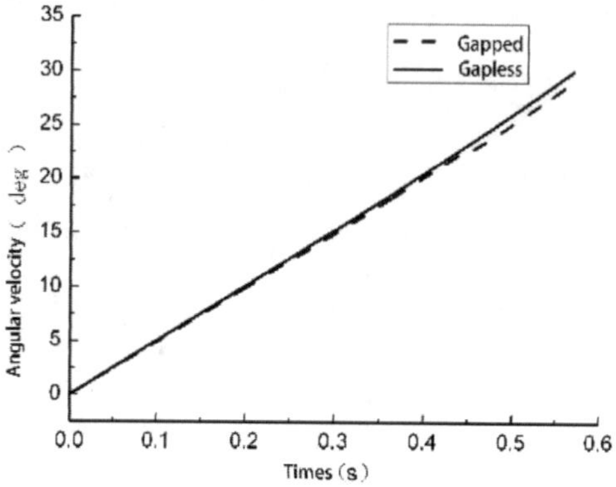

Figure 5. Angular velocity comparison curve. **Figure 6.** Angular displacement comparison curve.

It can be seen that, compared with the theoretical situation, when the hinge has a gap, the output angular velocity fluctuates, especially when the load changes, the collision between the hinges occurs, the fluctuation is relatively severe, the amplitude of the fluctuation is large, and due to the existence of the gap, Make the actual output angular velocity of the actuation system slightly lower than the theoretical value. The output angular displacement curve is basically the same in the two cases, but there is a small deviation. This is due to the existence of the gap. When the actuator has not really acted, the rudder shaft will move under the external action, resulting in a short time lag.

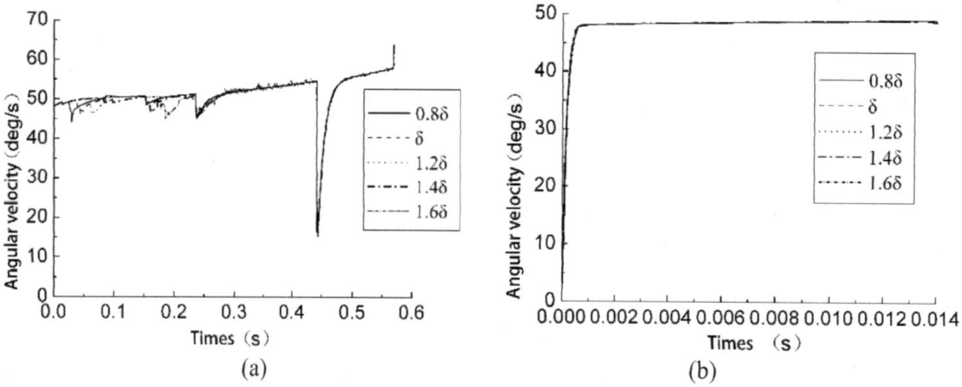

Figure 7. Numerical results with different clearance size: (a) Angular velocity curve; (b) Enlarged view.

(2) Under different clearances, the output angular velocity curve of the transmission mechanism is shown in Fig. 7, which describes the angular velocity change curve under 5 kinds of clearances. The load corresponds to working condition 4 in Fig. 4.

From Fig.7, it can be seen that as the gap increases, when the simulation starts, the appearance of the speed fluctuation peak is delayed in time. However, within the gap range of the object analyzed in this question, the effect of gap size on the degree of velocity fluctuation does not appear to be an obvious rule.

(3) Under the same friction coefficient and different clearances, the angular displacement curve output by the system is shown in Fig.8. The load corresponds to working condition 4 in Fig.4.

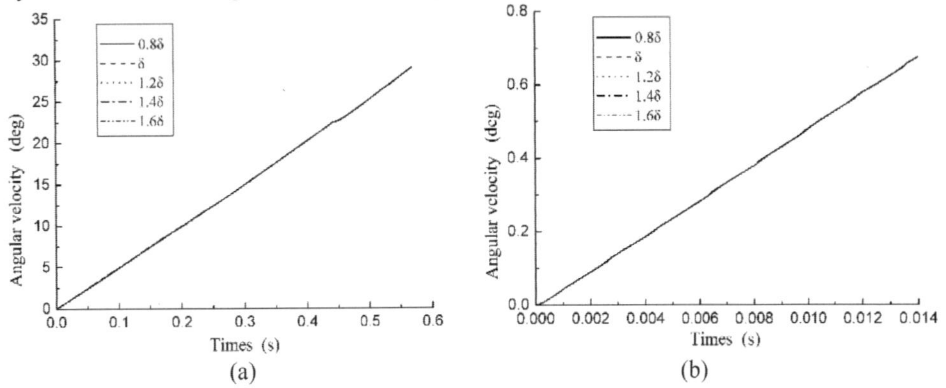

Figure 8. Numerical results with different clearance size: (a) Angular displacement curve; (b) Enlarged view.

It can be seen from Fig.8 that for the configuration shown in Fig.2, within the analyzed gap range, as the gap increases, the output angular displacement of the actuation system does not change significantly, and the output angular displacement the curves basically coincide.

5. The Influence of Friction Coefficient on Motion Characteristics

When analyzing the influence of friction coefficient on the motion characteristics, the gap at the hinge is fixed to the maximum gap given by the fit, that is, δ (DV_1, DV_2, DV_3 and DV_4 are: 19.987, 19.987, 50, 24.957, respectively). Then respectively analyze the movement characteristics of the transmission mechanism when the friction coefficient (DV_5) is 0.06, 0.08 and 0.1.

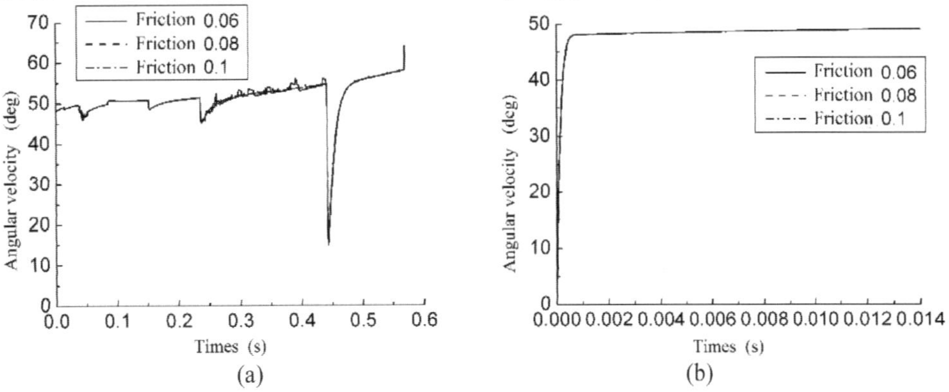

Figure 9. Numerical results with different clearance size: (a) Output angular velocity curve; (b) Enlarged view.

(1) Under different friction coefficients, the output angular velocity curve of the transmission mechanism is shown in Fig.9, which is the angular velocity change curve under three friction coefficients. The load corresponds to working condition 4 in Fig.4. It can be seen from Fig.9 that the friction coefficient will affect the fluctuation of the angular velocity. As the friction coefficient increases, the degree of fluctuation of the angular velocity curve decreases and the peak value of the fluctuation also decreases.

(2) Compare the output displacement of the transmission mechanism under the conditions of a fixed gap and different friction coefficients, as shown in Fig.10. The load corresponds to working condition 4 in Fig.4, and the displacement curve under different friction coefficients is shown in Fig.10. Under different friction coefficient conditions, the output angular displacement curves basically coincide. The output angular displacement of the dynamic system has less influence.

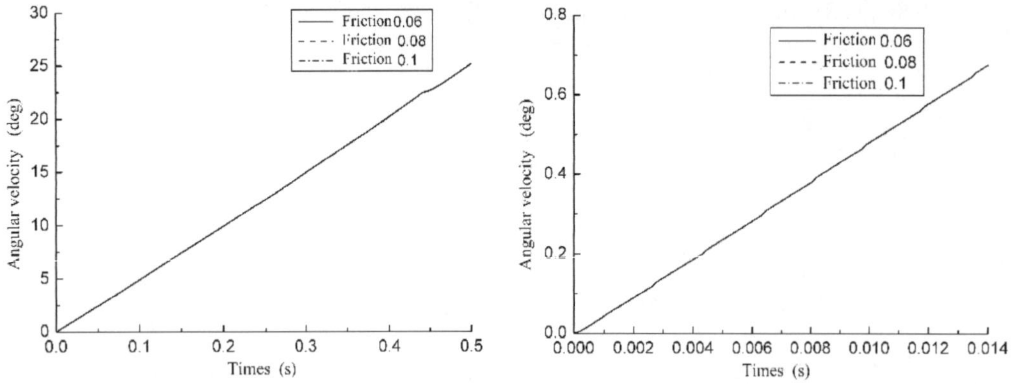

(a) (b)

Figure 10. Numerical results with different clearance size: (a) Output angular velocity curve; (b) Enlarged view.

6. Summary

Based on the theory of multi-body dynamics, this paper proposes a parametric optimization model modeling method. Taking the aileron rudder surface transmission mechanism of an aircraft as an example, the transmission mechanism configuration is optimized in size, which is suitable for multiple rudder surface configurations. Type size optimization provides an effective means. On this basis, considering the connection hinge gap and friction factors under the condition of mechanism rigidity, a multi-body dynamics simulation analysis was carried out, and the influence of 5 hinge gaps and 3 friction coefficients on the speed and output displacement of the rudder transmission mechanism was studied.

The research results show that: (1) Compared with the ideal state (no gap and no collision between rigid bodies), when the hinge has a gap, the output angular velocity fluctuates, especially when the load changes, the collision between the hinges will cause fluctuations. It is relatively severe, and the fluctuation amplitude is relatively large. (2) Under different gaps, as the gap increases, the appearance of the peak velocity fluctuations in the initial stage of the simulation will be delayed in time. (3) Under the same friction coefficient and different gaps, within the analyzed gap range, as the gap increases, the output angular displacement of the actuation system does not change significantly, and the output angular displacement curves basically coincide. (4) The coefficient of friction will affect the fluctuation of angular velocity. As the coefficient of friction increases, the degree of fluctuation of the angular velocity curve decreases and the peak value of the fluctuation decreases. (5) For a fixed gap, under different friction coefficient conditions, the output angular displacement curves basically coincide. Within the analyzed friction coefficient range, the friction coefficient has little effect on the output angular displacement of the actuating system.

Reference

[1] Jiang Hongchao, Liu Shirong, Zhang Botao. Inverse kinematics analysis of six degree of freedmodular manipulator [J]. Journal of Zhejiang University: Engineering Edition, 2010 (7): 1348-1354

[2] Chen Xiao, Deng pan, Anning, et al. Synchronous control technology of double cylinder propulsion marine electric actuator [J]. Marine engineering, 2018, 40 (5): 68-71, 76

[3] Xie Junchao, Xiao Qing, Hua Jing. Feasibility study of new marine electric actuator [J]. Ship science and technology, 2013, 35 (10): 78-81

[4] Yan Feifei, Chen Shengdong, Liu Yali. Design of ship electric actuator control system based on ADRC [J]. Ship science and technology, 2013, 35 (12): 88-91

[5] PAN Dafu. Modeling and control algorithm of electric actuator system [J]. Journal of Sichuan ordnance industry, 2013, 34 (9): 1-4

[6] Xie Zuiwei, Wu Xinyue. Collision simulation analysis based on ADAMS [C]. // Proceedings of the third China Annual Conference on CAE engineering analysis technology and 2007 National Advanced Symposium on Computer Aided Engineering (CAE) technology and application. Dalian: 2007:345-348

[7] LAN fengchong, Cai Zhihua, Chen Jiqing, et al. Biomechanical response and injury evaluation of chest and abdomen in automobile collision [J]. Journal of South China University of Technology (NATURAL SCIENCE EDITION), 2012, 40 (12): 70-78

[8] Zhou yinglu, Wang Zhiliang, Zhu Songqing, et al. Design of multi-channel actuator controller based on pca9685 [J]. Journal of Nanjing Institute of Engineering (NATURAL SCIENCE EDITION), 2017 (4): 26-31

[9] Zhang Puxing, Jia Qiuling. Design of multi-channel actuator controller for manipulator based on FPGA [J]. Micro special motor, 2011 (4): 73-75

[10] Xie Junchao, Xiao Qing, Hua Jing. Feasibility study on new marine electric actuator [J]. Ship science and technology, 2013-35 (10): 78-81

[11] Fu Yongling, Han Xu, Yang Rongrong, et al. Overview of design methods of electro-hydraulic actuators [J]. Journal of Beijing University of Aeronautics and Astronautics, 2017, 43 (10): 1939-1952

Research on Probability Analysis Technology of Circular Section Cabin Socket Structure Based on APDL

Long Yi*, Yuliang Li, Henggui Wu, Yukun Zhao and Maoping Zhong

Jiangxi Hongdu Aviation Industry Group Co., Ltd., Nanchang, China

*Corresponding author email: 34016@nchu.edu.cn

Abstract. In the strength design of sleeve joint structure, manufacturing tolerance and fit clearance are generally ignored. The load of connecting screw is usually calculated by the method of linear distribution. The contact force between screw and cabin is usually simulated by deterministic finite element method. The command flow of probability analysis for typical circular section socket structure is developed based on APDL. The uncertainty of manufacturing tolerance and fit clearance for socket structure is defined as random input variables, such as the cabin diameter, cabin hole position, and cabin aperture and so on. Monte Carlo method is used in probability analysis, and the input variables are random uniform distribution. The analysis results show that the transfer load of screws is redistributed under the probability analysis compared with the finite element analysis, and some screws bear larger load, while the nearby screws bear smaller load. At the same time, the load transferred by screw is not only affected by its own variables, but also by others. The probability analysis method can fully consider the influence of manufacturing tolerance of the structure, and evaluate the structural strength margin more accurately. It can not only prevent the harm caused by uncertainty of random manufacturing tolerance, but also prevent over design caused by conservative method.
Keywords: Sleeve structure; Uncertainty; Probability analysis; APDL.

1. Introduction

Sleeve connection has the advantages of simple structure, good openness and easy to process, which is widely used in the docking structure of circular section cabin with small force and sufficient stiffness[1]. In the preliminary design stage, the load of the connecting screws for socket structure is generally calculated by the method of linear distribution. The screw shear force, the cabin extrusion and shear stress are checked respectively. In the detailed design stage, the deterministic finite element method is generally used to simulate the contact relationship between the screw and the cabin. The stress state of the screw is obtained, as well as the local stress distribution of the cabin[2 ~ 5]. However, due to a large number of contact surfaces, the manufacturing tolerance and fit clearance are generally not considered in the calculation process[6 ~ 8].

When two sections are connected, due to the manufacturing tolerance and assembly clearance, the possible deviations include relative torsion, bending and dislocation. Aiming at the uncertainty of manufacturing tolerance and fit clearance of sleeve joint structure, the command flow of probability analysis is developed based on APDL, and the random uniform distribution model for clearance is adopted. This method can be used to evaluate the structural safety more accurately.

2. ANSYS Probabilistic Design System

The probabilistic design system (PDS) of ANSYS can be used to determine the influence of single or multiple random variables on the analysis results, and provides a set of practical tools to improve the efficiency of probabilistic design. We can graphically display the influence of an input variable on an output variable, and can easily add custom samples and additional analysis loops to obtain further analysis results. Figure 1 shows the process of PDS design analysis. The analysis file must be input independently, and the probability design database is independent of the model database.

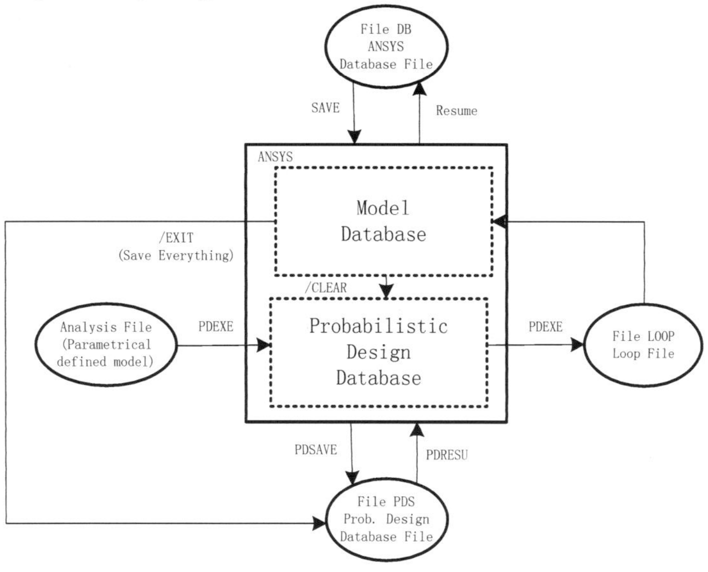

Figure 1. ANSYS Probabilistic design data flow.

3. Parametric Modeling

The typical sleeve structure of round section cabin is shown in figure 2. The connecting frames of two adjacent cabin sections are respectively processed as the inner and outer surfaces of the cylinder. The sleeve is alignmented with their mating surfaces, and then fixed with radial screws along the circumference.

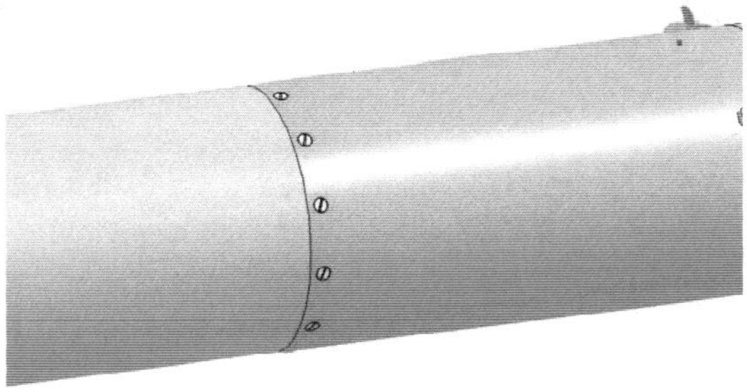

Figure 2. Structural model of cabin socket.

The load distribution of socket connection is complex, which is not only affected by the position and number of screws, but also related to the installation clearance, the stiffness of the connection frame and many other factors. When two sections are connected, due to the manufacturing tolerance and assembly clearance, the possible deviations include relative torsion, bending and dislocation. In view of the uncertainty of manufacturing tolerance and assembly clearance, this paper uses ANSYS parametric design language (APDL) to carry out parametric modeling of socket cabin. The connecting screw and the rear cabin are tight fit, which is simplified as the overall structure. The clearance fit is between the connecting screw and the front cabin, and between the front cabin and the rear cabin. The structure of each part is shown in figure 3 ~ figure5.

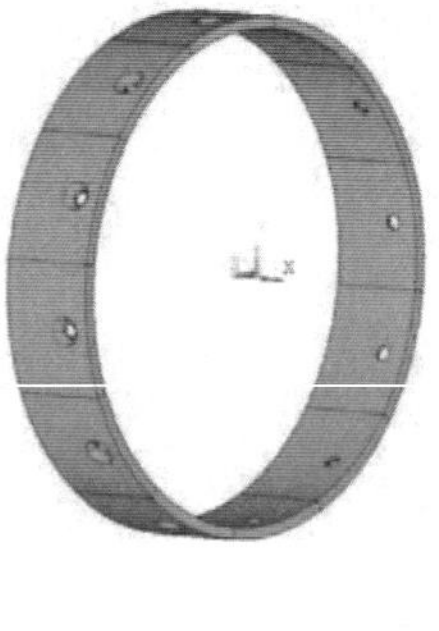

Figure 3. Front cabin structure.

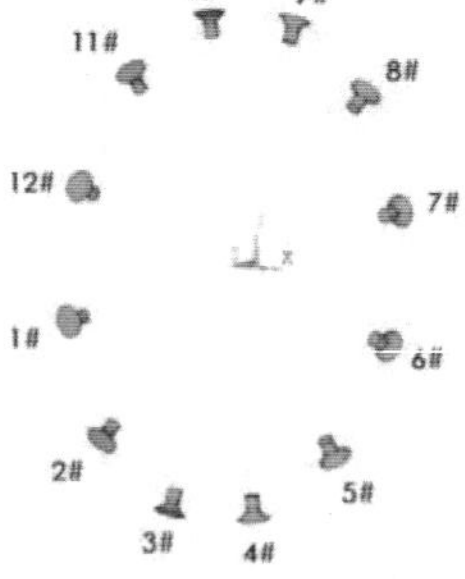

Figure 4. Connecting screw.

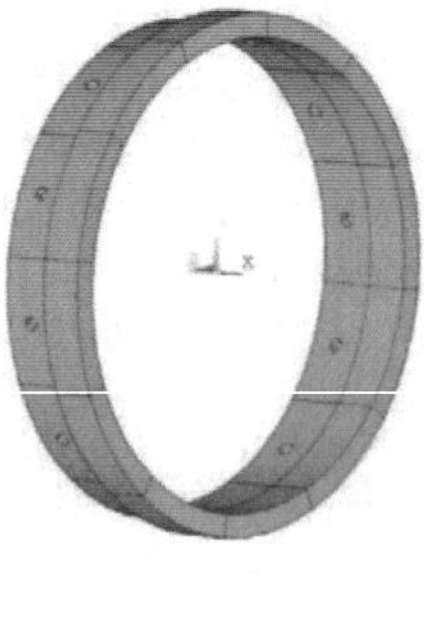

Figure 5. Rear cabin structure.

4. Results of Finite Element Analysis

The cabin material is defined as TC4 and the screw material is 30CrMnSiA. The finite element grid of the structure is divided. The contact relationship is defined between the mating surface of the front cabin and the rear cabin, and between the extrusion mating surface of the screw and the screw hole, as shown in figure 6. There are a total of 425000 nodes, 357000 solid elements and 49000 contact elements. At the same time, the geometric plane at the end of the rear cabin is used to impose fixed constraints. A concentrated mass element and coupling relationship are used to impose shear and bending loads at the end of the front cabin, as shown in figure 7.

Figure 6. Finite element model of cabin socket structure.

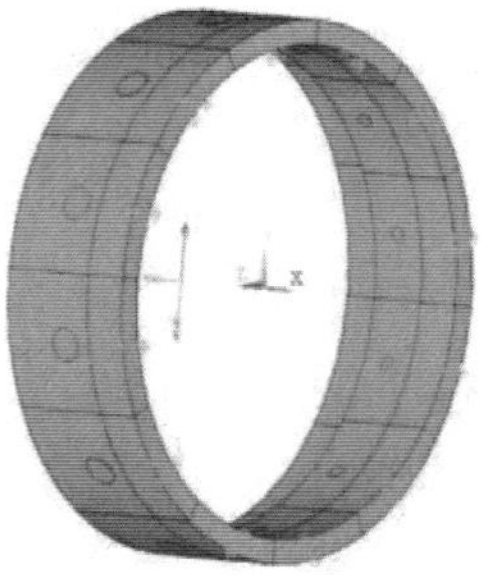

Figure 7. Constraints and loads.

The finite element analysis of the structure is carried out without considering the manufacturing tolerance. The equivalent stress of the structure is shown in figure 8 ~ figure 10. It can be seen that due to the influence of bending moment, the load of the screw presents a directional difference, and the stress at the hole edge also presents the same trend. The maximum stress of the screw is 867mpa under shear and bending load, and the maximum stress of the screw is 659mpa.

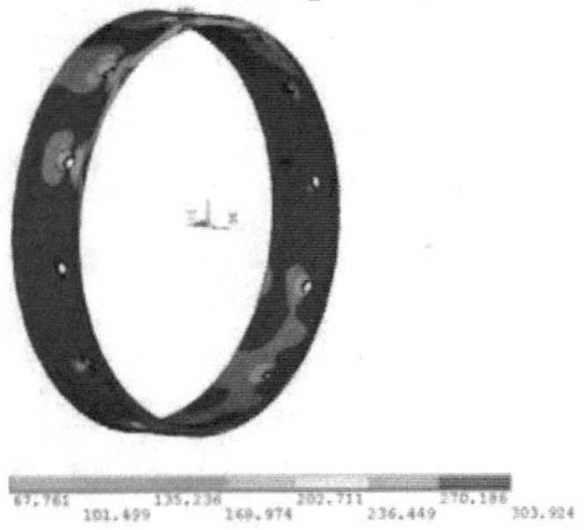

Figure 8. Front cabin stress.

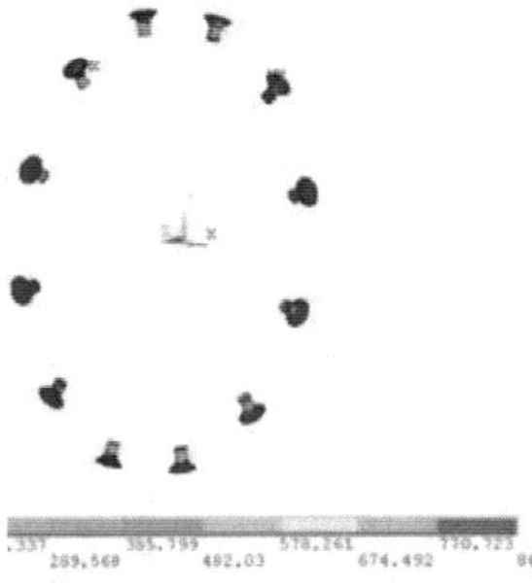

Figure 9. Connecting screw stress.

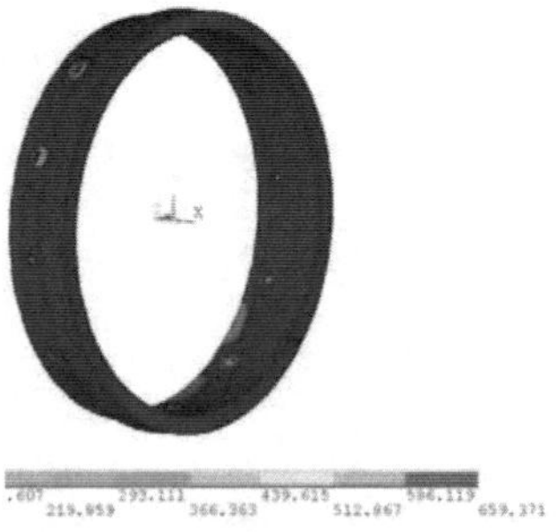

Figure 10. Rear cabin stress.

5. Probability Analysis Model

According to the requirements of manufacturing tolerance, cabin diameter, angle and position of screw hole in front and rear cabin, diameter of screw hole in front cabin are selected as random input variables, with a total of 65 items, as shown in table 1. The angle tolerance of each screw hole is the angular position deviation of the hole center along the circumferential direction of the cabin, and the position tolerance of each screw hole is the axial position deviation of the hole center along the cabin. According to the actual manufacturing tolerance range, each dimension tolerance is randomly and evenly distributed. Of course, we can also measure the actual size of the parts, and use the measured value as the input variable to establish the probability analysis model.

Table 1. Random variables of manufacturing tolerance.

Tolerance variable description	Variable name	initial value	Variable range	Unit
Outer diameter of front cabin	GC_wai_W	0	-0.1～0.1	mm
Inner diameter of front cabin	GC_wai_N	0.03	0～0.065	mm
Outer diameter of rear cabin	GC_nei_W	0	-0.1～0.1	mm
Middle diameter of rear cabin	GC_nei_Z	0.13	0.1～0.175	mm
Inner diameter of rear compartment	GC_nei_N	0	-0.1～0.1	mm
Front cabin No. 1 aperture	GC_Kong_DW1	0.015	0～0.03	mm
Front cabin No. 1 Hole location	GC_Kong_LW1	0	-0.1～0.1	mm
Rear cabin No. 1 Hole location	GC_Kong_LN1	0	-0.1～0.1	mm
Angle of No.1 hole in front cabin	GC_Kong_JW1	0	-5～5	'
Angle of No.1 hole in rear cabin	GC_Kong_JN1	0	-5～5	'
...	...		...	...
Front cabin No. 12 aperture	GC_Kong_DW12	0.015	0～0.03	mm
Front cabin No. 12 Hole location	GC_Kong_LW12	0	-0.1～0.1	mm
Rear cabin No. 12 Hole location	GC_Kong_LN12	0	-0.1～0.1	mm
Angle of No.12 hole in front cabin	GC_Kong_JW12	0	-5～5	'
Angle of No.12 hole in rear cabin	GC_Kong_JN12	0	-5～5	'

The maximum equivalent stress and maximum heading displacement of the whole model, as well as the load transferred by each screw, the maximum equivalent stress of screw, the maximum equivalent stress of hole edge and the maximum heading displacement of screw are taken as random output parameters, with a total of 62 items, as shown in table 2.

Table 2. Random output variables of probability analysis.

Variable description	Variable name	Unit
Maximum equivalent stress of full model	Sall	Mpa
Maximum course displacement of full model	Wall	mm
transfers load of No.1 screw	LiDing1	N
Maximum equivalent stress of No.1 screw	Sding1	Mpa
Maximum equivalent stress at the side of rear hatch hole at No.1 screw	Sxia1	Mpa
Maximum equivalent stress of front cabin hole at No.1 screw	Sshang1	Mpa
Maximum heading displacement of No.1 screw	Wding1	mm
...	...	...
transfers load of No.12 screw	LiDing12	N
Maximum equivalent stress of No.12 screw	Sding12	Mpa
Maximum equivalent stress at the side of rear hatch hole at No.12 screw	Sxia12	Mpa
Maximum equivalent stress of front cabin hole at No.12 screw	Sshang12	Mpa
Maximum heading displacement of No.12 screw	Wding12	mm

6. Probability Analysis Results

Monte Carlo method is used for probability analysis. In order to improve the calculation efficiency, the sample space of this paper is 60. In practical application, the sample size should be increased as much as possible. The calculated values of the random samples tolerance of the outer diameter for the front cabin are shown in figure 11, and other random input variables are not described. It can be seen that the sample distribution of input variables basically conforms to the trend of manufacturing tolerance distribution.

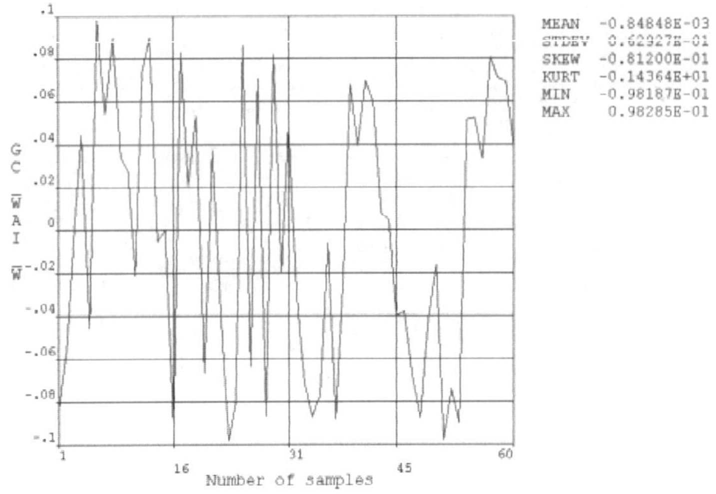

Figure 11. Random samples of tolerance of outer diameter of front cabin.

Because the load transferred by screw has a significant effect on the calculation results of hole edge stress, the calculation results of load transferred by screw are mainly analyzed. It can be seen from the comparison that the shear load transferred by No. 4 screw is the largest. The random distribution of its calculated value is shown in figure 12. Other random output variables are not described. It can be seen that the range of shear load transferred by No. 4 screw is large, and the maximum value of the corresponding 43 # sample is about 82% higher than the average value.

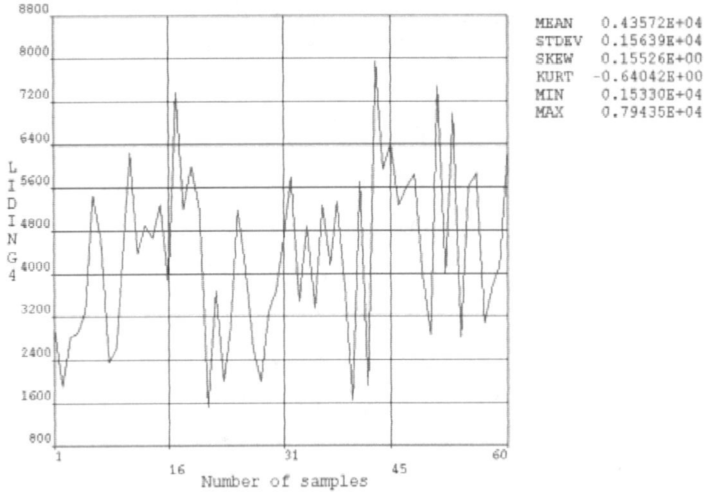

Figure 12. Calculation results of shear load transferred by No.4 screw.

The results between probability analysis (PA) and finite element analysis (FEA) of each screw of 43 samples are compared, as shown in figure 13. Using the finite element analysis method, the load symmetry and linearity of the screw are good, and the overall load value is low. Using probability analysis method, due to the random change of geometric size, the load of screws is redistributed. Individual screws bear larger load (No. 4 screw in the figure), while the nearby screws bear smaller load.

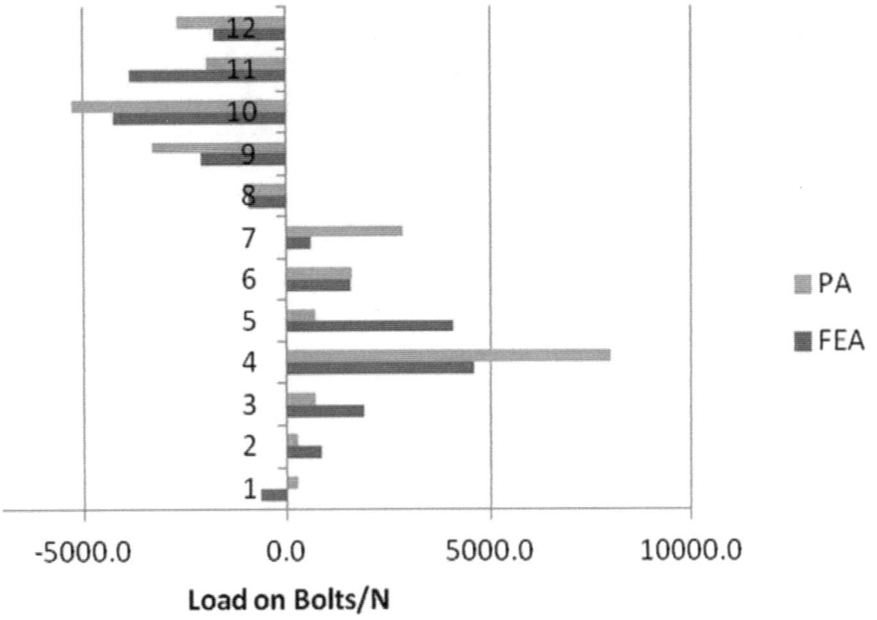

Figure 13. Screw load comparison.

The sensitivity correlation of each input variable to the shear load transferred by No. 4 screw is compared, as shown in figure 14. It can be seen that the six input variables which have significant influence on shear load transfer of No. 4 screw are: front hole tolerance of No. 4 screw, rear hole tolerance of No. 3 screw, rear hole tolerance of No. 5 screw, front hole angle tolerance of No. 9 screw, rear hole angle tolerance of No. 7 screw and rear hole angle tolerance of No. 10 screw. The sensitivity relationship among other variables is not detailed.

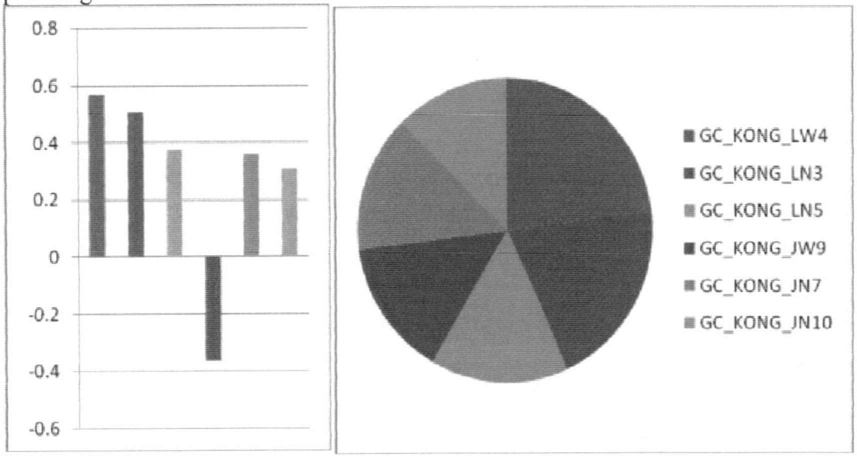

Figure 14. Sensitivity comparison of input variables to No. 4 screw.

The confidence was 95%, and the failure probability of No. 4 screw was calculated, as shown in figure 15. It can be seen that if the ultimate bearing capacity of screw is 7000N, the failure probability of screw is 6%. If the ultimate bearing capacity of screw is 7500N, the failure probability of screw is 2.8%. If the ultimate bearing capacity of the screw is 8000n, the screw is safe.

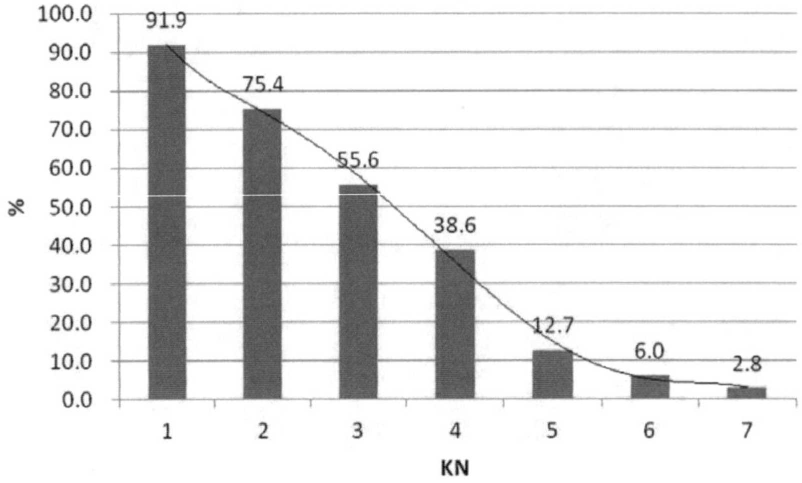

Figure 15. Screw failure probability.

7. Conclusion

In this paper, the command flow of probability analysis for typical circular section socket structure is developed based on APDL. The uncertainty of manufacturing tolerance and fit clearance for socket structure is defined as random input variables, such as the cabin diameter, cabin hole position, and cabin aperture and so on. The probability analysis technology is studied based on PDS of ANSYS. The analysis results show that, compared with the finite element analysis, the transfer load of screws is redistributed under the probability analysis, and individual screws bear larger load, while the nearby screws bear smaller load. At the same time, the load transferred by screw is not only affected by its own variables, but also by other screw variables. The probability analysis method can fully consider the influence of manufacturing tolerance on the bearing state of the structure, and evaluate the structural strength margin more accurately. It can not only prevent the harm caused by the uncertainty of random manufacturing tolerance, but also prevent the over design caused by conservative methods.

References

[1] Ge Jinyu 1986 *Structural design principle of winged missile* (Beijing: National Defense Industry Press)
[2] Yu Xudong 2007 *Missile modern structure design* (Beijing: National Defense Industry Press)
[3] Fan Donglin 2012 *Research on the localization of a certain type of air-to-air missile wedge* (*Aviation maintenance and engineering* vol 1) pp 74-76
[4] Fan Huitao 2013 *Air to air missile scheme design principle* (Beijing: Aviation Industry Press)
[5] Editorial board of China Aviation Materials Manual 2001 *China Aviation Materials Manual* (Beijing: China Standards Press)
[6] Game studio 2005 *APDL parametric finite element analysis technology and its application examples* (Beijing: China water resources and Hydropower Press)
[7] Compilation group of missile structural strength calculation manual 1978 *Missile structural strength calculation manual* (Beijing: National Defense Industry Press)
[8] Zhou Hong 2011 *Sensitivity analysis on strength reliability of a missile cabin connection structure* (*Solid rocket technology* vol 34) pp 768-771

Design and Experiment Study on Three-dimensional Rotating Ultrasonic Composite Generating Electrolysis Machining

Wanwan Cheng, Jing Li, Zhenghui Ge and Yongwei Zhu*

College of Mechanical Engineering, Yangzhou University, Yangzhou, China

*Corresponding author email: ywzhu@yzu.edu.cn

Abstract. Proposed a multi-dimensional rotating ultrasonic composite electrochemical generating machining method. Designed and built the shaft and radial three-dimensional rotary ultrasonic composite electrolytic generating machining system, analyzed and studied the machining mechanism according to the interaction relationship between rotary ultrasonic, mechanical and electrolysis.Selecting and optimizing the machining parameters for to machine plane by grinding, ultrasonic and rotary ultrasonic composite electrolytic machining. By the tests, the machining system is stable and reliable, and the rotary ultrasonic composite electrolytic generative machining technology could effectively improve the machining efficiency and accuracy. Adding 3μm-5μm radial ultrasonic vibration, carried on the rotating ultrasonic generating machining the plane and cutting the grooves, the technical advantages of two-dimensional rotary ultrasonic composite electrochemical machining are verified.
Keywords: Ultrasonic machining; Multi-dimensional rotating ultrasound; Composite electrochemical machining; System design; Experiment research.

1. Introduction

With the development of science and technology, the applications of various difficult-to-machine materials and parts with special-shaped surfaces have become more and more extensive. These materials usually have the mechanical properties of high hardness, high strength, and high brittleness. Traditional mechanical machining is difficult to meet the machining requirements, especially in terms of complex micro structure and good machining surface quality. The new non-traditional machining (NTM) technologies are very useful to machine these parts for the advantages such as not being affected by material hardness, small machining force, and low tool loss [1-3].

Electrochemical machining(ECM) and electric discharge machining(EDM) are effective machining ways for the conductive parts with special-shaped surfaces and difficult-to-machine materials. Ultrasonic machining(USM) is an effective machining method on hard and brittle materials , but the ultrasonic tool has severely worn and it is difficult to compensate online. In addition, EDM, ECM, and USM method are copy forming, and the electrode manufacturing is difficult[4-6]. So, it is important and urgent that study the composite machining technology keeping the advantages of single machining methods and quitting the disadvantages. Chemnitz University of Technology in Germany[7] found that machining efficiency of the ultrasonic-assisted machining is 40% higher than that of the single EDM by comparing the EDM tests with or without ultrasonic assist in the machining of micro-holes. Krefting D[8] used the three-dimensional surface roughness and surface fractal dimension to characterize the machining surface quality of SiCp/Al composites, the results show that the main factor affecting the surface roughness is the spindle speed, followed by vibration amplitude, cutting depth and feed rate. Deng Shihui[9] of Shandong University used the method of combining high-speed rotating spiral electrodes with ultrasonic vibration assisting to explore the influence on the milling

accuracy and processing quality of high-frequency pulse power micro electrodes, the spiral cylindrical micro electrodes with a diameter of 100μm are used in the thickness micro-grooves, and three-dimensional structures are milled on a 510μm stainless steel plate. Liu Zexiang[10] from Nanjing Agricultural University studied the composite machining of small holes by rotating ultrasonic electrolysis, and the results showed that both cathode rotation and cathode ultrasonic vibration are beneficial to the bubbles leaving the anode surface and promote the discharge of electrolyte, the rotating ultrasonic electrolytic composite machining has the higher material removal rate.
This paper designs and constructs the shaft and radial three-dimensional rotating ultrasonic composite ECM system, analyses and studies the machining mechanism. The main machining parameters are selected and optimized, comparative tests for the plane machining and the groove cutting are carried.

2. Design of Three-dimensional Rotating Ultrasonic Composite Generating ECM System

Figure 1(a) shows the diagram of the rotating ultrasonic composite generating electrolysis machining system. The machining system is composed of a displacement motion platform, a rotating ultrasonic vibration device, a pulse power supply of electrolysis, the ultrasonic generator and control system. The ultrasonic vibration system is composed of an ultrasonic transducer, an horn and a tool head. The tool head performs ultrasonic vibration while rotating at high speed, and the *Z*-direction can move vertically. The pulse electrolysis power supply and the ultrasonic generator provide the electric energy and ultrasonic energy required by the machining process.
The displacement motion platform drives the workpiece to move linearly in the *X* and *Y* directions. Figure 1(b) is the rotary ultrasonic electric spindle, it is composed of an ultrasonic transducer, an horn and a tool cathode, so that the tool cathode can perform ultrasonic vibration while rotating at a high speed, and the *Z*-direction feed movement at the same time. The rotary ultrasonic electric spindle rotates and the axial ultrasonic frequency vibrates, and the ultrasonic vibration device along the electrode radial direction is added in the *X* and *Y* directions, which can realize two-dimensional or three-dimensional ultrasonic composite generating ECM machining.

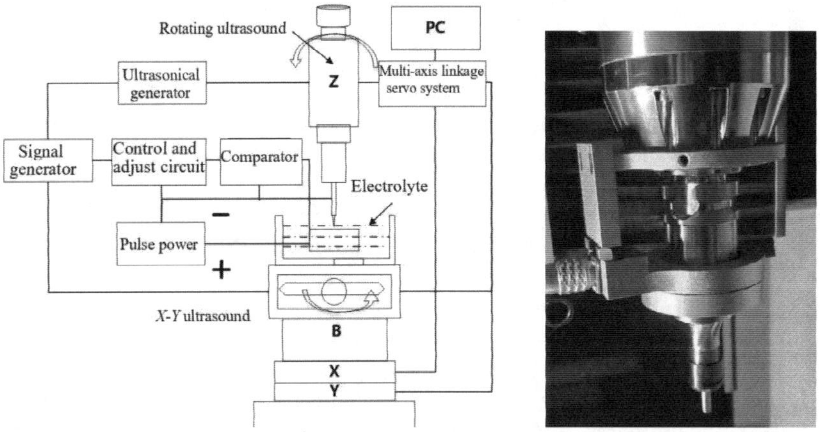

(a) Rotary ultrasonic composite ECM system Diagram (b) Rotary ultrasonic electric spindle

Figure 1. Diagram of the rotary ultrasonic composite generating ECM system.

Figure 2 is the control system of three-dimensional rotating ultrasonic composite generating ECM. The control system is composed of a control computer, an ultrasonic vibration device, a single-chip microcomputer, a pulse electrolysis power supply, three ultrasonic generators, three ultrasonic vibrate driving units, multi-axle movements control unit, three data complication interfaces.

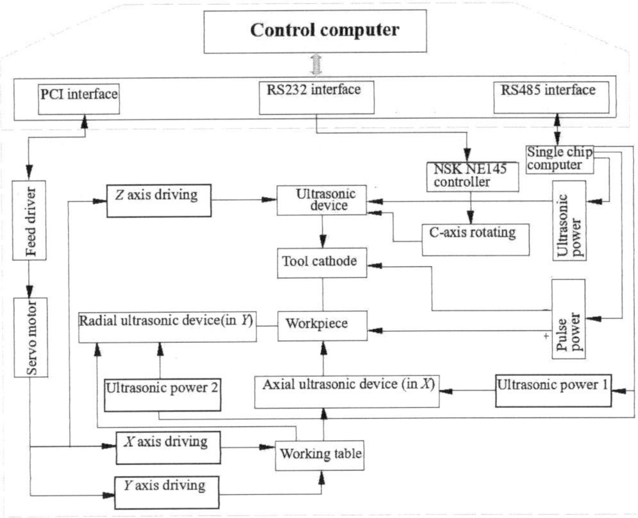

Figure 2. Diagram of three-dimensional rotating ultrasonic electrolysis control system.

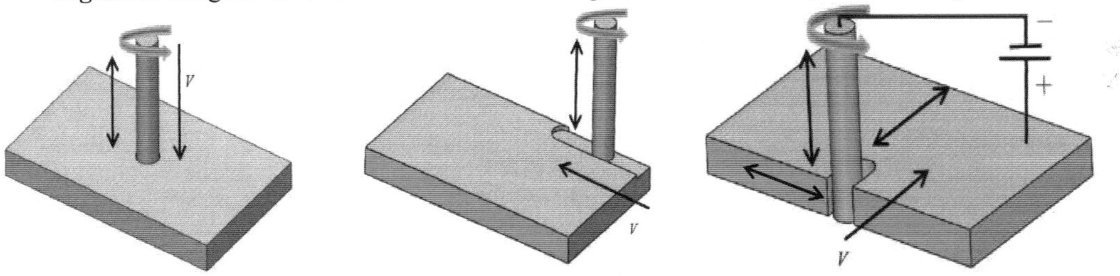

(a) Ultrasonic machining hole (b) Ultrasonic machining plane (c) 3-D ultrasonic machining

Figure 3. Scheme of ultrasonic vibrating and feed movements on the ultrasonic composite ECM.

Figure 3, the tool electrode rotates and does ultrasonic vibration relative to workpiece in the axial direction, and the radial ultrasonic can excite workpiece to vibrate in two directions, that is, two-dimensional (*Z* and *X*, *Z* and *Y*) relative to the tool, so the machining system can achieve the three-dimensional (Z, X, and Y) ultrasound. One-dimensional ultrasound is only suitable for copy forming machining, two-dimensional ultrasound can be used for cylindrical forming machining, and three-dimensional(3D) ultrasound can be used for the spatial curve generating machining.

The Figure 4 shows the principle of rotary ultrasonic composite generative ECM. The workpiece is connected to the positive electrode of the power source, the tool electrode is connected to the negative electrode of the power source, and the electrolytic solution is added to the machining area to realize electrolytic machining. At the same time, the tool electrode also vibrates axially with the ultrasonic frequency when the main shaft rotates.

The vibration displacement is shown in Figure 4. As there are the absolute abrasive particles fixed on the end face of the tool cathode, the conductive tool end can't conduct the workpiece during the ultrasonic vibration period, so, there is a gap $L_{\min}$ about the size of the absolute abrasive grain, it can avoid short-circuit discharge caused by the large electric field intensity as the gap between the workpiece and the electrode becomes too small. In a period of ultrasonic vibration, the abrasive particles on the tool end contact the workpiece to generate mechanical cutting force, which can remove the workpiece material. The machining way mainly rely on electrolysis and mechanical grinding of abrasive particles to remove materials. Rotation ultrasonic effects remove difficult-to-machine raised points and electrolytic passivation film. At the same time, the ultrasonic effect can update the electrolyte in real time, and remove the processed products in time, the ultrasonic effect have a polishing and shaping effect on the processed surface, and improve the quality of the processed surface.

Figure 5 shows the current waveform of the three states during the machining process. There will be occasional small spark discharges during the machining process. The main reason is that hydrogen bubbles and cavitation bubbles are generated in the electrolyte to reduce the conductivity. At the same time, the surface is machined. Due to the slow removal speed caused by electrolytic passivation, some local points gradually become convex points, which causes the distance between the workpiece and the tool to become smaller and the electric field strength becomes larger, the electrolyte would broken down easily and causes local spark discharge. This phenomenon is conducive to solving for the removal of local difficult-to-machining points, the amount of material removed is small and can be ignored. The current waveform of the three machining states in machined is shown in Figure 5.

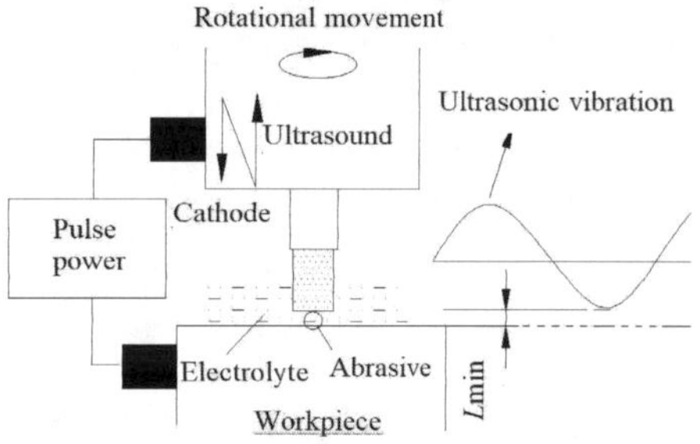

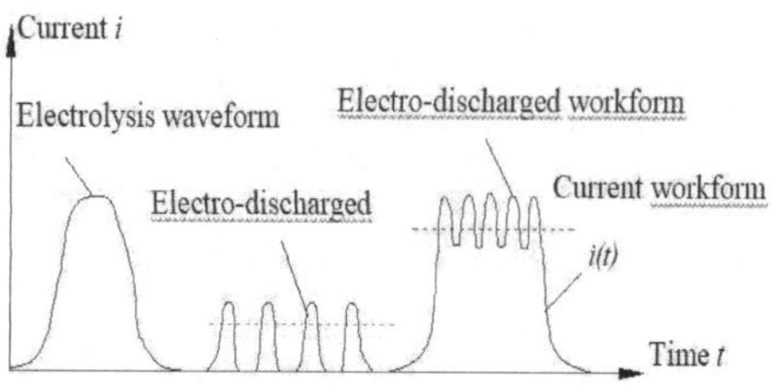

Figure 4. Diagram of rotary ultrasonic ECM. **Figure 5.** Current waveforms of the three states.

On the basis of the above machining methods, adding a computer digital control workbench to drive the workpiece to move on a specific trajectory relative to the cathode tool head, forming a rotary ultrasonic-assisted electrolytic generative machining method.

Figure 6 shows the principle of the rotating ultrasonic generative machining. The rod-shaped tool head with fixed abrasive grains rotates at high speed with the axially rotating ultrasonic device and simultaneously vibrates vertically in the vertical direction, thereby forming the required workpiece shape. Compared with ultrasonic composite copy machining, the generating machining way no longer need manufacturing complex shape electrodes, which will improve the machining flexibility greatly.

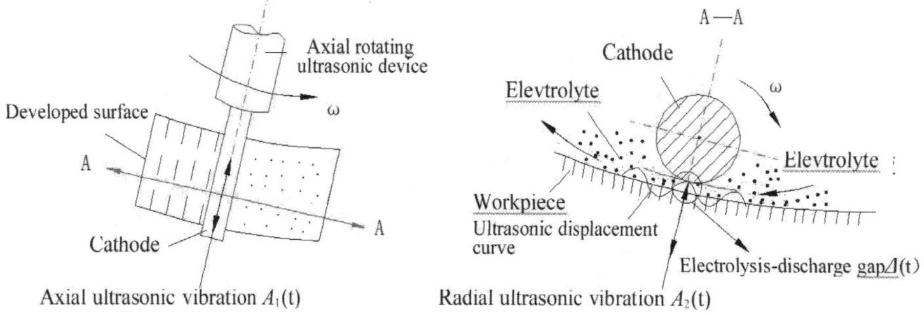

(a) Generative machining surface (b) Ultrasonic gap between cathode and workpiece

Figure 6. Diagram of rotating ultrasonic generative machining.

After the rotation movement is added the ultrasonic vibration tool cathode, the generative machining can be realized. Figure 6 is a diagram of the tool end surface generative machining. The shape of the machining depends on the generative motion trajectory and the inter-electrode gap.

When the feed speed is equal to the material removal rate, the processing state is the best and the efficiency is the highest; when the feed speed is slower than the material removal rate, the abrasive particles the mechanical cutting force is small; when the feed speed is faster than the speed of material removal, the mechanical cutting force of abrasive particles is relatively large, and the electrolytic effect is relatively weak. The forming error (machining accuracy) of the generative machining is mainly caused by the error of the generative feed motion trajectory and the inconsistency of the electrical machining gap.

3. Tests and Analyses of the Rotating Ultrasonic Generating ECM

Two-dimensional rotating ultrasonic composite evolution ECM is shown in Figure.7. The rotary ultrasonic composite ECM method uses a fixed abrasive cathode with an abrasive particle size of 100 mesh. The maximum spindle speed is 20000 r/min, the maximum ultrasonic vibrating amplitude is 20 μm, and the maximum output voltage of the power supply is 30 V, choosing 4V-6V for to the tests. The current and voltage signals can be measured by the digital storage oscilloscope in real time.

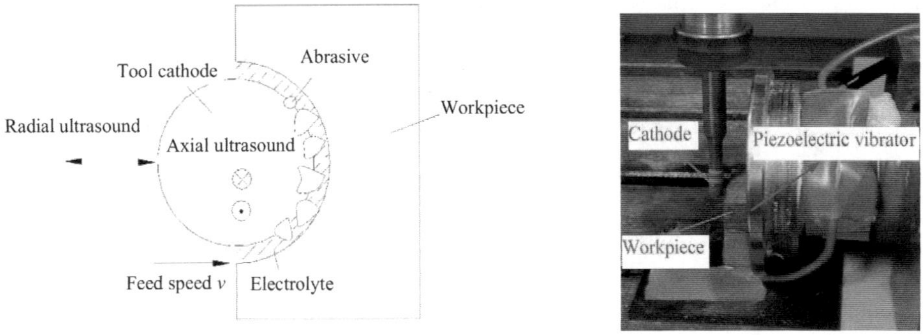

(a) Diagram of rotating ultrasonic composite ECM (b) Composite ECM device

Figure 7. Two-dimensional rotating ultrasonic composite evolution ECM.

The voltage and current during machining are displayed, measured and stored by a digital storage oscilloscope, which could be using for analysis and calculation the machining status. Using two channels of digital storage oscilloscope, the synchronization and the changing of the voltage and current can be displayed at the same time, so as to analyze and optimize the machining parameters in real time. Test parameters table is shown in Table 1.

Table 1 Test parameters.

Ultrasonic amplitude(μm)	Spindle speed(rpm)	Ultrasonic power(W)	Voltage amplitude (V)	Pulse frequency(Hz)
16	5000	150	4-6	5000

3.1. Comparative Test of End-face Composite Evolution ECM

The evolution feed speed is 15.0 mm/min, the grinding depth is 0.15mm, 5% sodium nitrate electrolyte, and other test parameters are shown in Table 1. Comparative tests are carried about the mechanical grinding, the ultrasonic grinding and the rotary ultrasonic grinding on high-speed steel workpieces. The machining area map can be seen in Figure 8(a). As the electrolysis effecting, the black film generates on the workpiece surface.

(a) Machining area (b) Mechanical grinding

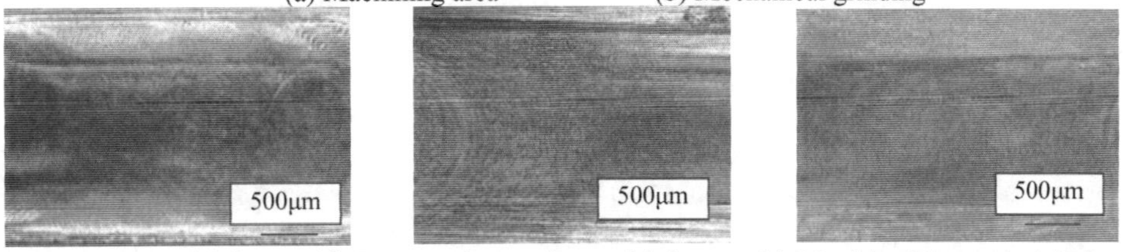

(c)Ultrasonic machining (d) Rotary ultrasonic machining (e) Machined by optimizing parameters

Figure 8. Comparison of machining area and test results.

Comparing Figure 8(b), Figure 8(c), Figure 8(d), cut-in and cut-out traces can be seen in the three machining methods. This is because the tools slowly accelerates and decelerates at the beginning and end of machining, and the cutting-in and cutting-out traces are inevitable.
There are the protruding marks in different degrees on the workpiece surface of the three machining ways, because the abrasive grinding speed of the outer ring of the tools is greater than the abrasive particle grinding speed of the inner ring, result in slow material removal speed in the intermediate machining area, and the machining scrap cannot also be discharged in time. The workpiece surface after the rotating ultrasonic composite evolution ECM in Figure 8(d), has relatively shallow protruding marks, because compared to the other two processing methods, electrolysis is added, and the workpiece material is removed by electrolysis and abrasive grinding, which weakens the end surface convex marks caused by inconsistent abrasive velocity.
In order to solve the problem of convex marks in the surface, the end face evolution machining method of rotating ultrasonic composite electrolysis machining is used to optimize the machining parameters. The machining material is cemented carbide YT15, the feed speed is adjusted to 5 mm/min, and the grinding amount is 0.10 mm. Figure8 (e) is the machining result diagram using optimizing parameters, it can be seen that the machining surface is relatively smoothed. The reason is that decreasing the feed speed and grinding depth can effectively solve the problem of convex marks.

3.2. Test of Two-dimensional Rotating Ultrasonic Composite Evolution ECM

In order to solve the electro-discharge sparks problem in cutting groove by ultrasonic composite evolution electrochemical machining, it is necessary to provide ultrasonic vibration in the radial direction. The radial ultrasonic vibration device can be seen in Figure 9, the transducer is composed of two ceramic sheets superimposed, and the ultrasonic amplitude is adjustable within 0 μm-7 μm.

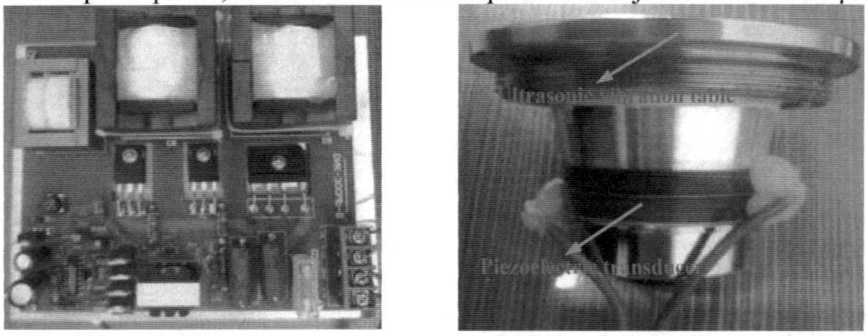

(a) Ultrasonic power (b) Piezoelectric vibration device

Figure 9. Radial ultrasonic vibration device.

The cemented carbide material YT15 sample is fixed to the end surface of the radial ultrasonic device. After the ultrasonic power is turned on, the YT15 sample does the ultrasonic vibration in the radial direction relative to the tool cathode, and the rotating tool cathode performs axial ultrasonic vibration, so that, two-dimensional ultrasonic vibration generates. The feed speed along the X direction is 1.0 mm/min, the radial ultrasonic amplitude is 5 μm, and 5% sodium nitrate electrolyte. Other machining parameters are shown in Table 1.
Figure 10 shows the principle of two-dimensional ultrasonic composite evolution ECM. The workpiece has radial ultrasonic vibration in the direction of cathode feeding. During the machining process, the gap between the workpiece and cathode undergoes periodic changes in the ultrasonic frequency, at same time, and the electrolyte can fully enter the machining gap to reduce the mechanical friction and the chance of spark discharge.
The cutting test result figure of the two-dimensional ultrasonic evolution ECM is shown in Figure 11. After the radial ultrasonic vibration is added, there is almost no spark discharge during the machining process, the electrolysis is strengthened, the machining efficiency becomes larger, and the machining depth is 1.5 mm.
The electrolysis current signal and the ultrasonic excitation signal of the rotating ultrasonic composite evolution electrochemical machining are shown in Figure 12.

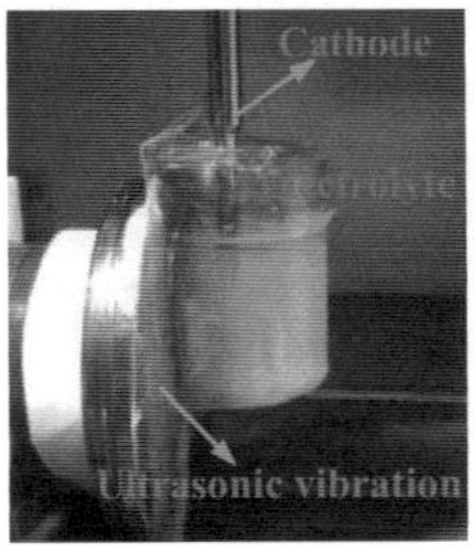

Figure 10. Ultrasonic machining device.

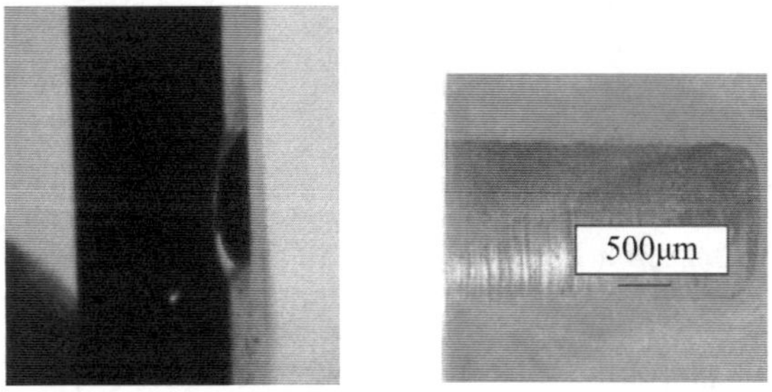

(a) Side of machined part (b) Front of machined part

Figure 11. Figure of two-dimensional composite ECM.

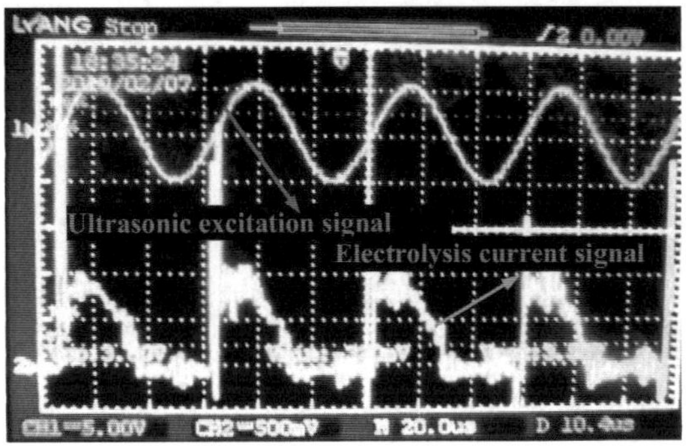

Figure 12. Electrolysis current signal and ultrasonic excitation signal of the composite ECM.

Two-dimensional ultrasonic vibration improves the renewal speed of the electrolyte in the machining gap and further improves the flow field conditions in the gap. Micro free abrasives are mixed in the electrolyte, and the fine polishing reduces the wear marks on the machined surface.

The pulse electrolysis voltage and the electrolysis current keep changing synchronously during the machining process. With the increase of the pulse voltage, the ECM current gradually increases, and the ECM effect will enhance, because the machined scrap is easily removed by the rotating ultrasonic vibration in time, the wear of tool electrodes will decrease, the material removal efficiency will increase, and the machining accuracy will be improved[11-12].

3.3. Cutting Groove Tests on Two-dimensional Rotary Ultrasonic Generating ECM

Using the same test parameters in table 1, the radial ultrasonic is added to the workpiece, the ultrasonic amplitude is set to 3μm, Figure 13 is the two-dimensional ultrasonic generating ECM.

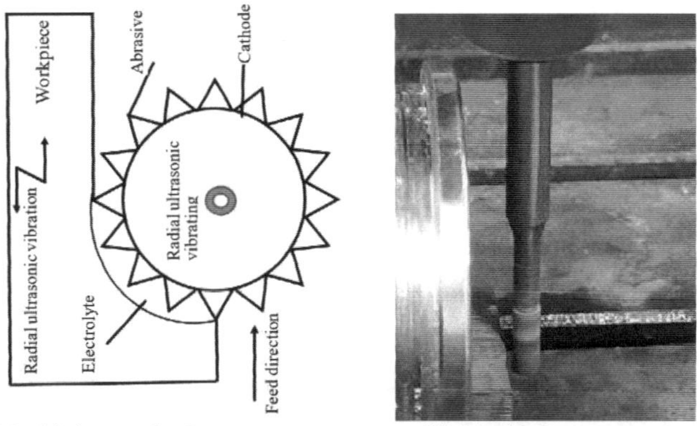

(a) Machining method (b) Machining area

Figure 13. Two-dimensional rotating ultrasonic composite generating ECM.

Figure14(a), Figure14(b) and Figure14(c) are the workpiece shapes machined in three methods.

(a) Rotary ultrasonic machining (b) Rotary ultrasonic ECM (c) Free abrasive rotary ultrasonic ECM

Figure 14. Machining effect diagram of three machining methods.

4. Conclusion

1) Three-dimensional(3D) ultrasonic composite generative ECM system is designed and constructed, the rotating ultrasonic composite ECM system's working is stable and reliable.
2) The convex marks on the workpiece surface of the rotary ultrasonic composite ECM are shallower and the surface quality is better; when the feed rate is optimized 5mm/min, the machining depth is less than 0.10mm, the problem of convex marks on the workpiece surface can be solve effectively.
3) The electric discharged sparks and the friction in the gap are serious, adding the radial ultrasonic vibration 5μm, ECM cutting can effectively reduce (eliminate) the discharge sparks and friction, and the electrolysis effect is significantly enhanced, the machining efficiency can be improved also.
4) Experiments on ceramic particle-reinforced metal matrix composites are carried out, a smooth edge groove was machining, verifying the technical advantage of two-dimensional ultrasonic vibration, the ultrasonic composite ECM accuracy become better under the condition of free abrasive particles.

Acknowledgment
This project is supported by National Natural Science Foundation of China (51775484, 52175438), authors give cordial thanks for the funding organization.

References

[1] CAO Fengguo. Ultrasonic machining[M].Beijing:Chemical Industry Press, 2014.
[2] ZHANG Jianhua, ZHANG Qinhe, JIA Zhixin, et al.Composite machining technology[M]. Beijing: Chemical Industry Press, 2014.
[3] XU Jiawen,YUN Naizhang, WANG Jianye. Electrochemical Machining Technique[M]. Beijing: National Defence Industry Press, 2008.
[4] XU Jiawen, ZHAO Jianshe, ZHU Yongwei, et al. New technologies for special machining of aircraft engine components[M]. Beijing: National Defence Industry Press, 2011.
[5] Khairy ABE. Assessment of some dynamic parameters for the ultrasonic machining process[J]. Wear, 1990, 137(2),pp:187-198.
[6] FENG Pingfa, Wang Jianjian, Zhang Jianfun, et al. Research Status and Future Prospects of Rotary Ultrasonic Machining of Hard and Brittle Materials[J]. Journal of Mechanical Engineering, 2017, 53(19), pp: 3-21.
[7] Zarepour H, Yeo SH, Tan PC, Aligiri E. A new approach for force measurement and workpiece clamping in micro-ultrasonic machining [J]. The International Journal of Advanced Manufacturing Technology, 2011, 53,pp:517–522.
[8] Krefting D, Mettin R, Lauterborn W, High speed observation of acoustic cavitation erosion in multibubble systems [J]. Ultrasonics Sonochemistry,2004,11(3-4), pp:119-123.
[9] DENG Shihui. Research on ultrasonic vibration-assisted micro-electrolytic drilling and milling [D], Shan Dong University,2019.
[10] LIU Zexiang, KANG Min, LI Shusheng. Flour Field Simulation of Small Holes Machined by the RUECM [J]. China Mechanical Engineering, 2015, 27(6), pp:748-753.
[11] ZHU Yongwei, WANG Zhanhe, FAN Zhongjun, et al. Micro ultrasonic machining technology and application of composite electrodeposition [J]. Journal of Mechanical Engineering, 2010, 46(3), pp:179-186.
[12] ZHAO Qingqing, DENG Zhengquan, ZHU Yongwei. Parameters On-line Measurement and Experiments of Ultrasonic Compound Electric Machining[J]. Journal of Mechanical Engineering, 2018, 54(17), pp:255-264.

Structural Design of Die-Casting Die for Industrial Cabinet Lock Handle Based on Pro/E

Fenfang Tan, Zhihang Tan* and Kuang Luo

School of Mechanical Engineering, Wuhan Polytechnic University, Wuhan, China

*Corresponding author email: tanzhihang@126.com

Abstract. This paper uses Pro/E and CAD software to carry out the design of die-casting mold structure for industrial cabinet lock handle.Firstly, the three-dimensional modeling, process analysis and calculation of the handle parts are carried out, and then the solid model is established.The die-casting machine is selected according to the clamping force. Finally, the opening and closing die is simulated, and the general assembly drawing of the die-casting die is drawn to ensure the cooperation with other parts without interference, which is of certain guidance to the actual production.
Keyword: Die-casting; Industrial cabinet lock handle; Process analysis; Pro/E; Mould design.

1. Introduction

In modern metal processing technology, pressure casting is a less special casting method without cutting. Looking around the world, the manufacturing level of die-casting molds in China is at a low level [1]. China's die-casting enterprises are generally small in scale, the comprehensive quality of enterprises still needs to be improved, the technical level is not in the international leading position, and the production efficiency is low. Efforts are still needed to participate in international competition, and the ability of independent research and development capabilities should be enhanced. In terms of high-grade molds, we cannot always rely on imported foreign advanced molds. This paper uses Pro / E and CAD software to carry out the die-casting die structure design of industrial cabinet lock handle.

2. Three-Dimensional Modeling and Process Analysis of Industrial Cabinet Lock Handle

Firstly, Pro/E and CAD software were used to model the industrial cabinet lock handle parts in three-dimensions [2], as shown in Figure 1.The material is 4-1 cast zinc, which has good casting properties.

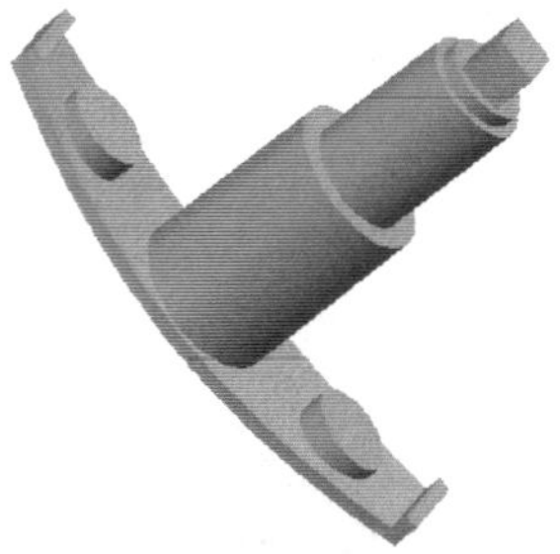

Figure 1. 3D drawing of industrial cabinet lock handle parts.

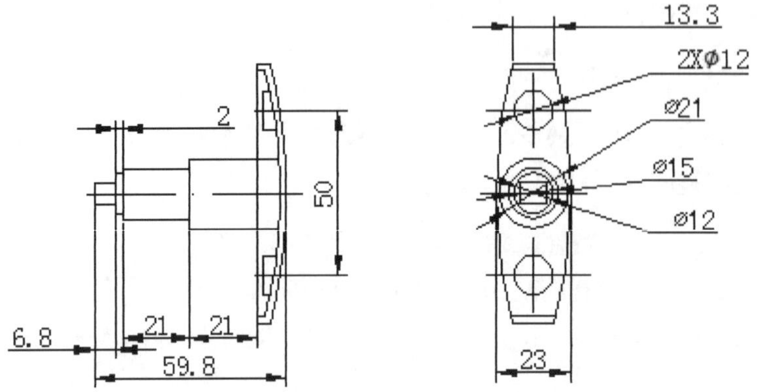

Figure 2. 2D part drawing of industrial cabinet lock handle.

Analyzing the dimensional accuracy requirements of Figure 1, it is planned to use a two-cavity die-casting mold for mass production, as shown in Figure 3.

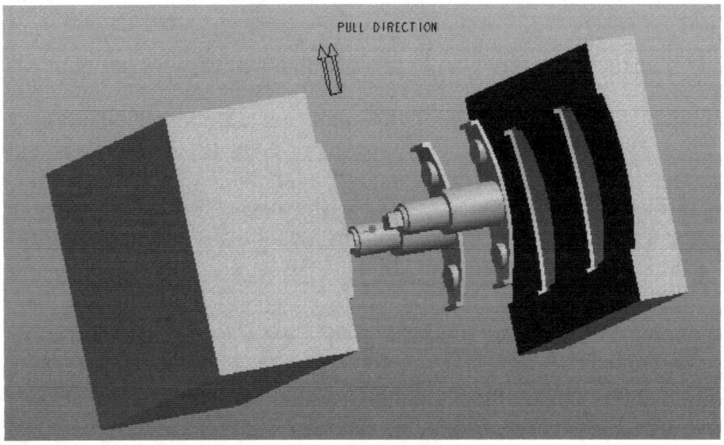

Figure 3. One mold and two cavities.

By calculation, the clamping force:

$$F = 1.25 \times 20MPa \times 8\text{cm} \times 96.2\text{cm}^2 = 19240KN \quad (1)$$

Then checking the literature [3], primary selection is the horizontal cold chamber die casting machine with model J11200.After the part is formed by die-casting, there are no obvious casting defects on the outer surface of the part. After die-casting, there is no subsequent processing, but only chrome plating on the surface of the part.

3. Die Structure Design

The liquid metal is injected from the gate sleeve at high speed and high pressure and enters the gating system to fill the mold cavity. After cooling and forming, the moving die part moves to the right and the moving and fixed die are separated, so that the ejection mechanism ejects the part from the driven die core to complete one-time die casting. This paper carries out the mold opening and closing simulation [4-5]. The mold opening state diagram of the die-casting mold for the industrial cabinet lock handle is shown in Figure 4, and the general assembly diagram of the die-casting mold is drawn, as shown in Figure 5, which ensures the cooperation with other parts without interference, and has certain guiding significance for the actual production.

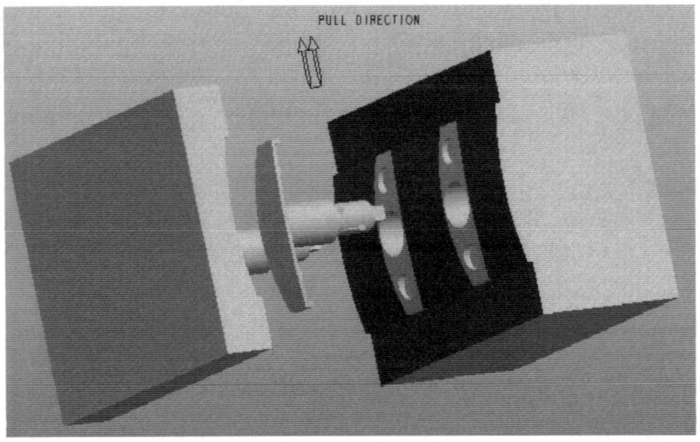

Figure 4. Die-casting die opening state diagram of industrial cabinet lock handle.

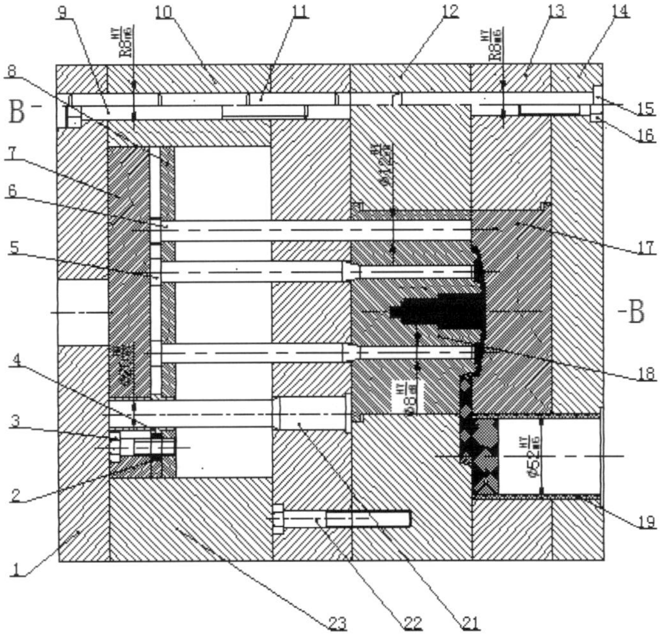

(a)

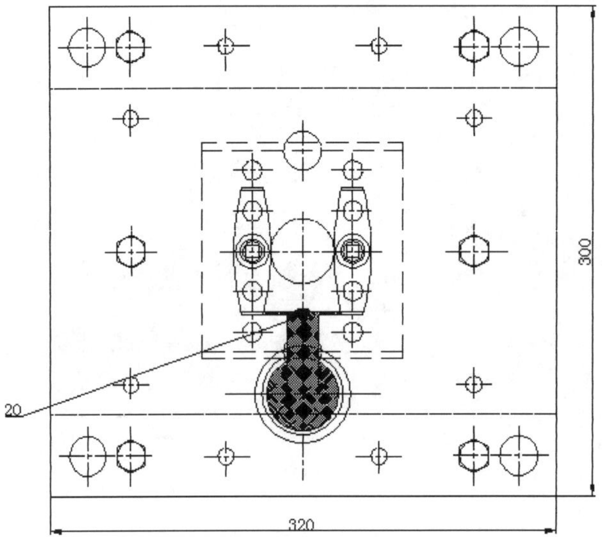

(b)

1- Movable template; 2- Limit washer; 3- Hexagon screw; 4- Push plate guide sleeve; 5- Push rod; 6-Reset rod; 7- Push plate; 8- Push rod fixing plate; 9- Hexagon screw 10-Cushion block; 11-Cylinder pin; 12-Fixed die set plate; 13-Moveable die set plate; 14-Fixed die plate; 15-Guide post; 16-Hexagon screw; 17-Fixed die core; 18-Moveable die Core; 19- Gating sleeve; 20- Shunt cone 21- push plate guide post 22- hex screw; 23- Support plate.

Figure 5. Assembly drawing of die-casting die for industrial cabinet lock handle.

4. Conclusions

This paper uses Pro/E and CAD software to design the mold structure of industrial cabinet lock handle. By considering the performance of the industrial cabinet lock handle material 4-1 cast zinc, the casting accuracy and dimensions, a series of designs are made for the molds, so that the designed parts have high dimensional accuracy and the molds can work properly and ensure the fit with other parts.

References

[1] Jianbo Zhou and Fuxiang Tian 2006 The status and development of die-casting [J] foundry equipment and technology pp 48-51

[2] Jinyou Li and Xianglin Zhang 2010 3D design and calculation guidance of die-casting mold M]. Chemical Industry Press pp 51-73

[3] Yusheng Gan 2008 Introduction and proficiency of professional skills of die-casting mold engineers [M]. Mechanical Industry Press pp 233-235

[4] Yanchen Tian and Baoshan Tian and Wenguang Wang 2005 Metal die-casting die design skills and examples [M]. Chemical Industry Press pp15-70

[5] Yao Huang and Yong Huang 2014 Practical teaching material for die-casting die design [M] Chemical Industry Press Second Edition pp 30-40

Research on Valveless Piezoelectric Micropump Based on COMSOL and 3D Printing Technology

Xinxin Wang, Shanshan Zhao, Xiaoxiao Yan*, Zhen Wang, Gang Tang, Liling Zhang, Bin Xu, Xiaozhen Deng and Yuwen Li

Jiangxi Province Key Laboratory of Precision Drive and Control, Nanchang Institute of Technology, Nanchang 330099, China

*Corresponding author email: qwwxiao@126.com

Abstract. In order to make the output flow of the valveless micropump larger and obtain continuous pumping capacity, a valveless piezoelectric micropump with a combination of V-shaped structure inlet and Y-shaped structure outlet is designed, which saves the design and installation of microvalve and makes the structure relatively simple and easy to process. In this paper, first, the operating principle of the valveless piezoelectric micro pump is analyzed; secondly, COMSOL simulation software was used to perform modal analysis on the drive and the appropriate vibration mode was obtained. The displacement trend of the piezoelectric vibrator was simulated and tested at the frequency of 40 Hz and different voltages; finally, 3D printing technology was used to produce an experimental prototype of the micro-pump, and t the specific size of the micro pump is 20 mm×20 mm×6 mm. The test results show that, driven by AC voltage, the discharge liquid of the valveless piezoelectric micropump can reach 658.8μl/min.

Keywords: Valveless; 3D printing; COMSOL; MATLAB.

1. Introduction

Piezoelectric-driven micropumps have the advantages of simple operation, portability, easy integration, and low power consumption. They are mainly applied to different degrees in biological monitoring, medical drug delivery, chemical analysis, and other fields. Piezoelectric micropumps are divided into valved and valveless. Valved micropumps have shortcomings such as large operating pressure loss, low transmission accuracy, and large volume, which severely restrict their further promotion and application. Valveless piezoelectric micropumps[1-3] can effectively solve the problems of valved piezoelectric micropumps, which makes domestic and foreign researchers very interested in this. Mohith Santhya[4] proposed the design, manufacture and testing of a new type of valveless amplifier piezoelectric actuator micropump for biomedical fluid transfer. K. Srinivasa Rao[5] proposed a piezoelectric micropump that uses PDMS as the pump body. However, the existing valveless piezoelectric pumps are relatively large in volume, inconvenient to carry, and have a relatively complicated inlet and outlet structure. When pumping macromolecular liquids, they are easy to remain and cause blockage, are inconvenient to clean, and are less likely to be reused.

In this paper, a valveless micropump whose inlet and outlet are at the bottom of the pump body, laying a foundation for the subsequent use of microneedles in medical continuous drug delivery. The structure uses the combined design of the liquid inlet and outlet ports of the V-shaped and Y-shaped structures, which saves the design and installation of the microvalve, and makes the structure relatively simple and easy to process.

2. Structural Design and Operating Principle of Piezoelectric Micro Pump

The structure of the micropump designed in this paper is shown in Figure 1. The valveless piezoelectric micropump is mainly composed of an inlet, an outlet, a piezoelectric sheet, a substrate, a PDMS film and a pump body.

The operating principle of the valveless piezoelectric micro pump is shown in Figure 2. When an AC voltage is added to this micro pump, the upward and downward movement of the piezoelectric vibrator will cause the volume and pressure of the cavity to change, thereby causing the liquid to flow. When the driver moves upward and deforms, the volume of the cavity of the micro pump increases. Because there is no stop valve at the inlet and outlet of the micropump, the liquid enters the pump cavity from the inlet and outlet together, but the inlet is a diffuser with a large end facing inward, and the liquid is easy to flow in. The inflow is large; the outlet is a constricted opening with the small end facing inward, which makes it difficult for liquid to flow in, and the inflow is small; When the driver moves downward and deforms, the volume of the cavity of the micropump becomes smaller, the outlet pressure is small, and the inlet pressure is high. In contrast, the outlet is easier to discharge more liquid. This micropump can realize the ability to pump liquid continuously.

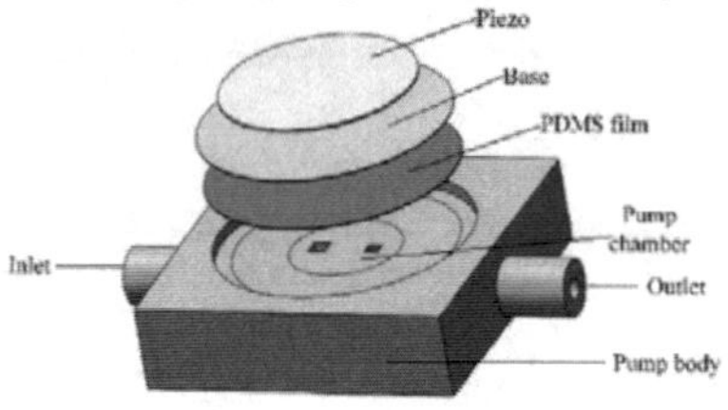

Figure 1. Exploded view of micro pump.

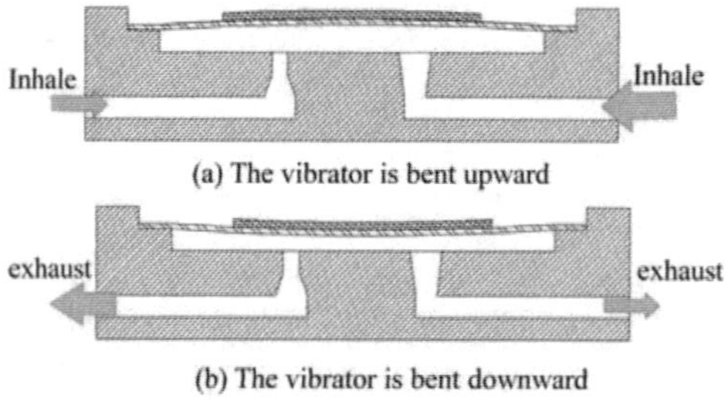

Figure 2. Sectional view of the working state of the micropump.

3. Simulation Analysis and Displacement Test of Piezoelectric Vibrator

3.1. Establishment of Simulation Model of Piezoelectric Vibrator

The model of the drive established in Comsol (COMSOL Inc. Stockholm, Sweden) software is shown in Figure 3. The driver can be divided into two upper and lower circles for modeling, where the upper circle is the piezoelectric sheet, the lower circle is the beryllium bronze substrate, and the lower surface is the fluid-solid coupling surface that is applied to the micropump analysis in contact with the fluid. When simulating the assembly of the micropump, the adhesion constraint is used on the periphery of the circle below. Constraints can be added in the subsequent analysis without affecting their integrity. The piezoelectric vibrator model is meshed, using a free tetrahedral mesh, and the mesh structure after division is shown in Figure 4. Add a fixed constraint to the periphery of the lower circle.

Figure 3. Piezoelectric vibrator simulation model.

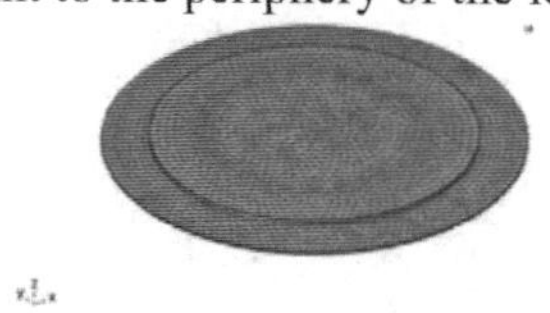

Figure 4. Piezoelectric vibrator meshing.

The vibrating displacement of the beryllium bronze substrate and the piezoelectric sheet of different thicknesses will produce large differences in deformation, then affect all aspects of the micropump. In the model simulation analysis, set the diameter of the beryllium bronze base to 18mm and the diameter of the piezoelectric sheet is 14 mm; the beryllium bronze and piezoelectric sheet of different thicknesses are selected, the AC voltage is 200 V, and the center displacement of the piezoelectric vibrator is shown in Table 1.

Table 1. Displacement of piezoelectric vibrator with different thickness.

Beryllium bronze (thickness μm)	PZT-5H (thickness μm)	Center displacement (μm)
200	100	31.62
200	200	22.18
100	200	37.7

It can be seen from the results of the simulation analysis that when using beryllium bronze substrate thickness is 100 μm and the thickness of PZT-5H is 200 μm, which is the most suitable for making piezoelectric vibrator.

3.2. Dynamic Simulation Analysis of Piezoelectric Vibrator

This section first conducts modal analysis, the first four vibration modes of the driver are solved, and the first four vibration frequencies are: 3919.2 Hz, 8196.1 Hz, 8197.1 Hz, 12829 Hz. As shown in Figure 5. It can be clearly seen that the center deformation of the driver is the largest in the first mode shape, and the deformation direction of the entire driver is the same, and the variation of the cavity formed is also large, which is in line with the micropump counter pressure requirements for electric vibrators.

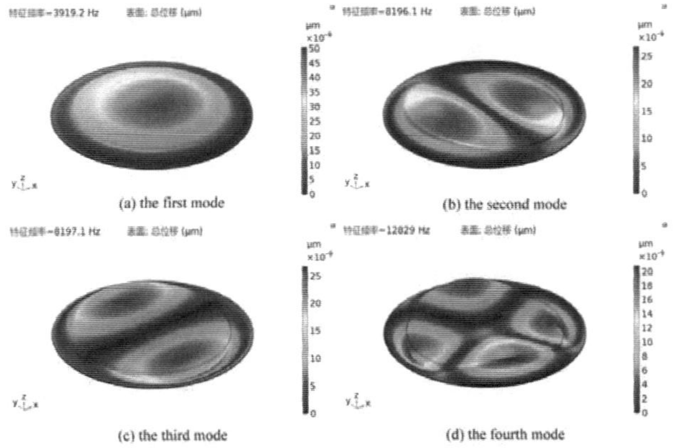

Figure 5. The first four modes of piezoelectric vibrator.

3.3. Piezoelectric Vibrator Displacement Test

The Microtrak™3 laser triangulation sensor (Shanghai Yinxu Mechanical and Electrical Equipment Co., Ltd., China) is used to measure the center displacement of the drive. The frequency of driving the valveless piezoelectric micropump is 40Hz, and the relationship between the deformation and displacement of the driver's center and the driver's added voltage in the simulation and experimental testing by MATLAB (MathWorks, Natick, Massachusetts, U.S.) software, as shown in Figure 6. From the results, the deformation displacement of the driver center point increases with the increase of the added voltage, showing a good linear relationship, and is approaching the result of Comsol simulation. When the driving voltage is 200 V, the displacement of the drive can reach 36.522 μm.

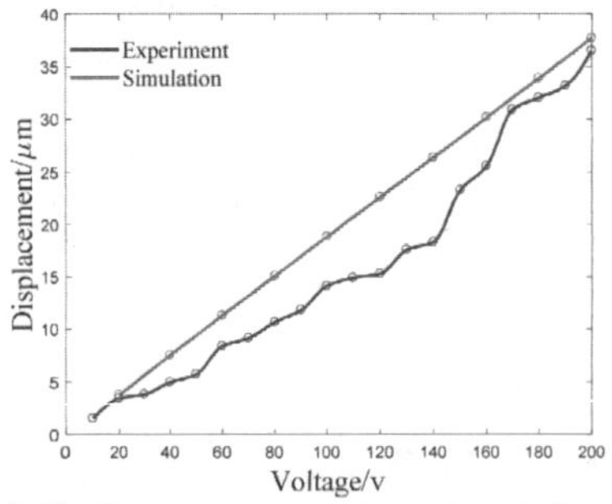

Figure 6. Comparison curve of displacement experiment and simulation of piezoelectric vibrator.

4. Prototype Production

3D printing with SLA technology is used to process the valve body of the valveless piezoelectric micropump. Design the pump model according to the required size in the 3D drawing software SolidWorks 2019 (SolidWorks, Massachusetts, America), as shown in Figure 10, then save the file in STL format and import it into the Materialise Magics 21.0 (Materialise, Belgium) software, and set the 3D printing parameters and import it into Aurora SLA350 printer (Jiguang Erwo Technology Co., Ltd., Shenzhen, China) is the main operation step. The pump cavity is round table size ϕ16 mm×ϕ8 mm×0.15 mm. The side length of the large end of the entrance is 1.246 mm, the side length of the small end is 1 mm, and the length is 2 mm. The side length of the large end of the outlet is 1.246 mm, the side length of the small end is 1 mm, the length of the upper part and the lower part are equal, both are 1mm. The physical points of modeling and final assembly are shown in Figure 7.

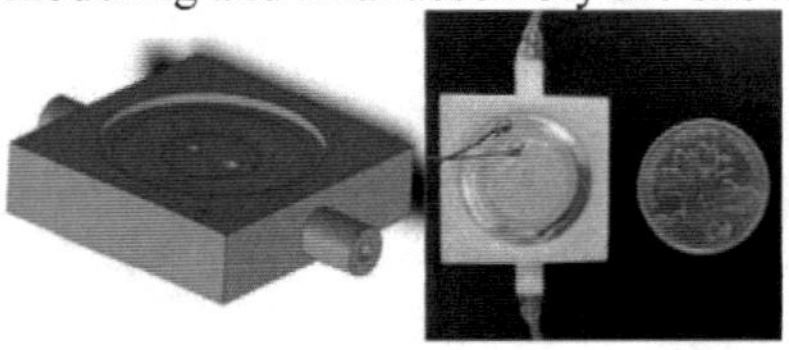

(a) (b)

Figure 7. Model and physical image (a) Overall model (b) Overall assembly physical object.

5. Discussion

According to the test method of Huang Liang[6] et al. of this research group, the test device mainly includes AC voltage source, petri dish, silicone tube, precision balance and valveless piezoelectric micro pump prototype. During the test, the ambient temperature was 25 °C, and the liquid was deionized water. Test method: Put the inlet hose into a petri dish with deionized water, the piezoelectric vibrator is energized to squeeze the pump cavity, and the liquid flows from the outlet into an empty petri dish placed on a precision balance. After the micropump has been running for a period of time, the liquid in the petri dish is measured, repeated several times, and finally the average volume is calculated.

When 200V voltage is applied to the driver and the frequency is 40Hz, the discharge flow rate from the outlet is 658.8μl/min. But the effect is not ideal, the main reasons are:

(1) The pump cavity is too high, the deformation generated by the driver when driven by AC voltage is too small.

(2) The air tightness of the inlet and outlet is not good, which leads to the leakage of the suction of the micropump.

Based on the above reasons, the height of the pump cavity will be reduced as much as possible in the future, and the piezoelectric drive will be improved or coupled with other drive methods. Optimize the size of the entrance and exit structure.

6. Conclusion

This article introduces a valveless piezoelectric micropump with a V-shaped inlet, a Y-shaped outlet, and a rectangular cavity for liquid transportation. The pump body is printed with off-white photosensitive resin material. Compared with the existing micro pump, It has the advantages of simple structure, easy realization, light weight, etc. The simulation and experimental analysis results show that the displacement test results of the piezoelectric vibrator are basically similar to simulation results, and the displacement is 36.522 μm at a 200V voltage and a frequency of 40Hz are applied to the driver, and the discharge flow rate from the outlet is 658.8μl/min. According to the experimental results, after analyzing the reasons, the structural size and driving mode of the valveless piezoelectric micropump will be optimized in the near future.

Acknowledgments

This work was partly supported by the National Natural Science Foundation of China (Nos. 62064008 and 51765044), and Natural Science Foundation of Jiangxi province (Nos. 20202BABL204041, 20202BABL204025, and GJJ201927). The authors are also grateful to their colleagues for their essential contributions to this work.

References

[1] Zhang J H, Wang Y, Huang J. Advances in valveless piezoelectric pump with cone-shaped tubes[J]. Chinese Journal of Mechanical Engineering, 2017, 30(4): 766-781.
[2] Huang J, Zou L, Li Z, et al. Development and performance comparison of valveless piezoelectric pumps with asymmetrical channels[J]. Sensors and Actuators A: Physical, 2020, 314: 112241.
[3] Huang J, Zou L, Tian P, et al. A valveless piezoelectric micropump based on projection micro litho stereo exposure technology[J]. Ieee Access, 2019, 7: 77340-77347.
[4] Santhya M, Upadhya A R, Panambur N K, et al. Performance analysis of a novel piezo actuated valveless micropump for biomedical application[C]//AIP Conference Proceedings. AIP Publishing LLC, 2020, 2236(1): 070002.
[5] Rao K S, Ganesh G V, Lakshmi G S, et al. Analysis of PDMS based MEMS device for drug delivery systems[J]. Microsystem Technologies, 2021, 27(3): 659-664.
[6] Huang L, Tang G, Hu M, et al. Study on double piezoelectric layers driven pump based on MEMS and 3D printing[C]//International Conference on Human Centered Computing. Springer, Cham, 2018: 571-579.

Angle Steel Tower Bolt Installation Detection Based on Simplified Attention Mechanism

Jinfeng Zhang[1,*], Tianzhong Zhang[1] and Yihua Lou[1]

[1] State Grid Anhui Electric Power Co.,LTD., Hefei 230022, Anhui, China

*Corresponding author email: 5952544@qq.com

Abstract. Incorporating position and channel attention into the image semantic segmentation network is an important strategy to improve the segmentation effect. However, there is a huge computational overhead in counting the correlation between any two positions or channels. To address the issue, this paper proposes a simplified attention mechanism network integrating position attention and channel attention. In addition, to further optimize the segmentation results, this paper adopts a new data-dependent upsampling method to expand the high-level feature map to the input size. Compared with the traditional attention network, our method reduces 32.6% of parameter and calculation amount, and obtains an average cross-comparison result of 95.42% on the dataset of angle steel tower bolt.
Keywords: Image semantic segmentation; Data-dependent upsampling; Simplified attention mechanism.

1. Introduction

Image semantic segmentation is a kind of intensive classification, and its goal is to assign the correct category label to each pixel in the image. Therefore, when semantic segmentation classifies a target, it is not only necessary to distinguish the type but also to identify the location of the target. Recent studies have shown that in order to accurately locate the target location, it is essential to capture long-range information [1-2]. Moreover, there are close associations between categories, and capturing the correlation between channels in the network is more conducive to pixel classification [3-5].

Capturing long-range information means that each pixel of the output feature map contains large-area input features. Since the convolution operation is a local operation, the output features can only contain the input features of the size of the convolution kernel. Therefore, in order to increase the scope of action, some methods [6] keep the size of the convolution kernel, continuously stacks the number of convolution layers, and increases network depth. Zagoruyko *et al.* [7] reduced the number of convolutional layers, expand the size of the convolution kernel, and increase the network width. Erancois [8] et al. used hole convolution to expand the size of the convolution kernel by ignoring part of the input features. These improved methods have optimized the segmentation performance to a corresponding extent, and achieved good segmentation results in specific segmentation scenarios, but the accompanying input feature loss and huge amount of calculation have affected the actual deployment of the strategy. DANet[9], GCNet[10] and other models were proposed. The implementation of non-local network can be divided into three steps: firstly, calculate the correlation between the query location feature on the feature map and all the location features, then use it as a weight to aggregate all features by weighted sum, and finally aggregate the features and the query of the corresponding location feature fusion.

On the other hand, the number of output channels of the semantic segmentation network is the same as the total number of categories, that is, each channel simulates one category. Methods proposed by Zhang *et al.* [3] showed that there is a close correlation between different categories, and capturing this correlation in the network is conducive to the accurate classification of pixels. In order to capture the inter-channel dependence, Niu *et al.* [5] and Fu *et al.* [9] applied the attention mechanism to the channel dimension. Like position attention, the realization process of channel attention can be divided into three steps: first calculate the inter-channel correlation, and then apply this correlation is used as the channel weight to aggregate the information of different channels, and finally the characteristics of the aggregated different channel information are merged with the corresponding channel characteristics. Similarly, position attention has the correlation between computing channels and is accompanied by huge computational overhead. Embedding position attention in the network will affect network performance.

Therefore, this paper aims to capture the global information with long-range dependencies while reducing the computational overhead associated with calculating the correlation, and proposes a simplified attention mechanism network. For position attention, we learn the correlation independent of all positions and utilize this correlation in a shared way. For channel attention, global average pooling and maximum operation are used to simulate category information, and then the inter-channel dependence is learned through the bottleneck fully connected layer. In addition, this paper adopts a new data-related up-sampling [11] method to expand the high-level feature map to the input size, which effectively improves the segmentation effect.

2. Method

2.1. Data Set and Implementation Details

Figure 1. Visualization of angle steel tower bolt installation data set.

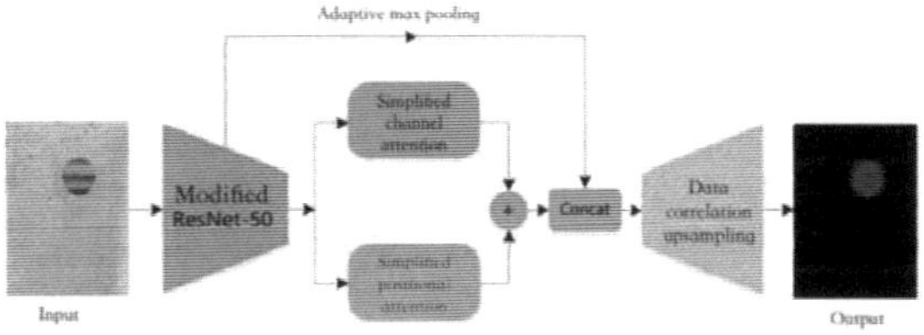

Figure 2. The overall structure of the model.

The angle steel tower bolt installation data set is a two-category test, which corresponds to two cases of incomplete installation of bolts (defective) and complete installation (non-defective). Among them, there are 123 defective pictures and 160 non-defective pictures. Figure 2 shows 4 original images and their corresponding labels (highly normalized).

2.2. Overview

The overall structure of the model is shown in Figure 2. The pre-trained ResNet-50 is used as the backbone network. In order to reduce the number of downsamplings and retain more detailed information, only the first three blocks of the residual network are used, and the step size of the third block is changed and set to 1. Since the image input to the network is normalized to 224×224, the width and height of the intermediate feature map obtained after the backbone network feature extraction is 28; then, the simplified position and channel attention two parallel modules capture the position And the global information that the channel depends on, and the respective extracted features are added and fused. The article uses a data-related up-sampling method to restore the final feature map to the input size to achieve pixel classification; in order to reduce the effective information lost by down-sampling, down-sampling high-resolution features through 1×1 convolution and adaptive maximum pooling the graph aggregates the characteristics of the attention module fusion.

2.3. Position Attention

Location attention captures global information with long-range dependencies by simulating the correlation between features. It focuses its attention on "where" to help accurately locate the target.

For the arbitrary position feature X_{pi} on the intermediate input feature map $X_{\mathrm{p}} \in R^{C\times H\times W}$, the corresponding output feature Y_{pi} after being mapped by the position attention can be expressed as

$$Y_{\mathrm{pi}} = X_{\mathrm{pi}} + W_2 \sum_{j=1}^{N} \frac{f_P(x_i,x_j)}{C(x)} (W_1 \cdot X_{\mathrm{pi}}), \tag{1}$$

where i and j represent the query position and the aggregate feature position respectively, $N = H \times W$, W_1 and W_2 represent linear conversion, 1×1 convolution is used to transform the number of channels in actual deployment, $f(x_i, x_j)$represents the correlation between the location features i and j which is normalized by $C(x)$. Semantic segmentation network usually uses the embedded Gaussian function to simulate the correlation between pixels [10], that is, $\frac{f(x_i,x_j)}{C(x)} = \frac{\exp(<W_4x_i,W_3x_j>)}{\sum_{j=1}^{N}\exp(<W_4x_i,W_3x_j>)}$.

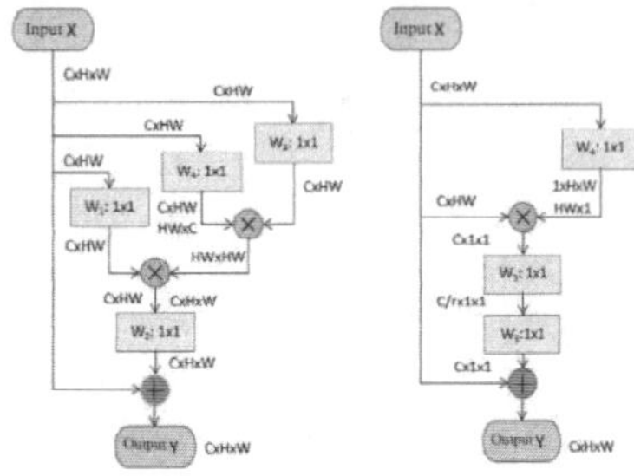
(a) position attention (b) Sharing attention

Figure 3. Position attention realization process.

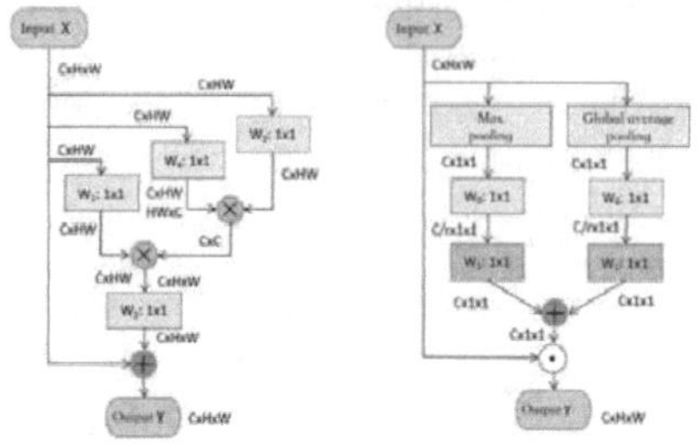
(a) Channel attention (b) Simplify channel attention

Figure 4. Channel attention realization process.

The process is shown in Figure 3(a). Further, this work moves the linear conversion operation W_1 after calculating the pixel correlation, and merges it with the conversion function W_2, namely

$$Y_{\mathrm{pi}} = X_{\mathrm{pi}} + W_3 \sum_{j=1}^{N} \frac{\exp(W_4x_j)}{\sum_{j=1}^{N}\exp(W_4x_j)} X_{\mathrm{pi}}. \tag{2}$$

This process further reduces the amount of calculation HWxC^2 to C^2. In addition, from the ResNet bottleneck structure, replacing the ordinary convolution with the dimensionality reduction-feature extraction-restoration method can further reduce the amount of calculation. Finally, the simplified attention is shown in Figure 3 (b).

2.4. Channel Attention

Channel attention captures long-range channel dependence by simulating inter-channel correlation. Different from position attention, channel attention focuses attention on "what" to strengthen classification. For the arbitrary position feature X_{ci} on the intermediate feature map $X_{\mathrm{c}} \in R^{C\times H\times W}$ input to the channel attention network, the corresponding output feature Y_{ci} after the channel attention network mapping can be expressed as

$$Y_{\mathrm{ci}} = X_{\mathrm{ci}} + W_2 \sum_{j=1}^{C} \frac{f_C(x_i,x_j)}{C(x)} (W_1 \cdot X_{\mathrm{ci}}). \tag{3}$$

The process is shown in Figure 4(a). It can be seen from Figure 3(a) and Figure 4(a) that when calculating channel correlation, the calculation amount of channel attention is the same as position attention, which is approximately $C^2 \times HW \times HW$. To reduce the amount of calculation, a simplified channel attention is developed as shown in Figure 4(b). It performs the operations of global average pooling and maximum pooling to simulate category information and learn the inter-channel dependence through two fully connected layers. For position attention, a bottleneck structure is set to further reduce the amount of calculation. The simplified channel attention can be expressed as

$$F = \sigma(W_1\,(W_0(F^c{}_{avg})) + W_1(W_0(F^c{}_{max}))), \tag{4}$$

where σ represents the Sigmoid activation function, W_0, W_1 represent two fully connected layers, r represents the attenuation rate, $W_0 \in \frac{C}{r} \times C$, $W_1 \in C \times \frac{C}{r}$, the calculation amount of the whole process is approximately $\left(HW + \frac{4}{r}\right)c^2$.

2.5. Data Related Upsampling

Bilinear interpolation is to perform two interpolations in the horizontal and vertical directions around the pixels to determine the target pixel. An important finding is that the pixels in the label are not independently distributed, and the pixels contain structural information. Therefore, the label $L_1 \in R^{H_1 \times W_1 \times C_1}$ can be almost losslessly compressed to $L_2 \in R^{H_2 \times W_2 \times C_2}$ and then decompressed to L_1, and then the reconstruction matrix W learned in the decompression process is used for upsampling, L_2 has the same dimension as the final aggregated feature map.

The label L_1 can be compressed to L_2 and can be divided into 4 stages. First, block L_1 into H_2 by W_2 sub-windows of r by r size, $r = \frac{H_1}{H_2} = \frac{W_1}{W_2}$, and then convert each sub-window into a vector $V = \{0,1\}^{N_1}$, $\{0,1\}$ means that the label is processed by one-hot encoding, $N_1 = r \times r \times C_1$, then the vector V is compressed into $Z \in R^{C_2}$, and finally horizontal and vertical Compress other sub-windows. For the third stage, this article uses a linear compression method as

$$Z = PV \quad V' = WZ. \tag{5}$$

In the above formula, $P \in R^{C_1 \times N_1}$ is used to compress V to Z, $W \in R^{N_1 \times C_2}$ is the decompression reconstruction matrix, and V' is the reconstruction vector of V. The gradient descent method is used to minimize the error of V' and V. Figure 5 shows the simple compression process when $r = 2$.

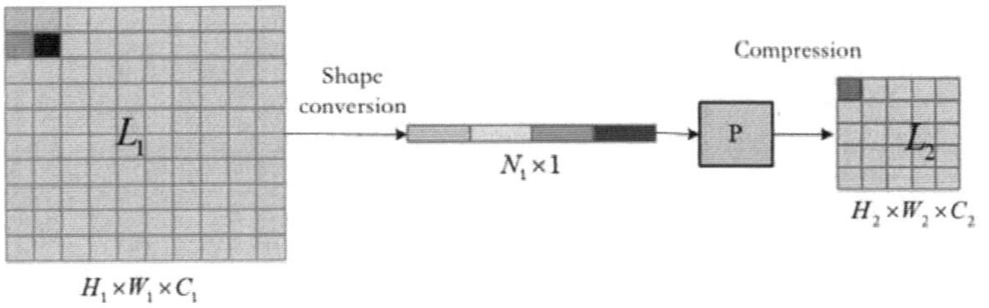

Figure 5. The process of compressing the label L_1 to the feature map size L_2 when $r = 2$.

3. Experiment

The training set is randomly extracted 75% of the dataset, and the remaining 25% is for the test set. HALCON software was adopted to randomly mirror and flip, crop 5% and rotate 200 data to enhance the training set images. As a result, the training set has 848 images and 71 are for tests. Moreove, the initial learning rate of the network is set to 0.05, the training is 50 cycles and it decays by half every 20 cycles, and the momentum is set to 0.93. The data set batch is 8, the attenuation factor is 0.0001, and the stochastic gradient descent algorithm is used as the optimizer to train the network.

3.1. Simplified Comparison of Attention Ablation Experiments

To verify the effectiveness of each module, this work conducts comprehensive ablation experiments. The experimental results are shown in Table 1. Here, the parameters (Params), calculations (FLOPs), and average intersection ratio (MIoU%) are used as evaluation indicators.

Table 1. Comparison of simplified attention ablation experiments.

Method	Params(M)	FLOPs(G)	MIoU(%)
Modified ResNet-50 (benchmark)	28.83	23.30	89.19
Datum + simplified position	29.36	23.30	93.08
Benchmark + simplified location + average pooling	29.87	23.31	94.76
Benchmark + simplified position + maximum pooling	29.87	23.31	93.64
Benchmark + simplified position + simplified channel (our model)	29.87	23.31	95.06

As shown in Table 1, this network has obtained an average intersection ratio result of 89.19% and only requires 28.83M parameters and 23.30 calculations. Experiment 2 integrates the simplified position attention module on the basis of the benchmark network, it can be seen that the average intersection ratio of the module increases by 3.89% without increasing the amount of calculation. This shows that the simplified attention module can not only improve the segmentation effect by capturing the global features with long-range dependencies, but also the shared correlation strategy will not increase the computational burden. It can be seen from experiments 3 and 4 that the integration of average pooling and maximum pooling operations can effectively improve the segmentation results without increasing the amount of model parameters and calculations. In Experiment 5, this method can achieve an increased average intersection ratio of 5.87%, compared with the benchmark network, and the amount of parameters and calculations only increased by 1.04M and 0.01G. This shows that the proposed model has a high degree of superiority for the angle steel tower bolt installation data set.

3.2. Hyperparameter Experiment Comparison

To obtain the best segmentation effect as much as possible, the article fine-tunes the learning rate and momentum hyperparameters. Figure 6 shows the change curve of the loss value under different initial learning rates when the momentum is 0.9. From the change curve of the loss value, it can be seen that when the learning rate is set to 0.05, the loss value is the smallest and converges quickly. Figure 7 shows the change curve of the loss value of different momentum when the learning rate is 0.05. It can be seen from the figure that when the momentum is 0.93, the loss value is the smallest and converges quickly.

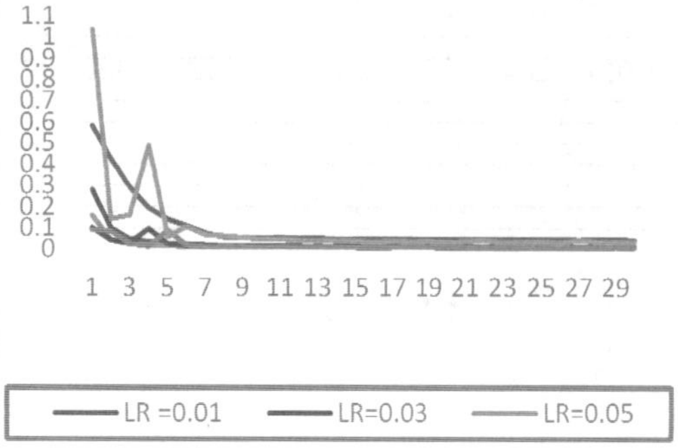

Figure 6. The change curve of the loss value of different learning rates.

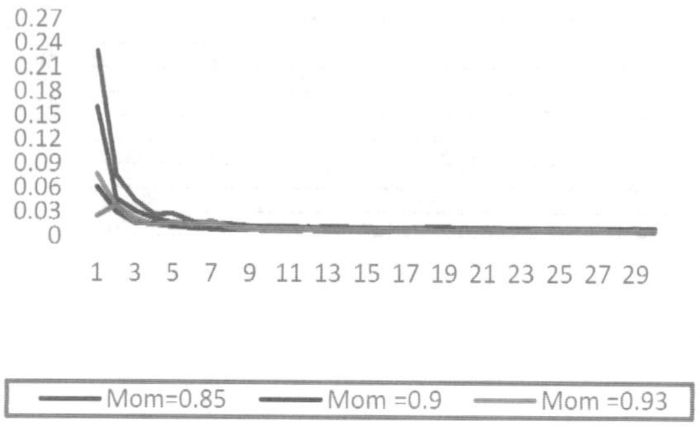

Figure 7. Variation curve of loss value for different momentum values.

3.3. Popular Method Experiment Comparison

The performance comparison with other popular methods are shown in Table 2. Deeplab V3+ takes advantages in the amount of calculation, but yields a huge gap of 19.01% for MIoU. SENet and NLNet only use channel or position attention to capture a single category or position correlation, so the segmentation effect has certain limitations. In addition, because NLNet needs to count the correlation between any pixels when calculating the correlation between features, the amount of parameters and calculations the amount has increased significantly. DANet uses parallel channel and position attention to simultaneously capture the correlation between channels and positions and achieves an average cross-combination ratio of 95.12% but needs a huge amount of parameters and calculation amount. The parameters of the proposed method basically do not affect the segmentation effect, while the amount of parameters and calculations are reduced by 32.6% and 32.3%, respectively. Therefore, it can be seen that the model proposed in this paper has higher superiority in the segmentation and detection of angle steel tower bolt installation data sets.

Table 2. Experimental comparison of popular methods.

Method	Params(M)	FLOPs(G)	MIoU(%)
Deeplab V3+	41.05	9.89	76.41
SENet	29.36	23.30	92.28
NLNet	35.92	28.86	93.79
DANet	42.99	34.41	95.12
Proposed model	28.97	23.30	95.42

3.4. Visible Experimental Prediction Result

The actual prediction results are shown in Figure 8(a). It can be seen that the network proposed in this article can not only achieve very good prediction results on edge smoothing targets, but also can be well predicted for unbalanced and disconnected targets. Therefore, it can be inferred that the network has a strong learning ability for the characteristics of the angle steel tower data set. In addition, from the angle steel tower data set training, test loss and average intersection ratio change curve in Figure 8(b), therefore the network can quickly learn defect features and produce better predictions. The article believes that certain fluctuations during testing are due to the lack of images in the test set.

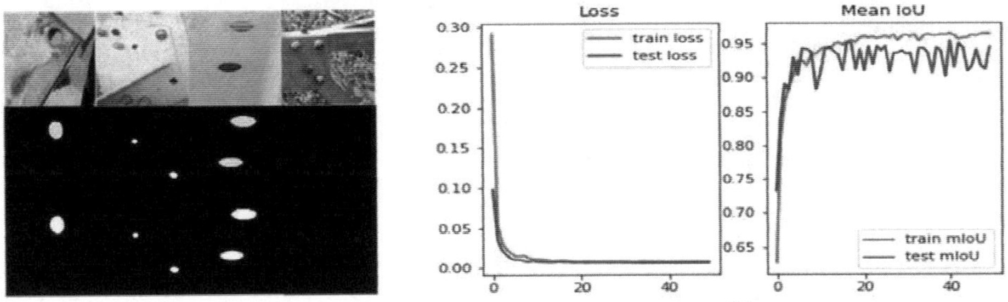

(a) (b)

Figure 8. (a) Visualization of the prediction results of the angle steel tower data set. (b) Loss during training and testing of angle steel tower data set and the change curve of average intersection ratio.

4. Conclusions

This paper proposes a simplified attention mechanism network: the modified ResNet-50 is used as the backbone network to extract features, and the simplified channel and location attention are two parallel modules to enhance feature learning. In addition, the article uses a data-dependent upsampling method to restore the feature map to the input size. From the experimental results of the angle steel tower bolt installation data set, it is reflected that the model proposed in this paper can not only segment the results very well, but also has the advantages of light weight.

Acknowledgement

This work was supported by the State Grid Anhui Electric Power Co., Ltd. (No. 52120019007G).

References

[1] MA S H, AN J B, YU B. Improved DeepLabv2 real-time image semantic segmentation algorithm. Computer Engineering and Applications, 2020, 56(18): 157-164.
[2] CHEN Y P, LI J, YAN S C, et al. A^2-net: Double attention networks. In advance in neural information processing systems, 2018, 352-361.
[3] ZHANG H, SHI Q P, ZHANG Z Y. Context encoding for semantic segmentation. Proceedings of the IEEE Conference on Computer Vision and Pattern Recognition, Los Alamitos: IEEE Computer Society Press, 2018, 7151-7160.
[4] HU J, SHEN L, SUN G. Squeeze-and-excitation networks. Proceedings of the IEEE Conference on Computer Vision and Pattern Recognition, Los Alamitos: IEEE Computer Society Press, 2018, 7132-7141.
[5] NIU R G. HMANet: Hybrid multiple attention network for semantic segmentation in aerial images. arXiv Preprint arXiv: 2001.02870(2020).
[6] FORREST N, MATTHEW W, ROSS B, et al. DenseNet: Implementing efficient ConvNet descriptor pyramids. arXiv Preprint arXiv: 1404.1869(2014).
[7] ZAGORUYKO S, KOMODAKIS N. Wide residual network. arXiv Preprint arXiv: 1605.07146(2016).
[8] ERANCOIS C. Xception: Deep learning with depthwise separable convolutions. Proceedings of the IEEE Conference on Computer Vision and Pattern Recognition, 2017,1800-1807.
[9] FU J, LIU J, TIAN H, et al. Dual attention network for scene segmentation. Proceedings of the IEEE Conference on Computer Vision and Pattern Recognition, 2019: 3141-3149.
[10] CAO Y, XU J R, HU H, et al. GCNet: Non-Local networks meet squeeze-excitation networks and beyond. Proceedings of IEEE International Conference on Computer Vision 2019, 1971-1980.
[11] TIAN Z, HE T, SHEN C, et al. Decoders matter for semantic segmentation: data-dependent decoding enables flexible feature aggregation. Proceedings of the IEEE Conference on Computer Vision and Pattern Recognition, 2019, 3126-3135.

Realization of Parameterized Serialization Design of Flanges in Chemical Equipment

Dongchen Liu[1], Junli Wang[2] and Yan Zhang[1,*]

[1] School of Computer and Communication Engineering, Liaoning Petrochemical University, Fushun, China
[2] R&D Department, Beijing Oulu Juxin Technology Company, Beijing, China

*Corresponding author email: 976802661@qq.com

Abstract. Flanges are important parts in the fields of machinery, chemicals, wind power, and sewage treatment. They are used in large quantities and have formed a series of products at present. In the paper, the secondary development of AutoCAD was adopted to parameterized series design, and 960 parts from 4 series flange (plate flat welding flange, neck flat welding flange, neck butt welding flange, socket welding flange) were drawn. The efficiency of part design has been significantly improved.
Keywords: Secondary development; Flange; Parameterization.

1. Introduction

Mechanical CAD design occupies a large proportion in the entire project. In the process of mechanical production, it is necessary to complete the production drawing of product model and size through AutoCAD. Many mass-produced products have a large drawing workload and complicated operations, or for a certain product, the user only uses a small part of the functions. Due to the many functions of the CAD software, the existence of other functions affects the user's development efficiency. Therefore, in order to improve the development efficiency of the parametric drawing system, it is necessary to carry out the secondary development of Auto CAD. In order to realize design automation and improve product market competitiveness, the relationship between current user needs and production products should be clear enough, and the designed drawing software will not have many problems [1-3]. This article selects four types of steel pipe flanges: plate flat welding flange, flat welding flange with neck, butt welding flange with neck, socket welding flange as the design objects , By entering the corresponding size, the parts can be drawn quickly, and the parameterized series design of steel pipe flanges can be realized.

2. Parametric Design of Flange

In the design of CAD systems, products of different models are often only different in size and have the same structure. When mapped to a geometric model, the geometric information is different but the topological information is the same. Therefore, the parametric design should reflect the topological structure of the parts, so as to ensure the consistency of the geometric topological relationship in the design process. Flange is a part that connects between shaft and shaft and is used for connection between pipe ends. According to national standards, there are series flange such as: integral flange, threaded flange, butt welding flange, flat welding flange with neck, socket welding flange with neck, etc. Flat welding flanges are suitable for the connection of carbon steel pipes with a nominal pressure of no more than 2.5MPa, and are mostly used when the medium is relatively mild. Butt welding flange is used for butt welding of flange and pipe. Its strength and rigidity are large, so it can withstand high temperature and high pressure environment. Socket welding flange is a flange whose one end is welded to a steel pipe and the other end is connected by bolts. It is often used in pipelines with DN⩽ 40. Flanges are divided into two types. The outer diameter of the steel pipe includes two series, A and B. The A series is the international general series (commonly known as the British pipe), and the B series is the domestically used series (commonly known as the metric pipe). The description and analysis of two-dimensional graphics can be divided into three parts: the topological relationship of the graphics, the coordinates of each point in the graphics, and the relationship between the structural parameters of the graphics (such as the length and width of the graphics). Figure 1 shows the structure of the PN10 butt-welded steel flange with neck. The geometric parameters of the figure are 19 point coordinates DP(x_0,y_0),P_1(x_1,y_1),P_2(x_2,y_2), P_3(x_3,y_3),…,P_{18}(x_{18},y_{18}). Known structural parameters are PN (nominal pressure rating), TYPE_NAME_A_B (flange type), N (flange neck part), A1/B1 (pipe outer diameter), S/R (flange neck part), L (Bolt hole diameter), F (sealing surface part), DN (nominal size), K (bolt hole center circle diameter), D (flange outer diameter), H1 (sealing surface part), C (flange thickness), H (Flange height). Establish the relationship between geometric parameters and structural parameters as shown in Table 1:

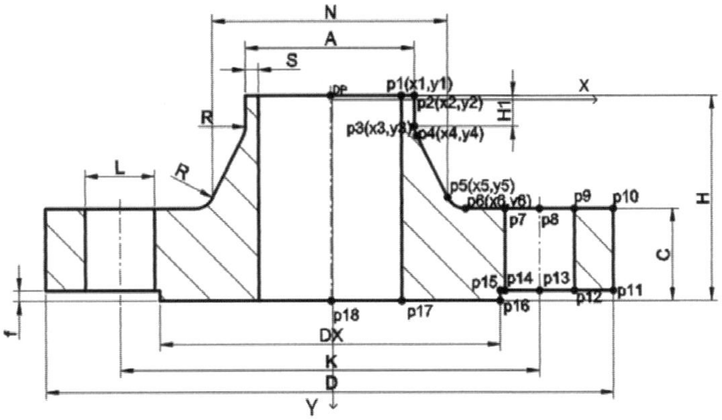

Figure 1. Structure diagram of butt-welded steel flange with neck.

Table 1. Parameter relationship of each coordinate point.

Coordinate points	X-axis	Y-axis	AutoLISP statement
DP	0	0	(setq dp (getpoint))
P1	A/2-S	0	P1(list (+ (car dp) (- (/ A 2) S)) (cadr dp))
P2	A/2	0	P2(list(+ (car dp) (/ A 2)) (cadr dp)))
P3	A/2	-H1	P3(list(+ (car dp) (/ A 2)) (-(cadr dp)H1)))
P4	A/2+R-Rcosα1	-(H1+Rsinα1)	P4(list(- (+(car dp)(/A 2)R))(*R cosα1) (-(+H1(cadr dp)(*R sinα1))))
P5	N/2+R- Rcosα_2	-(H-C-R+ Rsinα_2)	P5(list(- (+(car dp)(/N 2)R))(*R cosα2)) (-(+(-(+(cadr dp)H)C R)(* R sinα2)))
…	…	…	…
P18	0	-H	P18(list (car dp) (- (cadr dp)H)

3. Parameter Storage and Extraction in Flange Serialization Design

According to HG/T20592-2009, a butt-welded steel flange with neck is selected as the research object. Its axial interface structure is shown in Figure 1. The PN10 series flange can be manufactured according to different values of the nominal diameter DN (10mm-2000mm) 29 physical flanges of different specifications, see Table 2. DN refers to the nominal diameter of the pipe to which the flange is connected, not the inner or outer diameter of the flange, and the value of the nominal diameter DN of the pipe is generally close to the inner diameter of the pipe, but in order to unify the standard the nominal diameter is rounded [6-7]. Create a database text file as follows PN10.txt.

```
(No.  DN  A    B   D   K   L   n  Th   C   A   B   S   H1  R  H )
(1    10  17.2 14  90  60  14  4  M12  16  28  28  1.8  6  4  35 )
(2    15  21.3 18  95  65  14  4  M12  16  32  32  2    6  4  38 )
```

Starting from the second line, each space is defined as a value or table, the first value is the serial number, and the following items are: DN (nominal size), K (diameter of bolt hole center circle), L (bolt hole diameter)), n (number of bolt holes), H1 (sealing surface part), C (flange thickness), S, H1, R (structure size of flange neck part), H (flange height), these are in txt The file is stored in numerical form. DN is the key parameter on the user design interface. Data extraction is to extract the corresponding row information (1 10 17.2 14 90 60 14 4 M12 16 28 28 1.8 6 4 35) from the data file PN10.txt through the nominal diameter provided by the user (such as DN=10mm) and assign it to the table structure Variables such as list1. According to this table, the corresponding connection size can be extracted directly[8-10].

Table 2. Related parameters of PN10 butt welded steel pipe flange with neck

(I: Outer diameter of welding end, II: Connection size)

No.	DN(mm)	I		II				
		A	B	D(mm)	K(mm)	L(mm)	n	Th
1	10	17.2	14	90	60	14	4	M12
2	15	21.3	18	95	65	14	4	M12
3	20	26.9	25	105	75	14	4	M12
4	25	33.7	32	115	85	14	4	M12
…	…	…	…	…	…	…	…	…
29	2000	2032	20	2325	2230	48	48	M45×3

Table 3. Related parameters of PN10 butt welded steel pipe flange with neck.

(III :Flange thickness, IV: Flange neck part, V: Flange height)

No.	III C(mm)	IV N(mm)					V H(mm)
		A	B	S≥ (mm)	H1≈ (mm)	R≈ (mm)	
1	16	28	28	1.8	6	4	35
2	16	32	32	2	6	4	38
3	18	40	40	2.3	6	4	40
4	18	46	26	2.6	6	4	40
…	…	…	…	…	…	…	…
29	54	2072	2070	16	3	16	180

4. Flange Dialog Box Design and Parameter Transfer

The dialog box control language DCL is provided in the AutoCAD system to realize the dialog box design function. The information contained in the interface design of the flange dialog box includes: flange type option, PN type option, nominal diameter drop-down menu, as well as the input of specific related parameters, scale, angle, and insertion base point. Choosing the correct parameters can generate different types of 2D views or 3D solid models, but the premise is that you must choose type A or type B, and secondly, you need to select PN type, otherwise a dialog box will pop up and prompt "Please check whether you choose type A or type B ","No PN value selected". The nominal diameter list box contains DN (10mm~2000mm). When the user selects a DN as a parameter, the program will use this parameter as a keyword to traverse the data file, and stop traversing and calling until the corresponding keyword line number in the data file is found. This line of parameters. The program flow chart is shown in Figure 2. Start the dialog box program, the program reads the content of PN10.txt, and its nominal diameter option, PN type, etc. will be displayed in the dialog box. Fill in the scale, angle and insertion base point of the image and click the OK button at the bottom of the dialog box. The program selects one line of data in the txt, transfers the data to the program variables, draws and connects to generate a two-dimensional plan view as shown in Figure 3. The same is true for the three-dimensional map as shown in Figure 4.

5. Design Example of Flange

Through the secondary development of Auto lisp language, click the mechanical part in the custom command menu to call the flange part dialog box, select the AB type and PN model and the matching nominal diameter to get the flange part graphics, as shown in Figure 3. Shown are the results of the parametric serial design of plate flat welding flange, flat welding flange with neck, butt welding flange with neck, and socket welding flange. The four types of flanges are tested with A type PN16 nominal diameter DN=50mm and nominal diameter DN=20mm respectively. The reason why these two sizes are selected for testing is because PN16 plate butt welding flanges and neck flat welding flanges The nominal diameter range is 10mm~600mm, while the nominal diameter range of socket welded steel pipe flanges of the same model is 10mm~50mm, and the nominal diameter range of necked butt welded flanges is 10mm~2000mm.

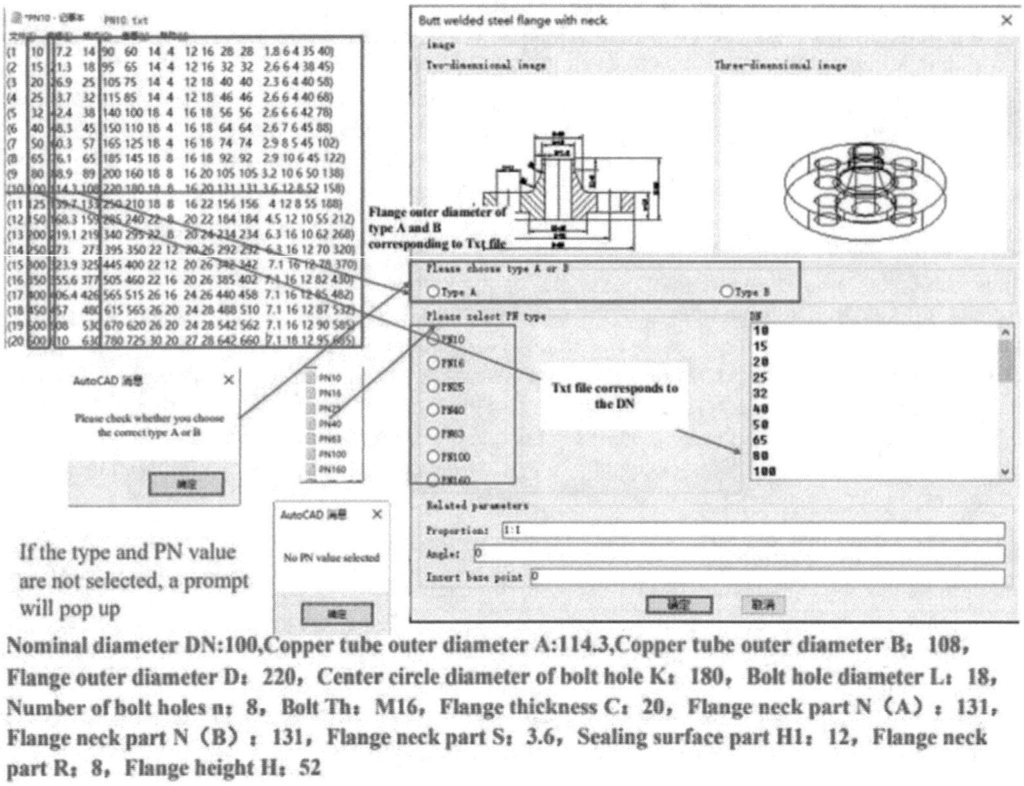

Figure 2. Dialog, data transfer process.

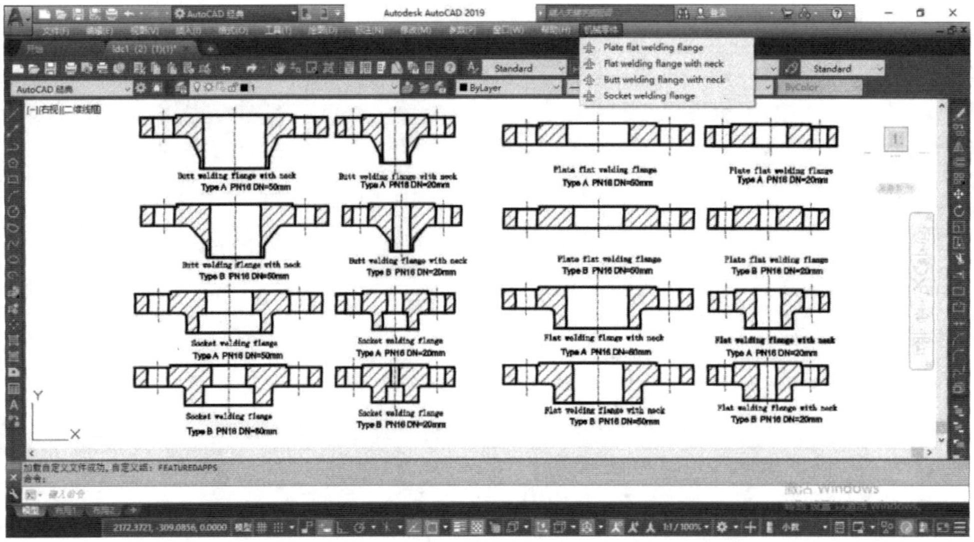

Figure 3. Two-dimensional view of parametric and serialized design of partial flanges.

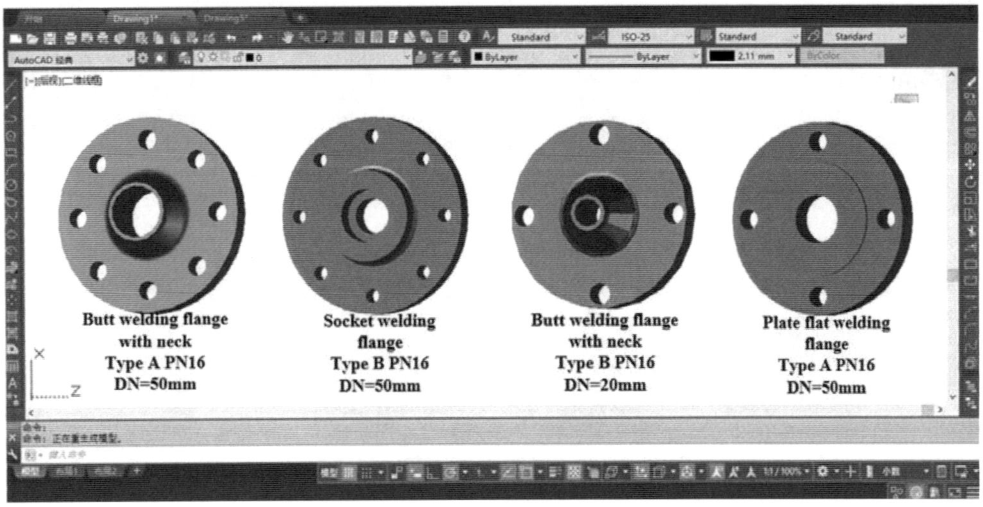

Figure 4. Generation of flange three-dimensional solid model.

6. Conclusions

Based on the secondary development of AutoCAD, this article introduces the process of the parameterized serialization design of the icon name and number of steel pipe flanges (PN series). Through the parametric analysis of the flange, determine the geometric parameters of the flange and the relationship between the parameters; through the serialization analysis, store and extract all the structural data and size information of the flange, and design the dialog box to complete with a visual interface Selection of flange parameter types. Finally, the parametric serial design of four types of steel flanges (960 series parts) was realized. Through the way of thinking in this article, other more complicated engineering drawing work can be realized without having to do a lot of repetitive work. This allows engineers and technicians to devote more energy to the innovative design of products and create better economic benefits for the enterprise.

References

[1] Bing Wan. Research on the application of AutoLISP language in computer-aided design[J]. Information and Computer (Theoretical Edition), 2019(3): 77-78, 81.

[2] Ziping Liu, Ping Xi, Ru Li. Research on the influence of model feature positioning technology on deformation in geometric modeling[J]. Development and Innovation of Mechanical and Electrical Products, 2017,30(05):82-84.

[3] Devaraj S. Malkapuram Ramakrishna, Singaravel B. Performance analysis of micro textured cutting insert design parameters on machining of Al-MMC in turning process[J]. International Journal of Lightweight Materials and Manufacture,2021,4(2).

[4] Huixia Wang, Decong Zheng, Hang Ding. The secondary development of the motor design system based on Auto LISP [J]. Agricultural Engineering, 2020, 3(3): 31-34.

[5] Lei Zhang. Realization of key technologies in the design of reducer based on Auto LISP[J]. Journal of Anhui University of Science and Technology (Natural Science Edition), 2018.38(2): 83-86.

[6] Huayong Zhou, Hongmei Zhang, Lulin Wang. Parametric drawing design based on AutoLISP language[J]. Machinery Manufacturing 56 volumes, 2018.5: 40-41.

[7] Haipeng Lan. The secondary development of drawing commands for the open sample drawing of the square-circle transition joint exhibition based on AutoLISP[J]. Information and Computer, 2020: 45-48.

[8] Yanru Liu, Feng Liu, Xiaochun Zheng, the secondary development of AutoCAD line type selection based on AutoLISP [J]. Coal Mine Machinery, 2017.38(7):158-159.

[9] Kehu Wang, Xiaohui Li, Jie Jia, Shaopeng Xie, Kunfei Lang. Secondary development of

reserved holes for positioning plates based on AutoCAD [J]. Land and Resources Exploration Center of Hebei Bureau of Geology and Mineral Resources. Precision Manufacturing and Automation, 2020(4): 38-41.
[10] Feng Feng, Fan Li, Tao Fu. Analysis of national standards for steel pipe flanges at home and abroad [J]. Standardization and Quality of Machinery Industry, 2017(7): 13-20.

Research and Development of Electric Power Lifting Truck

Zhao Yao[1], Hejia Li[1,*] and Liang Zhou[2]

[1] Non-commissioned Officer School of Army Academy of Armored Force, Changchun, Dalian University of Technology, Dalian, China
[2] 32381 Force

*Corresponding author email: 275039646@qq.com

Abstract. Equipment and materials are completely dependent on large machinery and human forklift, which is difficult to ensure the high efficiency of transportation. As a supplement to large transportation machinery and vehicles, it can make up for the lack of low transportation capacity of human-based transportation equipment, and solve the prominent problems such as backward material transportation equipment, poor road adaptability, time-consuming and laborious. The electric power assisted lifting truck to be developed can realize "one kilometer" of materials the transportation reserve with high travel efficiency has become the "range extender" of large vehicles. With the help of high and new technology, the R & D of electric power assisted lifting truck adopts the contractor guarantee mode to realize complementary advantages, resource sharing, joint development and win-win cooperation, so as to provide technical support for material support. The electric power assisted lifting truck is mainly composed of walking system, lifting system, control system and auxiliary system, which meets the needs of strong road trafficability, time-saving and labor-saving, simple operation, convenient maintenance, safety and reliability and low cost, and improves the rapid, accurate and convenient guarantee ability of materials.
Keywords: Electric power lifting truck; Walking system; Lifting system; Control system; Auxiliary system.

1. Introduction

In order to overcome the impact of road conditions and personnel physical endurance on the handling efficiency of combat readiness materials, the electric power assisted lifting carrier is used as a supplement to large transport machinery vehicles, make up for the lack of low transport capacity of human-based transport equipment, meet the needs of strong road trafficability, time-saving and labor-saving, simple operation, convenient maintenance, safety and reliability and low cost, and realize the "one kilometer" of combat readiness materials High efficiency transport reserve.

2. Research Status

Domestic well-known electric storage light and small truck manufacturers mainly include Noli Intelligent Equipment Co., Ltd., senmi (Shanghai) Industrial Equipment Co., Ltd. and nilang (Shanghai) Industrial Equipment Co., Ltd. Its light handling vehicle products can be divided into six categories: manual handling vehicle, walking (or semi-automatic) electric handling vehicle, pedal (or stand driving) electric handling vehicle, side leaning electric handling vehicle, counterweight forklift and high-altitude platform. Electric storage light and small truck belongs to industrial vehicle, which is a mechanical product of handling, storage and loading and unloading driven by internal combustion engine, battery or manpower. It can greatly save labor cost, improve labor efficiency, effectively avoid production accidents caused by human fatigue, and has broad application space[1]. However, in the complex environment and bad working conditions, if the common light handling vehicles in the market are used for the storage of goods handling and other materials, they will expose the lack of poor adaptability to the storage environment. The light truck commonly used in the market is mainly designed for the working environment with flat roads and ambient temperature in factories, warehouses, stations and yards, which can not adapt to harsh environmental conditions such as complex road conditions; The structural material is traditional structural steel, resulting in large weight of the whole machine, which is difficult to meet the requirements of portable and lightweight design; In order to achieve large lifting force, the hydraulic drive system is adopted, which has inherent defects such as oil leakage, temperature sensitivity, heavy structure and so on; The carrying platform has no anti-skid and anti falling design, which can not meet the needs of fast and stable transportation of dangerous goods in complex terrain in the field[2].

3. Overall Scheme

The electric power lifting truck to be developed is mainly composed of walking system, lifting system, control system and auxiliary system. The structural diagram of the electric power lifting truck is shown in Figure 1.

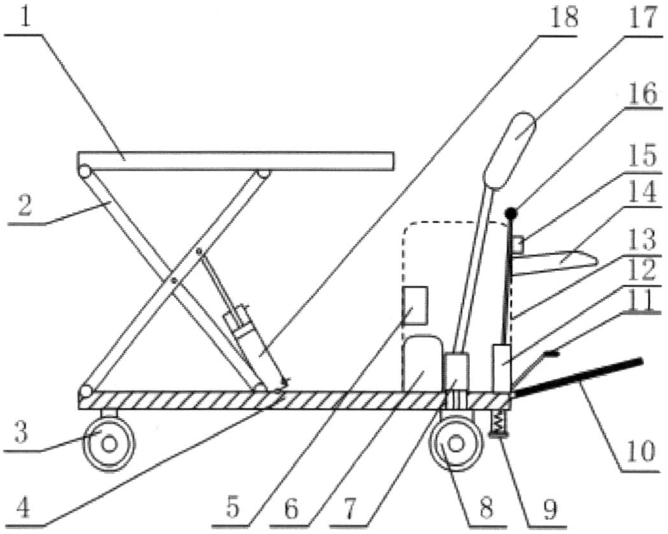

1.Carrying platform 2. Anti pinch fork lifting mechanism 3. Bearing wheel 4. Frame 5. Integrated control box 6. Lithium battery 7. Motor 8. Driving wheel 9. Safety brake 10. Operating pedal 11. Lifting pedal 12. Manual hydraulic pump 13. Chassis shield 14. Safety arm 15. Toolbox and parts box 16. Hydraulic rod 17. Handle head 18. Hydraulic cylinder

Figure 1. Structural diagram of electric power assisted lifting carrier.

Carrying platform: used to carry ammunition, equipment, oil and other materials, and can be used as a repair workbench.

Anti pinch fork lifting mechanism: used to lift the carrying platform.

Bearing wheel: used to support the vehicle body.

Frame: used to carry the whole vehicle parts.

Integrated control box: used to accommodate the electrical control elements of the whole vehicle.

Lithium battery: power supply for the whole vehicle.

Motor: provides power for the traveling system.

Driving wheel: as the power output device of the traveling system, it drives the carrier to move forward and backward.

Safety ground brake: lower the ground brake during lifting operation, which can effectively stabilize the vehicle body and prevent the vehicle body from shaking or moving. At the same time, it can prevent the wheel from local compression and deformation for a long time.

Operating pedal: operator's riding position.

Lifting pedal: an operating mechanism that manually drives the lifting mechanism to realize the lifting action.
Manual hydraulic pump: the power source of the hydraulic system.
Enclosure: used to accommodate precision components and control elements. It has the functions of dust prevention, protection and sound insulation.
Safety guard arm: ensure the stability of the operator's center of gravity, prevent falling and ensure safety.
Tool box and part box: hang on the lower part of the console at the rear of the vehicle body. It is convenient to take and place working parts, accessories and small tools during operation.
Hydraulic lever: control the switch of pressure relief valve to realize lifting conversion.
Handle head: a manual steering control handle integrating warning horn operation, emergency brake switch and other functions.
Hydraulic cylinder: the actuating element of the hydraulic system, which is used to push the scissor lifting mechanism to rise[3].

4. Main Technical Analysis

4.1. Research on Walking System

Wheel selection design. The tire material and structure are polyurethane solid tires. Compared with similar products, polyurethane solid tire has a series of advantages, such as good wear resistance, high tensile strength, high tear strength, impact resistance, resilience, aging resistance, shock absorption, water resistance, acid and alkali resistance, large bearing load, long service life and low price, which can meet the design requirements. In terms of wheel size, the tire diameter should be greater than the over ditch width. Empirical calculation can be carried out according to the set over sill height and over ditch width. Proper tire width can ensure that the truck can pass through the relatively soft road smoothly and ensure sufficient grip.
Research on drive axle and steering mechanism. It adopts rudder handle steering mechanism, floating drive axle and parallel design of double floating drive wheels. It can ensure that the driving wheel of the carrier can achieve favorable grip pressure and large grip area under different road conditions, so as to ensure flexible and labor-saving steering, firm ground and not easy to slip. The turning radius is small, and it can be flexible between material stacking in the warehouse[4].

4.2. Research on Lifting System

Mechanical structure: the lifting mechanism adopts anti pinch shear fork structure design. The utility model has the advantages of stable lifting, high stability, simple structure, reliable operation, etc. It is made of square steel, which can effectively save cost while ensuring sufficient strength. Check the length and section size of the boom and calculate according to the lifting height and load capacity.
Hydraulic power components: hand operated (foot operated) hydraulic pump. In case of light load or no-load, fast pedal is used to realize rapid height lifting. In case of heavy load, hand mechanism is used to realize greater lifting thrust by using the labor-saving principle of lever.
Hydraulic actuator: integrally cast single acting hydraulic cylinder. It can be purchased from the market, with simple structure, low price, reliable operation, firmness and durability.
Hydraulic control element: the oil circuit is equipped with adjustable constant flow valve to ensure the smooth and slow decline of the handling object under heavy load, and realize the rapid rise and fall of the height of the carrying platform under no-load or light load; The overflow valve is set inside to provide overload protection function, which can effectively avoid the damage of parts caused by excessive loading and instantaneous overload on bumpy roads and reduce the maintenance cost. All kinds of directional, pressure, speed and flow control valves have been standardized. They can be purchased flexibly according to the action demand. They have low price, stable performance, reliable operation and low cost[5].
Design and research of material carrying platform.
The size of the material carrying platform is designed according to the volume of the ammunition box and the carrying vehicle load, which is suitable for the whole box handling of various materials; The

size of ammunition loading platform and loading die are designed according to the volume of projectile, cartridge, fuse and handling vehicle load, which is suitable for unpacking and separate handling of ammunition. The carrying platform can also be used as a repair workbench. When repair work is required, lift the material carrying platform to an appropriate height through the lifting mechanism, and then carry out repair work for the workbench.

4.3. Research on Control System

Control terminal: fully consider the rudder handle head design of ergonomics, and each function button is convenient and easy to operate. The whistle button, down button, up button, travel button and electricity meter are moved to the handle head. The electricity detection is more intuitive, which is convenient to remind the operator to charge in time and protect the battery.
Power supply: lithium battery is used for power supply, which has the characteristics of large capacity, long service life, fast charging and maintenance free. In addition, the backup battery equipped with the vehicle can ensure the continuous operation of the electric power assisted lifting truck[6].

4.4. Research on Auxiliary Mechanism

Safety arm guard: the foldable safety arm guard and humanized arm guard structure are adopted to make the opening and closing operation of the arm guard convenient and labor-saving. It can improve the stability of operators during vehicle driving and prevent operators from falling off the driving pedal due to loss of center of gravity. When the protective arm is closed, it can only drive at low speed. After the protective arm is opened, it can drive at full speed to improve the safety of operators.
Safety ground brake: it is designed based on the self-locking principle of connecting rod mechanism and made of high-quality steel plate to ensure sufficient strength and service life. When in use, the carrier can be raised and fixed by stepping on the foot pedal, so as to ensure the stability of the vehicle chassis during lifting operation, effectively prevent the carrier from sliding and rolling during loading and unloading materials, and avoid safety accidents. At the same time, it can protect the wheel and avoid hub deformation caused by long-term unidirectional heavy pressure. The built-in spring can make the rubber foot pad close to the ground, further increasing the support stability.
Operating pedal: the operating pedal adopts folding design, which does not need to use the pedal or can be manually retracted during operation in a narrow space, so as to save the working space. The anti-skid design can avoid accidents of personnel and equipment caused by accidental sliding of operators when the vehicle is driving on rough roads or slopes. The cushioning design can eliminate the impact of the operator on the road surface during driving.
Design and Research on material fastening device: in order to avoid the handling object from bumping on the road or accidental sliding during climbing, a snap pin fastening device is designed to ensure the safety and stability of the handling object.
Emergency disconnect switch. Emergency power-off switch is set for emergency power-off and shutdown in case of system failure or out of control to avoid accidents.
Tool box and part box: hang on the lower part of the console at the rear of the vehicle body. It is convenient to take and place working parts, accessories and small tools during operation[7].

5. Conclusion

The electric power assisted lifting carrier can efficiently, safely and reliably complete the transportation of combat readiness materials under complex road conditions, fill the gap of automation and intelligence of combat readiness material transportation equipment of service support unit, and provide technical support for combat readiness material supply support. It has strong military and economic benefits, and has a broad application prospect.

References

[1] Wang Haibo, 2018. Research on energy recovery of lifting system of electric forklift [J]. Machine tool and hydraulic
[2] Nie Bo. Simulation Research on energy saving system of hydraulic lifting device of 2021 electric forklift [J]. Jiangsu University

[3] Guo Chaoyong 2015. A micro step control method of permanent magnet synchronous motor for sail driving mechanism [J]. Space control technology and application
[4] Jia Lei, 2016. Study on electromagnetic design and operation characteristics of wound rotor brushless doubly fed machine [D]. Huazhong University of science and technology
[5] Chen Butong, 2013. Energy consumption analysis and energy saving control technology research of forklift hydraulic system [J]. China Agricultural Machinery chemistry news
[6] Wu Hao, 2013. Research on energy recovery control strategy of parallel hybrid forklift [D]. Hefei University of technology
[7] Wu Ye 2016. Research on energy saving of forklift lifting system [D]. Taiyuan University of science and technology

Research on Design of the D-knotter Split Spindle

Jiayu Guo[1], Haitang Cen[2,*], Hao Tang[3], Zihan Shuang[4] and Risu Na[5]

[1] School of Mechanical Engineering, Inner Mongolia University of Technology, Hohhot, 010051, P.R. China

*Corresponding author email: 1054949591@qq.com

Abstract. The knotter is the core component of the square baler. The structure of the knotter is complex, the working load is large and the working environment is bad, which makes the knotter easily damaged. The imported D-knotter is widely used at present, once the parts are damaged, they often need to be disassembled and assembled as a whole, which takes a long time and seriously affects the efficiency of field operation. Aiming at the problem that the D-knotter assembly and its parts are inconvenient to replace after damage, this paper designs the split spindle of the knotter, and ensures the strength requirements of the spindle when working through an integral key connection. When replacing the knotter parts, the spindle is separated from the middle, and the knotter assembly is directly removed without removing other parts on the spindle, which obviously saves the replacement time of the damaged parts of the knotter. The statics and modal analysis show that the split spindle of the knotter can meet the strength requirements, and the spindle does not produce resonance.
Keywords: Knotter; Split spindle; Modal analysis.

1. Introduction

The square grass baler is a seasonally used mechanical product, especially in the vast areas of the north. The annual use of the square grass baler is generally between October and November [1-2]. At this time, it is the season of forage harvest, and the workload is large and the harvest time is tense. Because once there is heavy snow in the north, it will affect the normal harvest of forage, which requires the square grass baler to ensure high baling efficiency and stable product quality. The knotter is the core and the most prone to problems in the mechanical parts of the square baler. Once damaged, it will lead to the square baler can not work normally. Therefore, the quality stability of the knotter and the convenience of its replacement have higher requirements.

From the point of view of the quality stability of the knotter, the knotter designed and produced by domestic manufacturers is often poor in reliability. In the working process of the square baler, there is generally a problem that the service life of each part of the knotter assembly is not uniform. For example, the design life of the knotting jaw is 100,000 bundles. However, in the actual working process, when the bundle is 30000~40000 bundles, it will produce large abrasion and affect the normal release[3], so it is necessary to replace the parts. Many scholars have carried out a large number of studies on the mechanical properties and bearing analysis of the key components of the knotter[4-6], but these studies are carried out for the important parts of the knotter. In actual production, once the key components of the knotter of the square straw baler are damaged, they need to be replaced. The process of dismantling the knotter assembly is very complicated and takes a long time. In this paper, the integral spindle is divided into two parts and connected with an integral key to save the time of replacing the damaged parts of the knotter.

2. Main Spindle Design

2.1. Structure Design of the Main Spindle

The knotter is the core technical component of the baler. It mainly uses two structural forms: type C and type D. Among them, the D-knotter is widely used because of its low failure rate, fast knot forming action and compact structure. The power input of D-knotter is provided by the driving gear disc. It has a power input spindle and three power output routes. The power output forms are gear meshing and cam mechanism to drive other parts of the knotter, and the driving gear disc is driven by the main spindle. The existing D-knotter spindle is an integral type[7]. If the damaged knotter assembly is replaced, the dust cover, transmission sprocket, clutch, spindle frame and so on need to be removed one by one. The process is very cumbersome and takes a lot of time. Even if the individual parts in the knotter can be replaced without dismantling the spindle, the operation space is also very small.

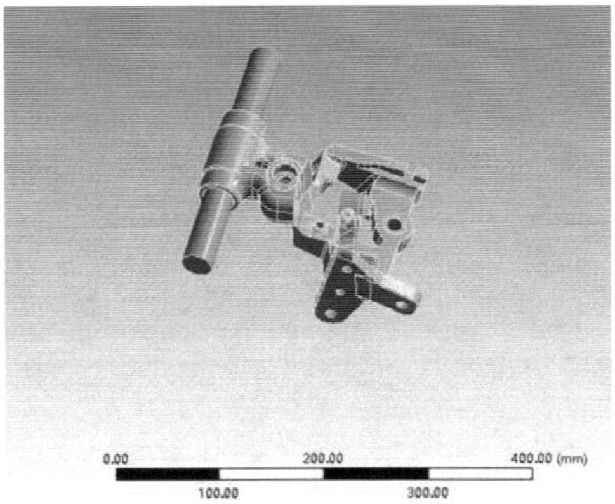

Figure 1. Drive spindle and bracket simplify assembly model.

In this paper, the main spindle is divided into two sections, and the separated two sections are connected by an integral key, but this design can not realize the positioning of the main spindle itself in the axis direction. It is necessary to apply a certain extrusion force to the main spindle with the help of external force to ensure the integrity of the two sections of main spindle. Therefore, the support bearings at both ends of the main shaft should be changed into tapered roller bearings that can bear both axial force and radial force, and mainly bear radial load. In addition, considering the weakening of the main spindle of the split design in strength, stiffness and vibration stability, a split shaft sleeve is designed at the split of the shaft as a reinforcement and also as an axial positioning component of the knotter gear disc and the frame body. The structural diagram of the shaft sleeve is shown in Figure 2.

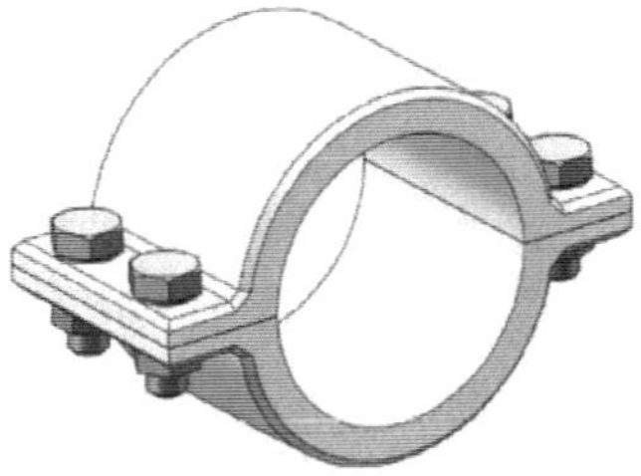

Figure 2. Shaft sleeve structure diagram of spindle split position.

When the split spindle needs to replace the parts of the knotter, it only needs to remove the bearing seats on both sides, and the spindle is disassembled as a whole. The spindle is separated from the middle, and the knotter assembly is directly removed without removing other parts on the spindle, so as to reduce the difficulty of replacing the damaged parts of the knotter, save time and improve the efficiency of the binding test. The structure of the improved knotter spindle has changed. In order to facilitate processing, the shape of the spindle is the optical spindle without a spindle shoulder, and the axial positioning of each component is realized by the shaft sleeve.

2.2. Calculation of Minimum Diameter of the Main Spindle

Since the main spindle of the knotter mainly bears torque and certain bending moment, the preliminary estimation of shaft diameter is carried out by reducing the allowable torsional shear stress. The torsional strength condition of the main spindle is:

$$\tau_r = \frac{T}{W_T} \approx \frac{9550000\frac{p}{n}}{0.2d^3} \leq [\tau_T] \tag{1}$$

In the above formula, τ_r is the torsional strength of the spindle, T is the torque on the main spindle and taken as T = 200N • m, W_T is the bending section modulus of the main spindle, n is the speed of the main spindle and taken as n = 90r / min, p is the transmitted power of the main spindle, p = 1.94Kw, d is the diameter of the main spindle, $[\tau_T]$ is the allowable shear stress.

According formula (1), the diameter of the main spindle is:

$$d \geq A_0 \sqrt[3]{\frac{p}{n}} \tag{2}$$

In formula (2), A_0 is the calculation constant, which is related to the material and bearing condition of the spindle, and takes A_0= 9550. According to the relevant data calculated above and the material parameters of 45 steel obtained by consulting the mechanical design handbook, it is substituted into the formula, and the preliminarily estimated minimum spindle diameter is d ≥ 28.4mm.

2.3. Extrusion Strength Check of Integral Key

The integral key connection is multi-gear, with strong bearing capacity, good neutrality and guidance, shallow tooth root, small stress concentration and small weakening of the shaft and part hole sleeve. The involute spline has the characteristics of high bearing capacity, compact structure, high precision and automatic centering[8]. Therefore, according to the actual use requirements, this paper selects an involute spline to connect the two parts of the main spindle. By referring to the basic dimension table of an involute spline in the mechanical design handbook, according to the actual working conditions and the calculation results of the minimum shaft diameter, the basic parameters of the selected involute spline are shown in Table 1.

Table 1. Basic parameters of the involute spline.

Modulus	Pressure Angle	Tooth Number	Pitch Circle Diameter	Working Length	Working Height of Tooth Side
1mm	30°	29	29mm	35mm	1mm

The main failure mode of integral key is that the working face is crushed (static connection) or the working face has excessively abrasion (dynamic connection)[9-10]. The integral key designed in this paper belongs to a dynamic connection, so the conditional strength is calculated according to the pressure on the working face. The calculation formula is:

$$p = \frac{2T}{\psi zhlD_m} \leq [p] \tag{3}$$

In formula (3), ψ is the uneven coefficient of load distribution and takes ψ= 0.8, z is the tooth number, h is full tooth height, l is spline joint length, D_m is average diameter. Refer to the mechanical design handbook and determine the allowable pressure of integral key connection as 10MPa ~ 20MPa according to the connection mode, use and manufacturing. Other parameters are brought in according to Table 1.

$$p = 16.99\text{MPa} \leq [p]$$

Meet the extrusion strength requirements of integral key, as shown in Figure 3

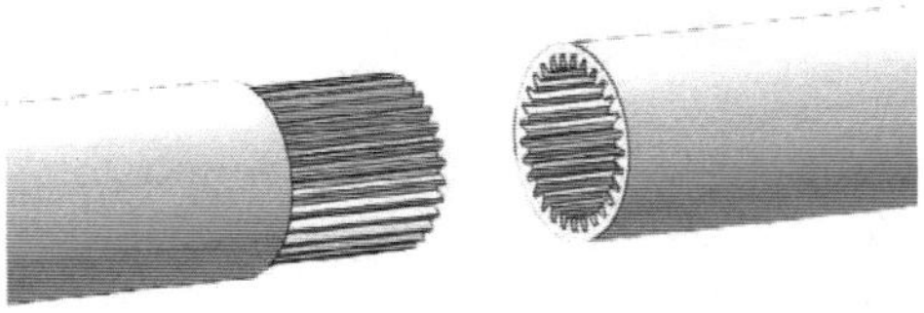

Figure 3. Spline connection model of the separated spindle.

3. Main Spindle Static Analysis

According to the above analysis, there are many parts on the knotter main shaft, and some parts have complex structures. To facilitate the static analysis of the spindle, this paper introduces each component on the spindle into ANSYS, defines the corresponding material, applies gravity load, calculates its gravity, and then converts the gravity of each component into external load and applies it to the spindle in the process of spindle analysis[11-13]. In order to facilitate the application of external load on the spindle, segment modeling is carried out in the spindle modeling process, and then the spindle is assembled. In this way, each part corresponds to a section of the spindle. When applying load, the corresponding force application surface can be directly selected. Figure 4 and Figure 5 are the segment modeling diagram of the spindle and the stress diagram of the spindle respectively.

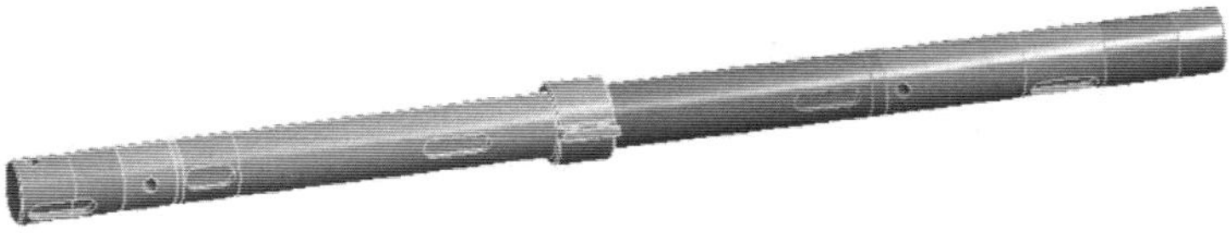

Figure 4. Piecewise establishment of spindle model.

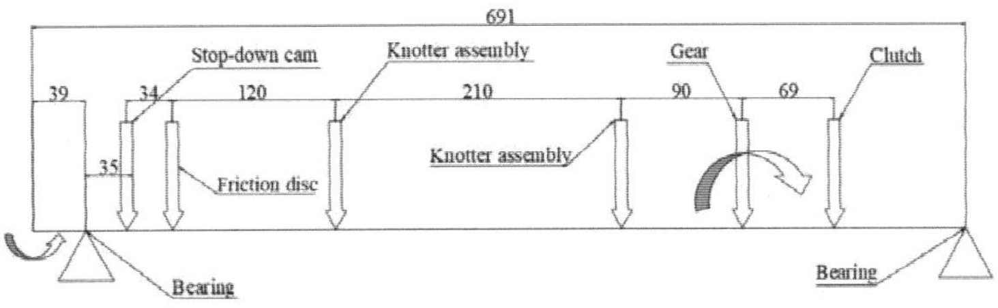

Figure 5. Force diagram of the main spindle.

With the help of the statics module of ANSYS, the geometric model is imported into ANSYS, then defining the material, and the external load, torque and its own gravity load transformed by the gravity of each component are applied. Since both ends of the main shaft are supported by bearings, the main shaft is regarded as a rigid constraint, then applying fixed constraints on the bearing mounting surface of the main shaft. The load application is shown in Figure 6.

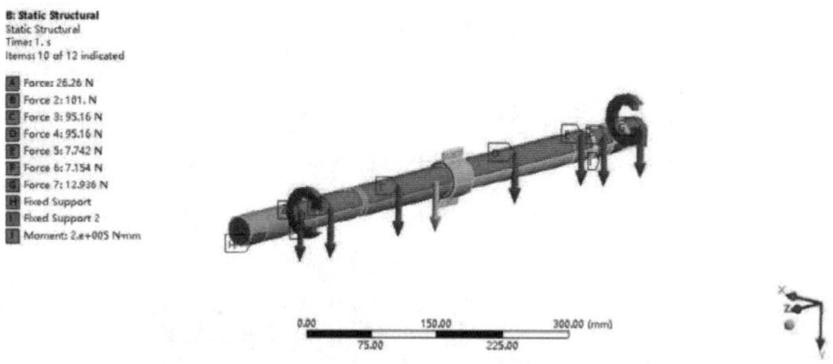

Figure 6. Spindle applied load diagram.

With the help of ANSYS analysis and calculation, the total deformation diagram and equivalent stress diagram of the main shaft are obtained as follows:

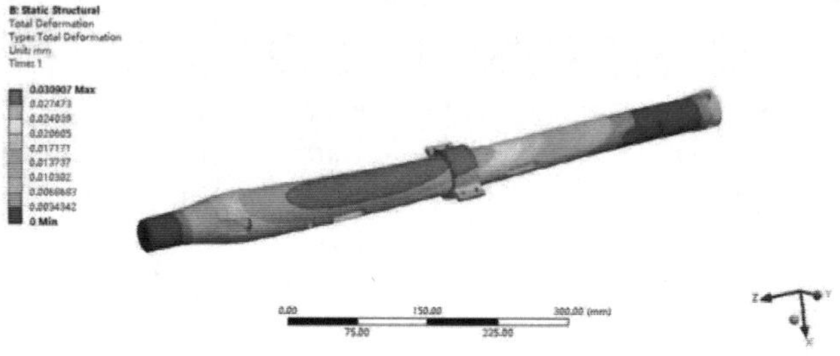

Figure 7. Total deformation diagram of the spindle.

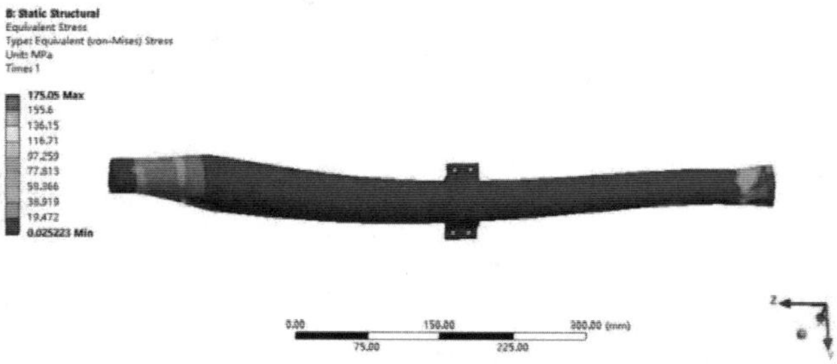

Figure 8. Equivalent stress diagram of the spindle.

It can be seen from Figure 7 that the maximum deformation of the main shaft is 0.03mm, which meets the use requirements.It can be seen from Figure 8 that the bearing supports on both sides of the main shaft have greater stress. Since one side of the main shaft is equipped with a crank (right end of the main shaft in Figure 8), the bearing can only be installed outside the crank, and the bearing on the other side of the main shaft is directly installed on the main shaft (left end of the main shaft in Figure 8). It can be seen from the figure that the maximum stress occurs at the bearing installation at the right end of the main shaft (on the bearing installation with crank), the maximum stress is 175.05Mpa, and the material of the shaft is 45 steel. Its yield limit is 430Mpa, meeting the strength requirements.

4. Results and Discussion

The knotter is a relatively precise mechanical product, so its working process requires that each part of the knotter can achieve very precise cooperation[14-17]. If there is a problem with the cooperation of each part, it is very likely to reduce the knot forming rate and cause collision damage to each part of the knotter. For the design of the main shaft, the main consideration is the coordination of the relative position between the needle brake, the bill hook and the clamp rope tray. Because if the main shaft vibrates greatly, it will lead to the change of the position of the knotter assembly, while the needle

brake driven by the crank and installed at the lower part of the bale chamber will hardly be affected by the vibration of the main shaft because it is connected outside the baling room. In this way, if the main shaft has a strong resonance, the coordination between the needle brake, the bill hook and the clamp rope tray will be affected. If it is light, it will not form a knot, and if it is heavy, it will collide between the needle brake and various parts of the frame. In addition, the main shaft will also be damaged by torsional fatigue and bending fatigue prematurely due to the rapid increase of dynamic stress. Based on the above description of the work engineering of the spindle and knotter, it is only a static analysis, which can not fully satisfied the design requirements of the spindle.
According to the knotter completing a knotting action in 0.67s, it can be obtained that the working frequency of the main shaft is 1.49Hz. In order to make the spindle have reasonable dynamic characteristics and avoid the resonance that may be caused during normal operation, it is necessary to conduct modal analysis on the spindle to avoid its resonance or vibration at a specific frequency. The geometric model of the spindle is imported into the modal analysis module of ANSYS for material definition and meshing. The load application is consistent with the above static analysis[18-20]. The default program-controlled analysis method is adopted, and the number of extended modes is set to 6. The natural frequencies and mode shapes of the first six modes of the spindle are obtained in Table 2.

Table 2. Natural frequencies and mode shapes of the first six modes of the spindle.

Module	Natural Frequency	Description of Vibration Mode
1	381.82Hz	The middle of the spindle swings along the Y-axis
2	406.61Hz	The middle of the spindle swings along the X-axis
3	1052.9Hz	The second-order torsion of the middle of the spindle in the YOZ-plane
4	1133.3Hz	The second-order torsion of the middle of the spindle in the XOZ-plane
5	2062.9Hz	The third-order torsion of the middle of the spindle in the YOZ-plane
6	2153.9Hz	The third-order torsion of the middle of the spindle in the XOZ-plane

The first six order modal analysis results are shown in Figure 9. It can be seen from the figure that the amplitude of each order of the spindle is relatively large. Therefore, once the spindle resonates in the working process, it will affect the normal work of the knotter. The vibration of the spindle is mainly concentrated in the middle of the spindle and the installation of the frame on both sides.

Journal of Physics: Conference Series **2218** (2022) 012077 doi:10.1088/1742-6596/2218/1/012077

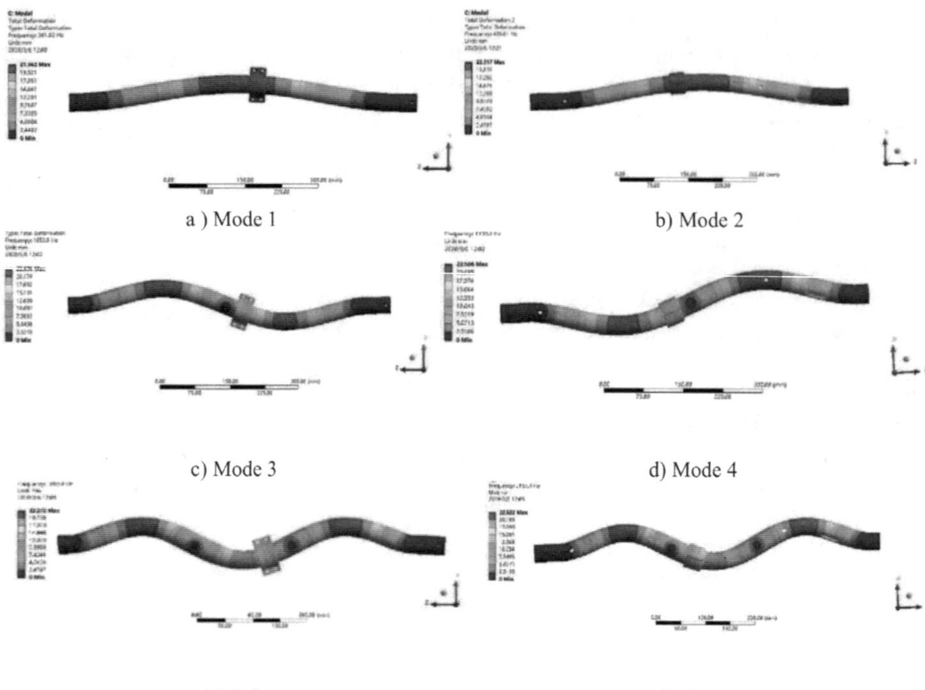

a) Mode 1 b) Mode 2

c) Mode 3 d) Mode 4

e) Mode 5 f) Mode 6

Figure 9. The first six modal analysis Results of the spindle.

In the first-order and second-order modes, it is mainly manifested as the swing of the middle part of the spindle along the Y-axis and X-axis, respectively. In the third-order and fourth-order modes, the main performance is the second order torsion in the middle of the spindle, which is located in the YOZ-planes and XOZ-planes respectively. In the fifth-order mode, the spindle has three twist swings, and the maximum amplitude is 22.27 mm. In the sixth-order mode, the spindle has three twist swings, and the maximum amplitude is 22.62 mm. From the results of the above modal analysis, it is concluded that the natural frequency of the knotter spindle is greater than 381.82 Hz, and the external load frequency of the spindle is 1.49 Hz, which is greatly different from the lowest natural frequency. Therefore, the spindle does not cause resonance when working normally, which meets the design requirements.

5. Conclusions

(1) Because of the problems of the existing D-knotter, such as the trouble in the replacement process and possible influence on the binding efficiency, the structural improvement of the main spindle is carried out. The original main spindle is divided into two parts and connected with an integral key, and the geometric dimensions such as spindle diameter are determined according to the load force analysis of the component.

(2) With the help of ANSYS, the static analysis of the improved spindle structure is carried out, and the total deformation diagram of the spindle and the equivalent stress diagram of the spindle are obtained. The maximum deformation of the spindle is 0.03 mm, and the maximum stress occurs at the installation of the right end bearing of the spindle. The maximum stress is 175.05 MPa, the material of the spindle is 45 steel, and the yield limit is 430 MPa, which all meet the strength requirements.

(3) According to the modal analysis of the improved spindle structure by ANSYS, the natural frequency of the knotter spindle is greater than 381.82 Hz, which is much larger than 1.49 Hz of the external load frequency of the spindle. Therefore, the spindle does not cause resonance when working normally, which meets the design requirements.

Acknowledgements
This work was financially supported by the Fund Project of Inner Mongolia Autonomous Region Science and Technology Plan Project (20190021).

References
[1] Wang X Y 2017 What kind of baler the market needs *Agricultural machinery market* p 30-31
[2] Li D P and Liu X Y 2002 Industrialization of forage and development of forage baler *Feed Wide Angle* p 10-12
[3] Yin J, Ji Z, Wang X X and Zhu H 2018 Fault analysis and surface shape improvement design of knotting mouth rope *Journal of Agricultural Machinery* p 82-89
[4] Liu Z Q 2009 Structural performance analysis of D-type knotter jaw *Hohhot: Inner Mongolia Agricultural University* p 118-124
[5] Li H T, Xiong Y and Chen L J, etc 2015 Wear analysis and improvement design of knotter cutting rope tripping mechanism *Journal of Agricultural Machinery* p 118-124
[6] Chen L J,Li C and Zhang A Q, etc 2015 Load test analysis of knotting jaws during straw baling *Journal of Agricultural Machinery* p 128-134
[7] Zhu H 2019 CNC machining and wear resistance analysis of the spindle hole of the double tooth drive knotter bracket *Jiangsu University* p 49-57
[8] Editorial Committee of Practical Manual for Mechanical Design 2013 *Practical Manual on Mechanical Design: 3rd Edition Beijing: Machinery Industry Press*
[9] Cheng G 2016 Reliability analysis of eccentric spline transmission rod *Science and technology outlook* p 171
[10] Pu L G,Chen G D and Wu L Y 2013 *Mechanical design: ninth edition Beijing: Higher Education Press*
[11] Fu S Y, Zheng B and Wen C 2021 Finite element analysis of engine connecting rod with different material properties *Mechanical design*
[12] Wang S F 2021 Research on transient dynamics analysis of gangue separation manipulator based on ANSYS *Coal technology*
[13] Bi F Q, Wang W L, Zhang L K, Chen J J, Jiang X, Zhang R B and Li Z 2021 Statics analysis of 6 DOF manipulator *Equipment machinery 15-18 + 27*
[14] Zhang A Q,Chen L J and HAN L j 2015 Synchronous acquiring system about information of bale knotter movement and cord tension *Transactions of the Chinese Society for Agricultural Machinery* p 31 -37
[15] Yin J J,Li S and Li Y M 2011 Kinematic simulation and time series analysis of D-knotter and its ancillary mechanisms. *Transactions of the Chinese Society for Agricultural Machinery* p 103-107
[16] Chen L j,Li C and Zhang A Q 2014 Reconstitution and motion simulation of D-hale knotter based on reverse engineering *Transactions of the Chinese Society for Agricultural Machinery* p 104-108
[17] Huang W 2009 Research on action and efficiency of D-knotter *Hohhot: Inner Mongolia Agricultural University* p 6-26
[18] O Z, Hu Z Q and Huang R H 2019 Optimization Design of Spindle Box Based on ANSYS *Machinery* p 61-66
[19] Ma H 2019 Modal analysis of split frame of graded crusher based on ANSYS *Proceedings of the National Coal Preparation Academic Exchange*
[20] Du J, Li J D, Zhao Z G, Meng Q K, Zhao Z K, Liang C, Zhao C and Liu C Y 2019 Modal analysis of disc mower frame based on ANSYS Workbench *Mechanical design*

Design of New Charging Pile Based on Pro/E

Fenfang Tan[1], Zhihang Tan[1,*], Song Wu[2], and Youping Peng[1]

[1] School of Mechanical Engineering, Wuhan Polytechnic University, Wuhan 430023, PR China
[2] Zhejiang Sany Equipment Limited Company, Zhejiang 313002, PR China

*Corresponding author email: tanzhihang@126.com

Abstract. This paper uses Pro/E, CAD and 3Ds max software to complete the modeling design of the new charging post.Firstly, 3D modeling, process analysis and calculation of the new charging pile parts were carried out. Then a solid model was established, according to the injection capacity, clamping force and injection pressure, the injection machine was initially selected, the cycle diagram of injection molding production process was drawn, and the mold opening and closing simulation was carried out. Finally, mold flow software was used to analyze the manufacturing process of the charging pile, virtual simulation of the problems of the injection machine during the injection process, and checking the feasibility of the design.A new type of charging pile is designed which has certain guiding significance for actual production.
Keyword: Injections; Park cartoon charging post; Process analysis; Pro/E; Mold Flow; Mould design.

1. Introduction

Through the study and analysis of the current situation of car battery charging piles at home and abroad, China's position in the world of charging cars is still at a low level, and there are many aspects to be improved [1][2][3] .At present, the shape of the charging post in the market, except for a few in the middle of the hole, curved, most of them are mainly square, the shape of a single, can not fully attract the eyes of users, hindering its large-scale development, the urgent need to design a vivid and dynamic new shape charging post, so that people can charge the car at the same time while strolling in the park again, can strengthen the popularity of new energy vehicles and can prompt people to exercise[4][5][6]. In this paper uses Pro/E, CAD, 3Ds max, and mold flow software to design a new type of new charging pile.

2. New Charging Post 3D Modeling and Process Analysis

First, Pro/E, CAD and 3Ds max software were used for 3D modeling of the new charging stake parts, as well as the rendering simulation of the appearance modeling, as shown in Figures 1, 2 and 3 .The material is FRPC/ABS, which has lower material cost and better performance, and can ensure the safety of the charging post well.

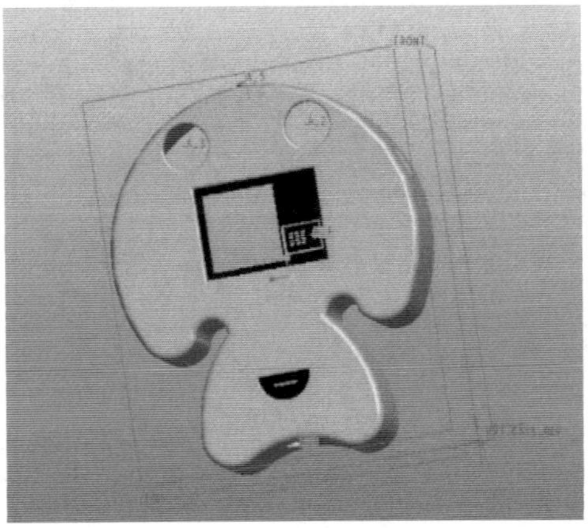

Figure 1. 3D part drawing of the new charging pile.

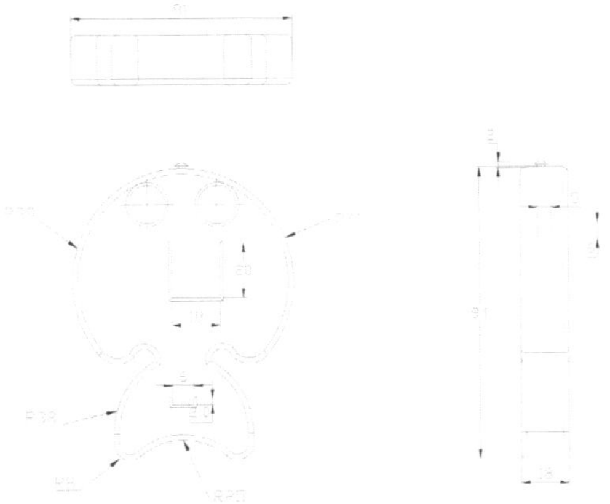

Figure 2. 2D part drawing of the new charging pile.

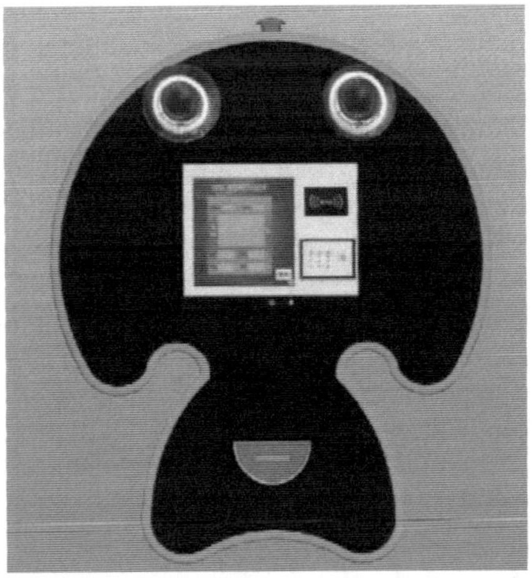

Figure 3. Rendering renderings of new charging pile.

The general principle of selecting parting surface is to ensure the quality of plastic parts, facilitate peeling and simplify the mold structure, as shown in Figure 4.

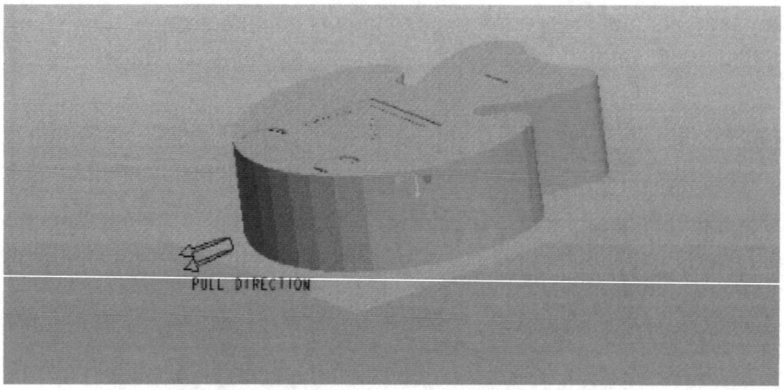

Figure 4. Parting surface of charging pile.

After calculation, the injection volume:

$$V = 0.8 \times \frac{\pi}{4} \times 10^2 \times 50 = 314\text{cm}^3 \tag{1}$$

Clamping force:

$$F = 314cm^3 \times 25MPa \times 15 \times 1.1 = 11775KN \tag{2}$$

Injection pressure:

$$P_1 = 132.6MPa \tag{3}$$

According to the main technical parameters, the primary model is SZ-320/1250 horizontal screw injection machine[7]. Where 320 represents the injection capacity of 320; the clamping force is $1250\text{x}10KN$, S represents plastic machinery, Z represents injection molding machine. After the processing of the charging pile shell is completed, the surface coating spray painting treatment operation is required. This paper uses the fluorine treatment surface layer technology, and the material is a fluorocarbon coating material.

3. Die Structure Design

The injection molding process varies depending on the model of the injection machine and is related to the action program of the injection machine [8] , the mold used and the process parameters, but the main steps are the same, mainly in six steps: (1) mold closing and locking, (2) injection device forward, (3) injection and holding pressure, (4) cooling and pre-plasticization of the part, (5) injection device backward, (6) mold opening and ejection of the part, the injection molding process is drawn Cycle diagram, as shown in Figure 5.This paper uses Pro/E to complete the 3D-DIMENSIONAL modeling of the cartoon charging pile in the park, the simulation of opening and closing molds is carried out. Die-casting die opening state diagram of the injection mold of the new charging pile is shown in Figure 6.

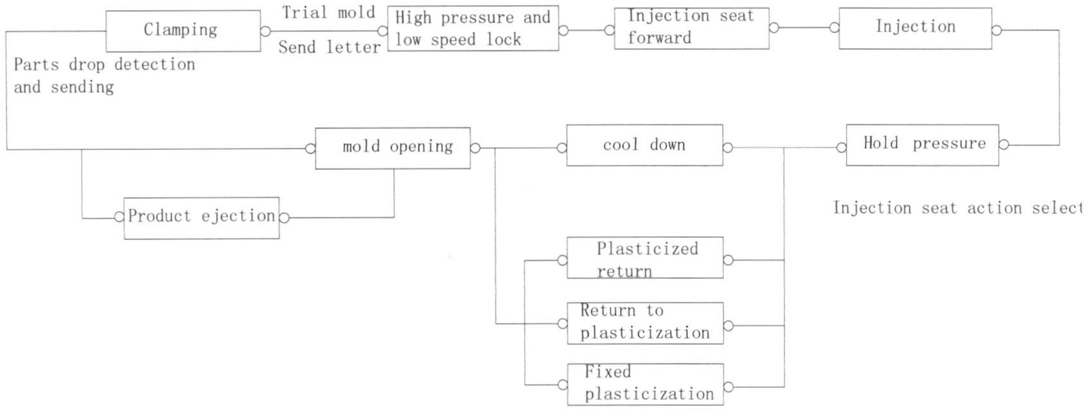

Figure 5. Production process cycle diagram of injection molding of new charging pile.

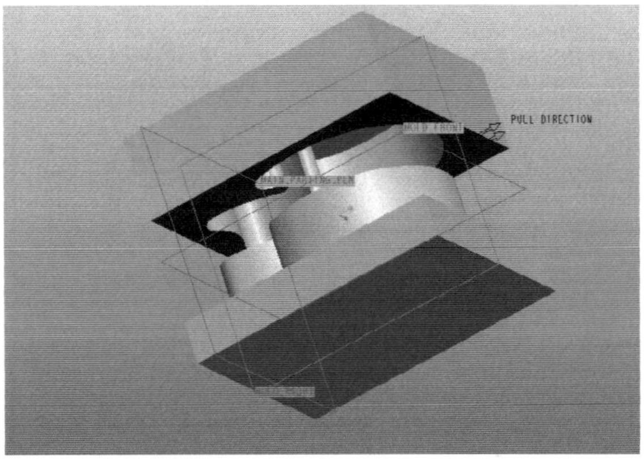

Figure 6. Die-casting die opening state diagram of the new charging pile.

4. Charging Pile Manufacturing Process Modal Flow Analysis

After the process parameters of the charging pile are analyzed, mold flow software is used to analyze the manufacturing process of the charging pile to virtually simulate the problems of the injection machine in the injection process and check the feasibility of the design.The basic process parameters such as filling time, leading edge temperature and clamping force are analyzed as shown in Figures 7, 8 and 9, and then the layout is rationalized in order to obtain the optimal design.The new type of charging pile designed in this paper applies curved surface design in appearance, which is more beautiful than before. There is also a big improvement in performance.

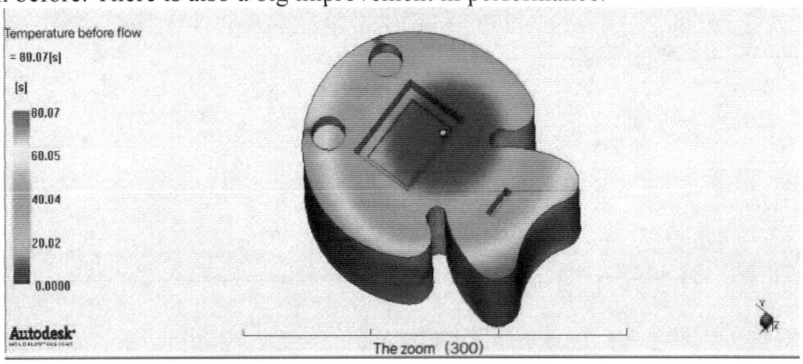

Figure 7. New charging pile filling time.

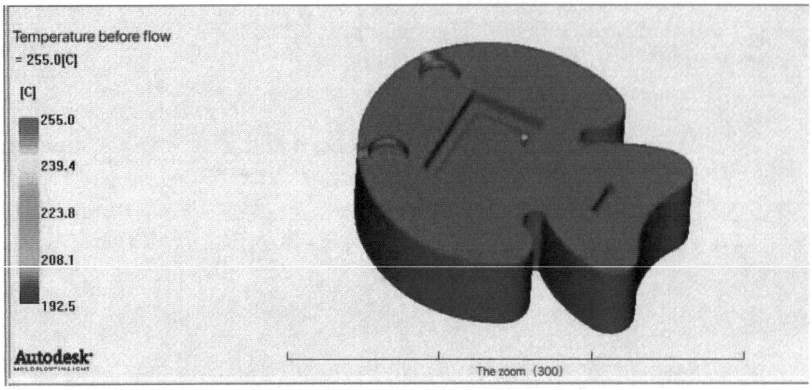

Figure 8. Flow front temperature.

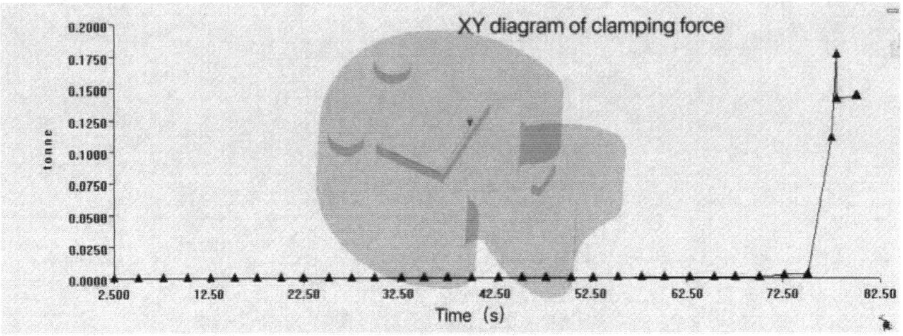

Figure 9. Clamping force.

5. Conclusions

This paper uses Pro/E, CAD and 3Ds max software to design the new charging pile, and then uses mold flow software to conduct mold flow analysis on the manufacturing process of the charging pile.This paper designs a new type of charging pile, as shown in Figure 10.The innovation points are: (1) This paper designs a vivid and dynamic cartoon modeling charging pile, which can greatly attract the user's eye and has a great breakthrough compared to the previous single design; (2) The charging pile adopts FRPC/ABS material as the outer shell of the charging pile. The cost of FRPC/ABS material is lower, but the resistance to deformation and impact is stronger than the commonly used PC, which can better ensure the safety of the charging pile; (3) Based on the cartoon shape, this article designs the two eyes of the charging pile to be used for lighting, which is more convenient for electric vehicle users to accurately find the charging position and facilitate their better operation of the product; (4) This article will use the charging pile The upper part is equipped with a solar light panel, which uses the principle of solar power generation to facilitate its use at night, which can save energy and strengthen environmental protection

Figure 10. Charging pile scene diagram.

References

[1] Mang Lu Xiaobing Zhou and Wei Zhang 2010 Research on the Development Status of Electric Vehicle Charging Facilities at Home and Abroad [J] Central China Power p 6
[2] Huaqing Liao 2016 Discussion on current Situation and Countermeasures of New Energy Vehicle Charging Facilities Construction at home and abroad [J] Company p 2
[3] Weifeng Yao, Junhua Zhao and Fushuan Wen 2015 Coordination planning of power distribution system and electric vehicle charging network [J] Automation of Electric Power Systems p 9
[4] Jiawei Chen Zezhen Liang and Xinmin Jin 2012 Electric Car Battery Pile design [J] Journal of Zhanjiang Normal University pp 107-113
[5] Chao Meng,Zhibin Li and Yan Ge 2013 Rural Electrification p 2
[6] Mengwei Lu 2016 Characteristic Design of Charging Pile Facilities in Scenic Spots of North Guangdong [J] Beauty and Times: City p 2
[7] Wenguang Wang, Baoshan Tian and Yanchen Tian 2004 Design techniques and examples of plastic injection molds [M] Chemical Industry Press pp 24-30
[8] Shanqin Wang 1997 Plastic injection molding process and equipment [M] China Light Industry Press pp 3-17

Research on the Development of Unmanned Underwater System Detection Technology

Yuqi Wu[1], Yuhan Deng[1], Longgang Zhang[1,*], Qidong Zhang[2] and Ran Bao[2]

[1] Yunnan Normal University, Kunming, 650500, China
[2] The 705th Research Institute of China State Shipbuilding Corporation, Kunming, 650118, China

*Corresponding author email: zhanglg_strong@whu.edu.cn

Abstract. In recent years, the number of unmanned systems in the world has been increasing. The emerging underwater platforms such as UUV are large in number, small in size, low in speed and low in noise. In the future, a variety of ocean combat missions will use a large number of UUV platforms. The existing methods for detecting underwater vehicles mainly include sonar, magnetometer. Based on this, by referring to a large number of literatures, a review of the research progress of underwater exploration at home and abroad in recent years. And compared with the United States, Russia and other countries unmanned system detection technology and equipment, analysis of their applicable conditions and advantages and disadvantages. This paper summarizes the technical difficulties of underwater detection methods and proposes corresponding solutions. At the same time, we forecast the development trend of underwater detection technology in the future.
Keywords: Underwater Unmanned System; Sonar Detection; Magnetic Detection; New Detection Technology.

1. Introduction

By consulting a large number of literatures, we know that sonar detection is one of the main technologies of antisubmarine detection technology. However, with the development of submarine noise reduction technology and the better and better performance of muffler, it is more and more difficult to detect enemy submarines by using traditional sonar detection technology. Therefore, more acoustic detection methods are the mainstream of our research. This paper briefly introduces the underwater electromagnetic detection and marine bioluminescence detection technology, briefly evaluates the application progress of underwater detection methods at home and abroad in recent years, and analyzes their applicable conditions, advantages and disadvantages. In the future, all countries will innovate their ideas, strive to innovate underwater detection technology and develop new underwater detection capabilities. The detection system is developing towards the direction of unmanned, integrated and intelligent.

2. Detection Method

2.1. Underwater Electromagnetic Detection Technology

Compared with acoustic field and other non-acoustic physical field, underwater electromagnetic field is an important source of exposure to underwater targets. Studies at home and abroad show that ship electromagnetic field is a signal source that can be used to detect and identify targets. At present, the United States, Russia and other countries have developed and installed submarine early warning and monitoring networks that include electrical and magnetic nodes for detection and attack, such as the Eisenhower submarine high-speed observation network, the Komor electromagnetic sea seal system, and the STL underwater warning system [1][2].

2.2. Underwater Exploration using Marine Bioluminescence

Submarines sailing in the ocean radiate sound waves outward, and the sound waves stimulate the fluorescence of microorganisms in the sea water, and the fluorescence track will automatically decay and disappear after about 15 minutes. If ultraviolet light is shone on the surface of the sea, the microorganisms in the surface of the sea can emit a visible light, which is closely related to the fluorescence properties of the microorganisms [3]. The principle of underwater exploration using Marine bioluminescence is shown in the Figure 1. The filter can detect microbial fluorescence containing submarine radiated sound wave information and filter out ultraviolet reflected light and scattered light. Under the excitation of seawater background noise and wind wave sound, microorganisms on the surface of seawater will also produce changes in fluorescence characteristics and content. The change law is the same as that of background noise characteristics, mainly DC or VLF components, which can also be filtered by filters [4].

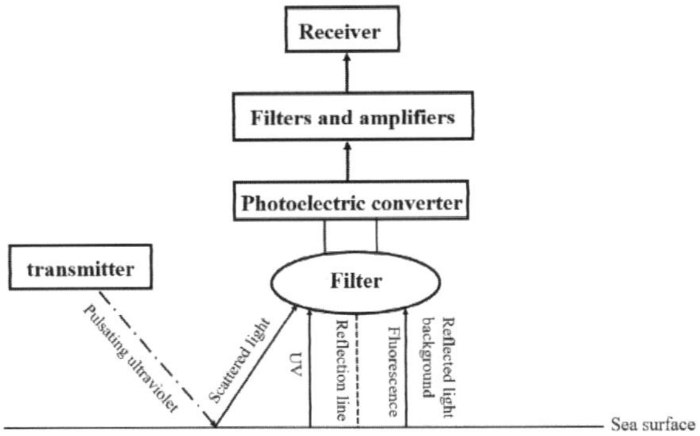

Figure 1. The principle of underwater exploration using Marine bioluminescence.

Figure 2. The new Edge Tech 2205 sonar system.

3. Underwater Detection System

3.1. The United States

The new Edge Tech 2205 is a modular, highly integrated sonar system that combines side scanning, shallow profile, and depth sounding functions. The Edge Tech 2205 is designed for ROVs, AUVs and other customizable vehicles. These modular units can collect side scan data images, shallow formation profile data and depth sounding data according to users' requirements, and collect and record data one by one or systematically. The new Edge Tech 2205 is shown in Figure 2.

The New Generation Wing-Body Fusion Underwater Glider (X-Ray) was developed with funding from the University of Washington Applied Physics Laboratory in the Office of Naval Research (ONR). The glider's innovative wing-body integrated configuration allows for high lift-to-drag ratios and large glide ratios while providing enough space for energy and payload (see Figure 3). With a wingspan of 6.1m and a glide speed of 1 to 2 kn, the X-Ray is designed to detect and track quiet submarines in shallow waters. The US military says the X-Ray can be deployed quickly and operated for months in a designated area, with a range of more than 1,000km [5] [6].

In the field of magnetic detection, G-882G Marine optical pump magnetometer (see Figure 4) developed by Geometrics company is a high sensitivity and high precision Marine magnetometer developed with the support of optical pump technology, which is applied in underwater submarine detection. Its high sensitivity (0.002 gamma) and high sampling rate (10 times per second) are highly valued by the Marine community [7].

The AN/AAR-37 infrared detector used by the U.S. Army P-3C can detect a temperature difference of 0.000 1° when the aircraft is flying at AN altitude of 1 50m -- 300m. Some models are also equipped with a forward-looking infrared detection system (FLIR, see Figure 5) on the nose, with a scanning range of +172°, up to 18° and down to -80°. The maximum detection range can be up to 37km, which can be manually controlled or automatically tracked. However, it can only be used as an auxiliary means, because it is greatly affected by Marine environmental factors, such as cloud, rain, fog and other weather can't be used, in addition, the sea temperature in a small range of differences will also affect the use of it [6].

The AN/BQQ-5 integrated sonar system is the most advanced attack nuclear submarine sonar system of the US Navy and one of the most advanced attack nuclear submarine sonar systems in the world. It adopts many of the most advanced sonar technologies and has become the most effective long-range detection equipment for the modern attack nuclear submarine. Its main function is to provide underwater target positioning data for torpedo launch, and at the same time, it undertakes the missions of underwater target detection, tracking, classification, recognition, noise direction finding, passive ranging, reconnaissance and underwater communication. An /BQQ-5 integrated sonar system is composed of nine independent sonars, including bow AN/BQS-13DNA sonar, AN/BQR-20 conformal array noise direction finding sonar, passive towed line array sonar, passive ranging wide aperture array sonar, target recognition sonar, AN/BQS-15 sounding sonar Lightning and collision avoidance sonar, AN/WLR-9A reconnaissance sonar, AN/WQC-5 communication sonar and AN/BQN-17 echo sounding sonar, etc [8]. An /BQQ-6 sonar system is a remote passive sonar system, which is modified and designed by AN/BQQ-5. The sonar system retains 73% of the configuration of the AN/BQQ-5 sonar system. The computer software system has been redesigned and the active sonar has been modified to improve the performance of the active sonar system.

Figure 3. The New Generation Wing-Body Fusion Underwater Glider (X-Ray).

Figure 4. Geometrics G-882G Marine Optically Pumped Magnetometer.

Figure 5. FLIR Marine stationary infrared thermal imager.

3.2. Russia

The MFK-540M submersible sonar system is equipped on the main battle nuclear submarines in service of the Russian Navy (as shown in Figure 5), including the Akula I, the Kan Saila, and the YK submarine. The system is mainly used to continuously monitor the surface and underwater conditions in the submarine's water area, detect, direct and track the target in a passive listening mode, which can be divided into wide band and narrow band, and the working frequency is audio frequency, low audio frequency and subaudio frequency. The MFK-5403M submersible sonar system array is shown in Figure 6[9].

Figure 7 shows the sonar system layout of Russia's newest conventional submarine, the Amur class. The bow array is conformal and covers most of the front of the bow. The base array extends from bow to stern to form a side array, and the vertical height of the side array is larger than the height of the conformal array. This design not only increases the bow receiving area, but also expands the observation range[10].

3.3. France

France's typical representatives in the field of unmanned craft mainly include ACSA's "Basil" and Sirehna's "Rodeur" unmanned high-speed gliding craft, among which the "Rodeur" surface unmanned craft can perform a variety of tasks, including mine anti-submarine, reconnaissance and surveillance, maritime security, Marine environment detection and investigation, etc. France "Allister" UUV is shown in Figure 8.

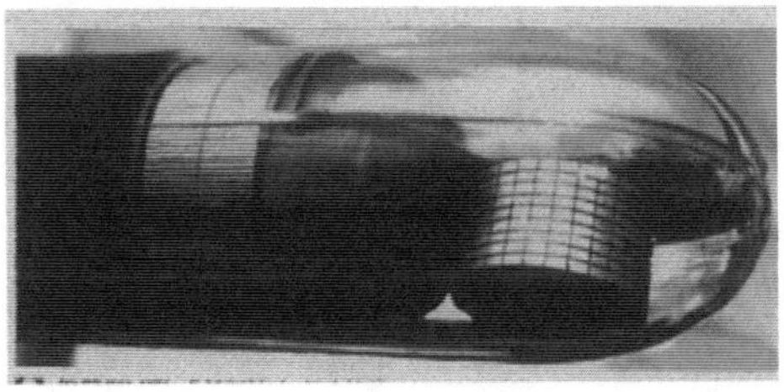

Figure 6. MFK-5403M submersible sonar system array.

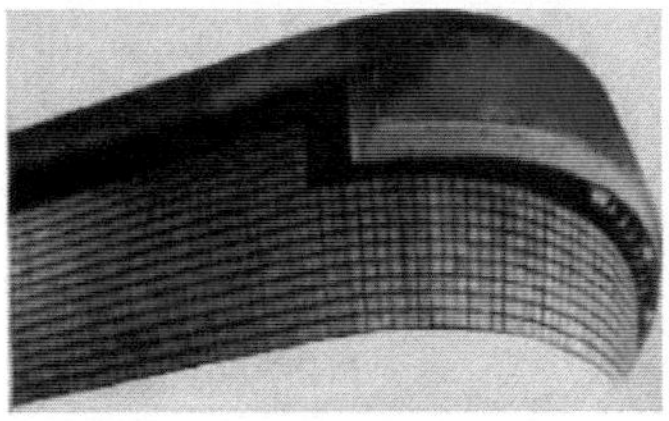

Figure 7. Large conformal array on the bow of the Amur class conventional submarine.

3.4. Germany

The German research Deep C has been delivered to the German Navy for testing. Deep C can be used for reconnaissance, communications, anti-mine and other missions. Deep C, an unmanned underwater reconnaissance vehicle developed by Germany Atlas Electronics, weighs 2.4t, has an endurance of 60 h, a maximum range of 400 km, a cruising speed of 4 knot, a maximum speed of 6 knot, a payload of 300 kg, and a maximum depth of 4000 m [11].

The German Class 214 submarine is highly integrated with various underwater acoustic equipment to form the Integrated Detection, Command and Weapon Engagement System (ISUS90). It includes passive ranging sonar, broadband active sonar (located in the command station envelope), With vertical aperture), reconnaissance detection and ranging sonar, bow columnar array sonar, head thunder avoidance sonar, side sonar, submersible towed sonar and multi-function processing and control system (see Figure 9) [11][12].

Figure 8. France "Allister" UUV.

Figure 9. Class 214 submarine.

3.5. Japan

Japan's underwater reconnaissance and early warning system is mainly composed of submarine units and underwater surveillance systems belonging to the Japan Maritime Self-Defense Force, which has a strong capability of reconnaissance and early warning against submersible and surface ships [13].

"Hyuga" battleship (see in Figure 10) is equipped with the most advanced detection equipment, mainly have OQ -21 sonar system in ship first cover sonar installation is OQS-XX low-frequency sonar, OQS adopted OXX low-frequency sonar QS-102 ships lower operating frequency, low-frequency sonar. So the range is much better than the original sonar, and the performance is much better in shallow water. The sonar, sonobuoy and anti-submarine intelligence processing device are combined to form the OQQ-21 shipborne sonar system [14][15].

3.6. India

When the submarine sails under water, it breaks the underwater equilibrium environment, and its tail will produce eddy current, which can be detected by detecting the change of optical characteristics of laser. Based on this principle, the study of eddy current detection under the principle of laser transmission characteristics was carried out at the University of Vellore in India. India developed a corresponding test system in 2013, as shown in Figure 11. The test system was installed on the anchor of an ocean research vessel, which was placed 42 m below the sea surface. The aft eddy was successfully detected by monitoring the A74 Marine acoustic research vessel.

Figure 10. "Hyuga" battleship.

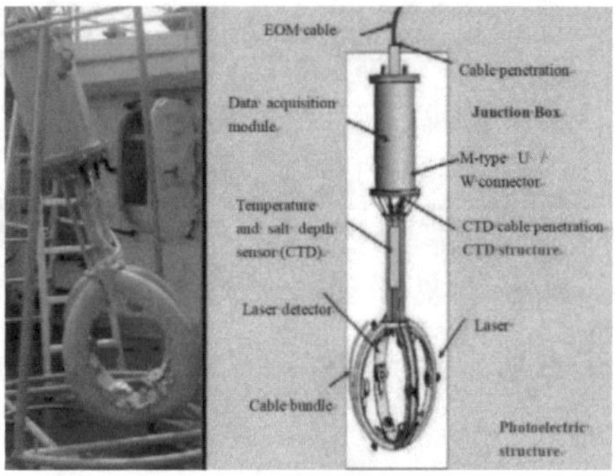

Figure 11. Laser test system developed in India.

3.7. China

Quantum radar technology, also known as the "killer" of stealth fighters, has important military application value in radar detection and imaging recognition. As early as 2016, China has completed research on quantum detection mechanism, target scattering characteristics and experimental verification of quantum detection principle, and completed target detection experiments in real atmospheric environment in the field. China's CETC publicly demonstrated a quantum radar (see in Figure12) as early as 2018, which at the time was the world's first truly military quantum radar. Quantum radar can effectively reduce the power consumption of the system, and can be applied to a variety of lightweight platforms. It can solve the bottleneck problems of traditional radar in the detection of low visible targets, survival under the conditions of electronic warfare, platform load limitation and many other aspects [16].
Besides, the Chinese Academy of Sciences developed an ultra-sensitive magnetic sensor (see in Figure 13) technology in 2017, known as the ultra-sensitive magnetic sensor, which can capture the weakest magnetic field from several kilometers away.

Figure 12. Quantum radar.

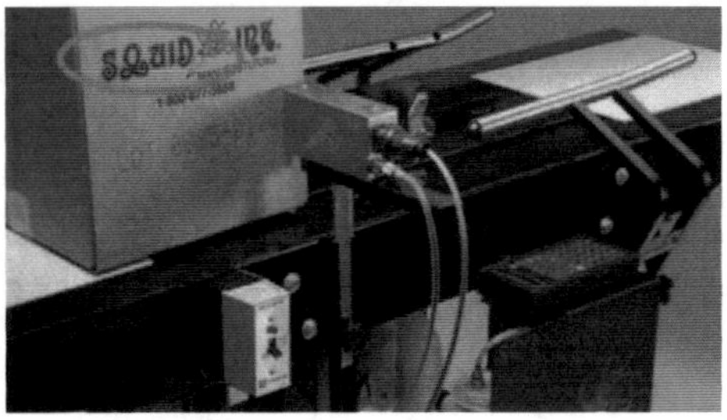

Figure 13. SQUID sensor in China.

4. Conclusion

Nowadays, the construction of sonar system presents the characteristics of openness, integration, digitization and modularization, and the application of sonar system is also very common. However, with the development of submarine noise reduction technology, the noise level of modern advanced submarines is very low, even at low speed near the sea and background noise level, using traditional sonar detection technology to detect enemy submarines has become more difficult, so the development of acoustic detection technology is the trend of the times. At present, the United States and other western countries have made considerable achievements in developing non acoustic submarine exploration technology. Using infrared detection method to detect the surface wave wake of underwater vehicle has been developed and applied in various detection equipment in western developed countries. The combination of radar processing technology and laser underwater transmission characteristics provides a new idea for underwater detection. The advantages of magnetic anomaly detection are strong recognition ability and good concealment. Due to its comprehensive advantages, western countries are strengthening the application in this field.

Acknowledgments
This work has been supported by Doctor Start Project of Yunnan Normal University (00900205020503127).

References

[1] Zhilin Lan and Jiabo Zhou 2008 Development of Unmanned Underwater Vehicles National Defense[J]. Science and Technology vol 02 pp 11-15.
[2] Large Displacement Unmanned Underwater Vehicle SteamingAhead. (2012-04-01) [2015-09-21].
[3] Shufu Li 2003 Underwater "secret agent" who turned the rivers and seas into the sea Underwater unmanned vehicle[J]. Shipborne device vol 04 pp 57-60.
[4] Jing Han and Zhao chaofang 2008 Application of ocean remote sensing technology in submarine detection [J]. Equipment environmental engineering vol 03 pp 67-70.1212.
[5] Yiling Cao and Minfang Wang 2001 Almighty underwater guard-unmanned underwater vehicle[J]. Flying Missile vol 04 pp 6-7.
[6] Huai Liu 2004 Unmanned underwater vehicle with broad application prospects Shipbuilding Industry[J]. Technical and Economic Information vol 12 pp 23-27.
[7] Long Zhang 2018 Performance analysis and application of g-882g marine magnetometer in marine engineering survey [J]. Automation and instrumentation vol 06 pp 135-138.
[8] Department of the Navy. Large Displacement Unmanned Underwater Vehicle Innovative Naval Prototype Technology[R]. ONRBAA Announcement, 2015.
[9] Yuelun Yang and Huiqing Zhu 2012 Research on the development of submarine sonar technology [J]. Acoustics and electronic engineering, vol 02 pp 40-44.
[10] Wei Dai, Wen Zhang and Li Zhang, Chunlei Xie and Junpeng Hui 2018 The development status and trend of foreign marine unmanned vehicles[J]. Ordnance Equipment Engineering Journal vol 39(07) pp 33-35.
[11] HUANG H, GAUNAURD G C 1997 Scattering of a plane acoustic wave by a spherical elastic shell near a free surface [J]. Int. J. Solids Structures vol 34 pp 1-602.
[12] German Navy to receive REMUS 100 AUVs from Hydroid[EB/OL]. (12-10-04)[2017-07-11].
[13] Mingtai Sun 1998 aviation antisubmarine introduction [M]. Beijing National Defense Industry Press pp 174-178 pp 183-185.
[14] Funnell, C. Jane's Underwater Warfare Systems[J]. 2010-2011.
[15] Brown P J. US satellites shadow China's submarines [N]. Asia Times. 2010-05-13.
[16] Reed A M, Milgram J H. Ship wakes and their radar images [J]. Annu Rev Fluid Mech, 2002, 34:469-502.

Electromechanical Actuator Design and Control Strategy for UAV Landing Gear

Tao Yang*, Jiufeng Wu and Xinlei Cao

China Academy of Aerospace Aerodynamics, Beijing, 100074, China

*Corresponding author email: sdyangtao@qq.com

Abstract. UAV is becoming more and more prevalent not only in the military field but also in the civilian field, whose flight safety and performance is of top concern. The landing gear system is undoubtedly a determining factor in UAV taking off and landing. In recent days, the all-electric trend in UAV design is not irreversible. As a result, the hydraulic actuator is replaced by EMA in the landing system. The electromechanical actuator for UAV landing system is designed in this paper. The detailed actuator system design is introduced in the beginning. And then communication protocol between the actuator system and the flight control computer is presented, serving as the basis of control strategy, which is the main content to study. The control strategy mainly deals with normal landing and emergency landing process. At last, the conclusion part summarizes its applications. The EMA devised in this paper and its corresponding control strategy provide a new idea in the landing system design of UAV.
Keywords: Electromechanical actuator; UAV; Landing gear.

1. Introduction

UAV (Unmanned Aerial Vehicle) is an aircraft which takes off and flies autonomously without manipulator. It can be used once or repeatedly by recovery technology [1]. And also, in order to complete different missions, it carries lethal weapons or non-lethal payloads. With revolution of international military affairs, development of national defence science and technology as well as application in the civilian field, UAV has successfully won a place in the aviation field because of its advantages such as outstanding mission performance, low development cost, low energy consumption, zero casualties and so on [2]. There is no denying that UAV enjoys a vast potential for future development. For military UAVs, the flight mission calls for long distance and long endurance, requiring higher performance of the airborne equipment and actuators [3]. As for large and medium-sized UAV, the taking off and landing is conducted by the ground sliding with the landing gear [4]. Traditionally, the landing gear system adopts the hydraulic source and utilizes the hydraulic actuator as its actuator [5]. With all-electric development process and continuously improving requirements, the hydraulic landing system constrains UAV for pursuing greater performance, as a consequence of its disadvantages such as a large amount of system accessories as well as increase of weight and fuel consumption [6]. It is urgent to develop an EMA for the landing system as a substitute. In this context, EMA (Electromechanical Actuator) is designed for UAV landing system in this paper. Dual redundancy technology is adopted in the actuator system design for higher reliability and better performance. In order to deal with all kinds of situations which the landing system will encounter during operation, the corresponding control strategy is also quite important.

2. Actuator System Design

The actuator system is made up of one controller and three actuators, namely, actuators for the nose gear, the left main gear and the right main gear respectively. The controller is the control core, and the actuator drives the landing gear to drop or retract [7]. In this part, working principle of the actuator system is introduced first. And then the controller design and the actuator design is presented one by one.

2.1. Working Principle

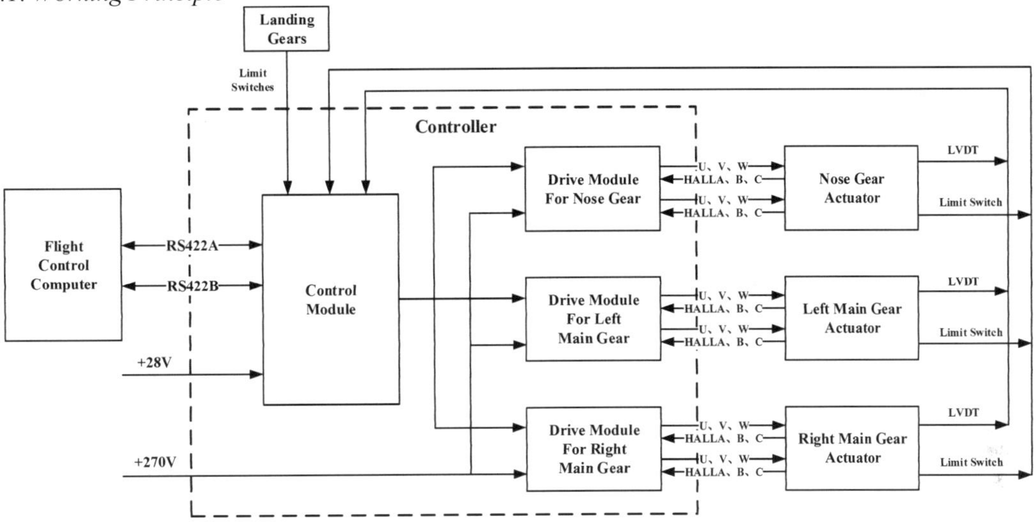

Figure 1. Block Diagram of Actuator System

As shown in Figure 1, the actuator system communicates with the flight control computer via two RS422 interfaces, which work together at the same time. One interface serves as the backup of the other, thus realizing dual redundancy. According to the command transmitted by the flight control computer, the actuator system reacts in real time, driving the landing gears to drop or retract while reporting the system status and faults back.

The controller design is in the form of dual redundancy. The control module and three drive modules for different landing gears are all main-backup design. There are two control cores working at the same time in the control module. One control core is used in the main channel connecting to the main RS422 port while the other is used in the backup channel connecting to the backup RS422 port. Two control cores don't communicate directly for information interaction between them are complex. As a result, there is a control management unit in charge of cooperation. The drive module for landing gear is also of dual redundancy, with the main drive channel and backup channel. The actuator is dual-redundant in the way which the motor is in the form of two windings and the feedback transducers are LVDT (Linear Variable Differential Transformer) and limit switches. LVDT indicates the real-time position and the limit switches show where the actuator reaches the desired position.

The actuator system can realize normal dropping or retracting and monitors the working status of landing gears in real time. When the flight control computer transmits dropping or landing command, the brake mounted coaxially with motor is unlocked. And then the actuator system conducts gear dropping or landing operation. When the indications show that the landing gear drops or retracts successfully, the actuator system stops operation and locks the brake. If there is any fault in process, the fault can be isolated via the fault protection and redundancy switching technology, which is applied in the actuator system. At the same time, the fault status reports back to the flight control computer.

2.2. Controller Design

The controller receives control command from the flight control computer via RS422 port. And then it instructs the drive module to drive the actuators, thus conducting dropping or retracting of different landing gears. The controller is mainly made up of two parts, the control part and the drive part as well as corresponding power supply. As shown in Figure 2, the controller design is divided into three kinds of boards, the control board, the drive board and the power board. The control module settles in the control board. There are three drive boards for nose gear, left main gear and right main gear respectively.

The power board offers power supply for the whole system. The power source is +28V used for the control part and +270V used for the drive part.

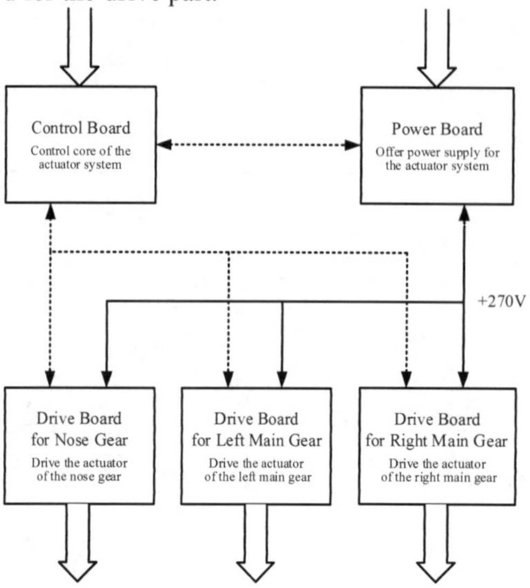

Figure 2. Hardware Design of Controller

2.2.1. Control Board. The control board conducts functions such as communication with the flight control computer, status monitoring, redundancy management, position feedback acquisition and so on. There are two control units in the control board, the main control unit and the backup control unit, which adopt the same hardware design. There are independent peripheral circuits and auxiliary power supplies. The main control unit communicates with the flight control system via main RS422 port with the backup unit via the backup RS422 port. Although data flows from two ports together, data in the main RS422 port is adopted prior to data in the backup RS422 port. Only under the circumstances that the main RS422 communication suffers from failure, the backup RS422 port becomes regular. In addition, the control board comprises the acquisition circuit of limit switch and feedback position, redundancy switching circuit, data storage circuit and so on.

2.2.2. Drive Board. The drive board is in charge of motor driving. In order to eliminate interference within the control board and the drive board, the electrical isolation is conducted by using the photoelectric coupler and isolated operational amplifier between the high voltage end and low voltage end. The drive boards for different actuators are the same, and there are two drive modules in the drive board for dual redundancy, namely the main drive module and the backup drive module. Each drive module adopts the integrated power drive module. The auxiliary power offers power supply independently. And both drive modules can enable or disable output by certain setup. The enabling status between two drive modules is interlocked. In this way, when the main drive module is at work, the backup drive module is standby, and vice versa.

2.3. Actuator Design

The actuator for landing gears is made up of double-winding motor, transmission mechanism, brake, limit switch and a linear displacement sensor. The double-winding motor is brushless DC motor, with two independent sets of windings and Hall sensors. The actuator needs to operate in linear motion for dropping or retracting of landing gears. On the basis of high reliability and small size, the ballscrew is selected as the main transmission mechanism. As a consequence of low speed requirement, a two-stage gear drive transmission is added between the ballscrew and the motor to enhance output thrust of the actuator.

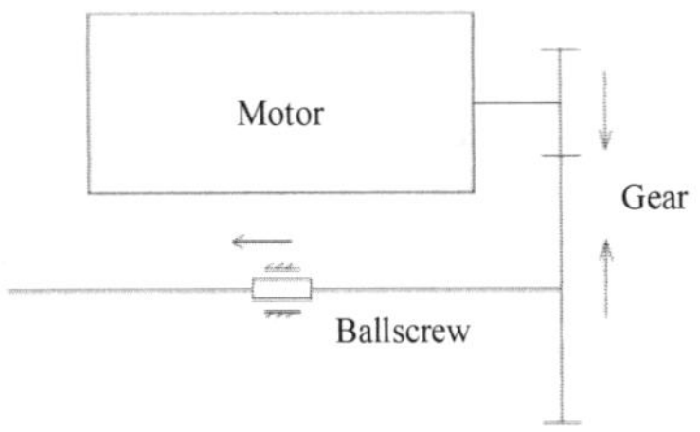

Figure 3. Schematic of Transmission Mechanism

As shown in Figure 3, the transmission mechanism works in the form that the motor drives the gear to rotate and then the gear drives the ballscrew to work. The brake which is locked when power is off locks the actuator when the landing gear drops or retracts successfully in order to ensure safety. The lock force is designed twice as much as nominal force of the actuator. The motor can work three times overload of the nominal force for a short time. Ever if the brake can't be unlocked, the actuator can still drop the landing gear successfully.

There are two limit switches in the landing gear system. One is mounted in the actuator. It can decide whether the landing gear is retracted successfully by detecting the position of screw nut. The other one is mounted in the landing gear to indicate when the landing gear drops successfully. The actuator position is indicated by the linear displacement sensor in real time, showing whether the landing gear is retracted or dropped successfully. The actuator works in the environment with strong vibration. As a result, the non-contact LVDT with long life is selected as the linear sensor in this design.

3. Communication Protocol

The actuator system communicates with the flight control computer via dual-redundant RS422 ports. The communication protocol is pretty important for information interaction between two devices, which has to contain various kinds of command, status and fault for control strategy of UAV landing gear [8]. The communication is in the periodical mode. The flight control computer and the actuator system transmits its data respectively in certain period. The communication protocol consists of two parts, namely, the downstream command and the upstream data.

3.1. Downstream Command

Data format of the downstream command is as follows: 1 bit start bit, 8 bits data bits, 1 bit stop bit and 1 bit old-even check bit with 115200 bps as the transmission baud rate. As listed in Table 1, the downstream command from the flight control computer begins with frame head which indicates it is a data frame for the actuator system. And then following it, it is the frame count. The downstream command mainly comprises landing command, authorized self-test command, emergency operation command and work mode setup. The detailed description of each command is listed in Table 1. At last, in order to validate the data, the checksum is added in the end of the data frame.

Table 1. Downstream command

No.	Item	Contents	Description
1	Frame Head	0xEB09	Indicating a data frame
2	Frame Count	0 ~ 255	Numbers of data package
3	Landing Command	0b00: No operation 0b01: Retract 0b10: Drop	Landing command for three landing gears
4	Authorized Self-test Command	0x00: No operation 0x01: Conduct Self-test	Decide whether the actuator system conducts self-test or not
5	Emergency Operation Command	0b00: No Operation 0b01: Retract	Emergency operation command for three landing gears
6	Work Mode Setup	0xAA: Flight Mode 0x55: Ground Mode	Work mode instruction for the actuator system
7	Checksum	0 ~ 255	Lower 8-bit of data Sum up

3.2. Upstream Data

Data format and transmission baud rate of the upstream data is same as that of downstream command. As listed in Table 2, there is also the frame head, frame count and checksum in the upstream data transmitted by the actuator system. The upstream data mainly includes status feedback of landing gears, redundancy status, fault status words, system status word, self-test status and emergency operation status.

Table 2. Upstream data

No.	Item	Contents	Description
1	Frame Head	0xEB09	Indicating a data frame
2	Frame Count	0 ~ 255	Numbers of data package
3	Nose Gear Status Feedback	0x00: No Operation 0x11: Retracting	Status indication of nose gear
4	Left Main Gear Status Feedback	0x12: Successful Retracting 0x13: Abnormal Retracting	Status indication of left main gear
5	Right Main Gear Status Feedback	0x21: Dropping 0x22: Successful Dropping 0x23: Abnormal Dropping	Status indication of right main gear
6	Redundancy Status	0b00: Main Channel 0b01: Backup Channel	Redundancy status of one control module and three drive modules
7	Fault Status Word#1	0: Normal 1: Fault	Control module fault and overcurrent fault of drive mode
8	Fault Status Word#2	0: Normal 1: Fault	Hall fault of drive mode
9	Fault Status Word#3	0b00: Normal 0b01: Secondary Fault 0b10: Primary Fault	Fault status of three landing gear actuator
10	System Status Word	COM Status &Work Mode	Status indication of actuator system
11	Self-test Status	0x00: No Operation 0x11: Successful Self-test 0x12: Self-test Failed	Self-test result of actuator system
12	Emergency Operation Status	0b00: No Operation 0b01: In Operation 0b10: Successful Operation 0b11: Operation Failed	Emergency operation status of three landing gear actuator
13	Checksum	0 ~ 255	Lower 8-bit of data Sum up

4. Control Strategy

The actuator system is a typical servo system. Once receiving command from the control computer, the actuator reacts as the control computer instructs it to. The control strategy for the UAV landing gear mainly studies on landing or dropping of landing gears. And there are the landing operation and the emergency operation. The landing operation is the normal landing or dropping process while the emergency operation deals with special cases when the normal process is not available [9].

4.1. Landing Operation

As shown in Figure 4, when the actuator system receives the dropping or landing command from the control computer continuously, the brake will be unlocked and the actuator commences operation with status reporting back to the control computer. Within 30 seconds, the controller in the actuator system judges whether the dropping or landing is successful or not according to the limit switch and LVDT. If the operation is conducted successfully, the controller reports the operation result. Otherwise, the controller reports the corresponding fault which leads to operation failure. No matter what happens after 30 seconds since the actuator begins to operate, the actuator will stop operation and the brake is locked at the same time, waiting for further instructions [10].

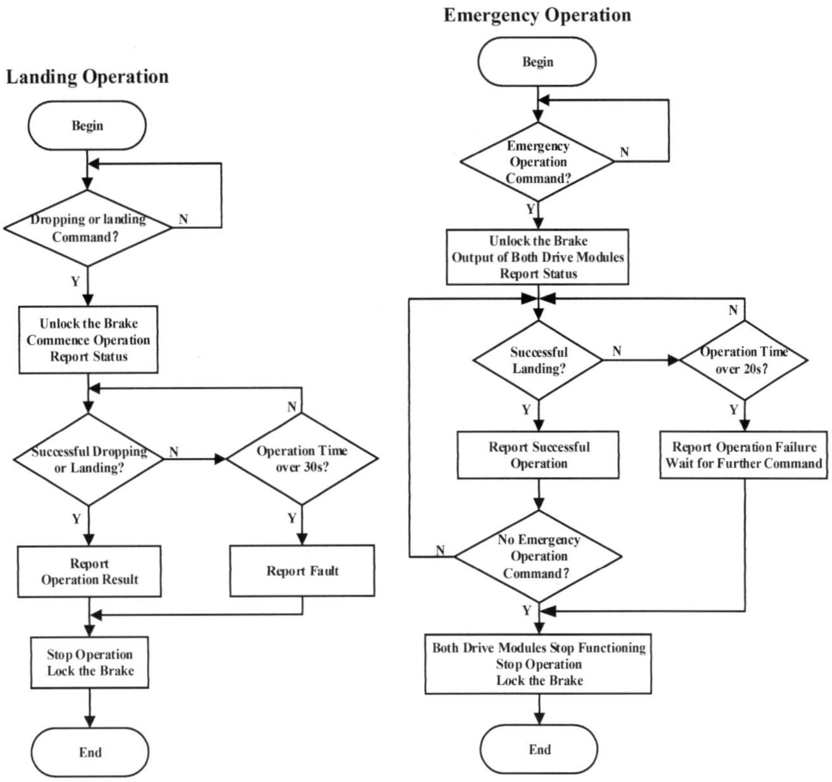

Figure 4. Flow Chart of Different Landing Operations

4.2. Emergency Operation

The emergency operation aims at emergency landing and its process is also shown in Figure 4. Emergency operation command for one certain actuator is initiated by the flight control computer on condition that it receives status of abnormal retracting, abnormal dropping or primary fault which lasts for more than 5 seconds [11]. And then the flight control computer takes action and transmits emergency operation command. Upon reception of emergency operation command from the control computer continuously, the brake will be unlocked and the main and backup drive module in the actuator system work together at the same time to secure stronger driving force to deal with unexpected landing or retracting process. The operation status is also sent back to the control computer. Within 20 seconds, the actuator controller system judges whether emergency landing is successful or not according to the limit switch and LVDT. If the emergency operation is completed, the controller reports the successful operation. At this time, only after the actuator system receives the no emergency operation command from the flight control computer, can it cut off output of both drive modules. The actuator will stop operation with brake locked. However, if the emergency operation fails, the controller reports operation failure. Operation is stopped and the brake is locked. The actuator system has to wait for downstream command and then conduct next operation.

4.3. Operation Comparison

As for normal landing operation, the actuator system plays a deciding role in the moment when the actuator stops operation, the initiative is taken by itself. What's more, if there are some undesirable results in normal landing operation, the flight control computer can turn to the emergency operation for help. In case of emergency, the coresponding operation is designed, which is the ultimate operation of the actuator system. As a result, to avoid landing failure led by the actuator system's misoperation, the decision-making right of stopping operation is given to the flight control computer.

5. Experimental Test

As shown in Fiugre 5, the electromechanical actuator designed in this paper is tested in the elastic force test bench to simulate all the working conditions which it may encounter with in practice and verify the control stategy. A testing system acts as the role of the flight control computer, transmitting command and conducting the landing operation. The normal landing operation is successfully completed every time when the actuator system receives the coresponding command. The tests is conducted 100 times and the success rate is 100%. As for the emergency operation, it is kind of destructive test. The test is only conducted 5 times. The actuator is able to overcome the difficulites which dose not encounter in the normal landing process. The result shows that the actuator system can work properly in case of emergency.

Figure 5. Experimental Test Setup

6. Conclusion

The hydraulic actuator is replaced by EMA in the landing system, with the process of all-electric UAV. According to the statistic data, the cost of the electromechanical actuator designed in this paper is reduced to 60% of the hydraulic actuator. At the same time, the dynamic performance is increased by 30%. Undoubtedly, there are a great many advantages in this upgrading, to list a few, low cost, high performance, high reliability, compact structure, easy maintenance, and so on. Thus, the safety of the flight mission with long distance and endurance is ensured. The dual-redundant EMA is designed, and the control strategy for landing gear is presented in this paper. Several ground tests are conducted as mentioned above in the experimental test part. Both the normal landing and the emergency landing process are simulated. The ground tests prove that the EMA designed in this paper is feasible and flexible. And the control strategy is appropriate in different working conditions. However, the real performance has to be verified further to see whether the EMA and control strategy can pass the practical flight tests.

References

[1] Tao Yujin, Li Peifeng 2014 *Development and key technology of UAV (*Aeronautical Manufacturing Technology vol 20*)* pp 34-39

[2] Gong Hongri and Tian Jin 2018 *Research on Organization and Application of Unmanned Combat System (Proceeding of the 6th China Conference on Command and Control* vol 4*)* pp 511-514

[3] Fan Bangkui, Mao Zexiao and Liu Jizong 2001 *Foreign UAVs* (Aviation Industry Press)

[4] Hong Nie and Xiaohui Wei 2008 *Key Technologies for Landing Gear of Large Civil Aircrafts (Journal of Nanjing University of Aeronautics& Astronautic* vol 40*)* pp 427-432

[5] Wei Jing, Ren Jie and Yuan Dongli 2014 *Modeling and Simulating of a UAV Hydraulic System (Electronic Design Engineering* vol 22*)* pp 8-11

[6] Xiaohui Wei and Hong Nie 2005 *Dynamic Analysis of Aircraft Landing Impact Using Landing region-based Model (Journal of Aircraft* vol 42*)* pp 1631-1637

[7] Dou Qingbo, Yang Wugang, Murangke and Shi Weiqi 2014 *The Design and Application of UAV Landing Test System* (*Mechanical Science and Technology for Aerospace Engineering* vol 33) pp 146-150

[8] Khapane P D 2003 *Simulation of Asymmetric Landing and Typical Ground Maneuvers for Large Transport Aircraft (Aerospace Science and Technology* vol 7*)* pp 611-619
[9] Yueping Guo, Yamamoto K J and Rob W 2006 *Stoker Experimental Study on Aircraft Landing Gear Noise (Journal of Aircraft* vol 43*)* pp 306-317
[10] Lazos B S 2002 *Mean Flow Features around the Inlinewheels of Four-wheel Landing Gear (AIAA Journal* vol 40*)* pp 193-198
[11] Yadav D 1995 *Landing Response of Aircraft with Optimal Antiskid Braking (Journal of sound and Vibration* vol 181*)* pp 401-406

Research and Development of a Real-time UAV Flight Visualization Simulation System

Yang Xiao[1,*], Shulin Dai[2], Chenfan Xiao[3] and Xinfeng Xu[1]

[1]Department of Mechanical Engineering, Southwest Petroleum University, Chengdu, China
[2]School of Automation Science and Electrical Engineering, Beihang University, Beijing, China
[3]Ant Group, Chengdu, China

*Corresponding author email: eastsun628@163.com

Abstract. On the premise of practical application, this paper describes the principle, technology and implementation methods of a real-time UAV flight visualization simulation system. The overall framework, hardware system composition, UAV flight model establishment, determination of flight control mode and control rate, design of 3D visual simulation system, management and processing of UAV flight state data are discussed, and finally, the actual application effect of the system is evaluated.
Keywords: UAV; Flight status; Flight simulation; Visualization; Data management.

1. Introduction

Unmanned aerial vehicle (UAV) are unmanned aircraft operated by radio remote control equipment and self-contained programmed control devices. In terms of civil use, the current application of UAV in aerial photography, traffic management, agricultural plant protection, surveying and mapping and other fields has greatly expanded the use of UAV itself. Whether controlling drones through remote controls or smartphone apps, they have the ability to reach the most remote areas, requiring little human support and little efforts, time and effort, this is one of the biggest reasons that drone are being adopted worldwide.

Small and medium-sized fixed-wing drones are heavy and wide used in civilian fields. For improving the development of small and medium-sized fixed-wing drones, beside to paying attention to the shape design, we should pay more attention to the research and development of flight control systems. The traditional research and development process of flight controllers is divided into object modeling, control law simulation, hardware design, software programming, integrated debugging[1]. Object modeling and control law simulation are based on the aerodynamic parameters of the UAV, establish a flight model and control model for engine characteristics and actuator characteristic parameters, then, the model was verified by the simulation software. Reference[2] establishes a non-linear six-degrees-freedom equation of motion, a guidance and control model with Matlab/ Simulink, the simulation results were analyzed by curves. Reference [3] similar to the former, the innovation point lies in the visual output of the simulator implements passed through Flightgear simulator. At present, most of the simulation process work on a computer platform. However, actual controllers used on drones are all embedded platforms. Considering the timeliness of the model simulation depends on the way the model is built, computing method and the operational power of the controller and the fixed-wing aircraft is operating during commissioning. Once a failure occurs, greater damage to the fuselage. So, it is urgent to add a more realistic flight control simulation and debugging method in the development process of the flight controller, make up for the deficiencies of the traditional debugging process.

To more truly verify the accuracy of the flight model with the control algorithm, the response efficiency of the hardware platform, the software program to the control signals, the compatibility between the software and the hardware, to reduce development costs, purpose of improving the commissioning efficiency, using visual simulation technology, the visual real-time flight simulation system is developed based on embedded controller ARM.

2. System Architecture

The whole system mainly includes flight simulation module and 3D view simulation module. The flight simulation module is built on the kernel STM32F103 microcontroller based on the ARM and on the associated peripheral circuits (The following referred to as the lower machine), complete the command control signal acquisition, flight state calculation and data transfer. The 3D visual simulation module receives the uploaded flight status information in real time, drive quick update of the 3D view module, realize the purpose of visualization. The system function is showed in Figure 1.

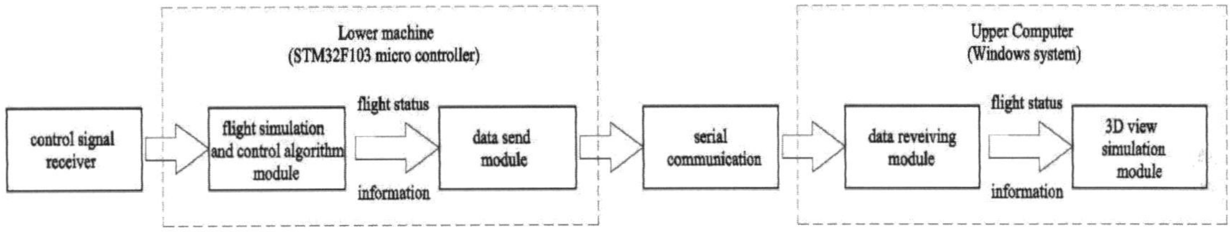

Figure 1. System Function.

3. Design and Implementation of the Flight Simulation Module

The flight simulation module is a core part of the system, flight models are mainly built using embedded microcontrollers, and under a given control signal and control law, UAV flight simulation was performed in real time using suitable numerical calculation methods, the continuous spatial position and posture information of the UAV are obtained. The functional structure is shown in Figure 2.

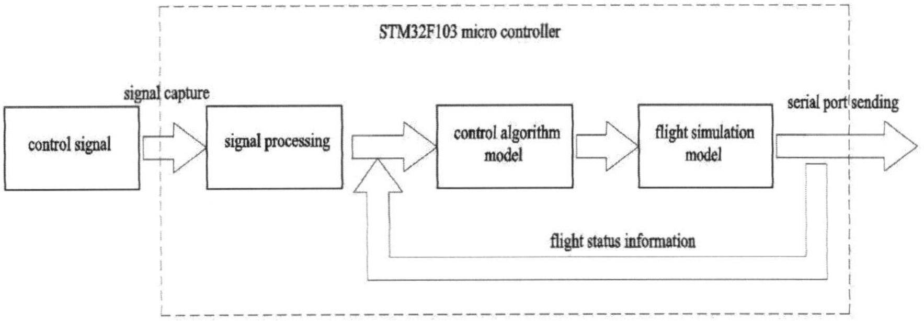

Figure 2. Module Function.

3.1. Hardware Design of the Flight Simulation Module

Considering the task requirements and functional requirements, the system uses Italian Semiconductor Inc. STM32F103 Series controllers serve as an embedded hardware platform. The series of microcontrollers include CPU, RAM, ROM, I/O Interface, Timer/Counter, UASRT, A/ D and D/ A converter. Its highest operating frequency is 72MHz, and can adjust the frequency of work. Single-cycle multiplication and hardware division can be achieved. The system utilizes its powerful operational capability and interface functions, realize the establishment and solution of flight simulation model and control model, control the signal acquisition and flight state data transmission.

3.2. Design of Flight Model and Control Law

3.2.1. Flight model establishment. The flight models established in this system include UAV dynamics model, kinematics mode, environmental model, engine model and steering machine model[4]. To simplify the equation reasonably, the following assumptions were made:

- Considering the drone flight altitude and mission radius, consider the Earth as an infinitely extended plane, ignoring the Earth's rotation and curvature;
- Considering the characteristics of the material and acting force of the UAV body, no deformation of the body, that is, the body as the ideal rigid body[5].

To ensure the fidelity of the UAV flight simulation, our system establishes the six degrees of freedom flight equations using the quaternions method as shown below:

$$\begin{cases} \mathrm{m}\dfrac{dV}{dt} = P\cos\alpha\cos\beta - X - mg\sin\theta \\ \mathrm{mV}\dfrac{d\theta}{dt} = P(\sin\alpha\cos\gamma_c + \cos\alpha\sin\beta\sin\gamma_c) + Y\cos\gamma_c - Z\sin\gamma_c - mg\cos\theta - mV\cos\theta \\ \dfrac{d\psi_c}{dt} = P\,(\sin\alpha\sin\gamma_c - \cos\alpha\sin\beta\sin\gamma_c) + Y\sin\gamma_c + Z\cos\gamma_c \end{cases} \quad (1)$$

$$\begin{cases} J_x\dfrac{dw_x}{dt} + (J_z - J_y)\omega_z\omega_y = M_x \\ J_y\dfrac{dw_y}{dt} + (J_x - J_z)\omega_x\omega_z = M_y \\ J_z\dfrac{dw_z}{dt} + (J_y - J_x)\omega_y\omega_x = M_z \end{cases} \quad (2)$$

$$\begin{cases} \dfrac{dq_0^*}{dt} = \dfrac{1}{2}(-\omega_x q_1 - \omega_y q_2 - \omega_z q_3) \\ \dfrac{dq_1^*}{dt} = \dfrac{1}{2}(\omega_x q_0 + \omega_z q_2 - \omega_y q_3) \\ \dfrac{dq_2^*}{dt} = \dfrac{1}{2}(\omega_y q_0 + \omega_x q_3 - \omega_z q_1) \\ \dfrac{dq_3^*}{dt} = \dfrac{1}{2}(\omega_z q_0 - \omega_y q_1 - \omega_x q_2) \end{cases} \quad (3)$$

$$\begin{cases} q_0 = \dfrac{q_0^*}{\sqrt{(q_0^*)^2+(q_1^*)^2+(q_2^*)^2+(q_3^*)^2}} \\ q_1 = \dfrac{q_1^*}{\sqrt{(q_0^*)^2+(q_1^*)^2+(q_2^*)^2+(q_3^*)^2}} \\ q_2 = \dfrac{q_2^*}{\sqrt{(q_0^*)^2+(q_1^*)^2+(q_2^*)^2+(q_3^*)^2}} \\ q_3 = \dfrac{q_3^*}{\sqrt{(q_0^*)^2+(q_1^*)^2+(q_2^*)^2+(q_3^*)^2}} \end{cases} \quad (4)$$

$$\begin{cases} \tan\psi = \dfrac{-2(q_1q_3+q_0q_2)}{q_0^2+q_1^2-q_2^2-q_3^2} \\ \tan\gamma = \dfrac{-2(q_2q_3+q_0q_1)}{q_0^2-q_1^2+q_2^2-q_3^2} \\ \sin\theta = 2(q_1q_2 + q_0q_3) \end{cases} \quad (5)$$

where q_0, q_1, q_2, q_3 are the quaternion, ω_x, ω_y, ω_z are the projection of angular velocity of rigid body on coordinate axis, ψ, γ, θ are yaw angle, inclination angle and pitch angles respectively.

In contrast to the system of equations of motion established by the directional cosine and Euler angle methods, the quaternionic method has outstanding advantages: Do not degenerate on any parameters, the number of parameters is only 4, the contact equation has only 1 [6].

The lower position machine uses a fourth-order Runge-Kutta method in the initial experiment to solve the system equations. The calculation results are smooth and highly realisti, but the computational rate is low. To improve the real-time nature of the computation, a second-order Runge-Kutta method is used later on the premise of ensuring smoothness and fidelity. The final results demonstrate that, in contrast to the Euler angle method, our proposed method combines the quaterniion method with the Runge-Kutta algorithm, the calculated time of the first-time pose information successfully reduce from about 60ms to 25ms, the computational rate was improved more than 2 times.
Both the engine model and the pneumatic data are determined by the specific UAV models, as known data. Procedures extract data from the corresponding interpolation table according to the atmospheric environment, flight status, throttle lever position and other states, interpolate current engine speed, thrust and aerodynamic coefficient.

3.2.2. Control mode and control law. The drone control mode includes manual control, instruction control, autonomous control. The system mainly adopts the instruction control mode. The control law used by the system is open-source, The control rules can be implanted according to the purpose of user development[7].The default control law is the classic PID Control law. The following are PID control law under three attitude angles of the command control mode.
1) Elevope angle holding and control loop closed loop control law:

$$\delta_e = P_\theta(\theta_r - \theta) + \int I_\theta(\theta_r - \theta)\,dt - D_\theta\omega_y \tag{6}$$

2) Roll angle holding and control loop closed-loop control law:

$$\begin{cases} \delta_a = P_\varphi(\varphi_r - \varphi) + \int I_\varphi(\varphi_r - \varphi)\,dt - D_\varphi\omega_x \\ \delta_r = D_r\omega_z \end{cases} \tag{7}$$

3) Yaw angle holding and control loop closed-loop control law:

$$\begin{cases} \varphi_r = P_\psi(\psi_r - \psi) \\ \delta_a = P_\varphi(\varphi_r - \varphi) + \int I_\varphi(\varphi_r - \varphi)\,dt - D_\varphi\omega_x \\ \delta_r = D_r\omega_z \end{cases} \tag{8}$$

where P, I, D are the proportional coefficients, integral and differential coefficients respectively. θ, r, ψ are the pitch angles, rolling angle and yaw angle respectively[8].

3.3. Programming of Flight Simulation Module

The Flight Simulation module is a multitask program, to ensure the reliability and real-time of the program operation, the operating system μC/OS-II is embedded in STM32F103, the microcontroller. The module mainly implements four functions: control signal acquisition, control model calculation, flight model calculation and serial port transmission of flight state data.

- Control signal acquisition task: Control signal is duty cycle from 5% to 10% adjustable 50Hz square wave signal. By configuring the control register of the microcontroller TIM(general timer), invokes its input capture function for pulse width acquisition.
- Control model calculation task: The program inputs the control amount and the flight state data at the previous moment as message, save the model calculation results in a preconfigured storage unit waiting for a call.
- The flight model computational task: Read the latest control model calculation results in the storage unit, input the flight mode, calculate current flight status, save the message, wait for the serial port transfer task extraction.
- Serial port transport task: The program calls the controller channel UASRT1 uploads the latest flight status data.

4. Design and Implementation of the UAV Flight 3D Visual Simulation Module

The 3D visual view simulation program is divided into three sub-module programs: the serial port communication module based on the multi-threaded and Windows customize message, the OpenGL3D scenario display module based on MFC and the UAV model establishment and import module.

4.1. The Serial Port Communication Based on Multi-thread and Windows Custom Messages

To ensure a higher asynchronous serial communication data transmission rate, at the same time, in order to ensure the flexible data interaction between the master and owner threads, the program builds on using shared storage for high-speed communication function between the threads by custom messages.

4.2. UAV Model Establishment and Lightweight

In the 3D visual simulation module, there are two problems for directly displaying 3D drone models in OpenGL:

1) Display the 3 D model with OpenGL code, the programming process is complex and the drone model quality is low;

2) It takes more time to load and update the models. Comprehensive consideration, the module uses 3 D software 3DMAX to build the drone model, ensure the quality of the model. At the same time, for the requirements of fluency display for the system, the drone model needs to be lightweight. We use Milkshape3D converts the model into a thin file format readable by the 3D View program[9].Figure 3 show the results of UAV model format transformation.

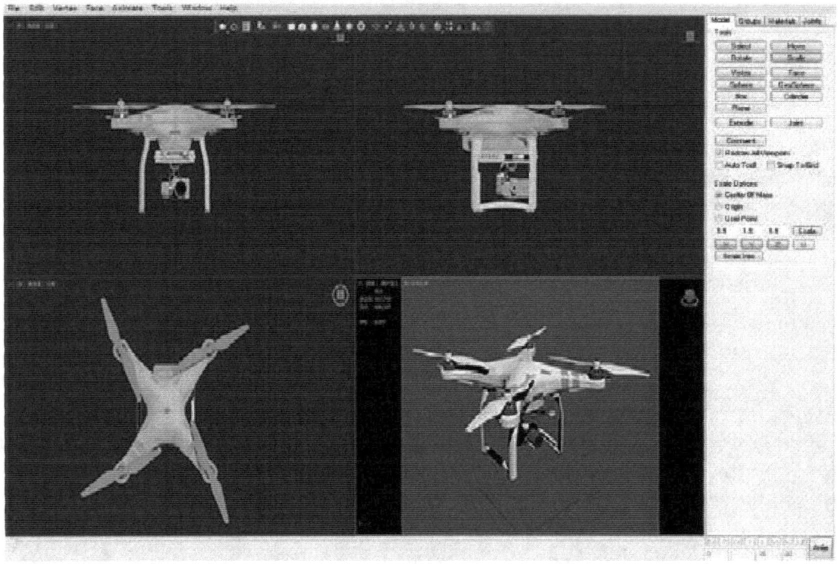

Figure 3. Modelling and Lightweight of the drone model.

4.3. The OpenGL 3D Scene Display Based on MFC

OpenGL is a high-performance 3D graphics processing system developed by SGI company, It is an open graphics software library, combined with Visual C++ programming, can implement relevant 3D objects computational and graphic algorithms, especially suitable for high-quality 3D dynamic display[10-11].

There are no window management functions in OpenGL library, so in Windows environment, build a 3D scene with OpenGL need to put window system with OpenGL. It can then be used to show and process the 3D scene[12]with OpenGL functions. MFC includes rich Windows application framework, Window and event management functions, so the combination of MFC and OpenGL can very well meet the requirements of the module.

When the 3D scene module display the image, the focus is on completing terrain and sky model building and flight state scenario-driven tasks. The specific implementation method is as follows:

The UAV model introduced in the 3D vision program will be moved according to the flight state data rules calculated by the flight simulation program. The motion of a 3D object in space can be described by its translation, rotation, and transformation matrix relations. Module Will UAV Flight state data is converted to the main thread OnDraw ( ). Translation function in the function glTranslate ( ), And the rotation function glRotate ( ), And the input parameter of the matrix transformation relation function, So as to realize the display and update of the UAV flight status[13].Figure 4 shows the flight of the 3D view program in the drone.

Figure. 4. The 3D visual display interface.

5. Management and Processing for the System Data

The system will have a lot of data generation and processing, with a module. Use the SQL Server database platform. SQL Server is a relational database management system that was developed by Microsoft Inc. SQL Server database has many advantages, making it the first in the database field and becoming the most popular database system. It graphics the user interface and makes system management and database management more intuitive and simple. Rich programming interface tools provide greater options for user programming and excellent scalability across multiple platforms from desktop computers running Windows to servers with large multiprocessors. Its support for Web technology makes it easy for users to publish data in the database to the Web page. It is easy to use, secure and available, fast performance, with an integrated development environment, running in a Windows operating system environment.

6. System Simulation Results

The 3D vision program contains functions to record flight state parameters in real time, these parameters record the results into the system's database for unified processing and management, to facilitate the later quantitative analysis of the data.

Figure 5 to figure 7 is the analysis curve of the results of the flight experiment, the traces of centroids of the UAV are shown separately, Speed change and attitude-angle response.

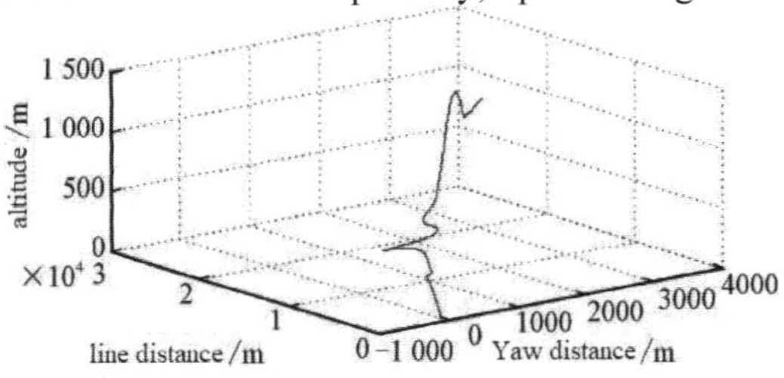

Figure 5. Track data curve.

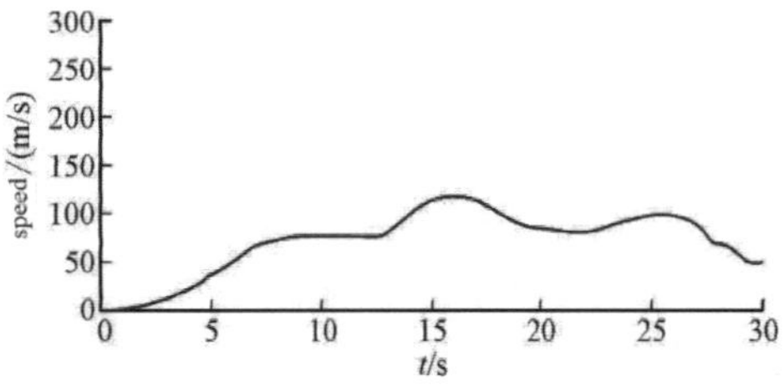

Figure 6. Speed response curve.

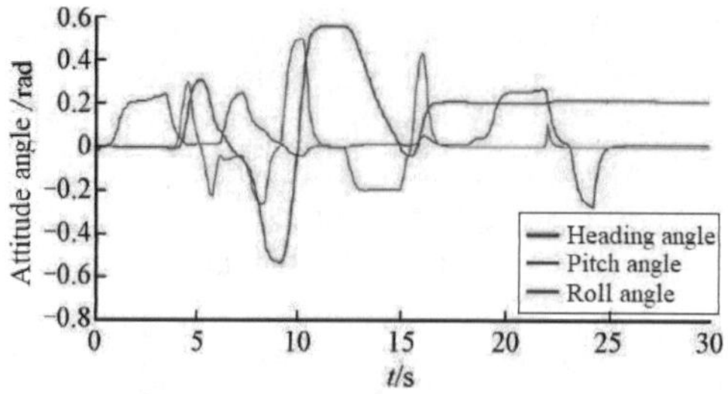

Figure 7. Posture-angle response curve.

Figure 8 is the main interface of the overall system for a real-time simulation of the UVA flight process.

Figure 8. Main interface of the system.

7. Conclusions

This is known from the experimental data analysis, it is feasible to realize the visual flight simulation of the UAV by combining with 3D visual simulation and embedded development based on ARM. The method discussed in this paper lays the technical foundation for making physical UAV flight controllers.

- To develop an open-source small and medium-sized fixed-wing UAV flight controller, The embedded UAV flight simulation system based on ARM is designed as a trantransition system for flight controller. The system implanted the traditional flight simulation process into an embedded microprocessor, the real-time flight state solution is realized.
- For dynamic, vividly show the results of the microcontroller calculating the UAV flight state, exploitation Developed a 3D OpenGL view simulation program that can be quickly update the drone flight data.
- The platform provided by this paper is also applicable for the flight control development of other unmanned aircraft vehicles. At the same time, build different flight models and control models, this system can also be expanded to other types of control methods and flight object.

Acknowledgments

This research is supported by the Open Research Project of the National Key Laboratory of Virtual Reality Technology and Systems VRLAB 2020C07 and the Major science and technology R&D project "civil large bypass ratio turbine engine unit delivery" of Sichuan Province (2019YFG0384).

References

[1] Yang Zhong, Shen Chunlin, Liu Jiufu. A UAV New technology for the virtual prototype design and verification of the flight controller [J]. Information and Control, 2005, 34 (1):81-85.
[2] Li Wenqiang, Peng Xuefeng, Zheng Zhiqiang. Six degrees freedom UAV simulation based on Simulink [J]. Systematic Simulation Journal, 2007, 19 (19): 4604-4606.
[3] Li Chao, Wang Jiangyun, Han Liang. Development of the flight simulation system of a fixed-wing aircraft based on Matlab [J]. Systematic Simulation Journal, 2013, 25 (8):1772-1777.
[4] Huang Hua, Xu Youping, Deng Zhiwu. A real-time visualization flight simulation system based on Flightgear simulator [J]. Systematic Simulation Journal, 2007, 19 (19): 4421-4423.
[5] Liu Xiongxiong, Zhang Weiguo, Huang Yijun, et al. Research on Advanced UAV Six Degree of Freedom Motion Modeling Technology [J]. Arrow and guidance paper, 2005, 25 (3):431-433.
[6] Yuan Zihuai, Qian Xingfang. Controlling flight mechanics and computer simulation [M]. Beijing: National Defense Industry Press, 2001: 38-80.
[7] Li Jun. Design and development of flight control software for small fixed-wing UAV [D]. Nanjing: Nanjing University of Aerospace Science and Astronautics, 2011.
[8] Hu Qing. Design of UAV flight control system for single-chip machine Based on STM32 [D]. Nanjing:Nanjing University of Aerospace Science and Astronautics, 2012.
[9] Ye Ge, Tian Zhaofeng, Yan Chuliang. Visualization of aircraft flight data based on OpenGL [J]. Aviation Journal, 2011, 32(6): 1050-1057.
[10] Payneb. Programming 3D Graphics with OpenGL [M] Houston: Rice University Press. , 2010: 325-362.
[11] Li Lirong, Shen Chunlin, Lu Yuping, et al. The 3D visual simulation system of ultra-low altitude/Intercept aircraft based on OpenGL [J]. Journal of Nanjing University of Aeronautics and Astronautics, 2005, 37 (3): 269-273.
[12] Hu Wei, Wang Hong. The OpenGL Programming in VC + +with MFC [J]. Computer application, 2001, 21 (8): 87-89.
[13] Li Haijun, Li Jiejie, Liu Xiao, et al. A visual simulation of the missile attack process Based on OpenGL [J]. Systematic Simulation Journal, 2004, 16 (3): 530-533.

Rolling Bearing Fault Diagnosis Based on PCA-ResNet

Xinjie Peng, Zhigang Wang*, Binbin Li, Long Qian and Bin Jiao

[1] School of Electrical Engineering, Shanghai Dianji University, Shanghai, China

*Corresponding author email: wangzg@sdju.edu.cn

Abstract. Aiming at the situation that there are many interference factors in the collected signals of rolling bearings working in harsh and complex environment. Propose A method of noise reduction based on principal component analysis (PCA) and fault diagnosis and classification based on residual neural network (ResNet). Experimental results show that this method can effectively complete the task of fault identification and classification, and is superior to LeNet-5, SVM, DNN and other common deep learning methods.

1. Introduction

Bearing is an important component of all kinds of mechanical and electrical equipment, which is widely used in various mechanical departments. However, bearing faults occur in use and have a certain randomness, which seriously affects the normal operation of mechanical equipment. Therefore, in-depth study of bearing fault mechanism and fault diagnosis and classification method has always been a hot issue for experts and scholars at home and abroad [1]. The vibration mechanism and common fault diagnosis methods of rolling bearings have been classified and counted by domestic scholars.[2]

Nowadays, with the development of AI and the improvement of computer computing power, human beings have entered the generation of big data. In pace with the data on the Internet more and more large, deep learning model is out of the advantages of large data processing, and is widely used in bearing faut diagnosis field. Literature[3] is put forward based on the short-time Fourier transform and convolution combined bearing fault diagnosis method of neural network, the recognition of different types of fault have high precision, It is a fault diagnosis method suitable for processing 'big data'. Literature[4] presents an improved random forest algorithm for faut diagnosis of motor bearing, which provides a new idea for fault diagnosis of motor bearing. By comparing the characteristics of variational mode decomposition (VMD) and empirical mode decomposition (EMD), and improving the multiscale permutation entropy, good results have also been achieved in the literature[5].

Although many fault diagnosis models have high recognition rate in bearing fault diagnosis, they fail to fully use the superiorities of deep learning in complex signal processing noise reduction, and the diagnosis accuracy will decrease with the deepening of the network[6]. Therefore, a fault diagnosis method based on the combination of PCA and ResNet was put forward. The original data were de-noised and then the residual network was used to diagnose and classify the data. The practicability and robustness of the method were proved by experiments.

2. The PCA-ResNet Model

2.1 Principal Component Analysis

Principal Component Analysis (PCA), It is a linear transformation.[7] The main idea of PCA is to map n-dimensional features to k-dimension, which is a new orthogonal feature, also known as the main component. It is a k-dimensional feature reconstructed from the original n-dimensional features. The job of PCA is to find a set of mutually orthogonal coordinate axes from the original room. The selection of new coordinate axes is closely related to the data itself. Among them, the first new coordinate axis selection is the direction with the largest variance in the original data, the second new coordinate axis selection is the one with the largest variance in the plane orthogonal to the first coordinate axis, and the third axis is the one with the largest variance in the plane orthogonal to the first and second axes. By analogy, n such coordinate axes can be obtained [8].

Its implementation method is as follows:

1)Demean, that is, each eigenvalue subtracts the respective mean;

2)Count assist variance matrix $\frac{1}{n}XX^T$;

3)The feature values and feature vectors of assist variance matrix $\frac{1}{n}XX^T$ are obtained by eigenvalue decomposition method

4)The feature values are sequenced from largest to smallest, the biggest k feature values are selected, and the homologous k feature vectors are respectively used for the purpose of row vectors to form the feature vector matrix P.

5)Transform the information entering a new room consist of k feature vectors, $Y = PX$.

2.2 Convolutional Neural Network

Convolutional neural network (CNN) is a feedforward neural network with deep structure. CNN has powerful data processing ability, and can directly realize feature extraction without extracting fault features. At the same time, CNN also has excellent representation learning ability. The structure diagram of LeNet-5 model, one of the commonly used CNN models is shown in Figure 1.

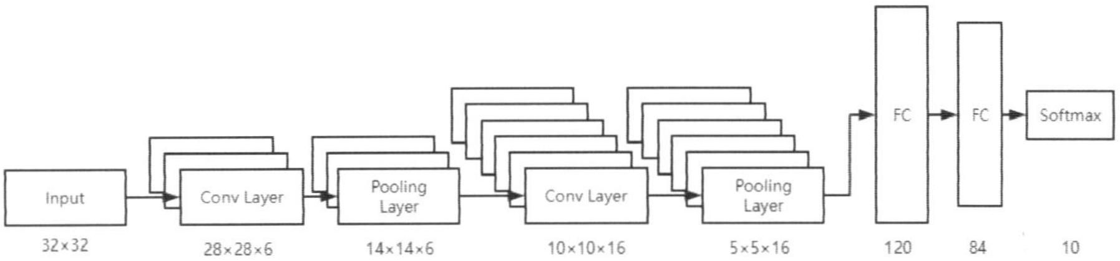

Figure 1. LeNet-5 model.

2.2.1 Convolution layer. The Convolution layer consists of several convolution units, and the parameters of Each convolution unit is iteratively optimized by one back-propagation algorithm to obtain the best parameters.

The convolution expression of the convolution kernel and the input data is:

$$S(i,j) = (I * K)(i,j) = \sum\sum I(i,j)K(m,n) + b \tag{1}$$

Where m and n are the dimensions of the nucleus, and b is the skew weight.

$$a = f(s(i,j)) \tag{2}$$

In which $f(.)$ to be the activation function, Relu function is selected in this paper.

2.2.2 Pooling layer. Pooling layer is used in CNN to reduce the size and improve the operation speed, which can also reduce the influence of interference and make each feature have better robustness. Max pooling and Average pooling are common, and Max pooling is more common in practical application. Its expression is as follows:

$$P_{max} = \max(f(s(i,j))) \tag{3}$$

$$P_{aveg} = \frac{1}{n}\sum f(s(i,j)) \tag{4}$$

Where n is the size of the nucleus.

2.2.3 Full connection layer and Softmax layer. Fully Connected layer (FC) plays a role of "classifier" in the whole CNN. The characteristic collected from the convolutional layer and the pooling layer are classified into a complete feature map by weight matrix.
Through Softmax layer, we can get the probability distribution problems of different types of current samples.

2.3 Residual Neural Network
Compared with traditional machine learning, the key features of deep learning lie in deeper network layers, nonlinear transformation (activation function), automatic feature extraction and feature transformation. Among them, nonlinear transformation is the key goal, which maps data to high-dimensional space to better complete "data classification". With the increasing depth of the network, more and more activation functions are introduced, and the data is mapped to a more discrete space. At this time, it is difficult to get the data back to the origin (identity transformation). In other words, the neural network mapping this data back to the origin requires far more computation than we can afford. The parameters can not be modified effectively, which greatly reduces the network performance.
The degradation phenomenon makes us reflect on the nonlinear transformation, which greatly improves the ability of data classification. However, with the increasing depth of the network, we have gone too far in the nonlinear transformation, and can not realize the linear transformation. Obviously, it is a good choice to add linear transformation branches to the neural network, and residual network (ResNet) was born in this environment.

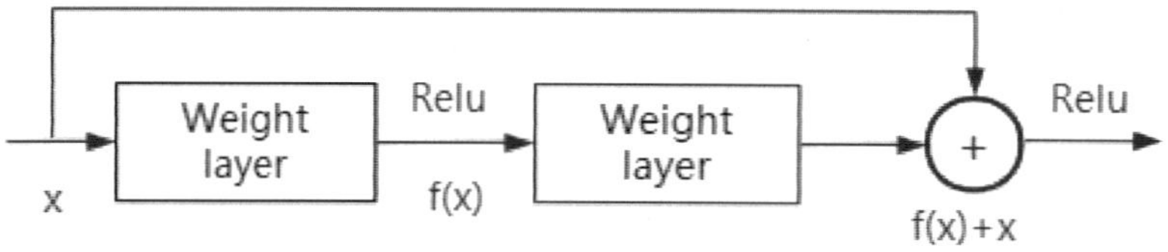

Figure 2. Residual block.

The residual block is shown in the figure 2: the original signal is x, and the residual network has two paths. The first main path is the output $f(x)$ obtained by forward propagation. The second shortcut superpositions the original signal x and the output $f(x)$, and import $f(x)+x$ into the Relu activation function to obtain the new output. By using the identity mapping in the residual network, the layers of the neural network are deep enough to ensure the normal propagation of the gradient, and effectively modify the subsequent network parameters, so as to improve the efficiency of the neural network.

3. Experimental Process

3.1 Experimental Steps
1) Using the bearing data of Case Western Reserve University to collect vibration signals under different bearing faults;
2) PCA denoising ratio is appropriately selected for denoising processing;
3) The pre-processed signals are randomly interpolated to form a two-dimensional atlas;
4) The two-dimensional feature atlas is divided into two parts according to proportion, one training set and one test set;
5) Initialize the weights in ResNet network, input the training set, and train ResNet network;
6) Input the test set into the trained model for fault diagnosis and recognition;
7) Analyze the results and compare them with other models.
The flow chart is as follows:

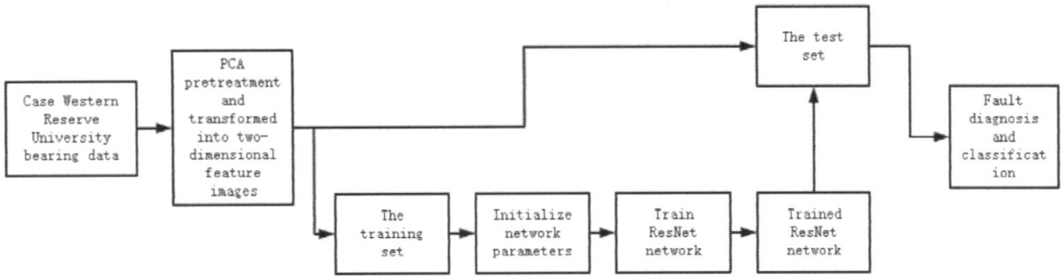

Figure 3. PCA-Resnet model flow chart.

3.2 Data Sources and Classification

In order to verify the effectiveness of the model, we choose to use the rolling bearing data provided by Case Western Reserve University to verify it, and divide it into four types: normal, inner ring wear, outer ring wear and ball wear, and there were 2500 sets of data for each fault, altogether 10000 sets of data. The method of random sampling is divided into two part which named test set and training set according to 30% and 70%, so there are 1750 groups of data in training set and 750 groups of data in test set, as shown in the table 1:

Table 1. Composition of experimental samples.

Sample	Training sample set	Text sample set	The label
Nor	1750	750	0
OR	1750	750	1
IR	1750	750	2
BB	1750	750	3

3.3 ResNet Model Construction

Tensor Flow, Caffe and Keras are the commonly used learning frameworks in convolutional light networks. Considering that this paper conducts training for multi-classification tasks, we choose Keras framework, which is a highly modular light network library. Written by pure living Python and with Tensorflow, Theano and CNTK as the back end, it can freely combine multiple modules of convolutional neural network with low operation difficulty.

Based on the classical Lenet-5 model, the ResNet model was constructed using Keras framework. The selection of each parameter is shown in Table 2.

Table 2. Residual network parameters.

Number	Parameter Name	Parameter setting	Number	Parameter Name	Parameter setting
1	Conv layer 1	$5\times5\times32,1$	6	Conv layer 5	$\begin{pmatrix}3\times3\times256\\3\times3\times256\end{pmatrix},2$
2	Pooling layer 1	$2\times2\times32,2$	7	Pooling layer 2	$2\times2\times256,2$
3	Conv layer 2	$\begin{pmatrix}3\times3\times32\\3\times3\times32\end{pmatrix},2$	8	FC	128
4	Conv layer 3	$\begin{pmatrix}3\times3\times64\\3\times3\times64\end{pmatrix},2$	9	Softmax	4
5	Conv layer 4	$\begin{pmatrix}3\times3\times128\\3\times3\times128\end{pmatrix},2$			

3.4 Analysis and Comparison of Experimental Results

The four features are successfully classified through operation, it can be seen from the feature classification diagram that the model can significantly classify faults, as shown in Figure 4

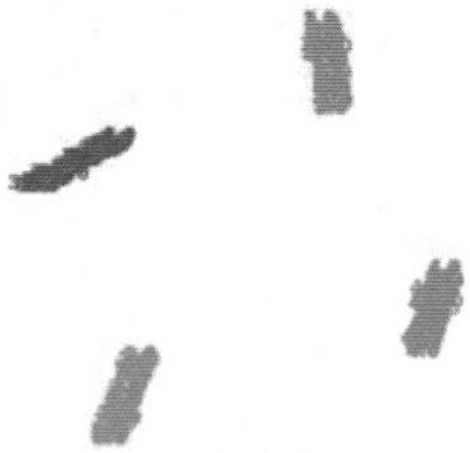

Figure 4. Feature classification diagram.

In order to verify that the model can accurately realize fault identification and classification, the classical convolutional neural network Lenet-5 model, DNN model and support vector machine (SVM) were used for comparison. As the control variable, 2500 groups of data were used uniformly, and they were divided into training set and test set according to the proportion of 70% and 30%. The experimental comparison results are shown in Table 3.

Table 3. Comparison of different experimental results.

Diagnostic model	Accuracy / %
PCA-ResNet	99.57
LeNet-5	86.91
DNN	98.84
SVM	97.54

As can be clearly seen from Table 3: The method studied in this paper can effectively extract fault features and identify fault states, with the highest average diagnostic accuracy of 99.57%. Compared with other algorithms, its feature recognition and diagnostic accuracy are significantly improved, proving the effectiveness of the model.

4. Conclusion

Using traditional deep learning methods, aiming at several common fault types of motor bearing, this paper proposes a bearing fault diagnosis model based on PCA-ResNet. Firstly, the principal component analysis (PCA) is used to reduce the noise of the collected data, and then the preprocessed data is divided into training part and testing part according to the proportion of 70% and 30%, and then it is identified and classified by residual neural network, which is realized in keras. Using the powerful feature extraction function of convolutional neural network, the task of high-precision bearing fault diagnosis and identification can be directly completed without extracting fault features alone, and the accuracy rate reaches 99.57% in experiments, the practicability of the model is proved.

Reference

[1] Sha Meiyu, LIU Liguo. Bearing Fault Diagnosis Technology Review Based on Vibration Signal [J]. Bearing,2015(09):59-63.
[2] Chen Qiangqiang, Dai Shaowu, Dai Hongde, NIE Zijian. Review of Rolling Bearing Fault Diagnosis Methods [J]. Instrument Technology,2019(09):1-4+42.
[3] Li Heng, Zhang Hydrogen, Qin Xianrong, Sun Yuantao. Bearing fault diagnosis method based on STFT and convolutional neural network [J]. Journal of vibration and shock,2018,37(19):124-131.
[4] Li Bing, Han Rui, He Yigang, Zhang Xiaoyi, Hou Jinbo. Proceedings of the csee,2020,40(04):1310-1319+1422.
[5] WANG XIAolong. Research on Rolling Bearing Fault Diagnosis Method based on Vibration Signal processing [D]. North China Electric Power University (Beijing),2017.
[6] Qu Jianling, Yu Lu, Yuan Tao, Tian Yanping, Gao Feng. Adaptive fault diagnosis algorithm for rolling bearings based on one-dimensional convolutional neural network [J]. Chinese journal of

scientific instrument,2018,39(07):134-143.
[7] Gu Yingkui, Cheng Zixin, Zhu Fanlong. Fault Feature Fusion Analysis of Rolling Bearing Based on Principal Component Analysis and Support Vector Machine [J]. China mechanical engineering,2015,26(20):2778-2783.
[8] Guo Weichao, Zhao Huaishan, Li Cheng, Li Yan, Tang Aofi. Enhanced Bearing Fault Diagnosis Method Based on Wavelet Packet Energy Spectrum and Principal Component Analysis [J]. Acta armamentarii,2019,40(11):2370-2377.

Research and Application of Display System of Component-Based Virtual Instrument for Land and Air Platforms

Hao Cai[1] **and Zheng Mao**[2,*]

[1,2] Department of Informatics, Beijing University of Technology, Beijing, China

*Corresponding author email: maozheng@bjut.edu.cn

Abstract. With the diversified combat missions in the new era and the development of industrialization, the demand for military and civilian equipment continues to increase, and the development and upgrade standards of land-air dual-use platforms must also move toward standardization. As the core of information visualization and control standardization, the component-based virtual instrument display platform is an important part of the integrated information display system of the land-air platform. The component display platform takes the optimization of the research and development, design and maintenance of the land and air mobile platform as the starting point, and at the same time meets the needs of wartime and meets the development trend of future informatization. It adopts engineering design ideas, hierarchical software layout and componentized ideas of man-machines. The concepts of interaction design are based on componentization techniques, serial communication, data dictionaries, pattern transformations, etc. A comprehensive information display platform aims to increase software productivity, reduce development costs and degree of coupling, and meet human-computer interaction as well as user-friendliness requirements.

1. Introduction

Land and air platform has the dual performance of land cross-country and air flight, mainly driving on the ground, can be flexible to shuttle the ground, with minimal cost and risk to transport soldiers or supplies to the destination When it encounters deep ditches, ponds, cliffs, minefields and other ground obstacles, or encounter ground fire blockade siege situation, it can take off vertically, quickly over the obstacles, or break through the fire blockade, and can hover in the air, select the landing site, land vertically and safely to complete the task of transportation, reconnaissance, rescue and logistics supply. Measured by current standards for conventional mobile vehicles and aircraft, land and air platforms with in situ takeoff capability are both capable of traveling on land and different from conventional transportation systems that are completely limited by their physical space, such as trains on tracks or cars on roads[1-2]. Amphibious platforms offer a significant expansion in operational reach and can greatly enhance the flexibility of flat-earth operations.

To cope with different task requirements, the layout design of the human-machine interaction interface of the land and air platform display terminal should be based on human visual characteristics as a reference, and the surface layout interval should not be too large to improve the efficiency of visual field search and delay the time of visual fatigue. But the majority of the display terminals are based on traditional electromechanical instruments, which are independently developed with the R&D team's project objectives. The aesthetics of the human-machine interface of the display terminals and the limitations of the display content greatly limit the amount of immediate information available, which is not conducive to the driver's judgement of the next control. What is urgently needed is a component-based software production method that facilitates agile development, low code redundancy, and easy maintenance and upgrading at a later stage, with low coupling, whose functional components can be reused and flexibly deployed[3] to the integrated information display terminals of land and air platforms. Based on the drawbacks and shortcomings of the above-mentioned issues, this paper proposes a component-based design of a display system for land and air platforms, which has positive reference value.

2. Component-based Virtual Instrument Display Platform

The integrated information display system of land and air platform adopts the design idea of building blocks, and divides the overall system architecture into two major parts, including data processing and management platform and integrated information display platform, and four modules, namely platform building block layer, platform support layer, software service layer and hardware support layer, in a hierarchical, open and low-coupling architecture. Using a software layered design, the layers are divided[4] and the architecture of the display platform is shown in Figure 1.

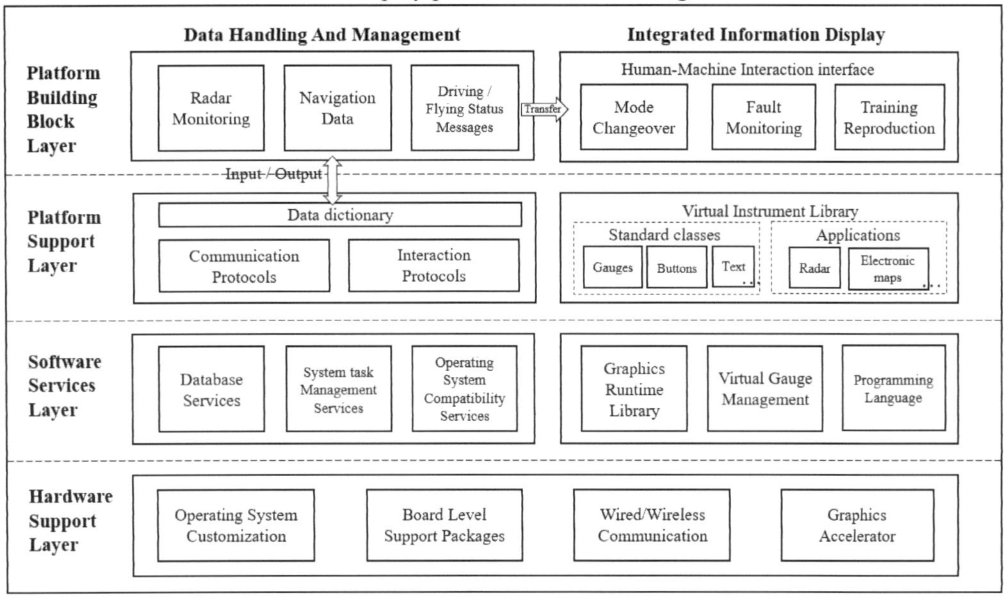

Figure 1. The architecture of display platform.

2.1. The Data Processing and Management Platform

Information acquisition, message parsing, data storage and reproduction of working conditions are the functions of the data processing and management platform. The platform is also built with a comprehensive electronic system as the core, which is connected to the measurement and control object through a signal conditioning-data acquisition card, serial/parallel port instrument or other equipment interface, and Interacting and parsing raw data messages with the computer operating system by means of serial communication.

The data is packaged in a uniform manner after the data has been collected by the industrial control equipment, and the packet format is: data header H, count bit C, identification bit S, data bits A, checksum T. The data dictionary completes the unified management of data, discriminates the validity and value range of data, and filters, extracts and classifies data. The platform abstracts and models the data protocols for external communication, establishes a generalised mathematical model and forms a dynamically configurable database file, which can be quickly adapted to various communication protocols and realises the separation of protocols and applications. The processing flow is shown in Figure 2.

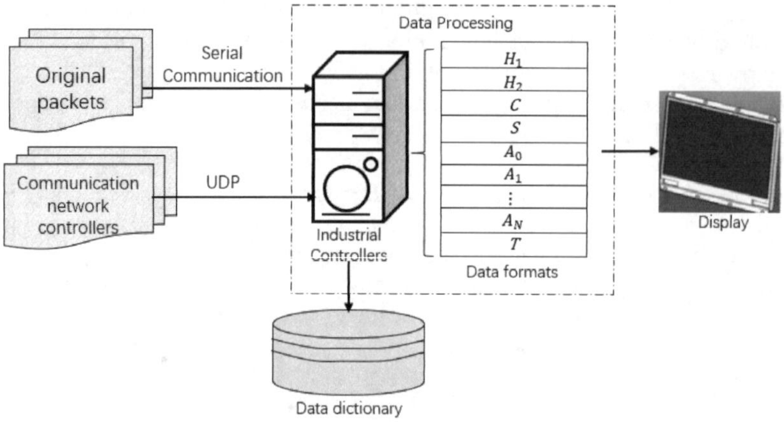

Figure 2. Data processing flow.

The stability and reliability of the software operation directly affects the performance of the hair reference collector[5], and the data processing method of the display platform implemented by the data dictionary in collaboration with the building blocks reduces the mutual interference between modules and further improves the flexibility and standardisation of the components.

2.2. Integrated Information Display Platform

The integrated information display platform is responsible for the display of the human-machine interface, the management of the virtual instrument library control library and the binding of the functional interfaces. The virtual instrument library is the core component of the display system architecture and the central module for the operation and human-computer interaction of the integrated information display platform. The display platform is an essential channel for users to obtain information about the vehicle platform and to interact with the platform on a human-machine basis. It will be the dynamic change of information in a timely manner for feedback output, it allows the operator to intuitively observe the various information of the platform, facilitating the judgment of their own movement trend, and effectively reduce the emergence of accidents.

This article focuses on embedding the idea of component-based design into the integrated information display platform, incorporating instrumentation technology and computer technology organically. While boosting the development efficiency, it coordinates the human-computer interaction interface with the unification of the components, and fits the users' usage habits to fulfill the actual requirements of modern diverse kinds of information display, in order to attain such characteristics as digitization and visualization of information.

3. The Idea of Componentisation

Component-based software development is the new generation of software development thinking and methodology[6]. Using the serial communication as the "communication link" and the serial address as the "resource number", the measurement and control object's data messages are packaged and sent to the information processing computer for data unpacking and calculation. The data dictionary serves as the "data storage and management warehouse" where all data is sorted and recorded, a process that can be considered as an entry operation. The process can be considered as an outbound operation, where data can be displayed on the interface by the "communication link" or conveyed to the platform for related operations according to the instructions given by the user for human-machine interaction. The allocation of serial port resources to data communication resources is shown in Table 1.

Table 1. Communication resource allocation table.

Cluster communication port	Communication links	Byte length (byte)	Transmission frequency (Hz)	Description
COM1	RS-485	22	5.0	Driving data
COM2	RS-232	58	5.0	Flight data
COM3	CAN	15	5.0	Other

In the current development of software technology, component-based programming has reached the binary level of software reuse, which follows the specifications of the unified component model and enables the mutual development of software components, as well as the effective control of component granularity[7]. Componentisation means that the original electromechanical instrumentation can be abandoned, with the aim of reducing the weight of the equipment and saving hardware resources, while at the same time providing access to information that is both rich and human-machine friendly. With the idea of component-based virtual instrumentation design and the construction of relevant standards and models, the most significant feature is the reconfigurability, which allows the functions of conventional instruments to be implemented by selecting the appropriate modules and auxiliary software, it allows for a clearer local and global situation of the overall platform[8]. Upon reaching a certain size, these virtual instruments with basic and special functions can be built up into standard and application components respectively, forming the core elements of the virtual instrument library. Once the virtual instrument library has been built, the interface of the display terminal can be designed without the hassle of the design process.

The establishment of a virtual instrument library is a crucial part of the project, which will enable the future reuse, expansion and maintenance of the project components. In addition, the industry standardisation of integrated information display terminals for land and air platforms can be further strengthened. The virtual instrument library needs to be developed, tested and reviewed by professionals or teams to meet the relevant criteria before it can be added to the library. The design must also be tailored to the user's habits, not only to improve the efficiency of access to information on the platform, but also to provide a powerful aid in the manipulation of land and air platforms for a variety of tasks.

4. Component-based Virtual Instrument Display Platform

The building blocks must be designed with the appropriate standards, such as identification, version, function, properties, interfaces and other relevant information. And with features such as independence, interface specification and extensibility as prerequisites for a standardised design. Specific design criteria are shown in Table 2.

Table 2. Component design standard.

Standards	Description
Name	It requires uniqueness and is character-based.
Identifier	Identifies the data as integer (int).
Versions	V***.***/Year.Month.Date.
Function	To achieve functional descriptions, with independent display functions required.
Attributes	Windows, layer, style, general, background, action, Widget1 (top, left, width, height, colour, style), ... , properties can be extended.
Runtime files	Runtime files (*.dll) containing interfaces, properties and state.
Interface	Contains methods for interacting with data and events, etc.
Configuration file	Configuration file *.xml for name, logo, version, run file and properties, etc.

The independence means that a building block is a unit that can be configured independently, favouring decoupling between modules in the software design process. The interface specifications implement the encapsulation of each building block to ensure the reliability of data communication and data interaction. The extensibility supports the reuse of building blocks to cope with changes in requirements or expansion of functionality.

5. Component Management

The components can be generated visually by any graphics editor, such as Qt, VC, OpenGL, and stored in a database as a dynamic link library. In Qt Designer, for example, it creates a component class in Qt's 2D drawing class, follows the component structure to complete the function code of each member and includes it in the related function events, then it connects the timer signal with the component update function to realize the dynamic refresh of the component display.

A virtual instrumentation library can be formed with a certain size of component design. When using it, it is only necessary to call the corresponding components from the library according to their type, achievable functions and display style preferences, and to analyse the requirements according to the functional implementation effect, if the component cannot fulfill the corresponding requirements and affects the normal working system of the information display platform. This is where development tools are used to update and extend the properties of the components in order to meet the constantly updated application standards of the information display platform. If all the current components satisfy the functional requirements, they can be arranged on the main interface and then stylistically modified and positioned for the human visual characteristics and the actual working environment. The information display platform is used to visually set the attributes and extend the functions of the components and to enable the layout and functional requirements of the human-computer interaction interface. The flow is shown in Figure 3.

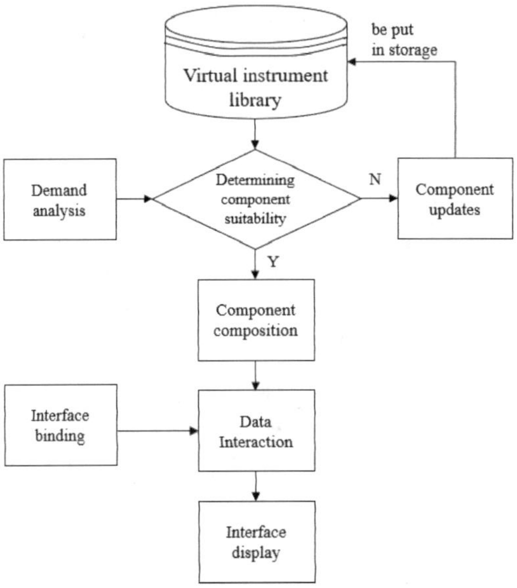

Figure 3. Component management process.

6. Virtual Gauge Interface Design

This paper is designed to complete the deployment of virtual gauges by means of component calls. The land and air platform needs to display various indicators in both "air" and "land" driving states, and the implementation of this function requires the invocation of button-like components as a channel for mode conversion, and the corresponding interface display through the judgment of the driving / flying state, so as to achieve the switching of the driving The switch between the car/flight display and the display of the multi-member working mode.

6.1. Component Operation

The display platform is taken over by the platform management task after the operating system has been loaded and the terminal hardware has been initialised. The platform management task completes the loading, configuration and data dictionary start-up of the graphics library. Component operation and display layers are shown in Figure 4.

Once the components have been designed and the virtual instrument library has been constructed, the components can be called up and the virtual instruments can be configured. The component models are differentiated by the different modes of travel of the land and air platforms.

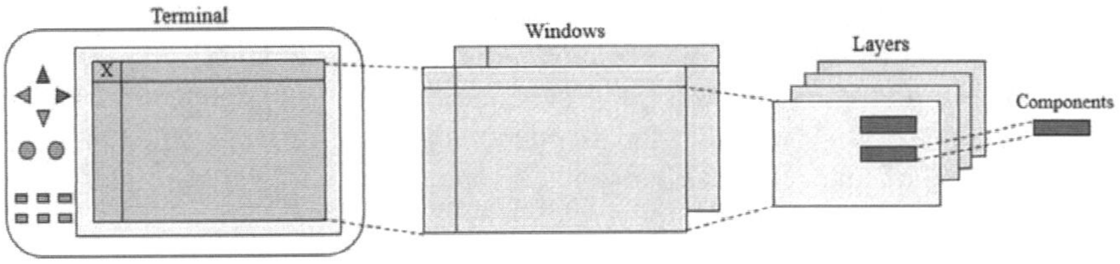

Figure 4. Component operation and display hierarchy.

6.2. Display Interface Layout

The layout of the interface is based on the visual characteristics of the human being. The horizontal dimensions of the instrument panel are larger than the vertical dimensions and are arranged in a left-to-right, top-down and clockwise direction. The layout is based on the principles of safety factors and driver ergonomics for monitoring vehicle working conditions during driving[9]. The layout of the interface should not be spaced too far apart to improve the efficiency of the visual field search and delay visual fatigue time. The placement of the components and their size are adjusted according to the priority of the data needs to be displayed in different marching modes.

The interface layout in driving mode places the two main parameter components, such as speed and engine speed, in the most easily observable position and increases the area occupied by the model. The parameter components such as water temperature, engine temperature and oil level are listed to the left and right of the driving interface. The high beam, low beam, left / right turn lights, warning lights and other indicator components are placed at the top, the number of kilometres driven and the Total mileage are placed on the right hand side of the interface, and the top right hand corner houses the driving / flying display interface switch button for mode switching operation The layout of the driving interface is shown in Figure 5.

Figure 5. The interface of driving.

Figure 6. The interface of flight.

The layout of the flight interface focuses on airspeed, vertical speed, vertical altitude, fuel level, engine temperature, heading, horizon and other parameters as the main display content. Heading and horizon are the most critical parameters to ensure normal flight conditions and are therefore prominently displayed in the centre of the interface to provide feedback on direction of travel and flight attitude. The airspeed and vertical speed parameters are positioned on either side of the interface and are also scaled up to allow the pilot to observe the dynamic changes in the platform's operating parameters. The layout of the flight interface is shown in Figure 6.

6.3. Mode Conversion

The integrated display system for land and air platforms uses a virtual push-button component as a changeover controller to change the human-machine interface between travelling/flying modes. Mode switching requirements include the demand for switching modes across platforms, display interface extensions and loading requirements. The initial state is the default mode of operation, which can be converted to the extended mode of operation and the co-operative mode of operation by means of the conversion controller, and there is a common functional partition between the three modes of operation that can be freely switched. This is shown in Figure 7.

The multi-modal collaborative way of working aims to build a safe working system for platforms in multi-task scenarios, supported by hardware and computer operating systems, with building blocks acting as changeover switches. The multi-modal collaborative way of working aims to build a safe working system for platforms in multi-task scenarios, supported by hardware and computer operating systems, with building blocks acting as changeover switches.

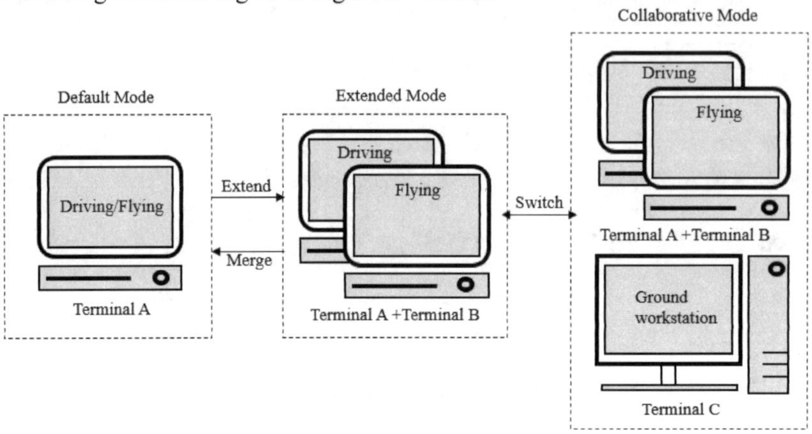

Figure 7. Component operation and display hierarchy.

The platform display terminal is responsible for the loading of the human-machine interface and the integration of the virtual instrument library, in accordance with the members' needs for access to information on the current platform's working conditions and for switching between platform display terminals. The data processing and management platform deals with the identification of driver usage patterns and the control of business processes. In order to cope with the functional requirements and working modes of the land and air platform under different mission scenarios, the default, extended and collaborative modes of the display terminal have been designed with corresponding configuration management databases to store the configuration files for each mode of the platform. The display platform management task judges and determines the operating mode of the land and air platform display terminals based on the dynamic commands received.

In the default mode, the platform operating member is authorised with the highest authority to independently perform platform function using and display mode switching to ensure their ability to use the platform display terminal building blocks. Extended working mode, i.e. extension of the original single terminal display to a dual one[10], allowing parallel display of driving/flying display modes and free switching between display terminals. In addition to the basic platform display, an extra ground condition monitoring workstation is added to support the ability of multiple members to work together while ensuring the safety of the driver and the land and air platform.

7. Conclusion

This paper proposes a virtual instrumentation display platform with a component-based design concept for the visualisation of information in different driving modes of land and air platforms to achieve effective output and display of measurement and control data. The transmission of data resources is made more efficient by the effective management of the communication model. At present, the initial application of the modular display platform technology has been successfully carried out on a land and air mobile platform, achieving specific functions with excellent human-machine friendliness.

With the rapid growth of industrial technology and the information industry, there is an increasing demand for display platforms that are more intelligent, but at the same time increase the cycle time and complexity of software development. A display platform with a component-based architecture not only assists the driver and workforce in their next steps, but also provides a strong guarantee of safe and stable operation. Information display and functional implementation of the display system is accomplished by creating a library of standard components and application components and assembling the parts according to rules during the development process, thus reducing the complexity of the software design, the readability and versatility of software and the productivity of hardware resources are improved, and reference standards for the independence, encapsulation and extensibility of components are established, ensuring that the development of current land and air display platforms is rule-based and has good prospects for development and application.

On the basis of meeting the platform display system can be developed successively, in the production engineering also need to solve the problem of refining the commonality of basic components, optimizing the framework of application components, etc. Secondly, constantly carry out the accumulation of components, realize the rapid construction of display terminals, and output the solution of multi-platform transplantation, in order to meet the future development trend of platform standardization and equipment efficiency.

References

[1] G. Pan and M. -S. Alouini, "*Flying Car Transportation System: Advances, Techniques and Challenges*"in IEEE Access,vol 9,pp.24586-24603,2021,doi: 10.1109/ACCESS.2021.3056798
[2] K. Rajashekara, Q. Wang and K. Matsuse, "*Flying Cars: Challenges and Propulsion Strategies*" in IEEE Electrification Magazine,vol.4,no.1,pp.46-57
[3] Yangyang Ge ,Bo Gao, Hao Feng. *Software design of component-based engine parameter collector[J]*. Information Technology and Informatization,2021(02):pp 49-51
[4] Ye Li, Yuan Gao. *Research on layered software architecture design and its application [J]*. Computer Knowledge and Technology, 2017, 13(13):pp 89-90
[5] Yangyang Ge, Jichao Wang. *Software design of serial bus communication based on building blocks[J]*. Information Communication, 2018(10):pp 196-197
[6] Szyperski C. *Component Software- Beyond Object Oriented Programming [M]* . Addision-Wesley,1997
[7] Xia Wang,. *Software reuse and software component technology [J]*. Information and Computer (Theory Edition),2018(10):pp 7-9
[8] M.Fandáková, K.Zábovská, B.Bučko, M. Zábovský. *Improvements of Computer Assisted Virtual Environment (CAVE)*. 2020 43rd International Convention on Information, Communication and Electronic Technology (MIPRO).IEEE,2020
[9] Han C and Shao X, "*Double signal double display intersection vehicle terminal system for parking navigation*" 2014 IEEE International Conference on System Science and Engineering (ICSSE), Shanghai, China, 2014,pp 103-107
[10] Hianzhen Jin, Wei Shang, Hongyun Chen. *An analysis of the design and optimization methods of the general interior arrangement [J]*. Automotive Parts,2019,21(11):pp 48-51

Study on the Division Method of Pipeline Robot Module

Haizhen Zhu*, Yan Yuan, Yongjun Lei and Xiaoru Li

School of Mechanical and Electrical Engineering, Xi 'an Technological University, Shaanxi Xi' an, China

*Corresponding author email: jenny_zhu9722@163.com

Abstract. The working environment in the pipeline is getting worse, and there are more and more functional needs for the pipeline robots. The traditional pipeline robot structure is fixed and single, and the rapid design and assembly cannot be realized. There are still problems of repeated design between different types of robots.In view of these problems, a top-down modular design method—combining of axiom design and IDEF(ICAM DEFinition method) functional modeling is proposed to divide the pipeline robot system architecture.This method uses the mapping relationship between function and structure, and extracts the general modules and special modules in different types of pipeline robot systems, which realizes the modular design of the non-modular pipeline robot, so that the module structure at its component level can be universally exchanged to form resource sharing.
Keywords: Pipeline robot; Functional requirement; Module division; Axiom Design; IDEF.

1. Introduction

Due to construction reasons, the pipeline has been underground or in a narrow place for a long time, and the internal environment is corroded, messy and closed. In order to avoid injury, the use of pipeline robot is introduced. However, the traditional pipeline robot has a single and fixed structure, which cannot quickly respond to the diversification of market demand. Modular design means to analyze the functions of products with different functions or products with the same functions but different properties and specifications within a certain scope, to disassemble the functional structure of products, and establish a series of functional modules, which are characterized by independent functions and structures. Then, the product needed can be obtained through the combination of various modules according to a certain configuration principle[1].Through modular design of pipeline robots[2,3], an innovative module division method was proposed, i.e., a method of combining axiomatic design and IDEF functional modeling to conduct module division for the pipeline robot system. This method can realize compatibility and interchangeability of components, form resource sharing, easy equipment updating, and rapidly satisfy user demands.

2. Determination of Module Division Method

The functions of pipeline robots must be realized through mechanical structure under the effect of the control system[4,5].By taking the pipeline maintenance robot as an example, structure decomposition was conducted via analytic hierarchy process, as shown in Figure 1. According to the structure decomposition, its bottom components are complex. If the bottom-up module clustering method is applied to module division, a huge computation burden will be involved, and the process is quite tedious. Therefore, a top-down design method that combination of axiom design and IDEF function modeling, was proposed to analyze the functions to be realized on the basis of the overall structure frame. Then modular design of the pipeline robot was studied from the angle of functions[6].The specific module division method is shown in Figure 2.

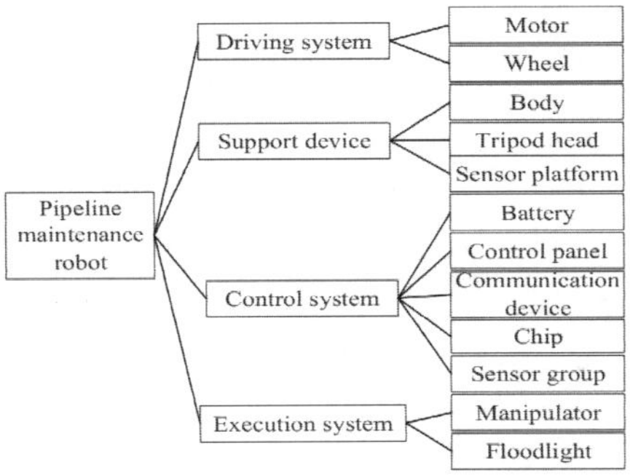

Figure 1. Structure Model of Pipeline Maintenance Robot

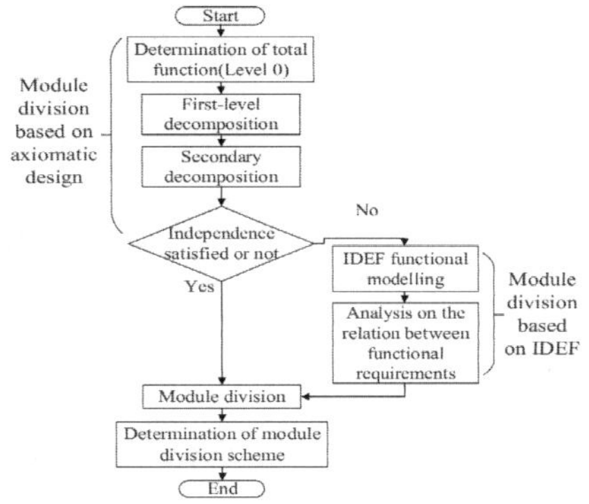

Figure 2. Division Method

3. Concept and Theory of Module Division Method

3.1. Axiomatic Design

The axiomatic design method can be described as the mapping and converting process between adjacent domains, as shown in Figure 3. Besides, independence axiom and minimum information axiom are adopted as the basis to verify the design effect[7-8].The "Z-shape" mapping process of axiomatic design can be expressed with the matrix form of Formula (1). In the formula, **[A]** is the product design matrix, indicating the relation of corresponding elements between FR and DP. In the matrix, 1 means that the two are associated; 0 means that the two are not associated. The requirements of independence axiom design of the axiomatic design can be met only when the design matrix is diagonal matrix or triangular matrix[9].

$$\{\boldsymbol{FR}\} = [\boldsymbol{A}]\{\boldsymbol{DP}\} \tag{1}$$

3.2. IDEF Functional Modeling

The mode of functional modeling in IDEF method adopts the graphical language to describe the relationships between various functional activities in the system abstractly and objectively by following the modularization criterion. The left arrows represent input, while the right arrows mean output, so box movement can be treated as data conversion. Upper and lower arrows indicate the environmental conditions to control the change and device support needed[10]. Figure 4 is the schematic diagram for layer-by-layer decomposition of IDEF functional model.

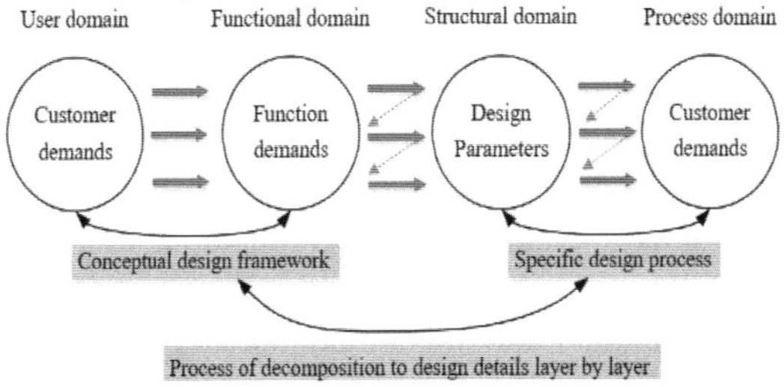

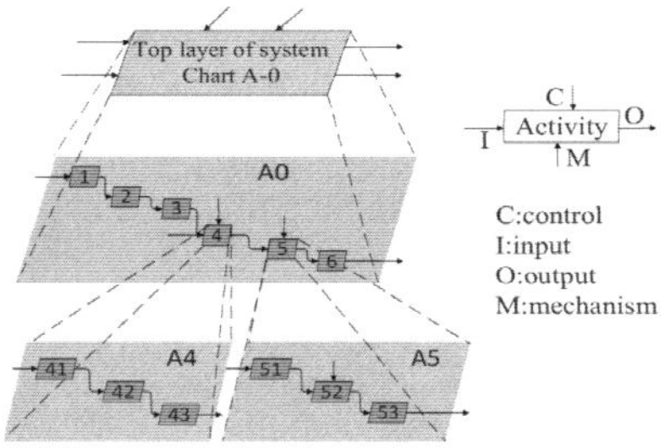

Figure 3. Axiomatic design process **Figure 4.** Results of Axiomatic Design Method Division

4. Module Division of Pipeline Robot System

4.1. Module Division Based on Axiomatic Design

4.1.1. The highest layer, Layer-0 decomposition. By taking the pipeline maintenance robot as an example, its fundamental aim is to realize the welding and dredging function in the pipeline. Hence, the functional requirement of the highest layer can be determined: pipeline maintenance. In other words, the FR_0-DP_0 design of the highest layer is: FR_0 = pipeline maintenance; DP_0 = modular pipeline maintenance robot.

4.1.2. Layer-1 decomposition. The basic functional requirement of pipeline robots is FR-pipeline maintenance, which can be further divided into 6 functional requirements: FR_1-Energy supply, FR_2-Moving, FR_3-Information collection, FR_4-Information processing, FR_5-User control, FR_6-Pipeline maintenance. The corresponding design parameters are: DP_1-energy supply component, DP_2-moving component, DP_3-acquisition information component, DP_4-information processing component, DP_5-control component, and DP_6-maintenance component. The design matrix between functional requirements and design parameters gained in first-level decomposition is a diagonal matrix.In the other words, the independence axiom is met. Therefore, the pipeline maintenance robot can be divided into 6 level-1 functional modules including energy supply, moving, information collection, information processing, user control and pipeline maintenance.

4.1.3. Layer-2 decomposition. In layer-2 decomposition, the layer-1 functional requirement of pipeline robots can be divided into the following layer-2 functional requirements, such as the energy supply function is decomposed into electric power supply function, energy conversion function and overtemperature protection function and so on. Control functions include input of command to control robots, image display, formatting and information recording. In view of the control and configuration, in-depth study will not be conducted, and they are divided into control functions. The most important maintenance functions can be decomposed into welding function and dredging function. Therefore, as shown in Table 1, "Z-shape" mapping and decomposition are further conducted for FR and DP at layer 2. The following is obtained:

Table 1. Layer-2 decomposition

xy	FR_{xy}	DP_{xy}	FR—DP design matrix
11	Power supply	Motor parts	$\begin{Bmatrix}FR_{11}\\FR_{12}\\FR_{13}\end{Bmatrix}=\begin{bmatrix}1&0&0\\0&1&0\\0&0&1\end{bmatrix}\begin{Bmatrix}DP_{11}\\DP_{12}\\DP_{13}\end{Bmatrix}$
12	Energy conversion	Spindle unit	
13	Over-temperature protection	Temperature detector	
21	Drive	Drive part	$\begin{Bmatrix}FR_{21}\\FR_{22}\\FR_{23}\\FR_{24}\end{Bmatrix}=\begin{bmatrix}1&0&0&0\\0&1&0&0\\0&0&1&0\\0&0&0&1\end{bmatrix}\begin{Bmatrix}DP_{21}\\DP_{22}\\DP_{23}\\DP_{24}\end{Bmatrix}$
22	Protection and support	Body parts	
23	Moving of maintenance device	Tripod head	
24	Moving of sensor	Sensor platform	
31	Collection of heat information	Heat source detection unit	$\begin{Bmatrix}FR_{31}\\FR_{32}\\FR_{33}\\FR_{34}\end{Bmatrix}=\begin{bmatrix}1&0&0&0\\0&1&0&0\\0&0&1&1\\0&0&1&1\end{bmatrix}\begin{Bmatrix}DP_{31}\\DP_{32}\\DP_{33}\\DP_{34}\end{Bmatrix}$
32	Collection of self-information	Position detection unit	
33	Collection of image information	Image collection unit	
34	Illumination	Lighting parts	
41	Information transmission	Information transmission unit	$\begin{Bmatrix}FR_{41}\\FR_{42}\end{Bmatrix}=\begin{bmatrix}1&0\\1&1\end{bmatrix}\begin{Bmatrix}DP_{41}\\DP_{42}\end{Bmatrix}$
42	Information collection	Information processing unit	
51	Control	Control panel component	
61	Welding	Welding assembly	$\begin{Bmatrix}FR_{61}\\FR_{62}\end{Bmatrix}=\begin{bmatrix}1&0\\0&1\end{bmatrix}\begin{Bmatrix}DP_{61}\\DP_{62}\end{Bmatrix}$
62	Dredging	Dredging parts	

According to Table 1, the design matrixes of information collection and information processing functions are full matrix and lower tridiagonal matrix, indicating that the functional structures under the same level and branch will influence each other, and that the independence in function and structure cannot be guaranteed. In view of the complexity of pipeline robots, the structure is decomposed into layer 2,i.e.,component level. Therefore, corresponding "Z-shape" function-structure mapping and decomposition are conducted for the remaining five parts, to realize module division, as shown in Figure 5.

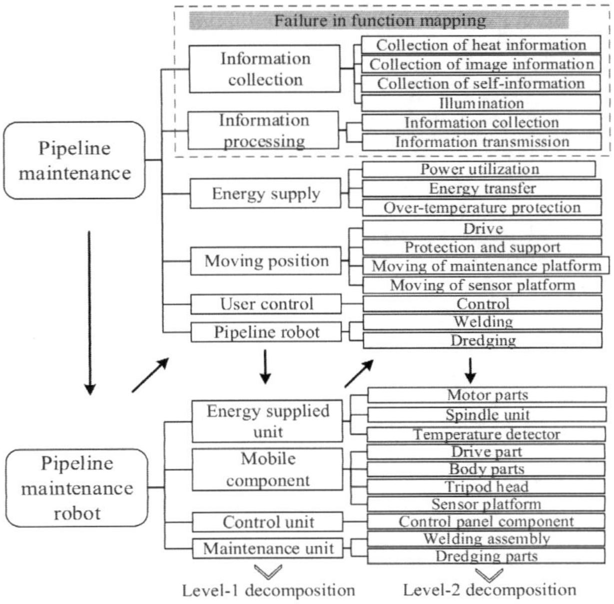

Figure 5. Results of Axiomatic Design Method Division

According to Figure 5, the structural components of layer-1 functional mapping are decomposed into several layer-2 modules through axiomatic design of pipeline maintenance robots: motor, battery, temperature detection, driver, body, tripod head, sensor platform, control, welding and dredging module.

4.2. Module Division Based on IDEF

In module division of pipeline maintenance robot based on axiomatic design, the illumination and camera shooting functions are associated, and the information transmission and information collection functions are also interactive, so the independence feature between modules is not satisfied. Therefore, the method of IDEF functional modelling should be further adopted, to analyze the internal relation between secondary functional activities of the two level-1 functions. If the relevancy between functional requirements is low and the mechanisms needed are different, then they can be classified into one functional module. Two models were established via the method of IDEF functional modelling to describe the input and output relation of information collection function, and the input and output relation of information processing function. By using software Visio 2013,A3 diagram was drawn to describe the former, as shown in Figure 6. and A4 diagram was drawn for another, as shown in Figure 7. The mechanisms required to achieve the second level of function can be found to be different, and therefore the functional structures with less correlation can be divided.

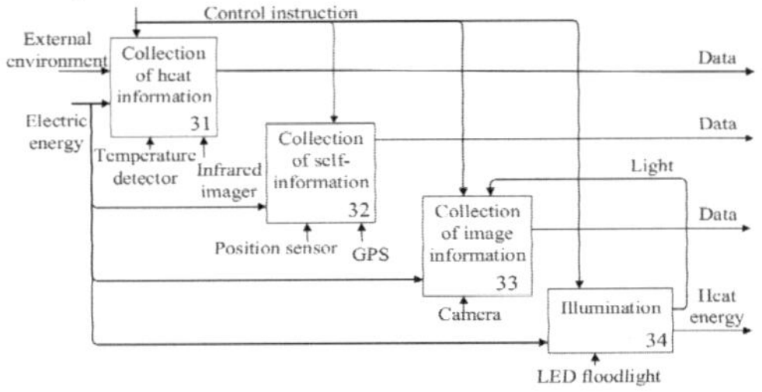

Figure 6. A3 Diagram of Acquisition Information Function

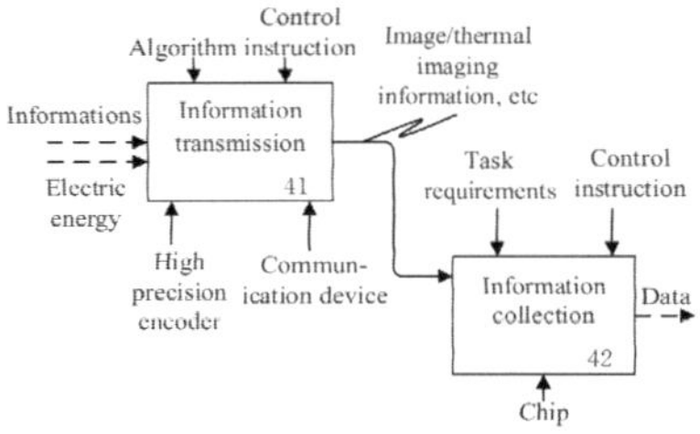

Figure 7. A4 Figure of The Information Processing Function

4.3. Extraction of Shared Functional Structures

Module division results of the complete pipeline maintenance robot system are shown in Figure 8.

According to the above module division method, complete module division was further conducted for the pipeline cleaning robot and pipeline inspection robot. Their results were put together for contrastive analysis. To simplify the process and provide convenience for module inquiry and management, general module and special module are extracted, to gain a modular pipeline robot system platform, as shown in Figure 9.

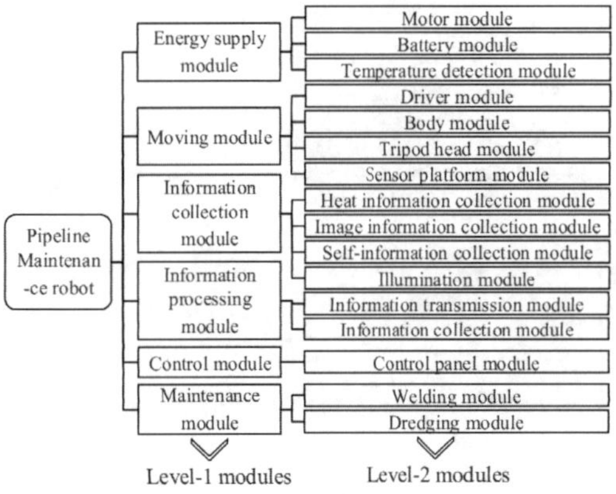

Figure 8. Full Module Partition Results

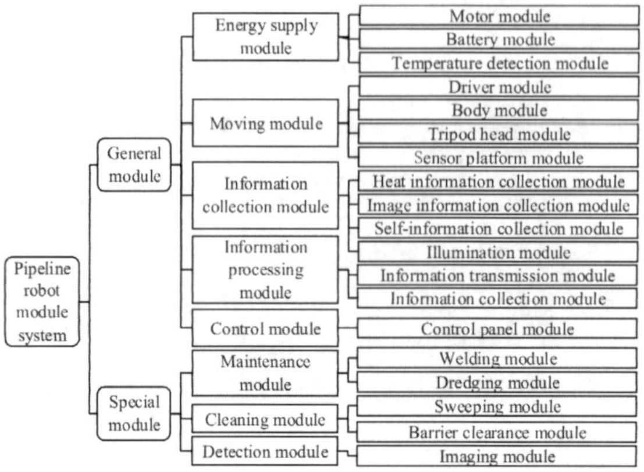

Figure 9. Pipeline Robot Module System

5. Conclusions

In allusion to the application environment of pipeline robots and changes & difficulties of functional requirements, a top-down module division method of combining axiomatic design and IDEF was applied for the pipeline robot system. The following advantages are presented:

(1) The module division method based on axiomatic design mainly combines the component collaboration and the degree of mutual influence, from the perspective of modules, and emphasizes the independence of functions between modules.

(2) The functional modeling method in the IDEF method can visualize the relationship between the functional modules and not only verify whether the results of the module partition based on the axiomatic design are reasonable, but also perfect for the unfinished partition modules.

(3) The two methods are combined to conduct module division, and simplify the division process. This method is more concise, visualized and efficient.

Acknowledgments

Project source: Basic Research Project of Shaanxi Province on Natural Science (2021JM-439)

References

[1] Song, Z., et al. (2020). "Modular Design of an Efficient Permanent Magnet Vernier Machine." IEEE Transactions on Magnetics PP(99): 1-1.

[2] Lin, H., et al. (2018). "Analysis and Control of the Permanent Magnet Synchronous Motor With Auxiliary Modular Design." IEEE Transactions on Magnetics PP(11): 1-6.

[3] Choi, H. R. and S. M. Ryew (2002). "Robotic system with active steering capability for internal inspection of urban gas pipelines." Mechatronics 12(5): 713-736.

[4] Kwon, Y. S. (2012). "Design and Motion Planning of a Two-Module Collaborative Indoor Pipeline Inspection Robot." IEEE Transactions on Robotics 28(3): 681-696.

[5] Rashid, M., et al. (2020). "Modeling of the in-pipe inspection robot: A comprehensive review." Ocean Engineering 203: 107206.

[6] Meng, et al. (2018). "Function Integration for Lightweight Chassis Based on Axiomatic Design and Design Structure Matrix." International Journal of Automotive Technology.

[7] Wang, et al. (2018). "A semantic model for axiomatic systems design." Proceedings of the Institution of Mechanical Engineers Part C Journal of Mechanical Engineering Science.

[8] Tang, D., et al. (2018). "Product Design as Integration of Axiomatic Design and Design

Structure Matrix." Matrix-based Product Design and Change Management.
[9] Wang, H., et al. (2018). "Model-based Axiomatic Design Approach for Complex Engineering Systems." Journal of Mechanical Engineering.
[10] Ang, C. L., et al. (1997). "A knowledge-based approach to the generation of IDEF0 models." International Journal of Production Research 35(5): 1385-1412.

Kinematic Modeling of 3P2R Welding Robot Based on D-H Parameters

Hongtang Chen[1], Haijun Ma[2], Hongli Jiang[3], Shenjin Lv[1], Yong Li[1] and Heshan Liu[4,*]

[1]School of Mechanical Engineering, University of Jinan, Ji Nan, China
[2]ShanDong Administration for Market Regulation, Ji Nan, China
[3]BinZhou Polytechnic, Bin Zhou, China
[4]School of Mechanical Engineering, Shan Dong University, Ji Nan, China

*Corresponding author email: liuheshan@sdu.edu.cn

Abstract. To meet the demand of different vehicle models welding task, as well as lower the cost of specialized welding facility, a kind of 3P2R arc welding robot has been developed and SolidWorks is used to set up the Three Dimensional (3D) solid model. The developed robot includes 3P2R body, controller, and a welding power supply. The direct kinematical model and inverse solution have been worked out basing on Denavit-Hartenberg (D-H) method, and the direct kinematical model and inverse solution are verified by experiment. This robot could also be used to burring or spray field with the actuator changed.
Keywords: Welding Robot; 3P2R; Denavit-Hartenberg method.

1. Introduction

For decades, robots have gained significant acceptance in manufacturing[1-3], especially in the automotive industry[4-6], but now automobile market is faced with the rapid development and the diversified purchase demands[7-10],so automobile production should be adjusted accordingly to meet the diversified purchase demands.

To meet the demand of different vehicle models welding task, as well as lower the cost of specialized welding facility, a kind of 3P2R arc welding robot has been developed and SolidWorks is used to set up the Three Dimensional solid model of the welding robot.

In this paper: the control system is described in Section 2. Section 3 highlights forward kinematics analysis. Section 4 describes the inverse kinematics. Finally, Section 5 provides the conclusions.

2. Control System

The robot controller is googoltech industrial 6-axis controller. The 6-axis controller is shown in Fig.1. The application software compiled by OtoStudio is running on the Windows 7.

Figure 1. Googoltech controller

3. Forward Kinematics Analysis

The 3P2R robot model set up by SolidWorks is shown in Fig.2, the welding torch positioning is controlled by the three prismatic(3P) joints and the welding torch orientation is controlled by the two revolute(2R) joints. Therefore, a finish welding could be performed for different vehicle models welding tasks.

Figure 2. 3P2R robot model

The following equations are used to describe the welding torch pose(orientation and position):

$$ {}^{i-1}T_i = T_{z,d} T_{z,\theta} T_{x,a} T_{x,\alpha} \tag{1} $$

$$ T_t = T_1 \times T_2 \times T_3 \times T_4 \times T_5 \times T_{tool} \tag{2} $$

$$ {}^0_6T = \begin{bmatrix} r_{11} & r_{12} & r_{13} & \mathrm{p_x} \\ r_{21} & r_{22} & r_{23} & \mathrm{p_y} \\ r_{31} & r_{32} & r_{33} & \mathrm{p_z} \end{bmatrix} \tag{3} $$

Where

$$ \mathrm{r}_{11} = \cos\theta_1 \tag{4} $$

$$ \mathrm{r}_{21} = \sin\theta_1 \tag{5} $$

$$ \mathrm{r}_{31} = 0 \tag{6} $$

$$ \mathrm{r}_{12} = -\sin\theta_1 \cos\theta_4 \tag{7} $$

$$ \mathrm{r}_{22} = \cos\theta_1 \cos\theta_4 \tag{8} $$

$$ \mathrm{r}_{32} = \sin\theta_4 \tag{9} $$

$$r_{13} = -\sin\theta_4 \sin\theta_1 \tag{10}$$

$$r_{23} = -\sin\theta_4 \cos\theta_1 \tag{11}$$

$$r_{33} = \cos\theta_4 \tag{12}$$

$$p_x = (a_2 - d_4)\cos\theta_1 - (d_5 \sin\theta_4 - d_2)\sin\theta_1 \tag{13}$$

$$p_y = (a_2 - d_4)\sin\theta_1 + (d_5 \sin\theta_4 - d_2)\cos\theta_1 \tag{14}$$

$$p_z = d_5 \cos\theta_4 - d_3 \tag{15}$$

The Denavit-Hartenberg parameters are as follows:

$$\begin{cases} \alpha_0 = 0 \\ \alpha_1 = \pi/2 \\ \alpha_2 = \pi/2 \\ \alpha_3 = -\pi/2 \\ \alpha_4 = -\pi/2 \end{cases} \tag{16}$$

$$a_0 = a_1 = a_3 = a_4 = 0 \tag{17}$$

$$d_1 = 0 \tag{18}$$

$$\begin{cases} \theta_1 = 0 \\ \theta_2 = \pi/2 \\ \theta_4 = \pi/2 \end{cases} \tag{19}$$

a2, d4 and d5 are constant.

4. The Inverse Kinematics

Given a desired welding torch pose(position and orientation), find the set of values for the joint variables which realizes the desired welding torch pose(position and orientation), is the inverse kinematics. In this paper, the inverse kinematics consists in deriving the joint variables ɵ1- ɵ 5, basing on the D-H parameters of 3P2R welding robot and the matrix T, the inverse kinematics of the 3P2R welding robot is solved by the analytical method of matrix inverse, and the correctness of the model is verified by the experiment.

5. Conclusion

To meet the demand of different vehicle models welding task, as well as lower the cost of specialized welding facility, a kind of 3P2R arc welding robot has been developed.

Denavit-Hartenberg (D-H) model is used to set up the robot joint coordinate system, and forward kinematics equation is derived by homogeneous coordinate transformation, based on the consideration of the kinematic constraint the forward kinematics model with the joint angle as the variable is obtained, by analyzing the solutions of the inverse kinematics model, the inverse kinematics model of 3P2R welding robot is solved by analytical method of matrix inverse, and the correctness of inverse kinematics is verified by the experiment. This robot can also be used to burring or spray field with the actuator changed.

Acknowledgements

This work was financially supported in part by Key R & D project of Shandong Province under Grant No. 2019JZZY010452.

References

[1] Liu YK, Zhang YM. Fusing machine algorithm with welder intelligence for adaptive welding robots. Journal of Manufacturing Processes, 2017,27:18-25.
[2] Wang XW, Zhou X, Xia ZL, Gu XS. A survey of welding robot intelligent path optimization. Journal of Manufacturing Processes, 2021, 63:14-23
[3] Zhang LJ, Bai QL, Ning J, et al.. A comparative study on the microstructure and properties of copper joint between MIG welding and laser-MIG hybrid welding. Materials & Design, 2016;110(15):35-50.
[4] Xu YL, Lv N, et al.. Welding seam tracking in robotic gas metal arc welding. Journal of Materials Processing Technology, 2017; 248:18-30.
[5] Wang XW. Three dimensional vision application in GTAW process modeling and control. International Journal of Advanced Manufacturing Technology, 2015; 80:1601-1611.
[6] Xiao RQ, Xu YL, et al.. A feature extraction algorithm based on improved Snake model for multi-pass seam tracking in robotic arc welding. Journal of Manufacturing Processes, 2021, 72 :48-60.
[7] Hu SJ, Ko J, et al.. Assembly system design and operations for product variety, CIRP Annals. 2011,60(2): 715-733.
[8] Putnik G, Sluga A, et al.. Scalability in manufacturing systems design and operation: State-of-the-art and future developments roadmap, CIRP Annals. 2013, 62(2):751-774.
[9] Stefania P., Nicola P, Lorenzo M.T., Anath F., et al.. Multi-robot spot-welding cells for car-body assembly: Design and motion planning. 2017, 44 : 97-116.
[10] Putnik G, Sluga A, et al.. Scalability in Manufacturing Systems Design and Operation: State-of-theart and Future Developments Roadmap. CIRP Annals.2013,62:751-774.

Structural Design and Analysis of Part of Manipulator

Yan Ma[1,*], Xingqiao Jiang[1], Jingxi Xu[2], Huang Liu[1] and Yue Yuan[1]

[1] Shenyang Academy of Instrumentation Science Co.,LTD, Shen yang, China
[2] Northeastern University, Shen yang, China

*Corresponding author email: MY9079cn@163.com

Abstract. In this paper, the manipulator can carry and grasp objects as the research object, manipulator is mainly composed of hands and motion mechanism. The adsorption type hand is mainly composed of suction cups and so on. The design object has three degrees of freedom, through the stepper motor and ball screw to complete the telescopic arm, through the cylinder to complete the lifting of the adsorption part, the workpiece from one coordinate adsorption and transport to another coordinate. Air suction hand structure design, hand calculation, cylinder control loop, cylinder working pressure, cylinder gas consumption analysisis presented.
Keywords: Manipulator; Stepper motor; Design.

1. Overview

Mechanical automation technology from the 20 s of this century, first started developing in the process of mechanical manufacturing cold mass production application, in order to adapt to the demand of the market after the '60 s and change, to strengthen the ability of mechanical manufacturing flexible rapid response to the market, began to build variability automated production systems, namely around flexible automation of computer technology. It is in the manufacturing system unchanged or minor changes, the information collected to process, analyze and judge the dicing saw or cutting process to achieve the desired operation or process and can automatically from manufacturing a part to manufacturing another different parts. [1-3].

The development of modern machinery manufacturing technology is mainly reflected in two directions: one is precision engineering technology, represented by the frontier part of ultra-precision machining, micro-machining and nano-technology, which will enter the era of micro-mechanical-electronic technology and micro-robot; The second is the high degree of automation of mechanical manufacturing, represented by the further development of CIMS and agile manufacturing. Green manufacturing is the embodiment of sustainable development of human society in manufacturing industry. In addition to technological innovation, all of this must rely on information technology, through computer simulation, simulation, to achieve [4-6].

In this paper, the manipulator can carry and grasp objects as the research object, manipulator is mainly composed of hands and motion mechanism. According to the object size, shape, material, weight and working environment requirements, there are several types of hand structure, supporting type, adsorption type and clamping type. The function of the motion mechanism is to enable the hand to complete various actions: movement, rotation and so on. The mechanism of telescopic, lifting and rotating motion, known as the degree of freedom of manipulator. The design object of this paper has three degrees of freedom, through the stepper motor and ball screw to complete the telescopic arm, through the cylinder to complete the lifting of the adsorption part, the workpiece from one coordinate adsorption and transport to another coordinate. Air suction hand structure design, hand calculation, part of the cylinder analysis, cylinder control loop, cylinder working pressure, cylinder consumption analysis.

2. Composition of Manipulator

Manipulator is mainly composed of actuator, drive system, control system and position detection device [2].
Hand, that is, the part in direct contact with the workpiece. Due to the different forms of contact with the workpiece, it can be divided into gripping hand and adsorption hand. This part adopts the adsorption type. The adsorption type hand is mainly composed of suction cups and so on. It relies on the adsorption force (such as the formation of negative pressure or electromagnetic force in the suction cup) to absorb objects. Wrist is connected hand and arm parts, adjust or change the position of the workpiece (that is, attitude). The main part of the manipulator including mechanical arm, mechanical claw and the fuselage device, which is the role of the mechanical arm positioning and support, so the mechanical arm in space should have at least three or more than three degrees of freedom, including but not limited to the front and back swing, about rotation and the pitch, to achieve through the transmission and drive system, Its positioning accuracy determines the precision of the manipulator. The mechanical arm is mounted on the fuselage device, and the movement is mainly dependent on the fuselage device. It can either walk on the track or be fixed in a specified position
Manipulator drive system is to drive the actuator movement of the transmission device. Commonly used hydraulic transmission, pneumatic transmission, pneumatic transmission, electric transmission, and mechanical transmission and other four forms. This design adopts mechanical transmission and pneumatic transmission.
Control system has mechanical control, electrical control, and jet control, hydraulic control and so on, generally common for the electrical control system.
Control manipulator actuator motion position, and the actual position of the actuator feedback to the control system at any time, and compared with the given position, and then adjust through the control system, so that the actuator with a certain accuracy to reach the set position.

3. Hand Design

3.1. Structural Analysis of Sucker Hand

The sucker hand is Is the use of flexible non-metallic materials under negative pressure formed by the suction plate to suck the workpiece; Suitable for carrying some thin irregular shape objects, compared with other manipulators, air suction hand has obvious advantages, small structure, will not damage and pollution of the workpiece surface. According to the method of forming negative pressure (or vacuum), air-suction hand can be divided into vacuum type, airflow negative pressure type and extrusion and exhaust type suction cup, all of which are applied to mechanical hands [13].
The sucke hand used in this design is shown in Figure 1.

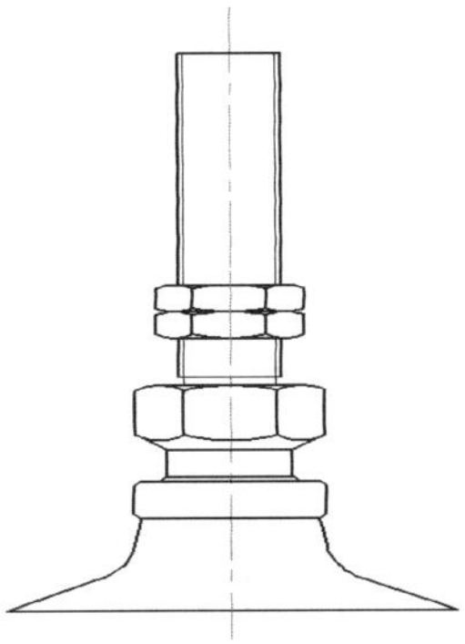

Figure 1. Air suction hand

3.2. Calculation of Hands

Intended service conditions: load W=10N

The suction force of vacuum sucker in the starting process is:

$$P = \frac{S}{K_1 K_2 K_3} . \Delta p \tag{1}$$

S -- area of the sucker(mm^2), $S = \frac{\pi}{4} D^2$, D is the diameter of the sucker;

K1 -- safety factor of adsorption workpiece, K1=1.2~2;

K2 -- Working condition coefficient, generally K2=1~3;

K3 -- azimuth coefficient, when the sucker vertical adsorption, $K_3 = 1/f$, F is the friction coefficient, rubber sucker adsorption metal material, f=0.5 ~ 0.8; When the suction cup is horizontal adsorption, take K_3=1;

Δp -- Pressure difference inside and outside the suction cup (MPa).

For vacuum sucker, the suction workpiece should be guaranteed to be sucked when it is started. In the process of work, negative pressure will be stable to ensure that the workpiece adsorption, will not fall. Micro-vacuum pump KVP04, vacuum degree -40kpa, sucker diameter 32mm, nitrile rubber material; K_1s is 1.2, K_2 is 1, f is 0.8, Δpis 0.04mpa,

$$P = \frac{\frac{\pi}{4} .32^2}{1.1 \cdot 1 \cdot \frac{1}{0.8}} \cdot 0.04$$

Then P = 21.45 N. Suction cups meet requirements.

4. Cylinder Part

4.1. Cylinder Control Loop

The design of cylinder control loop [7] is shown in FIG. 2. In Figure 2, 1 represents manual valve, 2 represents two-position five-way valve, 3, 4. One-way throttle valve, 5 represents travel valve.

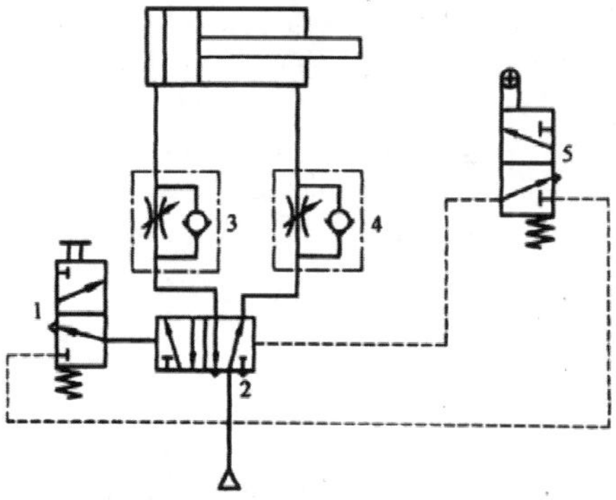

Figure 2. Cylinder control loop

4.2. Cylinder Working Pressure

The relationship between cylinder working pressure and load is shown in Table 1.

Table 1. Relationship between cylinder working pressure and load

Load F/N	<5000	5000~10000	10000~20000	20000~30000	30000~50000	>50000
Working pressure P /MPa	1.143	0.285	0.286	0.593	0.286	0.593

The cylinder working pressure is 0.4MPa. Primary cylinder MGPM40-150Z, cylinder diameter: 40mm, piston rod diameter 14mm, stroke 150mm, double acting cylinder.
The theoretical cylinder thrust is as follows:

$$F_1 = P.\frac{D^2 \cdot \pi}{4} \tag{2}$$

Theoretical cylinder tension:

$$F_2 = P.\frac{\left(D^2 - d^2\right) \cdot \pi}{4} \tag{3}$$

Type in the: F_1:--Theoretical cylinder thrust(N);
F_2:--Theoretical cylinder tension(N);
P:--Pressure of work(MPa);
D:--The cylinder diameter
D:--Piston rod diameter
Take the working pressure of 0.4mpa, bring into the cylinder diameter of 40mm, piston rod diameter of 14mm, can be obtained F1=502.7N, F2=441.1N;
In practical work, the efficiency of the cylinder should be taken into account. Generally, it can be calculated as 0.8, then F1=402.2N,F2=352.9N. Cylinder tension is much greater than the load, in line with the requirements.

4.3. Analysis of Cylinder Consumption

The piston speed of the MGPM40-150Z cylinder is 50mm/s~500mm/s, trips=150mm, Take working timet=3s, So the piston velocity is zero 50mm/s, Take the volumetric efficiency of the cylinder $\eta_v = 0.9$

$$q_1 = \frac{D^2\pi}{4} \cdot \frac{s}{t \cdot \eta_v} = \frac{0.04^2 \cdot \pi \cdot 0.15}{4 \cdot 3 \cdot 0.9} = 7 \times 10^{-5} \tag{4}$$

Is the air consumption of the cylinder when the load is lowered $q_1 = 7.0 \times 10^{-5} (m^3 / s)$

$$q_2 = \frac{\left(D^2 - d^2\right)\pi}{4} \cdot \frac{s}{t \cdot \eta_v} = \frac{\left(0.04^2 - 0.014^2\right) \cdot \pi \cdot 0.15}{4 \cdot 3 \cdot 0.9} = 6.1 \times 10^{-5} \tag{5}$$

Is the air consumption of the cylinder when lifting the load

$$q_2 = 6.1 \times 10^{-5} (m^3 / s)$$

With the development of modern control technology and micro-electronic control technology, the repeated positioning accuracy of manipulator will be improved, and its positioning accuracy will be enhanced. The manipulator will have a restructured mechanical structure, that is, modular. After assembling, the manipulator will start the required modules according to the actual operation needs, to achieve the corresponding functions, so that the same manipulator has different functions, increase its application range. In addition, with the popularity of low carbon concept, the development of manipulator will also be pollution-free, energy saving as the premise, such as using some new materials to manufacture non-lubricating components applied to pneumatic machinery hands, not only simplify the system, but also has a stable friction performance and a long life. In the next generation of industrial control technology, intelligent control will be more and more, the key is the application of sensors, will make the whole system into an intelligent closed-loop controllable system, will bring greater convenience to the equipment process, and will improve the overall accuracy.

5. Conclusion

Manipulator can imitate the human arm and hand some actions, can carry, grasp objects or operate tools, manipulator is mainly composed of hands and motion mechanism. The use of adsorption hand, mainly composed of sucker, etc., it is by adsorption force (such as the formation of negative pressure in the sucker or electromagnetic force) adsorption objects. The design object of this paper has three degrees of freedom, through the stepper motor and ball screw to complete the telescopic arm, through the cylinder to complete the lifting of the adsorption part, the workpiece from one coordinate adsorption and transport to another coordinate. The driving system of manipulator adopts mechanical transmission and pneumatic transmission. We through the photoelectric sensor on the position of the manipulator actual positioning, and the system given by the positioning comparison adjustment, to achieve the purpose of accurate control

Reference

[1] VerlA, ValenteA, MelkoteS, et al. *Robots inmachining*, CIRP Ann. 2019,68 (2)pp799–822
[2] Unai I, Imanol A, et al.*A practical and synchronized data acquisition network architecture for industrial robot predictive maintenance in manufacturing assembly lines*.Robotics and Computer–Integrated Manufacturing ,2022,74,102287pp1-9
[3] Abele E, Bauer J, Hemker T, et al. *Comparison and validation of implementations of a flexible joint multibody dynamics system model for an Industrial robot*. Journal of Manufacturing Science and Technology, 2011,4(1)pp 38-43
[4] Zhang Dongquan, E Mingcheng. *Manufacturing Equipment and Its Automation Technology*. Beijing: Science Press, 2007.2.
[5] ZHOU Boying. *Industrial Robot Design*. Beijing: China Machine Press, 1995
[6] WANG Hongyu. *Introduction to Stepper Motor Control Technology*. Shanghai: Tongji University Press, 1990.9.
[7] Chen Yan. *Hydraulic and Pneumatic Transmission* Beijing: China Railway Publishing House, 2017.2.

Research on Fuzzy Control Algorithm of Self-walking Platform for Maintenance of Circular Tunnel

Hao Liu[1], Wen Zhang[1], Zhenglong Wu[1], Tieying Guo[2,*], Dan Luo[2], Weiping Peng[2] and Tao Zhang[2]

[1]Sinohydro Bureau 7 Co., Ltd., Chengdu, China
[2]School of Power and Mechanical Engineering, Wuhan University, Wuhan, China

*Corresponding author email: 2020282080111@whu.edu.cn

Abstract. Self-walking platform is used to support subway tunnel maintenance and construction. However, the kinematics model and motion control of the subway tunnel self-propelled platform are different from that of the plane, it is a non-holonomic constrained system with serious sliding, strong coupling and highly nonlinear, and the pipeline environment is complex, so it is difficult to accurately model. In order to solve these differences, the kinematic characteristics of wheels in cylindrical workspace are discussed first. The geometric constraints of self-walking platform in circular tunnels are also analyzed. Secondly, considering the complexities of self-walking platform motion model, a control algorithm based on fuzzy logic is proposed. Finally, through MATLAB and ADAMS joint simulation, the scenarios for virtual platform moving in circular tunnels are simulated, which shows the effectiveness of the algorithm.
Keywords: Subway tunnel maintenance; Platform cars; Motion model; Fuzzy control.

1. Introduction

A new type of self-walking platform running directly on the curve wall has been designed and planned to use in subway tunnel maintenance and construction, which could effectively improve the safety of construction and speed up the progress of the project. Platform cars working in subway tunnel are similar to the wheeled pipeline robots, effective mobility and controllability are indispensable [1]. Therefore, the kinematic models and motion control of the platform are of great significance to operational safety and efficiency.

K lagnemma analyzed the interaction between wheel and ground on irregular ground and obtained a set of theoretical models, the kinematics model of the six-wheel suspended wheeled robots are analyzed using the theoretical model, and the motion control of the robot is realized [2,3]. According to the generalized high-order chain system framework. Zhangjun Song realized the smooth motion of the robot in the pipes or ducts by using the discontinuous control rate of the surface [4,5]. However, most of the existing commercial pipeline robots are controlled remotely by mobile control panel [6,7]. Therefore, in recent years, the research of pipeline robot mainly focuses on autonomous operation, which requires good motion control based on sensor information[8,9,10]. A novel simulation method combines the software of Matlab and Adams was proposed in this article to realize intelligent control of the platform.

2. Kinematic Model

2.1. Kinematic Characteristics of Wheel Purely Rolling on the Inner Surface of a Cylinder

In order to make the kinematics model not only general but also simple mathematical derivation, the following assumptions are made in the process of modeling [3]:

1) The wheel is simplified to a circular surface, and the contact with the wall of the pipe is point contact;
2) The robot is composed of a rigid body, and its wheels and pipe walls are non-deformable.

Based on the above assumptions, the wheel with cylindrical working space is modelled as a rolling disk, and these attitudes (α, β) is defined [11] according to the contact point O_3.And the attitude of the rolling wheel on the wall of a cylindrical workspace is defined on the tangent plane [12].

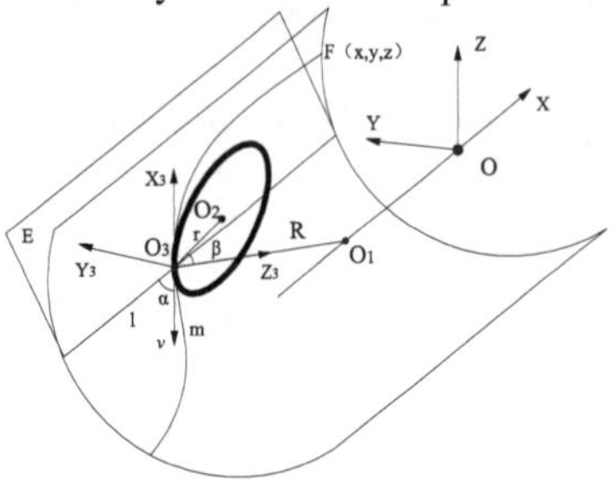

Figure 1. Frenet frame fixed to the contact point of the wheel wall.

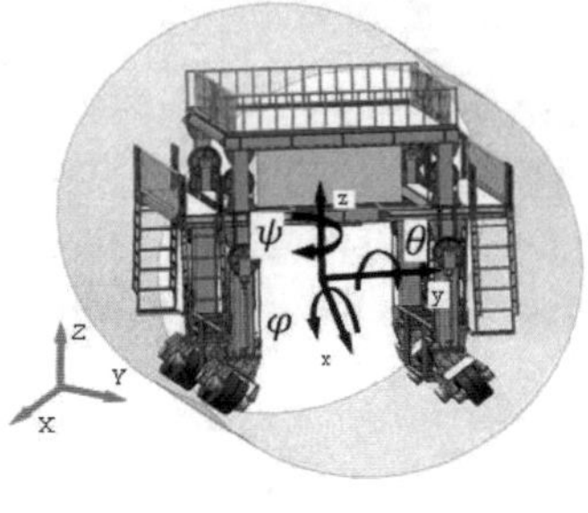

Figure 2. Moving coordinate system in the platform

When the wheel rolls purely on the cylinder with the angular velocity w, the set of contact points between the cylinder and the wheel is:

$$\begin{cases} x = -wtr\cos\alpha \\ y = R\cos\left(\dfrac{-wtr\sin\alpha}{R}\right) \\ z = R\sin\left(\dfrac{-wtr\sin\alpha}{R}\right) \end{cases} \tag{1}$$

where R is the cylinder radius and t is time; It can be expressed as the forming space curve.

$$\mathbf{p(t)} = \left[\text{-wtr}\cos\alpha \quad R\cos\left(\frac{-wtr\sin\alpha}{R}\right) \quad R\sin\left(\frac{-wtr\sin\alpha}{R}\right) \right] \tag{2}$$

The active coordinate system is established by the tangent vector, the main normal and the secondary normal of the wheel curve at point O_3. That is Frenet active coordinate system, as is shown in Figure 1. Coordinate transformation matrix is (Hartenberg-Denavit matrix) [13,14] as follow:

$$\mathbf{D}^{(ij)} = \begin{bmatrix} \mathbf{A}^{(ij)} & \vdots & \mathbf{d}_{ij}^{i} \\ \cdots & \vdots & \cdots \\ \mathbf{0} & \vdots & \mathbf{1} \end{bmatrix} \tag{3}$$

where $A^{(ij)} = e^{(i)} \cdot e^{(j)T} = [\tau\ n\ b]^T \cdot e^{(j)T}$, τ is unit tangent vector; n is unit principal normal vector; b is unit pair normal vector.

$$d_{ij}^{(i)} = \begin{bmatrix} -wtr\cos\alpha \\ R\cos\left[\dfrac{(-wtr\sin\alpha)}{R}\right] \\ R\sin\left[\dfrac{(-wtr\sin\alpha)}{R}\right] \end{bmatrix} \tag{4}$$

The coordinates of the wheel center in the moving coordinate system is $(0, rcos\beta, rsin\beta)$, Then the coordinates of the wheel center in the absolute coordinate system are as follows:

$$\begin{bmatrix} x \\ y \\ z \\ 1 \end{bmatrix} = D^{(ij)} \begin{bmatrix} 0 \\ r\cos\beta \\ r\sin\beta \\ 1 \end{bmatrix} = \begin{bmatrix} -r\sin\beta - wtr\cos\alpha \\ -r\cos\beta\left(\dfrac{wtr\sin\alpha}{R}\right) + r\sin\beta\cos\alpha\left(\dfrac{wtrs\alpha}{R}\right) + R\cos\left(\dfrac{wtr\sin\alpha}{R}\right) \\ r\cos\beta\sin\left(\dfrac{wtr\sin\alpha}{R}\right) + r\sin\beta\cos\alpha\cos\left(\dfrac{wtrs\alpha}{R}\right) - R\sin\left(\dfrac{wtr\sin\alpha}{R}\right) \\ 1 \end{bmatrix} \tag{5}$$

The speed of the wheel center can be obtained by differentiating the above equation, $v = (\dot{x}, \dot{y}, \dot{z})$; the velocity direction of the wheel center is perpendicular to the radius of the cylinder passing through the wheel center. The velocity of wheel center is as follows:

$$\mathbf{v} = \sqrt{\dot{x}^2 + \dot{y}^2 + \dot{z}^2}$$

$$= \frac{wr}{R}(-2r^2c^2\beta c^2\alpha + r^2c^2\beta - 2Rc\beta + 2Rrc\beta c^2\alpha + R^2 + r^2c^4\alpha c^2\beta + r^2c^2\alpha s^2\alpha)^{\frac{1}{2}} \tag{6}$$

2.2. Geometrical Constraints of Self-walking Platform in the Cylindrical Workspace

The motion of self-propelled platform in circular tunnel is a kind of three-dimensional motion. Using the method of robot, a dynamic coordinate system is fixed on the robot body, and the motion of dynamic coordinate system in absolute coordinate system is studied. As is shown in Figure 2.

The platform is symmetrical left and right, without steering wheel, and turns by differential speed of left and right wheels. And The coordinate transformation matrix is shown as follow:

$$\mathbf{K} = (\varphi, \theta, \delta) = \begin{bmatrix} \cos\varphi\cos\theta & \cos\varphi\sin\theta\sin\delta - \sin\varphi\cos\delta & \cos\varphi\sin\theta\sin\delta + \sin\varphi\sin\delta \\ \sin\varphi\cos\theta & \sin\varphi\sin\theta\sin\delta + \cos\varphi\cos\delta & \sin\varphi\sin\theta\cos\delta - \cos\varphi\sin\delta \\ -\sin\theta & \cos\theta\sin\delta & \cos\theta\cos\delta \end{bmatrix} \tag{7}$$

And φ is defined as roll angle, δ is defined as yaw angle and θ is defined as pitch angle. Therefore the coordinates of the rigin in absolute coordinates can be defined as vector

$$\mathbf{d} = [x_0, y_0, z_0]^T \tag{8}$$

The transformation matrix from the moving coordinate system to the absolute coordinate system is shown in the following diagram:

$$\mathbf{D} = \begin{bmatrix} \mathbf{K}(\boldsymbol{\varphi}, \boldsymbol{\theta}, \boldsymbol{\delta}) & \mathbf{d} \\ 0 & 1 \end{bmatrix} \tag{9}$$

It can be seen from the above that the coordinates of each wheel center are in the moving coordinate system, and the absolute coordinate system of each wheel center can be obtained through the transformation formula above. Meanwhile, the outer circle of each wheel can be regarded as four spatial curves. If the rotation velocity of the left wheel is w_l and that of the right wheel is w_r, then the vector of the first wheel in the moving coordinate system is expressed, and the expression of other wheels in the moving coordinate system is the same.

$$\mathrm{r}_1' = \left[-r\cos(w_l t) \quad -\frac{1}{2}w \quad \frac{1}{2}l + r\sin(w_l t)\right]^T \tag{10}$$

Similarly, the above vector can be converted to the form in absolute coordinates. At the contact point, the wheels and the duct or tunnel can be considered to be tangent, so there is a tangent vector perpendicular to the tunnel radius at this point. The tangent vector can be obtained by taking the derivative of the wheel circle formula with respect to t. The vector expression of the contact point radius in the absolute coordinate system is assumed to be:

$$\mathbf{R}=[0,y,z]^T \tag{11}$$

we have the following formula:

$$\frac{dr_i}{dt}\cdot\mathbf{R}=0 \tag{12}$$

And the contact points are on or within the round ducts or tunnels, which satisfies the following formula:

$$y^2+z^2\le R^2 \tag{13}$$

If any two parameters in x_0、y_0、z_0、φ、θ、δ are known, the remaining pose coordinates can be solved by combining the above equation.

3. Motion Control Algorithm Design

The motion control of the Platform cars in the subway tunnels is divided into two types, position control and attitude control. In this paper, we focus on attitude control. Therefore, we choose the platform-car's attitude Angle (roll Angle, pitch Angle and yaw Angle) in the tunnel as the control output, the angular velocity of the left and right wheels is selected as the control input. The motion control of Platform car in subway tunnels is a non-integrity constrained control problem, and there is a high degree of nonlinearity and coupling between its input and output. So, we choose the fuzzy control method to solve the above problems. Fuzzy control is a control method that imitates human thinking process and does not need accurate motion model of robot. At the same time, the nonlinear transcendental equation and wheel sideslip can be avoided [15,16]. When the Platform cars moves in the round ducts or tunnels, its attitude angle is coupled with each other due to the limitation of the round ducts or tunnels. In this paper, roll Angle and its change rate are selected as the main control basis.

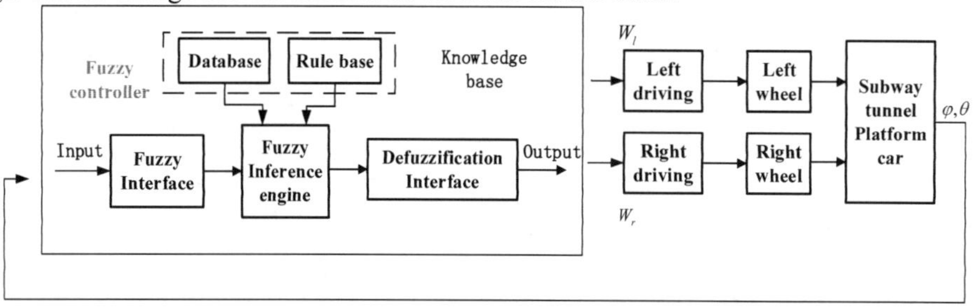

Figure 3. Schematic diagram of automatic walking control of robot on subway tunnel platform

Define the fuzzy control set: the fuzzy set of roll Angle φ, pitch Angle θ and the control output angular velocity u are $\{NB, NM, NS, Z0, PS, PM, PB\}$; Value range of roll angle is $[-10°\ 10°]$; Value range of pitch angle is $[-1°\ 1°]$; The value range of the difference between the two sides of the wheel speed u is$[-1\ 1]$. The domain of discourse of φ is$\{-10^\circ,-5^\circ,-1^\circ,0^\circ,1^\circ,5^\circ,10^\circ\}$. The domain of discourse of θ is$\{-1^\circ,-0.5^\circ,-0.1^\circ,0^\circ,0.1^\circ,0.5^\circ,1^\circ\}$. The domain of discourse of u is $\left\{-1,-\frac{2}{3},-\frac{1}{3},0,\frac{1}{3},\frac{2}{3},1\right\}$. With the quantization factor k_φ, k_θ, k_u, respectively mapping from the theory of things domain to the fuzzy domain [17]. Among them $k_\varphi=1$、$k_\theta=0.5$、$k_u=1.44$.The common triangle membership functions are used to realize the above mentioned fuzzification.

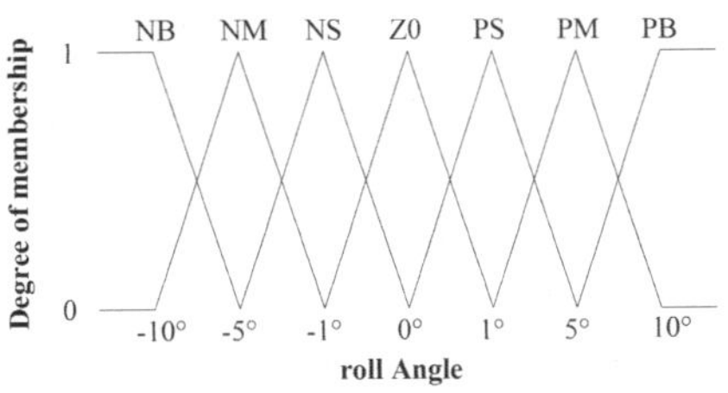

Figure 4. The membership function of robot roll Angle

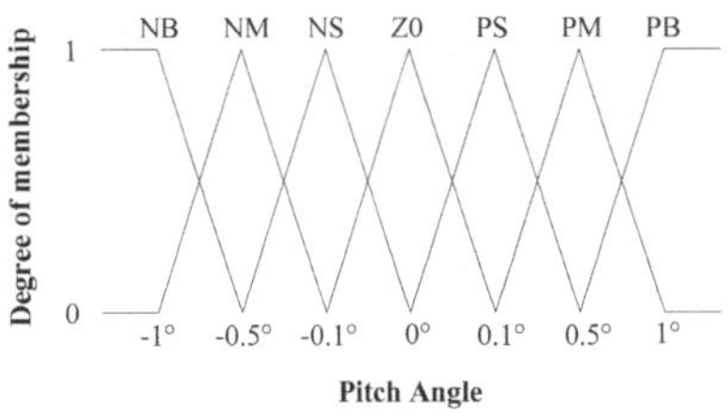

Figure 5. The membership function of rate of change of robot roll Angle change

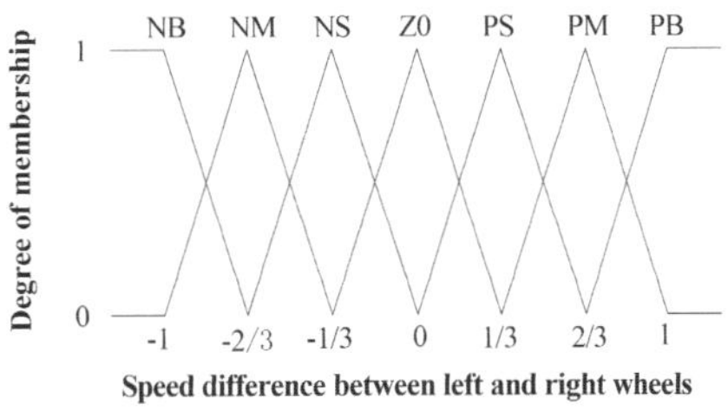

Figure 6. The membership function of rotation speed difference between left and right wheels of robot.

The fuzzy system designed by fuzzy reasoning belongs to the case of multi-input multi-rule fuzzy reasoning. *Mamdani* fuzzy implication and "max-min" synthesis operator are adopted [18]. According to the fuzzy relation of rule i in the rule, we can get:

$$R=U_{i=1}^{n} R_i = U_{i=1}^{49}(E1_i \times E2_j \times u_i) \tag{14}$$

According to the operation experience of human beings and the characteristics of subway tunnel environment, platform car parameters and motion equation, the following fuzzy rules are summarized.

Table 1. Fuzzy control rules

		E1						
		NB	NM	NS	ZO	PS	PM	PB
	NB	PB	PB	PM	ZO	PM	NB	PB
	NM	PB	PM	PS	ZO	PS	PM	PB
	NS	NM	NS	NS	ZO	PS	PM	PB
E2	ZO	NB	NM	NS	ZO	PS	PM	PB
	PS	NB	NM	NS	ZO	PS	PS	PM
	PM	NB	NM	NS	ZO	NS	NM	NB
	PB	ZB	NB	NM	ZO	NM	NB	NB

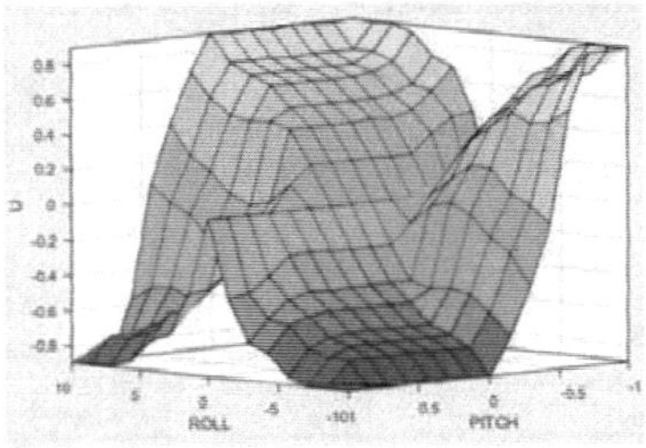

Figure 7. Characteristic curve of fuzzy controller of robot.

The process of transforming fuzzy quantity into clear quantity is called defuzzification. The commonly used methods include median method, maximum membership method and barycenter method. The process is the same as the fuzzification of legibility. The fuzzy control query table is established. In the real-time control, the legibility of the corresponding control quantity can be obtained from the query table according to the discrete value of the input variable and the actual control quantity can be obtained.

As can be seen from Fig.7, The controller is a nonlinear controller with nearly smooth space surface and continuous output surface, which is one of the remarkable characteristics of fuzzy control.

4. MATLAB and ADAMS Joint Control Simulation

The mechanical control system derived from Adams and the fuzzy controller established in the previous section were imported into Simulink. The simulation results of MATLAB and Adams verify the effectiveness of the algorithm. The Subway tunnel platform car is a multi-variable strongly coupled mechanical system, which is transformed into a linear multi-variable decoupling system during the joint simulation. The data interaction diagram between MATLAB and Adams is shown in Fig.8.

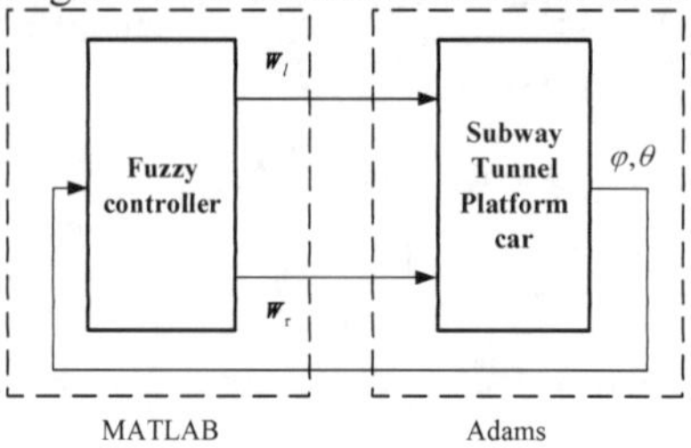

Figure 8. MATLAB and Adams data interaction diagram

4.1. Mechanical subsystem in Adams

The main material of the car is steel. The modulus of elasticity is $E = 2.07 \times 10^5 MPa$, and the density is $\rho = 7801 \text{kg} \cdot m^{-3}$. The material of wheels is rubber. The density is $\rho = 1.1 \text{ kg} \cdot m^{-3}$.The remaining relevant parameters are shown in table 2. The complete dynamic model in Adams is shown in Fig.9.The joint simulation control system structure is shown in Fig.10.

Table 2. Main geometric and dynamic parameters

parameter	numerical
Tunnel diameter(m)	6
Robot length(mm)	3300
Robot width(mm)	3200
Wheel diameter(mm)	553
Curve radius(m)	250
Initial angular velocity(rad/s)	1.44
coefficient of dynamic friction	0.3
Coefficient of static friction	0.8

Figure 9. Model of platform in ADAMS

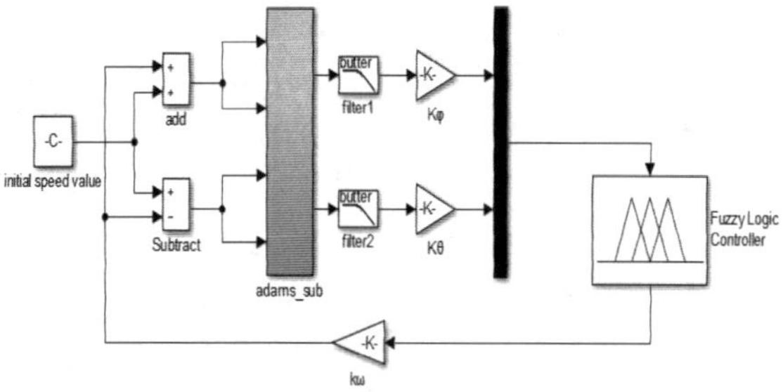

Figure 10. Joint simulation control system structure

4.2. Int Simulation Results

First, in Adams, we let platform car complete the simulation of bending in the round tunnel only at the initial angular velocity. In contrast to the above simulation, combined with the mechanical subsystem in Adams and the control system in MATLAB, the joint control simulation was carried out to verify the effectiveness of the designed algorithm, and the changes of roll Angle and pitch Angle of platform car are shown in Fig. 13 and 14. The angular velocity changes of the two sides of the platform car is shown in Fig15.

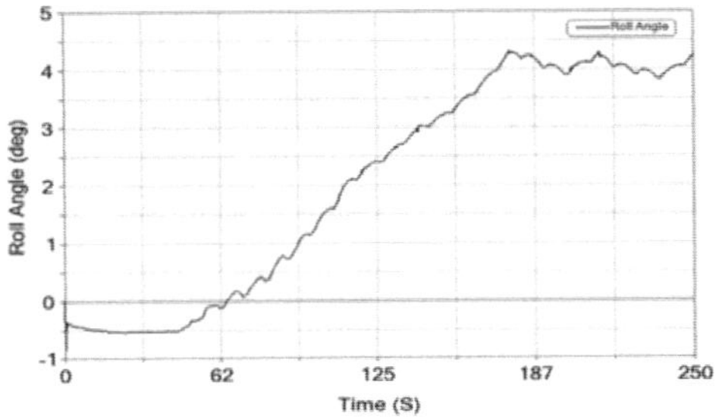

Figure 11. Roll Angle of platform without controller.

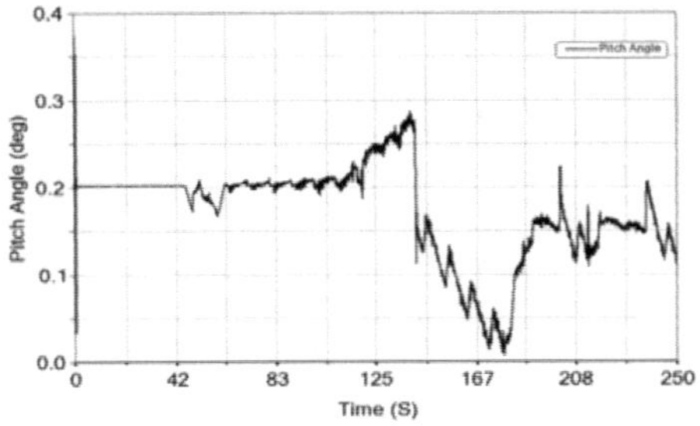

Figure 12. Pitch Angle of platform without controller.

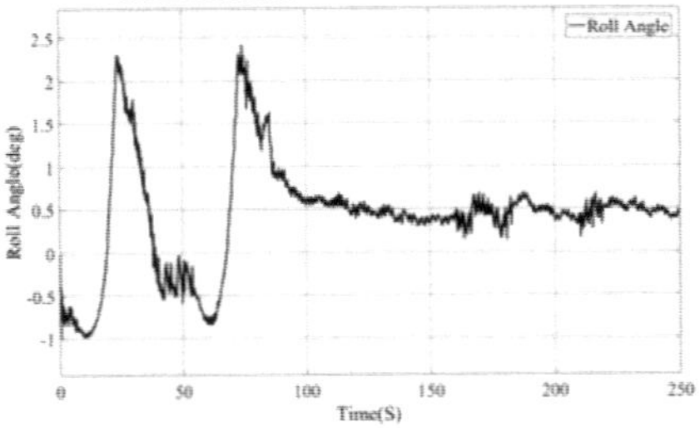

Figure 13. Roll Angle of platform car with controller.

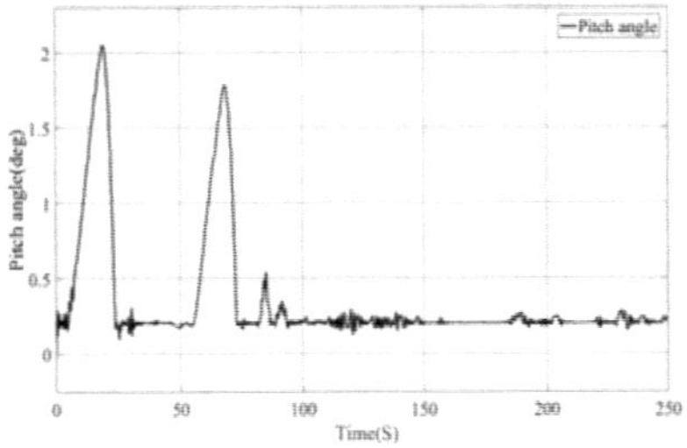

Figure 14. Pitch Angle of platform car with controller.

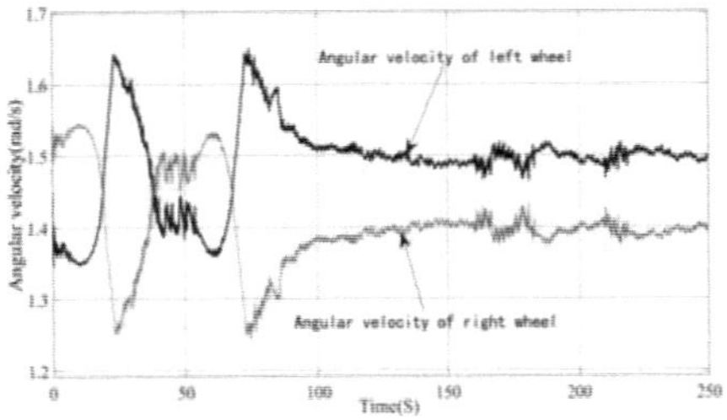

Figure 15. Diagram of wheel angular velocity change on both sides of platform car.

As shown in the Fig.11 and Fig.12.When the platform cars start to enter the curve with the initial driving angular velocity, the roll Angle of the robot increases sharply at 62 seconds. Then roll Angle gradually increased, when the simulation time to 180s, roll Angle eventually reached to around 4.2°. With the increase of roll Angle, it is more and more possible for the platform car to capsize in the subway tunnels. As shown in the Fig.13 and Fig 14. Under the control of the fuzzy controller, the rolling Angle and pitching Angle of the platform car gradually decrease after several rounds of regulation, with the rolling Angle gradually stabilizing at 0.5° and the pitching Angle gradually stabilizing at 0.2°. As shown in the Fig.15. It can be clearly seen that the platform car can adjust the attitude Angle in the tunnel by constantly adjusting the angular velocity of the left and right wheels.

5. Conclusions

The kinematic characteristics of single wheel rolling without sliding in subway tunnel are analyzed in this paper. Based on the analysis, the geometric constraint of platform car in subway tunnel is derived by using analytic geometry. Then the kinematic properties of Platform cars in tunnels are discussed. Due to the non-linearity of the whole model, a fuzzy intelligent control algorithm was designed to realize the automatic cruise of the control platform in the subway tunnels. Finally, a simplified dynamic model of the platform car was established in Adams, and a joint simulation with MATLAB was conducted to verify the effectiveness of the algorithm. The simulation results show that the control strategy can effectively ensure that the platform truck runs smoothly on the tunnel wall. The control strategy will be further improved based on the experimental data in the follow-up.

References

[1] Leiliu, Pingxiang, Yongji Wang, Trajectory tracking of a nonholonomic wheeled mobile robot [J]. Journal of Tsinghua university (natural science edition),2007, S2:1884-1889.

[2] Iagnemma K, Shinwoo K, Shibly H, et al. Online Terrain Parameter Estimation for Wheeled Mobile Robots with Application to Planetary Rovers[J]. IEEE Transactions on Robotics, 2004,20(5):921-927.

[3] Iagnemma K, Dubowsky S. Chapter 2 Rough Terrain Mobile Robot Modeling and Estimation[C]//Mobile Robots in Rough Terrain. Springer Berlin Heidelberg, 2004:17-50.

[4] Laiou M C, Astolfi A. Discontinuous Control of High-order Generalized Chained Systems[J]. SYSTEMS & CONTROL LETTERS, 1999,37(5):309-322.

[5] Ali S, Moosavian A, Alipour K. Tip-Over Stability of Suspended Wheeled Mobile Robots[C]// IEEE International Conference on Mechatronics and Automation (ICMA). Harbin: 2007:1356-1361.

[6] D.-H. Lee, H. Moon, and H. R. Choi, "Autonomous navigation of in-pipe working robot in unknown pipeline environment," in Proc. IEEE Int. Conf. RoboticsAutom.,2011, pp.1559–1564.

[7] L. A. Mateos, M. Sousa, and M. Vincze, "Dewalop-remote control for in-pipe robot," in Proc. 15th Int. Conf. Adv. Robotics, 2011, pp. 518–523.

[8] Song Z, Keng C, Yang D. Kinematic Analysis of Wheeled Mobile Robots in Cylindrical Workspaces [J]. Robots, 2006, 28(6): 636-641, 646. DOI:10.3321/j. issn: 1002-0446. 2006. 06. 015D.-H. Lee, H. Moon, and H. R. Choi, "Autonomous navigation of in-pipe working robot in unknown pipeline environment," in Proc. IEEE Int. Conf. RoboticsAutom.,2011, pp.1559–1564.

[9] Song Z, Ren H, Zhang J, et al. Kinematic Analysis and Motion Control of Wheeled Mobile Robots in Cylindrical Workspaces[J]. IEEE Transactions on Automation Science & Engineering, 2016, 13(2):1207-1214.

[10] M. Muramatsu, N. Namiki, R. Koyama, and Y. Suga, "Autonomous mobile robot in pipe for piping operations," in Proc. IEEE/RSJ Int. Conf. Intell.RobotsSyst.,2000, vol.3, pp.2166–2171, IEEE.

[11] S. Zhangjun and C. Ken, "Kinematic properties of wheeled mobile robots in round ducts/pipes," 2006.

[12] N. Chakraborty and A. Ghosal, "Kinematics of wheeled mobile robots on uneven terrain," Mechanism and Machine Theory, vol. 39, no. 12, pp. 1273–1287, 2004.

[13] Weber W. Automatic Generation of the Denavit-Hartenberg Convention Munich[C]// Robotics (ISR), 2010 41st International Symposium on and 2010 6th German Conference on Robotics (ROBOTIK). Germany: 2010:1-7.

[14] Atique M M U, Ahad M A R. Inverse Kinematics Solution for A 3DOF Robotic Structure Using Denavit-Hartenberg Convention[C]//Informatics, Electronics & Vision (ICIEV). Dhaka: 2014:1-5.

[15] Xiaohua Z, Hongjun C. Independent Wheel Drive and Fuzzy Control of Mobile Pipeline Robot with Vision[C]//Industrial Electronics Society, 2003. IECON '03. The 29th Annual Conference of the IEEE. 2003:2526-2530.

[16] Ding F, Jin L, Shouyong L, et al. Research on Obstacle Avoidance Based on Fuzzy Control for Inspection Robot in Oil Pipeline[C]//Information and Automation,2009. International Conference on ICIA '09. Zhuhai, Macau: 2009:189-19.

[17] Phu N D, Hung N N, Ahmadian A, et al. A New Fuzzy PID Control System Based on Fuzzy PID Controller and Fuzzy Control Process[J].INTERNATIONAL JOURNAL OF FUZZY SYSTEMS,2020,22(7): 2163-2187.

[18] Sun L H, Ma J, Yang B Q, et al. Fuzzy PID Design of Vehicle Attitude Control Systems: PROCEEDINGS OF THE 32ND 2020 CHINESE CONTROL AND DECISION CONFERENCE (CCDC 2020)[Z]. 32nd Chinese Control And Decision Conference (CCDC): 20201826-1830.

Experiment and Modelling of Triangular Holes Array by Non-traditional Machining Process

Panpan Wang, Yafeng He, Tao Yang, Bo Xu and Bing Long*

China Digital ECM Building Laboratory of Jiangsu Province, School of Aerospace and Mechanical Engineering /Flight College, Changzhou Institute of Technology, 213002, China

*Corresponding author email: longb@czu.cn

Abstract. Triangular holes array is a common structure of the specially shaped holes that can be used for various products in aerospace vehicles. These are commonly produced in the electrochemical machining (ECM) process. The high-quality and accuracy of a triangular grid are closely related to the processing parameters of cathode structure and sidewall insulation sleeve, power supply, electrolyte flow rate, feed speed, and machining gap. Simulations have been carried out to consider the influence of parameters including structure and dimension of the cathode, initial machining gap, feed speed, and current on the homogeneity of the flow field in the machining gap. Consequently, the modeling can support the design of the cathode. The paper is divided into two parts. Firstly, the tool cathode is simulated and fabricated. Secondly, the triangular holes array is produced by using the designed cathode tool. It was found that higher feed speed and lower machining gap increase the stability, accuracy, and efficiency of the machining process and the quality of the triangular holes array.
Keywords: Triangular holes array; Cathode design; Non-traditional machining.

1. Introduction

Various methods are used to fabricate the special-shaped holes pattern or surface texturing including laser beam machining [1, 2], electrochemical machining [3, 4], electrical discharge machining (EDM) [5], reactive ion etching [6], abrasive jet machining [7], chemical etching [8], photolithography [9], etc. ECM technology is applied for rapidly and flexibly producing these hole structures in metallic parts with good structural performance, like titanium alloy [10] and hard metals WC/Co [11], that is independent from the hardness and toughness of the material and without any mechanical and thermal impact [12, 13].

Grilles with special-shaped holes have been developed as lightweight and efficient structures, these have advantages in the improvement of shielding and radar absorbing properties. Consequently, these structures are considered perfect structures for intake and outlet design [14-16]. Some other surface texturing, like regular well-defined micro/nanopatterns on the surface, is greatly applied to medical, chemicals, telecommunication, data storage, optics, and microelectronics [17]. Functional properties of these patterns are characterized by friction and wear and wetting, especially its application to reduce friction and wear on carbide tools [18] or automotive engine parts [19].

The design of the cathode plays an important part in ECM technology because of its great impact on machining precision. Researchers around the world have performed various investigations corresponding to cathode tool design. Gu et al [20] have theoretically and experimentally investigated the cathode tool structure influence on the machining accuracy of blade fabrication in the electrochemical trepanning process. Zhao et al [4] have fabricated diamond holes on the metal grille.

The results show that the uniformity of electrolyte velocity distribution and the vibration feed of the cathode are helpful to significantly improve the machining localization and machining stability. Fang et. al studied the effects of the hollow hole shape and electrode size on the homogeneity of flow fields in the machining gap [21]. It was found that the uniformity of flow fields can be greatly increased by using electrodes with circular hollow holes compared to that with square hollow holes. The uniformity of flow fields, the machining stability, and accuracy increase with the hollow hole size increasing. The surface roughness, machining gap, and machining gap standard deviation of the square hole on titanium alloys are 0.8 μm, 286 μm, and 4.7 μm, respectively. It was fabricated by using an electrode with a 2.7-mm-diameter circular hollow. Factors affecting the accuracy of diamond-hole grilles have been optimized utilizing cathode sidewall insulation, pulse power supply, and passivating electrolyte by Zhao et al [3]. It presents that precise electrochemical machining, with precise synchronization of compound feed vibration and pulse power, offers better machining precision and process stability for diamond-hole grilles.

At present, studies have been published on the special hole-shaped pattern mainly focused on the optimization of the processing parameters and in the flow field during the creation of rectangular and circular holes, etc. Fabrication of triangular holes array was discussed in this paper. Cathode tool structure was designed to consider the influence of the shape of the electrolyte flow channel, applied voltage, machining gap, and machining feed rate on the quality of the creation of triangular holes array. The simulation of machining gap flow field under different parameters was analyzed by COMSOL software, and the flow field characteristics of electrochemical machining of triangular array holes were obtained. Parameters affecting the precision and reliability of ECM for triangular holes array were analyzed. This provides a theoretical basis for the experiment of a triangular hole's array formation. Consequently, triangular holes array on stainless steel was produced and the quality of these holes was analyzed.

2. Methods

2.1. Experimental

Taking into account the effect of main ECM process parameters, a designed cathode tool has been used to generate a triangular-shaped holes array on stainless steel workpiece. The electrochemical machining process consists of machining cells, control cells, and electrolyte circulation. The triangular cylinder cathode tool with a circular hollow of 5.5 mm inside length has been prepared by CNC lathe, CNC milling machine, and electric discharge machining as shown in Figure 1. The scanning and photographing of triangular holes array on the stainless steel surface was investigated by Olympus microscope.

Table 1. Processing parameters for experiments.

Parameter	Range
Workpiece material	304 stainless steel
Electrolyte	10% Aqueous $NaNO_3$
Electrolyte temperature	40 ℃
Inlet electrolyte pressure	0.5 MPa
Applied voltage	17 Volts
Cathode feed rate	0.4,0.5,0.6,0.7 mm/min
Cathode feed depth	2 mm
Interelectrode distance	5 mm
Initial machining gap	0.05 mm
Cathode inner/outer dimension	1.8/5.5 mm

The assembled cathode is composed of a triangular cylinder with an internal hollow and sidewall insulation sleeve. Tool electrode with epoxy resin-coated on its side surface was designed. During the ECM process, the anode workpiece is settled, and the tool cathode electrode is moved toward it. The electrolyte, supplied from the inner hollow of the cathode, is injected into the machining gap. This electrolyte flow is perpendicular to the workpiece that helps to take away the hydrogen bubbles and side-products from the processing area. When a voltage is applied across the electrodes, the material removal from the workpiece begins. Accompanied by the feeding of the cathode, the triangular holes array is generated eventually.

304 stainless steel is used as the anodic workpiece. It has been marked by significant properties, such as high strength, corrosion resistance, and good electrical conductivity. The electrolyte composition of 10% $NaNO_3$ aqueous solution is used. The hole diameter of the electrode, cathode tool length, and depth of machining is 1.8 mm, 20 mm, and 2 mm, respectively. The interelectrode gap is 5 mm. Table 1 shows details of experimental parameters. For conducting the experiments, four variables namely the shape of the cathode internal hole, size of the hollow, applied voltage, and the machining gap are considered in cathode design.

2.2. Model Description

Pro/E 5.0 and Comsol Multiphysics 5.4 are used to design the cathode tool structure and simulate the flow field distribution in the machining gap, respectively. In ECM, the simulation of flow and electric field in the machining gap is very complex. To simplify the electric model, we make three assumptions as follows: (1) in the machining gap, the electrolyte flow is regarded as an ideal turbulent and no compressible liquid; (2) according to Ohm's law, electrolyte conductivity and applied voltage determine current density; (3) the process is in the equilibrium state.

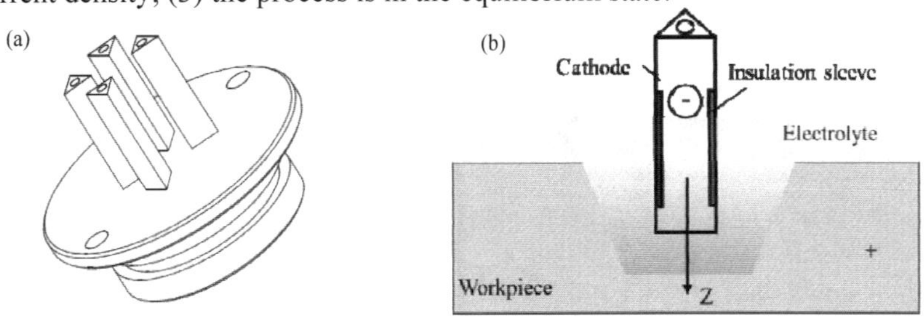

Figure 1. Schematic image of (a) structure of cathode tool and (b) machining zone.

Table 2. Simulation parameters for cathode design.

Parameter	Range
Electrolyte density	1×10^3 Kg/m^3
Electrolyte viscosity	0.0142 Pa.s
Inlet/Outlet electrolyte pressure	0.5/0 MPa
Conductivity of electrolyte	12 S/m
Cathode inner dimension	1.5,2,2.5 mm
Machining gap	0.05, 0.1, 0.15 mm
Applied voltage	12, 14, 16 Volts

Figure 1 shows geometrical shape of cathode tool and machining zone on the surface of workpiece. The sidewall insulation sleeve is used to prevent secondary corrosion of the workpiece. The fabrication of cathode tool is optimized based on the analysis of flow field simulation in the machining gap. Simulation parameters are given in Table 2. In the modeling, parameters such as shape and size of cathode hollow hole, applied voltage, feed rate, machining gap are considered.

3. Results and Discussions

3.1. Cathode Design

The dynamic behavior of the electrolyte flow is simulated in the machining gap. The flow velocity distribution at the machining gap is the most representative of the stability of the flow field during processing, and it can be seen intuitively that the more uniform the velocity distribution, the better the stability. Therefore, the flow rate and pressure distribution contour of the electrolyte are analyzed by Comsol Multiphysics 5.4. Through these simulations, the most suitable cathode hollow structure is selected to improve the stability of the triangular holes array formation processing.

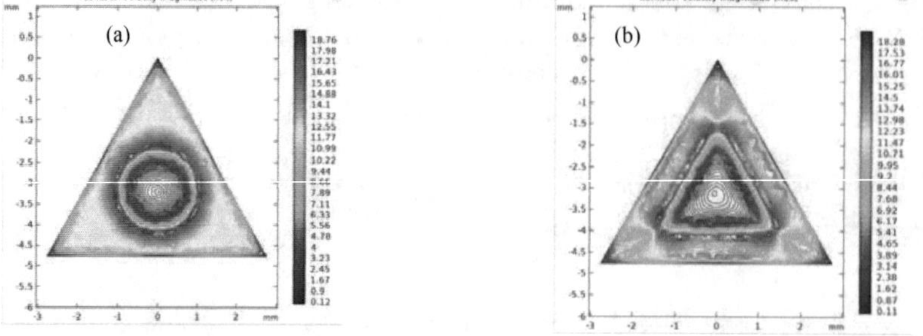

Figure 2. Contour of velocity magnitude in the machining gap corresponding to two types of cathode structure (a) in a circular hole and (b) in a triangular hole.

It can be seen from Figure 2 that for both the circular shape structure and the triangular shape structure, a certain degree of electrolyte shortage occurs at the corners of the triangle. The electrolyte-deficient area of the circular hole is significantly smaller than that of the triangular hole, and the flow of the triangular hole is disordered when the electrolyte enters the machining gap, so the electrolyte is not easily discharged, which may cause short-circuiting and reduce the stability of the machining process. The circular flow is uniformly distributed, which can ensure stability during machining.

3.2. Triangular Holes Array Formation

An experimental study was performed to discover the impact of ECM process parameters like feed speed and machining gap on the structure and surface roughness of triangular holes array on stainless steel surface using a designed tool cathode, as shown in Figure 1. Figure 3 shows machining fillet radius depending on cathode feed speed. As the feed speed increases, the machining fillet radius decreases. An increase in feed rate increases the homogeneity of the holes array and improves edge radius by reducing secondary corrosion. If the feed rate is too low, the material removal rate, side gap, and stray current effects will be increased.

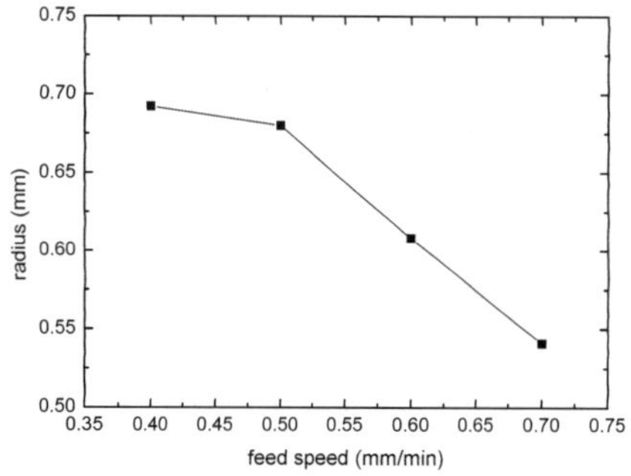

Figure 3. Machining fillet radius depending on cathode feed speed.

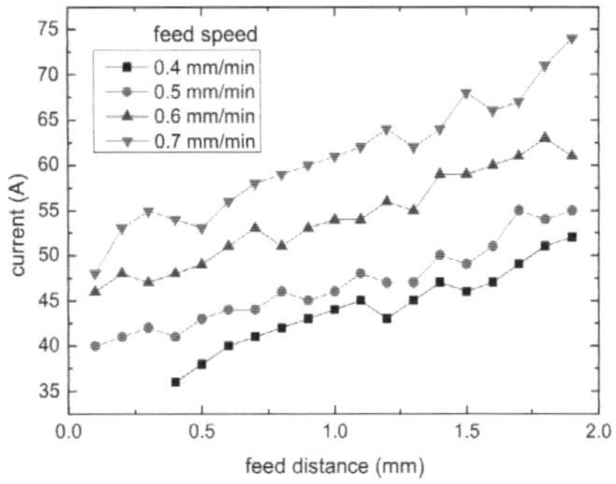

Figure 4. Current depending on cathode feed speed and feed distance.

Figure 4 represents the influence of cathode feed rate and feed distance on the current. Regardless of the feed rate, the current increases as the feed depth increase. The current at different feed depths increases as the feed rate increases. An increase in current leads to the increase of the material removal rate and the machining efficiency. As a result, the residence time of the cathode feeding decreases relatively. As the feed rate increases, the side dissolution area of the workpiece is reduced, and the taper of the hole also decreases. Therefore, an increase of the feed speed without short-circuiting occurring shortens the residence time, improves the machining efficiency the surface quality of the holes array.

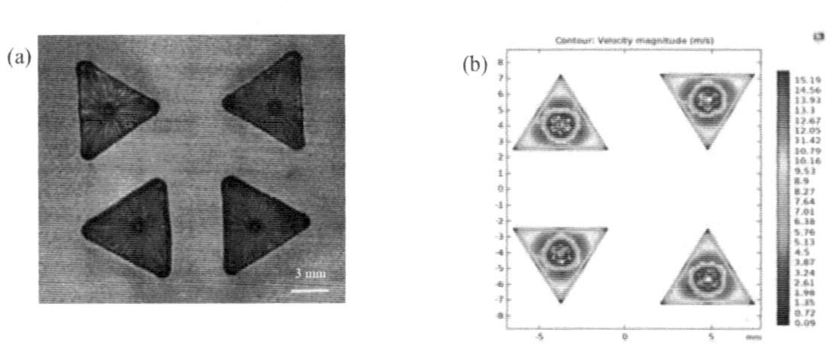

Figure 5. Images of triangular holes array (a) optical micrograph at feed speed of 0.7 mm/min and (b) contour of velocity distribution.

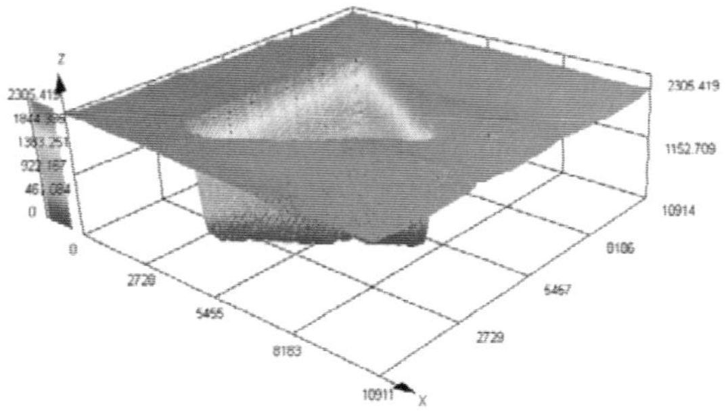

Figure 6. 3-D view of details of a triangular hole.

Figure 5 shows the optical image and contour of the velocity distribution of the triangular holes array in the machining gap. Figure 6 shows a 3-D view of a triangular hole. It is produced by combining specific parameters like the interelectrode gap of 5 mm, inlet electrolyte flow of 0.5 MPa, feed rate of 0.7 mm/min, and machining gap of 0.05 mm using a circular hollow of a diameter of 1.8 mm. The modeling is carried out under the same parametric conditions. The velocity distribution for different holes is almost the same. According to the velocity distribution contour, the electrolyte can be evenly distributed from the distribution chamber to each cathode. Due to controlled anodic dissolution and the proper parameter combination, almost all profiles of triangular holes array maintain the appropriate and uniform geometry. The produced depth of triangular holes array is almost the same because of the current density distribution evenly.

4. Conclusions
The paper describes the fabrication of triangular holes array in the electrochemical machining process using a single triangle-patterned cathode tool. The influence of main process parameters is experimentally examined and analyzed on the optimized structure of cathode design and surface roughness of triangular holes array. The following conclusions can be obtained:
(1) The effects of the hollow shape and dimensions of cathode on the uniformity of flow fields in the machining gap have been investigated. Cathode with circular hole enables the uniformity of the electrolyte flow and current density compared to that of a triangular hole.
(2) Lower machining gap, higher electrolyte flow rate, and higher cathode feed rate are recommended to produce high-quality triangular holes array using the designed cathode tool.
Experiments show the analyzing and reporting test results using the specific process parameters, like cathode tool profile, feed rate, and machining gap, during the fabrication of triangular holes array. The modeling is applied to optimize the machining process and validated by the experiment. However, the production of complex patterned cathode tool, and workpiece movement strategies will be further studied to produce various special-shaped holes array in different advanced applications.

Acknowledgments
This work was significantly supported by the Natural Science Foundation of Jiangsu Province (BK20161193), key R&D Projects of Jiangsu Province (BE2018067), Natural Science Foundation of the Jiangsu Higher Education Institutions of China (20KJB590001).

References
[1] Johan Meijer 2004 *Journal of Materials Processing Technology* vol 149 pp 2-17
[2] Kyoung Ho Ha and Se Won Lee 2015 *Journal of Micromechanics and Microengineering* vol 25 p 027001
[3] Jianshe Zhao and Feng Wang 2018 *Procedia CIRP* vol 68 pp 684-689
[4] Zhao Jianshe and Wang Feng 2016 *Chinese Journal of Aeronautics* vol 29 pp 1830-1839
[5] D. B. Meshram and Y. M. Puri 2016 *Journal of the Brazilian Society of Mechanical Sciences and Engineering* vol 39 pp 2593-2605
[6] Heon Lee and David B. Oberman 1996 *Journal of Electronic Materials* vol 25 pp 835-837
[7] Klocke F 2018 *Procedia CIRP* vol 68 pp 541-546
[8] Zhipeng Huang and Nadine Geyer 2011 *Advanced Materials* vol 23 pp 285-308
[9] Nan Cui and Xueyan Ding 2019 *IEEE Electron Device Letters* vol 40 pp 788-791
[10] Qu Ningsong and Fang Xiaolong 2013 *Chinese Journal of Aeronautics* vol 26 pp 224-229
[11] M M Lohrengel K P Rataj 2014 *Power Metallurgy* vol 57 pp 21-30
[12] Matthias Hackert-Oschätzchen and Raphael Paul 2015 *Procedia CIRP* vol 31 pp 197-202
[13] H Matthias Hackert-Oschätzchena and Gunnar Meichsner 2012 *Precision Engineering* Vol 36 pp 612-619
[14] Jianshe Zhao and Xiangli Zhang 2016 *The International Journal of Advanced Manufacturing Technology* vol 91 pp147-153
[15] Peng XL and Xu YM 2009 *Journal Composite Materials* vol 03 pp 162-168
[16] Jiang Shicai 2006 *Journal of Aeronautical Materials* vol 26 pp 196-198
[17] S Kunar and B Bhattacharyya 2018 *Materials and Manufacturing Processes* vol 34 pp 1-7
[18] Minghua Pang and Yongfang Nie 2018 *The International Journal of Advanced Manufacturing Technology* vol 99 pp 737-746
[19] Yin Bifeng and Gao Dashu 2018 *Journal of Tribology* vol 140 pp 061701
[20] Zhouzhi Gu and Weiguo Zhu 2019 *The International Journal of Advanced Manufacturing Technology* vol 100 pp 857-863
[21] XL Fang and Nm Qu 2013 *Journal of South China University of Technology* vol 41 pp 137-142

AUTHOR INDEX